Liaison Functions 2016

Core Programming Area at the 2016 AIChE Spring Meeting and 12th Global Congress on Process Safety

Houston, Texas, USA
10-14 April 2016

ISBN: 978-1-5108-2495-9

Printed from e-media with permission by:

Curran Associates, Inc.
57 Morehouse Lane
Red Hook, NY 12571

Some format issues inherent in the e-media version may also appear in this print version.

Printed by Curran Associates, Inc. (2016)

For permission requests, please contact AIChE
at the address below.

AIChE
120 Wall Street, FL 23
New York, NY 10005-4020

Phone: (800) 242-4363
Fax: (203) 775-5177

www.aiche.org

Additional copies of this publication are available from:

Curran Associates, Inc.
57 Morehouse Lane
Red Hook, NY 12571 USA
Phone: 845-758-0400
Fax: 845-758-2633
Email: curran@proceedings.com
Web: www.proceedings.com

TABLE OF CONTENTS

449794 WIC AIChE: 12 Steps for a Successful Career Transition

Abstract:

Career Transitions

Those in career transition may feel overwhelmed, fearful or lack purpose and are searching for motivating strategies to overcome self-limiting beliefs and formulate a plan to find their ideal career.

We will explore how to position seasoned professionals for success by providing 12 career transition strategies that will deliver the changes they are seeking in their career.

These holistic strategies will clarify how to get "unstuck", prepare for a successful career campaign, increase confidence, be empowered to take action, commit to goals, minimize fears and take action regardless, and ultimately create a realistic action plan to help gain the confidence, courage, and clarity necessary to make the career changes they seek.

Benefits include:

- Clarify "how" to survive a job transition
- Be prepared for a successful career campaign
- Develop richer relationships and increased confidence
- Be empowered to take action
- Be strongly committed to your goals
- Minimize fears and take action regardless
- Create a 12-step action plan

442035 Scaling Liquid Flow through Packed Beds from the Laboratory to the Commercial Plant

Abstract:

In 1952, Sabri Ergun published an article reconciling all the data pertaining to flow through packed columns at that time. The equation presented in that paper has since become known as the Ergun equation. It has been used with varying degrees of success since its publication. The Ergun equation has various problems. First, it uses a Reynolds number based on particle diameter, not fluid conduit diameter. Thus, the Reynolds number used in the Ergun equation does not describe fluid flow. Second, the Ergun equation was developed for spheres. However, myriad odd-shaped packing exists; thus, the Ergun equation must be modified when using non-spherical packing. That modification is generally encapsulated in a dimensionless "sphericity factor" … a cheat factor, in other words. Third, the Ergun equation violates geometric similarity, thus it cannot be used for scaling up or scaling down.

In this paper, we address these problems via Dimensional Analysis. We use Dimensional Analysis to formulate a new method for characterizing liquid flow through packed columns; i.e., for characterizing filtration. We also propose a new method for characterizing the packing material. Since our results obey the laws of similarity, our analysis can be used for up scaling and down scaling filtration units. This method is applicable to liquid flow through packed beds in chemical plants as well as oil flow through fractured rock.

Scale-up of Oscillatory Helically Baffled Reactors

Safaa Ahmed

Supervisors:
Dr. Anh Phan
Prof. Adam Harvey

➢ Outline

- ➢ Introduction
- ➢ Aim of study
- ➢ Devices and procedure
- ➢ Results
- ➢ Conclusions
- ➢ References

➢ Background

<u>**Oscillatory baffled Reactor (OBR)**</u>

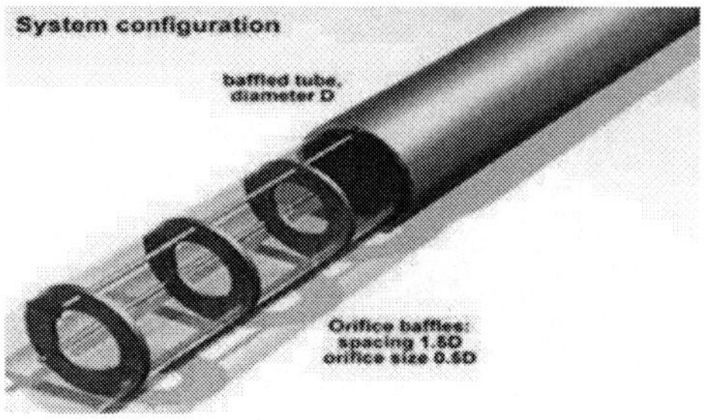

➢ Aim of study

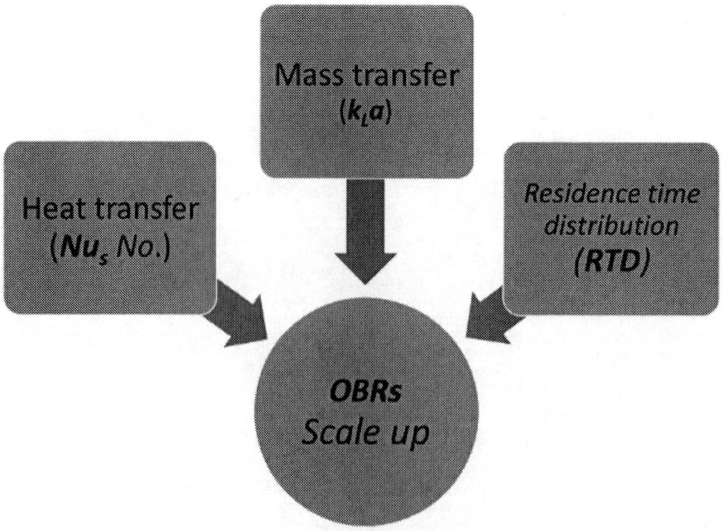

Residence time distribution (RTD)

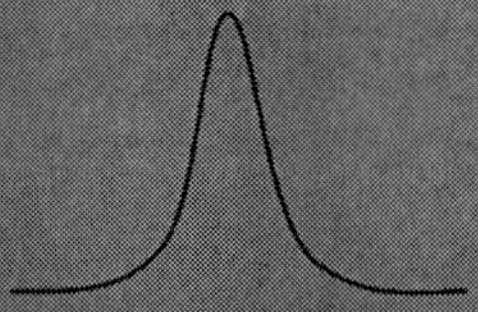

Tanks-in-series model

$$E(\theta) = [N(N\theta)^{N-1}/(N-1)!]\,e^{-N\theta}$$

- **Baffle designs:**

✓Orifice baffles

✓ Helical baffles

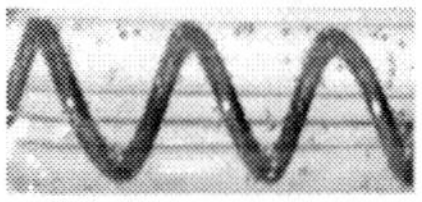

✓ Integral baffles

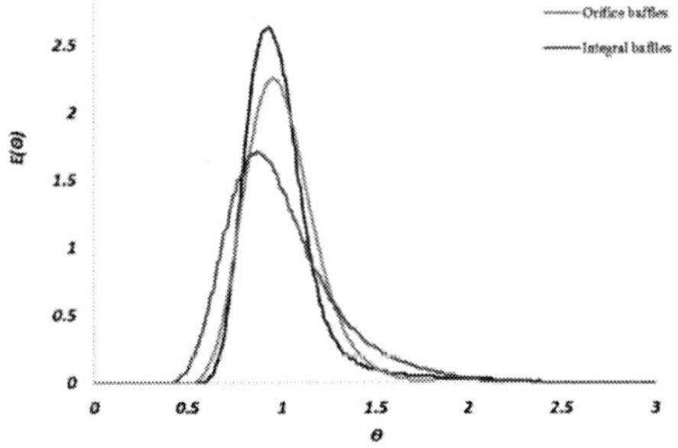

REN=100, F=3HZ, AND XO=4MM

➢ Devices and Methods

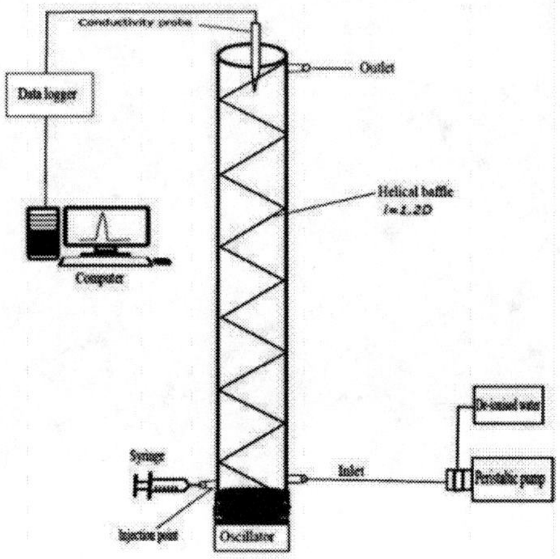

➢ Scale-up limitations

- **Geometric similarity**:
  - ✓ Baffle spacing *(l/D)=1.2*
- **Dynamic similarity:**
  - ✓ Strouhal Number (*Str*): 0.37, 0.27, 0.17
  - ✓ Velocity ration (ψ) = *Reo/Ren*

➢ OHBR scale-up results

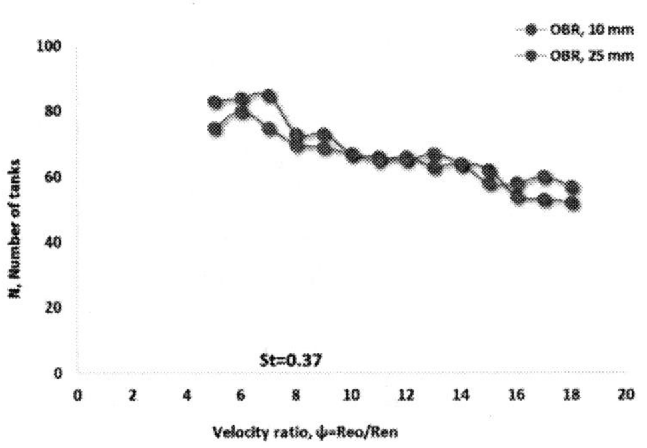

RTD behaviour at *Str*=0.37

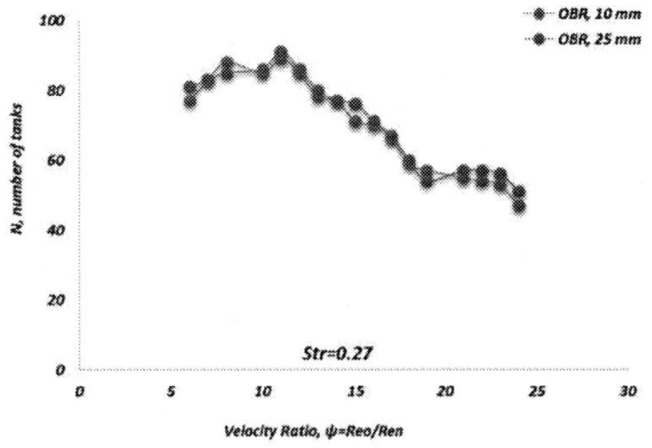

RTD behaviour at *Str*=0.27

➢ Scale up correlation

- $N = 25\, Str^{-0.3}\, \psi\, e^{-0.13\psi} + 30.2$

$$N = \frac{41.5 + aRe_n - bRe_n^{\;2}}{1.36}\left\{\left[\left(\psi + 1.8 - cRe_n^{\;-d}\right) + 1\right] e^{-0.35(\psi+1.8-cRe_n^{\;-d})}\right\} + 1 \qquad [1]$$

Comparison between the experimental data and correlation data

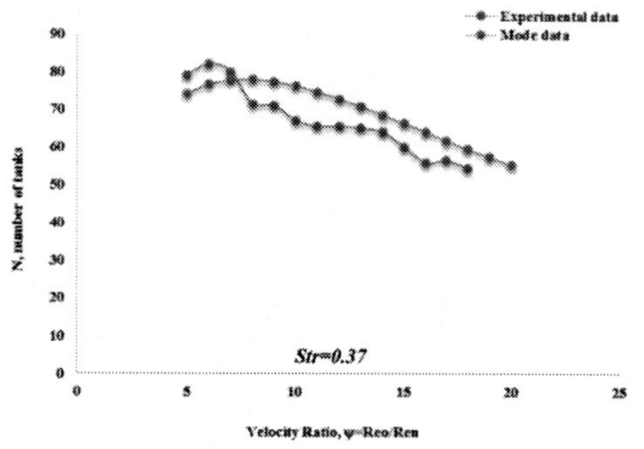

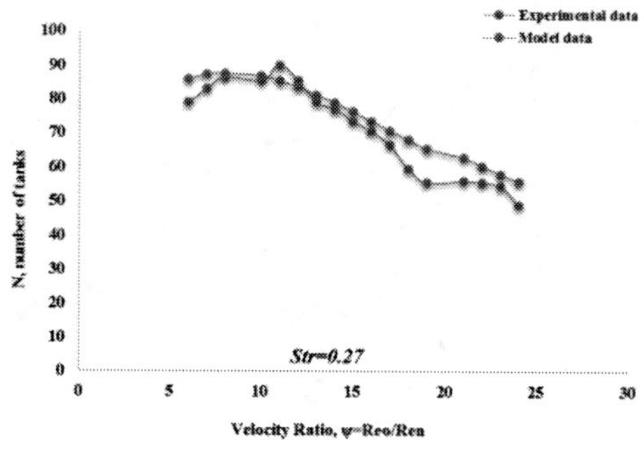

➢ Validation

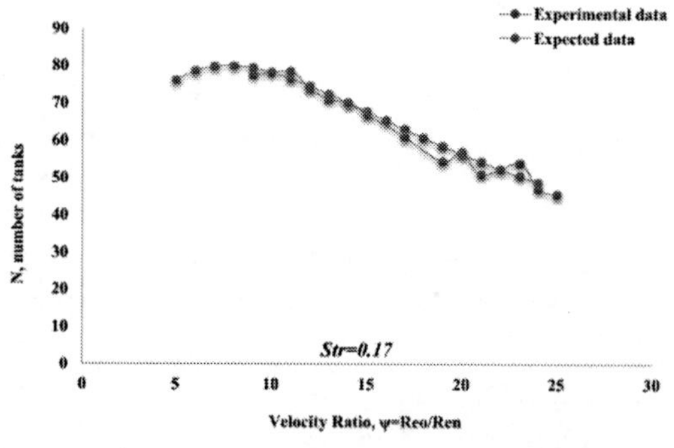

$R^2 = 0.98$

Visualisation of flow pattern inside meso-OBR

(PIV technique & CFD simulation)

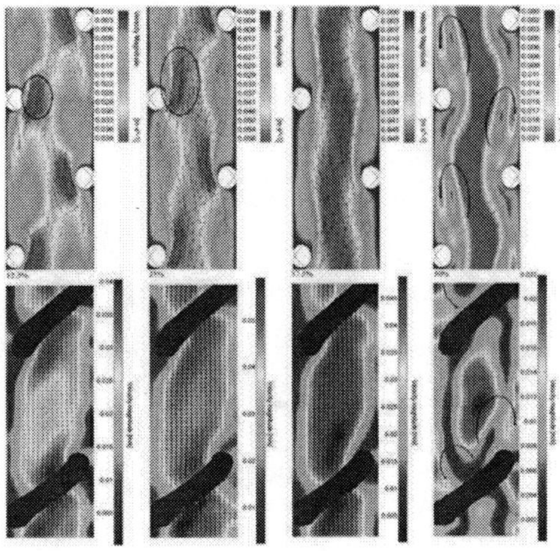

Oscillation conditions: *Ren = 0, Reo = 126, St = 0.2 (xo = 2 mm, f = 2 Hz)*

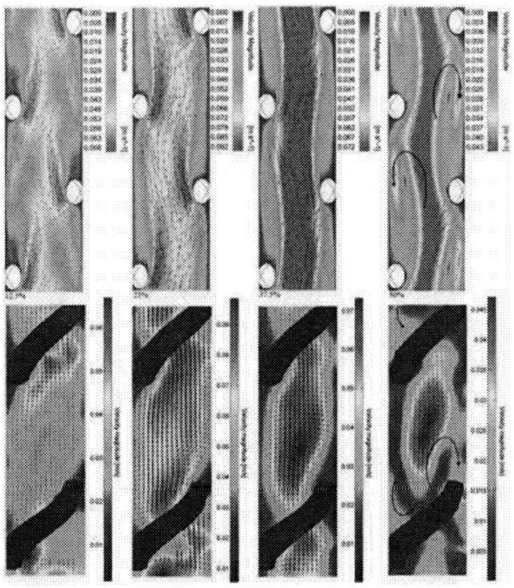

Oscillation conditions: *Ren = 0, Reo = 188, St= 0.13 (xo = 3 mm, f = 2 Hz)*

OBRs scale up based on Mass Transfer Coefficient

(Air-Water system)

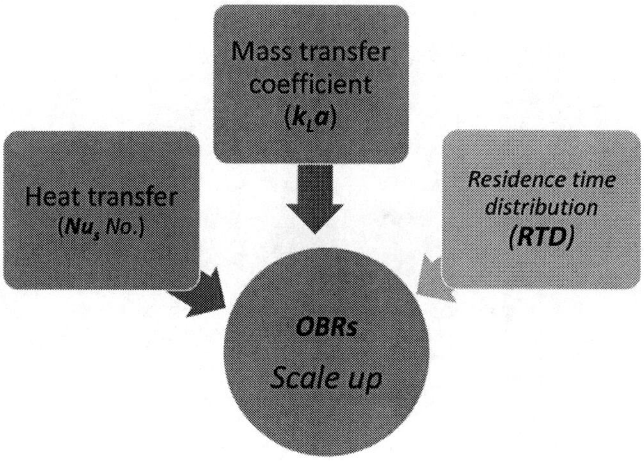

➢ Aim of this step

- Choose the best design of OBR (based on *kLa* results)
- Gas holdup calculation for each scale.
- Mass transfer coefficient (***kLa***) calculation for each scale.
- Determine a model of ***kLa*** for each scale.
- Compare the results of all scales.
- Validate the models.

➢ OBRs designs

✓Single-orifice design

✓ Helical design

✓ Smooth Periodic Constrictions design (***SPC***)

➢ *kLa* comparison (10 mm scales)

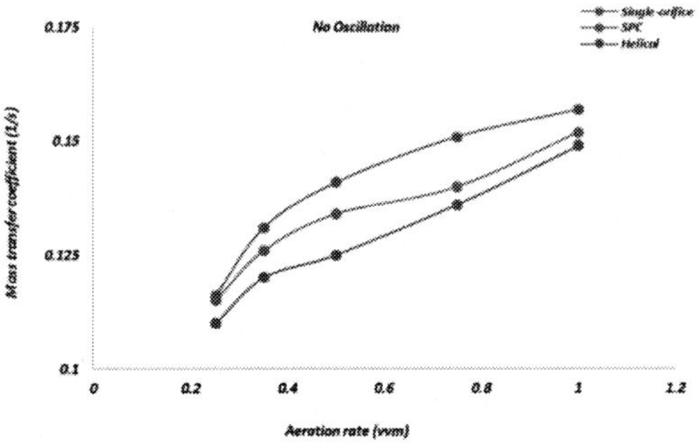

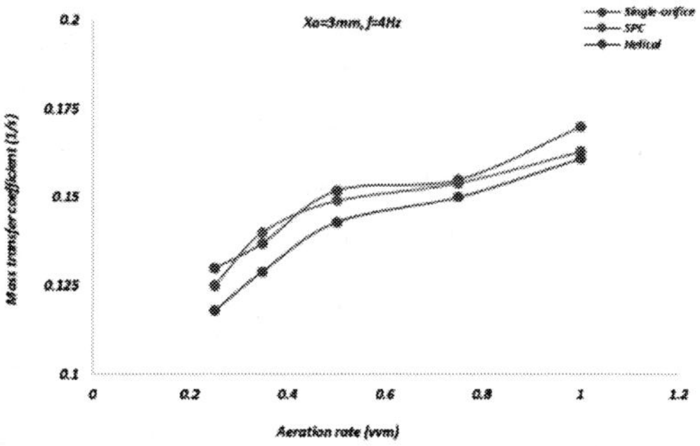

➢ **Baffles modifications**

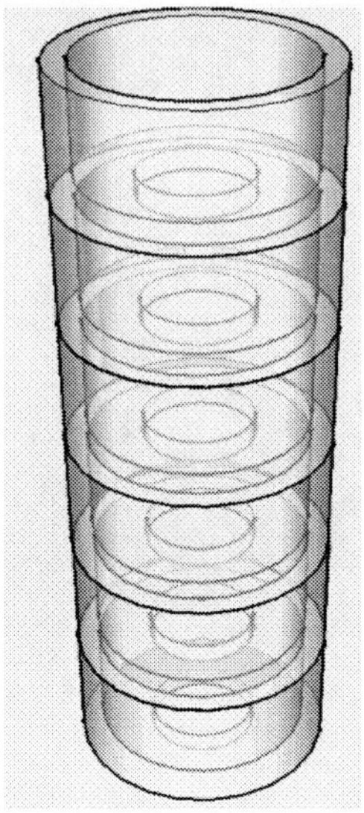

Single-orifice ⟶ Multi-orifice

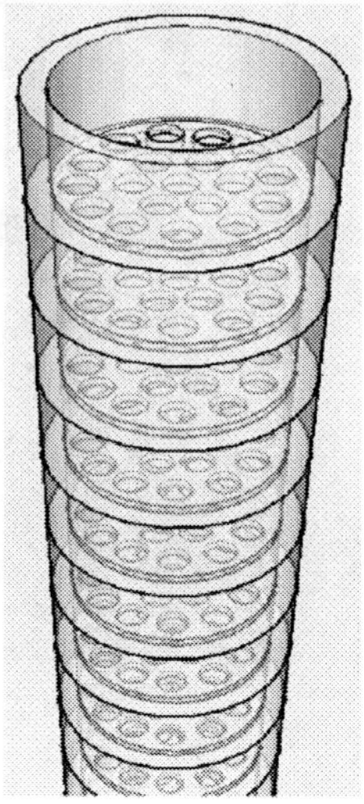

➢ Scale up limitations (Similarity)

1- Geometrical parameters:

- Open cross sectional area *(α)= 56%*
- Baffle spacing *(l/D)= 0.5*

2- Dynamic parameters:

- Maximum velocity *(2πfxo): 0 - 1000 mm/s*
- Aeration rate (*vvm*)*: 0.1-0.5 1/min*

➢ Conclusions

- The OBRs were scaled up successfully based on RTD .
- The scale-up correlation was developed.
- The scale-up correlation was validated.
- Flow pattern inside meso-OBR was visualised using PIV technique and Ansys software.
- A best design of OBR has been chosen and modified.

➢ References

[1] Stonestreet, P. and Van Der Veeken, P.M.J. (1999) 'The Effects of Oscillatory Flow and Bulk Flow Components on Residence Time Distribution in Baffled Tube Reactors', Chemical Engineering Research and Design, 77(8), pp. 671-684.

[2] Phan, A.N. and Harvey, A. (2010) 'Development and evaluation of novel designs of continuous mesoscale oscillatory baffled reactors', Chemical Engineering Journal, 159(1-3), pp. 212-219.

[3] Pereira, F.M., Sousa, D.Z., Alves, M.M., Mackley, M.R. and Reis, N.M. (2014) 'CO2Dissolution and Design Aspects of a Multiorifice Oscillatory Baffled Column', Industrial & Engineering Chemistry Research, 53(44), pp. 17303-17316.

[4] Smith, K.B. and Mackley, M.R. (2006) 'An Experimental Investigation into the Scale-up of Oscillatory Flow Mixing in Baffled Tubes', Chemical Engineering Research and Design, 84(11), pp. 1001-1011.

Thanks

Strategy for Designing a Successful Pilot Plant

Anil J. Mehta, The Dow Chemical Company, Freeport, TX, U.S.A.

Abstract

Testing a chemical or a petrochemical production process at a pilot scale is a vital step in scaling up the new process to commercial scale successfully. The author has extensive experience in scaling up new chemical and petrochemicals production processes successfully. The author presents an effective strategy and approach for designing and operating a pilot plant with examples to achieve the desired results. This could be an interactive discussion.

1. Introduction

One thing we all can count on in life is change. The planet and the world we live-in are in a state of change constantly. Our needs are always changing driven by technology changes or challenges such as climate change or other factors. To meet the ever-changing needs, new materials and processes to manufacture them are needed. Sometimes, the same products need to be made using different raw materials which also requires developing a new production process for the product. Thus, new manufacturing processes are always needed.

New materials and processes usually are invented in laboratories at a very small scale. If the product or the process is found promising from initial testing then the process needs be scaled up to produce the product in a much larger quantity at an affordable price to satisfy demand for the product.

Scaling-up a new manufacturing process from a small laboratory scale to a commercial scale requires designing and building a commercial plant, a highly complex job. Commercial manufacturing plants can be very expensive, costing anywhere from a few millions to billions of dollars, depending on the scale and complexity of the process.

2. Why test the new process at a pilot scale?

Scaling-up a new process from a laboratory scale to a commercial scale directly is very risky because many things can go wrong in the scale-up process. The new commercial plant may not work as intended, either not being able to meet product specification or operating at a rate far less than the design rate or both. The process industry is full of examples where a new process scaled-up failed to produce the quality product or failed to operate at a rate close enough to the design capacity. If the new plant fails to produce the intended product at an economical price, improvements to the plant are needed, requiring additional capital and time. In the worst case, the investment may have to completely written off, resulting in a total loss of capital which could be hundreds of millions of dollars and time.

Testing the new process at a pilot scale before commercializing minimizes such risks present in scaling up the process to commercial scale [1-4]. Pilot scale implies a scale much smaller than the desired commercial scale [5]. The pilot testing serves as an insurance for scaling up the new process to commercial scale because although the pilot testing is expensive, it is still much cheaper to invest in a pilot program than experience failure at the commercial scale.

3. Pilot testing strategy

Building a pilot plant is expensive; it can cost millions of dollars. Therefore, if the pilot testing can be done in an existing facility either in-house or at a contract manufacturer's facility, it can save the cost of building the new pilot plant. However, this depends on the type of the process to be scaled-up.

The new production process to be tested may be a continuous process, a batch process, a semi-continuous process or a combination of these. Batch processes tend to be simpler; hence, it is usually easier to find facilities for pilot testing a batch process. Complexity of a process increases as one goes from a batch process to a semi-continuous process, a continuous process and a continuous process with a recycle or even multiple recycle loops. The new process may be for making an existing product using a new synthesis route or using a different raw material e.g. producing ethanol from algae or it may be for a new product.

If the new process is for a new product, in addition to developing the production process, the product also has to be tested extensively for the intended end-use application and the market for the product developed. Launching a new product in the market brings additional and usually even bigger risk. Only a few of many products that get launched, gain sufficient acceptance in the market. Therefore, if the new production process is for a new product, it is best to minimize the pilot plant expenses by testing the process at an existing facility, either within or outside the company such as at a contract manufacturer's facility, until the product gains sufficient acceptance in the market. If a new process is for an existing product, is continuous, has one or more recycle loops, it will likely require designing and building a dedicated pilot plant as such customized pilot plants are hard to find in the market.

3.1 Before committing to pilot testing

A decision to test a process at a pilot scale is a major step in the process development. It requires committing significant capital and resources. Therefore, it is expected that the decision to test the process at a pilot scale follows rigorous testing of the process at a bench scale including safety and toxicology studies. It is expected that by this time all process technology options have been evaluated from techno-economic and safety aspects and there has been a rigorous internal debate for the technology choice before committing to the pilot testing.

3.2 Designing a successful pilot plant

3.2.1 Pilot plant objectives

There must be a clear understanding of the purpose of the pilot plant before designing it. Define in detail the goals for the pilot testing. Prepare a detailed list of the objectives and the data to be generated and how the data will be used – e.g. data on reaction rate, conversion, yield, selectivity, effect of flow regime, recycle, separation – distillation / extraction, product quality, process control, materials of construction etc. Ensure management input in this process to make sure nothing will be missed.

3.2.2 Scale-down to scale-up

One of the purposes of the pilot plant is to test at a much smaller scale the production process to be used at the commercial scale. Therefore, as far as possible, the pilot plant should have the same kind of key equipment and materials of construction that will be used in the commercial plant. It is often not necessary to use the same type of a pump, filter or a heat transfer equipment as in the commercial plant if such unit operations are well understood and not key part of the pilot testing.

The best way to ensure that the pilot plant will have the same kind of key equipment that will be used in the commercial plant is to first complete preliminary design of the desired commercial plant. This requires preparing material and energy balance for the commercial plant. Commercially available process simulation programs such as Aspen Plus or others can be used for the same.

Physical properties data form the backbone of a process simulation and preparing a process model will help identify missing physical properties data if any. Importance of accurate physical properties data can not be over-emphasized for scaling up a process successfully. Search literature extensively for the missing data. If the missing data can't be found in the literature then it needs to be generated within or outside the company.

Determine the type of equipment to be used in the plant and their preliminary size based on the material and energy balance for the process. The next step is to scale down the material and energy balance. At this point, a question arises – what should be the scale of the pilot plant?

3.2.3 What scale for the pilot plant?

The pilot plant should be of the smallest possible and workable scale while ensuring that it provides the high quality scale-up data required for designing the commercial plant, the scale-up data for various unit operations used in the plant.

The smallest possible size minimizes the pilot plant cost which includes cost of equipment, raw materials, waste generation and disposal. It also minimizes handling of raw materials, waste and potential exposure to hazardous chemicals, process inventory and the process safety risk.

Determining the minimum scale for the pilot plant requires determining the minimum scale for each unit operation to be tested such as reaction, distillation, extraction etc. In Dow, we can scale-up distillation columns from as small as a 2" diameter column [6]. In one of the process development project I led, we performed scale-up analysis for the reactor and determined that 1-liter reactor size was sufficient to generate the scale-up data for the commercial scale CSTR. However, this was an extremely corrosive process requiring a glass-lined reactor. The smallest glass lined reactor available was 1 gallon in size. Therefore, the pilot plant was designed based on the 1 gallon reactor size. The smallest size for piloting the epoxidation step required an order of magnitude larger feed rate.

Scale-up factors, the ratio of the commercial scale to the pilot scale, can vary over a very wide range depending on the type of unit operation and complexity of the process. For example, we scaled up successfully an absorber by a factor of over 1500 and we scaled up successfully improvements in a slow batch reaction from about a 100 grams scale to over a million pounds scale.

It is worthwhile to seek input from the subject matter experts for the unit operations being tested or used in the pilot plant such as reaction engineering, mixing, distillation, metallurgy. Pilot testing is expensive so it is imperative that the best expertise available within or outside the company be used to ensure that the correct decisions are made in designing the pilot plant.

3.2.4 Design for abnormal operation scenarios

Once the pilot plant scale is determined, the next step is to prepare material and energy balance for the pilot plant. Simulate various pilot plant operation cases such as start-up, shut-down, low throughput rate, high throughput rate etc. and ensure that the pilot plant is designed for these situations.

3.2.5 Testing materials of construction

If the process is corrosive, requiring corrosion resistant materials, make provision to test in the pilot plant the materials of construction you plan to use in the commercial plant and collect material performance data on long term exposure to the process.

3.2.6 Pre-construction review of the design

Once, key parts of the pilot plant design are completed, review the design with the larger team involved in the project, appropriate experts and supervision before ordering equipment and beginning construction of the pilot plant. It is much easier to make changes in the design at the design stage then after constructing the pilot plant. This will minimize potential for changes after pre-startup review and save the time in making the changes before the pilot plant can be started up, ultimately speeding up the process.

3.2.7 Process Safety

Most chemical processes involve dealing with hazardous materials. Exposure to a hazardous material can result in a serious safety incident. As scientists and engineers, it is our utmost responsibility to make the pilot plant as safe as possible. Safety must be an integral aspect of the design and operating discipline – how the pilot plant is started-up, operated and shut down. We can not depend upon personal protective equipment as they are the last layer of defense. Integrating safety in design builds the first layer of defense. Ensure input of process safety experts in the pilot plant design. Conduct process hazard analysis of the design. Identify worst case scenarios and design safe guards to protect against them.

Reactive chemicals testing should be done before even conducting laboratory scale experiments. Risk of a run-away reaction increases significantly as the scale of the process increases from a laboratory scale to a pilot scale. The surface area available for cooling from heat loss to the surrounding decreases rapidly per unit volume of the process materials. The highest temperature in a distillation column is usually found in the column's reboiler. If the liquid mixture in the reboiler is reactive, temperature for a potential run-away reaction should be determined and the reboiler should be operated well below the potential run-away reaction temperature. The run-away temperature is the temperature at which the rate of temperature rise exceeds the rate of cooling for the reboiler.

Reactive chemicals testing such as accelerated rate calorimetry can provide self-heating rate for the mixture. See the example in Figure 1 [7]. Cooling rate for the given equipment can be measured under real conditions as shown in Figure 2.

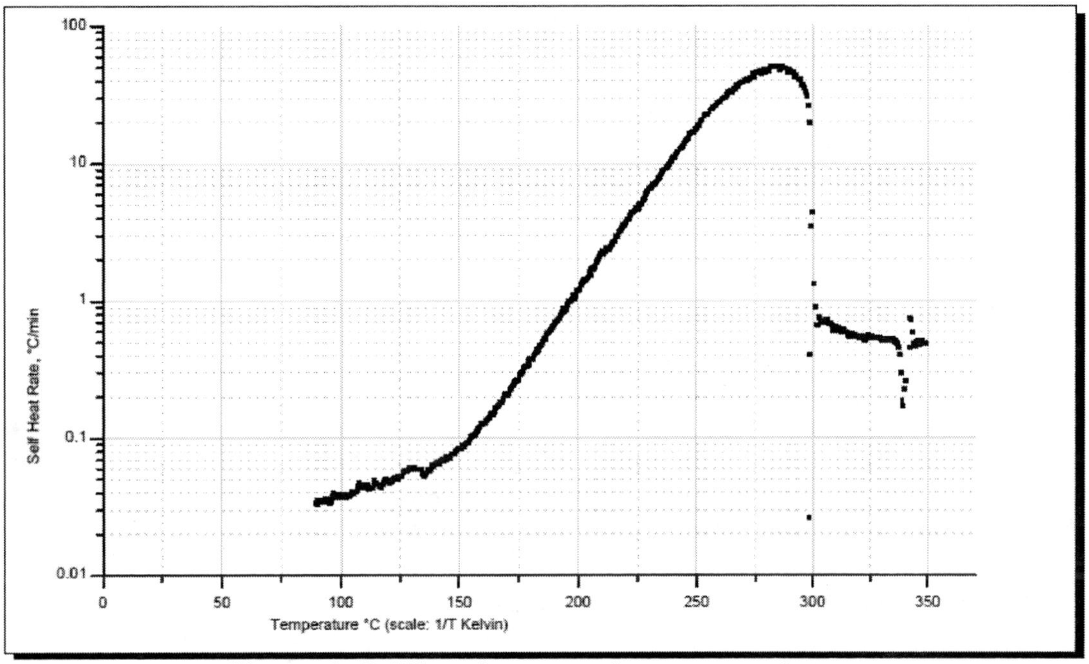

Figure 1: Self heating rate of a process mixture

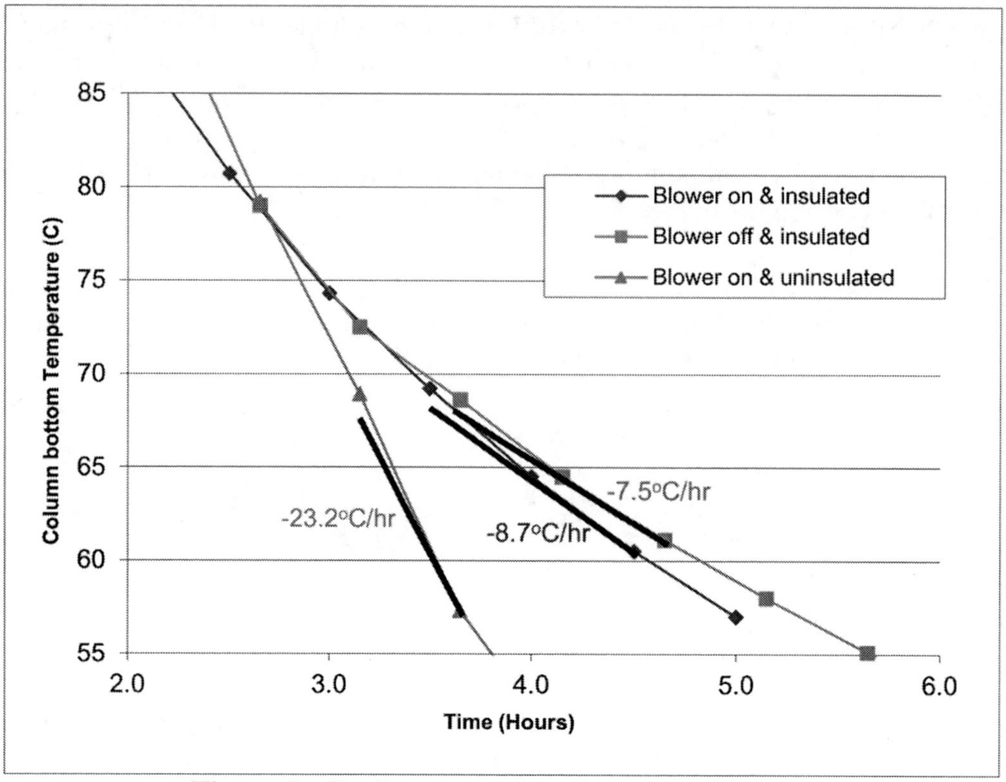

Figure 2: Cooling rate of a pilot plant reboiler

Safety must be a top priority, especially the safety of the people. When dealing with hazardous chemicals, do not take them lightly. Do not take shortcuts to meet the timeline. Conduct pre-startup review and correct deficiencies before starting up the pilot plant.

3.3 Process Samples Analysis and Data quality

One of the purposes of a pilot plant is to generate scale-up data required for designing the commercial plant, in addition to demonstrating the process at the pilot scale. Getting data on process streams composition at key locations in the process is critical.

Identify sampling points in the process during the design phase. Prepare a comprehensive process sampling and analysis plan describing all process sample points, sampling frequency and methods to be used for analyzing the samples. Set-up required analytical capability – set up the hardware such as gas chromatographs, titrators etc. Develop and validate the methods for analyzing the samples before starting up the pilot plant. Define the experiments to be conducted in the pilot plant and have an experiments plan ready.

Extract maximum possible information from each experiment. Each process sample is a result of a real physical phenomenon. Even a sample from a process upset or from a startup or a shutdown period presents an opportunity to learn about the process. It takes a lot of investment of time and resources to design, build and operate the pilot plant. Therefore, it is very important that the experiments are not wasted from lack of inadequate analytical capability.

High quality analysis of the process samples goes a long way in achieving the objectives of the pilot plant. The two graphs in Figure 3 show an example of a high quality analysis from one of the projects I led in the past. The two charts show sample composition accountability for two separate sample points in a continuous glycerine hydrochlorination pilot plant operating 24 hours a day, 7 days a week. The samples from both sample points were analyzed using the same instruments. Each data point in the graphs is a sum of the composition from 3 separate analyses – organic species analyzed on a gas chromatograph (GC), acid by titration and water by Karl-Fisher titration. The sum of the three needed to be 100 ± 5% for the analysis to be valid.

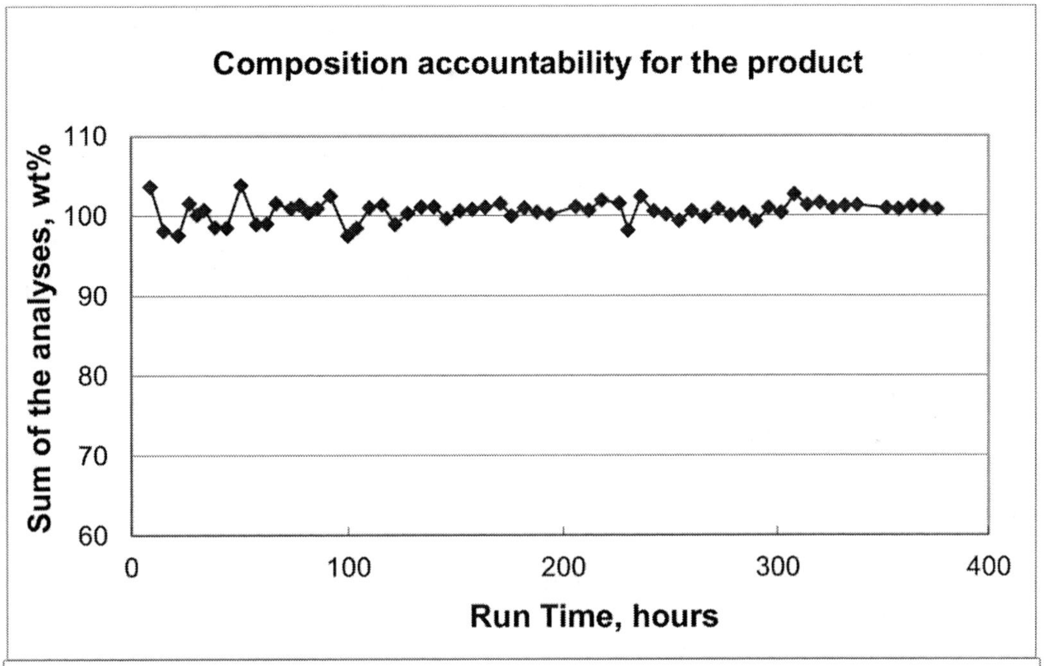

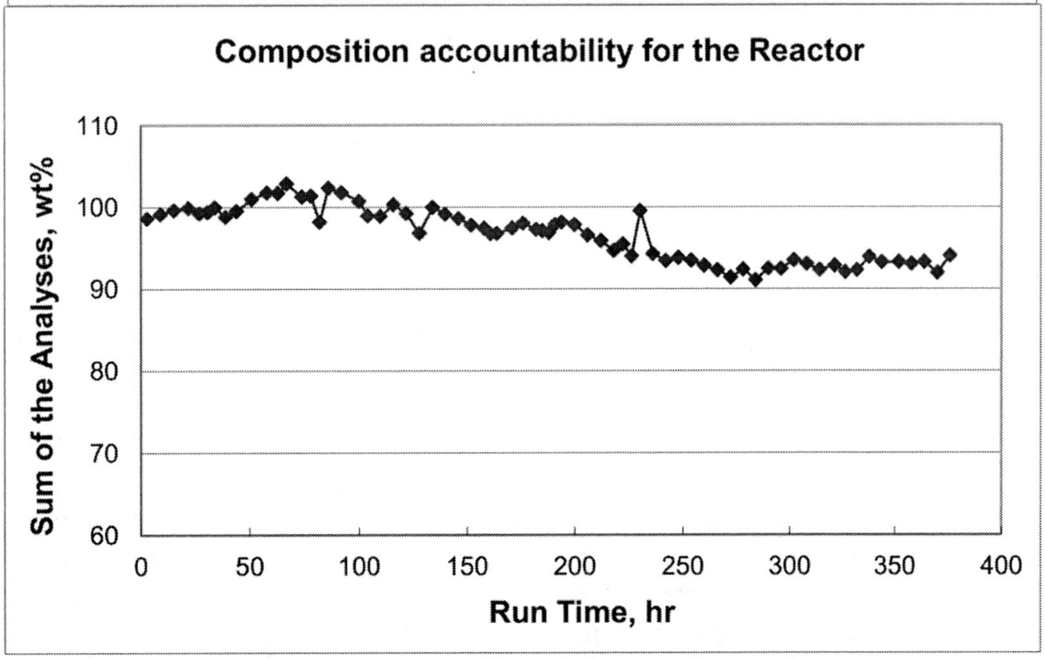

Figure 3: Accountability of the samples analyses

The first chart in Figure 3 which is for the product stream, shows that the sample composition obtained by using the three separate and independent analytical methods was usually within +/-2% for over 350 hours of continuous operation. The second chart which is for a reactor stream shows that the sample composition was within +/-2% for the first 100+ hours of operation and then it gradually decreased to 90-95% range. Excellent accountability for the first sample point confirmed that the three analytical instruments and the analysis methods worked well. The lower accountability later on for the second sample point could not result from the instruments but must have been a result of a real phenomenon. After further investigation, we concluded that the decrease in the accountability was a result of gradual build-up of heavier components in the process which did not elute from the GC column. Thus we were able to learn about the rate and extent of heavies build up in the process as a result of having the high quality analysis of the samples.

3.4 Staffing, training, start-up and operation

Determine in advance the number of persons required to operate the pilot plant, analyze process samples, dispose the waste etc. and staff the project accordingly. Train personnel on the process and samples analysis before starting up the plant. In the case of a complex process, a training program may be needed for the pilot plant operation. Start-up the pilot plant with water if possible and train personnel before introducing chemicals to the process. It may not always be possible to operate the plant with water first.

3.5 Data management- storage, retrieval and learning from the data

Plan ahead for data storage, set up data historian for the process and the analytical data for easy data retrieval at a later date. Analyze the data, learn from the data, refine the process model and the design of the commercial plant. Redo the techno-economic evaluation of the refined commercial process and decide accordingly. This is usually a major decision stage for the project on whether to commercialize the new process.

3.6 Document

It is very important to document in detail all significant aspects of project. Document design and construction of the pilot plant including the reasons for the choices made. Document the data generated, analysis of the data and learning. Document any model refinement achieved as a result of the new data and the results of the refined techno-economic evaluation of the process.

4. Conclusion

Pilot testing of a new production process is a vital step in scaling-up the process from a small laboratory scale to a commercial scale. There are many aspects of the process requiring attention. An effective strategy with examples has been presented in this paper for avoiding the pit-falls and testing the new process at the pilot scale successfully.

5. References

1. "Pilot plants" Kirk-Othmer Encyclopedia of chemical Technology (2005)
2. Ploufe, Patrick, Arturo Macchi and Dominique Roberge; "Chemical Process Development and Scale-up challenges" in Flow Chemistry, Vol. 1 – Fundamentals (2014)
3. Distillation: Equipment and Processes (2014), Chapter 10: Laboratory Distillation and Scale-up
4. Cusack, Roger; "Pilot Plants confirm Process Validity" Chemical Engineering, June 1998
5. Oxford Dictionary of Chemical Engineering (2014), P-282
6. From past consultations with Dow Chemical's distillation experts
7. From a test conducted in Dow Chemical's reactive chemical laboratory

441847 Design of Multiple Extractors for Optimizing Liquid-Liquid Extraction Processes

Abstract:

Separation processes involving liquid-liquid extraction cannot always be solved by a single, counter current extraction column. In some applications, two columns or a column with mixer/settler in series should be considered and may offer the best solution from a process optimization and/or economic standpoint. Fractional extraction is an obvious process whereby two columns in series can be considered in lieu of a single column. However, other processes such as columns with backwash, extractions with reaction, unfavorable density gradient and presence of solids can also involve more than a single column. In any of these types of processes, pilot plant testing is critical to the optimal design of the system. This paper covers a number of different applications where a single column was considered vs two extractors in series. The results from pilot plant testing are presented along with the eventual design for scale up to the commercial processes.

CARBONYL REMOVAL IN STEAM CRACKERS & FCC UNITS

The Merits of Carbonyl Pre-Treat (CPT) Technology

Dr. Sabah Kurukchi
and
Joseph Gondolfe
Janus Technology Solutions, LLC
Houston, TX (USA)

16AIChE
Spring Meeting, Houston, TX

JANUS

Presentation Outline

- **Statement of Problem Definition**
- **Technology Overview (SC Plus FCCU)**
- **Experimental Results from Absorption**
- **The Perfect Technology Solution for SC**
- **Experimental Results from L/L Extraction**
- **The Perfect Technology Solution for FCCU**
- **Conclusions**

16AIChE
Spring Meeting, Houston, TX

JANUS

Statement of Problem Definition
Oxygenates in Steam Crackers and FCC Units

- Hydrocarbons at high temperatures undergo decomposition to C and H_2
- In the presence of steam, some of the carbon converts catalytically at the operating temperature to CO and H_2 on the Steam Cracker radiant steel coil surfaces containing Ni and Cr.
- Similar reaction mechanism for the formation of CO & H_2 is true within the Fluid Catalytic Cracking Unit (FCCU) over its aluminosilicate catalysts.
- Oxygenates are catalytically formed in SC and FCC Units including organic acids, alcohols and carbonyls in concentrations ranging from 10-1000 ppm depending on the type of feedstock, catalyst, ratio of dispersion steam-to-hydrocarbon (feed) at operating temperature.

16AIChE
Spring Meeting, Houston, TX

JANUS

Technology Overview
SC Plus FCCU Key Operating Parameters

	FCC	DCC	SC
Reaction temp/°C	500–550	530–590	760–870
Reactor pressure/barg	1–3	1–2	1–0
Residence time/s	1–5	1–10	0.1–0.2
Catalyst/oil ratio (wt/wt)	4–8	10–15	–
Dilution steam (%)	1–3	5–30	30–80

16AIChE
Spring Meeting, Houston, TX

JANUS

Effects of Aldol Condensation in Caustic Services

Aldehydes and Ketones (acetone) present in Hydrocarbon streams polymerize by Aldol Condensation in the presence of caustic (acting as catalyst) in the formation of Red Oil polymer.

Red Oil causes fouling of Steam Cracker Caustic Tower and the spent caustic stream.

Red Oil causes fouling of FCC Caustic Contactor and Oxidizer

Red Oil Formation Leads to Fouling

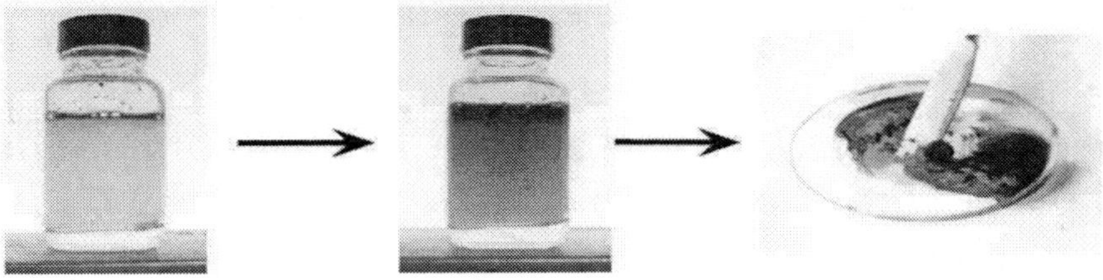

Yellow Oil → Red Oil → Sticky-Viscous Solid Resin

16AIChE Spring Meeting, Houston, TX

JANUS

Caustic Equipment Fouling by Aldol Polymer

Consequences of Red Oil Formation

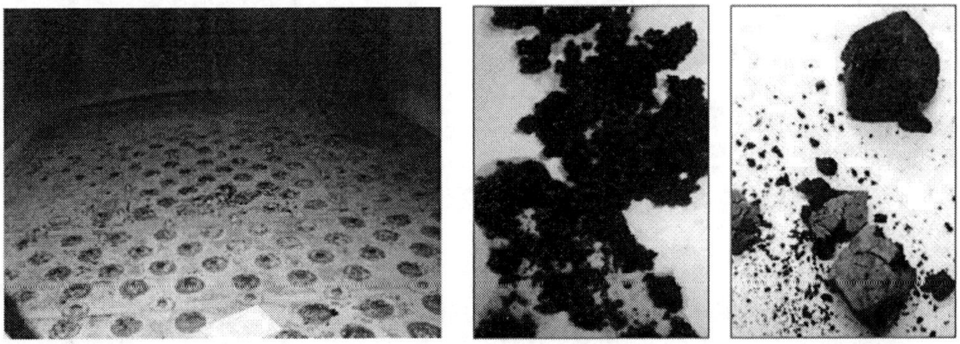

16AIChE Spring Meeting, Houston, TX

JANUS

Preventing Red Oil Polymer Formation in Caustic Towers

- Aldol Polymer (Red Oil) present technologies deal with treating the symptoms by:
  - Filtering the large polymer particles from the recirculating caustic solution
  - Additive(s) addition to disperse the formed polymer

The perfect solution is to prevent Red Oil formation by removing the Carbonyls upstream of the Caustic Contactors.

16AIChE
Spring Meeting, Houston, TX

JANUS

Mechanism of Aldol Condensation Reactions

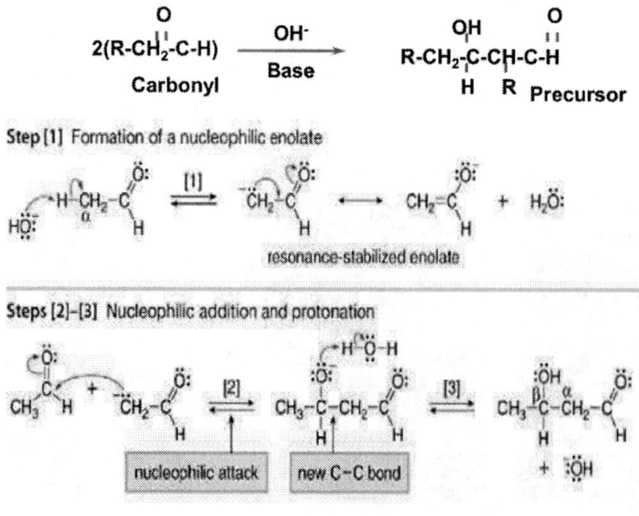

- In Step [1], the base removes a proton from the α carbon to form a **resonance-stabilized enolate.**
- In Step [2], the nucleophilic enolate attacks the electrophilic carbonyl carbon of another molecule of aldehyde, thus forming a new carbon–carbon bond. **This joins the α carbon of one aldehyde to the carbonyl carbon of a second aldehyde.**
- Protonation of the alkoxide in Step [3] forms the **β-hydroxy aldehyde.**

16AIChE
Spring Meeting, Houston, TX

JANUS

Typical CG Caustic Tower Feed Composition

Typical stream contaminants are:

- CO_2 200 - 400 mppm
- H_2S 20 -100 mppm
- Acetaldehyde 100 - 200 mppm
- Acetone 5 - 10 mppm

- Acetaldehyde is very reactive in the presence of caustic solution and react to form Aldol Polymer.
- Acetone is less reactive and present in very low concentrations; thus, cannot promote fouling tendency.

Feed to Caustic Tower		Liquid Cracker	Gas Cracker
Composition, mol %	Comp. MW	mol%	mol%
H2O	18.01	0.400	0.300
HYDROGEN	2.02	17.000	36.000
CO	28.01	0.100	0.054
METHANE	16.04	28.000	7.500
ACETYLN	26.04	0.500	0.166
ETHYLENE	28.05	29.100	32.800
ETHANE	30.07	7.000	21.000
MACETYLN	40.06	0.350	0.006
PROPDIEN	40.06	0.300	0.006
PROPYLEN	42.08	11.000	0.783
PROPANE	44.10	0.600	0.094
13BUTD	54.09	3.000	0.434
C4's	56.11	1.100	0.181
C5's	66.10	0.550	0.177
C6-C8 NONA	84.16	0.400	0.000
BENZENE	78.11	0.500	0.352
TOLUENE	92.14	0.040	0.044
CO2	44.01	0.02 to 0.04	0.02 to 0.04
H2S	34.08	0.002 to 0.01	0.002 to 0.01
Acetaldehyde	44.05	0.01 to 0.02	0.01 to 0.02
Acetone	58.08	0.0005-0.001	0.0005-0.001
Mole Wt		25-26	18.5-19
Flow Rate Ton/h		300-400	300-400
Pressure Bars		14-17	14-17

16 AIChE
Spring Meeting, Houston, TX

JANUS

Experimental Results from Absorption Metal Hydrogen Sulfite Solution Removes Carbonyl from Cracked Gas Stream

Extensive laboratory testing showed Metal Hydrogen Sulfite solution contacted with the cracked gas in absorption tower removes the Aldehydes and Ketones down to negligible ppm levels.

16 AIChE
Spring Meeting, Houston, TX

Carbonyl Removal Packed Absorber Rig Sketch

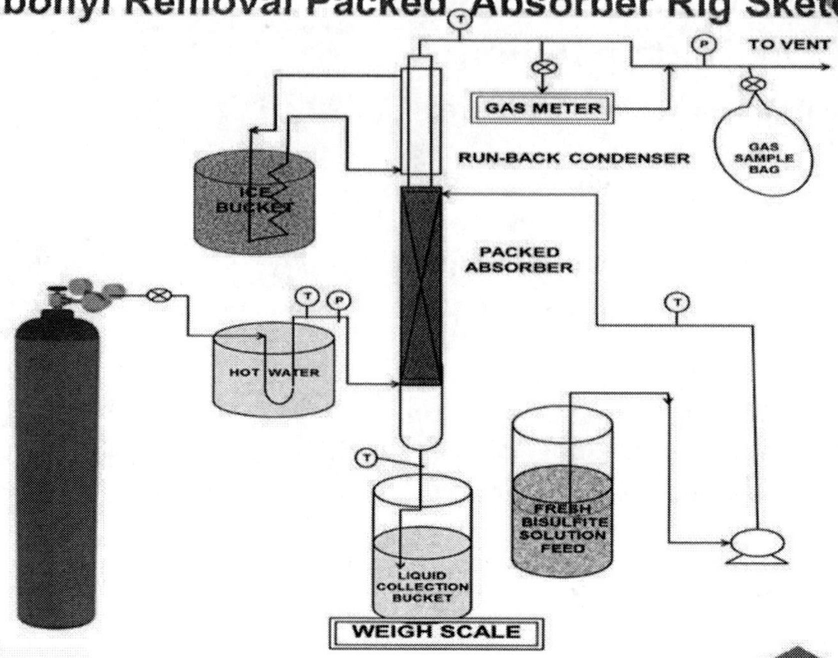

16 AIChE
Spring Meeting, Houston, TX

JANUS

Experimental Carbonyl Removal Absorber Results Summary

Run Set #	Metal Bisulfite wt%	Inlet Carbonyl Conc, wppm	Exit Carbonyl Conc, wppm
1	10	200	<1 (Detection Limit)
2	5	200	<1 (Detection Limit
3	1	200	<3

16 AIChE
Spring Meeting, Houston, TX

JANUS

Carbonyl Pre-Treat (CPT) Technology Fouling Prevention Advantages

- **Pretreating the cracked gas with Metal Bisulfite Solution removes >99% of the Carbonyl.**
- **The amine absorber and /or caustic tower treating the cracked gas will operate without fouling; thus less expensive tower internals can be used.**
- **The spent caustic will essentially contain sodium salts and dissolved hydrocarbons that can be cold stripped with nitrogen or methane to obtain hydrocarbon (HC) free spent caustic.**
- **HC free spent caustic is saleable to Pulp & Paper plants and Leather Tanneries.**
- **HC free spent caustic eliminates fouling in spent caustic treatments:**
  1. **WAO and Catalytic WAO**
  2. **Incineration**
  3. **Deep Neutralization - Stripping**

16AIChE
Spring Meeting, Houston, TX

JANUS

Conventional CG Caustic Treating Diagram

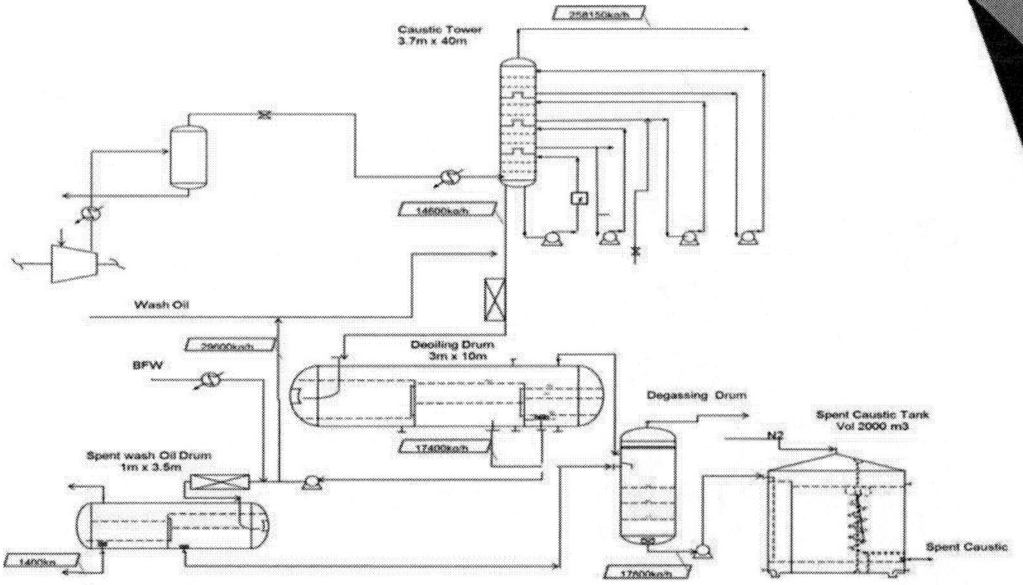

16AIChE
Spring Meeting, Houston, TX

JANUS

The Perfect Solution for SC Caustic Treating with CPT Technology

Eliminate:

- Deoiling Drum
- Spent Wash Oil Drum
- Use of Wash Oil

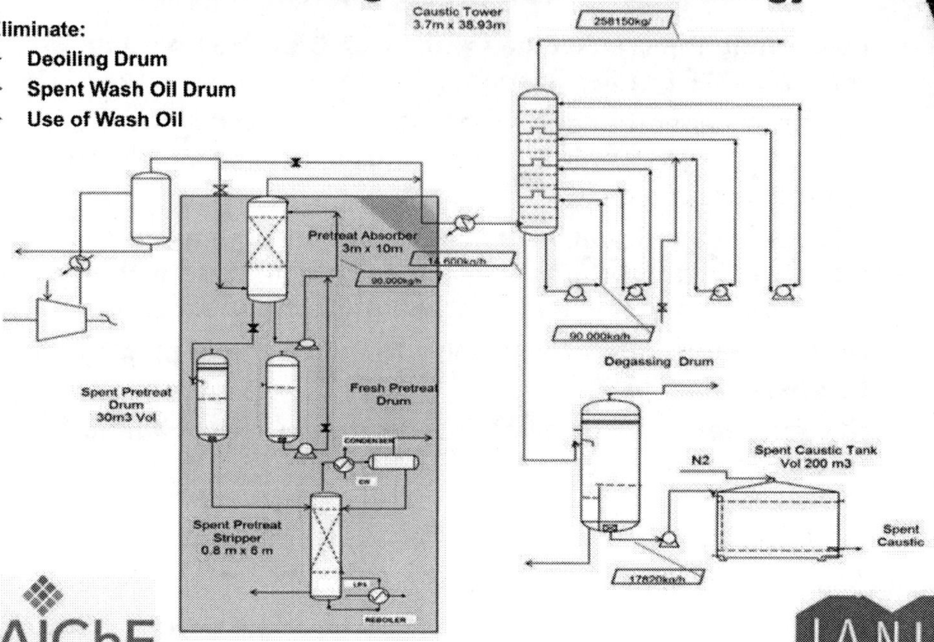

16AIChE
Spring Meeting, Houston, TX

JANUS

Experimental Results from L/L Extraction Metal Hydrogen Sulfite Solution Removes Carbonyl from FCC LPG Streams

Extensive laboratory testing showed Metal Hydrogen Sulfite solution contacted with the Liquid HC stream in L/L Extractors removes the Aldehydes and Ketones down to negligible levels.

16AIChE
Spring Meeting, Houston, TX

JANUS

Carbonyl Removal Packed L/L Extractor Rig Sketch

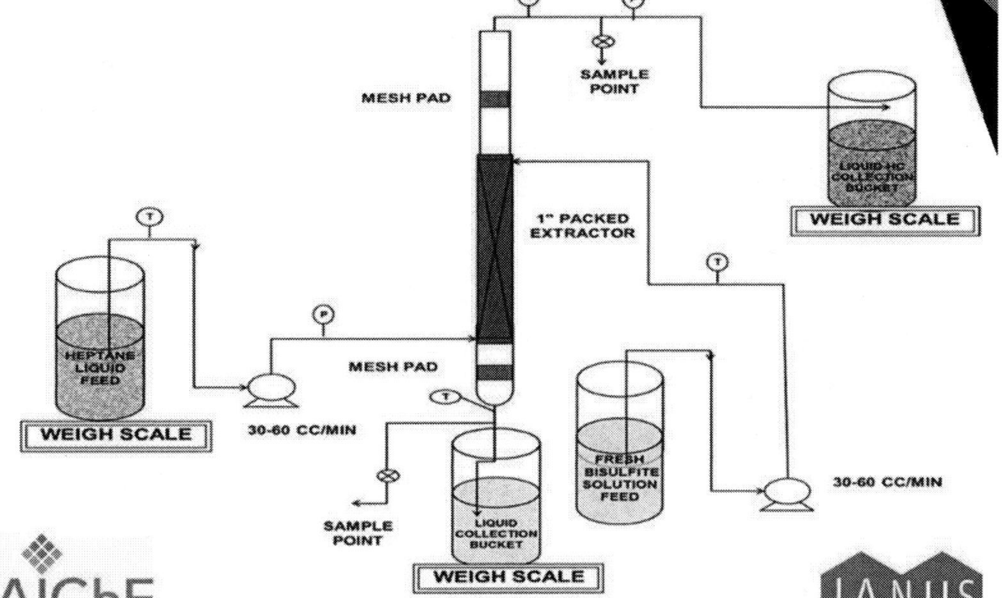

16AIChE Spring Meeting, Houston, TX

JANUS

Experimental Carbonyl Removal L/L Extractor Results Summary

	Wt% MHS Conc.	Inlet Acetaldehyde Conc. wppm	Inlet Acetone Conc. wppm	Outlet Acetaldehyde Conc. wppm	Outlet Acetone Conc. wppm
1	10	200	200	<0.1	0.5
2	10	506	496	<0.1	0.6
3	1	200	200	0.3	9
4	1	500	500	<0.1	10

16AIChE Spring Meeting, Houston, TX

JANUS

The Perfect Technology Solution for FCCU LPG Treating with Carbonyl Pretreat Unit

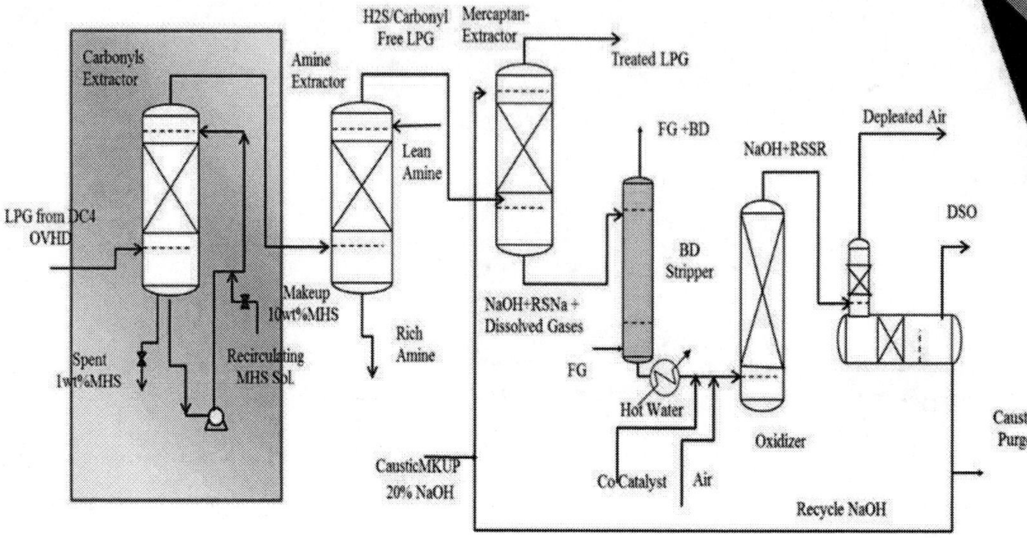

FCCU LPG Stream Carbonyl Pre-Treat (CPT) Fouling Prevention Advantages

- **Pretreating the LPG stream with Metal Bisulfite Solution removes >99% of the Carbonyl.**
- **The amine Extractor and /or caustic RSH Extractor treating the LPG stream will operate without fouling; thus more efficient sulfur Removal is achieved**
- **The treated LPG will be free of carbonyls and with lower sulfur content.**
- **Carbonyl free LPG will meet requirements of :**
  - **Alkylation Feed**
  - **Hydrogenation Feed**

16AIChE Spring Meeting, Houston, TX

JANUS

Conclusions

- Oxygenates are formed in both steam crackers and FCC units.
- Carbonyls in steam crackers and FCC Units once contacted with a caustic solution form RED OIL that foul both process equipment and render caustic and spent caustic streams to be "highly fouling" by their continued reactivity in solution.
- Removal of carbonyls from hydrocarbon streams can be achieved by washing with metal bisulfite solution.
- Innovative "Patent Pending" Cost Effective methods have been devised to totally remove the carbonyls from hydrocarbon streams upstream of caustic contacting equipment in an Carbonyl Absorber or Carbonyl L/L Extractor.
- CPT Technology provides a "perfect solution" to eliminate problematic Red Oil formation in Steam Crackers and FCC Units.

16AIChE Spring Meeting, Houston, TX

JANUS

CARBONYL REMOVAL IN STEAM CRACKERS & FCC UNITS

The Merits of Carbonyl Pre-Treat (CPT) Technology

Contact Information

JANUS

Dr. Sabah Kurukchi, VP & CTO
JANUS Technology Solutions, LLC

Tel (W): 713-7836128
Tel (M): 713-203-8746
Email: skurukchi@janusts.com
5820 Doliver Drive
Houston, Texas 77057-2455 (USA)

16AIChE Spring Meeting, Houston, TX

JANUS

444758 Combustion Pyrolysis of Light Hydrocarbons

Abstract:

Combustion pyrolysis of light hydrocarbons

Balamurali Nair, Pankaj Gautam, and Sreekanth Pannala

SABIC technology Center, Sugarland TX-77478

The growing abundance of natural gas resources and their competitive advantage with respect to other light paraffinic hydrocarbons make the production of chemical intermediates and other high value chemicals from natural gas an attractive option. At SABIC (Saudi Basic Industries Corporation), we are committed to efficient and optimal utilization of natural resources. One method to convert the natural gas resource to high value products is through combustion pyrolysis. Combustion pyrolysis involves partial oxidation of hydrocarbons to produce smaller unsaturated molecules which can be used as building blocks for commodity chemicals. In addition, combustion pyrolysis can also be used to upgrade low value off gases/ feed streams to produce valuable products instead of using these streams as fuel or for flaring. Although combustion pyrolysis technology is five decades old, it is not widely practiced commercially due to several reasons including operational and scalability challenges. The homogeneous nature of the reaction without catalysts, simple reactor design and favorable economics could make this process attractive in the coming decades. In this work, a brief review of the existing combustion pyrolysis technology followed by experimental results on a pilot unit will be discussed.

The performance of a two stage combustion pyrolysis pilot unit for the production of chemical intermediates acetylene and ethylene- from a variety of hydrocarbon feeds is elucidated here. Combustion pyrolysis uses oxy-fuel combustion for heat generation. The extreme temperatures prevalent in the combustor (2500-2800 C) make the process development challenging. The hot gases from the combustor are mixed with the reactor feed and fed into a reactor where pyrolysis reactions takes place. The important parameters which determine the performance of a combustion pyrolysis unit are the combustor firing rate, residence time, and type of feedstock. Some of the key aspects of this technology include feed flexibility and modularity. The process development challenges include the heat management, combustor material stability, and scalability to industrial sizes. The variation of acetylene and ethylene yield with firing rate, residence time and type of feedstock will be discussed. The results of piloting help to scaleup the process to commercial level.

444990 Utilizing Pilot Plants for Process Development to Bridge the Gap Between Laboratory and Industry

Abstract:

Pilot plants have historically been constructed to manage the risk of scaling up commercial plants using laboratory data or simulated models. Pilot plant operating data can be correlated to process specific scale-up factors to be used in modeling as a more robust tool for predicting commercial yields. The capital investment to construct and operate a pilot plant is substantially less than a commercial demonstration plant, which allows systematic operational changes to pinpoint the desirable operating ranges. This presentation will illustrate real-world examples of how to bridge the gap between laboratory testing and commercial plants, showing the need for a variety of technical expertise, relatively low capital investment, and site infrastructure.

445110 Energy and Hydrogen Systems REAL Time Optimization

Abstract:

Hydrogen management can have a significant effect on refinery utility supply through the integration with the energy system. Real-time optimization of hydrogen production in conjunction with the steam, power and fuels can yield significant savings opportunities for the refining industry.

Hydrogen is required in refining processes to produce low and ultra-low sulfur fuels. Hydrogen sources can be external or internal, of course. External sources are generally nearby third party industrial gas providers, who supply pure hydrogen streams, sometimes in exchange of fuel gas (FG), liquid petroleum gases (LPG) or natural gas (NG) for either or both feed and fuel. In some cases, impure hydrogen can also be provided by external sources, which can be purified in the refinery in the Pressure Swing Absorption (PSA) units, producing FG as a by-product.

Many refineries operate internal high purity hydrogen production units, generally based on gas reforming, using either NG or FG as feedstock. Gas reforming processes are endothermic, requiring FG for heating the reactors, and use steam as diluent of the feed and power to drive the feed charge and product compressors, which can be either electric or steam driven. The Steam Methane Reforming (SMRs) units usually produce steam and FG as byproducts.

When FG is used as a feedstock, critical decisions must be made about the best use of sources of the FG streams within the refinery units. When NG is used as a feedstock, many times it is considered a petrochemical feedstock rather than a fuel, having an incremental cost that could depend on its final use.

Any decision to import or internally produce hydrogen will affect not only the hydrogen system purity and availability, but also the FG system in addition; affecting the volumes of any externally purchased or internally supplied fuels (like NG or LPG).

Any attempt to reduce the costs from the utilities side (i.e., to optimize the energy system) should be based on accurate utility system models including steam, fuels, and power and calibrated with validated and consistent set of measurements. In order to consider the interaction of the hydrogen with the fuel system, a compositional model of the fuel must be included in the model. Solving and optimizing the energy and hydrogen balances at the same time is key to ensuring consistency with operations and constraints handling.

This paper describes how an integrated model optimizing the energy and hydrogen systems altogether has been implemented in real industrial environments.

444795 Optimization of Refrigeration and Storage Systems for Propane and Butane Export

Abstract:

The shale gas boom has changed the energy landscape, not only in the U.S. but also beyond our borders and is impacting World Markets. With the advent of horizontal drilling and "fracking" in North American Shale Plays, the U.S. has become a net exporter of natural gas liquids (NGL's), including propane and butane.

This presentation provides an overview of studies which compare options for refrigerating and storing propane and butane for loading on to ocean-going vessels for export from the U.S. The methods evaluated include using a single core-in-kettle chiller, phased chilling using multiple core-in-kettle chillers, an open loop refrigeration design and combining propane and butane chilling into a single refrigeration system. Storage arrangements such as off-site cavern storage and on-site refrigerated storage also have an impact on refrigeration system design, operation and economics. Very large amounts of power and energy are required for these chilling cycles and the options involved. This paper includes a summary of the optimal designs resulting from the studies, including an overview comparison of equipment costs, energy costs, and payback periods.

444569 Utilizing Vent Gases from Chemical Storage and Transfer to Produce Power and Heat

Abstract:

Rather than using a traditional thermal abatement methods for destruction of various chemicals that vent during storage or transfer, the Ener-Core PowerStation can utilize these gases to generate power and recover heat. The Ener-Core PowerStation's destruction efficiency is >99% and has virtually zero NOx emissions.

444065 Flowsheet Development and Simulation on the Use of Deep Eutectic Solvents for Carbon Dioxide Capturing

Abstract:

Capturing and restoring of carbon dioxide (CO_2) is getting growing attention as a potential method for regulating greenhouse gas emissions. Carbon dioxide content in natural gas can cause corrosion and other environmental issues. Low cost ionic liquid analogues, namely, deep eutectic solvents (DESs), have attracted more attention for use in a diversity of applications. DESs exhibit many favorable properties, such as availability, non-toxicity, biodegradability, recyclability, non-flammability, and low price. The use of ionic liquids for carbon dioxide capture is considered cost-effective because it requires less energy for solvent recovery compared to other conventional processes. In this work, an attempt to study the use of ionic liquids (ILs) for CO_2 capturing in typical absorption/stripper processes is considered. The attempt is tackled through simulation of a typical treatment system where ILs are introduced and circulated. The flowsheet simulation based approach implemented in Aspen HYSYS process simulators, is used to study the effect of temperature, pressure, adsorbent concentration, energy demand and flow rate on the CO_2 removal performance using the proposed system. The CO_2 content in natural gas is within the range of local industrial companies located in Qatar. The system is simulated using the most recent Acid Gas removal technology introduced in Aspen HYSYS simulation with closed loop circulation of the absorbent. The developed system is going to be validated against conventional amine system and experimental ILs data. The preliminary results showed a promising reduction on the required energy duty compared to Amine based treatment while higher purity is achieved.

440908 Chill More with Less and Properly Control Your Mechanical Refrigeration Unit (MRU) via Dynamic Simulation

Abstract:

Use Dynamic Simulation to Validate Your Process Control of a MRU Gas Processing Plant

There are many misconceptions about the closed refrigerant loop widely used in Mechanical Refrigeration Unit (MRU) to extract Natural Gas Liquid (NGL) from natural gas. Two common mistakes are: 1) How can I control the refrigerant compressor discharge pressure? 2) Why can't I control the liquid level in the receiver? The best way to answer these kinds of questions is to model the MRU in a dynamic process simulator. The dynamic simulation allows users to operate the virtual MRU plant without worrying about shutting down the real plant. From the dynamic simulation, users quickly learn that the compressor discharge pressure is actually controlled by the ambient temperature and/or the fans speed. The reason why the liquid level can't be controlled is because, 1) the total mass (not mass flow) is fixed in the closed loop, 2) the amount of mass in liquid form is controlled in the chiller and the economizer if any; thus the liquid level in the receiver is dependent on the chiller and economizer level.

Chill More with Less by Optimizing Set Points

Further, Dynamic Simulation is able to help to extract more NGL with less energy consumption. Most operators operate the plants with the fixed setpoints given by the plant equipment provider. However, this dynamic simulation demonstrates that it is possible to achieve a lower cold separator temperature while using the same electricity consumption or less by optimizing the setpoints.

Summary

Dynamic Simulation can be used to validate the process control during the design phase and to optimize the plant output while under operation.

Energy optimisation in natural gas processing plants

Leandro Labanca, Process Integration Ltd.
Yuhang Lou, Process Integration Ltd.

Abstract

Natural gas processing plants require a series of process units to remove contaminants such as water, carbon dioxide and hydrogen sulphide. These units have different process heating and cooling demands. Proper matching of the heating and cooling requirements leads to significant operating cost reductions. Excessive heat from process units can be recovered by producing steam, which may be used for satisfying heating requirements of other units or for power generation.

Systematic procedures for energy optimisation in natural gas processing plants are introduced and discussed during the paper. Novel software design tools with embedded state-of-the-art process integration technologies are presented for the simulation and optimisation of natural gas processing plants.

An industrial case study is presented to illustrate the benefits of performance optimisation both on the individual processes and on the overall steam system simultaneously. Eleven cost-effective projects were identified to save energy consumption in individual process units through enhanced heat recovery and process modifications. The influence and impact of the individual energy saving projects on the utility system was evaluated effectively with an integrated process / utility system model. Such a systematic and integrated energy optimisation approach significantly reduced utility consumption and improved the overall economics of the natural processing plant. Typical range of Internal Rate of Return (IRR) achieved for these energy saving projects were found to be between 25% - 180%.

The paper is extracted from a publication in GPA Europe's AGM and Technical Meeting, 2015 [7].

Introduction

Natural gas has been playing an increasingly important role in the past few decades as a clean and safe source of energy. Natural gas from underground has to be treated to remove impurities such as water, carbon dioxide and hydrogen sulphide before it is transported to end consumers. Separation of heavy hydrocarbons and sulphur recovery are often employed in natural gas processing plants to take advantage of the commercial value of by-products.

A typical natural gas processing plant consists of inlet facilities, an acid gas removal unit, a dehydration unit and other supporting process units. In all these process units, varieties of process heating and cooling are required. For example, in the acid gas removal unit, process heating is required at the bottom of the regeneration column, whilst the lean amine stream from the reboiler requires significant cooling before being sent to the absorption column. These process heating and cooling requirements are met by process-to-process heat exchangers or by heaters and coolers connected to site utilities.

If excess heat is available in an individual unit, steam could be produced to move heat across different process units and used for process heating or power generation purposes. In order to identify the potential for indirect heat recovery in the site, energy performance evaluation on a site-wide level, Total Site Analysis, is useful to review the steam balance and estimate the potential for power generation. Selection of mechanical drivers, including steam turbines, direct drive gas turbines or electric motors, has a direct impact on the ratio of heat and power supply. All these areas for utility system design should be carefully reviewed and evaluated systematically to achieve the best energy performance.

Process Integration Ltd. (PIL) has developed state-of-the-art software design tools, i-Heat™ and i-Steam™, for systematic energy optimisation of Heat Exchanger Networks (HENs) and utility systems, respectively. Advanced process integration technologies embedded in both software tools can be applied to effectively and efficiently synthesise a new process design or to identify and evaluate cost-effective retrofit options for an existing process. A recent heat integration and energy optimisation study for a natural gas processing plant using these novel tools is presented in this paper for demonstration purposes.

HEN design and optimisation with i-Heat™

Environmental policies and high energy prices have led companies to search for options to reduce their energy consumption as well as their carbon footprint. The optimal design of HENs may improve heat recovery and, hence, reduce the operating costs. Increasing heat recovery often requires an increase in the heat transfer area. Generally, there are economic trade-offs between the operating costs and the capital costs when designing and retrofitting HENs [1].

It is common to find options for process changes in a production process to unlock opportunities to reduce the operating and capital costs of the HEN. At this point, the user might consider reviewing the design and operation of other sections of the process before considering the HEN design further.

Across the year, the process is subjected to different operating schemes due to variations in the operation of other units or fluctuations in the ambient conditions. The HEN has to be designed to handle these variations and guarantee continuous operation of the unit [2].

To summarise, a good HEN design must achieve high energy recovery whilst balancing its capital investment. In addition, the HEN has to present flexibility to adapt to different operating scenarios [3]. Therefore, a HEN model needs to be set up to analyse different configurations and accurately evaluate its operating and capital costs.

1. Standard procedure for HEN design and optimisation

The following systematic procedure is used by PIL to design or retrofit a HEN.

- Step 1: Data collection

Base case and detailed scope of work are defined and agreed with the client. The relevant design or operating data from the process as well as the economic information for the study are collected. Different operating scenarios to be considered during the study are also agreed.

- Step 2: Stream data extraction

Stream's heating and cooling demands (duties, supply and target temperatures) are extracted from the collected documents for the process and entered in i-Heat™.

- Step 3: Establish HEN

Unit models such as heat exchanger, splitters and mixers are used construct new or existing HEN in i-Heat™.

- Step 4: Pinch Analysis

i-Heat™ performs pinch analysis [4] to identify energy targets for maximum heat recovery and minimum utility requirements at the selected minimum approach temperature for existing or new HENs.

- Step 5: Identify sources of inefficiency in the HEN

A suite of analysis tools are available in i-Heat™ to evaluate the energy performance of a HEN and identify the bottlenecks in the network. Sources of thermal inefficiencies, such as cross-pinch heat transfer, are identified and quantified automatically by the software tool.

- Step 6: HEN Optimisation

Operational optimisation of an existing HEN is carried out by adjusting operational variables, such as flows in stream-splits and duties in the exchangers, to achieve the minimum energy consumption without additional capital investment.

i-Heat™ also considers a combination of operational changes and structural modifications to automatically resolve the performance bottlenecks which limit energy recovery. Different retrofit designs are obtained based on maximum additional area and allowed locations of new heat exchangers defined by the user.

For a new design, i-Heat™ automatically synthesises cost-optimal heat exchanger networks whilst satisfying design, practical and operational constraints set by the user.

- Step 7: Economic analysis of retrofit options

Economic analysis is performed by i-Heat™ to quantify the trade-offs between the reduced energy costs and the additional capital investment. The user can compare the alternative retrofit designs based on overall costs to identify the best HEN configuration for the process.

Utility system design and optimisation with i-Steam™

Most production industrial sites possess significant heating and cooling requirements for their normal operation [5]. Most of the heating needs of various processes is satisfied by the same utility system which distributes steam around the site at several pressure levels. The steam system is also vital in meeting the power demands within the site [3]. Thus, the main aim of utility systems is to convert fuels to steam and electricity and satisfy process energy requirements within a production site.

There are two main scenarios to study and optimise site utility systems during the design stage: new design and retrofit design. A grassroots utility system may be needed for a site. Modifications to the configuration of an existing utility system might be required to reduce operating costs or to deal with changes in heat and power demands. At this stage, several decisions need to be addressed that will have a direct impact on the performance of the final design.

A number of configurations can be considered to satisfy the steam and power requirements of a site. For instance, boilers may be considered to generate high pressure steam, whilst steam turbines may be used to generate power through steam expansion. Alternatively, gas turbines may be employed to produce power whilst the exhaust gases could be sent through a Heat Recovery Steam Generator (HRSG) for steam production. In general, the efficiency of the utility systems may be increased with higher capital investment being considered. The user will need to identify the optimal configuration to satisfy the site requirements within the client's budget.

Process operating conditions change throughout the year for various reasons (e.g. start-up and shut-down of different units and planned maintenance cycles). In some regions, ambient conditions vary across the year or even across the day, which influences the operating conditions of processing units. Utility systems have to be prepared to cope with these variations to guarantee continuous operation of the site. The penalty for ensuring the integrity of the site operation is that the utility systems are generally over-designed and present significant redundancy in place. The redundancy also provides a number of degrees of freedom to improve the performance of the system [3].

Large and small process machines on a site need a mechanical driver (electric motor, steam turbine or other driving device) for their operation. The user should evaluate which is the best option for each particular driver taking into consideration that the selection of drivers will have a direct effect on the site requirements of steam and power.

To summarise, utility systems must be designed to cope with steam and power demands of the site whilst achieving the highest efficiencies with the lowest possible economic cost. To obtain the best configuration to accommodate the site needs and carry out a detailed economic evaluation of the diverse designs, an accurate optimised model should be developed [3].

1. Standard procedure for utility system design and optimisation

The following is the systematic procedure used by PIL to design a new utility system. A similar procedure is followed when analysing and optimising an existing utility system.

- Step 1: Data Collection

Detailed scope of work is defined and agreed with the client. A study basis is set which includes the definition of the base case. The relevant design or operating data and the economic information for the study are collected from the site.

- Step 2: Identify and understand utility demands

Hot and cold utility requirements are identified by reviewing the design documents of individual processes. Utility duties as well as the required temperatures for each utility requirement are extracted for energy performance analysis.

- Step 3: Definition of multiple scenarios

A number of scenarios considering various process operating conditions and seasonal ambient variations are defined to ensure the utility system can cope with different energy demands across the year.

- Step 4: Total Site Analysis

Utility demands are extracted from the processes and are used to construct the total site composite curves in i-Steam™. i-Steam™ contains sophisticated targeting tools to define the best energy performance for a site under a range of different operating scenarios. Total Site Analysis helps to identify heat and power co-generation targets and optimal steam header conditions.

- Step 5: Utility Systems Modelling

A comprehensive steam system model is produced in i-Steam™ to represent the major equipment items, such as boilers, steam turbines and gas turbines, and their interactions. Modelling of the steam network and the individual equipment items is based on thermodynamic principles and the overall heat and material balance of the system. Part-load performance models for key equipment are established [6].

- Step 6: Utility Systems Optimisation

Considering the multiple scenarios defined, operational optimisation is carried out by adjusting operational variables, such as steam distribution or steam header pressures, to explore the potential for performance improvement of the existing utility system. i-Steam™

automatically selects between electric motors and steam turbine drivers for all power demands within the utility system taking into consideration operating conditions and market prices.

Structural changes are also analysed in order to improve the flexibility of the system as well as to adjust site heat to power demand ratio. Finally, the overall steam network is verified for performance, flexibility and economics.

Energy optimisation studies in natural gas processing plants

PIL follows standard procedures to ensure high quality designs are produced in its energy optimisation studies. These procedures cover every aspect of the study from data extraction to process modelling and optimisation. Multiple layers of checking ensure that the energy study meets PIL's quality standards and every potential design improvement is captured during the heat integration and energy optimisation study.

- Step 1: Data collection

Obtain process design information from the client, including Basis of Design document, process description, Process Flow Diagrams (PFDs) and Heat and Material Balance (H&MB).

- Step 2: Data review and reconciliation

Review and cross check process design information from various sources. Discuss with the client any inconsistent data items and agree on priority for different data sources.

- Step 3: Process unit screening

Review and select those process units containing heat exchangers and therefore providing opportunities for design improvement.

- Step 4: Heat integration study for individual process units

Standard procedure for HEN design and optimisation is followed in the heat integration study for individual process units. As discussed in the previous sections, the key steps include stream data extraction, establish HEN in i-Heat™, pinch analysis, identify sources of inefficiency in the HEN, HEN optimisation and economic analysis of retrofit options.

- Step 5: Evaluate the potential for cross-process heat recovery

Stream data from multiple process units is analysed simultaneously to exploit the potential for cross-process heat recovery.

- Step 6: Energy optimisation for site-wide utility system

Standard procedure for utility system design and optimisation is followed in the energy optimisation for site-wide utility system. As discussed in the previous sections, the key steps include data collection, identify and understand utility demands, definition of multiple scenarios, total site analysis, utility systems modelling and optimisation in i-Steam™.

- Step 7: Economic evaluation of combined improvements

1) Verify the recommended retrofit option from each heat integration study for practicality and feasibility.

2) Establish new steam balance in i-Steam™ taking into consideration the recommended retrofit options from the heat integration studies.

3) Adjust the operation of the utility system to meet the new steam supply and demand.

4) Verify the operation of the utility system under different operating scenarios to guarantee the reliability of the site.

5) Evaluate the economic benefits with all recommendations combined.

Case study

PIL completed a heat integration and energy optimisation study for a new design of a natural gas processing plant. The natural gas processing plant imports steam at various pressure levels from a co-generation plant nearby. PIL was asked to identify opportunities to improve energy efficiency of the existing process design and exploit the interactions between the natural gas processing plant and the co-generation plant to minimise the overall life cycle cost.

- Step 1: Data collection

Various design information was provided by the client to support the heat integration and energy optimisation study.

- Step 2: Data review and reconciliation

Since data discrepancies were observed in some cases, a hierarchy of data sources was agreed between PIL and the client for the project to proceed further.

- Step 3: Process unit screening

PIL received the requested design information for twenty-three process units in the natural gas processing plant. After reviewing the process design data, PIL identified those process units containing heat exchangers and, therefore, providing opportunities for heat integration. Thirteen out of twenty-three units were selected for the energy optimisation study.

Three out of thirteen process units are discussed to demonstrate the benefits of heat integration for design and performance improvement using i-Heat™.

- Step 4: Heat integration study for individual process units

Unit 1

Unit 1 is used as an example to illustrate the standard procedure for a heat integration study. This unit has three hot streams and four cold streams. Process stream conditions and heat exchange duties were extracted from the design information and were entered in i-Heat™, as shown in Table 1. It was agreed with the client that heat loss during heat exchange was not considered as part of the study.

Table 1. Stream data summary for Unit 1

	No.	Stream	Segment	TS [°C]	TT [°C]	MCp [kW/°C]	DH [kW]	DTmin [°C]	HLoss [%]
1	1	Hot Stream 1	1	139.4	37.8	26.22	2666.1	Global	0.0000
2	2	Cold Stream 1	1	4.4	50	58.52	2666.1	Global	-
3	3	Cold Stream 2	1	80	90	186.54	1865.4	Global	-
4	4	Cold Stream 3	1	86.7	139.4	396.95	20950.2	Global	-
5	5	Hot Stream 2	1	110.6	60	385.51	19489.6	19	0.0000
6	6	Hot Stream 3	1	136.9	60	39.18	3012.2	19	0.0000
7	7	Cold Stream 4	1	137.2	147.2	245.2	2452	Global	-

Figure 1 shows the existing heat exchanger network design of Unit 1 in i-Heat™.

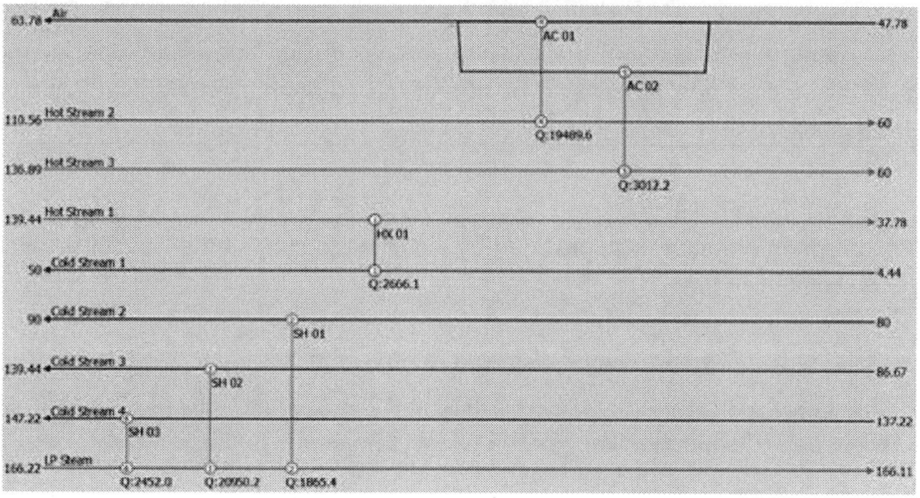

Figure 1. Existing heat exchanger network design for Unit 1. Duties (Q) are in kW and temperatures are in °C

There is one heat exchanger HX 01 for process heat recovery in this unit. Three cold process streams are heated by LP Steam. Two hot process streams are cooled down in air coolers to their target temperatures. The utility consumption of the existing design is summarised in Table 2.

Table 2. Utility consumption of the existing design (Unit 1)

Utility Type	Utility	Heat Exchanger	Duty (kW)
Hot	LP Steam	SH 01	1,865
		SH 02	20,950
		SH 03	2,452
		Total	25,267
Cold	Air	AC 01	19,490
		AC 02	3,012
		Total	22,502

Pinch analysis was carried out in i-Heat™ to identify the minimum energy targets for both hot and cold utilities. Temperature vs. Enthalpy segments for hot and cold process streams were combined to generate the hot and cold composite curves, as shown in Figure 2. The hot and cold temperature pinch points are represented by the red and blue dots in the diagram.

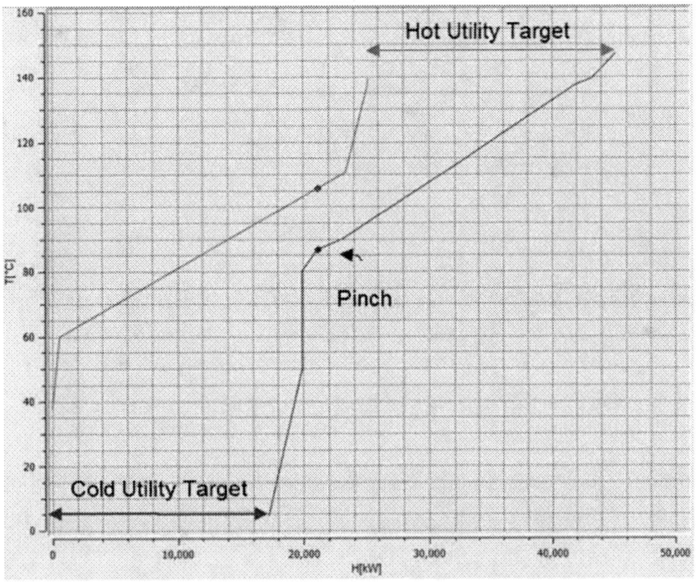

Figure 2. Composite curves for Unit 1

The global minimum approach temperature (minimum DT) at the thermodynamic pinch is set to 19°C. The corresponding energy targets were obtained from i-Heat™ and compared with current consumption, as shown in Table 3.

Table 3. Energy consumption and target for Unit 1 with minimum DT = 19°C

	Current Consumption	Energy Targets	Difference
Hot Utility Duty [kW[	**25,267**	**20,030**	**5,237**
Cold Utility Duty [kW]	**22,502**	**17,265**	**5,237**

By comparing the utility consumption of the existing design against the minimum energy target, thermodynamically (without any constraints), there is the opportunity to reduce hot and cold utility consumption by 5,237 kW.

Figure 3 shows the analysis of cross pinch heat transfer on the existing design. Heat exchangers that have cross pinch heat transfer are represented with an oblique line that crosses the pinch. Four heat exchangers were found to have cross pinch heat transfer, as shown in Figure 3. These heat exchangers are the source of thermodynamic inefficiency in the unit.

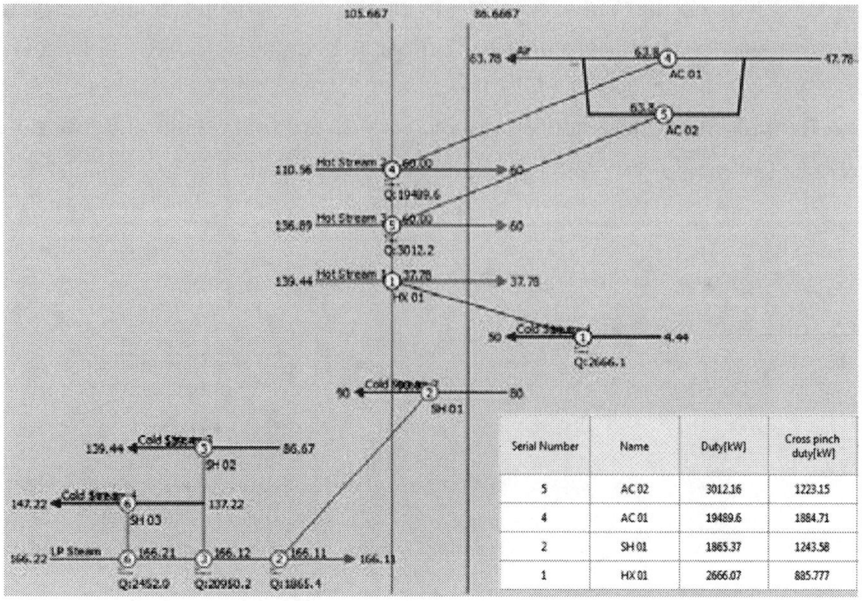

Serial Number	Name	Duty[kW]	Cross pinch duty[kW]
5	AC 02	3012.16	1223.15
4	AC 01	19489.6	1884.71
2	SH 01	1865.37	1243.58
1	HX 01	2666.07	885.777

Figure 3. Cross pinch heat transfer analysis for Unit 1
Duties (Q) are in kW and temperatures are in °C

Design improvements on this unit could be achieved by reducing cross pinch duties within the existing HEN. An energy saving project, PROJECT1, was proposed by PIL to reduce the cross pinch heat transfer duty in SH 01.

As shown in Figure 4, a new heat exchanger named 'New HX 01' was added in the design to recover heat between Hot Stream 2 and Cold Stream 2 which results in a reduction of LP Steam consumption by 1,865 kW. As a result, the LP steam heater SH 01 is no longer required and the duty of the air cooler AC 01 is reduced by the same amount.

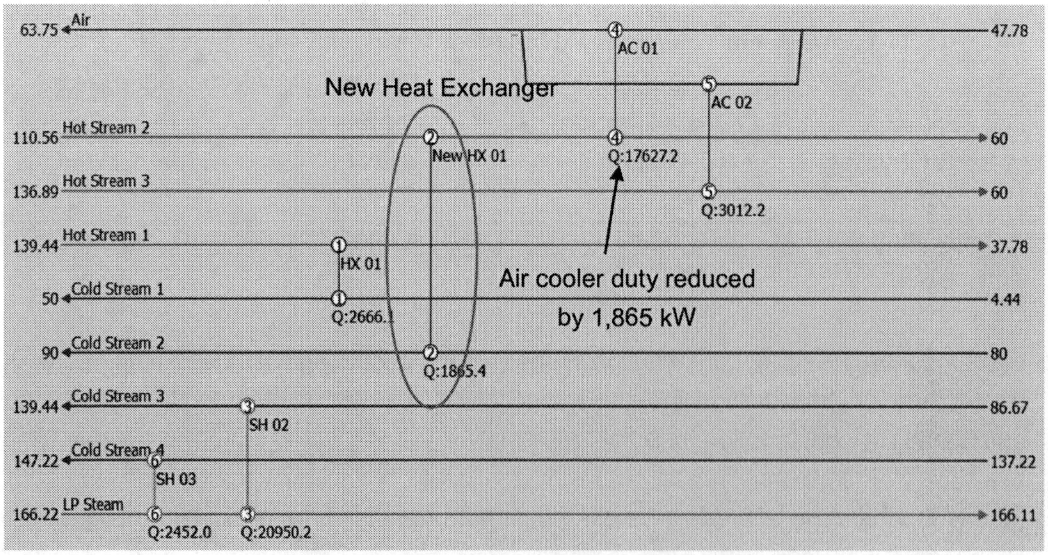

Figure 4. New heat exchanger network for Unit 1
Duties (Q) are in kW and temperatures are in °C

Process economic analysis was carried out to explore the trade-offs between the reduced operating cost and the extra capital investment for the additional heat exchanger. The changes in overall capital and operating costs between existing design and new design are summarised in Table 4. The addition of a new heat exchanger 'New HX 01' and removal of existing exchanger SH 01 require an increase in investment of 2.46 MM$. Energy savings of 1,865 kW of LP steam and air cooling duties are worth 0.62 MM$/yr. The project was found economically attractive with 20-year Net Present Value (NPV) of 6.51 MM$ and Internal Rate of Return (IRR) of 25.8%. PROJECT1 was recommended to the client for further investigation.

Table 4. Process economics comparison for PROJECT1

	Change in Installed Capital Costs (MM$)	Change in Operating Costs [1] (MM$/yr)
Single Train	+ 0.82	- 0.31
All Trains	+ 2.46	- 0.62

[1] Annual operating hours of 8760

Unit 2

A similar heat integration study was carried out for Unit 2. The existing heat exchanger network design is composed of 20 heat exchangers. Practical constraints, such as forbidden matches between process streams, were considered when looking for opportunities for further heat integration. The process streams and heat exchangers that do not allow structural changes were removed from the HEN model in i-Heat™. Figure 5 illustrates the simplified HEN composed of 6 heat exchangers.

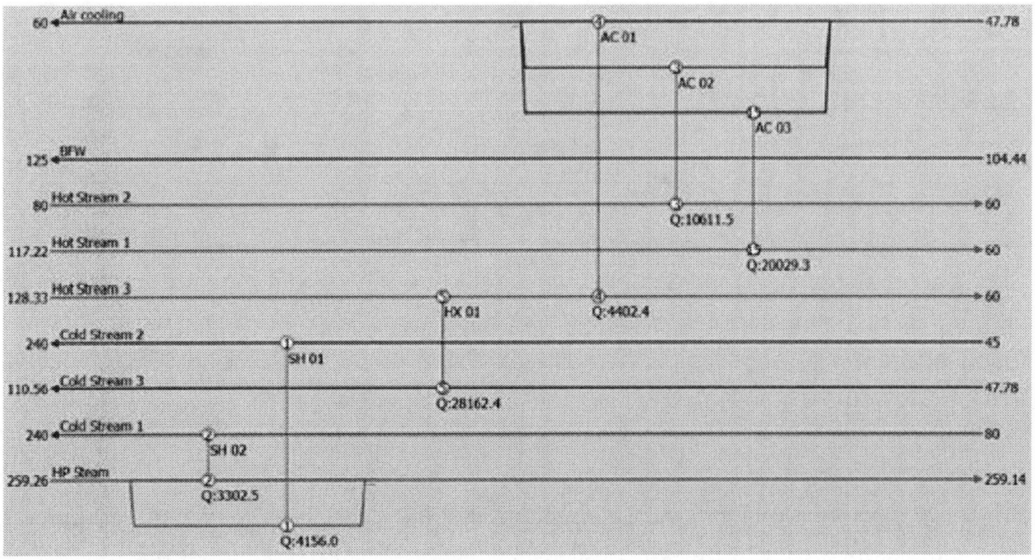

Figure 5. Simplified existing heat exchanger network design for Unit 2
Duties (Q) are in kW and temperatures are in °C

Cross pinch heat transfer analysis showed that steam heaters SH 01 and SH 02 have a cross pinch duty of 1,409 kW and 642 kW, respectively. In order to eliminate the cross pinch heat transfer in the existing design, two heat exchangers "New HX 01" and "New HX 02" were added in the heat exchanger network, as shown in Figure 6.

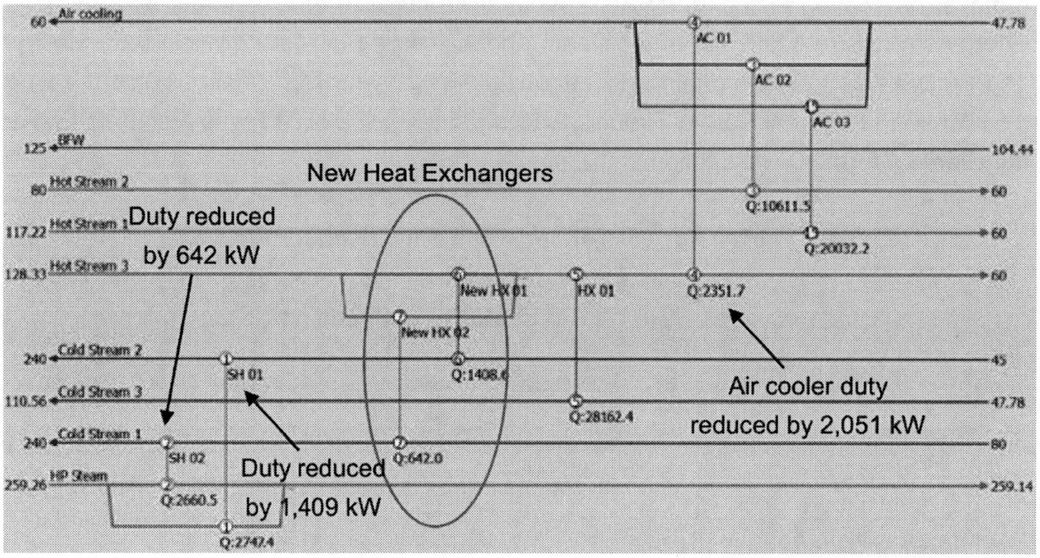

Figure 6. New heat exchanger network design for Unit 2
Duties (Q) are in kW and temperatures are in °C

The results summary of process economic analysis for the new heat exchanger network is summarised in Table 5.

Table 5. Process economics comparison for PROJECT2

	Change in Installed Capital Costs (MM$)	Change in Operating Costs [1] (MM$/yr)
Single Train	+ 0.65	- 0.554
All Trains	+ 3.92	- 2.77

[1] Annual operating hours of 8760.

Considering all trains, an increase in investment of 3.92 MM$ is required to modify the existing HEN. The associated energy savings are worth 2.77 MM$/yr. The project was found economically attractive with a 20-year NPV of 38.8 MM$ and IRR of 72.8%. PROJECT2 was recommended to the client for further investigation.

Unit 3

The initial pinch analysis for Unit 3 suggested that the existing HEN design had met the minimum energy targets. However, process simulation showed that further pre-heating the feed to the column could potentially reduce the energy consumption of the existing design. By using heat recovery to pre-heat the stream, LP steam consumption in the column reboiler was successfully reduced by 2,198 kW. The overall economic benefit of this design change is a reduction in operating costs of 0.37 MM$/yr. The required capital investment is increased by 0.20 MM$. Therefore, the project was economically attractive with a 20-year NPV of 6.0 MM$ and IRR of 188.7%. PROJECT3 was recommended to the client for further investigation.

The previous examples illustrate how heat integration analysis and optimisation can be carried out easily and efficiently with i-Heat™. Cost-effective energy saving projects can be quickly identified and evaluated. In total, eleven energy saving projects were identified and evaluated by PIL using i-Heat™. Seven projects were found to be cost-effective and recommended to the client for further feasibility evaluation.

- Step 5: Evaluate the potential for cross-process heat recovery in i-Heat™.

Following the heat integration studies of the individual process units, the potential for heat integration between process units was also considered in the energy optimisation study. Although it was possible to match heat sources and heat sinks across process units in the thermodynamic analysis, practical constraints ruled out options for feasible heat recovery.

- Step 6: Energy optimisation for site-wide utility system

The steam system and the nearby co-generation plant were analysed for energy optimisation. The steam system was designed to provide process heating and power for the natural gas processing plant. The existing steam system design was modelled in i-Steam™ for performance analysis. Figure 7 illustrates the various steam headers, process steam generators, steam turbine generator and steam consumers present in the system.

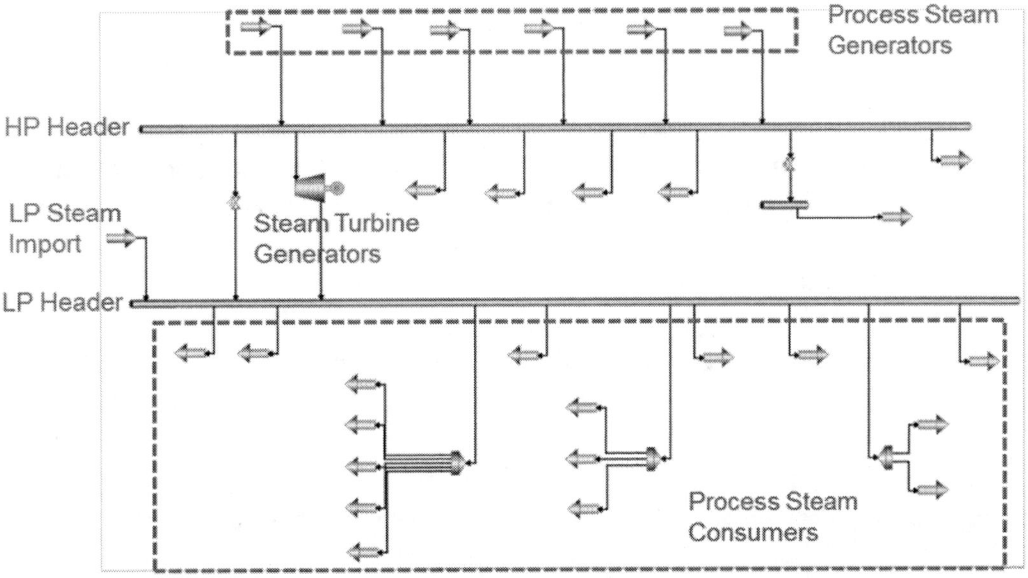

Figure 7. Steam system model in i-Steam™

The equipment items present in the nearby co-generation plant were also modelled in i-Steam™ to represent the power generation and LP steam export to the natural gas processing plant. The co-generation plant model was used to estimate the price of LP steam, as shown in Figure 8. Considering fixed steam flow rate through the HP turbine, if the LP steam export is reduced by 1t/h, the steam flow rate going through the condensing turbine is increased by 1t/h to generate additional power. The value of 1t/h LP steam was estimated as the economic benefit from extra power generation. Based on an electricity price of 0.07$/kWh, the value of LP steam was estimated at 10.87$/t. This LP steam price served as the basis for analysis and evaluation of process economics.

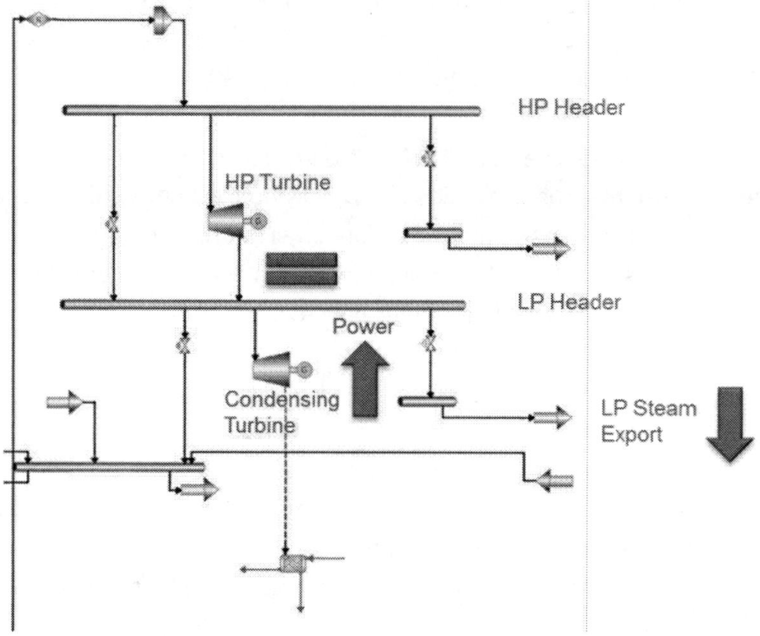

Figure 8. Co-generation plant model in i-Steam™

During the energy optimisation study, PIL also considered other options to improve the overall performance and economics of the natural gas processing plant. These options included driver selection as well as the optimisation of refrigeration systems. The integration of a chilled water system in the overall process was found to improve the process economics significantly.

- Step 7: Economic evaluation of combined improvements

Every cost-effective energy saving project was discussed with the client for further evaluation. A combined overall process model was developed to evaluate how the operation and configuration of the steam system and the co-generation plant were affected with the design improvements. The process economics of the combined process model was represented by the improvement of the NPV over the plant life of 20 years.

When considering all the energy saving projects that were accepted by the client, 20-year NPV of the natural gas processing plant is increased by 284.1 MM$. The IRR is 137.8%.

Conclusion

The case study illustrates how the advanced design tools for heat integration and energy optimisation, i-Heat™ and i-Steam™, can significantly improve the energy performance and process economics of a natural gas processing plant. Cost-effective projects can be identified by i-Heat™ to save energy consumption in individual process units through enhanced heat recovery. The influence and impact of the individual energy saving projects on the utility system can be evaluated effectively with the combined overall process model in i-Steam™. Such a systematic and integrated energy optimisation approach significantly reduces utility consumption and improves the overall economics of the processing plants.

The paper is extracted from a publication in GPA Europe's AGM and Technical Meeting, 2015 [7].

References

1. Douglas, J., 1988, Conceptual Design of Chemical Processes, McGraw-Hill.

2. Chen. L., Heat-Integrated Crude Oil Distillation Design, PhD thesis, Centre for Process Integration, University of Manchester, UK, 2008.

3. Smith, R., 2005, Chemical Process Design and Integration, John Wiley & Sons, Ltd., Chapter 23.

4. Kemp, I.C., 2007, Pinch Analysis and Process Integration: A User Guide on Process Integration for the Efficient Use of Energy, 2nd edition, Elsevier/Butterworth-Heinemann.

5. Varbanov, P., 2004, Optimisation and Synthesis of Process Utility Systems, University of Manchester Institute of Science and Technology (UMIST), Manchester.

6. Lou, Y., Comprehensive Modelling and Optimisation of Industrial Steam Systems, AIChE Spring Meeting, Austin, USA, 2015.

7. Zheng, X., Labanca, L., Energy Optimisation in Natural Gas Processing Plants, GPA Europe's AGM and Technical Meeting, London, UK, 2015.

445506 Reflecting on the Past Envisioning the Future: A Vision 2020 Primer

Abstract:

In the future, it is likely that some companies will have significantly better process safety performance than others. What separates the companies with great performance from those with just average performance? In Vision 20/20, five industry tenets of companies with great process performance were identified. During the Vision 20/20 work, it was recognized that there are four societal Themes that are also needed to have great process safety performance. This presentation relates seminal incidents in our industry to each of the tenets and themes. It is intended to be a reminder of our past learnings and a primer on Vision 20/20.

445505 Addressing Human Factors in Process Safety – Corporate and Plant Practices

Abstract:

Human factors engineering (HFE) is a scientific discipline concerned with (a) understanding interactions between human and non-human systems and (b) applying rigorous theory, principles, and methods to optimize human well-being and overall system performance. The field spans neuroscience, physiology, and cognitive science, among several other studies. This interdisciplinary nature provides HFE professionals unique opportunities in process safety management to prevent and mitigate dangerous mishaps. This talk discusses how companies can address HFE issues in the context of the Center for Chemical Process Safety *Risk Based Process Safety* framework. Critical lessons learned and best practices are offered for human factors implementation in capital project designs and existing process plants.

445504 Developing Power, Influence and Leadership for Successful Process Safety Culture

Abstract:

It is widely recognized that the development of strong safety cultures has the single greatest impact on the reduction of organizational accidents, a perspective shared by chemical process safety professionals. This, along with findings following disasters such as the loss of space shuttle Columbia, suggests that a top priority for all senior leaders, managers, and supervisors in the chemical processing industry should be to positively influence the shared beliefs, practices, and attitudes prevalent in the workplace. This session will identify key issues associated with this need for influence and will provide actionable insight regarding how attendees can develop the power they need and the techniques required to catalyze their organization toward a committed, risk aware mindset rather than mere regulatory compliance. Topics will include elements of organizational culture, power, group decision-making, and leadership framing.

444404 AIChE YPC: The Hope to Bring the New Development of Chemical Industry with the Millennial Generation

Abstract:

The millennial generation has become an important engine in the development of chemical industry in the decades to come. The advance in technology and computing has accompanied their growth in the past few decades. They are the group of people who are faithful fans of Facebook, iPhone and iPad. They have also become the major talent group of chemical industry and the majority of AIChE YPC members. The dynamic nature of the industry requires all the people to pursue professional development in order to develop new knowledge. Therefore this presentation will cover two aspects: it will provide the introduction of the characteristics of the millennial generation and their impact to the current chemical industry; in addition, it will bring the open discussion on the challenge and opportunity for YPs to take the leadership in the current and future development of chemical industry.

443215 WSA (Wet gas Sulfur Acid) Technology - A Competitive Solution for Sulfur Management

Abstract:

For more than 30 years Haldor Topsoe has designed plants for purification of sulfur containing waste gases derived from a wide variety of process industries. The WSA (Wet gas Sulfuric Acid) technology is efficient for recovering H_2S or other sulfur compounds in the form of commercial grade sulfuric acid. Moreover, the exothermic reaction energy generated when converting the sulfurous off-gas is efficiently recovered in the process as valuable superheated high pressure steam available for export.

The WSA technology is also characterized by high sulfur removal efficiency with more than 99% sulfur recovery, up to 99.99% when required, and is always in compliance with emission legislation. The process does not consume any chemicals nor will there be any solid or liquid effluents requiring additional treatment. This in combination with very low cooling water consumption makes WSA technology a very competitive solution for treatment of acid gases.

The WSA technology has proven feasible for treating sulfurous gases in many different applications, and today more than 140 units have been contracted worldwide handling a wide range of gas flows and with daily sulfuric acid production ranging from 4 to 1,140 MTPD.

The paper will explain the basic principles of the WSA technology and will include economical figures and the comparison with the typical alternatives for sulfur removal.

443717 Air-Quality Conscious Scheduling for Multiple Olefin Plant Start-Ups

Abstract:

The ground-level ozone is a pervasive air pollutant, which can be formed from flare emissions during chemical plant start-up operations. Especially in the concentrated industrial areas, the regional air quality (i.e., the ozone concentration) might be aggravated by simultaneously start-up emissions from multiple chemical plants. Thus, it is environmentally important for optimal scheduling of multi-plant start-ups under their manufacturing allowances to minimize the possible adverse air-quality impact. A systematic methodology on air-quality conscious scheduling for multi-plant start-ups has been developed. It demonstrates that multi-plant start-ups without any coordination could result in the significant air-quality impact. However, an optimal scheduling plan with several-hour tuning on the starting time of their start-up operations would significantly reduce about 80 % ozone increment. The study couples process dynamic simulation for industrial flare emissions with regional air-quality modeling together. It can provide valuable and quantitative supports for all relevant stakeholders, including environmental agencies, regional plants, and local communities.

444021 Integrated Waste Water Collection and Treatment System for Zero Waste Discharge Liquid Plant

Abstract:

Efficient design for collection of waste and effluent treatment is major challenge in EPC industry today. In this paper two cases of integration of waste water collection & treatment system are discussed. In first case, Power plant is considered and in other case LNG plant is taken. The sizing of waste water system and treatment of the effluents for zero waste discharge are compared for both of the cases. This paper throws light on significant savings by rightly sizing the collection and treatment system which majorly includes sumps, sump pumps, oil water separator, equalization tank, clarified water tank, and clarifier based on the intermittent and continuous waste from the plant. The flow sheets have been designed for power plant as well as LNG plant for integrated waste water collection & treatment system for zero waste discharge liquid plant. All the discharges like oil, waste water and chemicals are treated effectively and can be regenerated to use in the same plant. It is not only cost effective but also environment friendly.

Agenda

- What is the "RTR"
- "RTR" – Numerous Potential Impacts
- Long-Lead Compliance Projects Timing
- Roadmap to Success
- FCCU
  - Potential Effects and Solutions
- Delayed Coker:
  - Potential Effects and Solutions
- Flare Systems:
  - Potential Effects and Solutions
- Catalytic Reforming
  - Potential Effects and Solutions
- Benzene Fenceline Monitoring
- Conclusions

BURNS McDONNELL 2

What is the "RTR"?

- "Risk & Technology Review" rulemaking only applies to petroleum refineries
- New US EPA Clean Air Act rulemaking finalized Dec. 1, 2015
- Five key rules updated
  - NSPS J & NSPS Ja focus on criteria pollutants VOC, SOx, NOx, CO & PM
  - MACT Y, MACT CC & MACT UUU focus on Hazardous Air Pollutants (HAP)
- Almost every US refinery will be affected
- Complex new requirements (following years of actual emissions reductions)
- Potential high initial compliance costs for many refiners:
  - EPA estimated **$2 million** per refinery average across 143 US refineries
  - American Petroleum Institute estimated **$8 million** per refinery average
  - Actual costs will depend on many site-specific factors
- Most costs will occur in 2016 - 2018 for overall Jan. 30, 2019 deadline

BURNS & McDONNELL 3

Refinery Risk & Technology Review (RTR): numerous potential impacts

- New/revised air emission standards drive compliance projects & add new requirements to certain economic projects.
- Each refinery needs detailed technical evaluations to identify, evaluate & address potential RTR impacts, including:
  - **FCCUs** – do regen pollution controls meet new requirements at all times?
  - **Reformers** – do formerly "exempted" regen vents <5 psig meet new requirements?
  - **Delayed Cokers** – are modifications needed for >2 psig decoking vents to atmosphere?
  - **Flares** – flare gas recovery and/or control/instrumentation upgrades needed?
  - **Storage Tanks** – new floating roof retrofits or roof fitting modifications required?
  - **Startup/Shutdown** – do emissions comply after RTR eliminates SSM exemptions?
  - **Overall** – does refinery have roadmap to evaluate all impacts and meet all deadlines?

BURNS & McDONNELL 4

RTR Timeline: long-lead compliance projects may be short on time

- For existing refinery sources, comply "as expeditiously as practical, but no later than..."
  - **Startup/Shutdown/Malfunction** – "exemptions" eliminated on Jan. 30, 2016 & replaced by "general duty" requirements, then new SSM limits apply by Aug. 1, 2017
  - **Storage tank standards** – implement by April 29, 2016 (can request more time)
  - **FCC emissions testing** – new tests/frequency with first new tests by August 1, 2017
  - **Benzene fenceline monitoring** – start by Jan. 30, 2018
  - **All other requirements** – comply by Jan. 30, 2019
- Will EPA revise the rule? minor "corrections" were proposed & industry will request more changes (outcomes uncertain)

BURNS & McDONNELL 5

Fluid Catalytic Cracking Units: potential project impacts for normal operation

- **RTR focus** for FCCUs is on the catalyst regeneration vent, which involves complex equipment that is extremely expensive to upgrade or modify
- **RTR impacts** are highly dependent on unit-specific configuration & operation: many units are ready to comply but detailed review needed for each unit
- **Normal Operating Requirements**:
  - Previous limits & requirements were selectively made more stringent
  - New testing by Aug, 1, 2017 & most other requirements apply Jan. 30, 2019
- **Startup, Shutdown & Hot Standby Requirements:**
  - Previous SSM "exemptions" replaced with new requirements matrix
  - Examples: new >20 ft/sec cyclone velocity for metal HAP & new 1% excess O2 limit for organic HAP
  - Some changes effective immediately: complete transition by Aug. 1, 2017

BURNS & McDONNELL 6

Fluid Catalytic Cracking Units: Solutions

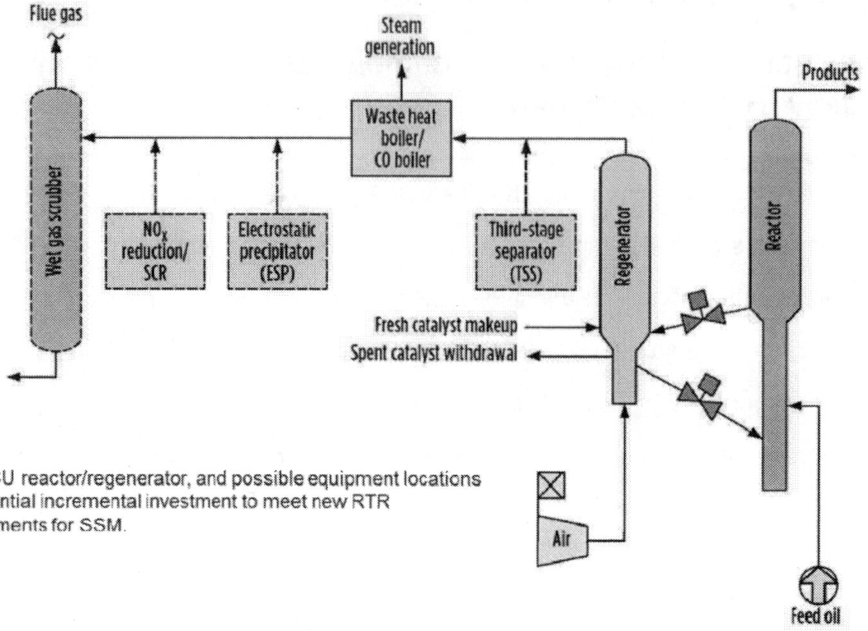

An FCCU reactor/regenerator, and possible equipment locations for potential incremental investment to meet new RTR requirements for SSM.

Fluid Catalytic Cracking Units – SSM

Third stage separators (TSS)	CO boiler (Partial burn)	Electrostatic precipitator (ESP)	Selective catalytic reduction (SCR)	Wet gas scrubber (WGS)	Comments:
				X	WGS assists with meeting PM, NO_x and SO_x requirements. CO met by complete combustion operation. Robust system.
	X				CO boiler assists with meeting CO. Evaluate incremental investment for TSS or ESP for meeting PM requirements on startup. SCR can assist with meeting NO_x.
		X			ESP assists in meeting PM but typically bypassed on startup. CO, NO_x and SO_x controlled by operating conditions in FCCU.
	X			X	CO boiler assists with meeting CO. WGS assists with meeting PM, NO_x and SO_x. Robust system.

Equipment exists

Equipment to consider as a minimum incremental investment

Delayed Coking Units: potential project impacts

- **Decoking vent pressure standards tightened for all delayed cokers:**
  - Old rule allowed drum venting as soon as pressure dropped to <5 psig outlet
  - New rule requires drum vapors to be processed or treated for organic HAP emission reduction down to <2 psig outlet
  - <2 psig limit applies to venting, draining, deheading & cutting
  - Additional technical requirement differences for "new" units versus "existing" units and for water overflow units versus other unit types
- **Emission reduction requirements unchanged:** all vapors that cannot meet <2 psig exemption must be recovered or treated to existing standards:
  - route to flare; or
  - reduce organic HAP by at least 98%; or
  - meet <20 ppmv Total Organic Carbon

Delayed Coking Units - Solutions

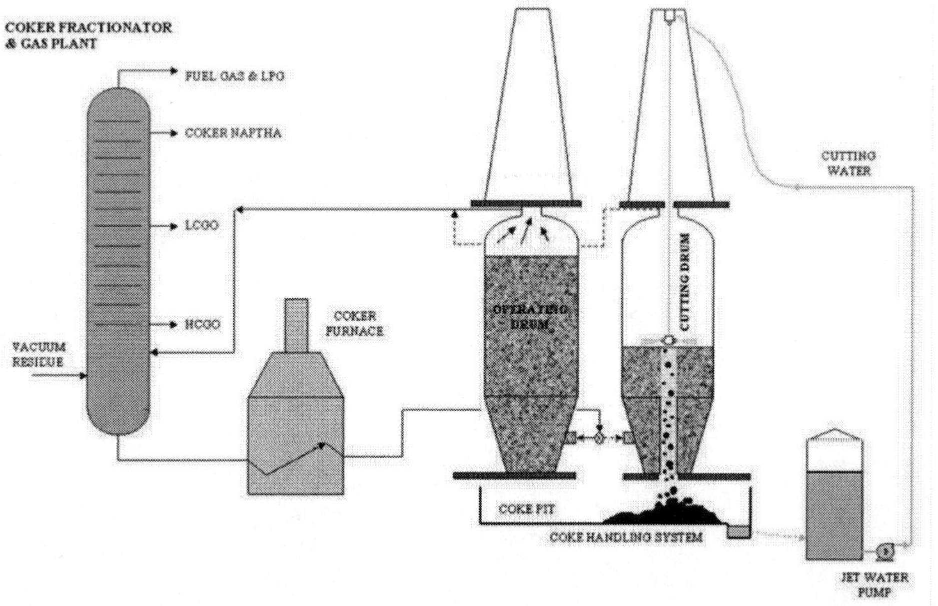

Source: Adams Project Managers, Inc.

Potential Mitigations

- **Achieving < 2 Psig Limit**
  - Use steam ejectors on blowdown drum
  - Use compressors on blowdown drum
  - Use compressors on coke drum overhead
  - Reduce pressure drop through blowdown system
  - Water filling in coke drums
  - Flare gas recovery compressor

Delayed Coking Units - Solutions

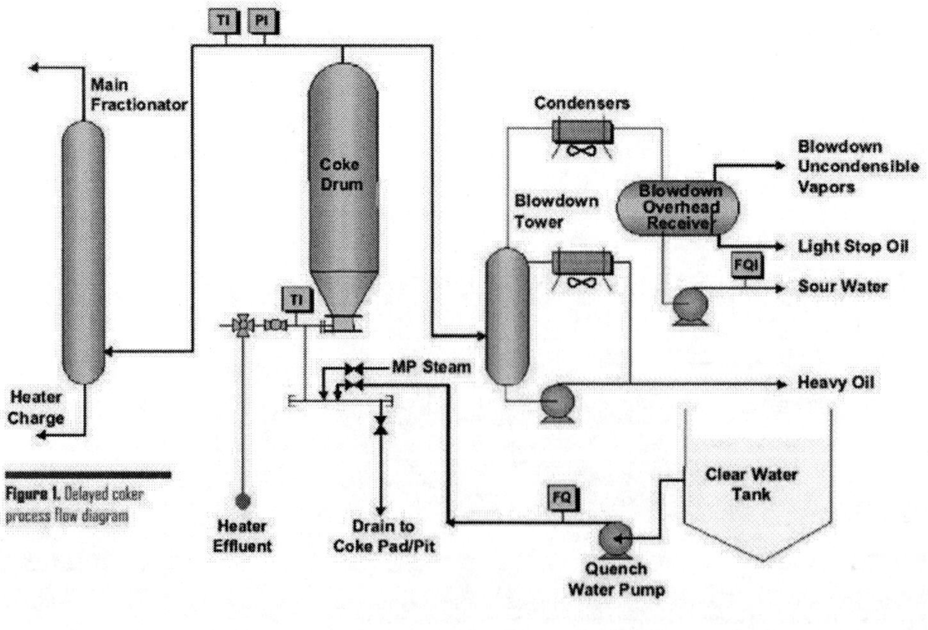

Figure 1. Delayed coker process flow diagram

Source: Lummus Technology, USA.

Design Considerations

- Can Coke Drums now achieve < 2 psig?
- Does achieving < 2 psig require more time which may affect capacity?
- What is the pressure drop through the blowdown system?
- Review increasing blowdown condensing capacity
- If add more motive force:
  - Ejectors – motive fluid; steam availability?
  - Compressors – Electrical capacity
- Main Fractionator – overhead cooling, compression and sour water handling.
- Consider a Flare Gas Recovery (FGR) system

Flare Systems: potential project impacts

- **RTR overhauls previous flare requirements**
- **New flare gas measurements required:**
  - Continuous volume flow measurement required for all "Combustion Zone" (CZ) contributing flows: vent gas, sweep, total steam, pre-mix, etc. (not pilot)
  - Quality of all flared gasses must be analyzed by instruments or sampling
- **Revised operating constraints for "normal" flare operation:**
  - New minimum Net Heating Value (NHV) ≥270 Btu/scf limit applies to all CZ flows
  - Tip velocity limits are similar to pre-RTR rules but apply to all CZ flows
  - NHV & velocity compliance calculations now required every 15-minute block
  - Startup/shutdown, excess fuel gas, PRV/PSV malfunctions, etc. regulated as "normal"
- **Emergency & high rates** >smokeless capacity subject to detailed new requirements
- **Bottom line**: RTR changes will drive detailed analysis & many flare projects

Flare System Solutions

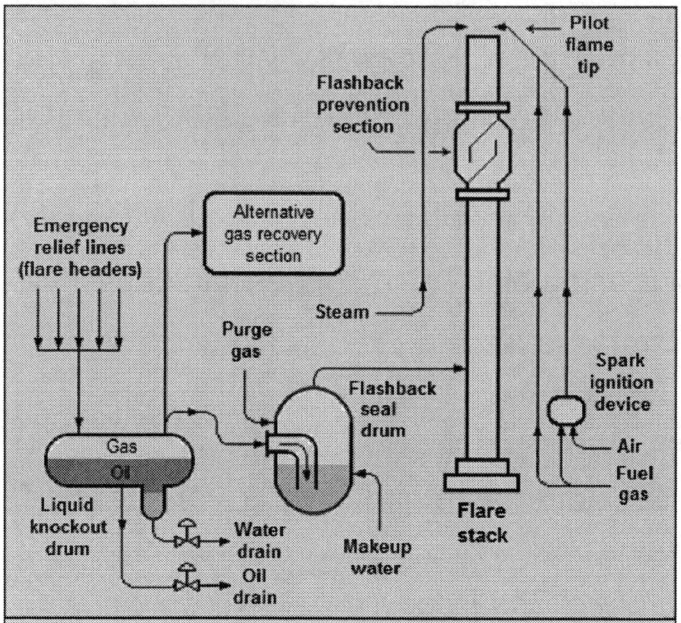

Diagram of an overall flare stack system in a petroleum refinery

Source: chemengineering.wikispaces.com.

Catalytic Reforming Units: potential project impacts

- **Scope of RTR changes for reformers:**
  - RTR only changes organic HAP requirements before catalyst burn-off
  - Inorganic HAP requirements for burn-off are unaffected by RTR
- **Old <5 psig vent "exemption" narrowed with more units regulated:**
  - Exemption now "clarified" to only apply to passive vessel depressurization
  - Any active purge vent (even if <5 psig) with sufficient pressure to route into emission treatment must now be treated to remove organic HAP
- **Organic HAP standards unchanged:** previously "exempted" <5 psig vents can no longer be vented unless routed to control: flare or other device to meet 98% HAP reduction or <20 ppmv TOC

Catalytic Reforming

Continuous Catalytic Regeneration (CCR) Units

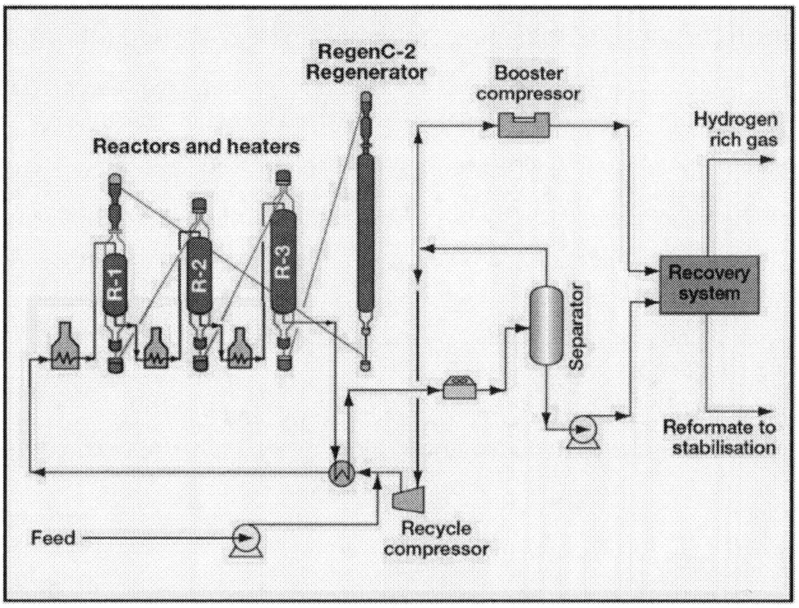

Source: Axens IFP Group Technologies.

BURNS & McDONNELL 17

Catalytic Reforming

Semi-Regenerative Units

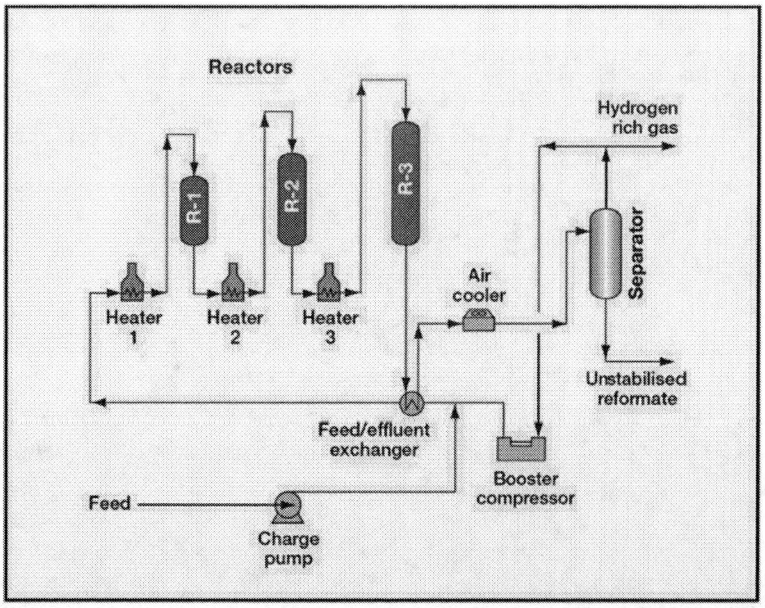

Source: Axens IFP Group Technologies.

BURNS & McDONNELL 18

Benzene Fenceline Monitoring Requirements

New program required to measure ambient benzene around refinery:

- Install 12 to 24 permanent fenceline/perimeter monitors (depending on refinery footprint) plus any background monitors to identify off-site benzene contributions
- Passive absorbent technology detects very low concentrations 24/7
- Meteorological station also required but can use existing/nearby met station
- Benzene data must be disclosed to public with associated new community perception & activist empowerment challenges
- Conduct root cause analysis if annual average measured benzene is >9 µg/m3 after discounting off-site contributions
- Multiple high measurements with related causes may lead to project requirements or increased compliance costs e.g., enhanced piping leak detection and repair program

Benzene Fenceline Monitoring Example

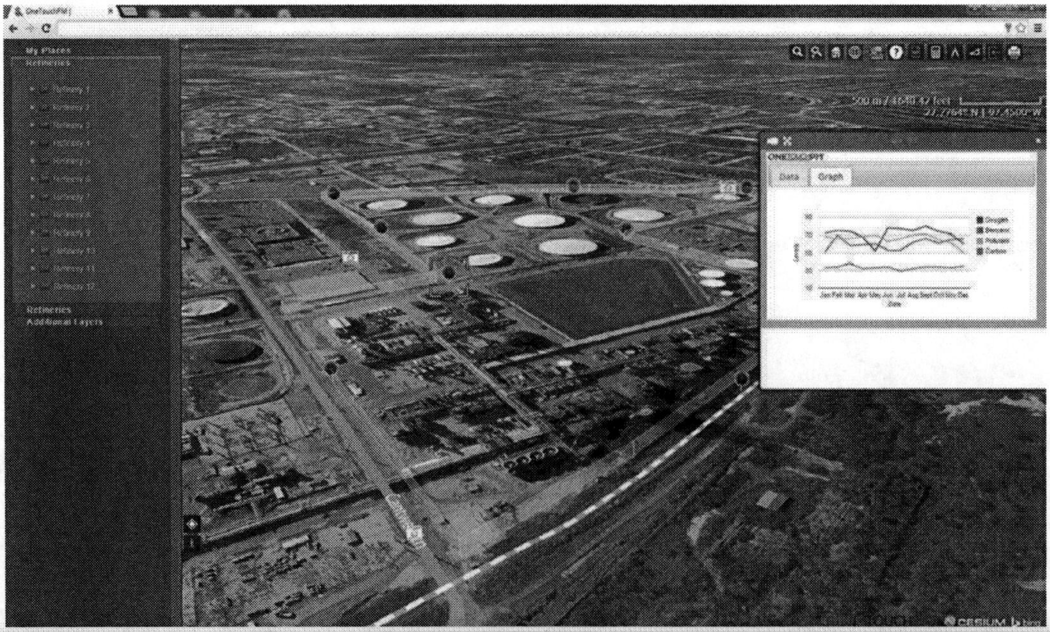

Other Impacts

- **Sulfur Plant**: new S/U & S/D requirements for purging gas
- **Storage Tanks**: some existing tanks may need new floating roofs, some existing floating roof fittings may need upgrades with April 29, 2016 deadline
- **Relief Valves:** new equipment requirements so that every release has: immediate detection, time/duration measurement, and operator notification
- **Maintenance Vapor Clearing:** new requirements apply even to small equipment clearing: affects procedures & compliance documentation
- **Onstream Analyzer Vents**: are now regulated process vents requiring vent-specific evaluations (many should meet flow/concentration limits)
- **Marine Vessel Loading**: certain loading was exempt from submerged-fill requirements but those exemptions were eliminated

Floating Roof Tanks, and Other Impacts:

External Floating Roof Tank (Pontoon Type)

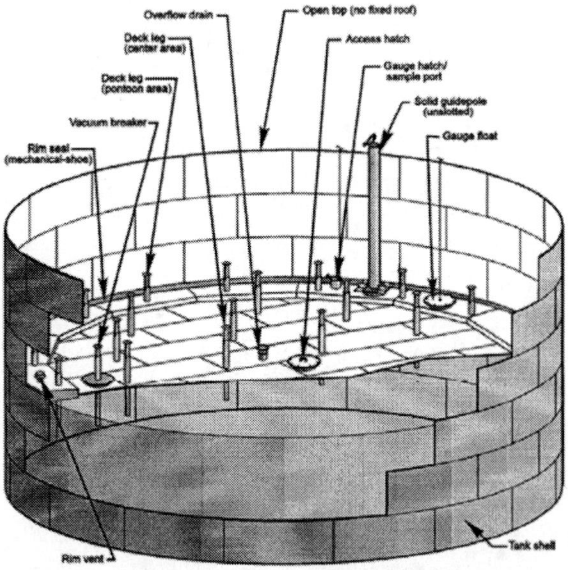

Source: US EPA, AP 42, November 2006.

Conclusions/Recommendations

- **Complexity:** is not reflected in slides, see paper for more detail
- **Priority**: plan for significant RTR compliance implementation costs with typical costs estimated in $2-8 million range per refinery
- **Scope**: the RTR can impact key refinery units and drive complex projects that may need coordination with turnarounds or other projects
- **Urgency**: the Jan 2019 overall compliance deadline combines with earlier deadlines for certain individual requirements to create schedule difficulty
- **Roadmap** is recommended to reflect priority, scope & urgency to chart course for successful compliance implementation with managed impacts
- **Team**: a purpose-built RTR implementation team is recommended to engage many necessary disciplines & stakeholders
- **Success** depends on site-specific RTR roadmap, implemented by an integrated RTR team, aligned with commercial priorities, and driven by compliance commitment

BURNS & McDONNELL 23

Contact Information

Blake Soyars, PE - Author/Presenter

Austin | 512-872-7150 | brsoyars@burnsmcd.com

Mark Lockhart, PE - Author/Presenter

Houston | 832-214-1975 | mlockhart@burnsmcd.com

Jim Corbit - Author

Houston | 832-214-2818 | jrcorbit@burnsmcd.com

Greg Neve - Author

Houston | 832-389-5746 | gneve@burnsmcd.com

BURNS & McDONNELL

CREATE AMAZING.

444421 AIChE Young Professionals Committee: The Mentor for Your Professional Development

Abstract:

Do you want to get mentor for your professional development? Young Professional (YP) members are a vital and active group of AIChE members. The Young Professionals Committee (YPC) works to increase participation among Young Professionals in AIChE benefits, services and volunteer opportunities. The committee serves as an intermediary between undergraduate students moving into the next phase of their professional or academic careers. This presentation will provide the introduction of YPC and all the available resources you can get for your professional development as an AIChE YP member.

445805 Entrepreneurs Guide to Starting a Company

Abstract:

This presentation will cover some of the key aspects of developing a successful startup based on the experience of two of the cofounders of DexMat. DexMat is a Houston-based company that is commercializing conductive carbon nanotube (CNT) technology for aerospace wire and cable applications. The company was founded in early 2015. DexMat was a finalist in the Rice Business Plan Competition and has secured SBIR funding from the Air Force to make lightweight cables for spacecraft. Topics that will be covered during the presentation include intellectual property management, fundraising, manufacturing scale-up, and developing corporate partnerships.

443347 The Pros, Cons, and Challenges of Consulting

Abstract:

The challenges of starting and maintaining a successful consulting business vary with the nature of the specialty, the target customers, long term plans for the business, resource availability, family support, resources required, and stage of career at which the business is started.

This presentation will review the development of a consulting business which has lasted 15 years and has morphed over time as different opportunities have developed and customer needs have changed. The business focuses on introductory chemical engineering training and TRIZ inventive problem solving training, two totally disconnected subjects that rarely overlap, both overlapping with AIChE activities. This presentation will review how these opportunities developed, which include lessons in identifying consulting business opportunities. Suggestions on some of the practical aspects of starting a consulting business and being flexible over the long haul will be presented.

445659 Sustainability Decisions in a Post COP21 Era

Abstract:

This presentation will summarize critical developments arising with and after the 2015 Paris Climate Conference which hosted the 21st Conference of the Parties (COP21) to review the United Nations Framework on Climate Change Conference, and the Sustainable Innovation Forum 2015. Additionally, the presentation will put into perspective the implications for US businesses and other stakeholders of the Obama Administration's commitments made at COP21 given the US political landscape, hostile congressional response and the 2016 election year, litigation regarding the EPA's GHG regulation, the state of the economy, price of oil, dynamic weather including drought and/or flooding exacerbated by sea level rise and increased storm surges, as well as the enforcement posture of federal agencies regarding permitting, compliance, and disclosures of material risk arising from threats to supply chain and other factors central to overall reliability and resilience.

16AIChE
Spring Meeting, Houston, TX

The Principle of Intrinsic Reality
Solving the World's Woes through Chemical Engineering and Economics

Jed Moyano Bellen, SCPM, PMP
Hyundai Engineering & Construction Co. Ltd.
9 Temasek Blvd., #27-02 Suntec Tower 2, Singapore 038989
jed.bellen@hdec.co.kr

Prepared for Presentation at
American Institute of Chemical Engineers
2016 Spring Meeting
11th Global Congress on Process Safety
Houston, Texas
April 10-14, 2016

UNPUBLISHED

The Principle of Intrinsic Reality Solving the World's Woes through Chemical Engineering and Economics

Jed Moyano Bellen, SCPM, PMP
Hyundai Engineering & Construction Co. Ltd.
9 Temasek Blvd., #27-02 Suntec Tower 2, Singapore 038989
jed.bellen@hdec.co.kr

Keywords: Principle of Intrinsic Reality, Chemical Engineering, Economics, Sustainability

Abstract

Human beings in profound rationality always seek for explanations of the phenomena around them. To understand these phenomena, human beings have developed systematized body of knowledge as a practical way of dealing with life's challenges. Among the systematized body of knowledge developed are chemical engineering and economics.

An avant-garde way of looking at each of these disciplines leads to the Principle of Intrinsic Reality. This principle relates the fundamental laws of the two disciplines to glean insights on the constraints imposed by conservation laws on economic systems at both the macro and micro levels.

It can be argued that if mass and energy in the universe are constant, then logically diminishing returns and marginal utility are bound to be observed. This logic is quite evident in a microorganism culture which follows the Monod growth model.

It can also be thought that conservation laws imply scarcity of resources. As such, it enjoins everyone to find optimal and efficient use of resources in the context of trade-offs, production decisions and sustainability. It even calls for the development of new technologies to produce more outputs from less inputs (i.e. shifting of production possibility frontier to the right). In this way, the Principle of Intrinsic Reality leads to the imperative for sustainable optimization.

Furthermore, it is deemed that the conservation laws can explain the phenomenon called "polarity of wealth." As mass and energy when harnessed create wealth, then the constancy of mass and energy implies boundary for wealth creation. In effect, wealth can neither be created nor destroyed but can only shift from one entity to another thereby creating polarity in wealth distribution. This assertion can be the key towards more equitable restructuring of economic systems to close income gap and thereby result to stable and resilient populace.

In conclusion, this paper envisions for a paradigm shift from traditional money-centric view of chemical engineering and economics to a synergistic convergence of disciplines which can powerfully tackle the world's woes and thereby effect sustainability.

1. The World in which We Live Today

The past few years saw unprecedented global challenges which have put into question existing economic models and scientific paradigms. Among these challenges are the dystopian global economy, climate change, environmental degradation and terrorism. The magnitude of these challenges has grown to a profound level that newer ways of thinking and perspectives become necessary. With that end in mind, this paper attempts to look at the fundamental laws of chemical engineering and economics in new ways to recognize our intrinsic reality from which these pressing challenges can be tackled at large.

2. The Principle of Intrinsic Reality

It is an observable fact that advancement in science increasingly blurs boundaries of various disciplines in tackling proactively the pressing global issues or problems. Within this purview, two seemingly diverging disciplines namely chemical engineering and economics can be taken together to glean insights on our intrinsic reality and what it means for our economic and ecological systems on a larger scale.

In the author's contention, intrinsic reality is grounded on the fundamental laws of chemical engineering and economics. These laws are the Conservation Laws (Chemical Engineering) and Law of Diminishing Returns (Economics). It can be argued that the Law of Diminishing Returns is observable because of the constraints imposed by the Conservation Laws. As such, this leads to the Principle of Intrinsic Reality which can be stated thus;

> *In any system, both at the macro- and micro-levels, the limiting factor for growth is the constancy of potential inputs.*

Potential inputs mean not the actual inputs to processes but the constant material or energy available which can be harnessed to create wealth. The reasoning or proof of this principle is presented in subsequent discussions.

2.1 Proof Based on the Monod Growth Model

The best illustration to understand the Principle of Intrinsic Reality is the Monod Growth Model. Monod in his 1949 seminal paper presented an empirical relation between exponential growth rate and concentration of essential nutrient. The data was fitted into hyperbolic equation:

$$R = R_K\left(\frac{C}{C_1 + C}\right)$$

In this equation, R is the growth rate during the exponential phase; R_K is the rate limit for increasing concentrations of C; C is the concentration of the nutrient; and C_1 is the concentration of nutrient at which the rate is half the maximum.

Figure 1: Graphical Representation of Monod Growth Model

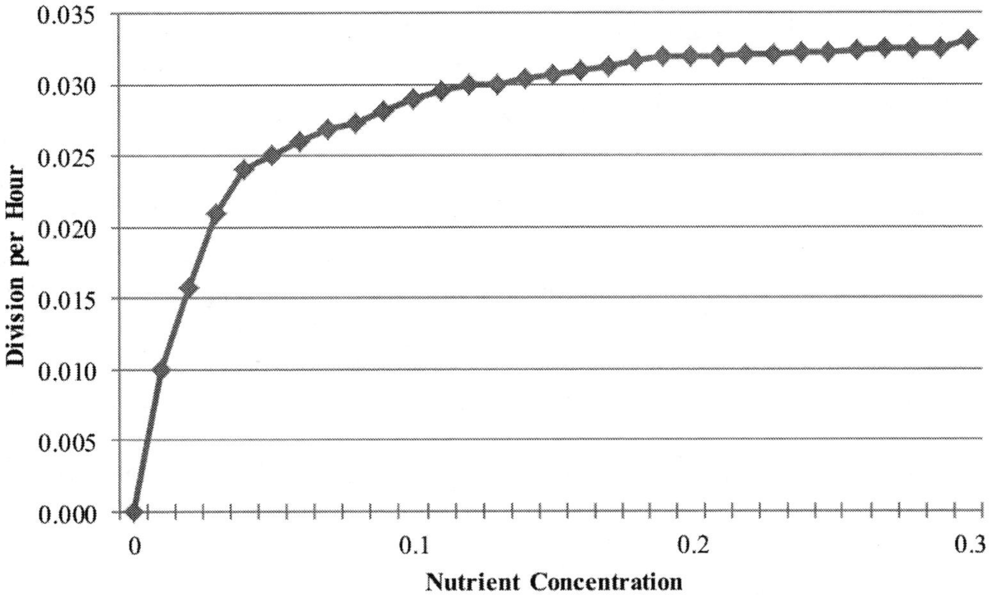

Monod (1949) further pointed out that the limiting factors commonly found are exhaustion of nutrients, accumulation of toxic metabolic products and changes in ion equilibrium.

The foregoing discussion has been observed at the micro-level with the bacterial culture. At the macro-level (i.e. the world at large), analogous phenomena are also observed to occur because of the same constraint imposed by the constant quantity of resources (i.e. material and energy). One example of these phenomena is that human beings cannot grow indefinitely. If they do, then how can the world sustain that growth with limited available resources (e.g. foods) to support it?

Another analogous phenomenon is in population growth. Thomas Malthus (1789) predicted that when population growth exceeds agricultural production, the population level is forced to return to subsistence-level conditions. This phenomenon as Malthus (1789) believed is due to the restriction of available resources supporting population growth, which others have claimed to be exploding as the earth becomes over populated.

In a surprising manner, a similar phenomenon can be observed in production processes in which the increase in a single factor of production assuming all other factors constant, results not to more marginal output but to a reduced one. This was attributed by classical economists, Thomas Malthus and David Ricardo, to the decreasing quality of inputs while neoclassical economists attributed this to the disruption in the entire production process as additional units of input are added to a fixed amount of capital. But in view of the Principle of Intrinsic Reality, the phenomenon is viewed intrinsic to the production system in which case infinite amounts of mass or energy cannot be processed with perfect efficiency. Hence, the law of conservation of mass and energy comes into the fore.

It can be deduced from the discussion that there is an underpinning intrinsic reality to the more obvious phenomena in the world from the micro-scale (bacterial growth) to the macro-scale (human growth and production processes).

2.2 Proof Based on Diminishing Marginal Utility

Another way to prove the Principle of Intrinsic Reality is through the diminishing marginal utility observable among humans and perhaps other organisms as well. Marginal utility, as defined in economics, is the gain from an increase or the loss from a decrease in the consumption of goods or services. The diminishing marginal utility is an observable phenomenon in which the utility or satisfaction an individual gains diminishes as he or she consumes more of the same unit of goods or services. Graphically, this can be represented as in Figure 2.

Figure 2: Graphical Representation of Diminishing Marginal Utility

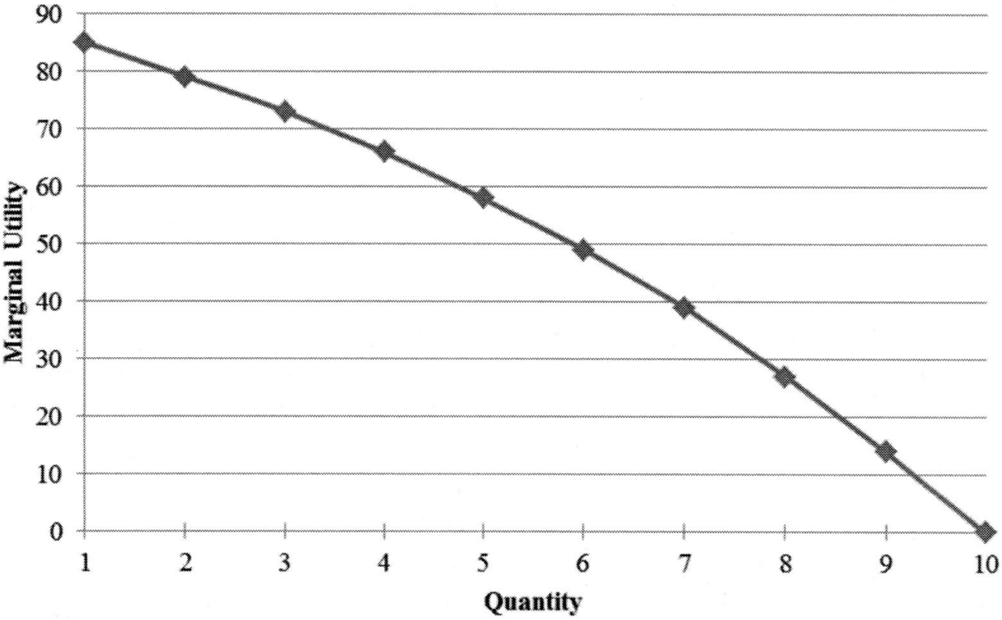

It can be deduced from this phenomenon and in support of the discussion in section 2.1 that human beings cannot grow indefinitely due to the fact that in our intrinsic reality, there is always a limit so as the constancy of available resources (i.e. mass and energy) can support our existence.

This means that in view of our intrinsic reality, we cannot have infinite greed otherwise an upsurge of unintended consequences are bound to happen (i.e. environmental degradation, terrorism and eventually demise of humanity).

Therefore, recognizing our intrinsic reality is the first step towards solving the world's woes and thereby resulting to sustainability and security.

2.3 Expanding the Production Possibility Curve through R&D

The caveat here is that the constant mass or energy (i.e. the limits) leads to trade-offs. Trade-offs is encountered when personal or business decisions are made. In making these decisions, one has to ensure that he or she becomes better off with his or her chosen option. We also realize in our life that we cannot get all that we want and we have to make trade-offs so long as it places us in a better or preferred position. This is more of a philosophical discourse, but it goes to show that the concepts of chemical engineering can be as useful in life and in business as it is in chemical processing.

To tackle these trade-offs, the Production Possibility Frontier in economics can be used. In view of the Principle of Intrinsic Reality where growth is limited by the fixed amount of material and energy, the Production Possibility Frontier needs to shift to the right as shown in Figure 3.

Figure 3. Graphical Representation of Expanding the Production Possibility Curve

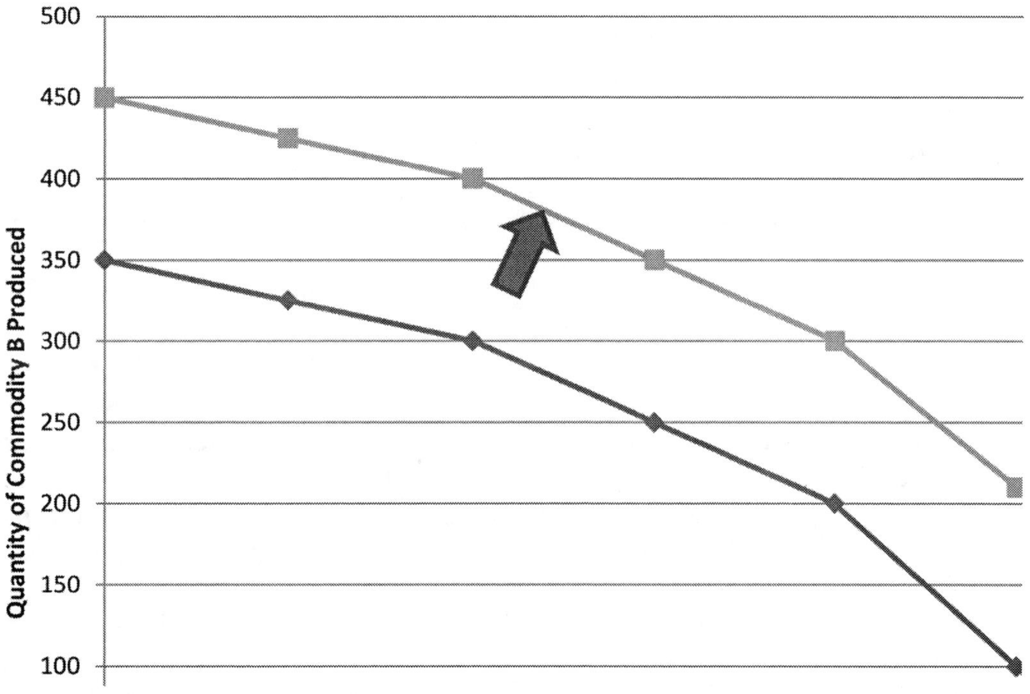

This shift has an implication on the practice of chemical engineering as Research and Development (R&D) is needed to achieve the said shift. R&D enables the economy to produce more of the two commodities (i.e. Commodity A and Commodity B) resulting in economic growth. The key challenge here to Chemical Engineering, particularly on Chemical Engineering R&D, is to look at the bigger picture where economic growth emanating from R&D output is pivoted on sustainability; considering its impact to the environment and the society at large.

3. Paradigm Shift in the Practice of Chemical Engineering

"Money" or "Profit Maximization" has been at the center of the practice of chemical engineering (Smith 2005). Various mathematical optimization methods are used to design systems, processes or components of processes with the end in mind of optimizing profits. This mental model however, will no longer work in our situation today as we face economic, ecological, social and spiritual crisis (Scharmer 2010).

To address this, Scharmer (2010) pointed out seven acupuncture points to shifting capitalism into a regenerative ecosystem economy. Within this purview, the practice of chemical engineering has a greater role to assume. For example, Scharmer (2010) suggested that the design of all production and consumption cycle be completely earth-to-earth, i.e. without necessitating for landfill and that it has to co-evolve with the natural ecosystem. In this regards, the way we do optimization should be revamped into sustainable optimization. By sustainable optimization, it means optimization of the entire system and adding in environmental as well as societal objectives to the equation.

In this definition, two things need to be considered. The first one is systems thinking which emphasizes the understanding of the entire system so the interaction among components will be known and unintended consequences mitigated (Senge 2006). This is a holistic approach to problem solving and finds enormous use in process design.

It should be noted however that the construction of processing plant, whether it is a greenfield or a brownfield project, is rooted on projected market demand. In some instances, economic volatility necessitates companies to experiment with multiple pathways to survive. In this circumstance and within the purview of systems thinking, design thinking comes into the fore. Design thinking is a collective principle emphasizing empathy for users, doing prototyping and tolerance for failure (Kolko 2015). This principle finds enormous use in chemical engineering designs such as in product development, process development, systems improvement and the like. This is also useful for simplifying or humanizing chemical engineering designs with the end in mind of contributing to sustainability. It also has to be noted that of paramount importance in using this tool is imagining the future with the view in mind of status quo and the likely outcome for sustainable development.

The second consideration is the inclusion of environmental and societal objective functions to the objective functions (economic criteria) of profit, cost, energy and yield (Edgar, et al. 2001) in optimization. The environmental objectives can be the impact or risk to environment and the societal objectives can be the impact or risk to society or common good.

These are just few examples of the needed paradigm shift in chemical engineering practice in consideration of the principle of intrinsic reality. But there can be more. A picture of the envisioned new paradigm for the practice of chemical engineering is shown in Figure 4.

Figure 4. New Paradigm in the Practice of Chemical Engineering

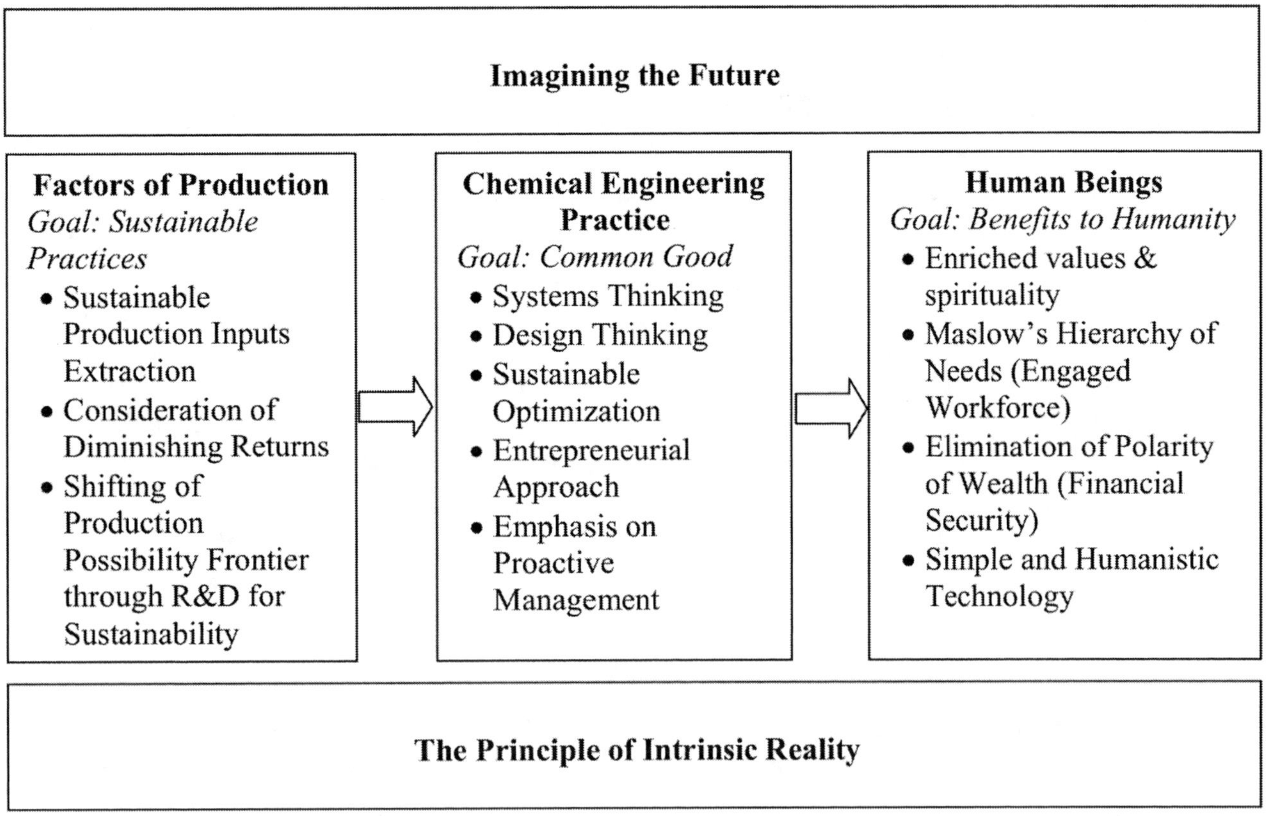

It should be realized that at the core of our economic systems is chemical engineering which can drive the economy into positive direction with the least unintended consequences. Related to that is capitalism in which upgrading of its operating systems can affect the practice of chemical engineering to a substantial degree. A discussion of this is provided in the following section.

4. Wealth Creation and Its Imperative for Sustainability

The development of this paper was inspired by the book "Freakonomics" in which the authors Steven D. Levitt and Steven J. Dubner (2005) asserted that economics can explain almost everything. It came to mind that if economics as a basic science can explain almost everything, then why not chemical engineering which was founded on chemistry, the central science?

Reflecting on it deeper, while explanations clear our doubts, we need more than that so we can tackle the world's woes. As such, this author thought that perhaps combining the two disciplines of chemical engineering and economics can be a potent force towards this end. How this can possibly be done is discussed in this section.

The starting point here is a review of the economic history of the world. In his book "A Farewell to Alms", Gregory Clark (2007) reviewed the economic history of the world in which he divided it into two periods; the period before 1800 and the period after. This is graphically represented in Figure 5.

Figure 5. Shares of Major Economies in World GDP through Time

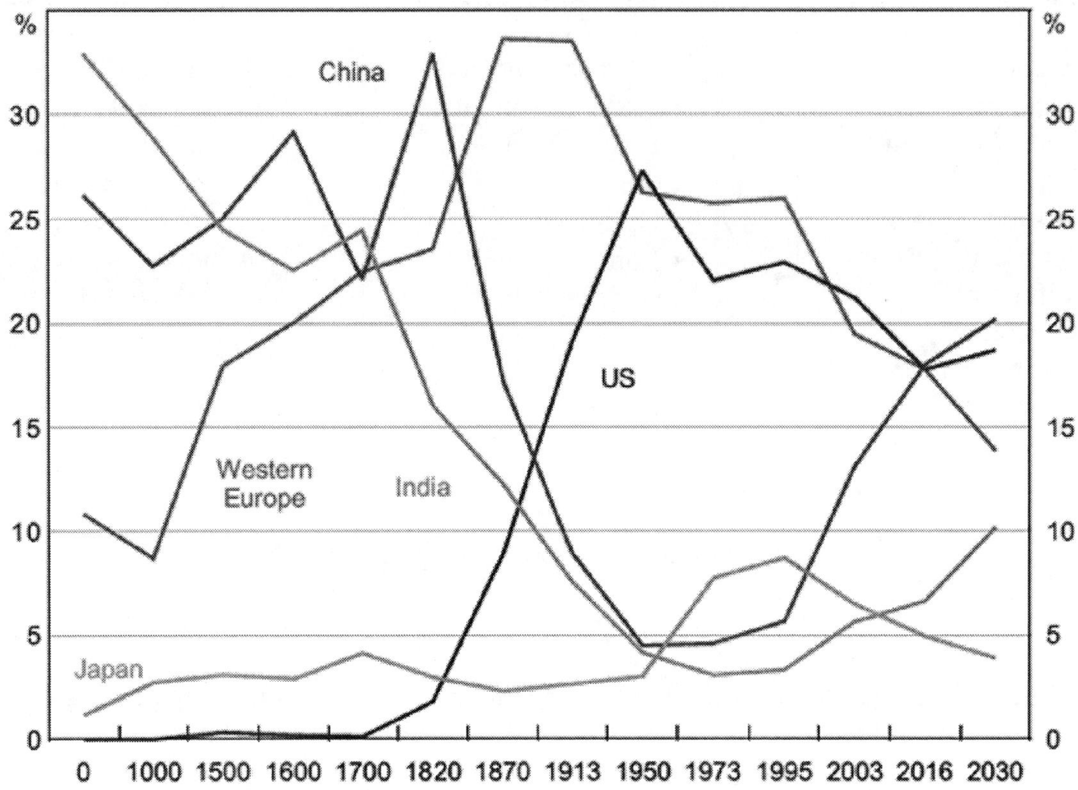

Note: Based on purchasing power parity (PPP) valuation of country GDP

Sources: International Monetary Fund (IMF), *World Economic Outlook* database, April 2011; Maddison (2006)

Thompson (2012) interpreted this graph by stating that "everything to the left of 1800 is an approximation of population distribution around the world and everything to the right of 1800 is a demonstration of productivity divergences around the world." In other words, everything to the left before Industrial Revolution depicts economic power by the sheer size of population and everything to the right after the Industrial Revolution shows the effect of mass production on country's economic output.

Clark (2007) on the other hand called the era before the industrial revolution, i.e. ca. 1800, as Malthusian era in which income per person varied across societies and epochs but growth is not observed. As Clark (2007) explained, this was due to a mechanism called Malthusian trap in which short-term increase in income through advancement in technology is dragged down by population growth. He further explicitly stated:

> "Economics as a discipline arose in the dying decades of the Malthusian era. Classical economics was a brilliantly successful description of this world. But the torrent of goods unleashed by the Industrial Revolution not only created extremes of wealth and poverty across nations, it also undermined the ability of economic theory to explain these differences."

In view of this statement, Figure 5 is reinterpreted in this paper. The contention of this author is that *polarity of wealth* can be gleaned from the graph. *Polarity of wealth* means the uneven concentration of wealth in the global economic system as some parts of the world are richer than others. The polarity can also be observed to shift as some richer parts of the world decline in economic power and some poorer parts of the world rise in economic wealth.

A historical example can be gleaned from Figure 5. India and China up until 1500 CE had 50 and 60 percent share of the world's economy, respectively. In 1 CE, China had a population of almost 60 million, while the area of the present day United States had a population of 680,000. In the late 18th century, the Industrial Revolution rendered countries with large population, just countries with the largest populations.

Looking at North America, it took them 1800 years to overtake China's economic output. However, by 1950, the US economic output was three times as great as China, even though its population is three times less. Comparing China with UK in 1913, the UK had a population of 45 million and China had 437 million but the economic output was almost the same. Analyzing what happened in Europe and later in the United States, Europe was on the lead even before 1800 (Nye 2012).

In 2010 according to data, China overtook Japan as the second biggest economy, though the numbers may not be strictly comparable. Japan's data are seasonally adjusted while those for China were not. Angus Maddison suggested that China and India were the biggest economies in the world for most of the past 2000 years. That is why economists find it a mystery why they fell so far behind and now they are flourishing ('China's economy' 2010). This may not be mysterious considering the *polarity of wealth* as was mentioned earlier and that this polarity can be argued to emanate from the Principle of Intrinsic Reality.

The reasoning here is that the world has constant mass and energy. Assuming that all these mass and energy can be harnessed to create wealth, it could mean that there is a limit for wealth creation. Theoretically, the limit can be expanded indefinitely through R&D but experience demonstrates that this was not achieved yet or getting anywhere near being achieved. As such, the constant mass and energy vis-à-vis the limit for wealth creation can be thought of as the reason for the polarity of wealth, i.e. wealth can only shift from one entity to another.

The implication of this assertion can be demonstrated by the fact that the 1% richest have accumulated more wealth than the rest of the world combined (Oxfam International 2016). The world has never seen 100% of its population to be wealthy. As such, time and again, equitable distribution of wealth is called for and mechanisms to achieve such_should be made a priority in government policies.

In a more concrete way, Scharmer (2010) suggested measures to address this growing inequality which is consistent with the mindset espoused by the Principle of Intrinsic Reality. These measures are as follows:

- Creation of a New Class of Common Property Rights

The creation of common property rights is intended to address the cost to society of private economic undertakings. For example, a private company is extracting natural resources from a certain place. While the private company can profit from that endeavor, the community in that place might suffer the undesirable consequences (e.g. flooding due to forest denudation). To ensure sustainability and institutionalize the voice of future generations in this case, a trust can be established which will own the rights and distribute funds (e.g. from cap-and-trade system for carbon emissions) to citizens (one share per person), communities and the government for investment in public infrastructure.

The trust is an effective means because it does not function like a company (driven by profits) or government (driven by special interest groups) but that it is accountable for long-term sustainability of the specific commons it manage for future generations.

- Creation of a Human Right to Basic Income

One way to ensure equitable distribution of wealth is to create a human right to basic income for every human being. This right when combined with free or affordable access to healthcare and education would create a playing field where everyone has a chance to pursue aspirations and dreams. In other words, it enables individuals to put their real creativity for the benefit of larger community. Another intention of this measure is to free up paid labor (a form of dependence) so that human beings can enhance their creativity and entrepreneurial skills.

- Inclusive Capitalism

To eradicate the *polarity of wealth*, capitalism should be made inclusive so that money and capital flows are directed to all sectors of the economic system. Currently, our money system leads to accumulation of oversupply of money and capital in high financial but low environmental and social gain areas. At the same time, there is an undersupply of money and capital in areas that address societal and community needs. It implies that there is too much money in areas where it is not needed but not enough on areas where it is needed. To address this imbalance, it is recommended to redesign the system so that the current extraction-based financial system is replaced by institutions which can support the real economy and thereby serve the ecology and community.

It should be noted that shifting of wealth is also observed in various geographical regions. When this occurs, geo-political instability can result. This goes to show that if we want lasting stability and security within the limits of our intrinsic reality, then we should not tolerate unabated greed. Otherwise, *polarity of wealth* will continue to persist and we cannot realize our full potential as individuals and collaborators for common good.

5. Way Forward...

Recognizing our intrinsic reality leads us to greater understanding of our very nature bounded by the constancy of mass and energy. It shed light on why growth stabilizes after addition of more inputs. Within this purview, trade-offs are also encountered which necessitates for R&D. R&D initiatives however should be underpinned on sustainability.

Looking at the practice of chemical engineering from the lens of intrinsic reality, the approach should be futuristic while considering relevant aspects related to factors of production and human beings.

The bigger picture shows that Engineers particularly Chemical Engineers are major players in driving the real economy. As Scharmer (2010) aptly puts it, "The essence of technology and knowledge creation, and the ultimate driving force of the economy, is the same: it is the human capacity to create." As such, if sustainability is to be achieved, Chemical Engineers have a greater responsibility to innovate for regenerative and social technologies. In this way, we can stem the polarity of wealth by providing affordable technology to enable disadvantaged individuals enhance their creativity, innovation and entrepreneurship. This can help them partake in the benefits of inclusive capitalism.

As demonstrated in this paper, the synergy emanating from the convergence of Chemical Engineering and Economics can powerfully tackle the world's woes. As such, there is a good reason for chemical engineers and economists alike to collaborate particularly on sustainability initiatives.

6. References

[1] 'China's economy overtakes Japan's in real terms' 2010, *The Economist*, 16 August, viewed 5 March 2016, http://www.economist.com/node/16834943

[2] Clark, G 2007, *A Farewell to Alms: A Brief Economic History of the World,* Princeton University Press, Princeton, NJ.

[3] Edgar, T, Himmelblau D & Lasdon, L 2001, *Optimization of Chemical Processes*, McGraw-Hill, New York.

[4] Huang, Y & Wang, B 2011, 'From the Asian Miracle to an Asian Century? Economic Transformation in the 2000s and Prospects for the 2010s', in H Gerard & J Kearns (eds), *Proceedings of the Reserve Bank of Australia annual conference*, Reserve Bank of Australia, Sydney, viewed 30 January 2016,

http://www.rba.gov.au/publications/confs/2011/huang-wang.html

[5] Kolko, J 2015, 'Design Thinking Comes of Age', *Harvard Business Review*, pp. 66-71, viewed 24 February 2016, https://hbr.org/2015/09/design-thinking-comes-of-age

[6] K.N.C. 2012, 'Mis-charting economic history: more 2,000 years in a single graphic', *The Economist blog*, web log post, 20 June, viewed 29 February 2016,

http://www.economist.com/blogs/graphicdetail/2012/06/mis-charting-economic-history

[7] Levitt, S & Dubner, S 2005, *Freakonomics: A Rogue Economist Explores the Hidden Side of Everything*, William Morrow, New York, NY.

[8] Malthus, T 1798, *An Essay on the Principle of Population*, J. Johnson in St. Paul's Church-yard, London.

[9] Maslow, A 1943, 'A Theory of Human Motivation', *Psychological Review*, vol. 50, pp. 370-396, viewed 24 February 2016, http://psychclassics.yorku.ca/Maslow/motivation.htm

[10] Monod, J 1949, 'The Growth of Bacterial Cultures', *Annual Review of Microbiology*, vol. 3, pp. 371-394, viewed 2 February 2016,

http://www.annualreviews.org/doi/abs/10.1146/annurev.mi.03.100149.002103

[11] Nye, J 2012, 'Fascinating new graph shows the 'economic history of the world since Jesus'', *MailOnline*, 23 June, viewed 15 November 2015,

http://www.dailymail.co.uk/news/article-2163610/Fascinating-new-graph-shows-economic-history-world-Jesus.html

[12] Oxfam International 2016, *An economy for the 1%, how privilege and power in the economy drive extreme inequality and how this can be stopped*, Oxford, viewed 5 March 2016,

https://www.oxfam.org/sites/www.oxfam.org/files/file_attachments/bp210-economy-one-percent-tax-havens-180116-en_0.pdf

[13] Samuelson, P & Nordhaus W 2009, *Economics*, McGraw-Hill Higher Education, New York, NY.

[14] Scharmer, O 2010, 'Seven Acupuncture Points for Shifting Capitalism to Create a Regenerative Ecosystem Economy', *Oxford Leadership Journal*, vol. 1, pp. 1-21, viewed 24 February 2016, http://www.oxfordleadership.com/journal/vol1_issue3/scharmer.pdf

[15] Senge, P 2006, *The Fifth Discipline: The Art & Practice of the Learning Organization*, Doubleday/Currency, New York.

[16] Smith, R 2005, *Chemical Process Design and Integration*, John Wiley & Sons Ltd., West Sussex, England.

[17] Thompson, D 2012, 'The economic history of the last 2,000 years in 1 little graph', *The Atlantic blog*, web log post, 19 June, viewed 29 February 2016, http://www.theatlantic.com/business/archive/2012/06/the-economic-history-of-the-last-2-000-years-in-1-little-graph/258676/

Acknowledgment

The author would like to thank Hyundai Engineering and Construction Co. Ltd. for the support in this endeavor. The author is also thankful to Ms. Sofia Ayu for meticulously editing this manuscript and Mr. Marcial N. Barroga for the poster design.

445030 Heavy Metal Detection in Treated Water Via a Fabric Nanocomposite Sensor

Abstract:

We present the conception of an electrochemical sensor for on-site/portable water analysis. This flexible sensor was developed to quantify the amount of heavy metals (in real-time), alleviating the wait time, large sample size, possible contamination, and expensive analytical equipment associated with current procedures for water analysis. This sensor is composed of a multiwalled carbon nanotube (MWNT) functionalized polymer mat, for a flexible yet conductive sensor. The MWNTs are then functionalized with a cyclo-oligomeric calixarene, which is shown to selectively form a supramolecular complex with metal ions. Upon complex formation, the charge carriers are drawn away from the carbon layer and the detection is prevalent at appropriate levels. The optimization of MWNT and calixarene functionalization, as well as the selectivity and sensitivity of the sensor will be discussed.

444086 Optimization of Layered Graphene Oxide Nanostructures through Controlling the Membrane Deposition Rate Towards Size-Sieving Separation

Abstract:

Single-layer graphene oxide (GO) flakes were coated on anodized aluminum oxide (AAO) substrate through a simple vacuum filtration process, the effect of the deposition rate on the nanostructures of obtained membrane was investigated though pure water flux and organic vapor diffusion, which indicated GO membrane assembled with slow deposition rate might have more ordered lamellar structure and, correspondingly, have smaller effective pore size. In contrast, under fast deposition rate, GO flakes assembled in a more random way. Therefore, through controlling the membrane deposition rate, the nanostructure of layered GO membrane could be optimized for size dependent separation. We found after thermal reduction, 100 nm GO membrane (prepared @ 0.02 nm/s) has effective pore size smaller than 6.1 Å, which exhibit great potential in n-hexane/2,2-dimethybutane (DMB) separation.

444376 Ozone Source Apportionments of Local Anthropogenic Emissions

Abstract:

Ozone is one of six principal air pollutants designed by US EPA. NOX and VOC emissions from anthropogenic sources are the most important factors to control ozone pollutant.

In this report, photochemical dispersion models are used to estimate the effect of ozone from local anthropogenic emission sources. EKMA-type diagrams are generated for NOX- and VOC- effects. Furthermore, APCA (Anthropogenic Precursor Culpability Assessment) technology is used to estimate ozone source apportionments.

EKMA-type diagrams show that 10-30% emission reductions could reduce ozone concentrations to meet standard. Further APCA results show the average ozone contributions from regions and emission categories. APCA results explain different effects for different monitoring stations.

444142 Robust Design of a Radiant Wall Burner for a High Temperature Environment Utilizing Computational Fluid Dynamics Simulations

Abstract:

Radiant Wall Burners are often utilized in Ethylene Cracking Furnaces. The conditions and fuel gases for a specific unit can vary considerably within the unit and across similar units. With burner air preheated to high temperatures, premixing the fuel can result in dangerous and destructive flashback, while complete nozzle-mixing of the fuel can result in burner instability at startup or turn-down conditions. These challenges can be solved through the generation of a robust radiant wall burner design, which is facilitated by executing Computational Fluid Dynamics simulations of many different geometric configurations.

444397 Flare Minimization and Emission Characterization during Start-Ups of Chilling Train in Ethylene Plants

Abstract:

Flaring is a crucial safety operation to protect chemical plant personnel and equipment during plant start-up. However, it generates huge amounts of off-spec products, and causes negative environmental and social impacts, as well as results in tremendous raw material and energy losses which could be saved to generate much more needed products. Thus, cost-effective start-up flare minimization strategies which is a double-win practice, simultaneously benefiting both the industrial and environmental sustainability, are becoming more and more important and attractive to the industries.

The chilling train is a very important section for the ethylene plants and accounts for more than 50% of the total start-up time and flare emissions of the whole ethylene plants. Within this section, the charge gas from the upper stream process is refrigerated to a very low temperature (lower than -160 °C) to separate hydrogen and methane. Meanwhile, the left charge gas is liquidized for cryogenic separation. The chilling train system consumes over 40% of the refrigerating capacity of an ethylene plant. During the chilling train start-up, different operational conditions with different compositions and flowrate of precooling process will affect both product separation efficiencies, start-up stability time and flare emission.

Traditional start-up scenarios are developed based on experience and improved by trial and error, which is inefficient and dangerous. In this study, a general methodology on flare minimization for chilling train start-up operations via rigorous dynamic simulation is presented. The methodology starts with setup and validation of chilling train steady-state and dynamic simulation models. The initial state of start-up in the real plant is achieved through systematically transforming from the validated dynamic model, and several operation conditions with different compositions and flowrate of precooling stream are examined to identify an optimal initial state of start-up based on the obtained initial state. Thereafter different flare minimization strategies for the chilling train start-up are conducted on the optimal initial state model. Also, emission source characterization under different flare minimization strategies for chilling train start-up is investigated. Deep insights of the start-up strategies, emission source distribution and dynamic emission profiles are provided. This study provides detailed technical support for both the industry and environmental agencies on evaluating and developing cost-effective flare minimization strategies in the future.

439811 Purification of Ibuprofen Enantiomers By a New Technique Combining Distillation and Crystallization

Abstract:

Stripping crystallization (SC) is a new separation technique which combines distillation and crystallization. Basically, SC is operated at a triple-point condition, in which the liquid is simultaneously vaporized and crystallized due to the three-phase equilibrium. In essence, the process is continued until the liquid phase is completely eliminated and only the pure solid crystals remain in the feed. Thus, crystal washing is not required since only pure solid crystals remain in the feed and no impurities are adhered on the crystal surfaces at the end of SC. In the previous research, SC has been successfully applied to separate p-xylene from the mixed xylenes by Shiau et al. (2005, 2008, 2013).

Ibuprofen, also called 2-(isobutylphenyl)-propionic acid, is a non-steroidal anti-inflammatory drug widely used to treat headaches and minor pains. Although it is known that S-ibuprofen is responsible for the anti-inflammatory effects while R-ibuprofen might cause some long-term side effects, ibuprofen is still sold as a racemic mixture in the market due to the complexity involved in the separation of enantiomers. A thermodynamic model is developed in this work to simulate the three-phase equilibrium conditions during the SC operation. The experiments show that seeding with ultrasound mixing or magnetic stirring can be efficiently employed to purify S-ibuprofen from ibuprofen enantiomers. The experimental results, including the final enantiomeric purity and recovery ratio of S-ibuprofen, are consistent with the simulation results predicted by the model.

References:

1. Shiau, L. D.*, Wen, C. C. and Lin, B. S. "Separation and purification of p-xylene from the mixture of m-xylene and p-xylene by distillative freezing", *Ind. Eng. Chem. Res*, **44(7)**, p2258-2265, 2005.

2. Shiau, L. D.*, Wen, C. C. and Lin, B. S. "Separation of p-xylene from the multicomponent xylene system by stripping crystallization", *AIChE Journal*, **54(1)**, p337-342, 2008.

3. Shiau, L. D.*; Liu, K. F. "Investigations into the effects of the cooling rate on stripping crystallization", *Ind. Eng. Chem. Res*, **52**, p1716-1722, 2013.

442557 Flow Field Simulation and Distribution Performance Analysis of a Gas Pipe Distributor in Fccu

Abstract:
A fluidized catalytic cracking (FCC) is an important petroleum refinery process, which, aided by highly selective zeolite catalysts, cracks large-molecule low-value heavy oil feedstock into small-molecule highvalue
products. When spend catalyst is fed into the regenerator for reactivating, it is fluidized by the gas distributor at the bottom of regenerator. The gas distributor is designed to uniformly distribute gas to maintain the fluidized state of the fluidized bed and to provide the reaction gas to guarantee the burning coke deposited on the spend catalyst. Therefore, the gas distribution performance of the gas distributor influences the quality of initial gas distribution directly. So the pipe distributor used in a industrial FCC unit, which is composed by main pipes and a number of branch pipes with many nozzle tubes, was investigated by means of flow field calculation. The gas distribution performance of the pipe distributor was analyzed based on the simulated flow field. The simulation result shows that there are a obvious nununiform
gas distribution from the different nozzle gas flow rate. The static pressure increases gradually along the flow direction in the nozzle tube while the axial flow rate decreases, which leads to the nozzle gas flow rate increasing gradually. Furthermore, an obvious bias flow occurs at the inlet of the nozzle tube, which influences the gas flowing into the nozzle. The pressure drops performance of nozzle and pressure varying in the branch pipe are analyzed comprehensively. Due to the nonuniform
gas distribution by the gas pipe distributor in the practical usage, the problems such as overlarge temperature difference among the fluidized bed layers in the regenerator have often happened, which will not only reduce burning coke efficiency and but also decrease activity of the regenerated catalyst. Meanwhile, the erosion of the distributor also occurs, and in consequence the function of the distributor fails, which even causes the nonscheduled shutdown of the FCC unit and leads to tremendous economic loss. The support is provided for optimizations of both structural parameters and operation parameters of the gas pipe distributor and improvement of the gas distribution performance. Furthermore, improvement suggestions were provided based on the above reasoning. The results are expected to be useful for the distributor design.

441535 Hydrogen Disposal Flare System Hazard Identification and Mitigation Analysis

Abstract:

A hazard assessment on a hydrogen disposal flare system was conducted in preparation for a hydrogen turbopump test at the Air Force Research Laboratory (AFRL), Aerospace Systems Directorate at Edwards AFB, CA. The test operations required the disposal of large quantities of liquid and gaseous hydrogen through a commercial flare system to prevent the formation of an explosive environment. As part of the assessment, the primary steps in the test operation process were evaluated for the formation of a potential hazard by looking at each associated failure mode and the resulting affect on personnel and facilities. Evaluation of each operational step was conducted in a Process Failure Modes and Effects Analysis (PFMEA) that identified the probability and severity of the identified effects and associated hazards. If the probability and severity were unacceptable, mitigating steps were applied until the overall hazard risks were reduced to acceptable levels. The recommendations of this assessment were incorporated into both the flare system design and the operational procedures. This paper outlines the results of the PFMEA which include identification of the effects/hazards, available risk-mitigating techniques, and implementation of these techniques used for the mitigation of the identified hazards. Based upon the findings significant design modification to the vender's commercial equipment was needed to comply with recent changes in the code. Most of the identified hydrogen flare system hazards/effects were categorized into four groups: internal explosions, obstructions, external explosions, and maintenance degradation. Many corrective actions were implemented to mitigate both the probability and severity of the identified hazards. In addition, this paper discusses the installation of oxygen monitoring systems on the flare system, determination of appropriate system purge requirements, installation of purge system status alarms, incorporation of a triple redundant flare ignition system with status indicators, and a number of other hazard risk mitigation techniques. This comprehensive description of the hazards and mitigation of a hydrogen disposal system is a plant is vital part to safe operation of both hydrogen commercial and aerospace facilities.
<align="center">

444020 Thermodynamic Modeling of Aqueous Lithium Sulfate and Its Solubility in the Presence of Na+, K+ and Mg2+ Sulfates

Abstract:

This work is an extension of the previous works done by our group towards development of a rigorous and comprehensive thermodynamic model for high salinity produced water that is essential for mass and heat balance calculations and simulation and design of novel treatment processes. Specifically this work presents a thermodynamic model that accurately represents all thermodynamic properties of aqueous lithium sulfate solution including the solubility of Li2SO4 in water and also in the presence of Na+–K+–Mg2+ sulfates. By explicit consideration of solution chemistry of each subsystem, symmetric electrolyte NRTL model is used to correlate available experimental data of all pertinent thermodynamic properties including activity coefficient, osmotic coefficient, excess enthalpy, and liquid heat capacity over the entire concentration range. This facilitates quantification of eNRTL binary molecule–molecule, molecule–electrolyte, and electrolyte–electrolyte interaction parameters. Also, solubility data are used to regress the thermodynamic constants required for modeling the liquid–solid equilibrium of lithium sulfate in water and in the presence of Na+–K+–Mg2+ sulfates. Displaying a significant prediction ability, the model accurately represents the thermodynamic behavior of the system from infinite dilution up to saturation and temperatures from 298.15 K to 573.15 K.

442068 Enhanced Furfural Production By an Innovative Bottom Dividing Wall Column with a Decanter Configuration

Abstract:

Enhanced Furfural Production by an Innovative Bottom Dividing Wall Column with a Decanter Configuration

Nguyen Van Duc Long, Le Cao Nhien, Le Quang Minh and Moonyong Lee†

School of Chemical Engineering, Yeungnam University, Gyeongsan 712-749, South Korea

Furfural, which are used to derive other industrial chemicals, has been identified as one of the major bio-based platform chemicals that can compete with petroleum-based chemicals. However, the current commercial furfural process has low yield and requires intensive energy. This paper reports the simulation, detailed design and optimization of an industrial process recovering furfural from lignocellulosic biomass, a renewable and inexpensive feedstock. The aqueous feed mixture from the acidic hydrolysis of biomass was fed to the separation process to recover furfural as the main product, along with methanol as valuable byproducts. The purification process was intensified by applying an innovative bottom dividing wall column with a decanter configuration (BDWC-D). The response surface methodology (RSM), which allows the interactions between variables to be identified and quantified, was used to optimize the BDWC-D structure. The predictions by the RSM showed satisfactory agreement with the rigorous simulation results. The heat pump technology and heat integration were then employed to improve the purification process. The results showed that the proposed sequence could save the overall energy requirements and total annual cost significantly as compared to the conventional sequence. This sequence can be employed for both grass-root and retrofit designs.

Keywords: Lignocellulosic Biomass; Furfural; Energy Efficiency Improvement; Dividing Wall Column; Response Surface Methodology

Acknowledgments

The authors wish to acknowledge the financial support from the R&D Convergence Program of NST (National Research Council of Science & Technology) of Republic of Korea and KITECH (Korea Institute of Industrial Technology) (ES150001).

*** Formatting Issue - Best Available Paper/Graphic ***

444278 Insight into the Mechanism of Toxic Chromium (VI) from Wastewater Using Magnetite Coated PINE Powder

Abstract:

INSIGHT INTO THE MECHANISM OF TOXIC CHROMIUM (VI) FROM WASTEWATER USING MAGNETITE COATED PINE POWDER

Pholosi A., Ofomaja A., Naidoo E.B.

Biosorption and Wastewater Treatment Research Laboratory, Department of Chemistry,

Faculty of Applied and Computer Sciences, Vaal University of Technology, P. Bag X021, Vanderbiljpark, 1900 South Africa.

INTRODUCTION

A knowledge of the interaction between pollutant molecules or ions and active sites on an adsorbent surface is an important phenomenon in many industrial processes such as catalysis, adsorption and other separation techniques. The use of lignocellulosic materials for Cr(VI) adsorption has gained interest over the years **(Miretzky and Cirelli, 2010)**. The adsorption properties of lignocellulosic materials has been attributed to acidic functional groups on their surfaces which can take part in ion-exchange, complexation, chelation and hydrogen bonding with pollutant species in solution **(Vithanage et al., 2015)**. Two mechanisms for lignocellulosic adsorption of Cr(VI) have been proposed; (i) an electrostatic mechanism in which at low pH, the protonated acidic functional groups on the biomaterial carrying a positive charge attracts the negatively charged Cr(VI) species **(Demirbas, 2005)** and (ii) the adsorption-coupled reduction mechanism in which Cr(VI) is reduced to Cr(III) either in solution (Direct reduction) or after being adsorbed (Indirect reduction) by electron-donor groups of the biomaterial that have lower reduction potential values than that of Cr(VI) **(Gao et al., 2008)**. The Cr(III) ions formed in solution may then form complexes with Cr(III)-binding groups on the biomaterial surface or released into bulk solution by repulsion between Cr(III) and other positively charged groups on the biomaterial **(Miretzky and Cirelli, 2010)**.

METHODS

Sodium hydroxide treated pine coated with magnetite (NTP-NC) was prepared by co-precipitation method. The effect of solution pH was performed in a batch system in which 0.5 g of the adsorbent was contacted in six 250 cm^3 beakers containing 75 cm^3 of 75 mg/dm^3 Cr(VI) solutions set at pH 1, 2, 4, 6, 8 and 10 at 26 °C and agitated at 200 rpm for 2 h and the clear solution analysed for Cr(V/I) left using UV-Vis spectrophotometry at 460 nm

RESULTS

The results of the effect of initial solution pH on the adsorption of Cr(VI) onto NTP-NC is shown in Fig. 1. To confirm the operating mechanism, the change in $[H^+]$ concentration at each initial pH, the oxidation/reduction potential before and after adsorption and the amounts of Total Cr, Cr(VI)and Cr(III) present in solution at each pH was monitored for NTP-NC and the results displayed in Fig1 . The change in $[H^+]$ concentration ($\Delta[H^+] = [H^+]_{final} - [H^+]_{initial}$) at the end of the adsorption process for each initial solution pH was recorded and plotted against the initial solution pH. The results showed that the change in $[H^+]$ concentrations at the end of the adsorption process where higher for lower initial solution pH's (where lower amounts of Cr was found in solution) than for higher initial solution pH's (where higher amounts of Cr was found in solution) and the change in $[H^+]$ concentration became small and almost constant after initial solution pH 4. This is consistent with the assumption that the removal of Cr from solution is associated with [H^+] ion consumption and that as the concentration of [H^+] ions reduces in solution, the Cr removal reduces. Cr(VI) ion in acidic solution has a positive redox potential value (above 1.3 V at standard conditions) and in the presence of an electron donor becomes unstable and can be reduced to Cr(III) **(Park et al., 2008)**. The results show that the initial ORP values (before adsorption) for the Cr solutions at pH's of 1, 2, 4, 6, 8, and 10 were 339, 280, 165, 55, -53 and -172 mV for NTP-NC respectively. The positive values of ORP at lower solution pH's indicates higher possibility for reducing Cr(VI), while the lower and negative ORP values indicates lower possibility for reducing Cr(VI). On addition of the adsorbents into the solutions, a change in ORP were observed . The values of ORP at the end of the adsorption process became 330, 268, 30, 9 -29 and -61 for NTP-NC respectively. A reduction in ORP at the end of the adsorption stage signifies that reduction has occurred in the system. **Yu et al. (2014)** showed the amount of Cr(VI) removed from aqueous solution in contact with zero-valent iron (nZVI) could be related to the change in ORP and pH of the solution. The authors observed a decrease in ORP from 400 to 100 mV and a corresponding 10 unit pH increase as Cr(VI) was removed from solution.

Finally, analysis of the prepared Cr solutions at the various solution pH's showed that the concentration of Total Cr obtained from ICP and that of Cr(VI) using UV-vis measurements were very similar, suggesting that Total Cr was almost equal to Cr(VI) in solution. After contacting the various solutions with the adsorbents, the concentration of Total Cr in the pH 1 solution was observed to have reduced to 21.56 mg/dm^3 for NTP-NC, while the Cr(VI) concentrations was 10.66 mg/dm^3 for NTP-NC. With solution pH 2, the concentration Cr(III) left in solution was lower than that of solution pH 1 for both samples. The reason is that at pH 1 there is an abundance of H^+ ions in solution which accumulates on the functional groups on the adsorbent making the surfaces of the adsorbent positively charged and repelling away the positively charged Cr(III) ions from the adsorbent surfaces **(Suksabye et al., 2009)**. At solution pH 2, it is believed that there is less accumulation of H^+ ions at the adsorbent surfaces which led to reduced electrostatic repulsion and increased adsorption of Cr(III) from solution. This increase in Cr(III) adsorption is also confirmed by the reduction in Total Cr in solution between solution pH 1 and 2. The concentrations of Cr(VI) left in solution at pH 2 was higher than at pH 1 for NTP while for NTP-NC the Cr(VI) concentrations were lower at pH 2. For solution pH's higher than 2 and up to 10, the residual Cr(III) in solution becomes very low and almost constant. This is consistent with the small change in [H^+] in solution which is almost constant at this pH range.

FTIR Evidence for the adsorption of Cr(III)

When Cr(III) is formed in solution via adsorption-coupled reduction mechanism it may undergo hydrolysis and/or complexation depending on the solution pH and the total Cr(III) concentration. Hydrolysis may lead to the formation of various Cr(III) species with an increase in solution pH which gives rise to a reduction in its solubility and increasing the chances of precipitation of the Cr(III) species. At solution pH 1 – 4, Cr^{3+} and $CrOH^{2+}$ are the major species which can be adsorbed by negatively charged ligands on the biomaterial surface. On the other hand, at solution pH range of 4 – 10, Cr(III) species (hydroxo complex), such as $CrOH^{2+}$, and $Cr(OH)_3$, are the dominant forms in the aqueous system and at high pH is readily transformed into soluble complex, . When NTP-NC was contacted with Cr(VI) solution for 2 hr, the final solution pH was increased to pH 2.53. The FTIR spectra for NTP-NC (Fig. 2) shows the presence of characteristic Fe-O peaks of bulk Fe_3O_4 at 553.26 cm^{-1}, 564.36 cm^{-1}, 570.21 cm^{-1} and 576 21 cm^{-1}. Functional groups associated with the pine includes the broad band at 3250.99 cm^{-1} representing –OH of cellulose, 2905.17 cm^{-1} representing C-H stretching, while the peaks at 1627.69 and 1602.87 cm^{-1} shows the COO^- and C=O bonds respectively. After 2 hr contact with Cr(VI) solution, the peak at 564.34 cm^{-1} was observed to have split into several new peaks (Fig. 3b (ii)). These emerging peaks are characteristic of Fe-O bonds of maghemite (γ-Fe_3O_4), for example the characteristics peaks of maghemite was observed at 595.21, 685.65 cm^{-1} and 582.70, 634.91 cm^{-1} suggesting a change in the iron oxide phase (Fe_3O_4 to γ-Fe_2O_3) due to the conversion of Cr(VI) to Cr(III) by magnetite nanoparticles. According to **Peterson et al. (1997)**, Cr(III)(hydr)oxide produced is precipitated on the magnetite surface. In this study, α-Cr_2O_3 peak were observed at 604.63 cm^{-1} indicating that Cr(III) was formed during the adsorption process. Finally it was also observed that the carboxylate group at 1635.55 cm^{-1} was shifted to 1637.73 cm^{-1} while a new peak was observed at 1737.72 cm^{-1} suggesting that complexation between Cr(III) and Carboxylate ions also occurred.

REFERENCE

Demirbas, A. Energy Source 27 (2005) 1449–1455.

Miretzky, P., Cerelli, A.F. A review. J. Hard. Mater. 180 (2010) 1-9.

Vithanage, M., Rajapaksha, A.U., Ahmad, M., Uchimiya, M., Dou, X., Alessi, D.S., Ok, Y.S. J. Environ. Manage. 151 (2015) 443-449.

D. Park, C.K. Ahn, Y.M. Kim, Y.-S. Yun, J.M. Park, J. Hazard. Mater. 160 (2008) 422–427.

P. Suksabye, A. Nakajima, P. Thiravetyan, Y. Baba, W. Nakbanpote.. J. Hazar. Mater. 161 (2009) 1103–1108.

R.F., Yu, F.H., Chi, W.P., Cheng, J.C. Chem. Eng. J. 255 (2014) 568–576

M.L. Peterson, A.F. White, G.E. Brown, G.A. Parks Environ. Sci. Technol. 31 (1997) 1573–1576.

Fig. 1 Effect of solution pH on the adsorption of Cr(VI)

*** Formatting Issue - Best Available Paper/Graphic ***

Fig, 2: FTIR of NTP-NC

443101 Thermodynamic Modeling of HNO_3-H_2so_4-H_2o Ternary System with Symmetric Electrolyte NRTL Model

Abstract:

Nitric acid concentration/sulfuric acid concentration (NAC/SAC) process has been widely used for concentrating dilute aqueous nitric acid and recovering spent sulfuric acid. Nitric acid is concentrated by employing sulfuric acid to bind water and break nitric acid-water azeotrope at approximately 68 wt. % nitric acid. To support heat and mass balance calculations and process simulation for NAC/SAC processes, we develop a comprehensive thermodynamic model for nitric acid-sulfuric acid-water ternary system based on previously published thermodynamic models of nitric acid-water [Wang et al., Fluid Phase Equilibria, (2015) (DOI: 10.1016/j.fluid.2015.04.015)] binary and sulfuric acid-water [Que et al., Journal of Chemical & Engineering Data, 56 (2011) 963-9772] binary systems developed with symmetric electrolyte NRTL (eNRTL) activity coefficient model. The present work takes into account one-step hydration of nitric acid molecules, two-step hydration of sulfuric acid molecules, and further hydration of hydronium ions. With up to three temperature coefficients for each eNRTL binary interaction parameter and up to two temperature coefficients for each chemical equilibrium constant, the ternary model is validated with boiling point data from 383.15 to 471.45 K and activities at 273 K. Furthermore, contour plots of boiling points, vapor phase composition, and specific heat capacity of the ternary system are generated for engineering use.

Keywords:

Nitric acid, Sulfuric acid, Electrolyte NRTL model, Vapor-liquid equilibrium, Boiling point, Heat capacity, Nitric acid concentration/sulfuric acid concentration (NAC/SAC)

444551 Continuously Adjustable, Molecular-Sieving "Gate" on 5A Zeolite for Distinguishing Small Organic Molecules By Size

Abstract:

Zeolites/molecular sieves with uniform, molecular-sized pores are important for many adsorption-based separation processes. Pore size gaps, however, exist in the current zeolite family. This leads to a great challenge of separating molecules with size differences at ~0.01 nm level. Here, we report a novel concept, pore misalignment, to form a continuously adjustable, molecular-sieving "gate" at the 5A zeolite pore entrance without sacrificing the internal capacity. Misalignment of the micropores of the alumina coating with the 5A zeolite pores was related with and facilely adjusted by the coating thickness. For the first time, organic molecules with sub-0.01 nm size differences were effectively distinguished via appropriate misalignment. This novel concept may have great potential to fill the pore size gaps of the zeolite family and realize size-selective adsorption separation.

444791 Dissolution and Precipitation of Minerals Under Different Alkali Concentrations in a Reactive CO_2 Batch System

Abstract:

The primary focus of this research is to study the dissolution and precipitation of minerals and salts from Berea sandstone and brine samples taken from the Great Salt Lake in a reactive CO_2 batch system. Various experiments were constructed to measure dissolution and precipitation reaction rates under different alkali concentrations. Changing the alkali concentrations of the Great Salt Lake brine samples produced changes in dissolution and precipitation of minerals from the Berea sandstone and the Great Salt Lake brine samples. Data from ICP-MS was used to quantify mineralogical changes in the samples and Micro-CT was used to measure the porosity change in the Berea sandstone samples. The results from these analytical instruments showed patterns of mineralogical dissolution and precipitation in the post-experimental fluid that correlate to the change in alkali concentrations.

*** Formatting Issue - Best Available Paper/Graphic ***

443365 Technical Reconstruction of Double Effect Distillation for Methanol Process

Abstract:

Abstract: Methanol is one of important basis raw materials of chemical industry and also an excellent environment-friendly fuel, which make it as a significant commodity in the global market. Distillation is an essential unit operation for purified production of methanol. Product quality, production cost, environmental impact and energy consumption are largely related to distillation process.

Our research effort, under this circumstance, has been devoted since 2002 to energetically optimize 4-column methanol distillation system that is widely adopted in methanol units from coal, with reduced cycling water requirement. This patented process technology saw its industrial application in Chinese largest single unit of 1800Kt/a of Baofeng Group located at North China, with the lowest steam consumption of 0.9/1 t(steam)/t(methanol) compared with normally 25% higher for both simulative case study and on-site operation. Methanol product is in accordance with O-M-232G and the recycling water is reduced significantly from 67 t(CW)/t(methanol) to 52 t(CW)/t(methanol).

The flow diagram of this advanced double effect distillation process is showed in Fig.1. Its advantages arise for limitations found in current units: (1) heat exchange network failed to fully recover energy causing large CW flow to remove extra heat; (2) for pressured column, bubble point reflux need larger reflux ratio than the sub-cooled; (3) the heat of distillate from pressured column was normally ignored; (4) the low concentration of methanol of feed inlet to recovery column leads to higher reflux radio, responsible for unsatisfied unit energy efficiency.

Fig.1 The double effect distillation process.

(Column1.Light ends column; Column 2.Pressured column;

Column 3.Atmospheric column; Column4.Recovery column)

As a solution, a new flow diagram is built up and showed in Fig.2. (1) HEN totally reorganized (not completely show); (2) sub-cooled reflux to pressured column caused 10% reduction to the reflux ratio, economizing steam consumption; (3) top product of pressured column give heat to crude methanol before feeding light ends column; (4) change product and by-product position of recovery column to give efficient unit energy consumption.

Fig.2 The reconstruction of distillation process.

(Column1.Light ends column; Column 2.Pressured column;

Column 3.Atmospheric column; Column4.Recovery column)

It is estimated that the production of methanol will be over 60 million tons in 2014. If this reconstruction could be applied in the whole methanol industry, 1.3 million tons of standard coal would have been saved. This is equivalent to reduce 5 million tons emission of carbon dioxide, so the environmental impact can be largely decreased. Field operation shows reconstruction perfect substitution worth to spread to the whole methanol industry, energetically efficient and low environmentally impacting.

444353 A Combined Rheology and Stability Study on Emulsified Acid for High Temperature Applications

Abstract:

HCl emulsion has been a popular formulation to stimulate carbonate reservoirs due to the various advantages that emulsified HCl has over ordinary HCl; including but not limited to deeper penetration and less corrosion damages. Emulsified HCl has even more potential advantages if utilized properly. The reaction kinetics, stability and rheology of emulsified HCl plays a significant role on achieving successful acidizing treatment.

The objective of this paper is to experimentally study the factors affecting the stability of emulsified HCl. The emulsion was prepared using 15wt% HCl and diesel with cationic emulsifier. The acid solution-to-diesel ratio was 70:30. The effect of stirring intensity, concentration of the emulsifier, addition of corrosion inhibitor, temperature, addition rate of acid phase to oil phase and mixing time of the emulsion have been studied. Stirring intensity was varied from 600 to 1500 rpm. Emulsifier dosage ranged between 5 to 20 gpt. Temperature used for heating, to mimic reservoir condition, was as high as 120°C.

The results showed that there is an optimum condition for emulsifier dosage and for mixing time of the emulsion. The optimum emulsifier dosage was found to be 10 gpt with 5 minutes of mixing time. As for the addition rate of acid phase to oil phase, 1.4 ml/min of drop-wise addition was found to give a very stable emulsion. The effect of stirring intensity on the emulsion stability was proportional, the higher the rpm speed the better the stability. Corrosion inhibitor, on the other hand, played a role as a destabilizer. With regards to heating temperature, it was found out that the higher the temperature, the faster the emulsion breaks.

The novelty of this study is the comprehensive approach in combining stablity and rheology of emulsified acid for high temperature applications. The particle size analysis and solution morphology help in better understanding HCl emulsion and how it works.

443503 A Computational Analysis of Fire Scene for Safety in a Compartment

Abstract:

This article details numerical prediction of fire spread carried out to study the distribution of carbon monoxide concentration, smoke dispersion, soot concentration and temperature in laboratory compartment, using a commercial fire modeling software tool. The heat release rate curve predicted from the simulation was compared with literature values carried out in a road tunnel. The time variations of carbon monoxide transportation, smoke concentration and temperature spread in compartment room are reported to evaluate the effect on humans. The soot concentration was described and used throughout this study as an indicator of safety in the compartment room. The effects of vent dimension and ventilation capacity were analyzed by smoke detector response times. This work provides improvements derived from reinterpretation of actual fire scene considering the ability of mechanical exhaust system and ventilation in compartment room to mitigate the hazard of fire.

444359 Development of Novel Emulsified Acid Using Waste Oil

Abstract:

In carbonate reservoirs, hydrochloric acid (HCl) is commonly used for stimulation purposes. The main disadvantage for using HCl is its high and rapid reactivity allowing for less control on acid reaction with formation. In addition, HCl may react with the tubing causing severe corrosion and damages. Several slow-reacting acid systems, though HCl-based, have been developed to retard the HCl reaction and control its reactivity to allow deeper penetration in carbonate reservoirs before it breaks out and reacts with formation rock. However, the selection of the hydrocarbon phase of emulsified acid is critical to the creation and stability of the emulsified acid system. Diesel oil, crude oil and solvents such as xylene have been reported to be successful in the field.

In this work, alternative and less expensive source of oil is developed to replace the current practices. Emulsified acid is a water-in-oil emulsion where oil is the continuous phase and acid is the dispersed one. Industrial and filtered waste oil was used to prepare the emulsified acid of this work. Extensive work has been carried out to study the chemical composition of the waste oil, namely SARA and to select the proper emulsifier. Emulsions were prepared using 15% HCl with emulsifier concentration ranging from 0.5 to 2.0 vol% and 70:30 of acid-to-oil ratio. Droplet size distributions were measured and used to benchmark with other emulsions. Emphasis was given to the stability of the emulsion especially at high temperatures. In order to apprehend the effect of corrosion inhibitor on the emulsion stability, its concentration was varied (up to 0.3 vol%). Later, rheological analyses of the waste oil emulsion were compared to the commonly used diesel-based emulsified acid.

Results showed the potential use of waste oil to prepare high temperature emulsified acid (up to 100˚C). The addition of corrosion inhibitor to the emulsified acid caused a negative effect on stability resulting in a less stable emulsion. On the other hand, addition of cationic emulsifier helped to formulate a more stable emulsion. Apart from that, the optimum concentration of the emulsifier was at 2.0 vol%. The rheological data confirms the stability of the new emulsified formulation at higher temperatures. This paper summarizes the findings of using waste oil emulsion and recommends it for field applications. The newly proposed emulsified acid introduces waste oil as a replacement of diesel.

The novelty of using waste oil will help to significantly reduce the expenses of formulating emulsified acid. Since waste oil is basically waste it has a good impact on the environment, that being said, instead of throwing it, the industry will be able to recycle it.

443764 Optimization of Heat Exchanger Network in Steam Assisted Gravity Drainage Oil Sands Surface Facilities

Abstract:

Steam assisted gravity drainage (SAGD) is the most viable and environmentally safe recovery technology for extracting heavy oil and bitumen. It is a thermal in-situ production process where parallel wells are drilled horizontally into an underground bitumen reservoir. Steam is produced at the central processing facility (CPF) and injected into the reservoir through the wells. The steam heats the bitumen to a point where gravity allows it to flow down to the lower well where the mixture of bitumen and water is then pumped to the surface. The water and bitumen are separated at the CPF. Typically, more than 90% of the energy consumed in producing a barrel of bitumen using SAGD is used to generate the steam that is injected into the bitumen reservoir. Fossil fuels is employed in the process, which causes GHG (greenhouse gas) emissions. The usage of hot and cold utility is reduced by increased heat exchange and optimization of the heat exchanger network, which results in reduction of GHG emissions. In this work, pinch analysis is performed to minimize energy consumption.

443855 Modeling the Thermal Behaviors of an Ultracapacitor Module

Abstract:

Ultracapacitors, also known as supercapacitors, have the potential to meet the increasing power requirements of the sustainable energy systems for the grid-scale energy storage systems and electrified vehicles. As compared to batteries, ultracapacitors offer a higher power density, higher efficiency, and longer shelf and cycle life. Because the performance of an ultracapacitor depends on temperature, it is important to calculate accurately the thermal behavior of a single ultracapacitor cell and the module composed of multiple ultracapacitor cells for the efficient and reliable systems integration of an ultracapacitor in grid and transport applications.

In this work, modeling is performed to study the effect of the layout configugation of the ultracapacitor cells on the thermal behaviors of an ultracapacitor module. The ultracapacitor module is subject to the charge/discharge cycling with constant-current between the nominal voltage of the ultracapacitor module and the half of this voltage. The charge/discharge current values examined are 50, 100, 150, and 200 A. The validation of the modeling approach is provided through the comparison of the modeling results with the experimental measurements.

443856 Modeling the Electrolyte Flow Behaviors of a Zinc/Bromine Flow Battery Module

Abstract:

The zinc/bromine (Zn/Br_2) flow battery is a promising rechargeable power source for energy storage systems because of its inherent chemical simplicity, high degree of reversibility at the electrodes, good energy density, and abundant low-cost materials. The Zn/Br_2 flow battery system, intended for the grid-scale energy storage system, is to be made in modular form. A Zn/Br_2 flow battery module is composed of the multiple stacks consisting of bipolar carbon electrodes. Since the stacks require electrolyte circulation on both the negative and positive sides, two plumbing loops are provided. The flow loops contain an electrolyte storage tank and a circulation pump. A heat exchanger may be included for temperature control in the flow loop. The design of the electrolyte circulation plumbing determines the distributions of electrolyte flow and pressure within the stacks and the efficiency of the stack is dependent on the flow distribution throughout the stack. The modeling of the distributions of electrolyte flow and pressure within the module is needed to optimize the plumbing for electrolyte circulation and thereby to improve the performance of the module. In this work, modeling is performed to study the effect of the layout configuration of the plumbing for electrolyte circulation on the distributions of electrolyte flow and pressure within a 50 kWh Zn/Br_2 flow battery module composed of four stacks. It is seen that the plumbing layout for electrolyte circulation can have a significant effect on the flow rate distribution of electrolyte through the stacks.

443974 Observer-Based Model Predictive Control for Time-Delay Systems

Abstract:

Model predictive control is one of the most widely used advanced process control technologies in industry.[1] It has many capabilities such as dealing with input/state constrains and uncertainties. Observer-based model predictive control was developed to estimate states information when the states are not measurable directly. Many chemical processes have time delays in either states or measurements, such as polymer production and some bioprocesses. [2] In this presentation, we are proposing an unknown input observer based model predictive control algorithm for time-delay systems. Unknown input observer is an observer design to eliminate certain types of unknown inputs in the system. Stability of unknown input observer-based model predictive control without time delay was provided by our previous study. Terminal cost and terminal region are designed for a class of non-linear time delay systems. The close-loop stability of the controller is provided. [3][4]

An example of bioreactor is provided which has uncertain time delays in measurements but with a known upper bound. Two measurements are provided which are online biomass and offline ethanol measurements. Model-plant mismatch is considered as unknown inputs in the system. State estimation and control are calculated using the observer and an optimization algorithm.

Reference:

[1] Marcus Reble et al., Model predictive control of constrained non-linear time-delay systems. IMA Journal of Mathematical Control and Information, 2010, 1-19

[2] Seung Cheol Jeong and PooGyeon Park, Constrained MPC algorithm for uncertainty time-varying systems with state delay, IEEE TRANSACTIONS ON AUTOMATIC CONTROL, 50, 2005, 257-262

[3] Mayuresh Kothare et al., Robust constrained model predictive control using linear matrix inequalities. Automatica, 32, 1996, 1361-1379

[4] Wook Hyun Kwon et al., A simple receding horizon control for state delayed systems and its stability criterion, Journal of Process Control, 13, 2003, 539-551

444044 Organic Solar Cells

Abstract:

This paper describes highlights the need of next-generation "organic solar cells". There is a continuing need for organic solar cells and arrays with very high specific power densities (1000-5000 kW/kg) for generating power in a new generation of ultra light space payloads example in NASA. An emerging technology with promise to meet these ambitious goals are organic solar cells based on very thin films of discrete layers. Theoretical efficiencies for multi junction solar cells are more ~70% as compared to ~30% for single junction devices. There is continuous development going on high mobility, band gap engineered organic donors and acceptors, as well as development of three junction devices and scaled up devices on lightweight flexible polymer substrates. The main interest in this paper is organic semi-conducting crystal: Pentacene, which has photo-conducting states. Organic solar cells manufactured today have low efficiency other than their stability issues. The efficiency of organic solar cells can be improved by either designing polymer materials that absorbs more of solar spectrum or by minimizing the charge collection efficiency by controlling the morphology of polymer films. Once fully developed, organic solar cells would revolutionary power the industry due to their flexibility and minimal weight

444055 A Refined Non-Random Two-Liquid Segment Activity Coefficient Model for Solubility Modeling

Abstract:

Accurate correlation and robust prediction for complex molecule solubility or solid-liquid equilibrium (SLE) in solvents and solvent mixtures have drawn much interest recently [Tanveer et al., *Chem. Eng. Prog.* **2014**, 110 (9), 37-47]. Solubility models such as the Non-Random Two Liquid Segment Activity Coefficient (NRTL-SAC) model [Chen and Song, *Ind. Eng. Chem. Res.* **2004**, 43, 8354-8362] are now routinely used in the industries. However, an important unfilled need is to model the solubility in oligomers/polymers/surfactants such as asphaltene nanoaggregates in petroleum fluids and drug molecules in lipid-based or solid dispersion formulation [Tung, *Org. Proc. Res. Dev.* **2013**, 17, 445-454]. This paper presents a refinement to the very successful NRTL-SAC model to better account for the contribution due to size differences of molecules. Specifically, the Flory-Huggins combinatorial entropy expression for conceptual segments in the model is substituted with the Staverman-Guggenheim expression for real molecules. We present the rationale for the refined model, the updated conceptual segment numbers for common solvents, and the newly identified conceptual segment numbers for common drug excipients. The model performance is illustrated by prediction of drug molecule solubility in various lipid solvents and solvent mixtures, compared to experimental data and previous work by other models. The refined NRTL-SAC model shows enhanced capability to correlate and predict drug molecule solubility in common solvents and excipients used in drug formulation.

443640 Preparation and Characterization of (*FURCRAEA spp*) as Green Support of Iron Nanostructured Catalyst

Abstract:

PREPARATION AND CHARACTERIZATION OF (*FURCRAEA spp*) AS "GREEN" SUPPORT OF IRON NANOSTRUCTURED CATALYST

Karen G. Bastidas G[1], Anamaria Barrera B[1], Cesar A. Sierra A[2], Hugo R. Zea R[3]

[1] Department of Chemical and Environmental Engineering. National University of Colombia (Bogota).

[2] Associate professor. Director of the Research Group of Macromolecules. National University of Colombia (Bogota).

[3] Associate professor. Director of the Research Group of Materials, Catalysis and Environment. National University of Colombia (Bogotá).

Abstract.

The availability of natural water resources has declined over time, becoming one of the biggest problems today. Environmental risk in Colombia's rivers has increase due to the widespread use of harmful chemicals in a variety of processes, ranging from illegal mineral extraction to industry wastewater, e.g. Medellin's river, due to its strategic location along a textile industrial corridor might be subject to dye contamination; the gold and platinum mining zones in Choco and Antioquia regions are being heavily affected by the use of mercury in the extraction processes. In 2009, 520 tons of mercury were imported by miners working in this area, amount that exceeds several times the mercury used by regulated industry in previous years (Ministry of Mines and Energy of Colombia, 2007). Not only the environmental risk is a concern but also the economic conditions of many low-income families whose livelihood in many cases depends exclusively of the of the mining.

Due to the issues raised above, the main goal of this study is to research, develop and implement innovative technologies in the synthesis of highly efficient, low cost and environmentally friendly iron based catalysts capable of degradate pollutants, ensuring operational stability. A native fiber (Fique, FURCRAEA spp) has been selected as a solid support for these catalysts due to its abundance, low cost and special physical and chemical characteristics.

The combination of cellulosic fibers with iron oxide nanoparticles could provide exceptional biodegradable composite materials for the treatment of organic dyes and heavy metals present in wastewater. Likewise, the use of iron particles immobilized on fique fibers allows an easy separation catalyst-treated water, turning the biocatalyst in highly recyclable and reusable material.

Fique fiber was initially characterized by determining the percentage of ashes, total lignin, soluble and insoluble lignin, cellulose, hemicellulose, total solids and humidity, obtaining 0.94, 22.58, 0.5, 22.8, 36.3, 27.5 0.701 and 2.1%, respectively. Additionally, the isoelectric point (pzc), acidic and basic sites were measured.

Then, the fiber was subjected to a pretreatment in order to expose more active sites which on time serve as nucleation points for the deposition of nanoparticles. This was done by means of an ultrasound-assisted procedure, varying the sonication time (30, 60 and 90 minutes). The effect of the sonication was followed by measuring the contents of lignin, cellulose and hemicellulose within the fiber after each time. 60 minutes of sonication was found to be most beneficial time since the amount of lignin and hemicellulose present in the fiber was reduced, while conversely the amount of crystalline cellulose exposed increases.

In order to improve the chemical ability of the nucleation sites to interact with iron and then to form stable and size controllable nanoparticles, a cationization procedure was performed on the fiber; this was achieved by the exposure of the fiber to acidic and basic conditions promoting cationization. Three cationization times (1, 2 and 3 hours) were studied and their effect on the fiber chemical surface was evaluated using FTIR-ATR and XRD analysis. The results showed 3 hours as the optimal cationizing time.

Subsequent to the cationization of the fiber surface, a procedure of ionic/cationic exchange was implemented by exposing the fiber to a solution of iron salts in order to replace the cations in the fiber surface and anchoring the iron nanoparticles. Different conditions were studied in order to synthetize zero valent iron nanoparticles (nZVI) and others with different oxidation states. Finally, to track the impregnation process on the fiber, samples from the precursor solution were taken at different times and its iron concentration was determined by atomic absorption spectrophotometry (AAS). Complementary characterization performed by X-ray diffraction, BET surface area and electronic microscopy, among others.

Key words: Natural support (fique), iron nanoparticles, bionanocatalyst.

444057 Explaining the Paradox of Asphaltene Precipitation with n-Alkanes

Abstract:

Explaining the paradox of asphaltene precipitation with n-alkanes

Md Rashedul Islam and Chau-Chyun Chen
Department of Chemical Engineering, Texas Tech University

Abstract
Paraffins, specifically alkanes, promotes asphaltene precipitation when they are blended with asphaltene-containing petroleum fluids. The alkane/oil volume ratio required for onset of asphaltene precipitation increases with the carbon number and yields a maxima. Experimental study indicates that the maxima lays at C7[1] or between C8 -- C10 [2]. However, precipitation yield decreases asymptotically with increase of the carbon number even if excessively high alkane/oil volume ratio is used [3]. For example, n-pentane be a better solvent for asphaltene than n-hexane in regard of both onset of precipitation and precipitation yield. However, high carbon number alkane beyond the maxima, n-hexadecane for instance, can be a better solvent at onset of precipitation point and a poorer solvent in regard of precipitation yield than n-pentadecane. A thermodynamic explanation for this paradox is necessarily important to find the compatibility of bitumen or crude oil with paraffinic hydrocarbons. In this work, this paradox is explained by the aggregation thermodynamics of asphaltene precipitation [4]. Both the enthalpy and entropy of mixing are investigated to explain the effect of paraffin carbon number on asphaltene precipitation. Athabasca and cold lake bitumen samples are studied.

References

[1] G. Hotier and M. Robin, "Action de divers diluants sur les produits pÃ©troliers lourds: mesure, interprÃ©tation et prÃ©vision de la floculation des asphalténes," Oil & Gas Science and Technology, vol. 38, pp. 101-120, 1983.
[2] I. A. Wiehe, H. W. Yarranton, K. Akbarzadeh, P. M. Rahimi, and A. Teclemariam, "The paradox of asphaltene precipitation with normal paraffins," Energy & fuels, vol. 19, pp. 1261-1267, 2005.
[3] D. L. Mitchell and J. G. Speight, "The solubility of asphaltenes in hydrocarbon solvents," Fuel, vol. 52, pp. 149-152, 1973.
[4] M. Wang, Y. Hao, M. R. Islam, and C. C. Chen, "Aggregation Thermodynamics for Asphaltene Precipitation," AIChE Journal, Submitted.

Using a Large Data Bank to Develop a Robust and Simple Model to Estimate the Saturation Pressure of Oil Reservoirs

By

Hassan Alkandari[1], Adel Elsharkawy[1], Adel Malallah, Osamah Alomair

Petroleum Engineering Department, College of Engineering & Petroleum
Kuwait University
Po Box 5969 Safat 13060-Kuwait

Abstract

Measurements of saturation pressure are crucial for all hydrocarbon reservoir fluids. Below the crude oil saturation pressure gas reaches a critical saturation; then, a two phase flow occurs and results in decreasing oil production and recovery. To maximize oil production and recovery, the reservoir pressure has to be maintained closer to the original saturation pressure. This pressure is normally measured using bottom-hole samples or surface recombined samples of oil and gas. Occasionally, the samples become unavailable and the pressure needs to be estimated using computational methods.

In this study, a large data bank of 231 crude oil compositions and saturation pressure measurements including literature data, unpublished data, and newly measured data were used to develop two empirical models to predict the saturation pressure of a variety of crude oils if the oil sample is unavailable or the experimental measurements are unreliable. The first proposed model utilizes the extended compositions of hydrocarbons up to the heptane plus fraction in addition to non-hydrocarbons. The second model uses the lumped compositions of light, intermediate, and heavy components in addition to non-hydrocarbon components as the input variables. The lumped model has the advantage of using fewer input parameters while maintaining the thermodynamics. The accuracy and validity of both models to calculate the saturation pressure for volatile oils, black oils, and heavy oils are presented using several compositional data. The models performance is also compared to the Peng-Robinson equation-of-state (PR-EOS), and the Soave-Redlich-Kwong equation-of-state (SRK-EOS) as well as all published methods that use compositions as input variables. The comparison indicates that the proposed models are much simpler and more accurate than the other computational methods. The proposed models treat the heptane-plus fraction as a single component, thus eliminating the splitting and characterizing the plus fractions and the binary interaction parameters needed for the EOS's calculations.

Key words: Saturation pressure (Ps), Bubble-point pressure (BPP), Equation of state (EOS), Correlations, and PVT Data

[1] Corresponding Authors: Hassan Alkandari , Hassan@hkandari.com, Adel Elsharkawy, adel.elsharkawy@ku.edu.kw

Introduction:

Reservoir fluid properties are an essential source of information for all calculations on hydrocarbon reserves. Accurate estimation of the pressure-volume-temperature (PVT) measurement is crucial. Of all the PVT properties, saturation pressure (Ps) is considered the most important parameter. It is well known that gas starts to flow once the reservoir pressure drops below the bubble point pressure and the gas saturation reaches a critical value. To minimize the gas flow and maximize the oil recovery, it is very important to maintain the reservoir pressure close to the original bubble point pressure.

Although the PVT properties are experimentally measured in laboratories and provide reliable results, this process is expensive, time consuming, and sometimes unavailable; therefore, measured field data (separator gas oil ratio, separator pressure, stock-tank oil specific gravity, and reservoir temperature) or reservoir fluid compositions are used to estimate the bubble point pressure.

There are two methods in the petroleum industry to calculate the saturation pressure of crude oil when experimental is unavailable; either using production data or compositional data.

Methods Based on Production Data (Composition is unavailable)

Numerous attempts have been made since the 1940s to develop correlations to estimate the saturation pressure from production data, such as producing gas oil ratio, oil gravity, gas gravity, reservoir pressure and temperature. Table 1 presents summary of all the methods that use production data to estimate the saturation pressure.

Methods based on Compositional data

Many authors have reported that the application of any correlation in other regions results in serious errors. These conclusions arise from the fact that regional crude oils represent a physical-chemical trend; therefore, all attempts to develop a general correlation using field measured parameters to estimate the saturation pressure from a large data bank have failed. For this reason, Elsharkawy (2003) made the first attempt to correlate the saturation pressure to the reservoir fluid composition instead of using the production data. In recent years, there have been several attempts to estimate the saturation pressure of oil reservoirs. These methods are reported in Table 1.

Elsharkawy (2003) presented a correlation of the bubble point pressure for the oil reservoir using the same input data needed for the equation-of-state (EOS) calculations, i.e., the compositional data and the reservoir temperature. Elsharkawy's model is based on the saturation pressure measurements of 60 crude oil samples from the Middle East and 75 collected from the literature. He demonstrates that the calculation of the saturation pressure depends on the type of EOS, the number of pseudo components, the method used to characterize the pseudo-components and the binary interaction number. Elsharkawy compared the performance of his model to the SRK-EOS and PR-EOS for black oils, volatile oils, heavy oils, and synthetic oils. He indicated that his model eliminates splitting the heavy plus fraction, characterizing the pseudo-components and inclusion of the binary interaction parameters needed for the EOS calculations.

Later, Tale et al. (2009) made the second attempt to model the saturation pressure of crude oil using 132 data samples published by Elsharkawy (2003). Tale's model, which is based on the alternating conditional expectation (ACE), has a set of 14 equations with 42 constants. The average absolute relative error (AARE) reported by Tale et al. indicates that the model did not perform well with the selected data. The ACE model has an AARE of 12.3% compared with AARE of 6.7% by Elsharkawy (2003) and 8.1% AARE for the SRK-EOS.

AlQuraishi (2009) attempted to model the saturation pressure using Elsharkawy's (2003) data. The model was developed using linear genetic programming. He correlated the saturation pressure to the formation temperature, methane content, and heptanes plus molecular weight. Thus, AlQuraishi's model neglected the effect of the intermediate and heptane plus content on the saturation pressure. The error analysis reported by AlQuraishi indicates that his model has a 5.8% error compared to 7.7% by Elsharkawy, 9.9% using the SRK-EOS, and 10.1% using the PR-EOS.

Bandyopadhyay and Sharma (2011) presented a model to predict the bubble point pressure, using temperature and fluid composition. They used 129 experimental and literature data originally published by Elsharkawy (2003) to develop their model. The model introduced a temperature interaction parameter that depends on compositions. The proposed model has a set of 24 coefficients and the estimated bubble point pressure is calculated as the sum of the series calculations. They reported an average absolute error (AAR) of 7.83% using their model compared to 8.3% using Elsharkawy's simple model (2003), 10.15% using SRK-EOS, and 10.58% using the PR-EOS.

Khamehchi et al. (2011) used an intelligent proxy simulator (IPS), which is based on an adaptive neural fuzzy inference system (ANFIS) and a structured artificial neural network (ANN) to predict the bubble point pressure and asphaltene onset pressure. The model uses a reservoir fluid composition and temperature to predict the Ps. It is not clear how the Ps is calculated from the proposed method, and they did not present the accuracy of the proposed method compared to the other published methods.

Farasat et al. (2013) presented a new mathematical approach to calculate the crude oil saturation pressure as a function of the temperature and the reservoir fluid composition. The model developed by Farasat et al used 130 experimental data, which were originally published by Elsharkawy (2003). Farasat used a support vector machine (SVM) to model the saturation pressure. They did not clearly demonstrate how to calculate the saturation pressure from the SVM model. Moreover, they did not indicate the overall average absolute deviation (AAD) for their model compared to other published models and both the PR-EOS and SRK-EOS for the 130 data.

Ahmadi et al. (2014) also presented an ANN technique called gene expression programming (GEP) to determine the bubble point pressure of the oil samples. Similar to Farasat et al. (2013), Ahmadi's model used most of the data previously published by Elsharkawy (2003) to construct a non-linear equation using reservoir fluid compositions and temperatures as input parameters. The proposed model is highly complicated and consists of 8 equations that have several constants compared to the one simple equation proposed by Elsharkawy (2003). They compared the performance of their proposed model to the SRK-EOS and PR-EOS; however, their paper does not mention how the heptanes plus fraction were treated and characterized for the EOS calculations. They reported an AAD of 4% using the proposed model compared to 8% using Elsharkawy's model (2003), 10.9% by PR-EOS, 10.5% by SRK-EOS, and 15% by Bandyopadhyay and Sharma (2011).

Gholami et al. (2014) estimated the bubble point pressure using a support vector regression SVR and a supervised learning algorithm SLT based on the statistical learning theory, and then the ACE was used. The model used reservoir fluid compositions and temperature to estimate the bubble point pressure. Gholami et al. used the same data originally published by Elsharkawy (2003). They presented three attempts based on genetic programming to correlate the bubble point pressure: SVR, ACE, and power-law committee machine (PLCM). The ACE-based method resulted in a set of three equations with a matrix of [8 by14] constants. Gholami et al. compared the accuracy of their model to the SRK-EOS and PR-EOS predictions reported by Elsharkawy (2003). Based on their AARD analysis, the PLCM model has the lowest errors followed by the SVR model, Elsharkawy's model, ACE model, PR-EOS, and SRK-EOS.

Lately, Jarrahian et al (2015) presented two empirical models to estimate the saturation pressure of black oils. The first model is based on production information. The second model is based on compositional information mostly from Elsharkawy (2003) and Heidaryan and Moghadasi (2011). They have not presented comparison of the accuracy of their model to Farasat (2013) model, Ahmadi (2014) model, and Gholami (2014) model.

This paper is concerned with the estimation of saturation pressure using compositional data. Therefore, the objective is to develop a robust yet simple model to estimate the saturation pressure for various types of reservoir oils using a larger data bank than that used by other published studies. The accuracy of the proposed models is compared to previously published models as well as SRK-EOS and PR-EOS.

Data bank

This study uses a large data set of 231 crude oil compositions and saturation pressure measurements. The data set consists of 130 data sets originally reported by Elsharkawy (2003); in addition to others collected from literature (Jacopy, 1958; Vogel, 1980; Coats, 1986; Droham, 1980; Dansh, 1992; Moharam, 1996; Rosenegger, 1999); 94 new data sets that have not been utilized before; as well as 7 newly measured samples at the Petroleum Fluid Research Center (PFRC) in Kuwait University. These data represents a wide range of compositions, such as volatile oils, black oils, and heavy oils. The data cover a pressure range of 537 to 5065 psia and a temperature range of 58 to 319 °F, crude oil compositions from C_1 to C_{7+}, non-hydrocarbons, such as N_2, H_2S, and CO_2, and the molecular weight and the specific gravity of the heptane plus fraction.

Data screening and classification

The compositions and the saturation pressure measurements in the data bank are subjected to a meticulous screening process to check the validity of the data. Our screening criteria uses the following steps: the sum of the molar compositions of all the components must add up to one, otherwise, the sample is eliminated. Samples with a missing molecular weight or specific gravity of the plus fraction are also eliminated.

After filtering, the data were reduced to 214 and divided into 179 measurements used to develop the new models, and 35 data sets are used for the blind test of the robustness of the proposed models. Statistical analysis of the input variables of the data set used in training and testing the extended and the lumped models are illustrated in Tables 2.

Several of the reported compositional data have extended analysis up to (C_{20+}). The following techniques are used for lumping or regrouping the extended analysis into single pseudo-fractions (C_{7+}). The mole fraction (X_i) of the individual components of the extended analysis is equal to the plus fraction (Ahmad, 2011).

$$X_{C_{7+}} = \sum_{i=7}^{n} X_i \qquad i=7, 8 \ldots n \qquad (1)$$

The average molecular weight (Mw) of the individual component of the extended analysis is equal to the molecular weight of the plus fraction.

$$Mw_{C_{7+}} = \sum_{i=7}^{n} \frac{X_i Mw_i}{X_{C_{7+}}} \qquad i=7, 8 \ldots n \qquad (2)$$

The density of the total plus-fraction can be calculated as follows:

$$\gamma_{C_{7+}} = \frac{X_{C_{7+}} Mw_{C_{7+}}}{\sum_{i=7}^{n} \frac{X_i Mw_i}{\gamma_i}} \qquad i=7, 8 \ldots n \qquad (3)$$

It is important to note the significance of the density of plus fraction because its density determines its paraffinic naphthenic aromatic (PNA) content.

Equation-of-state calculations

The EOS calculation of the saturation pressure is compared with the newly developed models as well as the other methods. The two EOS calculations considered in this paper are the ones most considered and widely used in the petroleum industry to evaluate the volumetric and phase behavior of reservoir fluid properties: the Soave-Redlich-Kwong EOS (SRK-EOS) and the Peng-Robinson EOS (PR-EOS). It should be noted that the accuracy of any EOS calculation of the saturation pressure largely depends on the number of pseudo-components and the splitting schemes used to divide the heptane plus fraction; as well as the characterization and calculation of the critical properties of the pseudo-components and the binary interaction parameters. Therefore, the two EOS calculations of the saturation pressure are not the same, even though the number of pseudo-components, characterization of the pseudo-components, and binary interaction parameters were the same.

Model development

As previously stated, the objective of this work is to provide the petroleum reservoir engineer with a simple yet accurate tool to estimate the saturation pressure for a variety of crude oils using compositional data when experimental is unavailable. Two models are presented in this study to calculate the saturation pressure of the crude oil samples. The first model uses all input parameters, such as the molar compositions of the hydrocarbon and non-hydrocarbon components similar to the EOS models. This model is referred to as the extended model. The second model uses fewer input parameters in which some of the hydrocarbon components have been lumped into a single group; this is designated as the lumped model. Both models are simpler than all published correlations, physically correct, and have an acceptable accuracy compared with the well-known EOS.

Figure 1 shows a correlation matrix of all the data available under consideration. This figure also indicates the relative importance of each of the input variables into the first proposed model. It agrees with most published correlations and the EOS that the methane content, heptane plus content and temperatures have the greatest impact on the saturation pressure followed by the intermediate content. The data understudy indicates that the non-hydrocarbon has the smallest impact. This result is simply a reflection of the molar contribution of each component comprising the crude oil composition.

Model I - Extended Model

The extended model uses 13 variables comprising the molar compositions of the pure components, heptane plus properties, and reservoir temperatures, which are similar to the input parameters for the EOS. Figure 2 shows the correlation coefficient graph of all the input parameters in the extended model. Thus, the elimination of the intermediate effect has a negative impact on the accuracy of the predictive model. Our proposed extended model has the following form:

$$Ps = 124.72N_2 + 17.57CO_2 + 22.55H_2S + 64.22C_1 - 9.80C_2 - 52.49C_3 + 6.16C_4 - 19.22C_5 - 23.63C_6$$

$$- 21.43C_{7+} + 435.31\gamma_{C7+} + 1.14Mw_{C7+} + 4.29T \qquad (4)$$

Where Ps is the saturation pressure in psia, T is the temperature in °F, and the hydrocarbon and non-hydrocarbon compositions are expressed in mole.

Model II - Lumped Model

The second model is much simpler and uses only 7 -variables (the lumped model). Several lumping schemes are adapted in the petroleum industry for phase behavior modeling and reservoir simulation of the petroleum reservoir fluids (Ahmed, 2011). Hydrocarbon components are grouped into a pseudo fraction based on their molecular weight (Whitson, 1980), the similarity of their physical-chemical properties (Lee, 1979), their carbon number, their true boiling point or their number of families based on the chemical structure (Becker et al., 2015). (Emera and Samara, 2005) proposed lumping the hydrocarbon components into volatile (C_1 & N_2), intermediate (CO_2, H_2S, C_2 to C_4) and heavy categories (C_{5+}) to calculate the minimum miscibility pressure. (Duan et al., 2013) proposed lumping several components into a single carbon number (SCN) with predetermined values for properties, such as critical pressure, temperature, and eccentric factor. Others have proposed optimal lumping schemes based on the objectives (Lin et al., 2008). In lumped model, $C_{1+}N_2$, C_2, CO_2, and H_2S, the intermediate (C_3 to C_6), heptanes plus (C_{7+}) and its properties, and temperature are the input variable. The correlation matrixes of the lumped components are provided in Figure 3. Even though the non-hydrocarbon has the least effect their contribution is important, especially the H_2S in the case of sour crudes. Both models presented by AlQuraishi (2009) and Bandyopadhyay and Sharma (2011) have eliminated the effect of H_2S.

The proposed model has the following form:

$$Ps = 82.115*(C_1+N2) - 11.635*C_2 + 39.158*CO_2 + 38.244*H_2S - 1.217*(C_3 — C_6) - 890.701*(C_{7+}/MW_{7+}*\gamma_{7+}) + 4.217*T - 1042 \quad (5)$$

Where Ps is the saturation pressure in psia; and T is the temperature in °F.

Result and discussion

Model accuracy

The accuracy of the two proposed models used to calculate the saturation pressures of the previously described crude oil samples is shown in Table 3. This table shows reports the average relative error, average absolute relative error standard deviation and the correlation coefficients for all the methods considered in this study. Table 3 also reports the accuracy of the PR-EOS, SRK-EOS, Elsharkawy's model, AlQuraishi's model, and Bandoypandyay and Sharma's model and Jarrahian model. The methods proposed by Gholami et al., Ahmadi et al., and Jarrahian et al are included in Table 3, even though they produce a unreasonable estimate of the saturation pressures, as explained in the appendix.

This table indicates that the extended and the lumped models have an (AARE) in the order of 8% and 8.5%, respectively. It should be noted that the complications in the Bandoypandyay and Sharma model did not improve the accuracy of the saturation pressure, because it had a large error. It is also noted that Jarrahian et al model produces an average error of 35%. Compared with the EOS, the lumped model has a few constants and eliminates the splitting, characterizing the pseudo-components and binary interaction parameters necessary for the EOS calculations. The lumped model as well as the extended model are very simple and require one-step calculations compared with all the other methods reported in Table 3.

A cross-plot of the calculated saturation pressure using the extended model based on the training data versus the measured data is illustrated in Figure 4A. This figure indicates that most of the data falls on the unit slope line, and there is no over-fit. This trend is supported by the high correlation of regression of the model (97.6%) presented in Table 3. Figure 4B shows a cross-plot of the testing data (28 from literature and 7 new measured in PFRC at Kuwait University). The extended model indicates an excellent agreement with all of the 28 data sets chosen at random, with the exception of a few data points at higher pressures. Additionally, the model has a perfect match for the 7 newly measured samples.

Figure 5A shows the performance of the lumped training-model for capturing the measured saturation pressure for most of the 179 samples. A cross-plot of the calculated saturation pressure using the lumped model based on the training data versus measured data is depicted in Figure 5B. This figure indicates that most of the data scatters falls to the unit slope line. The lumped model also has a high coefficient of regression of 97.4%, which is indicated in Table 3.

Models Validity

The accuracy and validity of the models for estimating the saturation pressure of various types of crude oil are reported in Table 4. This table illustrates the capability of the newly proposed models to predict the saturation pressure for volatile oils, black oils, and heavy oils. Table 4; provides the compositions, properties of the heptane plus fraction, temperature, and measured saturation pressures of the samples. Additionally, the calculated saturation pressure using the PR-EOS, SRK-EOS, Elsharkawy's model (2003), AlQuraishi's model (2009), and Bandoypandhyay and Sharma's model (2011), Jarrahian et al (2015); as well as the extended and lumped models are provided in Table 4. It is clear from this table that the proposed models have a wide range of applications regardless of crude oil types. We have also included information regarding the 7 newly measured data; at the PFRC, compositional, and saturation pressure. At the bottom of Table 4, we have provided the unpublished data of the compositions; and the saturation pressures at low temperatures. It is well known that the saturation pressure of a crude sample is measured at low temperatures in addition to the reservoir temperature as a method to check the quality of the sample.

Figure 6A reports the error of all the methods considered in this paper to predict the saturation pressure for volatile oils (heptane plus content below 30%). Both the extended and the lumped models have an AARE of nearly 4% followed by Bandoypandhyay and Sharma's model (2011); and Elsharkawy's model (2003); the SRK-EOS has nearly twice the AARE of the newly proposed model. AlQuraishi's model (2009) has an AARE of 12% because the model is based on methane and neglects the effect of heavy and intermediate components, whose contribution is highly significant in the reservoir and production performance of volatile crudes. Jarrhaine et al (2015) model is not suitable for volatile oils. It shows a 35% error in estimated saturation pressure.

The performance of the extended and lumped models compared to the various methods for the black oil samples (35< C_{7+} <55%) is provided in Figure 6B. The newly proposed models have an AARE close to 10%; followed by Bandoypandyay and Sharma's model 14%, the SRK-EOS and PR-EOS 16%, and Elsharkawy 17%. AlQuraishi's model has an AARE of 24%. Jarrhain et al (2015) model shows a 33% error in estimated saturation pressure even though the authors claim that the model is suitable for black oils.

Figure 6C shows the error level of all the methods as well as the proposed models for the newly measured crude oil samples at Kuwait University using the PFRC. Again, Jarrhain et al (2015) model show a 40% error in estimated saturation pressure.

The performance of all published models at low temperature is illustrated in Figure 6D. The low-temperature calculation indicates that Bandoypandhyay and Sharma's model (2011) has accuracy problems and is not valid at low temperatures because the authors proposed a temperature-composition related coefficient. Also, Jarrhain et al (2015) model is not suitable at low temperature. It shows a 40% error in estimated saturation pressure.

The ability of the proposed models to capture the thermodynamic is thoroughly checked against its independent variables. The validity of the model is compared to the experimentally measured data as well as the predictions by other methods used to capture the physics of increasing the saturation pressure as a function of temperature is illustrated in Figure 7. Two main points are observed here; first, the model matches the measured saturation pressures much better than the other methods. Second, it follows the physical trend predicted by the EOS's. Figure 8 shows the model performance for capturing the effect of changing the saturation pressure as a function of the molecular weight of the heptane plus fraction compared with the SRK-EOS and PR-EOS. The extended model follows the trend of the PR-EOS in increasing the saturation pressure. Elsharkawy's model (2003) was unable to predict the correct trend as predicted by the EOS's. The model sensitivity in capturing the change in the saturation pressure as a function of increasing the methane content is shown in Figure 9. The proposed model as well predicts the correct trend of increasing the saturation pressure when increasing the methane content.

Conclusion

In this paper, two models are presented to estimate the saturation pressure of a variety of crude oils when measures are unreliable or samples are unavailable. The two models are developed using compositional data and saturation pressure measurements from a large data bank comprising literature data, unpublished data, and newly measured and reported data of 231 crude oil samples. The input parameters in the first model are the extended hydrocarbon and non-hydrocarbon molar percentages and reservoir temperatures. The second model uses lumped hydrocarbon fractions of intermediate, and heptane plus fractions in addition to its properties; and the reservoir temperature as input parameters. The accuracy of the two models has been tested and compared to that of the SRK-EOS and the PR-EOS; as well as published correlations that use compositional data as the input parameters. The comparison indicates that the two models are much more accurate than all the published methods. The validity of both models has also been tested using various types of crude oils samples and compared to experimentally measured saturation pressures and estimates using the EOS as well as the published correlations. The proposed models are physically correct and much simpler than all the published methods. The models do not require tedious computations and eliminate splitting of the plus fraction, characterization of the pseudo-components; as well as the binary interaction parameters that are required for the EOS calculations.

Acknowledgement

The authors thank the General Facility Lab at Kuwait University for their cooperation and conducting the compositional and saturation pressure measurements of the 7 new samples described in this paper using Petroleum Fluid Research Center (PFRC) project no. GE 01107.

References

Agarwal, R.K., Li, Y.K., Nghiem, L., 1990. A regression technique with dynamic parameter selection for phase – behavior matching. SPE Res. Eng. February, 115 – 120.

Ahmadi, M. A., Zendehboudi, S, James, L., Elkamel, A., Dusseault, M., Chatsiz, I., Lohi, A., 2014, New Tools to determine Bubble Point Pressure of Crude Oils: Experimental and Modeling Study, *Journal of Petroleum Science and Engineering, v.123, 204-216.*

Ahmed, T., 2011, Splitting and lumping schemes of the plus fraction, working guide to vapour-liquid phase equilibria calculations, Chapter 9, Hydrocarbon Phase Behaviour, Gulf Publishing Co. 111-126.

Al-Gathe, A., Abd-el Fattah, Kh., El-Mtwally, K., 2014, New artificial intelligent approach for bubble point pressure, Int. J innovation and Sci. Research, Vol.4, No.2, July, 121-135.

Al-Marhoun, M.A. 1985. Pressure-Volume-Temperature Correlations for Saudi Crude Oils. Presented at the SPE Middle East Oil Technical Conference and Exhibition, Bahrain, 11–14 March. SPE-13718.

Almehaideb, R.A. 1992. Improved PVT Correlations for UAE Crude Oils. Presented at the Middle East Oil Show and Conference, Bahrain, 15-18 March. SPE-37691.

Al-Najjar, H.S., Al-Soof, N.B.A., and Al-Khalisy, K.M. 1988. Correlations for Bubble-Point Pressures, Gas Oil Ratios and Formation Volume Factors for Iraqi Crude Oils. *Journal of Petroleum Research* (June 1988): 13.

AlQuraishi, A. A., 2009, Determination of crude oil saturation pressure using linear genetic programing, Energy & Fuel, 23, 884-887.

Al-Shammasi, A.A. 2001. A Review of Bubblepoint Pressure and Oil Formation Volume Factor Correlations. *SPE Res. Eval. & Eng.* 4 (2): 146-160. SPE71302.

Asgarpour, S., McLauchlin, L., Wong, D. et al. 1989. Pressure-Volume-Temperature Correlations for Western Canadian Gases and Oils. *J Can Pet Tech.* 28 (4): 103.

Asoodeh, M., Bagheripour, P.,2012, Estimation of bubble point pressure from *PVT* data using a power-law committee with intelligent systems, *J. Pet. Sci. and Eng., Vol. 90–91, July 2012, Pages 1-11.*

Bandyopadhyay, P., Sharma, A., 2011, Development of a new semi analytical model for prediction of bubble point pressure of crude oils, *Journal of Petroleum Science and Engineering, Volume 78, Issues 3–4, September 2011, Pages 719-731.*

Becker, P. J., Celse, B., Guillaume, D., Dulot, H., Costa, V., 2015, Hydrotreatment modelling for variety of VGD feedstocks: a continuous lumping approach, Fuel, 139, 133-143.

Coats, K.H., and Smart, G.T., 1986, Application of regression base EOS PVT Program to laboratory data, SPERE, May 277-299.

Dansh, A. Xu, D. H., Todd, A. C., 1992 A grouping method to optimize reservoir description for compositional simulation of gas injection process, SPERE, Aug. 343-348.

De Ghetto, G. and Villa, M. 1994. Reliability Analysis on PVT Correlations. SPE-28904, Presented at the European Petroleum Conference, London, United Kingdom, 25-27 October.

Dindoruk, B. and Christman, P.G. 2001. PVT Properties and Viscosity Correlations for Gulf of Mexico Oils. Presented at the SPE Annual Technical Conference and Exhibition, New Orleans, 30 September-3 October. SPE-71633

Dokla, M.E. and Osman, M.E. 1992. Correlation of PVT Properties for UAE (United Arab Emirates) Crudes. *SPE Form Eval.* (March 1992): 41.

Droham, J.K., 1988, Enhancing the evaluation of PVT data, OSEA paper presented the 7th Offshore South East Asia, Feb. 2-5, Singapore.

Duan, J., Wang, W., Liu, H., Cong, J., 2013, modelling the characterization of the plus fraction by using continuous distribution function, Fluid Phase Equilibria, 345, 1-10.

Elam, F.M. 1957. *Prediction of Bubble Point Pressures and Formation Volume Factors from Field Data.* MS thesis, University of Texas at Austin, Austin, Texas.

Elmabrouk, S., Zekri, A., Shirif E., 2014, The prediction of bubble point pressure and bubble point oil formation volume factor in absence of PVT analysis, PET. Sci. & Tech., 32, 1168-1174.

Elsharkawy, A. M., 2003, An empirical model for estimating the saturation pressures of crude oils, Journal of Petroleum Science and Engineering 38 (2003) 55-77.

Elsharkawy, A.M. and Alikhan, A.A. 1997. Correlations for predicting solution gas/oil ratio, oil formation volume factor, and undersaturated oil compressibility. *J. Pet. Sci. Eng.* 17 (3–4): 291-302.

Emera, M., Samara, H., 2005, Use of genetic algorithm to estimate CO2-oil minimum miscibility pressure- a key parameter in design of CO2 miscible flood, J. Pet. Sci. Eng., Vol. 46, issue 1-2, 37-52.

Farasat, A., Shokrollahi, A., Arabloo, M., Gharagheizi, F., Mohammadi, A., 2013, toward an intelligent approach for determination of saturation pressure of crude oil, Fuel Processing Technology 115 (2013) 201-214.

Frashad, F., LeBlanc, J.L., Garber, J.D. et al. 1996. Empirical PVT Correlations for Colombian Crude Oils. SPE-36105 Presented at the SPE Latin American and Caribbean Petroleum Engineering Conference, Port of Spain, Trinidad and Tobago, 23–26 April. -MS.

Gharbi, R.B.C., Elsharkawy, A.M., 1999, neural network model for estimating the PVT properties of Meddle East crude oils, SPE Reservoir Eval. Eng. 2 (3), 255-265.

Gholami, A., Asoodeh, M., Bagheripour, P., 2014 How committee machine with SVR and ACE estimates bubble point pressure of crudes, *Fluid Phase Equilibria, Volume 382, 25 November 2014, Pages 139-149.*

Glasø, Ø. 1980, Generalized Pressure-Volume-Temperature Correlations. *J Pet Tech.* 32 (5): 785-795.

Heidaryan, E., Moghadasi, J., 2011, Crude saturation pressure through imperial correlation, Thermodynamics 2011, Athens, 31 August- 3 September 2011.

Hasan, M., 1993, PVT Correlations for Indonesian Crude Oils. *Proc.* 22nd Annual Convention of the Indonesian Petroleum Association, Indonesia, Vol.1, IPA93-2.1-070.

Jarrhian, A., Moghadsi, J., Heidaryan, E., 2015, Empirical estimating of black oil bubblepoint (saturation) pressure, J. Petroleum Science and engineering, (126), 69-77.

Jacopy R.H., Berry V.J.,1958, A method for predicting the pressure maintenance for reservoir producing volatile oil, AIME 213, 59-65.

Kartoatmodjo, T., Schmidt, Z. 1994, Large Data Bank Improves Crude Physical Property Correlations. *Oil Gas J.* 92 (27): 51–55.

Kartoatmodjo, T.R.S. and Schmidt, Z. 1991, *New Correlations for Crude Oil Physical Properties*, SPE 23556.

Khairy, M., El-Tayeb, S., Hamdallah, M., 1998. PVT Correlations Developed for Egyptian Crudes. *Oil Gas J.* 96 (18): 114.

Khamekhcki, E., Behvandi, R., Rashid, F., 2011, Prediction of bubble point pressure & asphaltene onset pressure during CO2 injection using ANN & ANFIS models, Pet. Sci. & Tech. Vol. 1 No.2, 35-45.

Labedi, R.M. 1982, *PVT Correlations of the African Crudes*, PhD thesis, CSM.

Labedi, R.M. 1990, Use of Production Data to Estimate the Saturation Pressure, Solution GOR, and Chemical Composition of Reservoir Fluids, SPE-21164-MS. Presented at the SPE Latin America Petroleum Engineering Conference, Rio de Janeiro, Brazil, 14-19 October,

Lasater, J.A. 1958, Bubble Point Pressure Correlations. *J Pet. Tech.*10 (5): 65–67. SPE-957-G.

Levitan, L.L. and Murtha, M. 1999. New Correlations Estimate Pb, FVF. *Oil Gas J.* 97 (10): 70.

Lee, S., 1979, Experimental and theoretical studies on the fluid properties required for simulation of thermal process, Paper SPE 8393.

Li, Y.K., Nghiem, L.X., Siu, A., 1985. Phase behavior computations for reservoir fluids: effect of pseudo-components on the phase diagrams and simulation results. J. Can. Pet. Technol. Nov. – Dec., 29 – 36

Lin, B., Leibovici, C. F., Jorgensen, S. B., 2008, optimal component lumping: problem formulation and solution techniques, Computer & Chem. Eng., 32, 1167-2272.

Macary, S.M. and El-Batanoney, M.H. 1992, Derivation of PVT Correlations for the Gulf of Suez Crude Oils. *Proc. of the* 11th EGPC Pet. Exp. & Prod. Conference, Vol. 1, 374, Cairo, Egypt,

Moharam, H.M., 1995. Prediction of viscosity of heavy petroleum fractions and crude oil using a corresponding states method. Ind. Eng. Chem. Res. 34, 4140 – 4144.

Moradi, B., Malekzadeh, E., Amani, M., Boukadi, F.H., Kharat, R., 2101, Bubblepoint pressure imperial correlation, SPE 132756.

Obomanu, D.A. and Okpobiri, G.A. 1987, Correlating the PVT Properties of Nigerian Crudes. *J. Energy Res. Tech.* 109 (4): 214-217.

Okoduwa, I. G., Ikiensikimama, S.S., 2010, Bubble point pressure correlation for Niger Delta Crude Oils, SPE 136968.

Omar, M.I. and Todd, A.C. 1993. Development of New Modified Black Oil Correlations for Malaysian Crudes. SPE-25338 Presented at the SPE Asia Pacific Oil and Gas Conference, Singapore, 8-10 February.

Owolabi, O.O. 1984. *Reservoir Fluid Properties of Alaskan Crudes*. MS thesis, University of Alaska, Fairbanks, Alaska (May 1984).

Pedersen, K.S., Thomassen, P., Fredenslund, Aa., 1988a. Characterization of gas condensate mixtures. AIChE Spring National Meeting, March 6 – 10, New Orleans, LA.

Peng, D. y., Robinson. D. B., 1976. A new two-constant equation of state , Ind. Eng. Chem. Fund. 15, 59-63.

Petrosky, G.E. Jr. 1990. *PVT Correlations for Gulf of Mexico Crude Oils*. MS thesis, University of Southwestern Louisiana, Lafayette, Louisiana.

Petrosky, G.E. Jr. and Farshad, F. 1998. Pressure-Volume-Temperature Correlations for Gulf of Mexico Crude Oils. *SPE Res Eval & Eng*.1 (5): 416-420. SPE-51395.

Riemens, W.G., Schutle, A.M., Jong, L.N.J., 1988. Birba field PVT variations along the hydrocarbon column and confirmatory field tests. J. Pet. Technol. Jan., 83 – 88

Rosengger, L., Wu., R.1990, Integrated PVT data characterization- lessons from four case histories, J. Can. Per. Tech., 38, (13), (paper 97-05)

Shojaei, M., Bahrami, F., Barati, P., Riahi, S., 2014, Adaptive neuro-fuzzy approach for reservoir oil bubble point pressure estimation, *J. of Natural Gas Sci. and Eng., Volume 20, September 2014, Pages 214-220.*

Soave, G. 1972. Equilibrium constants from a modified Redlich-Kwong equation of state, Chem. Eng. Sci. 27, 1197-1203.

Standing, M.B. 1947. A Pressure-Volume-Temperature Correlation for Mixtures of California Oils and Gases. *API Drilling and Production Practice* (1947): 275-287.

Tale, M., Alquraishi, A., Al Mushaigeh, E. 2009, New model for crude oil saturation pressure prediction using alternation conditional expectation, Offshore Mediterranean Conf. & Exhibit. Italy 25-27, 2009.

Talebi, R., Ghiasi, M. Talebi, H., Mohammadyian, M., Zendehboudi,S., Arabloo, M., Bahadori, A., 2014, Application of soft computing approaches for modelling saturation pressure of reservoir oil, *J. Nat. Gas Sci. and Eng., Vol. 20, Sep. 2014, 8-15.*

Vazquez, M., Beggs, H. D. 1980, Correlations for Fluid Physical Property Prediction. *J Pet. Tech.* 32 (6): 968-970. Vazquez, M.E. 1976. *Correlations for Fluid Physical Property Prediction,* MS thesis, University of Tulsa.

Velarde, J., Blasingame, T.A., and McCain Jr., W.D. 1997. Correlation of Black Oil Properties at Pressures below Bubble Point Pressure - A New Approach. Presented at the Annual Technical Meeting of CIM, Calgary, Alberta, 8–11 June. PETSOC 97-93.

Vogel, J.R., Yarborough, L. 1980, the effect of nitrogen on phase behavior and phase properties of reservoir fluids, SPE 8815.

Whitson, C. H., 1980, Characterizing hydrocarbon plus fraction, SPEJ, 24 (4), 683-694.

Williams, C.A., Zana, E.N., 1980. Use of the Peng-Robinson equation of state to predict hydrocarbon phase behavior and miscibility for fluid displacement. SPE 8817, SPE/DOE on Enhanced Oil Recovery, April 20–23, Tulsa, OK.

Appendix

In this appendix, sample calculations are presented to show the deficiency of some of the published methods.

Input parameters:

N_2	CO_2	H_2S	C_1	C_2	C_3	C_4	C_5	C_6	C_{7+}	GR.C7+	MWC7+	T, °F	Ps, psia
0.65	0.02	0	45.02	12.45	8.93	6.03	3.02	1.44	22.44	0.81	184	140	3002

Ahmadi et al (2014)

Ahmadi et al. proposed an 8 steps calculation (copied from their paper) to estimate the BPP as follows:

$$BPP = A \times Mw_{C7+} - \frac{B}{C_{7+}} + C \times Vol./Inter. + D \times SG_{C7+} + \frac{E}{SG_{C7+}} + \frac{F}{Mw_{C7+}} + T \times 505.967 + \frac{2.45789 \times 10^6}{T} + \frac{G}{Vol./Inter.} \quad (5)$$

$$A = 1488.46 - SG_{C7+} \times 1087.64 - \frac{601.512}{SG_{C7+}} + 1670.58 + T \times 0.40771 + \frac{6345.73}{T} + \frac{8.72863}{Vol./Inter.} \quad (6)$$

$$B = \frac{4132.53}{Vol./Inter.} + \frac{6.1977 \times 10^7}{Mw_{C7+}} + T \times 3181.39 + \frac{1.06393 \times 10^8}{T} \quad (7)$$

$$C = \frac{20964.7}{C_{7+}} - \frac{1.76162 \times 10^6}{Mw_{C7+}} - T \times 19.1963 - Mw_{C7+} \times 21.5427 + \frac{12851.5}{SG_{C7+}} - \frac{712613}{T} - 3477.49 + SG_{C7+} \times 9532.73 \quad (8)$$

$$D = 1.05345 \times 10^{+06} - \frac{1.50748 \times 10^8}{Mw_{C7+}} - T \times 676.048 - \frac{1.20252 \times 10^7}{T} - \frac{12178.2}{Vol./Inter.} \quad (9)$$

$$E = 641357 - \frac{1.02891 \times 10^8}{Mw_{C7+}} - T \times 301.034 - \frac{1.91446 \times 10^6}{T} - \frac{3873.62}{Vol./Inter.} \quad (10)$$

$$F = 2.51471 \times 10^{+08} + T \times 25332.5 + \frac{3.47893 \times 10^8}{T} - \frac{244573}{Vol./Inter.} \quad (11)$$

$$G = 20125.6 - T \times 23.9271 - \frac{703193}{T} \quad (12)$$

Ahmadi et al. define the Vol. /Inter. as the mole percentage ratio of the volatile (C_1& N_2) to the intermediate components (C_2 to C_6, CO_2 and H_2S). Application of the above equations from their paper using the input parameters results in the following:

Vol./Inter.	A	B	C	D	E	F	G	BPP
1.43211	1643.9	1545062	-271.127	45122.71	23642.47	2.57E+08	11753	1,794,123

The calculated saturation pressure (BBP) from Eq.5 is 1,794,123 psia which is unreasonable. This indicates a deficiency in the model.

Gholami et al (2014)

Gholami et al. presented 3 steps calculation (copied from their paper) to estimate the BPP as follows:

$$Tr(M) = \sum_{i=0}^{5} B_i M^i \quad (15)$$

Application of Eq. 15 resulted in the following

N_2	0.00584
CO_2	0.060917
H_2S	0.050636
C_1	-0.47658
C_2	-0.41547
C_3	-0.45083
C_4	-0.02003
C_5	0.068769
C_6	0.273921
C_{7+}	1.814523
SG.C7+	-0.0197
MWC7+	37.03525
T., F	-0.11139

Application of Eq. 16

$$Tr(P_b) = Tr(N_2) + Tr(CO_2) + Tr(H_2S) + Tr(C_1) + Tr(C_2) + Tr(C_3) + Tr(C_4) + Tr(C_5) + Tr(C_6) + Tr(C_{7+}) + Tr(SG_{C7+}) + Tr(MW_{C7+}) + Tr(T) \quad (16)$$

Resulted that Tr(Pb) = 37.8159

$$P_b = (57.0840387741 \times (Tr(P_b))^2) + (1166.9102304199 \times (Tr(P_b))) + 2207.3573204358 \quad (17)$$

Application of Eq. 17 resulted in saturation pressure (P_b) = 127,967 psia, which is unreasonable, compared to the measured one is 3002 psia. This indicates that there is a deficiency in the model.

Jarrahian et al (2015)

Jarrahian et al. proposed the following Eq. (copied from their paper) to estimate the saturation pressure:

$$P_{sat} = A_0 \exp\Big(A_1 z_{N_2} + A_2 z_{CO_2} + A_3 z_{H_2S} + A_4 z_{C_1}^{A_5} + A_6 \sum_{2}^{6} z_{C_i} + A_7 z_{C_{7+}}^{A_8} \gamma_{C_{7+}}^{A_9} Mw_{C_{7+}}^{A_{10}} + A_{11}(T - 459.67)\Big) \quad (2)$$

Application of Eq. 2 resulted in saturation pressure (Ps) = 1812 psia, compared to the measured one is 3002 psia. This indicates that model has inaccuracy problem. This has been the pattern in all the crude oil samples shown in Table 4 and figures 6A through 6D.

Table 1- Summary of attempts to calculate the saturation pressure of crude oils

Method based on production information	Method based on compositional data
Standing, 1947; Elam, 1957; Lasater, 1958; Vasquez, 1976; Vasquez and Beggs, 1980; GlasØ, 1980; Labedi, 1982&1992; Owolabi, 1984; Al-Marhoun, 1985; Obomanu and Okpobiri, 1987; Asgarpour et al., 1989; Al-Najar et al., 1988; Dokla and Osman, 1990; Petrosky, 1990; Kartoatmodjo and Schmidt, 1991&1994; Macary and El-Batanoney, 1992; Omar and Todd, 1993; Hasan, 1993; De Ghetto and Villa, 1994; Farashad et al., 1996; Almehaidaeb, 1997; Elsharkawy, 1997; Velarde et al., 1997; Kairy et al., 1998; Petrosky and Farshad, 1988; Al-shammasi, 2001; Gharbi & Elsharkawy 1999 Levitan and Murtha, 1999; Dindrouk and Christman, 2001 Moradi et al., 2010 Okoduwa and Ikiensikimama 2010 Asoodeh and Bagheripour, 2012 Shojaei et al., 2014 Talebi et al., 2014 Elmabrouk et al., 2014 Kanu and Ikiensikimama, 2014 Jarrahian et al 2015	Elsharkawy , 2003 Tale et al., 2009 AlQuraishi, 2009 Bandyopadhyay and Sharma, 2011 Khamehchi et al., 2011 Farasat et al., 2013 Ahmadi et al., 2014 Gholami et al. 2014 Jarrahian et al 2015

Table 2- Description of the data used to develop and test the extended model

Input	Training model (179 Data)			Testing (28 Data)			Testing (7 new experimental)		
	Min	Ave	Max	Min	Ave	Max	Min	Ave	Max
N_2	0.00	0.35	1.67	0.00	0.27	1.67	0.10	0.26	0.35
CO_2	0.00	1.27	11.37	0.00	1.17	11.37	0.14	0.41	0.83
H_2S	0.00	0.15	3.68	0.00	0.15	3.22	0.00	0.00	0.00
C_1	13.16	33.86	74.18	19.50	34.79	73.36	20.64	28.25	42.79
C_2	3.36	8.07	13.71	4.64	8.26	11.63	6.69	7.88	10.82
C_3	0.89	6.59	11.87	2.46	6.76	12.03	6.05	6.59	7.10
C_4	0.95	4.46	8.4	1.66	4.41	6.58	4.03	5.05	5.93
C_5	0.40	3.21	5.95	1.00	3.11	5.23	3.36	4.33	5.52
C_6	0.00	3.14	6.37	0.99	3.15	5.46	2.88	3.74	4.33
C_{7+}	10.72	38.89	57.73	11.18	37.88	51.22	27.61	43.49	51.31
$SG.C_{7+}$	0.74	0.87	0.959	0.70	0.86	0.93	0.85	0.89	0.92
MWC_{7+}	134	240	368.9	105	223	292	216	262	298
T, ° F	58	156	319	58	165	261	121	157	230
Ps, psia	537	2198	5065	1025	2307	5000	1227	1779	3344

Table 3: Error analysis of the proposed models in comparison with all the investigated methods

Method	ARE	AARE	SD	R^2%
PR-EOS	6.24%	10.19%	1028	97.9
SRK-EOS	-3.12%	10.36%	1024	96.7
Elsharkawy (2003)	0.85%	8.57%	1015	96.4
Alquraishi (2009)	-6.06%	10.16%	1088	96.7
Bandyopadhyay (2011)	8.05%	12.28%	1054	96.8
Farasat (2013)	NA	NA	NA	NA
Gholami et al. (2014)	?	?	?	?
Ahmadi et al. (2014)	?	?	?	?
Jarrahian et al (2015)	35.52	35.56	983	95.2
New Extended Model	-0.37%	8.20%	1030	97.6
New Lumped Model	-0.50%	8.46%	1036	97.4

ARE = Average Relative Error

AARE = Average Absolute Relative Error

SD = Standard Deviation

R^2 = Correlation Coefficient

NA = Calculation was not possible, because the weight and bias were not provided.

? = The method produces unreasonable results as indicated in the appendix.

Table 4- Comparing model accuracy for calculations of saturation pressures for various crude oils

Volatile-Oil-Samples (C7+<30)																						
Fluid Source	index	N2	CO2	H2S	C1	C2	C3	C4	C5	C6	C7+	GR.C7+	MWC7+	T, °F	Ps, psi	SRK	PR	Elsh	Alq	Bandy	Jarr	Extd	Lump
Rosenegger	7	0.03	8.39	0	47.4	10.3	6.12	4.2	2.89	2.05	18.6	0.83	180	295	4000	3933	3879	4460	4420	4165	3317	4139	4178
Vogel-80	13	0.65	0.02	0	45	12.5	8.93	6.03	3.02	1.44	22.4	0.81	184	140	3002	2624	2548	2742	2895	2643	1812	3019	2996
Dansh-92	18	0	0	0	46.8	8.77	7.44	4.01	2.56	4.02	26.4	0.77	158	212	2941	2550	2594	3248	3617	3269	1944	3244	3374
	31	0.44	0.38	0	49.1	7.6	6.13	3.84	2.52	2	28	0.84	231	199	3739	3522	3507	3418	3951	3508	2309	3625	3644
Middle East	45	0.03	0.97	0	41.6	10.8	7.52	4.5	3	2.43	29.1	0.85	208	241	3069	3109	2979	3080	3513	3138	1962	3127	3140
Coats-86	51	0.55	1.03	0	36.5	9.93	8.85	6	3.78	3.56	30.4	0.84	200	234	2746	2537	2472	2714	2945	2643	1682	2777	2720

Black-Oil -Samples (35<C7+<55)																						
Fluid Source	index	N2	CO2	H2S	C1	C2	C3	C4	C5	C6	C7+	GR.C7+	MWC7+	T, °F	Ps, psi	SRK	PR	Elsh	Alq	Bandy	Jarr	Extd	Lump
Coats-86	73	1.64	0.08	0	28.4	7.16	10.5	8.4	3.82	4.05	36	0.84	252	131	1694	1674	1618	1615	1738	1644	1084	1713	1713
Middle East	88	0.08	1.37	0.69	36	8.67	5.94	2.97	1.5	3.1	39.7	0.9	274	168	2505	2710	2318	2247	2608	2321	1476	2441	2445
Middle East	114	0.29	0.48	0	28.4	8.29	7.38	5.06	3.42	4.41	42.3	0.88	252	133	1632	1672	1477	1544	1748	1517	984	1587	1598
Middle East	122	0	1.25	0	33.4	9.24	6.07	2.4	1.62	2.78	43.3	0.85	252	230	2110	2313	2278	2424	2737	2438	1410	2388	2412
Middle East	134	0	0.17	0	28.6	8.63	7.43	3.81	2.61	4.11	44.7	0.86	249	134	1548	1551	1429	1477	1764	1433	937	1503	1567
Middle East	146	0.44	0.83	0	27.8	7.54	6.9	4.61	3.1	3.11	45.7	0.9	272	133	1565	1799	1505	1558	1730	1528	1002	1602	1590
moharam-96	149	0.21	0.34	0	20	7.93	8	6.6	5.87	5.08	45.9	0.86	230	235	900	1353	1290	1531	1514	1225	773	1303	1294
Middle East	162	0.35	0.47	0	26.5	7.71	6.05	4.03	4.08	4.22	46.6	0.89	253	135	1500	1613	1380	1427	1631	1391	923	1478	1456
Middle East	176	0.12	0.51	0	28.8	7.91	6.46	3.17	2.07	3.25	47.7	0.88	250	134	1615	1665	1476	1574	1783	1502	981	1571	1614
Middle East	194	0.21	0.15	0	27.8	7.68	6.19	3.41	1.81	2.84	49.9	0.86	228	134	1590	1532	1377	1545	1680	1426	923	1466	1493
Rosenegger	207	0.37	0.02	0	17.3	6.11	8.23	5.47	3.76	3.96	54.8	0.85	242	156	537	938	888	914	1064	836	558	661	731

Newly-Measured Kuwait University Laboratory																					
Fluid Source	index	N2	CO2	H2S	C1	C2	C3	C4	C5	C6	C7+	GR.C7+	MWC7+	T, °F	Ps, psi	SRK	PR	Elsh	Alq	Bandy	Jarr	Extd	Lump
KU-1	209	0.33	0.22	0	25.6	6.87	6.39	5.61	4.68	4.33	46	0.879	222	133	1595	1360	1530	1442	1514	1394	860	1357	1337
KU-2	210	0.35	0.47	0	26.5	7.71	6.05	4.03	4.08	4.22	46.6	0.893	253	135	1500	1400	1620	1427	1631	1391	924	1478	1456
KU-3	211	0.27	0.74	0	23.1	7.2	6.45	5.93	4.66	4.16	47.5	0.906	290	138	1283	1140	1360	1284	1448	1296	800	1287	1219
KU-4	212	0.22	0.26	0	20.6	6.69	6.65	5.11	5.52	3.6	51.3	0.919	298	171	1227	1120	1300	1188	1415	1170	742	1176	1131
KU-5	213	0.19	0.14	0	27.8	8.37	6.88	5.28	4.33	3.85	43.2	0.89	278	121	1524	1300	1560	1402	1659	1409	928	1534	1489
KU-6	214	0.33	0.24	0	31.3	7.48	6.58	4.86	3.7	3.17	42.3	0.9	274	171	1978	1800	2200	1975	2242	1985	1224	2060	2026
KU-7	215	0.1	0.83	0	42.8	10.8	7.1	4.51	3.36	2.88	27.6	0.85	216	230	3344	3070	3400	3083	3568	3139	2003	3213	3200

Low-Temperature Oil Samples (59°F<T<71°F)																							
Fluid Source	index	N2	CO2	H2S	C1	C2	C3	C4	C5	C6	C7+	GR.C7+	MWC7+	T, °F	Ps, psi	SRK	PR	Elsh	Alq	Bandy	Jarr	Extd	Lump
Middle East	101	0	1.12	0.21	27	10.8	9.23	4.97	2.62	3.33	40.8	0.88	249	71	1247	1090	1033	1040	1235	564	808	1148	1205
Middle East	117	0.36	0.17	0	27.2	8.93	8.6	6.07	3.71	2.55	42.4	0.88	271	68	1130	1028	1245	1094	1248	766	820	1239	1228
Middle East	135	0	0.17	0	28.6	8.63	7.43	3.81	2.61	4.11	44.7	0.86	249	70	1270	1055	1255	1130	1311	783	826	1229	1297
Middle East	138	0.14	0.54	0	29.4	7.96	6.19	3.96	3.25	3.84	44.7	0.9	290	59	1265	1065	1350	1191	1301	880	894	1408	1390
Middle East	147	0.44	0.83	0	27.8	7.54	6.9	4.61	3.1	3.11	45.7	0.9	272	67	1320	1125	1400	1201	1268	921	879	1319	1312
Middle East	173	0.06	0.81	0	31.3	7.42	5.68	2.82	1.63	2.64	47.6	0.89	270	68	1460	1215	1535	1441	1463	1026	975	1548	1576
Middle East	193	0.16	0.3	0	24.7	7.51	6.92	4.76	2.65	3.43	49.6	0.88	239	71	1050	975	1156	1008	1107	701	714	958	986

Elsh: Elsharkawy (2003)
Extd: Extended model

Alq: AlQuraishi (2009)
Lump: Lumped model

Band: Bandyopadhyay and Sharma (2011)

Jarr: Jarrahian et. al (2015)

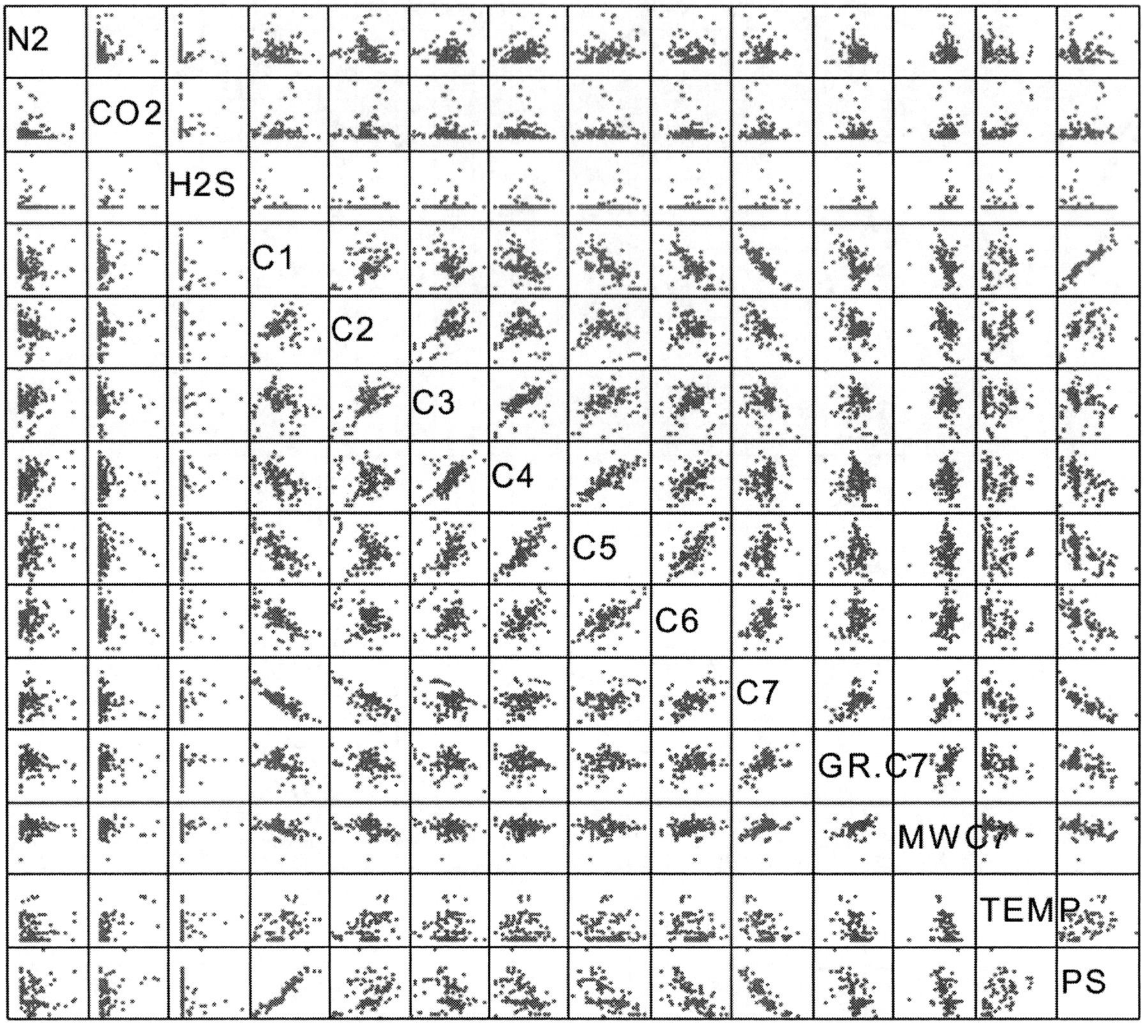

Figure 1: Correlation matrix for all the variables in the oil data bank

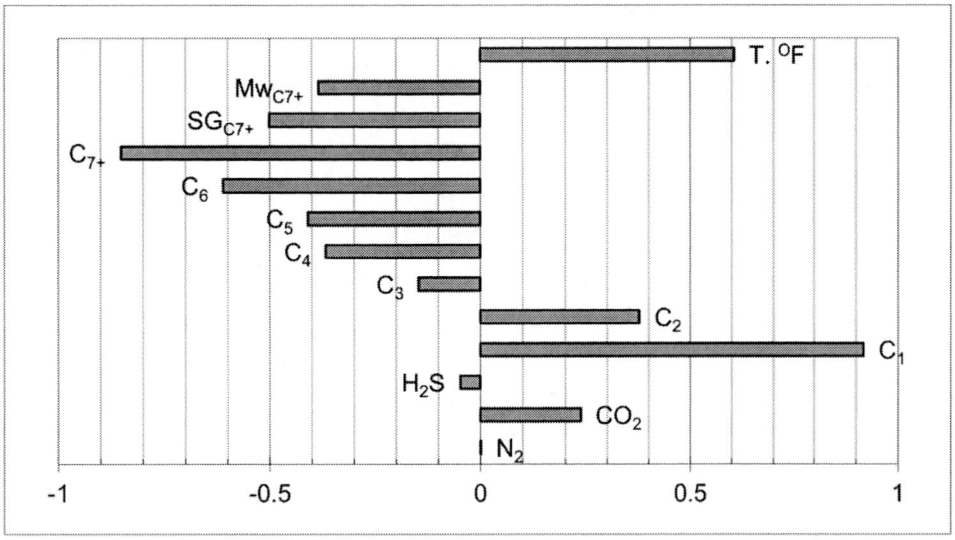

Figure 2: Correlation coefficient for all the variables in the training model (179 data)

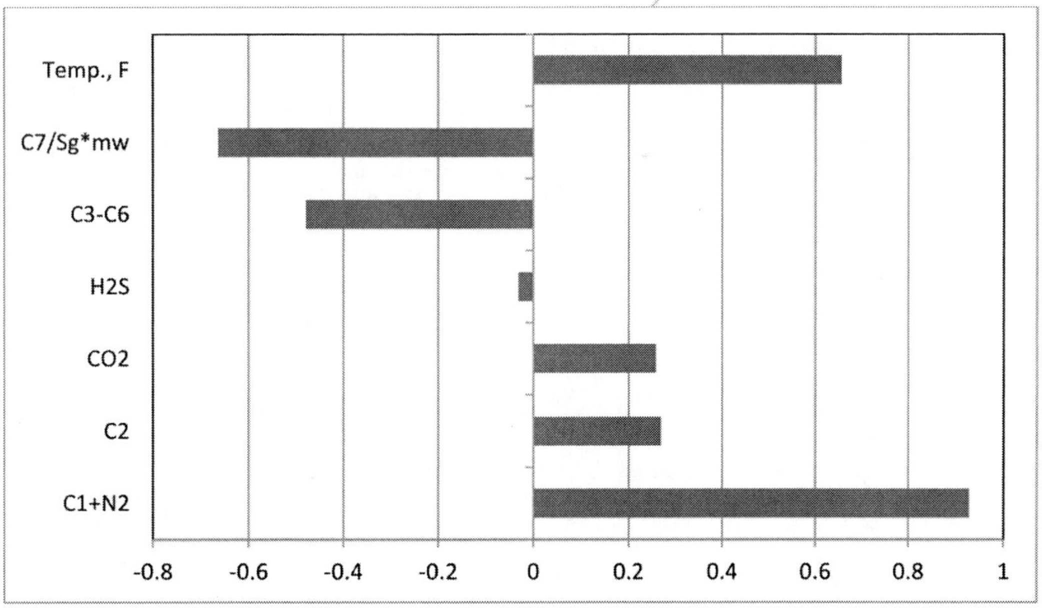

Figure 3: Correlation matrix for the lumped model (179 data)

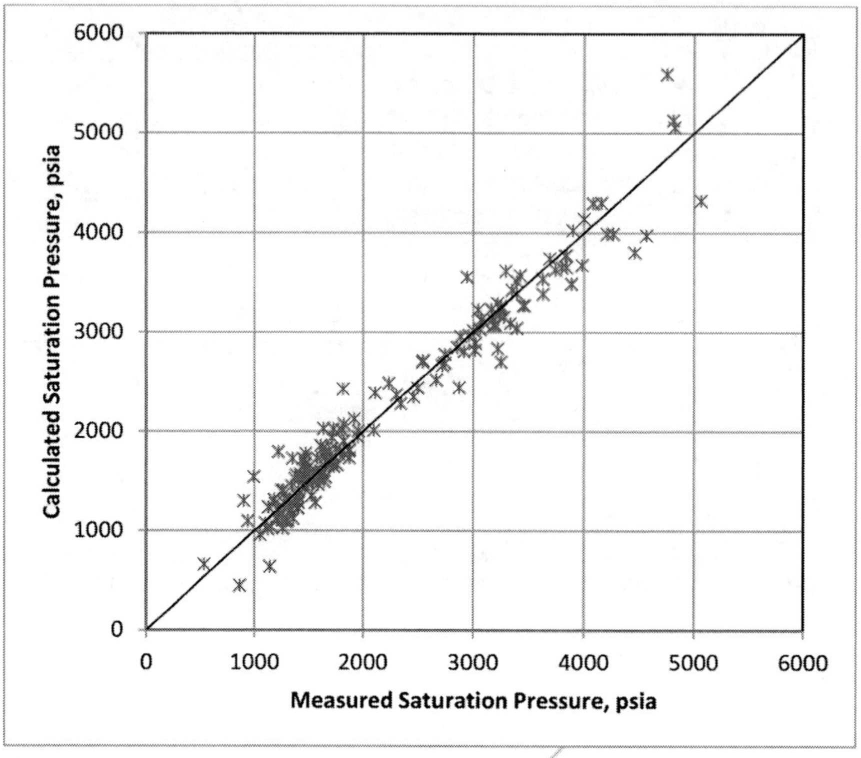

Figure 4A: Cross plot of the extended model using 179 training data

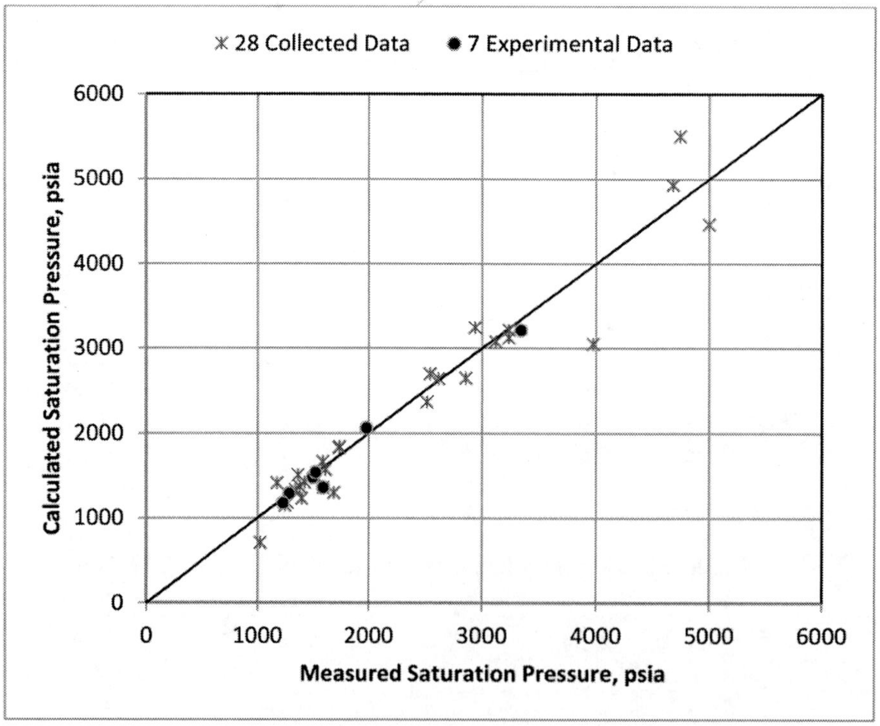

Figure 4B: Cross-plot of the extended model using 35-testing data

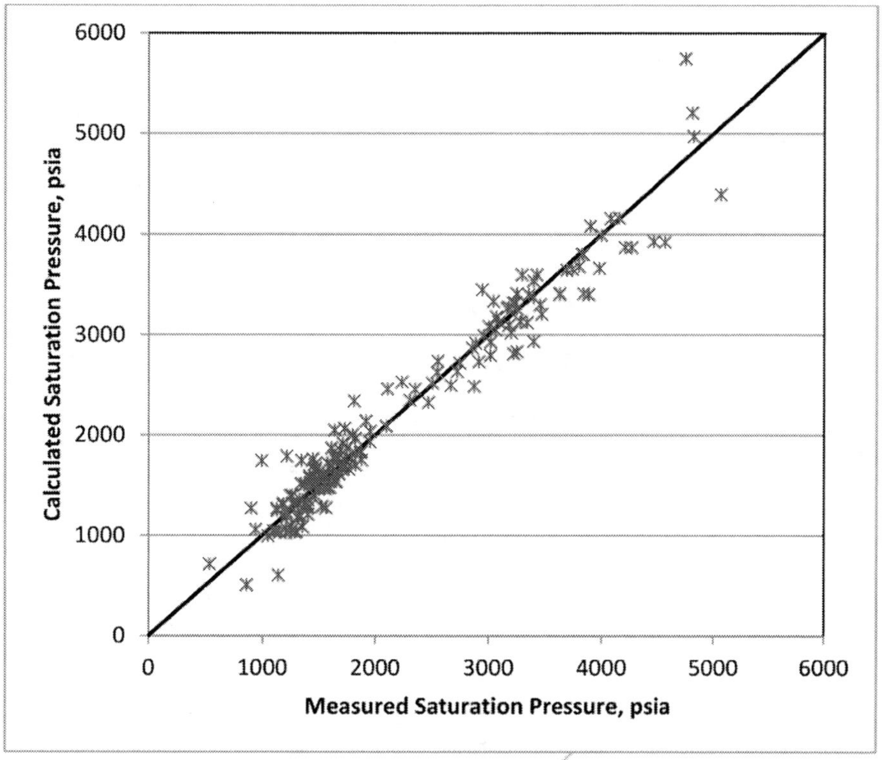

Figure 5A: Cross-plot of the lumped model using 179 training data

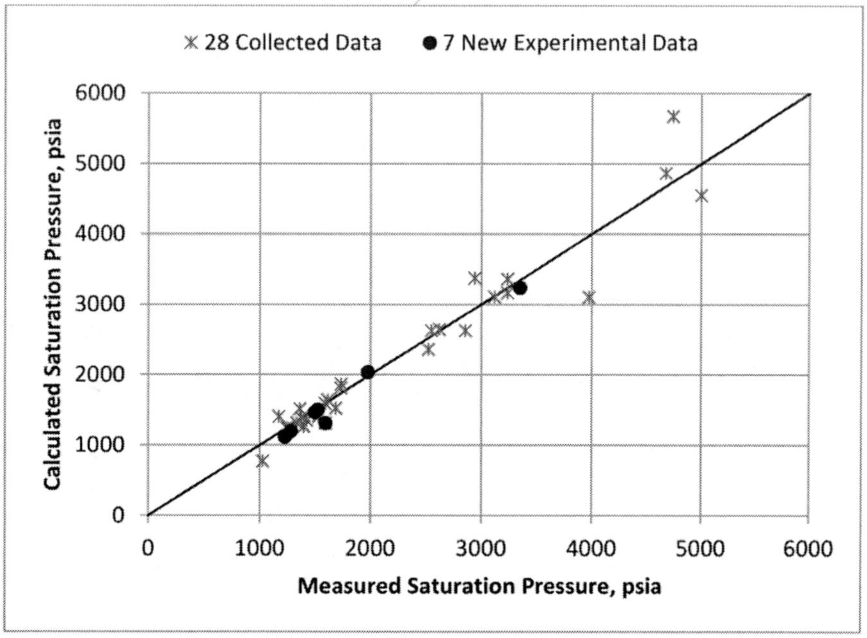

Figure 5B: Cross-plot of testing data for the lumped model

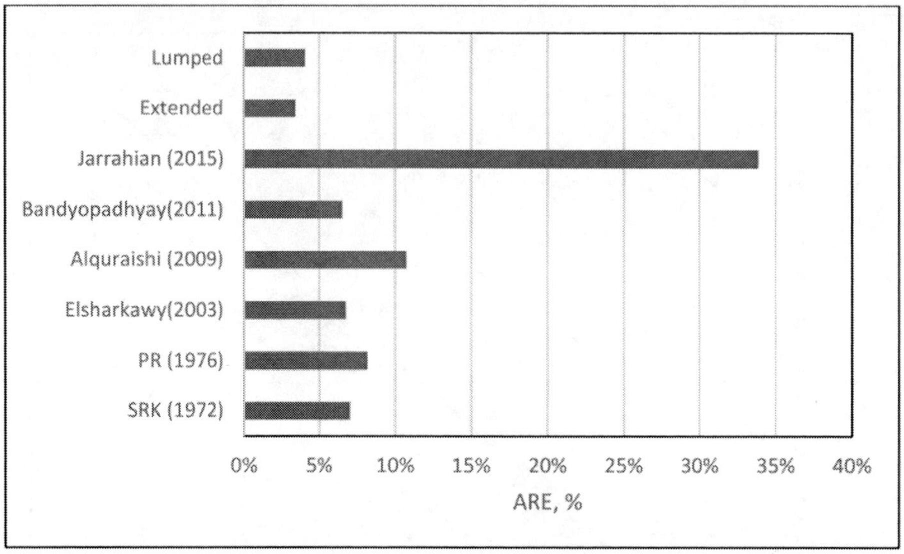

Figure 6A: Comparison of all the models' accuracy for the volatile oil samples

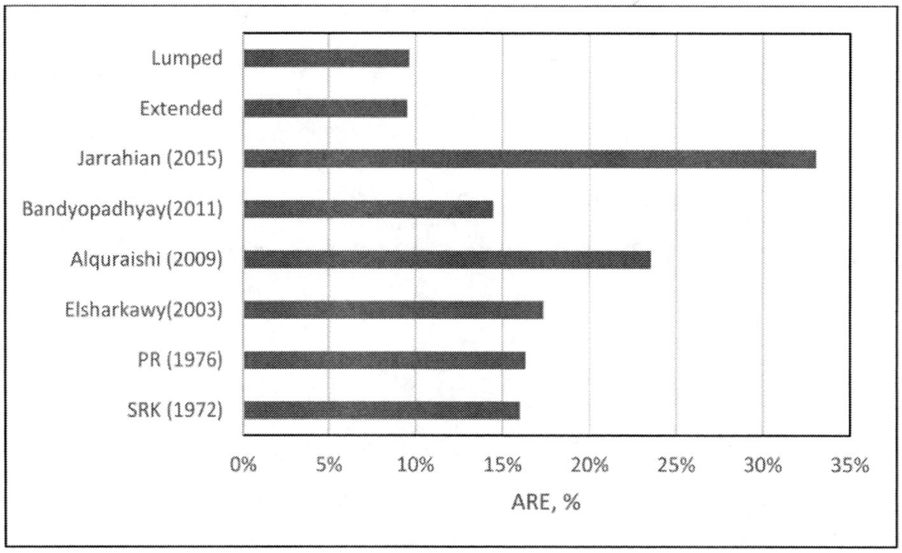

Figure 6B: Comparison of all the models' accuracy for the black oil samples

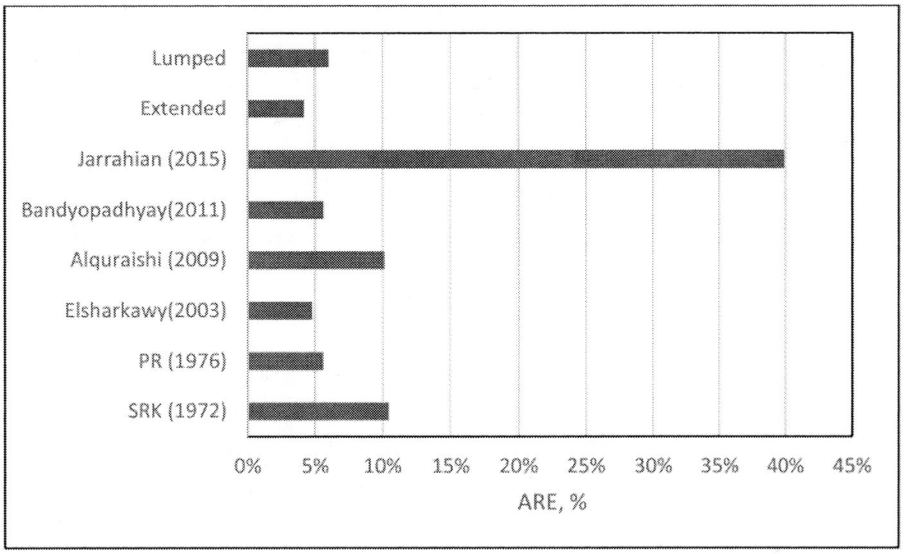

Figure 6C: Comparison of all the models' accuracy for the newly measured samples at KU

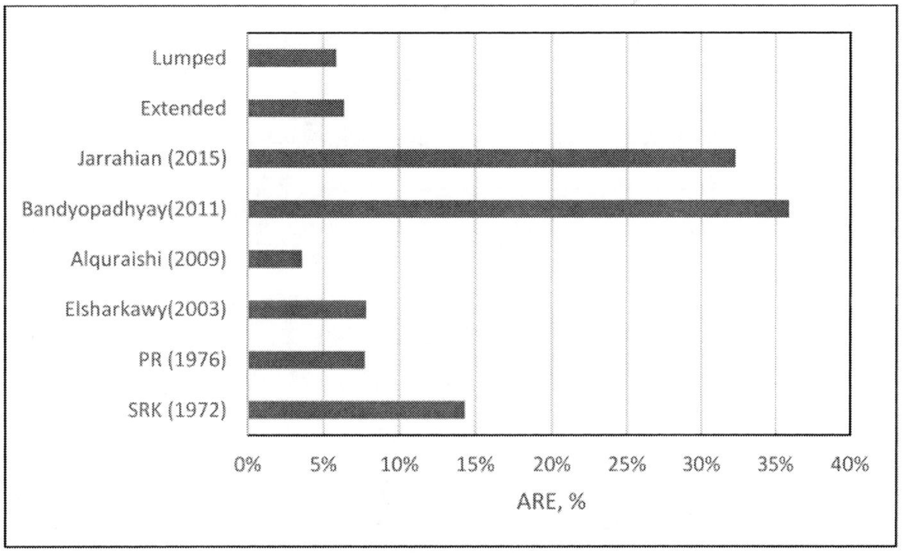

Figure 6D: Comparison of all the models' accuracy for low-temperature oil samples

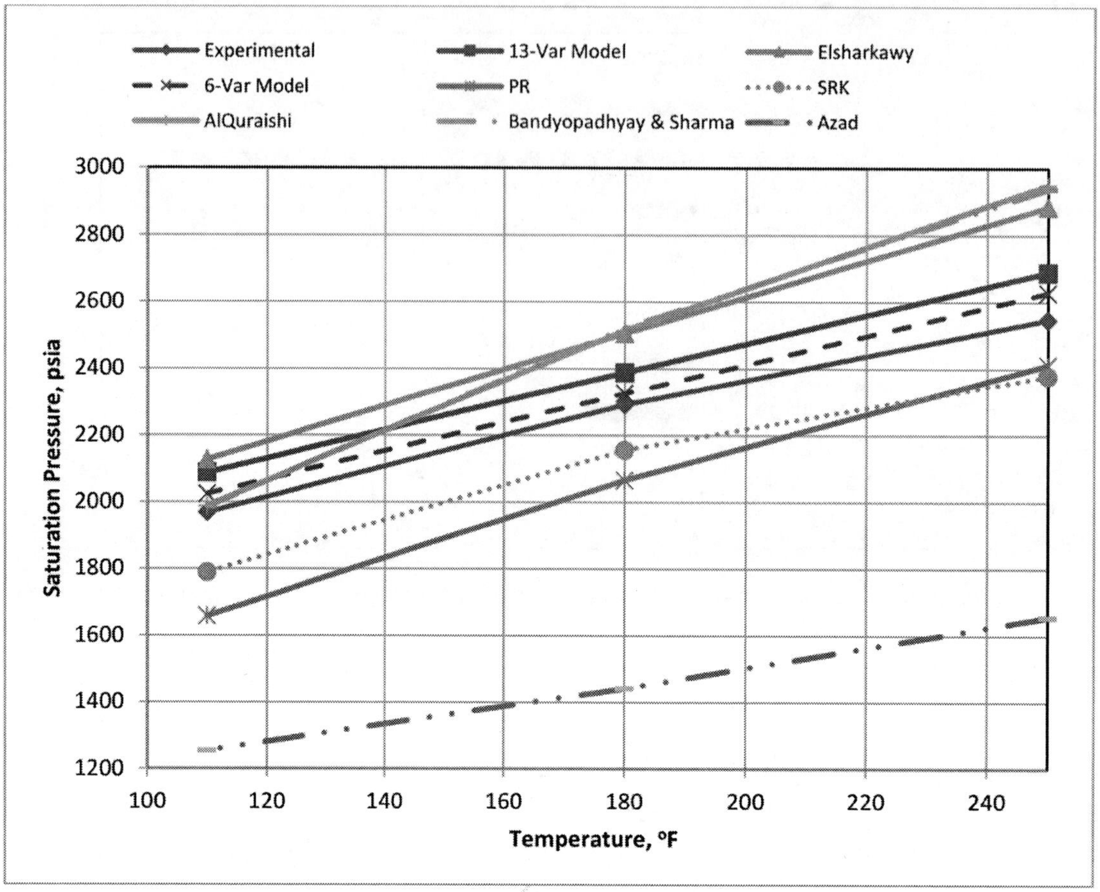

Figure 7: Changes in saturation pressure as a function of temperature for the different models (Sample #70)

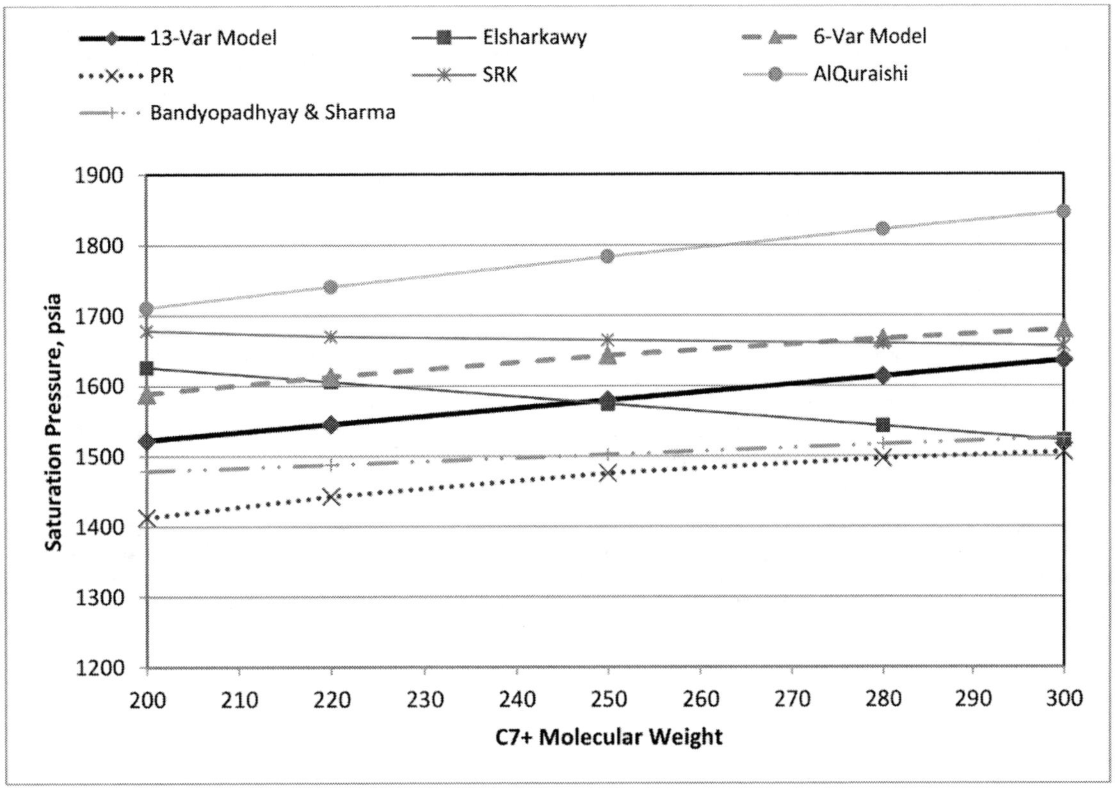

Figure 8: Comparing model performances: Changes in BPP as a function of C_{7+} molecular weight at 134°F (Sample #176)

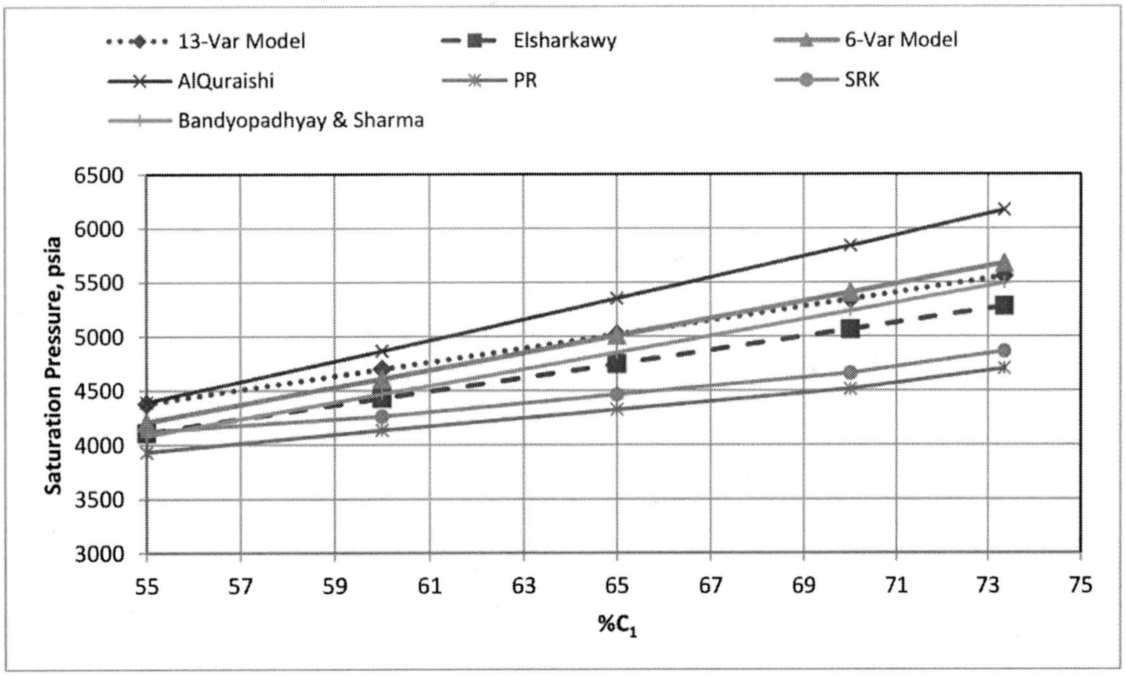

Figure 9: Sensitivity of saturation pressure to methane content at 212°F (sample #2)

444834 Stability and Rheology of Emulsified Acid in Waste Oil: Effect of Emulsifier Concentration, Mixing Speed, Temperature and Droplet Size

Abstract:

Acid in oil emulsion is used in the industry for stimulation purposes. In particular, HCl as the acid phase and diesel as the oil. Emulsifying HCl has numerous benefits over regular HCl. Perhaps the main reason is due to deeper penetration. Apart from that, less corrosion damages are caused since the external phase is hydrocarbon. Several studies showed the success of replacing diesel as an external phase with other hydrocarbon oil, such as crude oil and xylene.

This work utilizes the extra hydrocarbon left unused –or sometimes dumped- from refineries. This unused hydrocarbon is referred to as waste oil. First the chemical composition of waste oil was studied using HPLC, SARA fractionation as well as FTIR. This helped in selecting the proper emulsifier agent for the waste oil. The HCl-waste-oil emulsion was prepared using 15wt% HCl with a ratio of 70:30 of acid-to-oil. Extensive work was done to ensure reproducibility of data. This study was done to experimentally investigate the effects of distinct important variables on thermal stability and apparent viscosity of the HCl-waste-oil emulsion. These variables are; emulsifier concentration, mixing speed and the addition of corrosion inhibitor. All experiments were conducted at a high temperature 120°C. To assess the quality of the HCl-waste-oil emulsion, droplet size was measured as a function of temperature (up to 100 °C).

The results showed that the HCl-waste-oil emulsion is a shear-thinning fluid. Power-law model was applicable seamlessly to all of the apparent viscosity data for all of measured temperatures. The optimum conditions were found to be; 0.5vol% and 600 rpm for emulsifier concentration and mixing speed, respectively. Apart from that, the emulsions prepared with corrosion inhibitor were less stable compared to the ones without it. However, the addition of corrosion inhibitor tended to generate monodispersed emulsion but bigger size droplets for temperatures 25°C and 50°C and polydispersed droplets for higher temperatures. Droplet size measurements showed an average size of 1.47 to 3.09 μm. Overall, the lab results showed a promising potential for the HCl-waste-oil emulsion to be used in the field.

This work takes into account; reducing the budget of acidizing job and at the same time environmental concerns. It makes use of the unwanted hydrocarbon from refiners. And, it saves the environment by using the waste oil instead of dumping it.

443836 Oxidative Dehydrogenation of n-Butane on Carbon Catalysts

Abstract:

Butenes and butadiene, which are one of the important industrial chemical products to produce synthetic rubber , are sythesized traditionally by oxidative dehydrogenation (ODH) of n-butane over various mixed metal oxides. Metal-free carbon nanomaterials has recently received much attention by researchers as a catalyst for the ODH reaction due to its environmental acceptability, corrosion resistance, and unique surface properties. For the reason, the reaction system was modelled in this study for the sythesis of butene and butadiene. For the modelling of the reaction system, the catalytic performance of several carbon catalysts, such as graphite and activated carbon, were initially tested. Then, the reaction mechanisms and kinetics were developed based on experimental results. Lastly, the mathematical modeling of the reaction system was optimized to maximize an overall yield of the main products. This approach allowed us to develop the optimized reaction system more efficiently.

This research was respectfully supported by Engineering Development Research Center (EDRC) funded by the Ministry of Trade, Industry & Energy (MOTIE). (No.N0000990)

443837 Blowdown Dynamic Modeling and Analysis of Pressure and Temperature

Abstract:

Vessel blowdown is one of the safety operation procedures that discharges flammable gas or liquid to flare system during emergency situation such as fire. When blowdown occurs, the liquefaction and freezing effect cause a pressurized vessel to be brittle, exposing it to possible hazard situation. Also the blowdown systems are recommended to be accordance with API(American Petroleum Institute) 521 code. API 521 requires pressure in vessel should be half of the initial pressure or be lower than 100 psig in 15 minutes after commencing blowdown in fire case. Therefore, accurate estimation over the change of temperature and pressure of the vessel during blowdown period is crucial for the design of blowdown system. For the reason, researchers have developed a few dynamic blowdown models so far. However, the model was found to be rather inaccurate under certain conditions. Therefore a new dynamic blowdown model was developed in this work. For the accurate analysis, the model algorithm considers both vapour and liquid phase in non-equilibrium condition that the temperatures of vapour and liquid phases are different. Futhermore, heat transfer from the vessel wall to discharge gas at either laminal or turbulent flow in the vessel was considered in the model to increase the accuracy of blowdown estimation. For thermodynamics, both Peng-Robinson and Soave-Redlich-Kwong equations of state were applied to the model, and the results were compared each other. Finally, the simulation results were compared with Haque et al's experimental and steady-state simulation results.

444364 Molecular Simulation of Controlling Ions and Their Role in Low Salinity IOR for Carbonate Reservoirs

Abstract:

Carbonate geology represents around 50% of hydrocarbon-bearing reservoirs in the world and more than 70% in some areas such as the Middle East. Water flooding has not been very efficient in carbonate reservoirs due mainly to the low water wetness. Modifying the ionic composition and salinity of the injected seawater demonstrated that the oil recovery from these carbonate reservoirs could be improved. Although the underlying mechanisms are still a subject of debate, the improved oil recovery has been mostly attributed to the increased water-wetness of the rock through the interactions of specific ions like Ca^{2+}, Mg^{2+}, SO_4^{2-}, and Cl^- , which are present in seawater with the rock surface under favorable conditions

The study of hydrated gas-phase metal ions provides a link between intrinsic chemistry of the isolated ion and its chemistry in solution. The structures and binding enthalpies of a variety of gas-phase and solution mono- and divalent ion hydrates containing up to 12 water molecules have been studied computationally. A DFT and molecular dynamics simulations were employed to study hydrated structures of the form $[\mathbf{M}(H_2O)_n]^{2+}.mH_2O$ for calcium and magnesium and $[\mathbf{M}(H_2O)_n]^{+}.mH_2O$ for sodium, in which there are *n* water molecules in the first coordination shell and *m* water molecules in the second coordination shell. The hydration, dehydration, and interactions of metal ions with Cl^{-1} and SO^{-2}_4 have been studied to better understand the mechanism of low salinity IOR. Density functional (DFT) theory using the B3LYP functional, and 6-311++g** basis set has been used. The charge of the metal ion decreases monotonically as the number of water molecules increases in the complex. Extensive calculations of the energetic contribution of individual structural characteristics as well as the energetic trends in binding energy are developed, which will lead to better understanding of how low salinity water IOR works.

Keywords: Hydration, carbonate reservoir, DFT, binding energy, Low Salinity

447119 Development R&D of Manoir Industries and Emphasis on Coating

Abstract:

The Senior VP Technical Support & Projects will present the development R&D of Manoir Industries and especially the coating (it is our lattest development) as well the services we can provide such as metallurgy expertise, advice, convector cleaning, site welding, welding supervision in order to increase the furnace productivity. So it is 60% R&D and 40% services. Manoir Industries manufactures cast components for the petrochemical market and develops high temperature alloys such as the Manaurites.

442925 Identifying Gas Liquid Contactor Performances for Optimized Natural Gas Desacidification Amines Units - from Process Calculations to Experimental Strategy

Abstract:

Amine units are widely used for gas sweetening. The absorber performances of these reactive absorption processes are typically mass transfer limited. The proper selection of gas liquid contactor is thus critical to optimize the size of towers, the solvent flowrate and the duty of the reboiler. This work proposes an approach to identify the most suitable packings for a given application.

The case-study selected here to illustrate the approach deals specifically with a CO_2removal application where the sour gas is at 7 % vol, and the sweet gas specification is 2% vol. The gas flowrate is fixed at 0,18 MMSCFD with a solvent flowrate fixed at the minimum of the thermodynamic constraints (via equilibrium curves). An in-house rate-based simulator with accessible forms of mass transfer correlations, based on a double film model, is used to calculate the absorber height and diameter which are needed to reach the sweet gas specifications. A reference structured packing is compared with standard valve trays.

The calculations show that the bed height required with the first contactor is considerably higher than with the second one while the column diameter is reduced. In this case the use of standard valve trays is more attractive for cost optimization. Interestingly, because mass transfer is accelerated, the absorption rate with the structured packing is found to be thermodynamically limited due to a more severe temperature increase which reduces the driving force in the bottom section of the absorber.

The study determined that the optimum design for the investigated case is a compromise between a full trayed tower, that generally imposes a larger diameter, and a packed tower where the higher mass transfer forces result in a temperature rise in the bottom of the tower which in turn imposes a height increase and a much higher solvent flowrate. . To determine the mass transfer characteristics of a gas-liquid contactor providing this optimum, a sensitivity analysis has been conducted. Mass transfer parameters, i.e., liquid and gas side mass transfer coefficients and interfacial area, were varied by a factor 2 relative to the ones obtained with the reference structured packing. Calculations show that one should use a contactor that reduces liquid phase resistance, specifically, the mass transfer conductance $k_L a_e$ should be increased by at least 50%. This target could be reduced to +30% if one could simultaneously reduce the gas phase conductance $k_G a_e$ by 20%. This result can be explained by the link between mass and heat transfer (Chilton-Colburn analogy) whereby reducing $k_G a_e$decreases the temperature level which in turn thermodynamically increases the mass transfer driving force.

The next step will be to review the literature more extensively to make a short list of pre-selected random and structured packings that best meet the targets. Experimental tests will then be run on IFPEN facilities to measure flooding limits and mass transfer parameters in a small 150 mm diameter column. This step is required to confirm packing's behavior, considering the disparity of the values reported by different authors in the literature. The last step is to consolidate the existing correlations for the most promising packing, and mitigate differences between various authors by performing characterization tests on larger diameter test columns. The final correlations will be implemented in our in-house rate based simulator for industrial use.

The present approach can be applied to determine the required mass transfer characteristics for optimizing other gas treatment applications. However, the conclusions reached here as regards to mass transfer targets and the problematic temperature levels, are valid only to the present case –study, and are not directly applicable to other gas sweetening applications (selectivity, LNG, ...). For example, in a case where H_2S and CO_2 are treated simultaneously the mass transfer limitations could be very different.

443784 Measurement and Correlation of the Mutual Solubility of Diisopropylamine + Water and Triethylamine + Water Systems at High Pressure

Abstract:

The mutual solubility data of diisopropylamine + water and triethylamine + water systems were measured at the pressure up to 100 MPa and the temperature range was 280.2~316.2 K. Experiments were performed in a high-pressure autoclave with a moveable piston. The experimental data were correlated satisfactorily with NRTL activity coefficient model and the binary parameters were obtained.

444161 Novel Microwave Synthesis Carbon Dots from PINE Bark and Its Application for Heavy Metal Sensing

Monday, April 11, 2016
Exhibit Hall E (George R. Brown)

Saheed Sanni, Chemistry, Vaal University of Technology, Vanderbijlpark, South Africa, Puthalapattu ReddyPrasad, Chemistry, Vaal University of Technology, Vanderbiljpark, South Africa and Augustine E. Ofomaja, Chemistry, Vaal university of Technology, Vanderbiljpark, WV, South Africa

NOVEL MICROWAVE SYNTHESIS CARBON DOTS FROM PINE BARK AND ITS APPLICATION FOR HEAVY METAL SENSING

Sanni S.O., ReddyPrasad P., Ofomaja A.E.

Biosorption and Waste water Treatment Research Laboratory, *Department of Chemistry, Faculty of Applied and Computer Sciences, Vaal University of Technology, P. Bag X021, Vanderbijlpark-1900, South Africa.*

INTRODUCTION

Microwave pyrolysis for the production of carbon has the advantages of being cheap, produces high yield of carbon and is absolutely a green synthetic method [1]. In recent times, the use of microwave pyrolysis have been extended to the preparation of carbon dots. Carbon nanodots (C-dots) falls under the class of carbon nanoparticles with size below 10 nm[2]. C-dots are carbon materials which processes high water solubility and strong fluorescence depending on their size, edge, shape, surface ligands, and defects. Since C-dots preparation can be achieved from difference carbonaceous sources and by varying a range of parameters during their synthesis, the properties of the final product may vary depending on these factors. C-dots have sp^2 characteristics and possess several functional groups on their surfaces that makes them highly hydrophilic and easily functionalized with various organic, polymeric, inorganic, or biological spices[3]. Several synthesis methods that have been explored includes electrochemical synthesis, combustion/thermal/hydrothermal/acidic oxidation, supported synthesis, microwave/ultrasonic, arc discharge, laser ablation/passivation, and plasma treatment[4]. C-dots processes excellent such as water dispersibility, chemical stability and photostability, ease of modification, excitation-dependent multicolor emission, low toxicity, and good cell permeability, great efforts have been paid to their potential applications in biosensing, bioimaging, drug delivery, and other biological related aspects.

METHODS

The carbon dots (C-dots) with green luminescence was synthesized from microwave assisted pyrolysis of pine bark waste our carbon source at 800, 1000 1200 W for 1 Hr. The dots produced were characterized using FTIR, TGA, XRD, UV-Vis and Raman spectroscopy. The dots were applied for heavy metal sensing.

RESULTS AND DISCUSSION

The functional groups present on the C-dots (**Fig.1**) includes -OH stretching band mode at 3413 cm^{-1}, C-H stretching at 2934 and 2845 cm^{-1}, stretching peaks of C=O, C=C at 1731, 1604 and 1437 cm^{-1}. The XRD pattern (**Fig 2a &b**) of produced C-dots depicts sharp peak at 20.9° and 26.8°, which is highly crystalline cores and possess amorphous nature of C-dots. This peculiar peak is ascribed to 002 of the graphitic carbon, which in line with inter-planar spacing of 3.77 Å with higher spacing than bulk graphite at 3.34 Å. The thermogravimetric and differential thermal analysis display (**Fig 3**) that the C-dots possess thermal stability more than 100 °C and mass losses occurs afterwards. The thermal analysis depicts 3 endothermic peaks at 331, 542 and 735 °C, which are ascribed to oxygenated groups such as carboxyl, hydroxyl, carbonyl and phenols. The optical properties of the synthesize C-dots were explored using UV-vis absorption and photoluminescence (PL) spectroscopy. The UV-vis spectra display (**Fig 4**)intense absorption peak in the range of 260 to 290 nm, which is ascribed to π-π^* transition of the aromatic sp2 domain and also presence of multiple polyaromatic chromophores. A shoulder peak at 330 nm which is ascribed to n-π^* transitions of C=O and structural defects of the C-dots. PL spectrum of the C-dots (**Fig 5a**)exhibits an intense maximum emission peak at 432 nm upon excitation at 330 nm, which also shows characteristic yellowish green color under sunlight and intense blue coloration UV lamp (365 nm). The PL properties of as prepared C-dots was varied at (**Fig5b**) different excitation wavelength from 300 to 405 nm. There was red-shifting in emission peak as the excitation peak was increased from 300 to 405 nm and also simultaneously decrease in fluorescent intensity, which is ascribed to quantum effect and surface defects. This behavior of the prepared C-dots is dependent on excitation-fluorescence, which is accordance with other report [13]. The quantum yield of synthesized C-dots was measured to be 17% at maximum excitation at 330 nm

Fig.6 shows the Raman spectrum of the PB-GQDs, which has been usually used to confirm the quality of the prepared PB-GQDs. There is two major Raman peaks at 1370 cm^{-1} (D band), 1580 cm-1 (G band). The peak of G band corresponds to the Sp^2 bonded carbon atoms in a two-dimensional (2D) hexagonal lattice and the peak of G band is associated with the vibrations of carbon atoms with dangling bonds in the termination plane of tangled graphite or glassy carbon. The D/G ratio band ratio is ID/IG: 0.68 is confirm the formation of graphene like structure from synthesized pine bark.

Fluorescence detection of Cu^{2+} ion

The detection of Cu^{2+} was performed at room temperature in phosphate buffer (PBS) solution pH 4.3. 1.0 mL of PB-CQDs solution was added into 0.5 mL of PBS buffer, followed by the addition of 1.0 mL different concentrations of Cu^{2+}. The fluorescence emission spectra were recorded after reaction for 5 min at room temperature (~28 °C).

Selectivity and sensitivity studies

Moreover sensitivity and selectivity is additional significant parameter to assess the performance of the sensing system. Consequently, we examined the fluorescence intensity changes in the presence of representative metal ions such as including Cu^{2+}, Hg^{2+}, Ca^{2+}, Al^{3+}, Ba^{2+}, Mn^{2+}, Ni^{2+}, Pb^{2+}, Co^{2+}, Zn^{2+}, Fe^{3+}, $Cs^{2+}$and Mg^{2+}, of each at a concentration of 15.0 µg mL^{-1} were reacted with a PB-CQDs solution. Fig. 7 A displays the relative change in PL intensity of PB-CQDs in the presence of various metal ions. Fe^{2+}, Mn^{2+} and Hg^{2+} ions cause the slight PL changes (defined as the relative change of PL intensity in 60–80% as compared to blank) which can be attributed to the nonspecific interactions between the carboxylic and/or amine groups and the metal ions. Among these metal ions, Cu^{2+} displays the strongest PL quenching effect on PB-CQDs, assigning to the special coordination between the Cu^{2+} ions and the phenolic hydroxyl and/or amine groups of PB-CQDs which has been widely used for the detection of Cu^{2+} ions or colored reactions in traditional organic chemistry. Such a specific fluorescence quenching effect may originate from the strong interactions between Cu^{2+} ions and the surface groups of PB-CQDs which transfer the photoelectrons from PB-CQDs to Cu^{2+} ions. These results clearly demonstrate that the PB-CQDs-based Cu^{2+} sensor is highly selective to Cu^{2+} over the other metal ions. Fig.8. Fluorescence responses of CCDs to the addition of different metal ions under acetate buffer (pH 4.3). The concentration of each metal ion is 15 µg $mL^{-1.}$ F_0 and F correspond to the fluorescence intensities of PB-CQDs at emission wavelength of 440 nm excited at 320 nm in the absence and presence of metal ions, respectively.

As shown in Fig. 3, the fluorescence intensity of PB-CQDs gradually decreased with the Cu^{2+} concentration. The fluorescence intensity displayed liner responses with the Cu^{2+} concentrations, with two linear ranges of 0.0□14.0 µg mL^{-1} and 15.0□25.0 µg mL^{-1} (Fig.3B). The linear equations are $F/F0$ =0.9827□0.8485C_{Cu2+} (R = 0.993) and $F/F0$ = 0.7565–0.0033 C_{u2+} (R = 0.996), respectively.

REFERENCES

[1] W. B. Lu, X. Y. Qin, S. Liu, G. H. Chang, Y. W. Zhang, Y. L. Luo, A. M. Asiri, A. O. Al-Youbi, X. P. Sun, Economical, green synthesis of fluorescent carbon nanoparticles and their use as probes for sensitive and selective detection of mercury(II) ions, Anal. Chem. **2012**, 84, 5351-5357.

[2] J. Jiang, Y. He, S. Li, H. Cui, Amino acids as the source for producing carbon nanodots: microwave assisted one-step synthesis, intrinsic photoluminescence property and intense chemiluminescence enhancement. Chem Commun **2012**, 48, 9634–6.

[3] Y. Wang, A. Hu, Carbon quantum dots: synthesis, properties and applications, J. Mater. Chem. C **2014,** 2, 6921–6939.

[4] C. Liu, P. Zhang, X. Zhai, F. Tian, W. Li, J. Yang, et al., Nano-carrier for gene delivery and bioimaging based on carbon dots with PEI-passivation enhanced fluorescence, Biomater. **2012,** 33, 3604-13.

Figures:

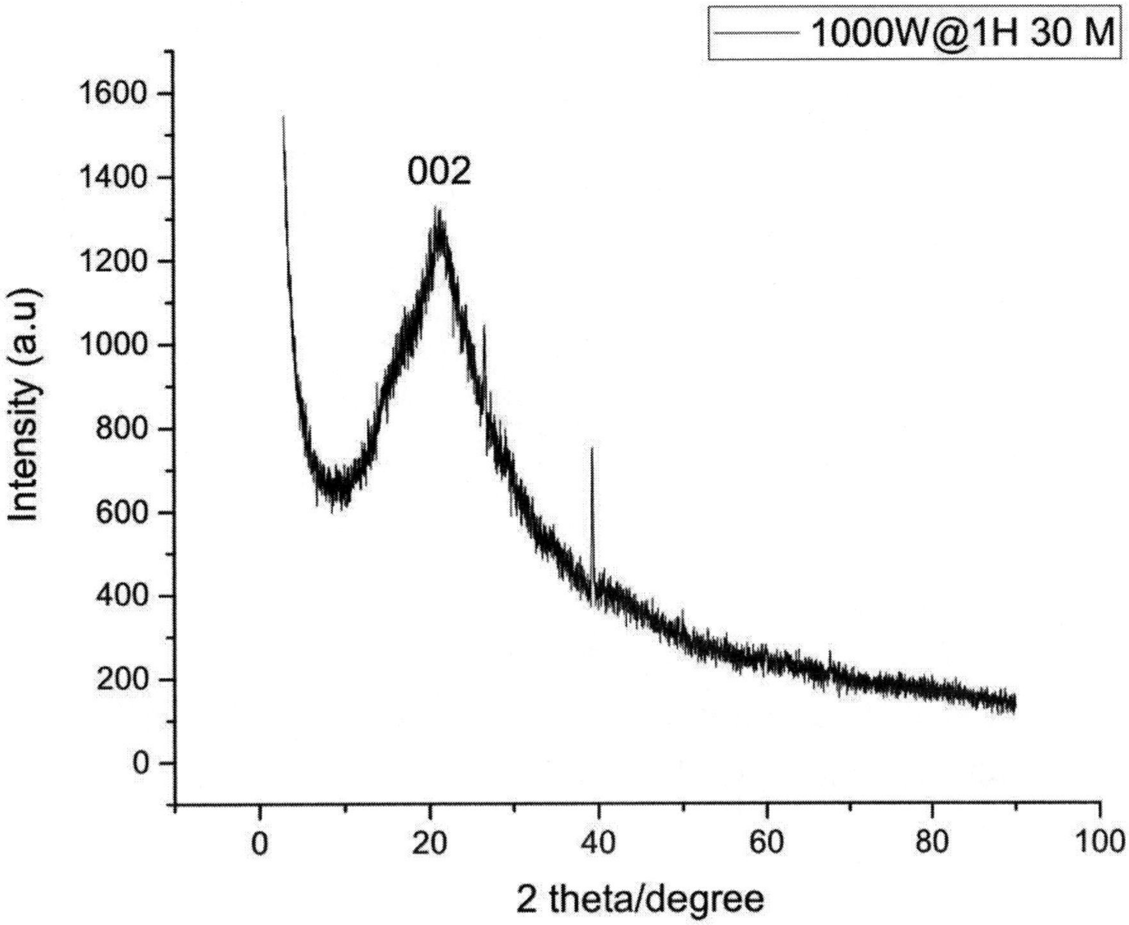
1000W@1H 30 M
1600
1400
1200
1000
800
600
400
200
0
002
Intensity (a.u)
0
20
40
60
80
100
2 theta/degree

Fig 1. FTIR spectra of as-prepared C-dots from Pine Bark

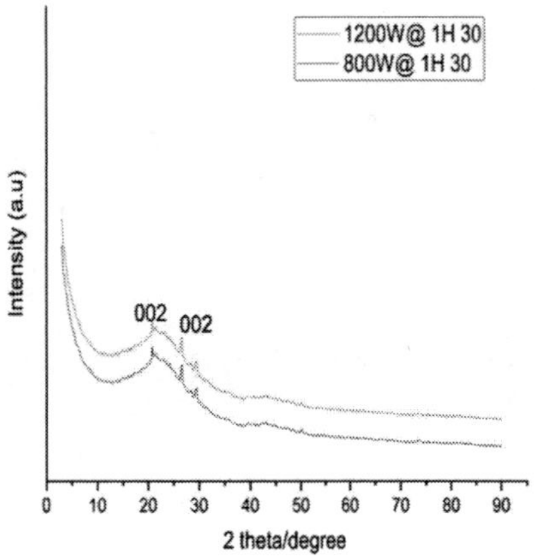

(a)

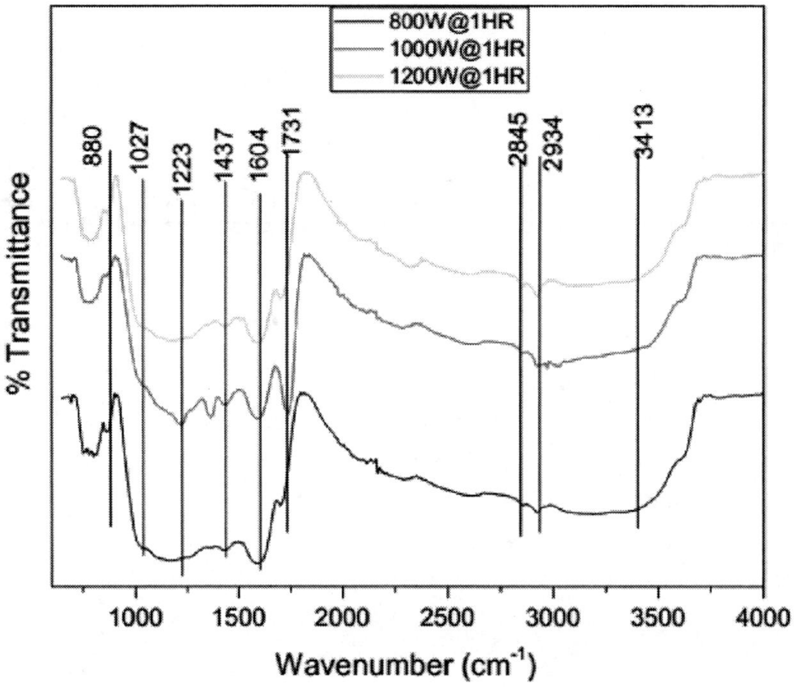

(b)

Fig. 2a & b. XRD pattern of C-dots.

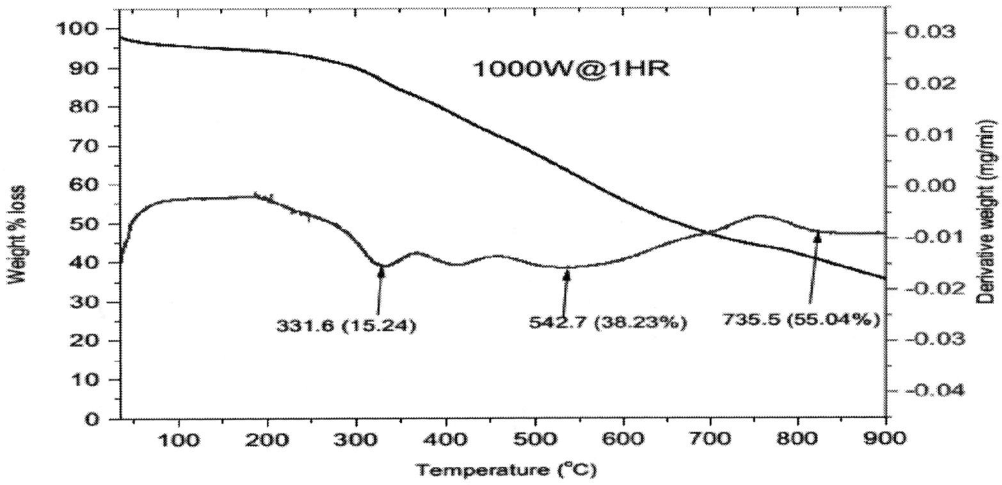

Fig 3. Thermal curves for C-dots under N_2 atmosphere.

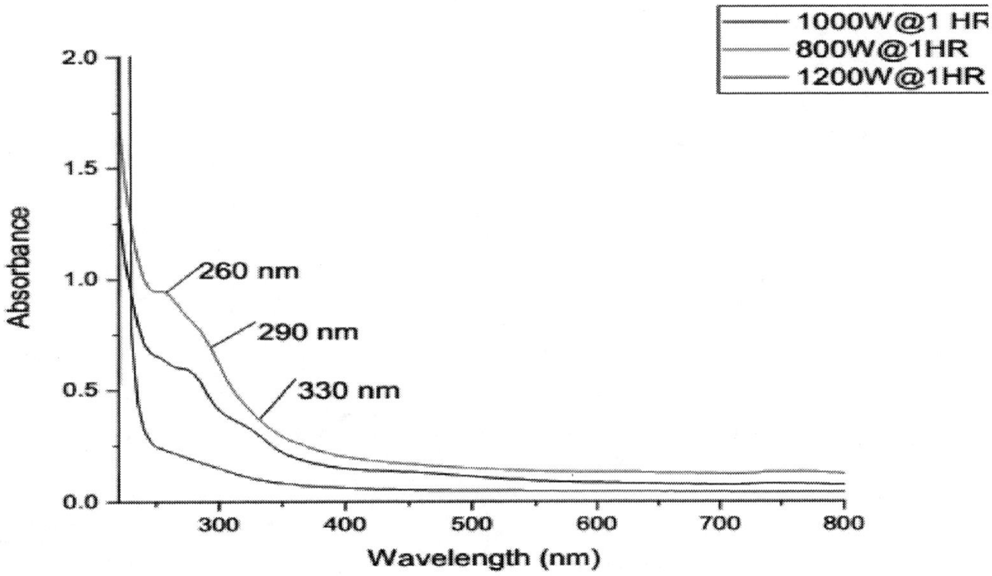

Fig 4. UV-vis absorption of C-dots.

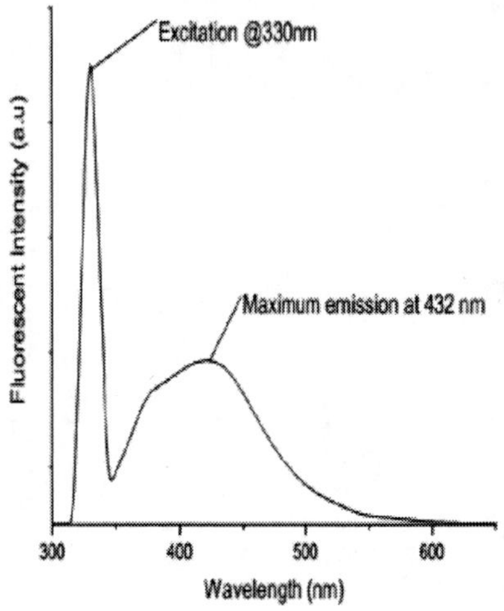

(a)

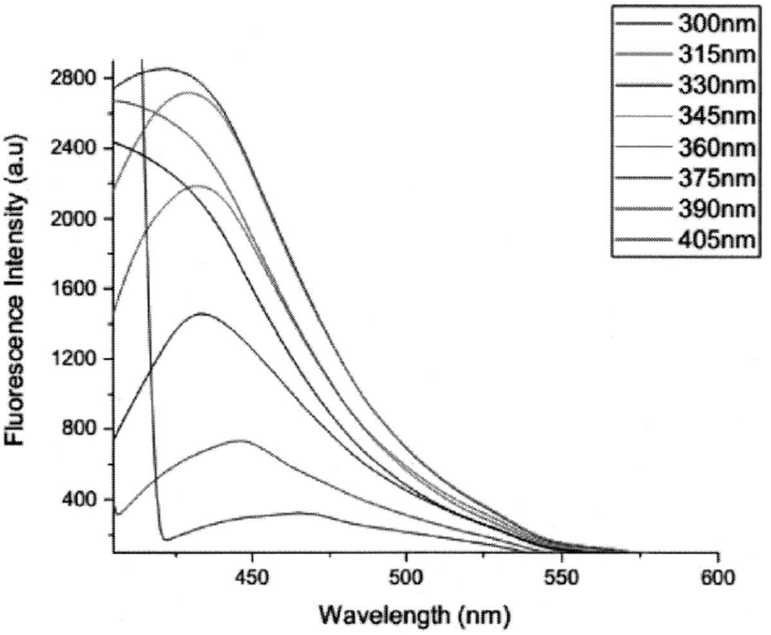

(b)

Fig 5a. PL excitation and emission wavelength (b) PL emission spectra at different excitation wavelength from 300 to 405 nm in 15 nm increments.

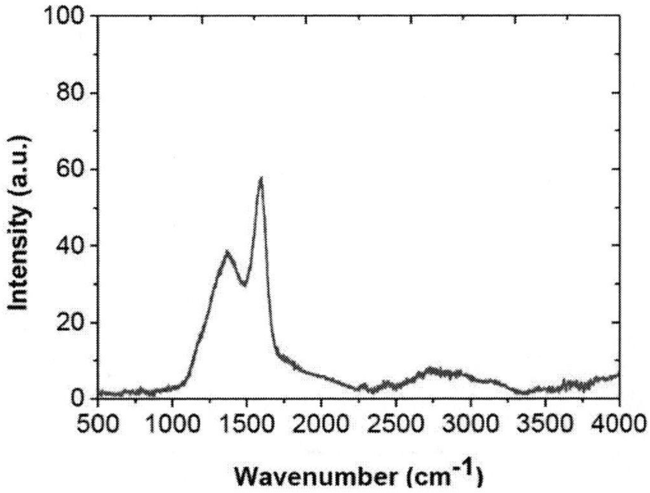

Fig. 6. Raman Spectra of PB-CQDs

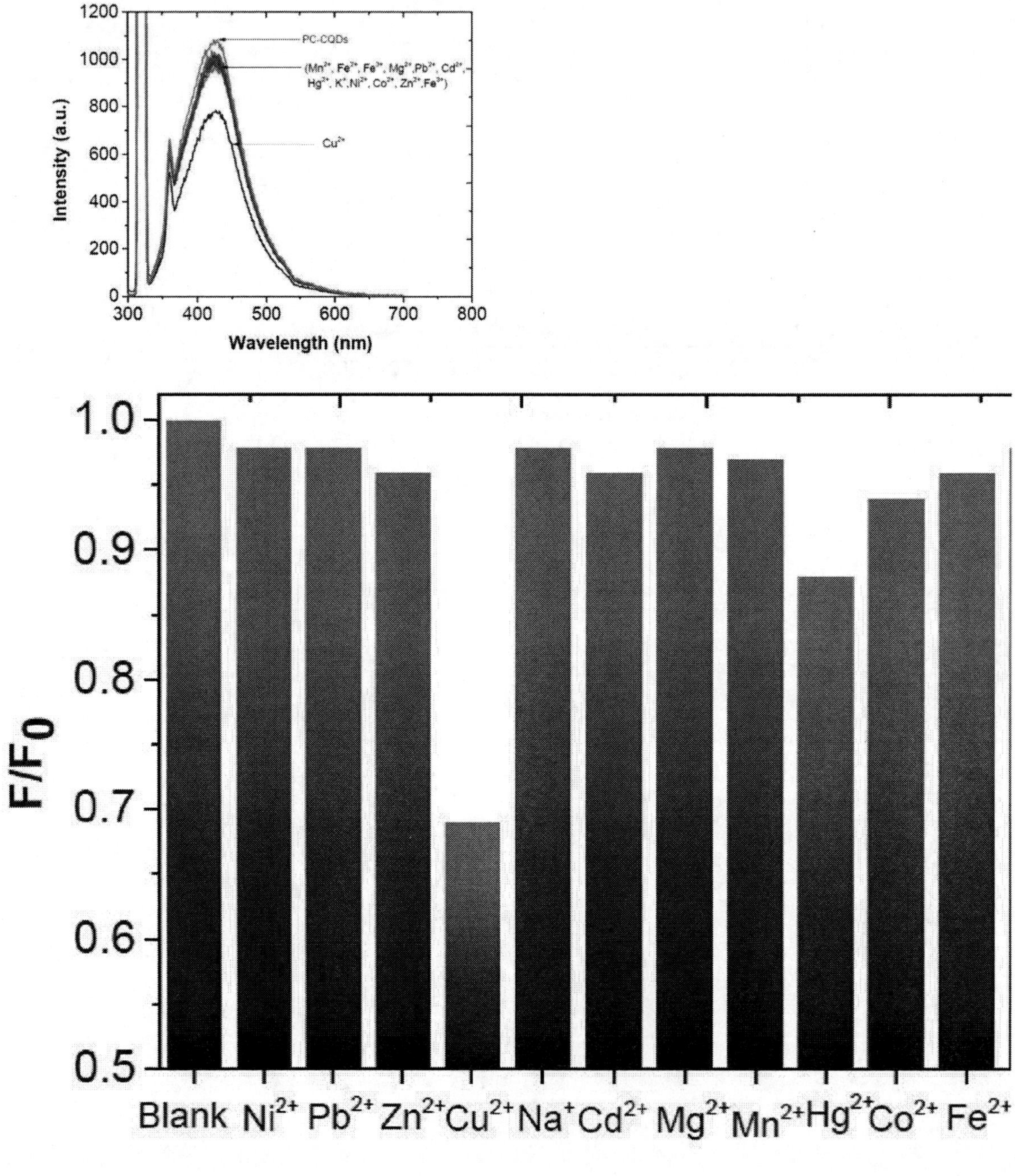

Fig.7. Fluorescence responses of CCDs to the addition of different metal ions under acetate buffer (pH 4.3).

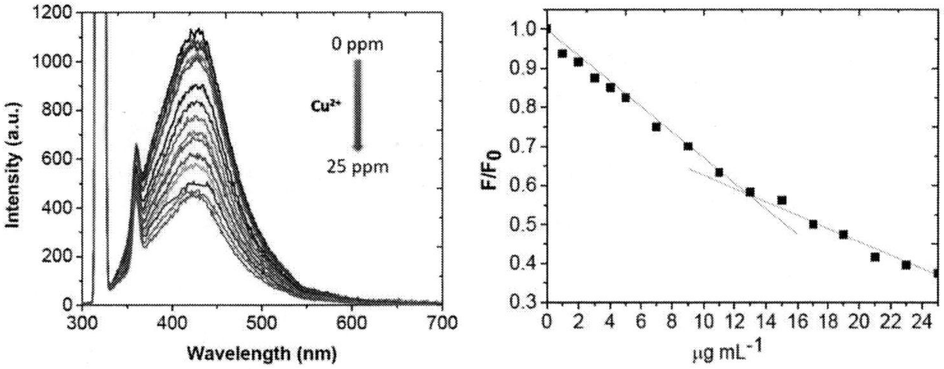

Figure 8. Fluorescence titrations of the y-CDs with Cu^{2+} from 0 to 25 μg mL^{-1}. (Influence of Cu^{2+} concentration on the fluorescence intensity of PB-CQDs (top to bottom: 0, 1, 2, 3, 4, 5, 7, 9, 11, 13, 15, 17, 19, 21, 23 and 25.0 μg mL^{-1})

443095 Experimental and Numerical Study of Transport Phenomenon of Falling Film Flow with Different Heating Methods

Abstract:

Falling film flow is a common phenomenon in chemical industry. It is widely used in chemical instruments, such as structured packed columns and falling film evaporators. An experiment was implemented to study the mass transfer phenomenon of binary-mixture (ethanol-water) falling film flows with different heating methods (uniform or concentrated heating). In order to study the detailed flow and mass transfer behaviors, a 2-D multiphase computational fluid dynamics (CFD) model based on volume of fluid (VOF) was developed. This model divides mass transfer into two parts: one is caused by the vapor-liquid equilibrium (VLE) and the other is caused by the heat flux. Results from simulation are inconsistent with the experimental data. Both experiments and simulations show that the ethanol concentration of the vapor with concentrated heating is 1.5% higher than the one with uniform heating.

443774 Thermodynamic Anaysis of Dehydration of n-Propanol Using Heterogeneous Azeotropic Distillation

Abstract:

Heterogeneous azeotropic distillation is a widely popular process in chemical process industries for the separation of binary azeotropic mixtures into their constituent pure components. It involves the addition of an entrainer which is partially miscible and forms a heterogeneous azeotrope preferably with one of the original components. This is a powerful technique which combines several physical phenomena such as presence of azeotropes and liquid-liquid immiscibilities, and is used widely in industries to enhance difficult or intractable distillation separations. For the design of heterogeneous azeotropic distillation columns, a detailed study of the heterogeneous region is crucial to plan the distillation sequence, which requires experimental vapor-liquid-liquid equilibrium (VLLE) data.

The present study focuses on the investigation of the heterogeneous region of the well known industrially available azeotropic mixture n-Propanol-Water system using Propyl acetate as entrainer. The VLE data of the system was generated by means of modified Othmer still and the still is modified using a probe type ultrasonic homogenizer to account for the heterogeneity. Isobaric vapor- liquid- equilibria and vapor - liquid - liquid - equilibria data at atmospheric pressure have been obtained for the ternary system n-Propanol-water-propyl acetate and the constituent binary systems. The equilibrium data has been correlated using NRTL, UNIQUAC and UNIFAC models and compared with the experimental data.

444269 Functionalization of OH-Terminated Nanoparticles By Cyclic Azasilanes

Abstract:

Colloidal suspensions of functionalized nanoparticles have been suggested for enhanced oil recovery, formation damage remediation, flow assurance, and water treatment operations. Surface modified nanoparticles can also be used as tracers in reservoir studies, active media in thermal management fluids, and fillers in high performance elastomers. Conventional functionalization of nanoparticles with alkoxysilanes and chlorosilanes is limited by needs for catalytic amount of water for the reaction to proceed, requirement of multiple hours to fully saturate the surface, and release of by-products upon reaction completion. Monolayer deposition with a high density of functional groups represents a particularly difficult challenge.

Our ongoing studies indicate that application of cyclic azasilanes that have a strained ring structure with adjacent Si and N centers provides highly efficient functionalization of hydroxylated silica, alumina, magnetite, and diamond nanoparticles which is catalyst-free and by-product-free. The data on synthesis and characterization of functionalized nanoparticles as well as possible reaction mechanism of functionalization will be discussed.

443776 Study of Ionic Liquids As Entrainers for the Extractive Distillation of *N*-Propanol and *N*-Propyl Acetate Binary Azeotropic Mixture and the Isobaric VLE Studies of the System

Abstract:

Distillation is a widely accepted technique for the separation of binary liquid mixtures. Separation by simple distillation is not possible for binary mixtures that form azeotropes. One of the commonly used methods to separate azeotropes is extractive distillation. It is the process of adding an entrainer that can alter the relative volatility of the mixture and thereby breaking the azeotropic point. Ionic liquids on account of their physiochemical properties as well as the recyclability have been used as a preferred replacement for the conventional entrainers. Extractive distillation with ionic liquids has the advantages of both liquid solvent (easy operation) and solid salt(high separation ability). The corrosion and dissolution problems by using inorganic salts can be rectified by these non-volatile, chemically stable, high electrical conductive molten salts.

The azeotropic system under consideration in the present study is *n*-propanol - *n*-propyl acetate mixture. These compounds which constitute the products/reactants of esterification reaction of n-propanol form a minimum boiling azeotrope at 94.7°C. *n*-propyl acetate has been included in a list of volatile acetates as an ecologically benign CFC-free solvent for aerosol sprays.

Vapour liquid equilibrium(VLE) measurements of *n*-propanol and *n*-propyl acetate are performed at atmospheric pressure in a modified Othmer still which is a continuous distillation still in which vapour phase is recirculated. The most suitable ionic liquids for shifting of azeotropic composition of the system are selected using COSMO-RS model with suitable combinations of anions and cations. The isobaric VLE of the system with [EMIM][HSO_4] is obtained and compared with the existing data. The influence of alkyl chain length on the performance of ionic liquids is also analysed with regard to the previous studies. The effect of polarity is the most important factor on the behaviour of ionic liquid solvation in alcohols. Experimental VLE data were correlated using modified UNIFAC model.

444280 Copper-Phyllosilicate Core-Shell Nanocatalysts with Balanced Active Sites for Carbon-Oxygen Hydrogenolysis Reactions

Abstract:

Copper-Phyllosilicate Core-shell Nanocatalysts with Balanced Active Sites for Carbon-Oxygen Hydrogenolysis Reactions

Hairong Yue, Weimao Fang, Changmin Liu, Jiankang Ying

College of Chemical Engineering, Sichuan University, Chengdu, China

Email: hryue@scu.edu.cn

Abstract

Hydrogenolysis of carbon-oxygen (C¨CO) bonds (e.g., esters, ethers, furfural, and CO_2) has emerged as a versatile synthetic tool in organic methodology, as it could produce a variety of products (e.g., chemicals, fuels, and polymers). Copper-based catalysts have been intensively explored for hydrogenation reactions as the copper sites account for the selective hydrogenation of carbon-oxygen bonds and relatively inactive for the hydrogenolysis of carbon-carbon bonds. Work on understanding the active sites of copper catalysts for hydrogenation reactions indicted that both Cu^0 and Cu^+ species were crucial to the activity of Cu-based catalysts. However, researchers were not able to establish a relationship between activity and Cu^0/Cu^+ active species since the conventional Cu-based catalysts is the deactivation by metal particle growth and unstable surface Cu^0 and Cu^+ active species in the strong reductive H_2 and oxidizing carbon-oxygen atmosphere.

We renctly present on efficient approaches for fabrication of a series of copper-phyllosilicate core-shell nanocatalysts and nanoreactors with balanced and stable Cu^0 and Cu^+ active species. We chose the hydrogenation of dimethyl oxalate, ethylene cyclicarbonate and as probe reactions for a better understanding of properties of active sites and the structure-activity relationship. The results indicated a synergy between copper and the oxide components and the balanced surface Cu^0 and Cu^+ species can greatly improve the catalytic C-O hydrogenolysis performance of the catalysts. These nanocatalysts with balanced and stable Cu^0 and Cu^+ active species, confinement effect, intrinsic high surface of Cu^0 and Cu^+ and unique tunable tubular morphology, has potential applications in the high-temperature hydrogenation reactions.

Keywords: core-shell nanocatalysts, Hydrogenation reaction; copper catalyst

References

1. Yue, H.; Zhao, Y.; Ma, X.; Gong, J.: Ethylene glycol: properties, synthesis, and applications. *Chem Soc Rev* 2012, 41, 4218-4244.

2. Gong, J.; Yue, H.; Zhao, Y.; Zhao, S.; Zhao, L.; Lv, J.; Wang, S.; Ma, X.: Synthesis of Ethanol via Syngas on Cu/SiO2 Catalysts with Balanced Cu^0-Cu^+ Sites. *J Am Chem Soc* 2012, 134, 13922-13925.

3. Zhao, S.; Yue, H.; Zhao, Y.; Wang, B.; Geng, Y.; Lv, J.; Wang, S.; Gong, J.; Ma, X.: Chemoselective synthesis of ethanol via hydrogenation of dimethyl oxalate on Cu/SiO_2: Enhanced stability with boron dopant. *J Catal* 2013, 297, 142-150.

4. Yue, H.; Ma, X.; Gong, J.: An Alternative Synthetic Approach for Efficient Catalytic Conversion of Syngas to Ethanol. *Acc Chem Res* 2014, 47(5):1483-1492

5. Yue, H.; Zhao, Y.; Zhao, L.; Lv, J.; Wang, S.; Gong, J.; Ma, X.: Hydrogenation of dimethyl oxalate to ethylene glycol on a Cu/SiO2/cordierite monolithic catalyst: Enhanced internal mass transfer and stability. *AIChE J* 2012, 58, 2798-2809.

6. Yue, H.; Zhao, Y.; Zhao, S.; Wang, B.; Ma, X.; Gong, J.: A copper-phyllosilicate core-sheath nanoreactor for carbon¨Coxygen hydrogenolysis reactions. *Nat Commun* 2013, 4, 2339.

443819 Improving High Shear Rate Friction Reduction Behavior of Polyacrylamide-Based Fracturing Fluids Using Xanthan Gum

Abstract:

The main component of slickwater in fracturing low permeability formations is a high molar mass polymer such as polyacrylamide (PAM), which acts as a drag reduction agent in turbulent flow. Polyacrylamide does not degrade in the environment and also suffers from mechanical degradation at high turbulent flow rates. Five different concentrations of PAM as well as mixtures of polyacrylamide/xanthan were prepared to investigate the possibility of improving shear stability of PAM. Drag reduction measurements were performed in a flow loop and friction factors at different flow rates were calculated using rheological parameters. Based on the observations, Xanthan (XG) is not as good a friction reducer as PAM, even at three-fold higher concentration. XG slightly improved the shear stability in mixed polymer solutions at low concentrations. But at high concentrations, XG significantly improves the friction reduction behavior of PAM at nonionic and ionic environments.

Stimuli-responsive Rubber in Oil and Gas Applications

Rostyslav Dolog, Sankaran Murugesan, and Valery Khabashesku

Baker Hughes, 14990 Yorktown Plaza Dr, Houston, Texas, 77040, USA

ABSTRACT

A problem that exists today in the oil & gas industry is the loss of mechanical properties by reactive rubber seals due to swelling, and the limited space in downhole well applications to accommodate larger swellable packing elements.

This paper presents cost-effective, reinforcing, stimuli-responsive fillers, such as certain classes of fly ash, for use in rubber composite materials in various oil and gas applications. Apart from cost effectiveness, these fillers can also enable stimuli-responsive effects, as well as improved mechanical properties (e.g. modulus) for rubber composites when swelled in water. Fly ash (Class C) is a common component of Portland cement which was used in the literature to achieve a stimuli-responsive effect when added to Hydrogenated Nitrile Butadiene Rubber (HNBR) as filler. Exposure to water brines can cause a filler to promote an increase in pH which enables the hydrolysis of HNBR nitrile groups to carboxyl groups, thus creating an ionomer. This process is followed by neutralization of the new ionomer with metal ions, and formation of ionic crosslinks, further improving mechanical properties of the rubber.

This paper demonstrates that incorporation of fly ash as a reinforcing filler allows development of swellable seals that can retain, or even increase, their modulus and stiffness upon swelling. These characteristics allow avoiding the loss of mechanical properties by seals due to swelling and also eliminating the need for large swellable packing elements that occupy additional space downhole.

KEYWORDS: Elastomers, stimuli-responsive, fly ash, composites

INTRODUCTION

There is a growing demand for high performance elastomers in the oil & gas industry. Rubber compounds are used for various sealing applications, packer elements, stators for pumps, etc. In general, rubber formulations are comprised of elastomer matrix, curing system, antioxidants, plasticizers, along with reinforcing filler, which provides the compounds with their strength. The reinforcing filler is usually represented by carbon black and, to a lesser extent, silica. Along with other factors, filler mechanical properties and concentration generally define hardness, modulus and tensile strength of the elastomer compound. Throughout the years, many different materials have been tried as alternative fillers for rubber composites, including nanomaterials (CNTs, graphene, nanofibers), recycled rubber particles, and ash/fly ash. In the case of ordinary carbon black, the filler serves not only for reinforcement, but also as a cost reducer, because raw elastomer is often a more expensive component. Mechanical properties of the composites are affected by the aspect ratio of the filler, quality of the dispersion, and interface between blend components. The latter determines the effectiveness of load transfer from polymer matrix to the reinforcing filler. High performance compounds can be designed by incorporating well dispersed fine particles with a high aspect ratio and good compatibility with the polymer matrix.

In this context, the fly ash is a cheaper, low reinforcing filler alternative to carbon black, which will help reduce cost of the compounds for "non-demanding" sealing applications requiring large amounts of raw materials. Fly ash is produced in large quantities as a byproduct of coal burned at power plants, and can be considered as a recycled material. Fly ash costs several cents per pound, while carbon black, depending on the grade, is usually one order of magnitude more expensive. Hemmings et al[1] demonstrated that fly ash can be used as filler for polymeric composites with enhanced mechanical properties. Adoption of this technology in the oil and gas industry could change the design approach for reactive rubber compounds. Further functionalization of fly ash through a cost-efficient process could ensure compatibility with the rubber matrix, while promoting filler dispersion and ensuring an efficient load transfer. Such functionalization's include, but are not limited to, surface modification with fatty acids, dicarboxylic acids or phosphoric acid. The reasoning for surface modification is to attach functional groups with aliphatic chains to fly ash particles to make them compatible with aliphatic rubber chains, or to attach acidic groups on their surface that can interact with nitrile groups of HNBR. Initial attempts towards chemical functionalization of fly ash have been made by Seal et al for treatment of offshore oil spills.[2]

Apart from cost effectiveness, composites with fly ash can show stimuli-responsive properties. It has been shown that adding Portland cement to the HNBR matrix improves its mechanical properties in the swollen state by water uptake.[3] Exposure to water causes Portland cement to grow in volume and mass, and promotes an increase in pH which enables the hydrolysis of HNBR nitrile groups to carboxyl groups, thus creating an ionomer. This process is followed by neutralization of the new ionomer with metal ions and formation of ionic crosslinks, further improving mechanical properties of the rubber. Fly ash (Class C) is a common component of Portland cement, which contains a large variety of metal oxides, and is capable of increasing the

---

[1] Filler Comprising Fly Ash For Use In Polymer Composites. Hemmings et.al., US 6,916,863 B2, July 2005.

[2] Hydrophobic High Surface Area Zeolites Derived from Fly Ash for Oil Spill Remediation, Seal et.al., Environ. Sci. Technol., 2013, 47 (11), pp 5843–5850

[3] Stimuli-Responsive Cement-Reinforced Rubber, Musso et. al., ACS Appl Mater Interfaces. 2014 May 14;6(9):6962-8.

pH of the water solution by hydrolysis, which has a potential to achieve a stimuli-responsive effect when added to HNBR as filler.

Degradation of tensile properties is a common problem for swellable packers and other sealing elements. By selectively using certain classes of fly ash as effective reinforcing filler, one can develop swellable seals that can retain or even increase their modulus and stiffness upon swelling. Therefore, there is a potential to minimize or even avoid the loss of mechanical properties by water swellable reactive packing elements due to swelling. This provides an opportunity to decrease the size of these elements, which are often represented by rather long sections, and hence occupy a significant amount of space downhole. In this paper we have studied the reinforcing effects of fly ash fillers with HNBR compounds as well as stimuli responsive properties of these compounds.

EXPERIMENTAL SECTION

Rubber compounds T1, T2, T3, and T4 are elastomers based on a custom-built peroxide curing system. Formulation F2 was prepared by adding 15 wt. % of fly ash class C to compound F1, which is a commercial swellable NBR sulfur-cured compound.

T series compounds were prepared using a C.W. Brabender® internal mixer operated at 60 RPM for 20 minutes at ~160°F and then molded into sheets and cured at 340°F for 30 minutes in a hot press. Compound F2 was prepared by mixing fly ash with compound F1 in the Brabender mixer for 6 minutes at 60 RPM at ~100°F. Compound F1 is a commercial water swellable formulation that was used as received. F series compounds were cured at 300°F for 60 minutes in a hot press.

Tensile properties of rubber samples were tested using dog bone specimens, according to the ISO37 standard, at room temperature and at 350°F using an MTS® testing system.

Swelling tests in 3% KCl brine and LVT 200 oil were performed to assess the effect of fillers on the chemical resistance of elastomers. Standard one-inch rubber buttons were tested at 200°F at ambient pressure for 6 days. Thermal properties were studied using TGA Q50 and DSC Q2000 instruments.

To analyze the stimuli-responsive effect introduced by fly ash fillers, rubber buttons were aged for 10 days in 3% KCl in the oven at 200°F and then removed from the solution, dried for 30 minutes in the oven and then for 3 hours at room temperature. Samples were later tested in the MTS testing system in compression mode.

Viscoelastic properties of stimuli-responsive samples, based on water swellable compounds, were accessed using the RSA G2 dynamic mechanical analysis (DMA) instrument. This experiment was conducted in a temperature ramp mode of 2°C/minute at 1Hz frequency and 0.2% amplitude strain. A Nicolet 6700 Fourier Transform Infrared (FTIR) spectrometer was used to confirm neutralization of carboxylic groups by metal cations.

RESULTS AND DISCUSSION

The reinforcing properties of fly ash were evaluated at room temperature (Table 1) and at 350°F (Table 2) and compared with N550 carbon black. Sample T3, reinforced with fly ash, demonstrated inferior tensile properties and hardness compared to a compound filled with carbon black. However, at room temperature, fly ash was capable of delivering a reinforcing effect to the rubber compound without sacrificing its elasticity, e.g. the ultimate elongation of elastomer compound remained virtually unchanged after incorporation of 50 parts per hundred rubber (PHR) of fly ash, while its modulus and tensile strength have increased. Therefore, potentially, one can substitute carbon black with high loadings of fly ash in certain applications, thus further decreasing the cost of the rubber product. The summary of mechanical properties is provided in Table 1, E_x denotes modulus of a compound at certain strain (x), i.e. stress at strain.

Sample	N550 CB (PHR)	Fly ash Class F (PHR)	Tensile strength (psi)	E_{25} (psi)	E_{50} (psi)	E_{100} (psi)	Elongatio n (%)	Hardness Shore A
T1	0	0	829	71	97	121	472	48
T2	50	0	3402	165	273	744	290	68
T3	0	50	1176	94	127	166	467	54

Table 1. Room temperature tensile properties and hardness of elastomer samples

The relatively low increase in modulus and hardness of T3 compared to T1 at room temperature can be explained by a low aspect ratio of fly ash and weak interactions between blend components. The latter problem can be solved by functionalization of fly ash particles.

The general data trends from the high temperature tensile tests closely resemble the results of the tensile test conducted at room temperature. However, at high temperature, rubber specimens break at much lower elongation, which leads to a great decrease in the measured tensile strength. This behavior is typical for elastomers and limits their application to a certain maximum service temperature and is also due to the fact that high temperature tests are conducted at conditions that push the limits of the material. The modulus values at 15% (E_{15}) and 20% (E_{20}) strain are not provided herein, because half of the tested specimens were broken before they reached such strains.

The superiority of carbon black as an efficient filler became even more pronounced at high temperature. However, incorporation of fly ash doubled the tensile strength and increased the elongation at break by one-third when compared to the unfilled elastomer.

Sample	E_5 (psi)	E_{10} (psi)	E_{15} (psi)	E_{20} (psi)	Tensile strength (psi)	Elongation (%)
T1	31	46			91	23
T2	61	130	194	261	832	39
T3	42	73	98	119	173	30

Table 2. Summary of high temperature (350˚F) tensile data for elastomer samples

Thermal properties of samples T1-T3 were assessed with Differential Scanning Calorimetry (DSC) and Thermogravimetric analysis (TGA). TGA allows evaluating thermal stability of samples and to verify the actual composition of a sample from the ash content. The results of the DSC and TGA analysis are summarized in Table 3. During TGA, samples were heated to 600°C at 10°C/minute rate in a nitrogen (N_2) environment, and subsequently heated at 10°C/min to 600 °C in air to burn off carbon-based fillers. This allows decomposing of organic matter first by applying heat, and then burn the remaining carbon-based material in air atmosphere.

The initial degradation of the polymer started at 200°C and progressed to complete decomposition of aliphatic chains in the temperature range of 420-500°C with a highest rate of decomposition at ~460°C. Sample T2 incurred the second rapid mass loss after 600°C due to burning of the carbon black, which comprised most of the ash content at 600°C in a N_2 environment. The DSC experiment demonstrated the glass transition temperature of the elastomer remained largely independent of the type of filler - the respective DSC heating scan was conducted with a 10˚C/minute temperature ramp after precooling the sample to -50˚C.

Sample	W_{600} (%)	W_{800} (%)	$T_{deg}1$, ˚C	$T_{deg}2$, ˚C	T_g, ˚C
T1	2.4	2.4	452.8	-	-29
T2	29.8	2.7	458.7	616.5	-29
T3	32.8	32.7	447.3	-	-28

Table 3. TGA/ DSC results for elastomer samples

Table 3 contains ash contents (W) at 600°C and 800°C, degradation temperatures (T_{deg}) which represent the temperature of the highest rate of decomposition during TGA experiment, and glass transition temperature (T_g) of the samples, as determined by DSC scan.

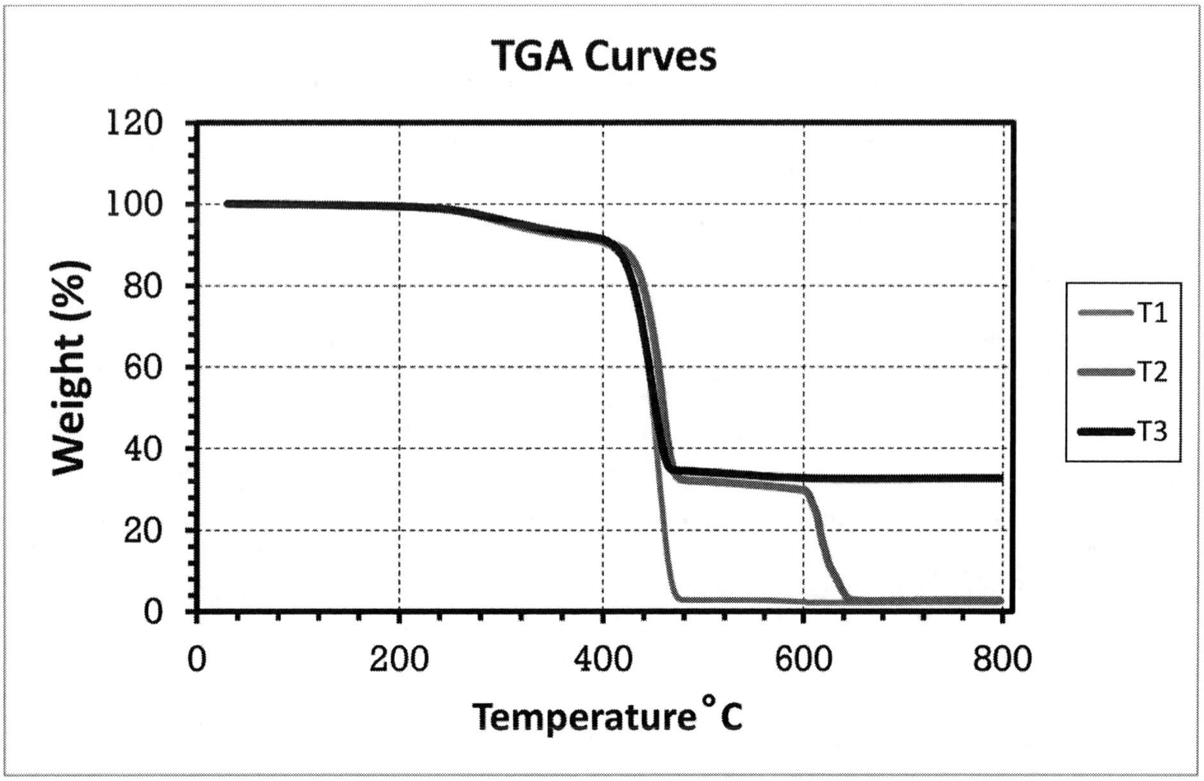

Figure 1. TGA Curves for samples T1, T2, and T3

Sample \ Testing fluid	3% KCl brine	LVT 200	Specific gravity
T1	0.6	13.0	0.96
T2	0.3	11.6	1.13
T3	0.8	9.1	1.18

Table 4. Volume increase (%) of elastomer samples due to swelling for 6 days at 200°F

Musso et. al. developed a stimuli-responsive HNBR composite reinforced with Portland cement that was increasing its stiffness and modulus when swelled in water[3]. Fly ash (Class C) is a common component of Portland cement, which contains the components necessary to achieve a similar effect. Therefore, one can avoid the loss of mechanical properties by water swellable packing element seals upon activation. To test this hypothesis, we studied the effect of aging in water brines on the properties of elastomers with fly ash fillers.

Rubber samples were molded and cured into standard size rubber buttons (cylinders of ~0.25" height and ~1" diameter). These buttons were aged for 10 days in 3% KCl brine in the oven at 200°F, removed from the solution, dried for 30 minutes in the oven and for 3 hours at room temperature. The swelling of both compounds was close to ~1% by volume (Table 4). Samples were later tested in an MTS machine in compression mode. Each button was compressed to 70% strain, buttons remained intact during the experiment, and visually all appeared to have a certain level of compression set at the end of the experiment. The results of the experiments are summarized in Table 5 and Figure 2. Only one button was tested for each sample. The provided compression modulus values (C_n) represent values of stress at a percentage strain (n), e.g. C_5 is stress at 5% compression strain.

Compound T4 was reinforced by fly ash Class C, which contains larger weight fraction of CaO than fly ash Class F (~25% vs ~10-15% respectively) and is expected to produce a higher pH and closely resemble cement that was used to prepare stimuli-responsive rubber in the original paper by Musso et.al.[3] The results demonstrate that exposure to water (3% KCl) at high temperatures can improve the modulus for the compounds filled with fly ash Class C (T4), while the control sample (T1) showed the opposite behavior. After aging, the compressive modulus at 50% strain (e.g. stress at 50% strain) of control sample T1 decreased by 19%, as compared to the unaged sample, while compressive strength of sample T4, reinforced by fly ash, increased by 15%. The difference in modulus values is even more drastic at 70% compression strain (-35% and 31%, respectively). This is in agreement with the hypothesis stated earlier.

Sample	Filler	Compressive modulus (psi)				
		C_5	C_{10}	C_{25}	C_{50}	C_{70}
T1	None	19	42	127	468	1810
T1 swollen	None	16	36	109	380	1185
T4	Fly Ash class C (50 PHR)	21	46	134	434	1628
T4 swollen	Fly Ash class C (50 PHR)	29	57	153	501	2125

Table 5. Summary of compressive modulus properties after 3% KCl soaking and 200°F drying

To further investigate the stimuli-responsive property of fly ash filler, the latter was incorporated into a water swellable compound, which provided for significantly more interaction of filler with water to occur. Sample F2 comprises unmodified commercial water swellable NBR formulation F1 mixed with 15 wt.% of fly ash Cass C. Sample F2 was processed into a 0.04" slab, then a rectangular sample was cut from the slab and swollen in 3% KCl solution until 100% volume increase. The F2 sample was then dried overnight in the oven, which lead to de-swelling of the sample due to solvent evaporation. Mechanical properties of the F2 sample were assessed with DMA prior to the swelling test, in a swollen state, and after drying - the data summary is shown on Figure 3.

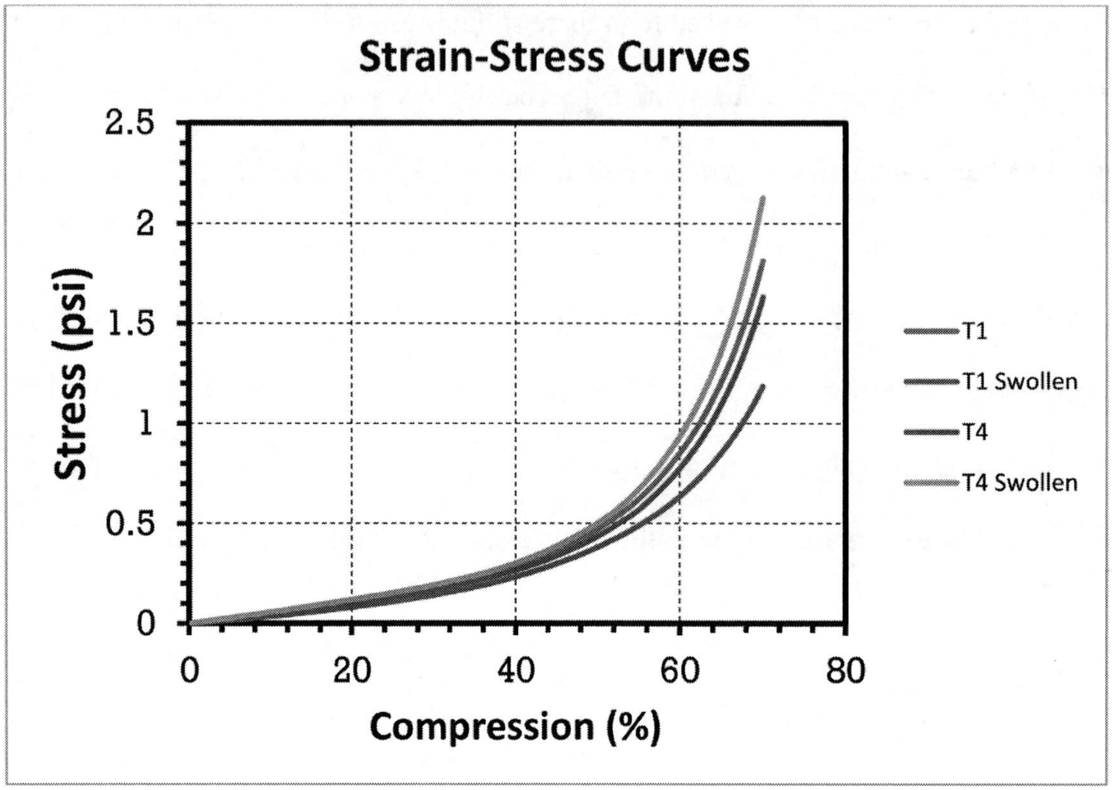

Figure 2. Stress-strain curves

The storage modulus of sample F2 after swelling and subsequent drying was about an order of magnitude higher than that of original sample F2 measured prior to swelling experiment. The storage modulus of the dried F2 sample was also less sensitive to the increasing temperature, contrary to the weakening original sample F2. The increase in the storage modulus of swollen F2 with temperature is simply due to the drying of the sample. The obtained result demonstrated that fly ash can essentially act as a stimuli-responsive material.

The significant improvement in the modulus is believed to be due to the neutralization of the swellable component of the commercial F1, and potential hydrolysis with subsequent neutralization of NBR elastomer. This is confirmed by the shift in T_g to a lower temperature range after drying as illustrated in Figure 3.

Carboxylic groups were detected in the original sample F2 prior to swelling, hence, they cannot be attributed solely to hydrolyzed HNBR, but rather to the super-absorbent polymer present in water swellable rubber compound (see Figure 4). FTIR confirms neutralization of carboxylic groups because a significant increase of the peak near ~1540 cm^{-1} and a lowering of peak intensity at ~1720 cm^{-1} have been observed. This neutralization is believed to promote the formation of ionic clusters in the compound with fly ash considered to be the source of cations. The neutralization of carboxylic groups could as well be attributed to K^+ cations present in the KCl brine solution, however, in that case, the respective peak in FTIR spectra would appear at higher wave numbers, approaching ~1600 cm^{-1}.

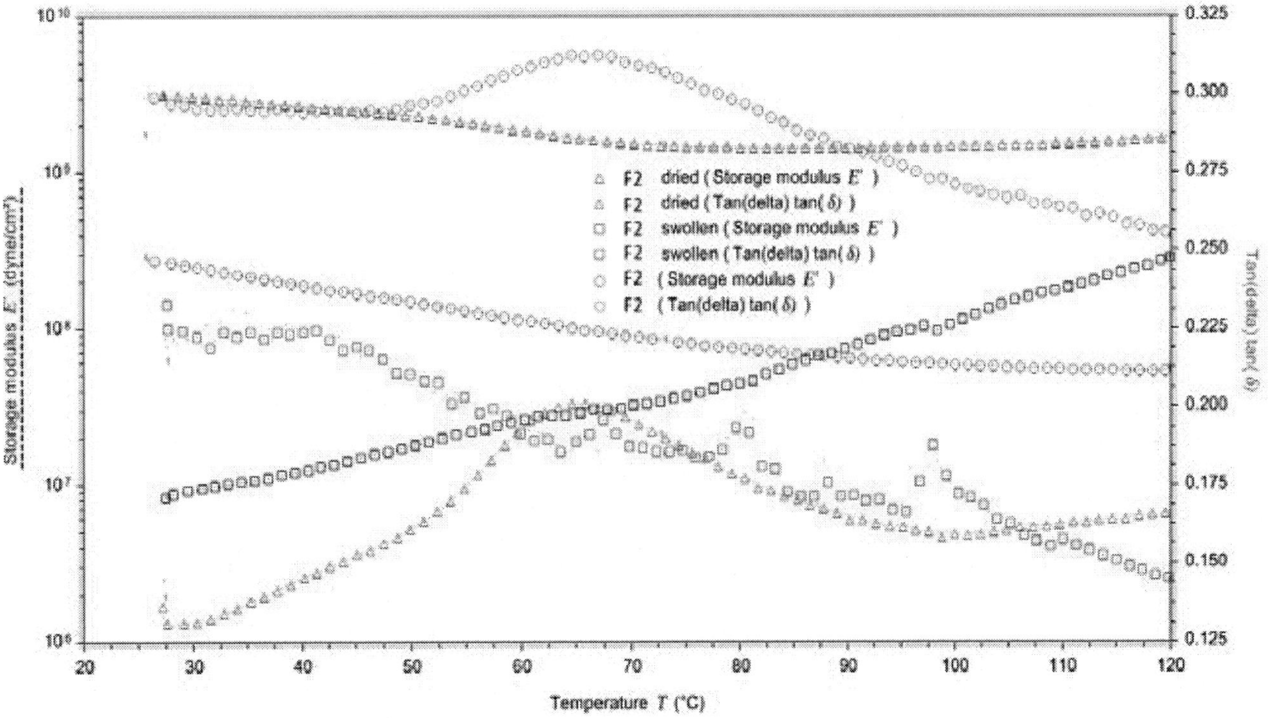

Figure 3. DMA data on F2 sample

It is important to mention that sample F2 exhibited a certain degree of brittleness after drying, which is undesired in rubbers, but this can be managed by incorporation of common ionomer plasticizers.

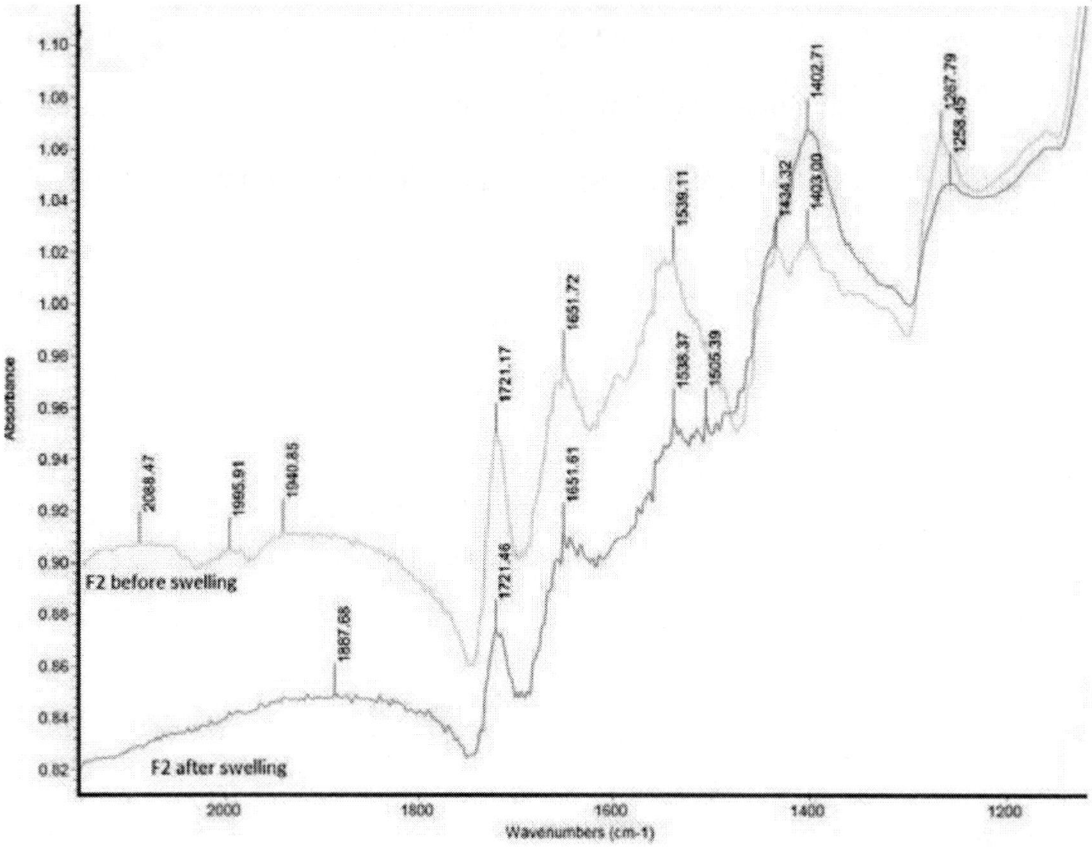

Figure 4. FTIR spectra of F2 sample taken before and after swelling

Conclusions

Fly ash demonstrated a great potential for consideration as cost effective, low reinforcing filler for downhole oil and gas applications. Although the reinforcing properties of fly ash are inferior to carbon black, incorporation of fly ash can improve the mechanical properties of HNBR elastomer compared to the unfilled raw crosslinked elastomer at both room temperature and at 325°F.

Furthermore, composites of HNBR with fly ash Class C demonstrated a stimuli-responsive effect by significantly increasing their modulus upon swelling in water brine. This effect was even more pronounced in composites of fly ash with swellable water compounds, which provided an easy access of filler to water upon swelling. A ten-fold increase in modulus was achieved in such fly ash reinforced sample, after swelling in 3% KCl water brine and drying in the oven overnight. The FTIR confirmed neutralization of carboxylic groups present in the investigated system, which contributed to overall improvement of mechanical properties of the compound upon swelling.

444093 New Development on Hoist Scheduling with the Relaxation of Cyclic Precondition

Abstract:

Multi-Stage Material Handling (MSMH) processes are designed to manufacture materials requiring a series of treatments and in industry, they are broadly used due to their capability to handle large work orders and compatibility with materials with different recipes. A typical example is electroplating process. A hoist handles the transportation of each job among units (stages) according to its processing recipe. The hoist movement plan is predetermined through optimization and this plan is repeated to perform continuity. Obviously, the ultimate goal is to maximize the productivity within each movement plan, which is equivalent to minimize the cycle time to finish one work order. The optimization process is generally called Cyclic Hoist Scheduling (CHS), since every movement plan is a cycle due to its repeatability. This precondition in term of hoist operation and unit usage is general for most cases. However, it could block the really optimal results.

In this study, this precondition has been relaxed for multi-capacity units, thus the unit usage can be more flexible. Thus, the real optimality can be found through CHS. To deal with the cycle connection issue from the relaxation of the precondition, a new module Cycle Connection Identification (CCI) is proposed. Obtained hoist schedule will be checked. Once a schedule with cycle connection problems is identified, a new cycle will be scheduled forcing the smooth connection with the first cycle. The objective of scheduling a new cycle is to match the beginning of the first cycle and the ending of the second cycle. If the two-cycle group is still with cycle connection issue, more cycles will be scheduled till the cycle group is free from cycle connection problems. Each cycle is called a sub-cycle and the constructed cycle group without cycle connection problems is called a full cycle. Note that in each sub-cycle, the work order and all transportation operations are completed.

In this paper, hoist scheduling models with/without such a precondition have been developed and proposed as mixed integer linear programming. The performance of the two models is tested through case studies and the difference is compared and demonstrated in detail.

444310 Formal Modeling of Accidents and Automatic Generation of Tutoring Scenarios for Intelligent Safety Training of Process Operators

Abstract:

With the emphasis of accident-free operations and prevention of catastrophic events, safety-training demands for the chemical plants have been on the increase globally. However, the research results on modeling of accidents and involving human operators and the advanced theory and technology for (safety) operator training have been shared limitedly among domain experts and commercial developers. Even with the recent trend of simultaneously delivering the operator training system with the construction of a new process plant, in commercial operator training systems (OTS), safety training used to implement quite limited and obvious scenarios (and the burden was almost up to the trainer), partially due to the difficulty in modeling propagations of probable accidents. By the regulatory requirements or proactive investment by companies, effective scenarios for training of safety and abnormal situation management are necessary to be improved in the aspects of the size of volume as well as the training depth. Thus, considering the emergence of enhanced virtual reality technologies and inexpensive gadgets like HMD, supporting the immersive training environment, the future efficiency of the training heavily relies on the quantity and quality of safety scenarios, which would be essential in guiding the entire training process.

In this research we first overview a government-sponsored research project developing an enhanced safety training system for chemical process operators, using haptic devices in addition to enhanced audio and visual equipment. The initial version of the system runs with a fixed set of training scenarios developed for predetermined accident scenarios. However, with a limited set of training scenarios, the trainees simply memorize and rote learn the steps, required to be mastered to proactively solve the problems with an advent of possible accidents. In addition to that, without formal modeling, the system cannot give timely feedbacks and the performance of the trainee cannot be measured in real-time. Then to enhance the contents and usefulness of the training, the knowledge of process operations is collected and systematically stored as a form of knowledge base, in addition to the accident scenarios obtained by executing Fault Tree Analysis and HAZOP study.

By using a formal modeling language like UML and Petri nets, the accident scenarios and operators' problem-solving tasks are modeled hierarchically and classified into distinguishable subtasks. Then a new training scenario is generated on-line through automatic synthesis based on planning using the knowledge base. Safety training scenarios are generated intelligently for the tutoring of a single operator's job function as well as effective collaboration among a group of operators for real-scale malfunction diagnosis and intervention at the initial propagation stage of an accident. As case study, safety trainings for city gas governing stations are executed based on systematic scenario generation and analyzed for improved effectiveness and field applicability of the system. Reduction of time and cost in the training scenario production and real-time customized training based on new operator's level of knowledge and skill are also discussed in the framework of intelligent tutoring systems.

Experimental Study on Najas Minor of Producing Bio-Oil

Abstract: Due to the diminishing of petroleum reserves and the environmental degradation caused by consumption of fossil energy , many attentions have been paid to the application of biomass energy which is acceptable as an abundant renewable energy resource. Najas minor grows in dense clusters with highly branched stems of which a small fragment could propagate into a thrive one.In this study, we researched the properties of Najas minor to make biofuels .Firstly,the najas minor after heavy metal ions degradation were fed into pyrolyser to prepare biofuels. Significant differences of the product was observed compared with only Najas minor pyrolysed.

KEY WORDS: Najas Minor; catalytic pyrolysis; bio-oil

1.Introduction

Energy demands and environmental problems caused by fossil energy have been a matter of growing concern.Bio-oil has been considered to be a promising way for producing renewable bio-fuels or chemicals[1-3].In recent years,many countries pay attention to the importance of biomass that is an abundant renewable energy resource.Biomass can be defined as any organic material mainly consisting of carbon, hydrogen,oxygen, nitrogen, and some other components in small proportions[4].Although the biomass has been studied for nearly a century,most studies mainly concentrated in Lemna minor, Lemna gibba, Spirodela polyrhiza and Landoltia punctata.We still know little about the biological characteristics and potential application value of other varieties of biomass[5].Najas minor,a submersed herb,grows in dense clusters with highly branched stems of which a small fragment could propagate into a thrive one.Najas Minor is native to Europe, Asia and North Africa from the Netherlands to Morocco east to Japan and the Philippines,including China,Siberia, Central Asia, Iran, Turkey, Ukraine, Germany, France Italy and a host of other Countries.

Catalysts used for catalytic cracking are basically divided into the following categories: molecular sieve, alkali metal salt, metal oxide, etc.Different catalysts have different catalytic effects.The use of metal nanoparticles has shown to be very important in recent industrial applications[6].The application of precious metal nanometer materials in catalysis,which has caused the extensive concern of researchers in recent 30 years,has become a hot research topic in the field of catalysis.Gold nanoparticles have received a great deal of attentions due to their special optical and electrical performances, catalytic properties and remarkable quantum size effect, surface effect etc[7]. The synthesis of metal nanoparticles can be classified broadly into wet and dry methods[8]. The cost of production is relatively high both materially and environmentally. There is an unequivocal need to develop more cost-effective and environmental benign methods to alternate these existing methods[9].Researchers in this field turn to use organisms to produce inorganic nanostructures extracellularly[10].

Although there has been a large volume of work on the synthesis of gold nanoparticles[11-13], and obtainning bi-oil by abundant biomass, there are relatively fewer attempts to combine these two methods.In this work, gold nanoparticles were obtained by aqueous chloroauric acid solution with the of Najas Minor at room temperature. The direct pyrolysis and catalytic pyrolysis experiments of Najas Minor were carried out to observe the effects of catalysts on pyrolysis process of Najas Minor.

2. Materials and Methods

2.1. Experimental samples

The Najas Minor collected from Tianjin city and the solutions of Au(III) were used in this study.The stock solutions of Au(III) were prepared by dissolving analytical grade chloroauric acid

($AuCl_3 \cdot HCl \cdot 3H_2O$, Kewei Reagent, China) in deionized water.

2.2.Experimental apparatus and procedure

Pyrolyzer with GCMS,shown in Fig. 1,is mainly used for analyzing the composition of polymer.Pyrolyzer with GCMS includes three parts :quartz pyrolysis tube,pyrolyzer and GCMS.

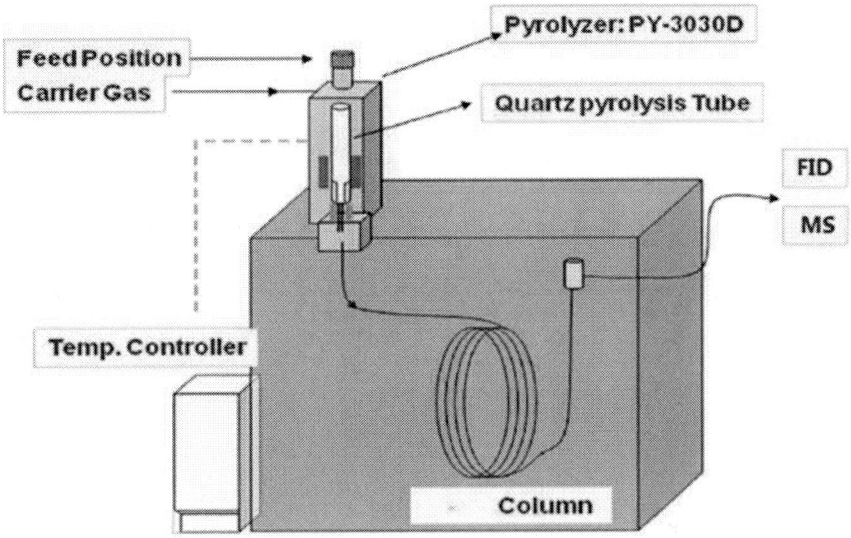

Fig. 1 Diagram of Pyrolyzer with GCMS

The quartz pyrolysis tube (shown in Fig.2) length is 12cm.The sample end diameter is 5 mm,and the pyrolysis products end diameter is 3mm.This special structure is beneficial to the fixation of the reactant in the tube.The top and bottom of the reactants are both supported by quartz wool to prevent reactants blowed by the carrier gas ,or fine particles into the chromatographic column.All pyrolysis experiments were carried out on the EGA/PY-3030D pyrolyzer(Frontier Lab,Japan) .The samples were filled in the special quartz pyrolysis tube to reduce the amount of valuable catalyst and reduce the cost of research work.

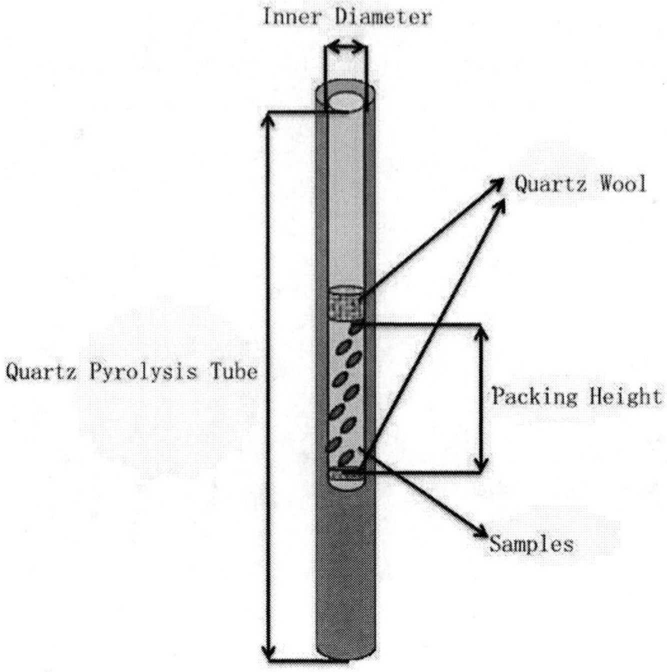

Fig. 2 Diagram of Quartz Pyrolysis Tube

Approximately 0.5g Najas Minors were used in each experiment.The samples were impregnated in 50ml deionized water and 50ml solution of chloroauric acid(0.0793mmol/L) for a night,followed by drying at 80℃ until the weight didn't change.Each experiment used 2.5mg Najas Minors as reactants.Each sample was heated, under the control of heating controller, from ambient temperature to 600 ℃ repeatly and kept at 600 ℃ for 2 h.

3. Results and discussion

3.1.MS Analysis of the Products during the Pyrolysis

Under the same pyrolysis conditions, bio-oil obtained by direct pyrolysis and catalytic pyrolysis were analyzed by GCMS,and the main chemical compounds were shown in Table1 and Table2.

Table 1 Indentification and quantification of main chemical compounds by direct pyrolysis

NO.	peak area%	Formula	Compound
1	12.67	$C_3H_6O_2$	2-Propanone, 1-hydroxy-
2	3.67	C_7H_8	Toluene
3	3.28	C_7H_8O	p-Cresol
4	3.81	$C_5H_{10}O$	Pentanal
5	3.53	C_8H_8O	Benzofuran, 2,3-dihydro-
6	2.17	C_8H_8N	Indole
7	5.93	$C_{20}H_{40}O$	3,7,11,15-Tetramethyl-2-hexadecen-1-ol
8	2.52	$C_{22}H_{42}O_2$	Phytol, acetate
9	5.25	$C_{16}H_{32}O_2$	n-Hexadecanoic acid
10	8.99	$C_{18}H_{30}O_2$	9,12,15-Octadecatrienoic acid, (Z,Z,Z)-

Table 2Indentification and quantification of main chemical compounds by catalytic pyrolysis

NO.	peak area%	Formula	Compound
1	6.01	C_7H_8O	p-Cresol
2	1.75	$C_7H_7NO_2$	3-Hydroxypyridine monoacetate
3	4.58	C_8H_7N	Indole
4	3.05	$C_{10}H9N$	1-Phenyl-1-cyclopropanecarbonitrile
5	3.90	$C_{12}H_{22}O_{11}$	.alpha.-D-Glucopyranose, 4-O-.beta.-D-galactopyranosyl-
6	8.04	$C_{20}H_{40}0$	3,7,11,15-Tetramethyl-2-hexadecen-1-ol
7	2.82	$C_{20}H_{40}$	2-Hexadecene, 3,7,11,15-tetramethyl-, [R-[R*,R*-(E)]]-
8	3.99	$C_{15}H_{28}$	1-Pentadecyne
9	2.12	$C_{15}H_{28}$	3,7,11-Trimethyl-2,4-dodecadiene
10	8.75	$C_{16}H_{32}\,0_2$	n-Hexadecanoic acid

3.2 discussing

According to the chemical category and common characteristics, the products were divided into hydrocarbons, alcohols, acids, ketones, nitrogen compounds and other compounds.Nitrogen compounds detected by GC/MS analysis ,mainly come from the pyrolysis of amino acids in Najas Minor.Table 3 showed the distribution of pyrolysis products.As shown in the table,the hydrocarbons ,alcohols and nitrogen compounds in biological oil of catalytic pyrolysis were significantly increased compared with direct pyrolysis.The content of hydrocarbons increased from 12.4% to 34.6%.The content of alcohols increased from 9.1% to 16.6%.The gold nanoparticles loaded in surface of Najas Minor, was conducive to the generation of hydrocarbon and alcohol in pyrolysis.By analyzing the GC/MS results, the main hydrocarbons in bio-oil by catalytic pyrolysis were long chain alkanes,and olefins.The main components in gasoline and diesel is hydrocarbons. As a result, whether the biomass can be effective as future power fuel oil, depends largely on the hydrocarbon content of bio-oil after further treatment.

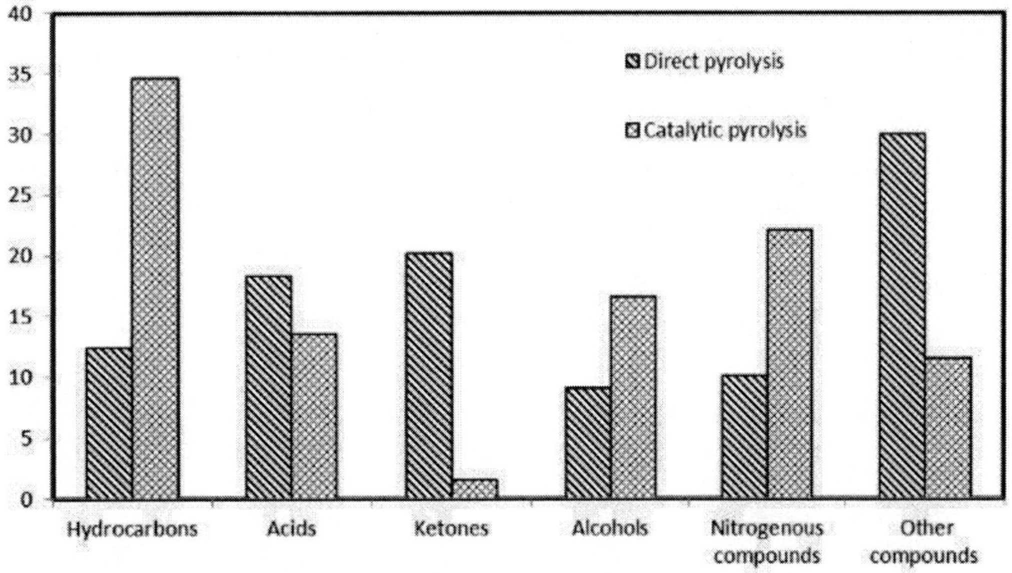

Fig.3 Distribution of pyrolysis products

4.Conclusion

Biomass was explored as a potential biomass energy resource. The experimental study on the direct and catalytic pyrolysis of Najas Minor had been presented, using Pyrolyzer with GCMS,to observe the catalytic effects of gold nanoparticles.In this study,remarkable catalytic effects for pyrolysis were observed.The pyrolysis of Najas Minor with gold nanoparticles,provided more hydrocarbons, alcohols and nitrogen compounds.

参考文献：

[1]Gong F, Yang Z, Hong C, et al. Selective conversion of bio-oil to light olefins: Controlling catalytic cracking for maximum olefins[J]. Bioresource Technology, 2011, 102(19):9247-54.

[2]Zhang Z, Bi P, Jiang P, et al. Production of gasoline fraction from bio-oil under atmospheric conditions by an integrated catalytic transformation process[J]. Energy, 2015, 90:1922-1930.

[3]Hui S, Yuqi D, Jinzhao D, et al. Transesterification of sunflower oil to biodiesel on ZrO2 supported La2O3 catalyst.[J]. Bioresource Technology, 2010, 101(3):953–958.

[4]Yaman S. Pyrolysis of biomass to produce fuels and chemical feedstocks[J]. Cheminform, 2004, 35(31):651-671.

[5]Hejný S. Ellias Landolt The family of Lemnaceae—A monographic study—Vol.1 biosystematic investigations in the family of Duckweeds (Lemnaceae) (vol. 2)[J]. Folia Geobotanica Et Phytotaxonomica, 1993, 28(1):50-50.

[6]Gardea-Torresdey J L, Tiemann K J, Gamez G, et al. Gold Nanoparticles Obtained by Bio-precipitation from Gold(III) Solutions[J]. Journal of Nanoparticle Research, 1999, 1(3):397-404.

[7]Marie-Christine D, Didier A. Gold nanoparticles: assembly, supramolecular chemistry, quantum-size-related properties, and applications toward biology, catalysis, and nanotechnology.[J]. Chemical Reviews, 2004, 35(16):293-346.

[8]Liu B ,, Xie J ,, Lee J Y, et al. Optimization of high-yield biological synthesis of single-crystalline gold nanoplates.[J]. Journal of Physical Chemistry B, 2005, 109(45):15256-63.

[9]Raveendran P, Fu J, Wallen S L. Completely "Green" Synthesis and Stabilization of Metal Nanoparticles[J]. Journal of the American Chemical Society Jacs, 2003, 125(46):págs. 13940-13941.

[10]Absar Ahmad , Satyajyoti Senapati , MIK, et al. Extracellular Biosynthesis of Monodisperse Gold Nanoparticles by a Novel Extremophilic Actinomycete, Thermomonospora sp.[J]. Langmuir, 2003, 19(8):3550-3553.

[11]Selvakannan P, Mandal S, Phadtare S, et al. Water-dispersible tryptophan-protected gold nanoparticles prepared by the spontaneous reduction of aqueous chloroaurate ions by the amino acid[J]. Journal of Colloid & Interface Science, 2004, 269(1):97-102.

[12]Gittins D I, Caruso F. Biological and Physical Applications of Water-Based Metal Nanoparticles Synthesised in Organic Solution[J]. Chemphyschem, 2002, 3(1):110－113.

[13]And A G, Murphy C J. Seed-Mediated Synthesis of Gold Nanorods: Role of the Size and Nature of the Seed[J]. Chemistry of Materials, 2004, 16(19):3633-3640.

COMPARATIVE ANALYSIS OF MOLECULAR STRUCTURE IDENTIFIABILTY BASED ON SIGNATURES AND DESCRIPTORS

Zelimir F. Kurtanjek
University of Zagreb, Croatia

Abstract

This work is focused on molecule computer representation aimed for molecule predictive model (QSAR) and inverse modelling. Two different approaches are compared, i.e. the molecule structure coding based on graph theory (Faulon et al.[1] extended valence) and the chemical based descriptors[2]. Applied are 1-3rd level molecule graph fragments and 630 2D molecular descriptors. The molecule test sets are alkanes, alkenes, acetones, aromatics, organic acids and halogenated hydrocarbons in the range of C_1-C_{12}, selection of biomolecules (including aminoacids) and a set of binary ionic liquids (cations: imidazole, pyridinium, quinolinium, ammonium, phosphonium). The comparative comparison of molecule description based prediction potential are statistically evaluated by octanol-water (PWA) partition coefficients (for hydrocarbons), viscosity of ionic liquids and biological impact on acetycholinesterase inhibition. The evaluation is based on two distinct levels. On the first level the comparative analysis based on internal and external correlation between "scores" in multidimensional spaces of the separate first, second and third order valance spaces, and the molecular signature space. Firstly, the potential of information content applicable for inverse modeling is tested on ability for molecule classification by use of linear discrimination (LDA) based on the singular value decomposition (SVD) of corresponding molecular fragments and descriptors. For chemical-biological property inverse modelling proposed is application of constrained genetic algorithm (GA). Tested are linear structure multivariate (chemometric) models such as principal component regression and partial least squares and nonlinear decision trees (random forest). Applied are singular value decompositions for reduction of the molecule signature and the descriptor spaces up to the significance level of p=0.05. The selected level of significance is presumed to be in coherence with experimental data of molecule physical and chemical properties. The main focus is on extraction of the main molecular fragments (partial graphs extracted from molecular signatures) and specific molecule descriptors. Importance of application of open source software modules provided by Bioclipse[3] and R software packeges[4] for computational-statistical numerical algorithms is emphasized.

Key words: Molecule signatures, descriptors, inverse modelling, decision tree

Introduction

Advancement of high-throughput experiments have lead to „big data" databases applied in pharmaceuticals, medicine, new material science, catalysis, and new industrial processes. The role of chemoinformatics as a science for computer encoding of molecule properties and algorithms for data mining and inference are the essential steps in the development. Computer encoding of molecular properties has been a part of chemical engineering from the beginning of process design. It is mostly based on molecule group contributions, but has been extended by application of molecular descriptors (1-2-3 D), molecule signatures, and application of molecular quantum mechanics. Modeling is faced with the problem of „high dimension curse". For example, about 5 000 various molecular descriptors can be determined for each molecule, although most of them are highly correlated. It results in an ill-conditioned problem of modelling with large uncertainties in parameter and poor model prediction accuracy outside modelling training sets. Great improvement has been achieved by application by use of model ensembles, like random forest of many individual decision trees. These models are highly flexible, can be trained on discontinuous and nonlinear functional relations, can encompass various data formats like numerical and categorical variables, and due to large number of randomly selected individual predictors in principle do not have tendency to overfitting. However, when the object of modelling is extraction of important molecular features, due to autocorrelation between descriptors random forest, produced lists of importance are large sets of poorly distinguishable variables. The focus of this work is to apply feature regularization for optimization and reduction of list of variable importance. Compared are two methods for structure encoding, molecular descriptors and molecular signatures of first and second order.

Modeling and Factor Regularization

In this work are selected two distinct data sets of organic molecules from open source data bases (J. Sangster, 1989) and ionic liquids ILs (The UFT/ Merck and NIST, 2015), and lists of the selected molecules are given in Table 1. As target properties for organic molecules are selected the corresponding molecular classes and partition octanol-water (POW) coefficients, and dynamic viscosity in the range of 30-80 ^{0}C of ILs is selected The molecules are coded into SMILES and used as input lists into software for evaluation of molecular signatures. Most popular open source software tools are DRAGON (R. Todeschini, et al., 2009) and PADEL (Chun Wei Yap, 2010). Here are used 1-2 D descriptors generated by PADEL. For the same list of molecules are determined first and second order molecular signatures or extended valances (J.P. Faulon, 2003). The signatures are two dimensional graph based structural molecular fragments. The vertices are atoms and the edges are corresponding bonds (lengths or heights), and are here only used of the first and second level. A molecule is characterized by a complete set of signatures as input data for modelling. Since the models are based on large number of variables (molecular features) possibly lager than number of samples (molecules), usually denoted as p>>n models. They are intrinsically „ill conditioned" leading to good predictions on train data but very poor predictions on test data. Also, produced list of variable importance are very dense "clusters are not distinguished" with large uncertainties (broad confidence intervals). This makes the problem of inverse modelling aimed to predict a set of candidate molecules for given set of target properties, such as in a case of high dimension combinatorial task for selection of ILs cations and anions, very unreliable. In order to reduce model dimensionality and improve robustness and parsimonious of important molecular features here is applied modelling with feature regularization. The model is based on ensemble of random trees (CART models, L. Breiman et al. 1998) with included feature regularization.

Data sets of organic molecules	**Range of number of atoms**				
Molecular classes	C	H	O	Cl	F – I - Br
Alkanes	5-12	12-30			
Alkenes	4-9	6-18			
Acetones	6-24	6-26			
Halogenated hydrocarbons	3-10	6-22	1		
Organic acids	1-16	2-32	2		
Esters	4-11	6-22	2		
Aromatics	1-12	1-21		3	2 – 2 - 1
Data sets of ionic liquids (ILs)					
Cations	**Anions**				
1,3-Dimethylimidazolium (+)	PF_6 (-) Cl (-) CH_3COO (-)				
1-Methylpyridinium (+)	BF_4 (-) $MeSO_4$ (-) $EtSO_4$ (-)				
1,1-Dimethylpyrrolidinium (+)	Tf_2N (-) CF_3SO_3 (-)				

Table 1. The molecular groups of the organic and ionic liquid molecules selected for random forest models and extraction of the most important molecular features. The properties for the organics are obtained from J. Sangster (1989) and for the ionic liquids from Merck and NIST data bases (2015). Molecular descriptors and signatures of ILs ions are recorded separately and adjoined as a single record.

The modelling is formulated as an optimization (functional and multivariable minimization) problem with two parts of the objective function (T. Hastie et al., 2009).

$$\min F_{object.}(X, T, \omega) = SSE(y, \hat{y}) + \phi(T, \omega) \tag{1}$$

The first part is the sum of square errors (SSE) based on difference between observed y and prediction $\hat{y}$ based on input data by an ensemble T of interrelated CART models (L. Breiman et al. 1998), The second part is a selected metrics ϕ for the model complexity depending the sum of leaf probabilities for population of individual CART models and their total number N_T.

$$\phi(T, \omega) = \gamma \times N_T + \frac{1}{2} \times \lambda \times \sum_i^{N_T} \omega_i^2 \tag{2}$$

The ensemble output prediction is a sum of predictions of individual T_i tree CART models:

$$\hat{y} = \sum_i^{N_T} T_i(X) \tag{3}$$

A very efficient open source code Extreme Gradient Boosting XGB (T. Chen, 2016) is available in R language (R Developm-ent Core Team, v. 3.2.3, 2015). The optimization algorithm is consecutive, starting with first CART model which is appended with a next tree boosted by errors in predictions of previous tree. At each step optimization is solved by linearization of the objective function to quadratic terms. Number and structure of trees as well as the parameters for regularization need to be found by trial and error and minimal number of trees by folding and cross validation error on the training set. The critical step is validation on untrained set to have acceptable model error and avoid model overfitting with minimal model complexity.

Results and Discussion

Tested are models for prediction of the seven molecular classes (Table 1.) based on molecular descriptors and molecular signatures (first and second level). Results for molecular descriptor feature importance are shown on Fig. 1. The effect of regularization is very pronounced. Out of hundreds of descriptors for the molecular classification needed are only few, the most important being BCUTw.1h. On contrary, when an ensemble of independent CART trees is applied, i.e. random forest model, level of accuracy of prediction is the same, but the list of variable importance is densely populated with numerous descriptors.

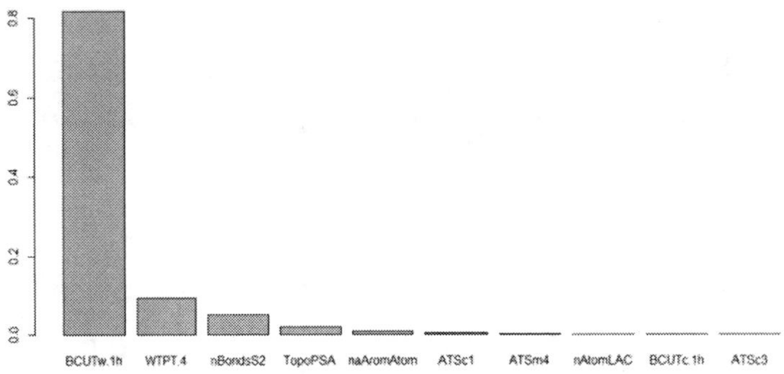

Fig 1. Variable importance molecular descriptor factors for the inverse modeling of the selected molecular classes. The first five descriptors are: BCUTw.1h (based on a weighted version of the Burden matrix which takes into account both the connectivity as well as atomic properties of a molecule), WTPT.4 (Randić's weighted path), nBondS2 (bond count), and naAromAtom (number of aromatic atoms).

The same classification problem was tested by the first and second order molecular signatures. The important signatures are depicted on Fig. 2-3. Accuracy of prediction on trained sets of molecules compared to descriptors is about the same, though the importance factors are not as distinctive as in case of the descriptors.

significantly better when molecular descriptors are applied. The

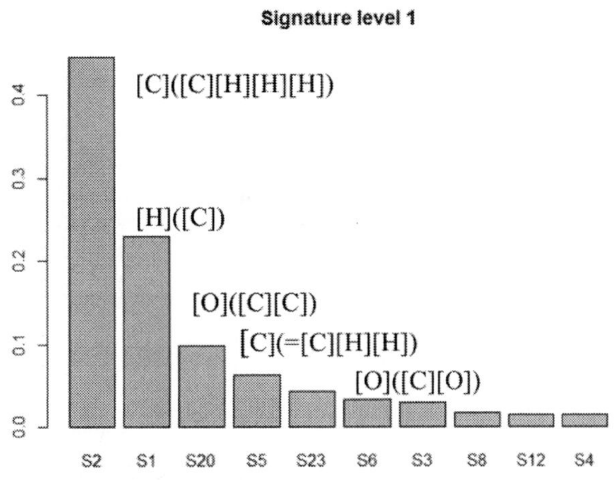

Fig 2. Relative importance of the individual first level signatures for inverse prediction of the selected molecular classes. The first five most important signatures are coded according to SMILES (Simplified Molecular Input Line Entry System).

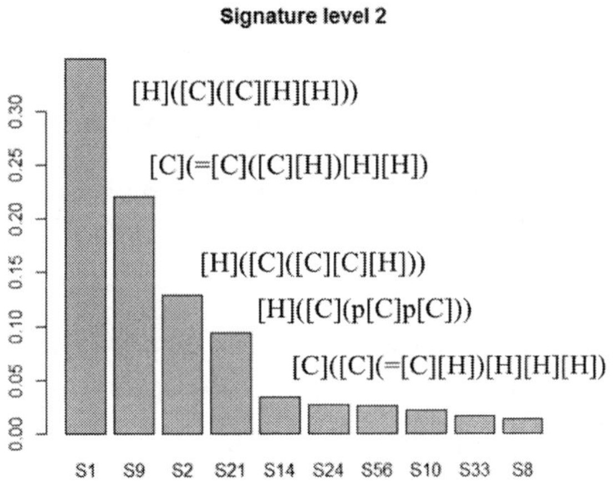

Fig 3. Relative importance of the individual second level signatures for inverse prediction for the selected molecular classes. The first five most important signatures are coded according to SMILES (Simplified Molecular Input Line Entry System).

However, model accuracy on untrained set of molecules, depicted in Fig. 4, is significantly lower for the model with molecular descriptors compared to signatures. The average error with descriptors is about 5% while for signatures about 11%. The "box and whiskers" plots are obtained by error evaluation of ten randomly generated untrained sets.

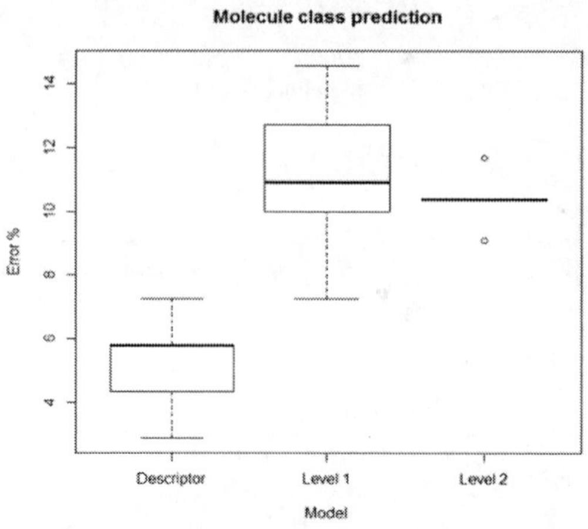

Fig 4. "Box and whiskers" plot for the model validation (relative errors) for inverse prediction of the molecular classes based on the set of untrained molecular descriptors and the first and second level of molecular signatures.

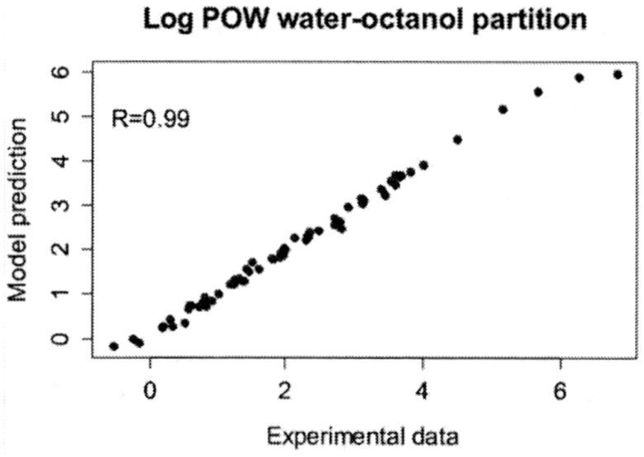

Fig 5. Predictions of the partition octanol-water coefficients POW, for the specified molecular classes of organic molecules, are based on the random forest model with the molecular descriptors.

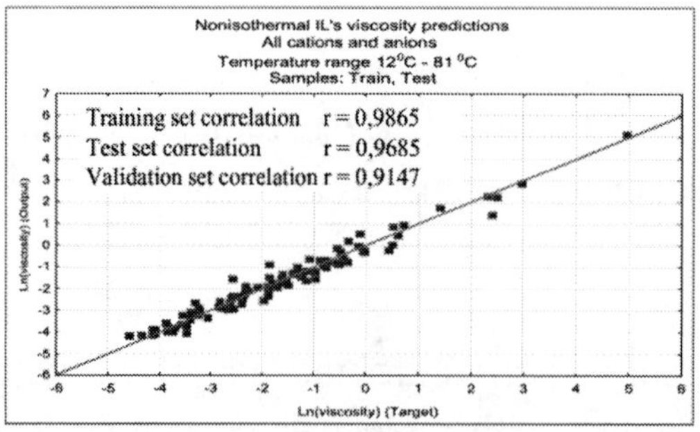

Fig 6. Predictions of dynamic viscosities for the specified groups of ionic liquids (ILs) based on the random forest model with the molecular descriptors. The viscosities cover the temperature range from 12 – 81 ^{0}C and five orders of magnitude.

In Fig. 4 can be observed that dispersion of errors (quantiles) for the second level molecular signatures is negligible, though two outliers are present.

For prediction of molecular properties applied are random forest models based on molecular descriptors. For all of the organic molecules, included in the seven classes, evaluated are octanol-water partition coefficients POW, shown in Fig. 5. The range of POW values is very high, on log scale fro 0 to 6, but the obtained accuracy of the model predictions is high (Pearson correlation R=0.99). The same approach is applied for prediction of ionic liquid viscosity data in temperature range from 12 to 81 ^{0}C. Since temperature effects on viscosity are intrinsically nonlinear, they represent a modelling challenge. However, the decision trees are very efficient for accounting nonlinearities, though nonlinearity is reflected in non-orthogonal decision planes, rendering trees with significantly increased number of branches. For the range of 6 order of magnitude the accuracy of prediction on trained set is R = 0.98 and on untrained set R = 0.91.

Conclusions

Compared are ensemble decision trees models for molecular feature class and property prediction based on extensive list of molecular descriptors and first and second level of extended valances or signatures.

The aim was to test effects of regularization on ensemble of CART decision tree models. Proven is effectiveness of the extreme gradient boosting algorithm in cases with extensive list of (molecular) features.

Obtained is increased accuracy on untrained sets of data, and more importantly, obtained are parsimonious lists for ranking of variable importance lists. This enables further significant reduction in development of robust and simple decision tree models.

The reduction in number of molecular features applied for decision tree models may greatly simplify the challenging problem of inverse modeling based on molecular descriptors and/or signatures.

References

1. J. Sangster, J. Phys. Chem. Ref. Data, **18**(3)(1989) 1111-1227
2. The UFT/ Merck Ionic Liquids Biological Effects Database, www.il-eco.uft.uni-bremen.de
3. NIST, Ionic Liquids Database Standard Reference Database #147, <ilthermo.bouder.-nist.gov /ILThermo/mainmenu.uix>
4. R. Todeschini and V.Consonni: "Molecular Descriptors for Chemoinformatics", (2 volumes), WILEY-VCH,Weinheim (Germany) 2009.
5. Chun Wei Yap, PaDEL-Descriptor: An Open Source Software to Calculate Molecular Descriptors and Fingerprints, J. Comp. Chem., (2010):**32**: 1466-1474.

6. J.-L. Faulon, D.P. Visco Jr., R.S. Pophale, The signature molecular descriptor. 1. Extended valence sequences vs. topological indices in QSAR and QSPR studies, J. Chem. Inf. Comput. Sci. **43** (2003) 707–720.
7. J.-L. Faulon, C.J. Churchwell, J.D.P. Visco, The signature molecular descriptor. 2. Enumerating molecules from their extended valence sequence, J. Chem. Inf. Comput. Sci. **43** (2003) 721–734.
8. L. Breiman, J. H. Friedman, R. A. Olshen, C.J. Stone, „Classification and Regression Trees", CRC, Boca Raton, 1998.
9. T. Hastie, R. Tibshirani, J. Friedman, The Elements of Statistical Learning, Data Mining, Inference and Prediction, Springer Verlag, Second Edition, New York, 2009.
10. T. Chen, T. He and M. Benesty (2016). xgboost: Extreme Gradient Boosting. R package version 0.4-3. https://CRAN.R-project.org/package=xgboost
11. R Development Core Team, , R: A language and environment for statistical computing. R, Vienna, Austria,www.R-project.org 2015.

444478 Biogas Reforming with Addition of Carbon Dioxide to Syngas over Ni-Based Bimodal Pore Catalyst

Abstract:

Methane dry reforming of (MDR) has been widely studied due to its important applications in producing of syngas ($CO + H_2$), removal of two greenhouse gases, and upgrading of biogas (mainly composed of CH_2 and CO_2) into value added chemicals. Compared to the noble metal catalysts, nickel-based catalysts are suitable for industrial scaled-up MDR due to their high activity and low cost. However, nickel-based catalysts have been encountering a significant challenge that is the active metal sintering and coke deposition, leading to the catalyst deactivation and then the poor stability. Nevertheless, it has been reported that the deactivation of nickel catalysts can be suppressed by adding promoters like strong Lewis bases with the enhancement of chemisorb CO_2, such as MgO and CaO, and like lanthanide elements with the capacity of oxygen storage and release, such as CeO_2 and La_2O_3. Al_2O_3 is a commonly applied catalyst carrier own to its good pore size dispersion, high specific surface area and high mechanic strength.

Herein, we report that the bimodal NiCeMgAl catalysts with different NiO-loading were synthesized by the refluxed co-precipitation method and were evaluated for MDR performance by using real biogas from anaerobic digestion of food wastes. The sample containing 15wt% NiO-loading (Ni_{15}CeMgAl) was found to be active enough at 750 °C with a high CH_4 conversion of around 90%. Ni_{15}CeMgAl catalysts kept their bimodal porosity after reduction and MDR process. The evolution of the Ni_{15}CeMgAl catalyst before and after MDR process was investigated by SEM, TEM, XRD, and TGA techniques. The correlation between the structure evolution and catalytic performance change was also discussed.

444994 Multiscale Simulation of Proteins Motion in the Microchannel

Abstract:

Protein adsorption is an important phenomenon, which plays a prominent role in chemical, industrial and medical areas. In this work, the dynamic behavior of the egg-white lysozyme has been investigated in a micro-channel with a multiscale modeling approach. The Molecular Dynamic simulation, free energy computation and Langevion Dynamics (LD) simulations have been combined to study lysozyme adsorption process on Au (111) surface. The gold surface has strong attraction to proteins. The protein-surface adsorption or desorption free energies is calculated. From the case studies of LD simulations, the surface has effects on the protein dynamic adsorption process. By the comparison of LD, MD and the experiments, the time step is concluded as an important factor to study protein adsorption behavior at the water-surface interface. Fine time step at the scale of 1 picosecond is necessary to provide enough accuracy to resolve proteins' dynamics trajectories.

444827 Syngas to Ethylene Glycol: A Technology Overview

Abstract:

SABIC is the world largest MEG producer petrochemical company and most of its MEG is being produced through the hydrolysis of EO. Ethylene oxide (EO) is a simplest form of epoxides and due to its molecular structure, acts as a versatile intermediate for a variety of chemicals such as; monoethylene glycol (MEG), di- and triethylene glycols, ethoxylates, ethanol amines, and glycol ethers. The growing demand of MEG owing to its downstream utilization in polyester industry has prompted the research of an efficient, cost effective, and environmentally benign EO production process.

The major feedstock for MEG are ethylene and syngas. The heavy dependence of MEG producers on ethylene and its scarcity in some parts of the world promoted the need of alternate feedstock for highly demanded MEG. The technological developments utilizing syngas as a feedstock will be discussed along with the commercial analysis and future growth in EO/EG business.

445029 Electrospun Nanocomposite Fibers for the Removal of Heavy Metals in Water

Abstract:

Uniform electrospun nanofibers with small diameters were successfully prepared by systematically optimizing the electrospinning parameters including polymer concentration, salt loading, applied voltage, tip-to-collector distance, and flow rate. These parameters are shown to affect the fiber morphology and the addition of iron (III) nitrate salt was found to be able to reduce the fiber size. Due to the conductive nature of iron salt in polymer solution, even a relative lower voltage of 6 kV could generate a stable fluid jet and yield the nanofibers. The thermal properties of the composite fibers were investigated by thermal gravimetric analysis (TGA) and differential scanning calorimetry (DSC). Rheological studies were also conducted. The composite nanofibers were further carbonized to carbon nanofibers. Applications of the nanofibers in environment remediation have shown promise.

445004 Grand Challenges in Future Energy and Chemicals: Climate Change and Sustainability As Drivers for New Technology and Innovation

Abstract:

Global demand for energy will double by 2050, driven by increasing population and affluence especially in emerging economies. While fossil fuels will continue to underpin the energy supply for the near future, concern over impact on climate change continues to grow, driving interest in expanded use of renewables and development of carbon capture and sequestration or re-use solutions. In addition to CO_2, water use and air emissions are critical issues to be addressed in future and current energy supply, as is energy storage for effective use of renewables. This presentation will provide an overview of what industry is doing today to address these issues, including the factors and metrics used in decision making.

444825 Significance of Energy Intensity and GHG Emission Reduction Towards Sustainability of a Petrochemical Plant

Abstract:

Sustainability is an important pillar in today's petrochemical industry for improving process efficacy by reducing specific energy consumption, greenhouse gas (GHG) emissions, and wastewater. SABIC (Saudi Basic Industries Corporation) like other giant petrochemical companies has set stringent sustainability targets and investing huge resources to transform its plants from the present position to the first quartile in each category.

SABIC is the world largest producer of ethylene oxide and ethylene glycol. For each ton of glycol produced, 0.2 to 0.5 tons of CO_2 is emitted as GHG. The significance of sustainability "in general" while the methodology adopted to reduce energy intensity, carbon footprint, and wastewater "in particular", will be discussed thoroughly in our talk.

444591 Lehigh University: Crude Lights End

Abstract:

A Light Ends Separation unit was required to process 607.08 kmol/h of light hydrocarbon material from the Pipestill column in a large refinery. Ethane, propane, isobutane, n-butane, mixed pentanes, and heavy naphtha components were separated by a series of distillation columns to meet given product purities. Conventional distillation columns, absorption columns, and divided wall columns were considered as potential separation methods of these components. An optimum scheme was chosen that included one absorption column, two conventional distillation columns, and a divided wall column, meeting all product purity specifications. Dynamic simulations of step changes in crude stock flow rate and composition, in Aspen Dynamics, were performed on separation schemes optimized in Aspen Plus on a Total Annualized Cost (TAC) basis. The plant was able to appropriately handle the given step changes, converging to a new steady state in a maximum of 11 hours and 12 hours for changes in feed flow rate and feed composition steps, respectively.

445006 University of Houston: Design of a Propylene Glycol Plant

Abstract:

Propylene is the second most common petrochemical feedstock in the chemical processing industry. It is the building block for a variety of products, such as polypropylene, propylene oxide, and propylene glycol. The propylene glycol market is currently growing at eight percent per annum and exceeds two million tons per year. The production capacity of propylene glycol in the world reached 1.6 million metric tons (3.5 billion pounds) in 2010. Among the many manufacturers are Dow, Lyondell, Sinopec, Shell, BASF, Repsol, Huntsman, and others. Dow Chemical has approximately half of the world production. The goal of this project is to design a plant that will produce 90,000 metric tons/yr of propylene glycol: 79,200 metric tons/yr will be propylene glycol (MPG) and 10,800 metric tons/yr will be dipropylene glycol (DPG). This presentation includes the design of the plant and the process economics of the plant.

12TH GLOBAL CONGRESS ON PROCESS SAFETY

LPS 50th LOSS PREVENTION SYMPOSIUM

Siting Bulk Hydrogen Storage Systems: Advances in Risk-Informed Siting Procedures in NFPA Codes

Carl Rivkin, P.E.
National Renewable Energy Laboratory
15013 Denver West Parkway
Golden, CO 80401
Carl.rivkin@nrel.gov

Martin Gresho, P.E.
FP2 Fire, Inc.
282 Ronnie Road
Golden, CO 80403
marty@fp2fire.com

Dr. Christine LaFleur, P.E.
Sandia National Laboratories
PO Box 5800 MS0748
Albuquerque, NM 87185
aclafle@sandia.gov

Alice Muna
Sandia National Laboratories
PO Box 5800 MS0748
Albuquerque, NM 87185
amuna@sandia.gov

Prepared for Presentation at
American Institute of Chemical Engineers
2016 Spring Meeting
12th Global Congress on Process Safety
Houston, Texas
April 11-23, 2016

Siting Bulk Hydrogen Storage Systems: Advances in Risk-Informed Siting Procedures in NFPA Codes

Carl Rivkin, P.E.
National Renewable Energy Laboratory
15013 Denver West Parkway
Golden, CO 80401
Carl.rivkin@nrel.gov

Martin Gresho, P.E.
FP2 Fire, Inc.
282 Ronnie Road
Golden, CO 80403
marty@fp2fire.com

Dr. Christine LaFleur, P.E.
Sandia National Laboratories
PO Box 5800 MS0748
Albuquerque, NM 87185
aclafle@sandia.gov

Keywords: hydrogen, compressed gas, safety distance, fuel cell vehicle, storage

Abstract

From 2007-2009, the NFPA 55 Industrial and Medical Gas Technical Committee and the NFPA 2 Hydrogen Technology Technical Committee developed revised separation distances for the bulk gaseous hydrogen storage systems based on an analytical process that integrated both Computational Fluid Dynamic (CFD) modeling to evaluate the consequences of hydrogen releases and probabilistic risk analysis to include the frequency of hydrogen releases, the harm criteria associated with these releases, and the acceptable risk level for the separation distances calculated as a function of these inputs. An NFPA 55 and 2 joint Hydrogen Task Group was formed in 2014 to accomplish the following tasks:

-Develop revised separation distances for bulk liquefied hydrogen storage based on a risk-informed methodology parallel to the process used in the previous update of the gaseous requirements.

-Evaluate hydrogen system safety measures for both gaseous and liquefied hydrogen storage and develop a system for separation distance reduction based on additional layers of safety or new safety measures.

-Evaluate the risk criteria used to develop the separation distances for bulk gaseous hydrogen storage and make recommendations for changes in these distances based on the revised criteria.

This paper summarizes this work including the risk-informed process for developing separation distances for bulk liquefied hydrogen storage which has not previously been addressed in NFPA codes.

1 Introduction

Hydrogen-utilizing technologies have been a part of the industrial work place for decades. [1] These technologies require bulk hydrogen storage systems. In the last twenty years there has been an effort to develop Fuel Cell Electric Vehicles (FCEVs) that require retail Hydrogen Fueling Stations (HFS). Moving hydrogen storage from the controlled industrial environment, where personnel exposure is typically limited, to the retail environment, where the general public is exposed, has triggered a review of the NFPA hydrogen storage requirements. The objective of this review is to establish risk-informed safety requirements for hydrogen storage which are based on scientific analysis of the physical behavior of unintended hydrogen releases as well as the probabilistic analysis of release frequencies. The risk threshold used to inform this analysis is based on risks incurred at conventional gasoline fueling stations which are deemed acceptable to the general public.

The process for developing risk-informed requirements for bulk hydrogen systems in NFPA codes began in 2007 and produced a set of risk-informed gaseous separation distances in the 2010 edition of *NFPA 55 Compressed Gases and Cryogenic Fluids Code* [2] that were also extracted into the 2011 edition of *NFPA 2 Hydrogen Technologies Code[3]*. The basis for these distances was extensively documented in Sandia Technical Report 2009-0874, *Analysis to Support Risk-Informed Separation Distances for Hydrogen Codes and Standards*. [4] The separation distances are the minimum distance between any point on the bulk hydrogen storage system and a defined set of exposures. Typically, the most restrictive exposures are the lot line and openings in nearby buildings. For the 2016 versions of NFPA 55 and NFPA 2, the exposures were sorted into three groups based on the risk criteria used to develop the separation distance. The significance of separation distances is that most HFS will be located in urban areas where there is limited space for siting storage systems. Many HFS will be located at existing gasoline retail facilities where space is also limited.

The separation distances for bulk liquefied hydrogen storage systems were not evaluated during the gaseous requirements revisions. The current requirements are based on distances developed by NASA in the 1960's and incorporated into Occupational Safety and Health Administration (OSHA) requirements. The technical bases for the existing requirements were not documented. In 2014, NFPA created a new NFPA Hydrogen Storage Task Group (2014 NFPA Task Group) to develop bulk liquefied hydrogen storage system separation distances based on scientific analysis and revisit the separation distances for bulk gaseous systems. This analysis was to include a methodology for converting redundant or extra safety measures into reductions in separation distances.

This paper documents progress to date on the efforts to develop science based separation distances for bulk liquefied hydrogen storage and revise the bulk gaseous separation distances based on information derived from project deployment. There is ongoing work that will likely produce further revised requirements and flexibility in bulk hydrogen storage beyond those described in this paper.

2 Historical Separation Distances

A risk-informed approach for selecting the leak diameters was utilized to establish the separation distances in the 2011 edition of NFPA 2 and the 2010 edition of NFPA 55. The risk-informed approach included three considerations: the frequency of leakage for typical hydrogen facilities, the cumulative frequency of system leakage, and the risk from leakage events for the example facilities. All three are dependent upon the establishment of example facilities and the generation of hydrogen-specific component leakage frequencies. Facility descriptions were provided for four typical facilities operating at different pressures.

Hydrogen-specific component leak frequencies were generated as a function of leak size for several hydrogen components using Bayesian analysis. Hydrogen leakage data was provided by members of NFPA Task Group for use in this analysis. The hydrogen-specific data was utilized to generate system leakage estimates for a 20.7 MPa and 103.4 MPa facility. For a 0.1% piping leak size, the system leakage frequency is 3 x 10^{-2}/yr. and 6 x 10^{-2}/yr. for the 20.7 MPa and 103.4 MPa systems, respectively. The higher leakage frequency for the 103.4 MPa system is due to the contribution from the additional compressor and high pressure storage modules. These values are comparative to the fire frequency in gasoline stations (7.4 x 10^{-2}/yr. [7]) and suggest that a 0.1% leak would be expected during the lifetime of these facilities. To reduce the potential for significant consequences to a person at the site boundary due to expected accidents, larger and less frequent leak sizes of at least 1% should be used as the basis for separation distances.

The cumulative probability for different leak sizes was then calculated to determine what range of leaks represents the most likely leak sizes. When evaluated on a cumulative distribution basis, leaks equal or less than 0.1% of the component flow area was estimated to represent 95% of the system leakage frequency. Basing separation distances on this size of leak would ensure that they cover the majority of possible leakage events. However, the use of a larger leak size was recommended to the NFPA Task Group based on consideration of the actual system leakage frequencies discussed above and uncertainty in the component and system leak frequency analysis.

The risk resulting from different size leaks was also evaluated for four standard gas storage configurations. The risk evaluation indicates that the use of 0.1% of the component flow area as the basis for determining separation distances results in risk estimates that significantly exceed the 2 x 10^{-5} fatality/yr. risk guideline selected by the NFPA Task Group for both the 20.7 MPa (3000 psig) and 103.4 MPa (15000 psig) example systems.

On the other hand, use of a leak size equal to 1% to 10% of the component flow area results in risk estimates that are slightly above, but reasonably close to the risk guideline. Based on this input, the NFPA Task Group decided that the separation distances specified in NFPA 2 and NFPA 55 would be based on a leak size equal to 3% of the largest flow area downstream of a gaseous storage system greater than 11.3 m^3 (400 scf).

More details on this entire process can be found in [4].

3 Bulk Liquefied Hydrogen Storage

An NFPA Hydrogen Storage Task Group was formed in 2014 to develop a scientific basis for the separation distances for bulk liquefied hydrogen storage systems. The Task Group had the following objectives:

- Define a representative bulk liquefied hydrogen storage system.
- Perform a risk analysis on the system to determine the release scenarios with the highest risk.
- Evaluate the risk criteria for each of the highest risk release scenarios by using a scientific model that predicts the extent of specific hydrogen concentrations.
- Acquire leak frequency data from the discharge points from the highest risk release scenarios
- Perform an integrated analysis of releases and leak frequencies to estimate the system separation distances using the risk threshold of 2 x 10^{-5} fatalities /yr.

The Task Group decided to use the liquefied hydrogen system described in *CGA P-28 OSHA Process Safety Management and EPA Risk Management Plan Guidance Document for Bulk Liquid Hydrogen Systems* 2009 (CGA P-28) [5] as the representative system. The representative system contains a liquid hydrogen storage tank, a pressure buildup circuit and economizer, vaporizer, and pump. CGA P-28 also contains a representative hazard and operability study (HAZOP) that was performed on the representative bulk liquefied hydrogen storage system. The process was divided into seven segments, called nodes:

1. Delivery trailer, hose, and fill line to the storage vessel
2. Storage tank
3. Pressure buildup circuit and economizer
4. Hydrogen line—liquid through the vaporizer to the final line
5. Hydrogen pump
6. Vent stacks
7. General items

In CGA P-28, each node was examined based on deviations from normal operating parameters that could have potentially hazardous consequences. The Task Group started with this division of the system and analyzed each node deviation to come up with event likelihood and hazard severity. Event likelihood was the annual probability that the deviation may occur for each node, with the highest level expecting the deviation to occur once per year or more and the lowest level expected to occur once per 10,000 years, documented in Table 1. Hazard severity classified the consequence from the node deviation. The hazard severity ranged from negligible to catastrophic, documented in Table 2. The Task Group estimated the event likelihood and severity based on industry experience and subject matter expertise and had to come to an agreement on the levels given to each node deviation.

Table 1. Event Likelihood Classification Table

Level	Annual Probability	Probability Description
1	Frequent > 1.0	Expected to occur once per year or more frequently.
2	Reasonably probable 1.0 to 0.1	Expected to occur once per 10 years.
3	Occasional 0.01 to 0.1	Expected to occur once per 100 years.
4	Remote 0.001 to 0.01	Expected to occur once per 1000 years.
5	Extremely remote 0.0001 to 0.001	Expected to occur once per 10,000 years.
6	Improbable < 0.0001	Expected to occur less than once per 10,000 years. Extremely unlikely to occur.

Table 2. Hazard Severity Classification Table

Level	Description	Potential Consequences
1	Catastrophic	May cause fatality to non-associated members of the public.
2	Critical	May cause severe injury to non-associated members of the public, fatality or serious injury to works of the public, fatality or serious injury to workers of persons conducting business at a refueling site or significant damage to equipment/facilities.
3	Marginal	May cause minor injury, or minor system damage.
4	Negligible	Will not result in injury or system damage.

A few assumptions were made for this analysis. First, the task group assumed an NFPA 55 and CGA H-5 code complaint system. The scenarios would only have a single failure, and cascading failures would not be analyzed. The current setback distances did not provide any credit during this analysis and tornadoes were not included in the natural disaster considerations. Based on the level of likelihood and severity, the node deviation was given a risk ranking, documented in Table 3.

Table 3. Risk Ranking Table

		Likelihood					
		1	2	3	4	5	6
Severity	1	1	1	1	2	3	4
	2	1	1	2	3	3	4
	3	2	2	3	3	4	4
	4	4	4	4	4	4	4
1: High Risk 2: Moderate Risk 3: Low Risk 4: Routine Risk							

After the Task Group had prioritized all 102 node deviations, the risk ranked results were reviewed and discussed. Two scenarios were added, based on Task Group feedback, which analyzed high and normal flow from a trailer vent stack. In total, the Task Group chose nine scenarios with the highest risk to analyze further to define separation distances to exposures. The liquid hydrogen storage exposures are similar to gaseous hydrogen storage exposures and are grouped in Table 4.

Table 4. Bulk Liquid Hydrogen Storage System Exposures

Group 1 Exposures
(a) Lot lines
(b) Air intakes (HVAC, compressors, other)
(c) Operable openings in buildings and structures
(d) Ignition sources such as open flames and welding
Group 2 Exposures
(a) Exposed persons other than those servicing the system
(b) Parked cars
Group 3 Exposures
(a) Buildings of noncombustible non-fire-rated construction
(b) Buildings of combustible construction
(c) Flammable gas storage systems above or below ground
(d) Hazardous materials storage systems above or below ground
(f) Ordinary combustibles
(g) Unopenable openings in buildings and structures
(h) Encroachment by overhead utilities
(i) Piping containing other hazardous materials

(j) Flammable gas metering and regulating stations

(k) Inlet to underground sewers[1]

1. Exposure not included in gaseous hydrogen storage separation distances

The Task Group's high risk release scenarios are shown in Table 5. Note that the first six scenarios listed occur during liquid hydrogen (LH2) transfer operations from a tanker truck to the bulk LH2 storage tank. The other three scenarios occur during normal system operations. The two scenarios that the Task Group added are included in this list. For each of the scenarios, the Task Group decided upon the inputs to liquid hydrogen models and the risk criteria for analysis. Sandia developed a cryogenic hydrogen integral jet and plume model, called COLDPLUME, which is intended to be used to predict release characteristics from various real-world hydrogen storage states. [7] Sandia is also developing a network flow model which accounts for internal system flows from the storage vessel to the release plane, called NETFLOW, which will be suitable for liquid and multiphase cryogen flows. [8] These models are currently under development and experimental validation. Once modeling results are available, the hydrogen exposure risk criteria used for the bulk gaseous separation distances (discussed in the following section) will be used to develop the new liquid separation distance table.

Table 5. High Risk Liquid Hydrogen Release Scenarios

	HAZOP Number and Description		**Modeling Notes**	**Separation Distance Driver**
Release scenarios during liquid transfer to bulk storage tank	1.18	High flow of gaseous hydrogen from trailer vent stack due to venting excess pressure after LH2 transfer	Sandia's NETFLOW and COLDPLUME models will characterize the temperature and concentration from the trailer vent stack. The task group would like to model an estimate of up to 50 kg of cold, gaseous hydrogen.	
	1.19	Normal flow from trailer vent stack due to venting excess pressure after LH2 transfer	Sandia's NETFLOW and COLDPLUME models will characterize the temperature and concentration from the trailer vent stack. The task group would like to model an estimate of up to 50 kg of cold, gaseous hydrogen.	Modeling results will be used to calculate separation distance from air intakes and overhead utilities because this is highest frequency release and release point is closest proximity to these exposures.
	1.6	High flow from line rupture, valve or component failure during transfer process	Sandia's NETFLOW and COLDPLUME models will characterize the temperature and concentration from the release to the air. A model is needed to characterize pooling and evaporation effects.	
	1.4	High temperature due to external fire causes high flow venting through tank vent stack	Sandia's NETFLOW and COLDPLUME models will characterize the temperature and concentration from the trailer vent stack. This is the worst case venting scenario from the pressure relief device to a high ambient temperature.	
	1.8	Reverse flow during transfer process caused by human error and pressure mismanagement	Sandia's NETFLOW and COLDPLUME models will characterize the temperature and concentration from the release to the air. A model is needed to	

			characterize pooling and evaporation effects.	
	1.16	Loss of containment from external impacts, consider all causes	The task group postulated a vehicle crash into the tanker during transfer. This causes a transfer hose rupture with NETFLOW and COLDPLUME models characterizing the release to the air. A model is needed to characterize pooling and evaporation effects.	
Release scenarios during normal system operation	4.15	Loss of containment from pipe leading from tank to vaporizer or vaporizer itself caused by thermal cycles or ice falling from vaporizers	Sandia's NETFLOW and COLDPLUME models will characterize the temperature and concentration from the release to the air. A model is needed to characterize pooling and evaporation effects.	Modeling results of hydrogen concentration plume and heat flux from a subsequent fire will be used for all other separation distance exposures because this is the highest risk priority during normal operations.
	6.15	Misdirected flow caused by operator error resulting in large low level release of cold gaseous hydrogen through bottom drain valve of vent stack during normal tank venting process	Sandia's NETFLOW model will capture how much gas is going out of the drain vent and vent stack. Sandia's COLDPLUME model can characterize the temperature and concentrations from the releases to the air. A model is needed to characterize pooling and evaporation effects.	
	2.1	High pressure because of a leak in inner vessel allowing hydrogen into the vacuum area	Sandia's NETFLOW and COLDPLUME models can characterize the flow out of the casing vent (a large hole instead of a pipe orifice).	

4 Bulk Gaseous Hydrogen Storage

As described in Section 2, the initial gaseous hydrogen setback distances were based on a few key risk criteria. The 2014 NFPA Task Group decided to revisit three of the risk criteria: the leak area as a percent of the pipe diameter, the incident heat flux threshold, and the hydrogen concentration threshold.

The cumulative probability for different leak sizes was calculated to determine what range of leaks represents the most likely leak sizes. The system leakage frequency corresponds to the largest internal pipe downstream of the highest pressure source in the system. The results of this analysis indicated that leaks less than 0.1 percent of the component flow areas represent 95 percent of the leakage frequency for the example systems, however the risk resulting from this small leak size significantly exceeded the 2×10^{-5}/yr. risk guideline set by the Task Group. At the same time, the use of a leak size between 1 percent and 10 percent of the component flow area results in risk estimates that are reasonably close to the risk guideline. [2]

Based on the results of both the system leakage frequency evaluation and the associated risk assessment, the Task Group decided that adjusting to a diameter of 1 percent value, instead of a 3 percent, would remove excess conservatism from the input assumption to the model. The 1 percent value still accounts for 95 percent of the leakage frequency from the example systems and does not exceed the 2×10^{-5}/year risk guideline established in the previous analysis. This results in more permissive separation distance requirements with no change in risk.

The Task Group also reviewed the heat flux values and determined that the use of a "no harm" criterion (1.6 kW/m^2) was overly conservative. This heat flux assumes exposed persons will not take protective actions, such as moving away from the fire scene. The task group deemed it reasonable to assume that exposed personnel will relocate away from a fire scene within a few minutes and therefore the "no harm" criteria is not appropriate for establishing separation distances. Exposures that were analyzed based on this heat flux value were updated to reflect the harm distance for a radiation heat flux level of 4.7 kW/m^2. The Task Group decided to not change the three other heat flux values used in the previous revision of the separation distances in 2009.

The Task Group reviewed the hydrogen concentration threshold. Based on work done at Sandia National Laboratories Combustion Research Facility, the Group concluded that there would not be sustained ignition at hydrogen concentrations of 8% or less. [6] There could be localized hydrogen ignition that would not develop into sustained combustion. Table 6 shows the effect on separation distances of the different leak areas, heat flux harm criteria at different system pressures, without the changes to the risk criteria.

Table 6. Harm Distances for Leak Areas, Harm Criteria, and Pressures [2]

Harm Criteria	Harm Distance (Leak Area)			
	>0.10 to 1.72 MPa (>15 to 250 psig)	**>1.72 to 20.68 MPa (>250 to 3000 psig)**	**>20.68 to 51.71 MPa (>3000 to 7500 psig)**	**>51.71 to 103.43 MPa (>7500 to 15000 psig)**
Un-ignited jet concentration - 4% mole fraction of hydrogen	31.2 m (20% Area) 22.1 m (10% Area) 15.7 m (5% Area) 12.1 m (3% Area) 7.0 m (1% Area)	36.1 m (20% Area) 25.6 m (10% Area) 18.1 m (5% Area) 14.0 m (3% Area) 8.1 m (1% Area)	22.6 m (20% Area) 16.0 m (10% Area) 11.3 m (5% Area) 8.8 m (3% Area) 5.0 m (1% Area)	26.8 m (20% Area) 19.0 m (10% Area) 13.4 m (5% Area) 10.4 m (3% Area) 6.0 m (1% Area)
Radiation heat flux level of 1.6 kW/m^2 (500 Btu/hr-ft^2)	23.4 m (20% Area) 15.9 m (10% Area) 10.7 m (5% Area) 7.9 m (3% Area) 4.1 m (1% Area)	28.1 m (20% Area) 19.0 m (10% Area) 12.8 m (5% Area) 9.5 m (3% Area) 4.8 m (1% Area)	16.6 m (20% Area) 11.2 m (10% Area) 7.8 m (5% Area) 5.5 m (3% Area) 2.6 m (1% Area)	20.5 m (20% Area) 13.8 m (10% Area) 9.6 m (5% Area) 6.8 m (3% Area) 3.3 m (1% Area)
Radiation heat flux level of 4.7 kW/m^2 (1500 Btu/hr-ft^2)	17.0 m (20% Area) 11.6 m (10% Area) 7.9 m (5% Area) 5.9 m (3% Area) 3.1 m (1% Area)	20.2 m (20% Area) 13.8 m (10% Area) 9.4 m (5% Area) 7.0 m (3% Area) 3.7 m (1% Area)	12.2 m (20% Area) 8.2 m (10% Area) 5.5 m (5% Area) 4.1 m (3% Area) 2.1 m (1% Area)	14.9 m (20% Area) 10.0 m (10% Area) 6.7 m (5% Area) 5.1 m (3% Area) 2.6 m (1% Area)
Greater of radiation heat flux level of 25237 W/m^2 or visible flame length[1]	13.0 m (20% Area) 9.2 m (10% Area) 6.5 m (5% Area) 5.0 m (3% Area) 2.9 m (1% Area)	15.0 m (20% Area) 10.6 m (10% Area) 7.5 m (5% Area) 5.8 m (3% Area) 3.4 m (1% Area)	9.4 m (20% Area) 6.7 m (10% Area) 4.7 m (5% Area) 3.6 m (3% Area) 2.1 m (1% Area)	11.1 m (20% Area) 7.9 m (10% Area) 5.6 m (5% Area) 4.3 m (3% Area) 2.5 m (1% Area)

Greater of radiation heat flux level of 20000 W/m^2 or <u>visible flame length</u>[1]	13.0 m (20% Area) 9.2 m (10% Area) 6.5 m (5% Area) 5.0 m (3% Area) 2.9 m (1% Area)	15.0 m (20% Area) 10.6 m (10% Area) 7.5 m (5% Area) 5.8 m (3% Area) 3.4 m (1% Area)	9.4 m (20% Area) 6.7 m (10% Area) 4.7 m (5% Area) 3.6 m (3% Area) 2.1 m (1% Area)	11.1 m (20% Area) 7.9 m (10% Area) 5.6 m (5% Area) 4.3 m (3% Area) 2.5 m (1% Area)

[1]The largest harm distances are predicted for the visible flame length.

Table 7 shows the revised separation distances for the four storage pressure ranges based on the task group's recommended changes in risk criteria. The safety distances in the table are also rounded to the nearest whole number and multiplied by a 1.5 safety factor. The significance of these potential changes can be shown in the high pressure column for the Group 1 Exposures. The reduction in separation distances from 34 feet to 16 feet for bulk gaseous systems would allow these systems to be placed at existing gasoline fueling stations and other space constrained locations. This expansion in the population of locations that could house hydrogen storage systems and therefore HFS would further the deployment of hydrogen fueling infrastructure to support FCEVs.

Table 7. Draft Updated Values to 2016 NFPA 2 and NFPA 55 Tables with 1.5 Safety Factor

Exposures		Separation Distance			
		>0.10 to 1.7 MPa (>15 to 250 psig)	>1.7 to 20.7 MPa (>250 to 3000 psig)	>20.7 to 51.7 MPa (>3000 to 7500 psig)	51.7 to 103.4 MPa (7500 to 15000 psig)
Group 1	Existing	12 m (40 ft)	14 m (46 ft)	9 m (29 ft)	10 m (34 ft)
	Proposed New	**5 m (16 ft)**	**6 m (20 ft)**	**4 m (13 ft)**	**5 m (16 ft)**
Group 2	Existing	6 m (20 ft)	7 m (24 ft)	4 m (13 ft)	5 m (16 ft)
	Proposed New	**5 m (16 ft)**	**6 m (20 ft)**	**3 m (10 ft)**	**4 m (13 ft)**
Group 3	Existing	5 m (17 ft)	6 m (19 ft)	4 m (12 ft)	4 m (14 ft)
	Proposed New	**4 m (13 ft)**	**5 m (16 ft)**	**3 m (10 ft)**	**4 m (13 ft)**

Group 1 Exposures include: lot lines, air intakes, operable openings in buildings and structures, and ignition sources. Group 1 separation distances are based on the higher value of radiation heat flux of 4.7kW/m^2 or the unignited jet concentration decay distance of 8% hydrogen volume fraction concentration. In this instance, the separation distance is higher for the concentration value than the heat flux value.

Group 2 Exposures include: parked cars, exposed persons other than those servicing the system. Group 2 separation distances are based on the higher value of the incident radiation heat flux of 4.7kW/m^2 exposure to employees for a maximum of 3 minutes or the visible flame length.

Group 3 Exposures includes everything else (ex: buildings of combustible construction, ordinary combustibles, openings in buildings and structures, etc.). Group 3 separation distances are based on the higher value of the radiant heat flux for non-combustible equipment of 25.2 kW/m^2 or the visible flame length.

5 Future Work

The NFPA Hydrogen Storage Task Group has several more tasks to complete its objective of developing a scientific basis for hydrogen storage separation distances and a methodology for converting safety measures into reductions in separation distances.

These tasks include:

- Completion of the liquefied hydrogen release analysis.
- Evaluation of liquefied hydrogen leak frequencies.
- Integration of leak frequencies and release analysis into a science-based set of bulk liquefied hydrogen separation distances.
- Creating a risk based methodology to convert both liquefied hydrogen and gaseous hydrogen storage safety measures into separation distance reductions.

6 Conclusion

There has been significant progress made in developing science-based justifications for bulk hydrogen storage separation distances. These changes in bulk hydrogen storage will enable a much broader deployment of HFS in support of HFCEVs.

The initial effort in the 2007-2009 timeframe produced bulk gaseous separation distances based on a risk-informed analytical approach that was well documented, peer-reviewed, and repeatable. The effort started in 2014 has led to a proposed a revised set of separation distances for bulk gaseous storage systems. In the 2016-2017 timeframe, a revised set of separation distances for bulk liquefied hydrogen storage systems will be developed.

7 References

[1] Austin, G.T., Shreve's Chemical Process Industries Fifth Edition, 1984 McGraw Hill, New York.

[2] *NFPA 55 Compressed Gases and Cryogenic Fluids Code*, 2010 NFPA, Quincy, MA.

[3] *NFPA 2 Hydrogen Technologies Code*, 2011 NFPA, Quincy, MA.

[4] J. LaChance, W. Houf, B. Middleton, and L. Fluer. Sandia Technical Report 2009-0874 *Analysis to Support Risk-Informed Separation Distances for Hydrogen Codes and Standards.* Sandia National Laboratories 2009.

[5] *CGA P-28 OSHA Process Safety Management and EPA Risk Management Plan Guidance Document for Bulk Liquid Hydrogen Systems*, CGA 2009, Chantilly, VA.

[6] R.W. Schefer, G.H. Evans, J. Zhang, A.J. Ruggles, and R.Greif. "Ignitability limits for combustion of unintended hydrogen releases: Experimental and theoretical result", International Journal of Hydrogen Energy 36 2011.

[7] W. Houf, R. Schefer, "Analytical experimental investigation of small-scale unintended releases of hydrogen," International Journal of Hydrogen Energy, Vol. 33, 2008, pp. 1435-44

[8] Winters WS, a New Approach to Modeling Fluid/Gas Flows in Networks. SAND2001-8422, Sandia National Laboratories, July, 2001.

Is a Two-Inch Hole Adequate for a Siting Study?

Jeffrey D. Marx
Quest Consultants Inc.
Norman, OK
jdm@questconsult.com

Antonino Nicotra
Bechtel Oil, Gas & Chemicals, Inc.
Houston, TX
anicotra@bechtel.com

Prepared for Presentation at
American Institute of Chemical Engineers
2016 Spring Meeting
12th Global Congress on Process Safety
Houston, Texas
April 11-13, 2016

Is a Two-Inch Hole Adequate for a Siting Study?

Jeffrey D. Marx
Quest Consultants Inc.
Norman, OK
jdm@questconsult.com

Antonino Nicotra
Bechtel Oil, Gas & Chemicals, Inc.
Houston, TX
anicotra@bechtel.com

Keywords: Building siting, API RP 752, two-inch hole, 2-inch hole, consequence analysis, QRA

Abstract

For several decades now, many consequence-based siting assessments (e.g., API RP 752) have relied upon a two-inch hole to represent what is termed a "credible release size" or a "maximum credible event" (MCE). Alternatively, a study may evaluate siting by use of a quantitative risk analysis (QRA), which would apply a full range of hole sizes. Due to QRA complexity and expense, the MCE analysis has often been the preferred approach for siting studies in the US, and the two-inch hole has been one of the more common choices for the MCE. Hole size choices range from small leaks (1/4-inch hole) through one-, two-, four-, and even six-inch holes. Why a two-inch hole? Does a two-inch hole provide a sufficiently accurate qualitative evaluation of risk? When a single hole size is chosen for a given purpose in a consequence analysis, that hole size should not be chosen based solely on engineering judgement, but should be based on more defensible information. This paper seeks to explain the basis for selecting the two-inch hole, including a literature review, concepts from consequence modeling, data from failure frequencies, and engineering judgement. Examples comparing the use of a single hole size (with emphasis on the 2-inch hole) to QRA results will also demonstrate some of the potential benefits and problems of an MCE analysis.

1 Introduction

Siting studies have always been a means to understand the potential impacts that could be imposed by hazardous materials on plant workers, the public, process equipment, or the environment. The most common siting study in the process safety industry is the building siting study, meant to ensure that the potential impacts to persons in buildings result in an acceptably low risk, or are mitigated to the point that the risk to those persons is acceptably low.

While hazard modeling technology and siting study methodology have progressed over the years, it is still a significant resource investment to evaluate siting issues using fully risk-based techniques such as quantitative risk analysis (QRA). Because of this, siting studies are often accomplished with simplifying approaches, such as a consequence-only study. Within this realm of simplifying methodologies, the choice of a given hole size stands out as the one with the most influence on the study's results. Other modeling parameters are also important, but the hole size's qualitative link to probability (providing an insight into risk) makes it the singular choice for defining the magnitude of hazards which are to be considered.

A general list of acute hazards at or near petrochemical facilities includes the following:

- Exposure to a flash fire (an ignited flammable gas cloud)
- Exposure to thermal radiation from a jet (torch) fire, pool fire, or fireball
- Exposure to overpressure following a vapor cloud explosion (VCE) or physical explosion
- Exposure to a toxic gas

These hazards have been evaluated in many consequence and risk analysis studies for persons who are outdoors, for persons indoors (e.g., workers in an occupied building at a petrochemical facility), and for equipment or property within the facility. The effects on persons or equipment is a matter of choosing the hazard impact levels, or endpoints, that are applied in consequence modeling.

1.1 Building Siting

The American Petroleum Institute (API) has published a Recommended Practice (API RP 752) titled *Management of Hazards Associated with the Location of Process Plant Buildings* [1]. It is the recognized standard for building siting assessments, and outlines a methodology for conducting occupied building siting assessments. API RP 752 provides guidance for identifying hazards that could affect the process plant buildings and/or occupants of those buildings, and for managing risks related to those hazards. API RP 752 addresses three categories of hazards:

- Explosion
- Fire
- Toxic material

Evaluation of these hazards is as varied as the number of analysts. Explosions are traditionally the focus of building siting; fires have often been addressed through spacing alone; toxic gas exposure has often been simplified to be a function of the concentration outside a building. While most process plant buildings will provide occupant protection in the event of an external flash fire and provide shielding against thermal radiation, the evaluation of hazards should account for what levels of protection are afforded by the type of building and any mitigation measures in place. For example, where flammable or toxic gases are present in the facility, dispersion analyses should check for infiltration to determine occupant impacts, taking into account the building construction and any air exchange system.

As the potential hazards are better understood, other aspects of the analysis are given consideration. The subject matter of this paper – the selection of release hole size – is arguably the most important variable when the analysis is not a full QRA. Hole size selection can define the general level of risk (larger, high consequence releases are significantly less frequent than small, low consequence leaks) and it can define an event magnitude that has an assumed probability, leading to a qualitative measure of risk. Selection of a maximum credible event (MCE), based on a specific release hole size, then forms an important basis for the building siting evaluation.

2 Methodology

The methodology that forms the background for this paper is consistent with API RP 752 and provides a four stage process that allows the study to stop at any stage by proceeding only as far as necessary to demonstrate that occupants of the subject buildings will be adequately protected. Similar methodologies can be applied for siting studies that involve offsite populations or onsite equipment.

Stage 1 is a screening analysis, serving to identify potential hazard sources and determine which buildings should be included in the analysis. This involves a review of the process to determine the potential hazards that are inherent to the materials being handled. A decision concerning the inclusion of buildings is then based on criteria found in API RP 752. The guidance provided in API RP 752 regarding building inclusion can be summarized by the principle that all buildings intended for occupancy are to be included in the analysis.

Stage 2 is a worst-case consequence analysis, and serves to evaluate the potential consequences of the hazards identified in Stage 1. The siting assessment can be stopped after Stage 2 if it has been shown that adequate protection against worst-case scenarios is provided for building occupants through building spacing, building construction, process controls, or other engineering methods.

Stage 3 is a qualitative risk analysis. This combines consequence analysis and estimates of accident frequencies to arrive at a qualitative measure of risk posed to each building's occupants. This is accomplished by modeling the MCE, often defined by a hole size that is deemed the maximum credible hole size. The intent is to determine if the risk to building occupants is below a qualitative risk acceptability criterion. Depending on the results of this stage, determination of acceptability may be qualitative or semi-quantitative. If the risk is found to be acceptable, the analysis can end after Stage 3.

Stage 4 is a quantitative risk analysis (QRA), a much more rigorous process than previous stages. In this stage, hazard footprints are first generated for the full range of release sizes, locations, and weather conditions. Next, the frequency of occurrence of each hazard footprint is determined using historical equipment failure rates, conditional probabilities, and site-specific weather data. All calculated hazard zones, with their respective probabilities and originating locations, are combined to produce contours which demonstrate the frequency with which an event originating in the facility could adversely impact an occupied building and its occupants due to vapor cloud explosions, fires, and toxic or flammable vapor clouds. If the level of risk is acceptable, the analysis is complete. If not, mitigation measures that reduce event consequences or frequencies (or both) are implemented. Stage 4 can then be repeated, with modifications of the event consequences and/or probabilities based on mitigation measures to determine the risk.

3 The Maximum Credible Event

API RP 752 defines the maximum credible event (MCE) as:

> A hypothetical explosion, fire, or toxic material release event that has the potential maximum consequence to the occupants of the building under consideration from among the major scenarios evaluated. The major scenarios are realistic and have a reasonable probability of occurrence considering the chemicals, inventories, equipment and piping design, operating conditions, fuel reactivity, process unit geometry, industry incident history, and other factors. Each building may have its own set of MCEs for potential explosion, fire, or toxic material release impacts.

The use of the word 'credible' is always problematic because it is inherently subjective. Based on the guidance given in RP 752 regarding the MCE, the selection of MCEs can be done in several ways.

- Those events that have a 'reasonable' probability of occurrence
- Those events that have the largest consequences on building occupants
- Those events that have historical precedence
- Some combination of the above (or other) factors

An investigation into the types of MCE selections (as available in open literature) shows that there is a great deal of variance in how MCEs are defined and applied in siting studies. Some notable examples are:

> In "*Application of Qualitative and Quantitative Risk Analysis Techniques to Building Siting Studies*" [2], the authors present a methodology that integrates a corporate standard of a two-inch hole as the largest credible event. The methodology subsequently applies a quantitative risk analysis (QRA) approach if occupied building impacts due to the two-inch holes are found.

> In "*Evaluation of Permanent Buildings at an Existing Upstream Oil and Gas Business*" [3], the authors apply a methodology found in a corporate engineering technical practice that specifies two-inch holes and full-bore pipe ruptures as the events to be evaluated in a building siting study. The case study used these consequence analyses to evaluate varying magnitudes of hazards and then transition to a risk-based evaluation which used the probability of each of the hole sizes in the analysis.

> The Center for Chemical Process Safety (CCPS) has published a book titled "*Guidelines for Fire Protection in Chemical, Petrochemical, and Hydrocarbon Processing Facilities*" [4]. Four release categories for fire hazard analysis studies are presented. Small, medium, large, and rupture categories are defined by hole sizes of 3-10 mm, 10-50 mm, 50-150 mm, and full-bore ruptures, respectively. Discussion around these hole sizes also points out that the large and rupture

releases are not always the worst-case events due to event durations and dosage effects that should be accounted for.

The paper "*Methodology for selection hole sizes for consequence analysis studies*" [5] further explores the issue of event duration relative to hole size. Their case studies show that for fires "an intermediate hole size provides the combination of intensity and duration that results in the maximum thermal effect." While the worst case hole size was found to vary based on the effect being evaluated, the process system properties, and the distance from the release point, a two-inch (50 mm) hole could be taken to be representative of the worst case for many of the cases evaluated. A hole size around the two-inch size was also found to be the worst case for toxic effects in their analysis.

"*A Frequency Based Approach to Hole Size Selection for Consequence Analysis*" [6] presents a methodology which selects the MCE based on a likelihood of occurrence approach for specific equipment groups. This results in a variable hole size for the MCE, based on the extent of the equipment group that is selected. While this method does provide a "leveling" effect for release probability, it is dependent on the subjective definition and delineation of equipment groups and results in larger MCE hole sizes for larger (more complex) equipment groups.

In the CCPS book "*Guidelines for Evaluating Process Plant Buildings for External Explosions, Fires, and Toxic Releases*" [7], general guidelines and technical details are presented for complying with API RP 752. The MCE approach offered by this text leads an analyst to consider many possible events, screening out those that are not deemed credible, and selecting an MCE for each building as the event that has the maximum consequences at that building.

The Canadian Standards Association (CSA) standard Z276, *Liquefied natural gas (LNG) – Production, storage, and handling* [8], specifies design spills which define the potential extent of hazard zones that are not allowed to impact the public. Z276 defines, among other (smaller) release scenarios, that the design spill should be "The maximum flow of LNG that could be spilled from … the largest branch connection, 60 mm (2 in) or smaller…." This effectively defines a 2-inch hole as the siting criterion for Canadian LNG plants.

These various approaches to MCE selection demonstrate the difficulty in determining the most technically sound approach. Due to the history of using the two-inch hole and its frequent occurrence in siting studies, further investigation into why it has been, and might continue to be used is warranted.

4 A Frequency Argument

While equipment failure frequencies are typically brought into a siting analysis only if the analysis has moved to a risk-based approach, they are sometimes integrated into a study as a comparative evaluation of various hole sizes. Whether for QRA or comparative purposes failure frequencies must be appropriate for the system being evaluated, and must be statistically valid. There are only a few failure frequency databases that meet these criteria. The most notable, and most popular for use by risk analysts, is the Hydrocarbon Releases Database (HCRD) [9]. This database provides 20+ years of incident data, which are all releases of hydrocarbon, as well as the data for numbers of equipment-years of service. Because it provides detailed information for both failures and the equipment numbers from which failures can originate (the numerator and denominator for calculating a failure rate) it is the most consistent database from which to take data.

Use of the HCRD is generally straightforward, although some manipulation is necessary in order to apply the data to accidental release scenarios [10]. For the purposes of this paper, an evaluation was made, using the HCRD, to determine the overall frequency distribution for the two-inch hole. The analysis took on the following steps:

1. For each equipment type, gather the data for number of equipment-years in service.
2. For each equipment type, gather the data for all hydrocarbon release incidents associated with that equipment type.
3. Remove all incidents which occurred under very low pressure (as these are typically maintenance events with little consequence).
4. Calculate a total release frequency for all incidents involving that equipment type.
5. Sort the remaining release incident data by release hole size.
6. Create a cumulative distribution curve of hole size versus fraction of the total release frequency. An example cumulative distribution curve is shown in Figure 1.

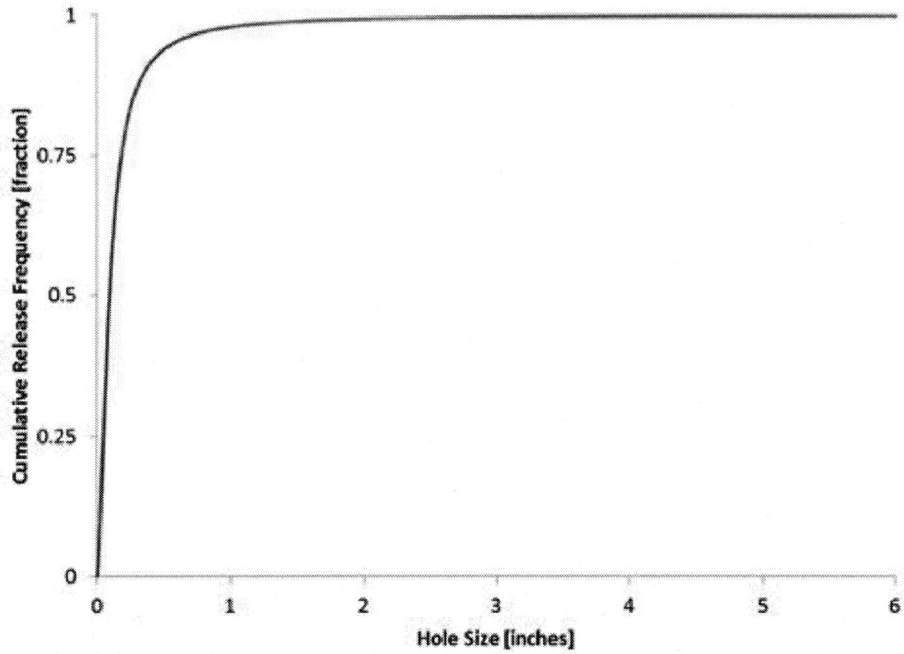

Figure 1. Example Cumulative Distribution Curve

With this approach, any hole size or hole size range can be chosen to evaluate its contribution to the overall failure rate. Taking this approach for the two-inch hole, the following conclusions were reached:

- For small equipment such as flange connections (gaskets), instrument connections, and small diameter pipe ($\leq$ 2 inches), hole sizes up to two inches represent 100% of the failure rate.
- For all other equipment (piping, compressors, pumps, heat exchangers, valves, filters, pressure vessels, storage tanks, and pig launchers/receivers), hole sizes up to two inches represent 95.46% of the total failure rate. This value is an aggregate average over all equipment types, not including the small equipment mentioned above.

Thus, it can be seen that the two-inch hole, if taken to represent all release hole diameters up to and including two inches, accounts for over 95% of the probability of hydrocarbon release events. This conclusion is reached from the evaluation of an offshore process systems database; events that are not well represented in this database (e.g., onshore storage tank overfill) may need to be included in a set of credible events.

5 Example Cases

To further illustrate how the two-inch hole may be an acceptable choice for an MCE, consider two example hydrocarbon facilities.

1) A small gas plant – This facility processes sour natural gas from a gathering system, removing contaminants such as water, carbon dioxide, and hydrogen sulfide (H_2S). It also separates out a stream of heavier hydrocarbons, or natural gas liquids (NGLs), which are separated, stabilized, and stored on site to be exported via truck. The remaining natural gas is compressed and sent out by pipeline.
2) A liquefied petroleum gas (LPG) transfer depot – This facility receives LPG (primarily propane) by pipeline and rail car. Propane is stored in pressure vessels on site and exported via tank truck to serve local markets. Besides propane and a mercaptan addition system for odorization (as needed), the facility does not handle, transport, or store any other hazardous materials.

The layout in Figure 2 represents this example small gas plant. Figure 3 shows the LPG facility used in this study. Both of these facilities are small, relative to many other hydrocarbon processing facilities. However, the principles applied to, and results obtained from, these smaller plants will apply to larger hydrocarbon processing facilities.

5.1 Screening Analysis

In the example gas plant facility, the control room, warehouse, and office are the three occupied permanent buildings on site. All three buildings are small pre-engineered metal buildings with similar sizes. As stated in the facility description, toxic and flammable hazards exist; therefore, the analysis will need to evaluate fires, explosions, flammable gas infiltration, and toxic gas infiltration.

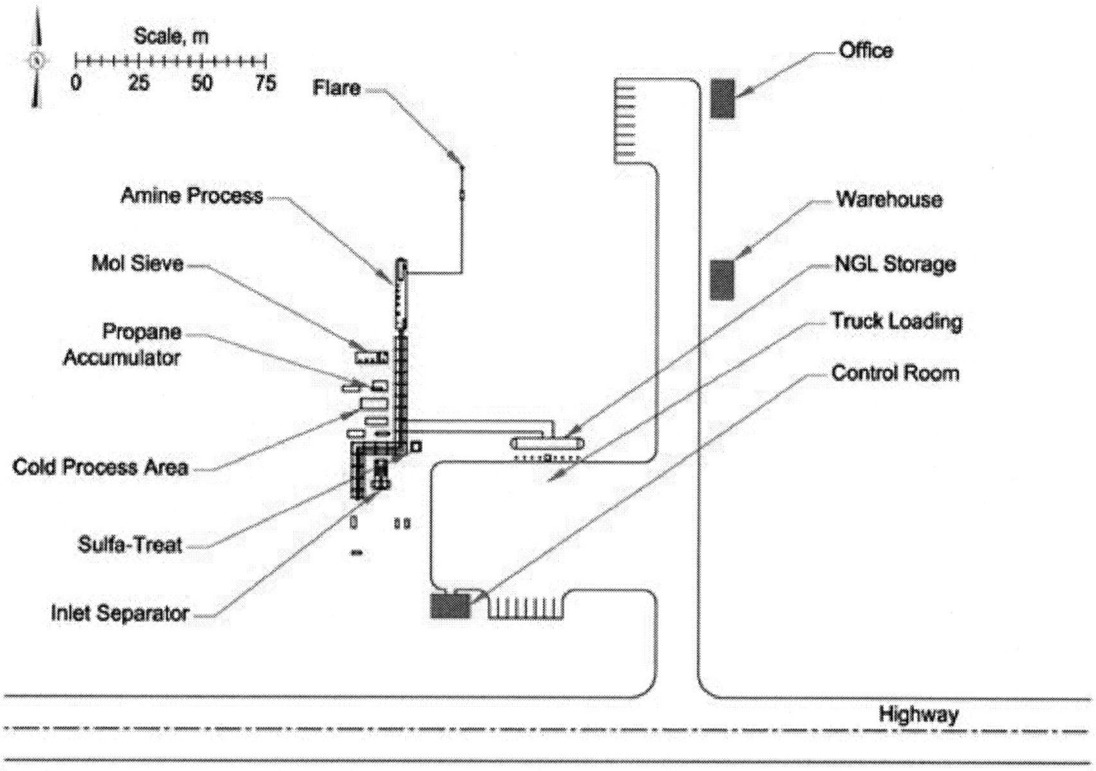

Figure 2. Example Gas Plant Facility Layout

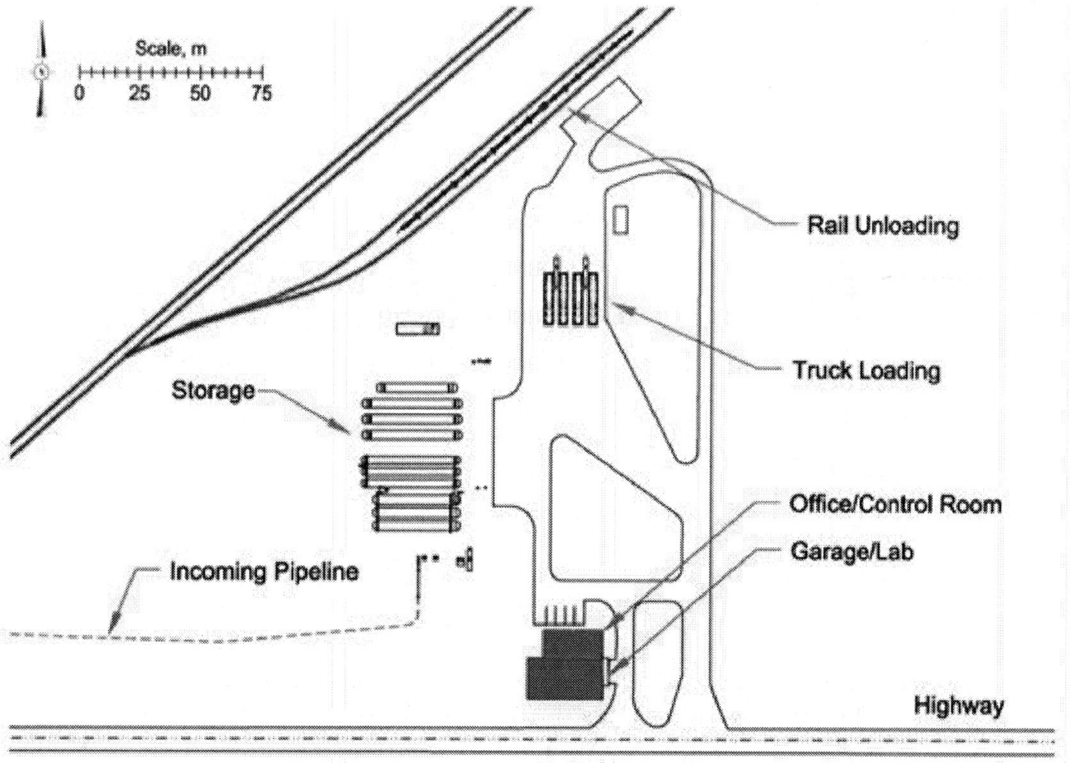

Figure 3. Example LPG Facility Layout

The LPG facility includes only two occupied buildings: a control room/office and garage/warehouse. While these two buildings are physically connected, they will be treated separately for the purposes of a building siting assessment. The site's hazards include fires, explosions, and flammable gas infiltration.

5.2 Consequence Analysis

The inherent hazards associated with these example facilities are well understood. A review of the process shows that there are multiple failure case scenarios that could result in fire, flammable cloud infiltration, toxic cloud infiltration, or explosion hazards that might impact the occupied buildings. A set of failure case scenarios was selected to represent each example facility – eight scenarios were chosen for the gas plant, and ten for the LPG plant (some additional scenarios were modeled to account for emergency shutdown systems and portions of the piping that are buried).

As part of the consequence analysis, locations for potential VCEs are selected to model representative explosion events. Locations within and around the facility that are confined or include congestion are referred to as potential explosion sites, or PESs. As the amount of confinement or degree of obstruction increases, so does the potential strength of the blast wave that could be created by a vapor cloud explosion within that area. This type of analysis is covered in more detail in other papers [11,12]. The explosion model used in the assessment is the QMEFS model [13]. All other consequence calculations were accomplished with CANARY by Quest® consequence analysis software.

The endpoints chosen for use in this study are listed in Table 1. The endpoints represent the threshold of fatality to building occupants when the building is a pre-engineered small metal building.

Table 1. Hazard Endpoints for Building Occupants

Hazard	Endpoint for Threshold of Fatality	Reference Number
Overpressure	1.13 psi	[14]
Radiation	35 kW/m^2	[15]
H_2S Infiltration	100 ppm†	[16]
Flammable Infiltration	Lower Flammable Limit (LFL)	[1]

† This is the emergency response planning guidelines level 3 (ERPG-3) value, as defined by the American Industrial Hygenists Association

At this point in the siting study, the assessment typically evaluates whether the potential impacts to the occupied building(s) are acceptable by comparison of worst-case and MCE impacts. For the purposes of this paper, both example facilities were taken through a QRA. Within that process, the release hole sizes that would have been assigned as worst-case (usually pipe ruptures) and MCE (two-inch hole) were included, so that a comparison could be made between the consequence-based results and QRA results.

As an example of consequence-based results, consider a range of hole sizes (1/4 inch through 6-inch pipe rupture) for the liquid discharge line of the propane storage tank at the LPG facility. Figure 4 demonstrates the hazards extents for these release scenarios, as a function of hole size, for fire (immediate jet fire) and flammable infiltration. The hazard magnitude values are the maximum found at that hole size, considering all weather conditions and release orientations.

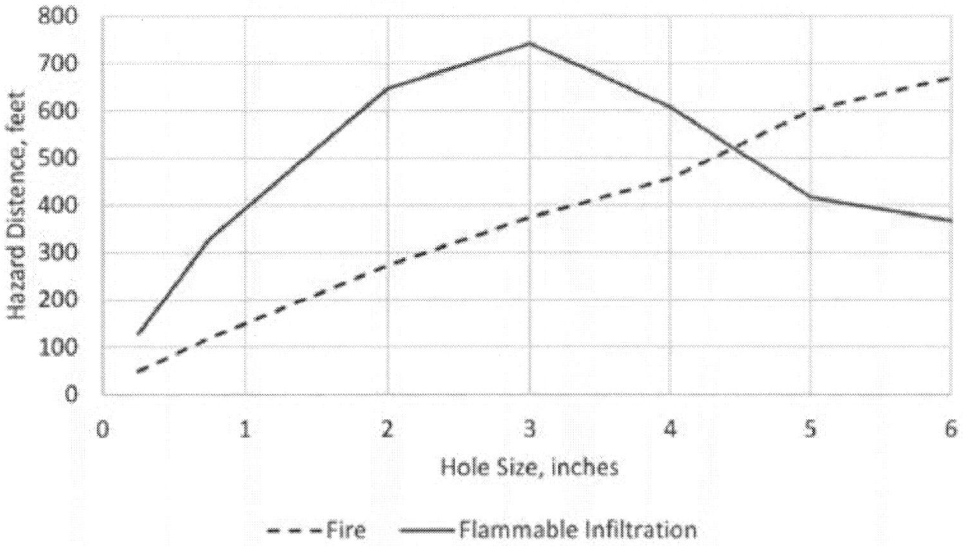

Figure 4. Effect of Hole Size on Maximum Hazard Magnitude

For this analysis the MCE was defined as a release up to a two-inch diameter hole in process equipment or piping. Due to dosage effects (in flammable and toxic infiltration, and occasionally radiation hazards), the worst case is not always the largest hole size of a selection set [5]. By setting the two-inch hole as a maximum, the MCE incorporates some element of frequency, but allowing smaller holes that may produce a larger hazard incorporates a maximum impact effect. Vulnerability zones for the various hazards, as defined by the MCE scenarios, are shown in Figure 5 for the gas plant and Figure 6 for the LPG facility.

5.3 Quantitative Risk Analysis

The risk that an individual inside a building is potentially exposed to a fatal hazard, as a result of events that could originate in the facility, can be represented by a numerical measure. This numerical measure represents the chance, or probability, that an individual will be exposed to a fatal hazard during a year-long period. For example, within the context of this example siting study, a value of 1.0 x 10^{-4} per year (or 10^{-4} in shorthand notation) represents one chance in 10,000 per year of being fatally affected inside a building by a release originating in the facility.

The risks due to all possible unique accidents in each example facility were combined to produce measures of risk for building occupants. The expansion of the failure cases to a full set of unique accidents involved the following parameters:

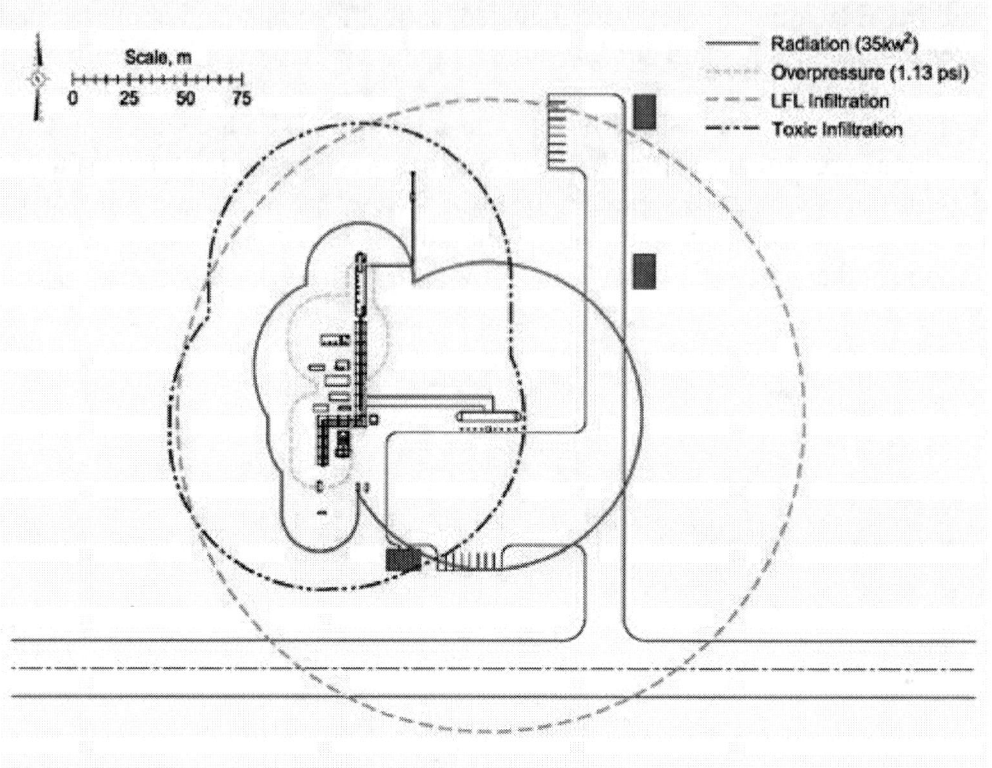

Figure 5. Gas Plant - Composite Vulnerability Zones for MCE Hazards

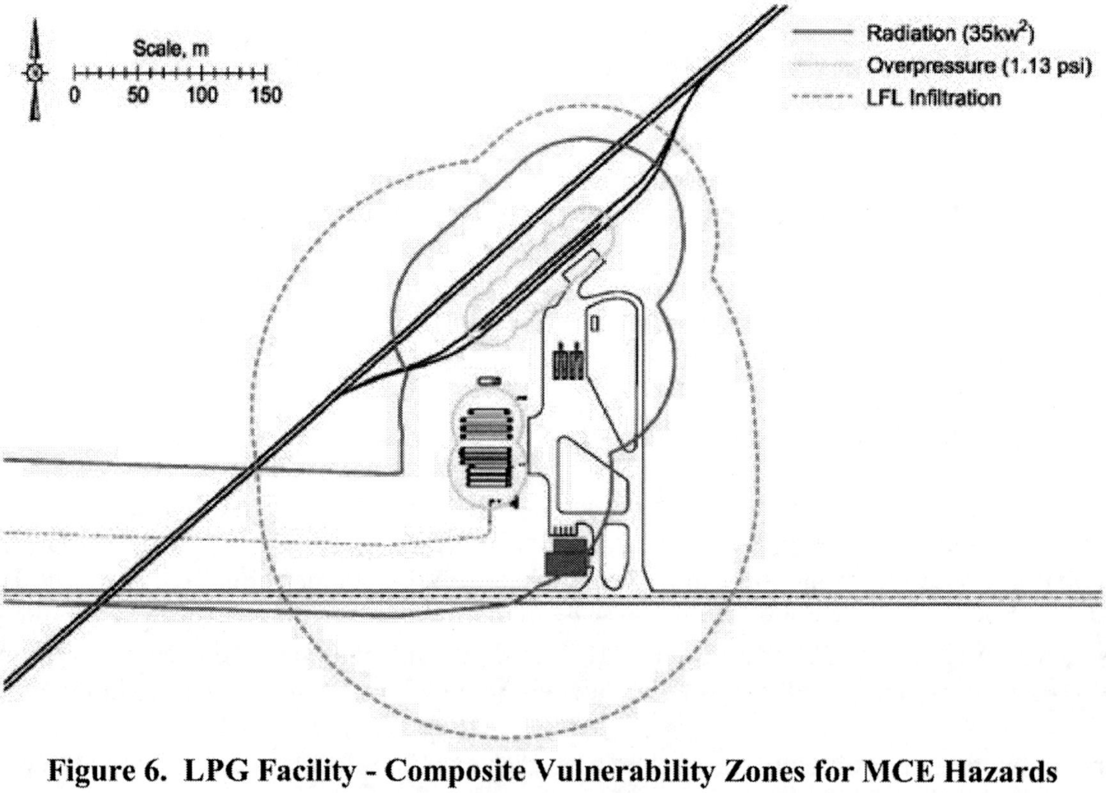

Figure 6. LPG Facility - Composite Vulnerability Zones for MCE Hazards

- 5 release hole sizes for each failure case scenario
- 2 release orientations (horizontal, upward)
- 21 wind speed and atmospheric stability combinations; 64 wind directions
- Fire radiation, vapor dispersion, and vapor cloud explosion hazards

The hazard impacts were computed and an event frequency was established for most of the above combinations (some combinations do not exist; for example, fire radiation models are not sensitive to atmospheric stability, so it is not a variable, and 6 wind speeds were applied instead of the 21 combinations). The event frequencies were calculated based on the Hydrocarbon Releases Database (HCRD), a comprehensive failure rate database [9], and the application of conditional probabilities. For any one unique scenario, the frequency of a release is obtained by following these steps:

- For the set of equipment described by the failure case (an equipment group which can be described by similar event consequences following an accidental release), the various pieces of equipment from which releases could occur are counted.
- The number of each type of equipment is multiplied by its failure frequency from the HCRD. The failure frequencies are summed for all equipment types involved in the failure case. Failure frequencies are developed for each release hole size, so the end result is a release frequency for each of the five hole sizes.
- The conditional probability for release orientation (horizontal=0.5, upward=0.5) is applied.
- The conditional probability for ignition timing is applied. Ignition is broken down into three possibilities: immediate ignition, delayed ignition, or no ignition.
  - Immediate ignition (jet fire, pool fire) is determined by the release rate and material type.
  - Delayed ignition (flash fire, vapor cloud explosion, jet fire, pool fire) is determined by the potential ignition sources encountered by the flammable vapor cloud.
  - The probability for non-ignition (toxic gas exposure, infiltration hazards) is the remainder after immediate and delayed ignition are accounted for.
- The conditional probabilities associated with weather conditions are applied. These include wind speed, atmospheric stability, and wind direction.
- Any conditional probabilities associated with a fractional time-of use for specific failure case scenarios, as applicable, are applied.

The total risk to building occupants is the summation of the potential impacts, at their respective event frequencies, for all combinations of failure case scenarios, hole sizes, release orientations, ignition timing, and weather conditions. The predicted risk represents the potential fatality for building occupants due to radiation from a fire, the effects of a vapor cloud explosion, flammable gas infiltration to the flammable limit (with ignition assumed), and toxic gas infiltration to the ERPG-3 concentration endpoint. The total risk is graphically presented in the form of location specific individual risk (LSIR) contours. The overall risk results are shown in Figure 7 for the gas plant and Figure 8 for the LPG facility. (Note that the LSIR contours are only valid where buildings exist or may be placed.)

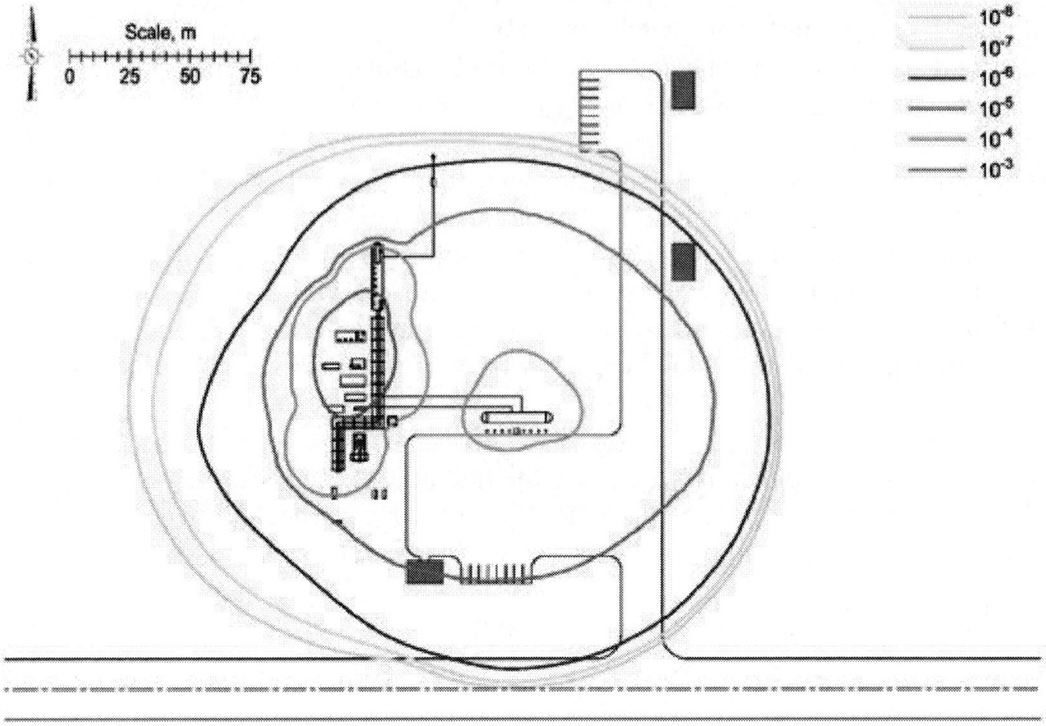

Figure 7. Gas Plant – Overall Risk Contours for Building Occupants

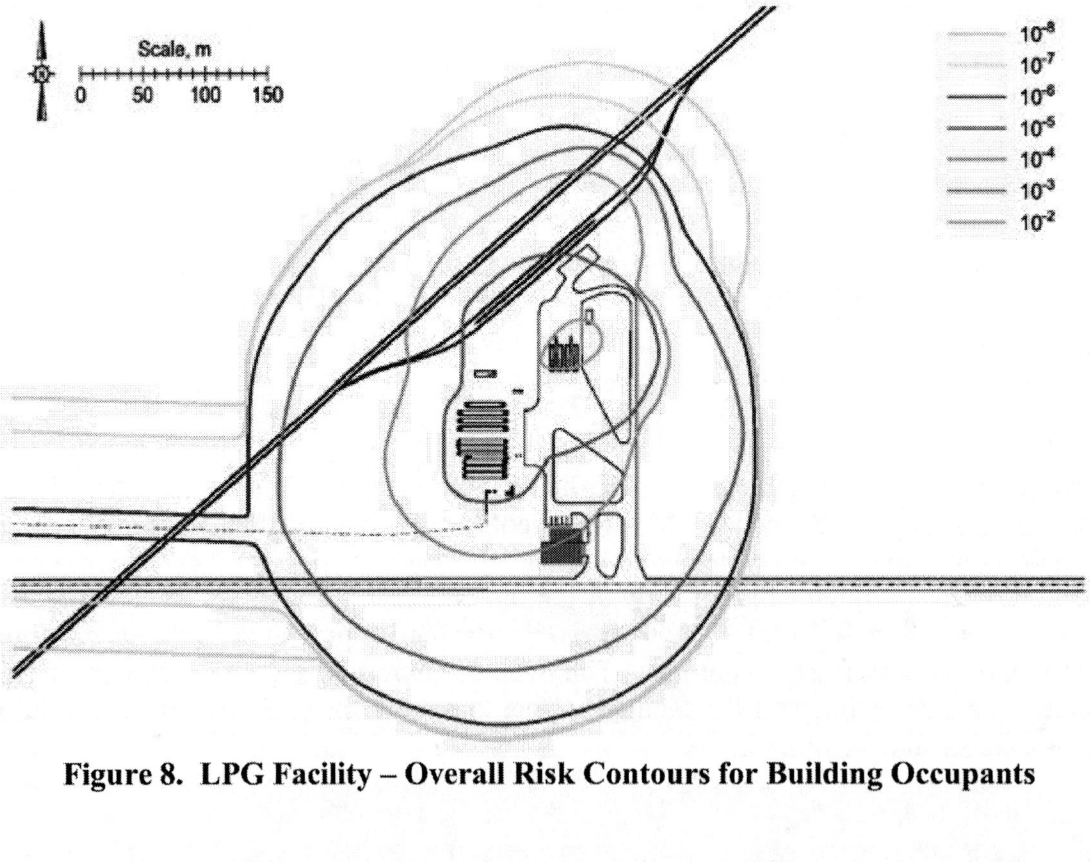

Figure 8. LPG Facility – Overall Risk Contours for Building Occupants

Because the risk contours shown in Figures 7 and 8 are defined based on impacts to occupied buildings, they are only valid at a location where a building exists (or could exist). For the control room in the example gas plant facility, the annual risk of fatality is predicted to be approximately 1.0×10^{-5} per year. For the control room in the example LPG facility, the annual risk of fatality is predicted to be approximately 1.0×10^{-4} per year. These risk levels assume continuous occupancy of the building(s), meaning that they are occupied 24 hours per day, 365 days per year.

6 Comparison

When the vulnerability zone results presented in Figures 5 and 6 are compared to the risk analysis results in Figures 7 and 8, several key findings can be extracted.

- For both facilities, the explosion overpressure vulnerability zones are within the overall 1.0×10^{-4} per year LSIR contour (those associated with all hazards). For these examples, the MCE vulnerability zones for explosions are small, but not expected to be any larger for less probable events due to the limited size of the PESs. This is generally true for most facilities.
- The fire radiation vulnerability zones following two-inch releases (as defined by the 35 kW/m^2 level) are approximately equal to the 1.0×10^{-5} per year risk contour at the gas plant. For the LPG facility, the MCE fire radiation vulnerability zones are approximately equal to the 1.0×10^{-4} per year risk contour, with the exception of the area surrounding the incoming pipeline.
- For the gas plant, the MCE flammable infiltration vulnerability zones extend well beyond the 1.0×10^{-6} per year risk contour. For the LPG facility, the MCE flammable infiltration vulnerability zones are approximately equal to the 1.0×10^{-6} per year risk contour.

7 Conclusions

The analysis presented in this paper sought to investigate the application of the two-inch hole to siting studies, with emphasis on building siting.

A review of published work on this subject found that the selection of hole size has been inconsistent, including some methods which select a variable hole size, some which use a fixed size, and some which leave the selection to events deemed credible. However, a more common choice was found to be the two-inch hole.

When evaluated according to release frequency, the two-inch hole was found to represent about 95% of all hydrocarbon release scenarios. This suggests that the two-inch hole is a reasonable selection as an upper bound of release hole sizes, and should be a reasonable choice for the MCE. This applies to general process system accidental releases and is not meant to exclude other events at a particular site that may be deemed credible.

In two example studies, the MCE vulnerability zones were compared to LSIR contours. Using an acceptable risk tolerance value for process plant personnel of 1.0×10^{-4} per year [17], vulnerability zones can be evaluated for suitability in siting studies. Based on this comparison for the example facilities, it was found that:

- When using an MCE defined by releases up to a two-inch hole, the fire radiation vulnerability zone compares well to the overall 1.0×10^{-4} per year LSIR contour for persons in buildings.
- MCE vulnerability zones for explosions are expected to be larger in extent than the 1.0×10^{-4} per year risk of fatality due to explosions, largely following the general observation that explosion impacts are limited by the volume of confined and congested space. Thus, the MCE explosion vulnerability zones can be taken to be representative of explosion risk.
- The flammable vapor dispersion and infiltration risk may not be well represented by MCE vulnerability zones – the vulnerability zones are aligned more with the 1.0×10^{-6} per year LSIR contour than the 1.0×10^{-4} per year contours.
- Toxic vapor dispersion and infiltration hazards may not be well represented; however, this will be a site-specific results that is highly dependent on the quantity and concentrations of toxic chemicals at the site.

As a generalization, the two-inch hole as an MCE selection appears to be a reasonable choice for representation of risk to occupied buildings. If a building is located outside of the MCE fire and explosion vulnerability zones, its risk will likely be acceptable. The exception to this is locations where there may be a high probability of flammable or toxic infiltration impact. This finding is certainly dependent on the size of the facility, the types of hazardous materials processed or stored, and the parameters applied to a siting analysis.

In conclusion, the general trend indicates that the two-inch hole can be used as a reasonable approximation of the risk to process plant building occupants: if a building is outside of the MCE vulnerability zones (as defined by releases up to two inches in diameter), the risk to occupants is generally found to be acceptable.

8 References

[1] API (2009), *Management of Hazards Associated with Location of Process Plant Permanent Buildings*. API Recommended Practice 752, American Petroleum Institute, 1220 L Street, Northwest, Washington, D.C. 20005, Third Edition, December 2009.

[2] Cornwell, J.B., Marx, J.D., and Lee, W.W. (1998), "Application of Qualitative and Quantitative Risk Analysis Techniques to Building Siting Studies". Process Plant Safety Symposium, Houston, TX, October 26-27, 1998.

[3] Aspengren, M., Whipple, T., Vralsted, D., Raghunathan, V., and Pitblado, R. (2009), "Evaluation of Permanent Buildings at an Existing Upstream Oil and Gas Business". 5th Global Congress on Process Safety, Tampa, FL, April 26-30, 2009.

[4] CCPS (2003), *Guidelines for Fire Protection in Chemical, Petrochemical, and Hydrocarbon Processing Facilities*. Center for Chemical Process Safety of the American Institute of Chemical Engineers, 3 Park Avenue, New York, New York, 10016-5991. (ISBN 0-8169-0898-2)

[5] Sanguino, O. and Hissong, D.W. (2013), "Methodology for Selecting Hole Sizes for Consequence Studies". Journal of Loss Prevention in the Process Industries 26 (2013) 483-488.

[6] Stahl, M.W. and Kenady, K. (2011), "A Frequency Based Approach to Hole Size Selection for Consequence Analysis". AIChE 7th Global Congress on Process Safety, Chicago, IL, March 13-16, 2011.

[7] CCPS (2012), *Guidelines for Evaluating Process Plant Buildings for External Explosions, Fires, and Toxic Releases, Second Edition*. Center for Chemical Process Safety of the American Institute of Chemical Engineers, Published by John Wiley & Sons, Inc., 111 River Street, Hoboken, New Jersey, 07030. (ISBN 978-0-470-64367-9)

[8] CSA Z276 (2015), Liquefied natural gas (LNG) — Production, storage, and handling. Canadian Standards Association, 178 Rexdale Boulevard, Toronto, Ontario, M9W 1R3, January 2015.

[9] HSE HCRD (2012), Hydrocarbon Releases System, Health and Safety Executive. https://www.hse.gov.uk/hcr3

[10] Bolsover A., (2013), "A Public Leak Frequency Dataset for Upstream and Downstream Quantitative Risk Assessment". AIChE 9th Global Congress on Process Safety, San Antonio, TX, April 28 – May 2, 2013.

[11] Marx, J.D., and K.M. Werts (2013), *The Application of Pressure-Impulse Curves in a Blast Exceedance Analysis*, Journal of Loss Prevention in the Process Industries, Volume 26, May 2013, pp 478-482.

[12] Marx, J.D., and Ishii, B. (2015), "A Comprehensive Approach to API RP 752 and 753 Building Siting Studies." Mary Kay O'Connor Process Safety Center International Symposium, College Station, TX, October 27-29, 2015.

[13] Melton, T.A. and J.D. Marx (2009), "Estimating Flame Speeds for Use with the BST Blast Curves." *Process Safety Progress*, Vol. 28, No. 1, March 2009, pp 5-10

[14] DOD (2009), *"Approved Methods and Algorithms for DOD Risk-Based Explosives Siting"*. Technical Paper No. 14, Revision 4. Department of Defense Explosives Safety Board, Alexandria, VA, July 21, 2009.

[15] TNO (1999), *Guidelines for Quantitative Risk Assessment* (First Edition), the Purple Book. CPR 18E, the Netherlands Organization for Applied Scientific Research, Committee for the Prevention of Disasters, the Hague, Netherlands, 1999.

[16] AIHA (2013), American Industrial Hygienists Association, 2013 Emergency Response Planning Guidelines, https://www.aiha.org/get-involved/AIHAGuidelineFoundation/EmergencyResponsePlanningGuidelines/Documents/2013ERPGValues.pdf , 2013.

[17] CCPS (2009), *Guidelines for Developing Quantitative Safety Risk Criteria*. Center for Chemical Process Safety of the American Institute of Chemical Engineers. Published by John Wiley & Sons, Inc., Hoboken, New Jersey, 2009. (ISBN 978-0-470-26140-8)

12TH GLOBAL CONGRESS ON PROCESS SAFETY

LPS 50th LOSS PREVENTION SYMPOSIUM

Retrofit of Non-Structural Components on Buildings Subjected to Blast Loading

David A. Holgado, M.S., P.E.
ABS Consulting
Advanced Engineering Division
140 Heimer Rd., Suite 300
San Antonio, TX 78232
dholgado@absconsulting.com

Jerry Collinsworth, M.S.
ABS Consulting
Advanced Engineering Division
jcollinsworth@absconsulting.com

Jeff Goodman
ABS Consulting
Advanced Engineering Division
jgoodman@absconsulting.com

UNPUBLISHED

Prepared for Presentation at
American Institute of Chemical Engineers
2016 Spring Meeting
12th Global Congress on Process Safety
Houston, Texas
April 10-14, 2016

Retrofit of Non-Structural Components on Buildings Subjected to Blast Loading

David A. Holgado, M.S., P.E.
ABS Consulting
Advanced Engineering Division
140 Heimer Rd., Suite 300
San Antonio, TX 78232
dholgado@absconsulting.com

Jerry Collinsworth, M.S.
ABS Consulting
jcollinsworth@absconsulting.com

Jeff Goodman
ABS Consulting
jgoodman@absconsulting.com

Keywords: Hazard Identification and Risk Analysis, Safe Design

Abstract

Structures in chemical and petrochemical facilities are often located in areas that may be subjected to blast loading. Occupied buildings typically have non-structural components located along the interior of the exterior walls and roof such as windows, doors, wall mounted AC units, lights, furniture, storage racks, hanging equipment, and loose articles.

Occupants of buildings subjected to accidental explosions may be injured from glass fragments and interior non-structural items becoming projectiles and impacting building occupants. As a pressure wave from a blast impacts the exterior of a building, the wall and roof components are rapidly accelerated inward. Equipment or contents mounted on or in contact with the exterior façade are also accelerated and may be dislodged and projected as debris. Items anchored to the ceiling structure can be thrown vertically from the initial forward deflection of the supporting member or break free from their supports and become falling debris hazards.

Therefore, evaluation and mitigation of non-structural debris for buildings subjected to blast load is important to further mitigate the potential hazards to personnel occupying these buildings. This paper provides design retrofit recommendations based on accident investigation experience at chemical and refining facilities and engineered solutions for typical hazards commonly observed at these facilities.

1. Introduction

In blast related events, injuries may result from glazing fragment hazards and interior articles that become projectiles and impact building occupants. Hazards associated with non-structural components occur when the pressure wave from an explosion impacts the exterior surfaces of a building. The wall and roof structural components and the attached interior components are rapidly accelerated inward. After reaching peak inward displacement the structural members spring back, creating a dynamically applied reaction force on the connection between the structural components (SC) and non-structural components (NSC). Connections that are too weak to resist these reactions break free from the loaded surfaces and the NSC are propelled into the occupied spaces of the building.

Non-structural hazards result from items of notable weight attached to or in contact with exterior walls or ceiling. Common examples of these items are bookcases, filing cabinets, storage racks, wall mounted fixtures, overhead lights and air conditioning diffusers. As a general rule most items mounted overhead, or located 48-inches or higher above the floor, can pose a hazard to occupants in the event of an accidental explosion.

Figure 1 depicts a typical vertical cross-section of a building with non-structural components secured to exterior structural components. An example of the non-structural hazards that can result even at lower building damage levels is provided below in Figure 2 [1].

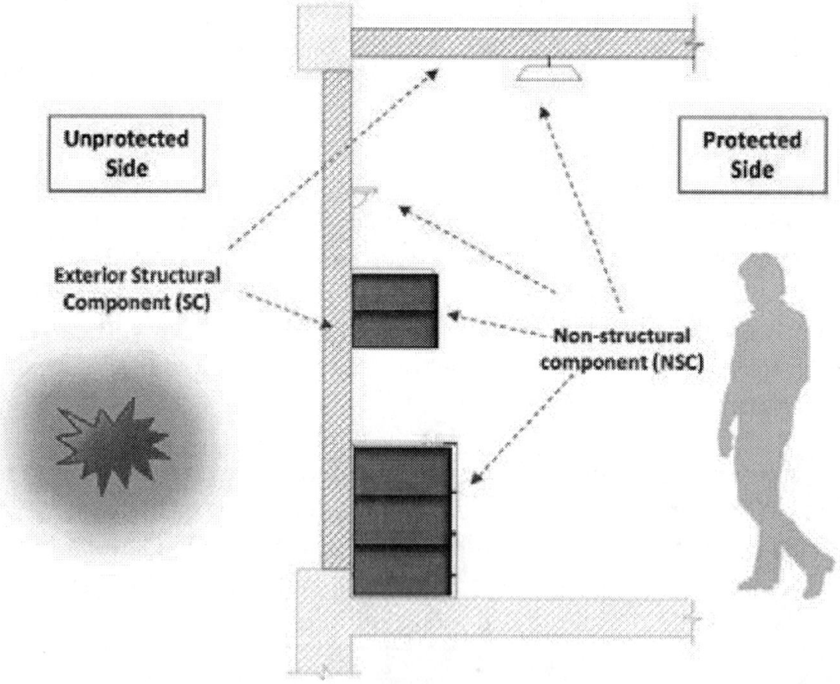

Figure 1. Non-Structural Hazards Adjacent to Exterior Building Surfaces

Figure 2. Lightly Damaged Building and Associated Non-Structural Hazards [1]

The response of non-structural components (NSC) to blast loading is associated with the response of the attached structural component (SC) in order to determine the design basis acceleration for supporting the NSC weight. This paper summarizes the basis of the methodology developed for this purpose.

Note that the upgrade solutions developed in this paper are not intended to prevent damage to the fixture itself (i.e. internal/external mechanical or electronic damage) but rather to mitigate the hazards to occupants. Also, window glazing and door response to blast loads are not included but can also present non-structural hazards. They should be addressed on a case by case basis for each building.

2. Design Methodology

2.1 Structural Response Prediction

Structural components subjected to blast loading are commonly analyzed utilizing the Single Degree of Freedom (SDOF) method [2]. The SDOF solution is obtained using simple numerical integration methods and the Single-Degree-of-Freedom Blast Effects Design Spreadsheet SBEDS [3] is one of the publically available tools.

Closed form solutions of the SDOF response are difficult to obtain because of material non-linearity and variation of loading. However, an analytical solution of the SDOF response has been achieved considering certain assumptions and simplifications as described in the following sections. The typical representation of SDOF system is depicted in Figure 3.

Although the peak deflection of SDOF elasto-plastic system could be predicted using simplified charts, such as the one shown in Figure 4, [2], the charts predicting the acceleration at the moment of maximum deflection, which is the design basis acceleration for restraining the non-structural fixture, are not available.

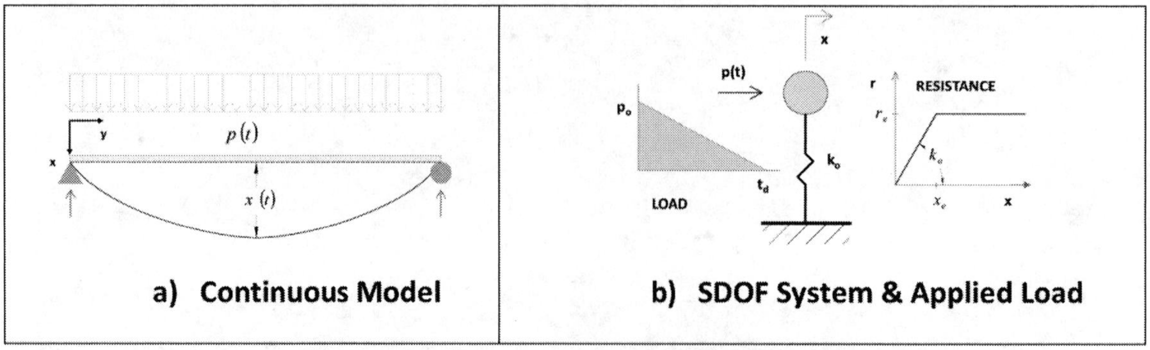

Figure 3. Component and SDOF Model

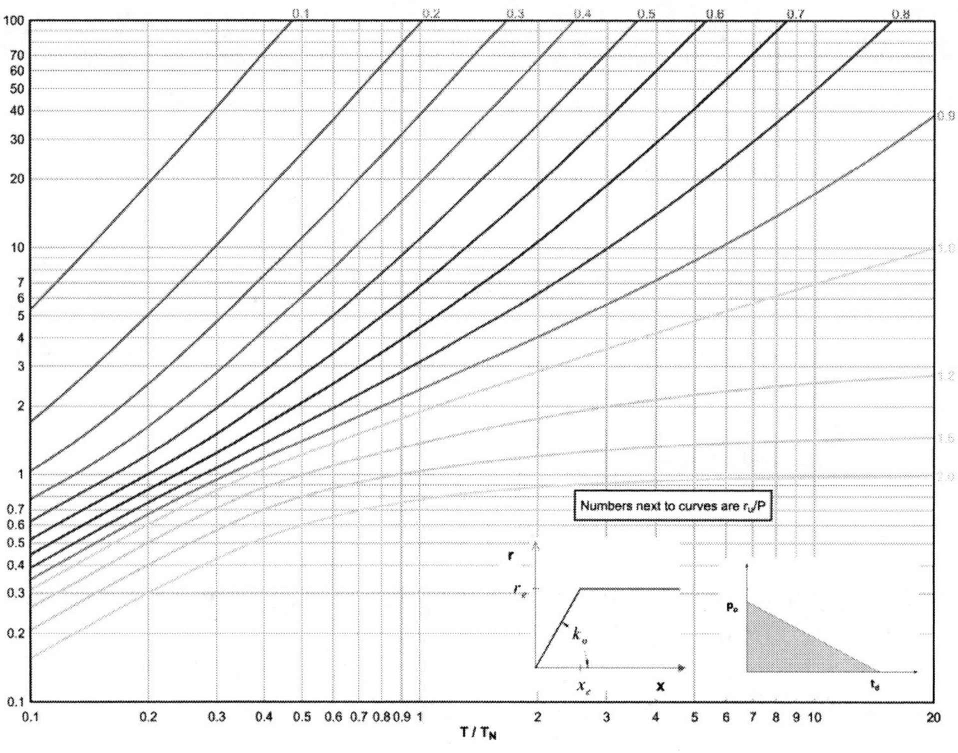

Figure 4. Non-Dimensional Maximum Deflection Response of SDOF [1]

2.2 Assumptions and Approximations

- Mass is uniformly distributed along the span of the component.
- The assumed shape function, $\phi(y)$, is applicable for deflection, velocity and acceleration.
- The applied blast load is assumed to be right triangular with an instantaneous rise to peak pressure and a linear decay to ambient as illustrated Figure 3 b).
- The resistance function follows a bi-linear configuration as shown in Figure 3 b).

- The component has adequate shear strength to transfer the reaction forces during the component response.
- For components having resistance other than bilinear due to support constraints (e.g. Fixed-Simple, Fixed-Fixed, etc.), the approach described herein is still applicable. The equivalent elastic stiffness, k_{EQ}, should be computed that results in the resistance deflection function having equal strain energy.

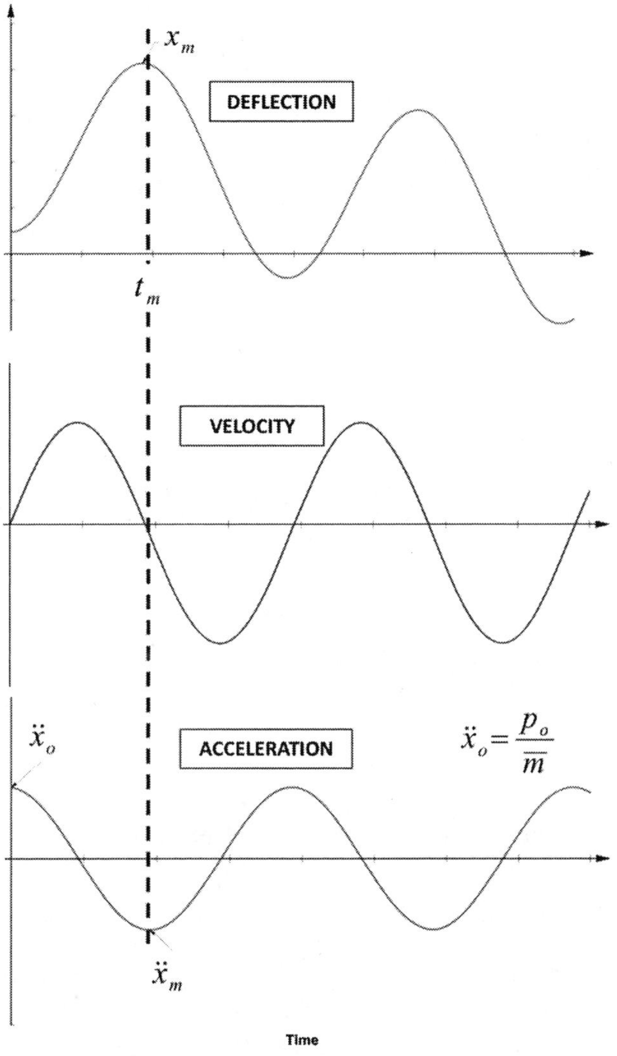

Figure 5. SDOF Dynamic Response

2.3 Component Response Prediction

2.3.1 Deflection Response

The analytical solution is implemented in Visual Basic code and the SC ductility is plotted for the SDOF system as a function of the ratio of the load duration to the natural period (t_d/T_n) and the dynamic load factor (r_e/p_o) = [0.1, 0.2.. 1.5, 2.0] and then, the non-dimensional plot is generated as shown in Figure 6. The computed solution in Figure 6 is identical to plots provided for simple SDOF solutions as shown previously in Figure 4.

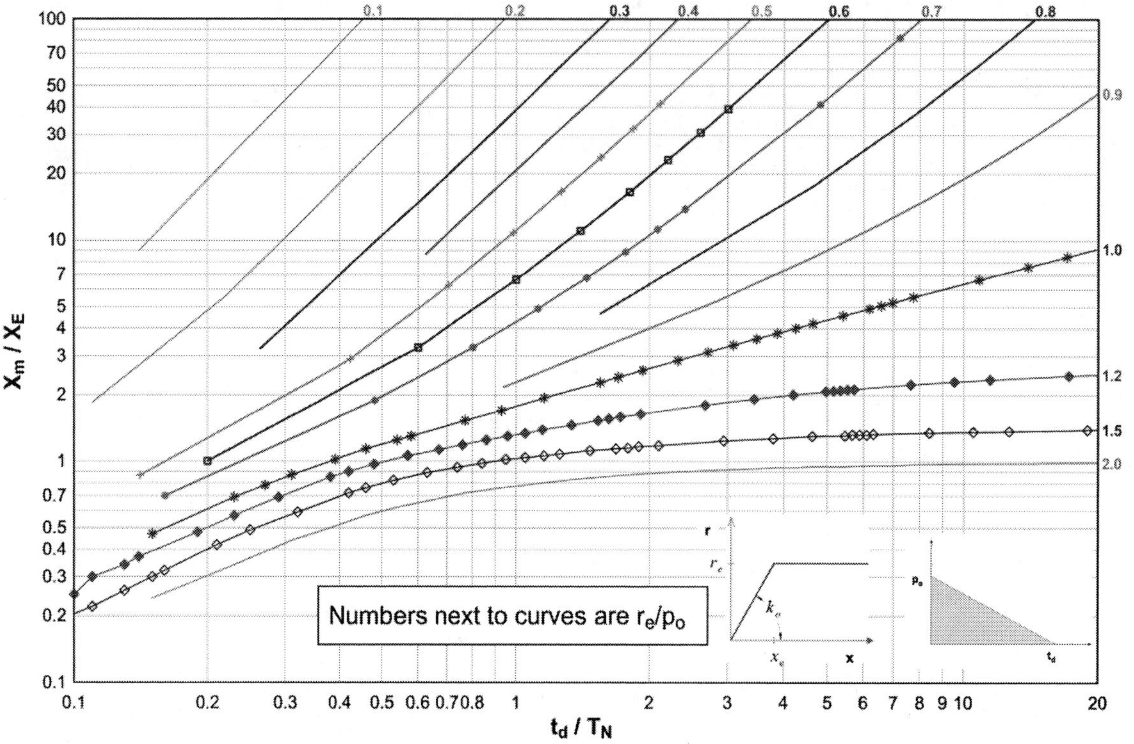

Figure 6. Non-Dimensional Maximum Deflection Response of SDOF

2.3.2 Acceleration Response

In a manner similar to the prediction of peak deflection described in previous section, the design basis acceleration (occurring at the time when the maximum deflection of the SC is achieved) is plotted also as a function of ratio of the load duration to the natural period (t_d/T_n) and the dynamic load factor (r_e/p_o) for $r_u \leq p_o$ and $r_u \geq p_o$ as shown in Figure 7 & Figure 8, respectively. Figure 7 & Figure 8 may be utilized to determine the design basis acceleration for retrofitting the non-structural fixture attachment to the supporting SC.

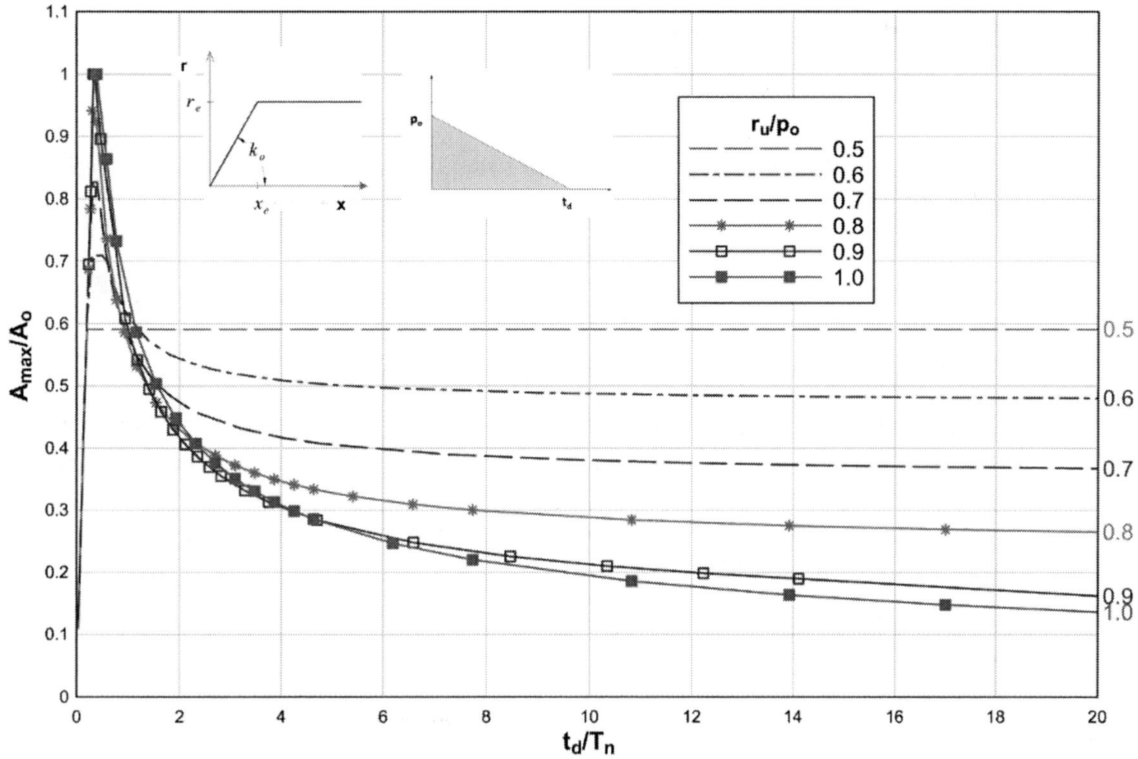

Figure 7. Non-Dimensional Acceleration Response for $r_u \leq p_o$ of SDOF

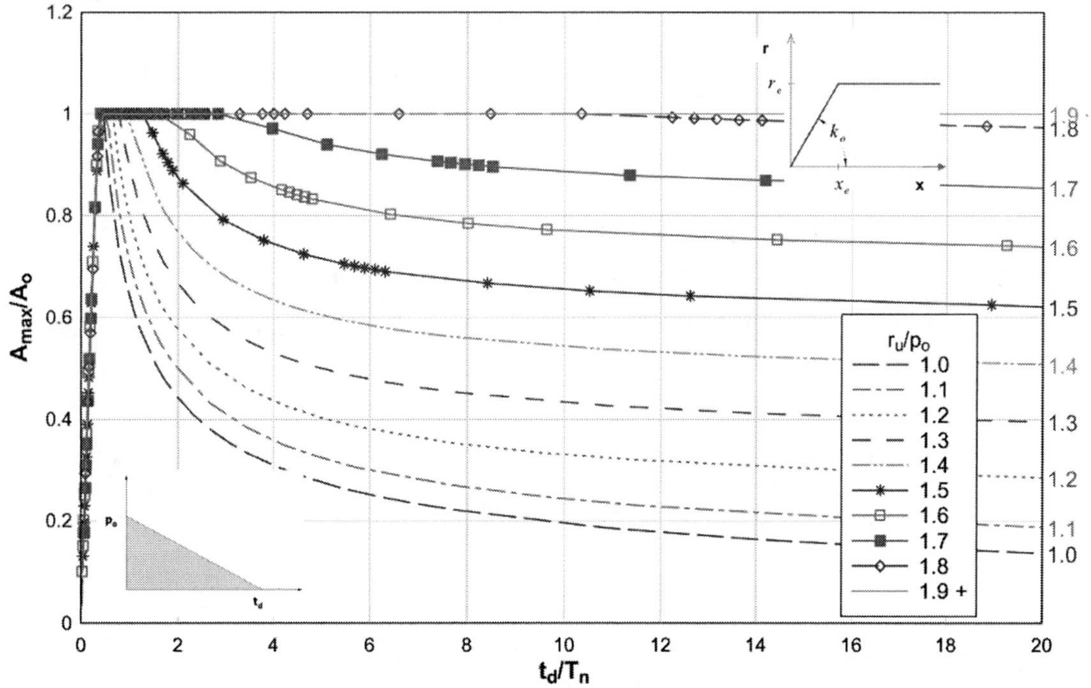

Figure 8. Non-Dimensional Acceleration Response for $r_u \geq p_o$ of SDOF

3. Retrofit Design Techniques

3.1 Assumptions and Limitations

- Structural components should be limited to Medium Response or lower as defined in Appendix 5.B of ASCE [5]; this component response allows certain permanent deformation during blast loading but it is still repairable.
- The building must be designed and constructed according to modern conventional construction practices, within the last 50 years using wood, steel, masonry or concrete. This excludes temporary buildings or trailers.
- The mass of the fixture is insignificant compared with the mass of the structural component and the fixture behaves as a rigid body during dynamic response of the structural component. In other words, there is no interaction effect during the response of the structural component and non-structural fixture as computed using the corresponding SDOF approach.
- The fixture is located near the location of the maximum response of the structural component (i.e. midspan).
- The fixture is in contact with the structural component prior blast loading (i.e. no impact effect).
- The fixture is not attached to cladding but is attached to a secondary framing component (e.g. purlin, girt, joist, etc.) or primary framing component (e.g. column, beam, Concrete Masonry (CMU) wall, etc.).
- The connection between the fixture and structural component is assumed to be rigid.
- The upgrade solutions developed herein are not intended to prevent damage to the fixture itself (i.e. internal/external mechanical or electronic damage) but rather to mitigate the hazards to occupants.
- Window glazing and door response to blast loads are not included but can also present non-structural hazards. They should be addressed on a case by case basis for each building

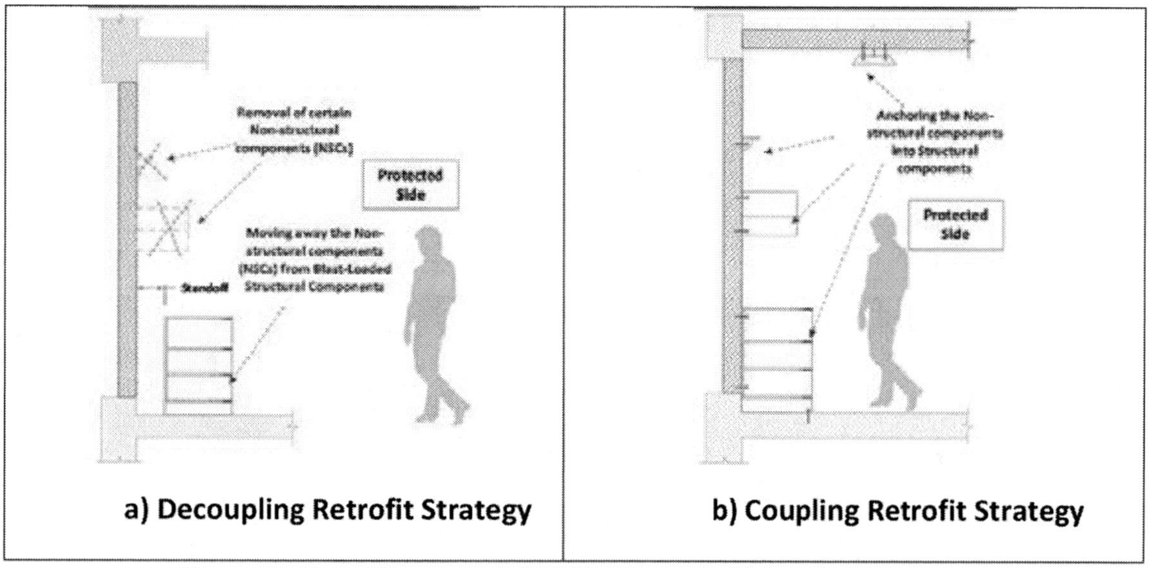

Figure 9 Non-Structural (NSC) Hazards Retrofit Strategies

3.2 Decoupling Retrofit Strategy

Of the two retrofit options, decoupling is the simpler one to implement. This technique is based on the concept of providing sufficient distance for all heavy fixtures from the exterior building surfaces (i.e. walls and roofs), as shown in Figure 9 a). A safe offset is computed based on the maximum inward displacement of the structural component along the exterior surface as shown in Figure 10. Thus, the maximum displacement of the primary or secondary framing system is computed for the applied blast load. Figure 11 depicts an example of maximum deflection of component under a set of applied blast loads. The non-structural fixture is then separated from the structure by sufficient distance to accommodate the total deflection of the supporting SC.

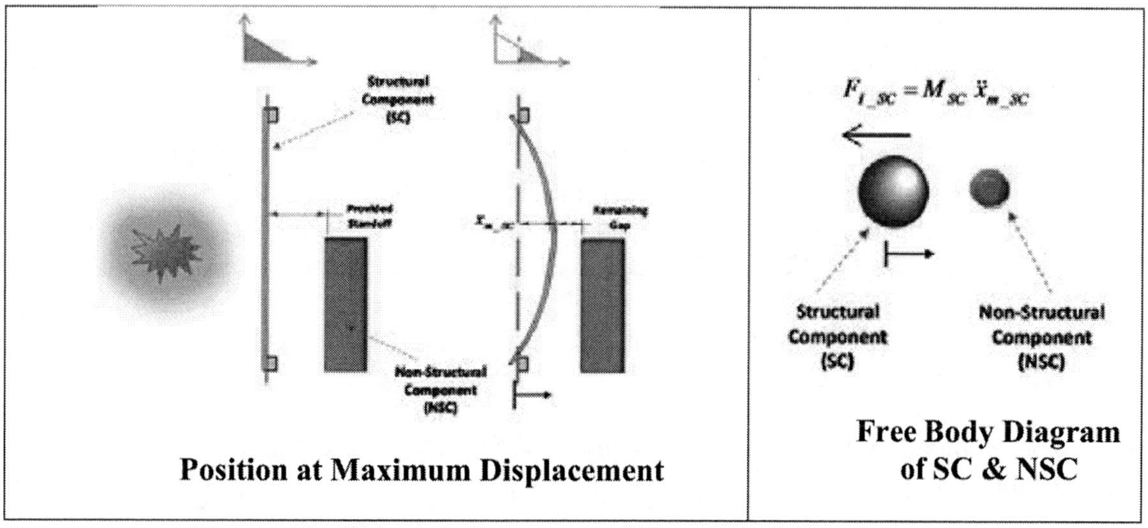

Figure 10. Dynamic Response of Decoupled Components

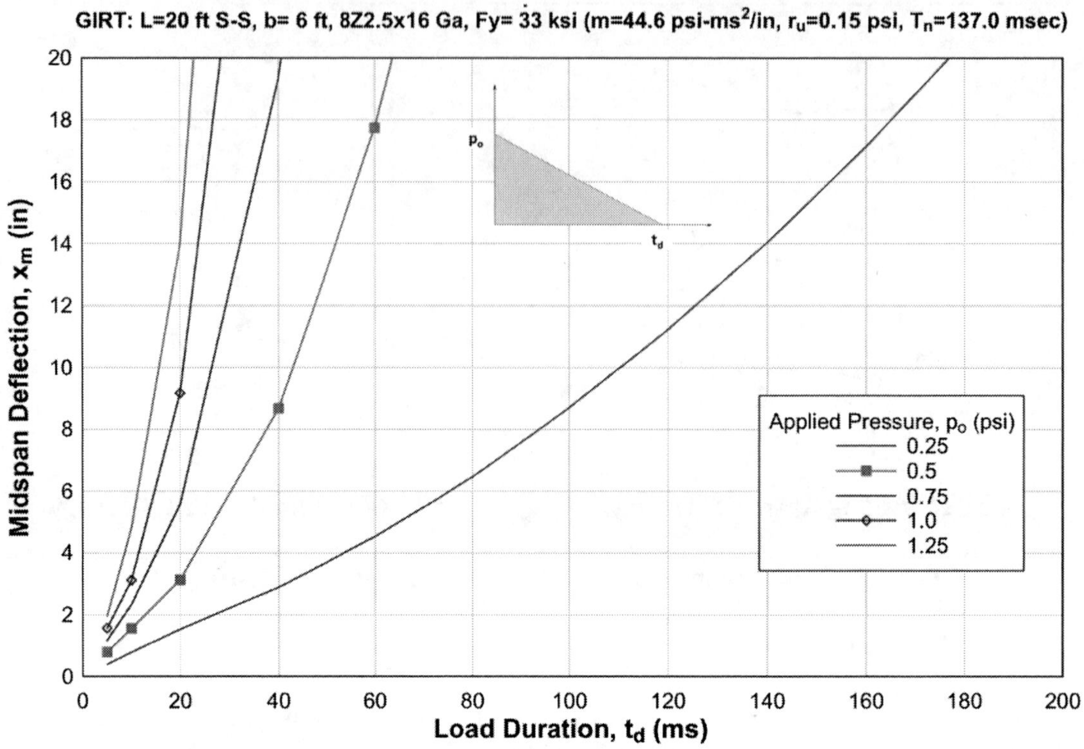

Figure 11. Maximum Deflection Response of Structural Component (SC), x_{m_SC} (example)

3.3 Coupling Retrofit Option

Since the decoupling retrofit technique is based on providing sufficient standoff distance between fixtures and loaded exterior surfaces, the availability of sufficient room is critical and in many cases that may not be a feasible retrofit option. In those instances, the coupling retrofit technique will need to be evaluated for implementation. Figure 9 b) depicts the general concept of the coupling option retrofit.

As the coupling retrofit relies on the fasteners between the SC and the non-structural fixture it is imperative the connection be sufficient to withstand the imparted forces at the time of peak displacement and the onset of structural rebound. The forces used to develop the connection requirements are based on the mass of the fixture and the acceleration of the wall at the point of maximum inward displacement, as depicted in Figure 12. Several assumptions are made in order to compute the acceleration of the fixture which are described in Section 2.3.2.

Figure 13 depicts an example of design basis acceleration of component under a set of applied blast loads

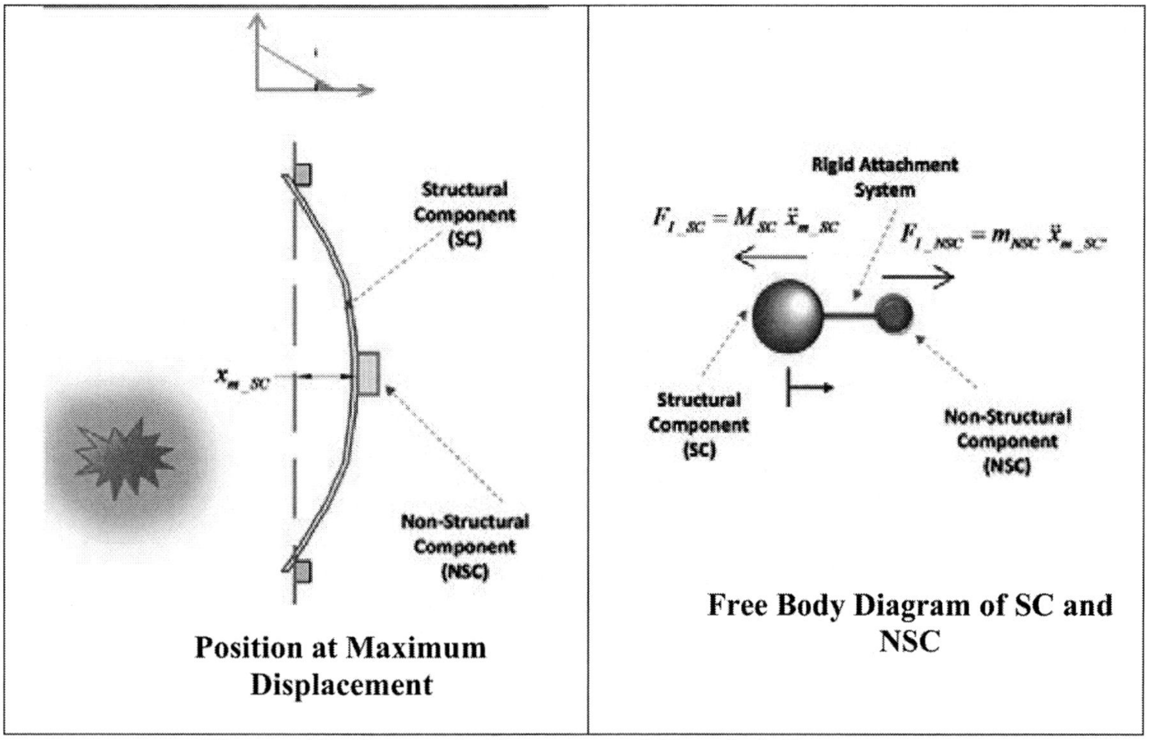

Figure 12. Dynamic Response of Coupled Components

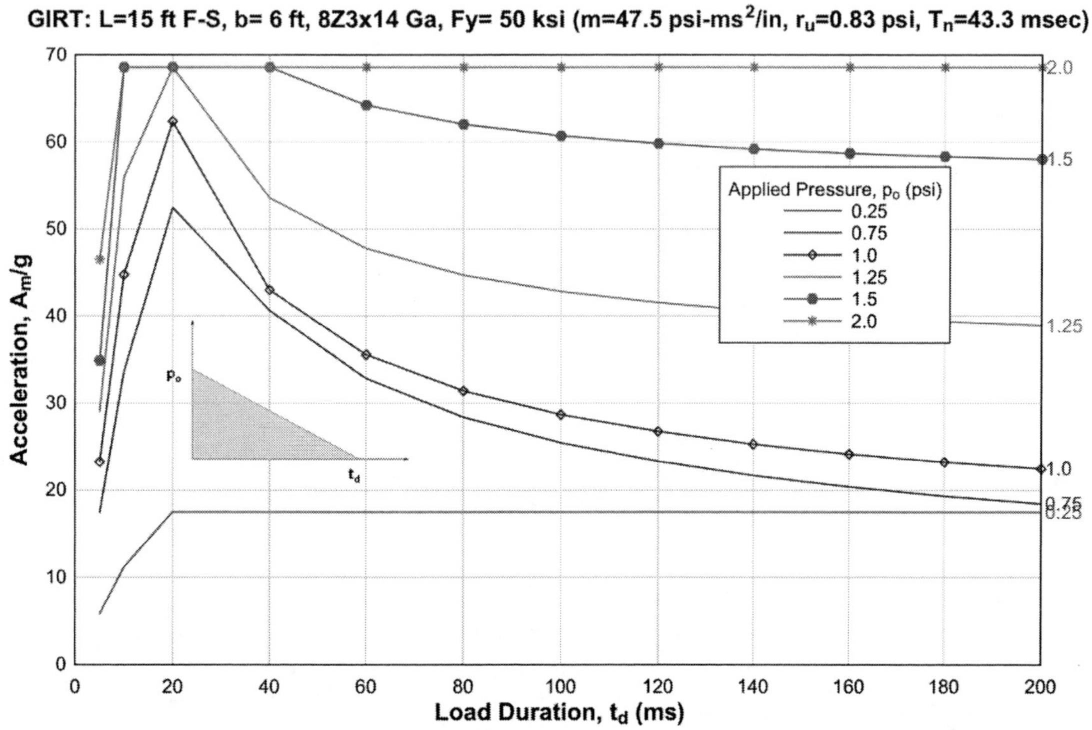

Figure 13. Acceleration of Structural Component (SC), $\ddot{x}_{m\_SC}$ (example)

3.3.1 Computation of Required Anchorage Forces

The required anchorage forces on NSC are based on the predicted acceleration of the structural response (at maximum displacement), as described in Section 2.2, and the mass of the attached fixture. Table 1 shows in example of peak acceleration of components per type of building construction.

Some common NSC to be retrofitted are identified including their corresponding weight as shown in Table 2.

Table 1. Maximum Acceleration of Structural Components (example)

Surface	Substrate	Amax (g)
Wall	R CMU	3.2
Wall	PEM	139.2
Roof	PEM	129.9
Wall	Steel	43.2
Roof	Steel	40.0
Roof	OWSJ	14.1
Wall	Wood	4.6
Roof	Wood	10.3

Table 2. Required Anchorage Forces per NSC (example)

			Required Anchorage Force, Fu (lb)		
Fixture	Weight (lb)	Surface	On Steel	On CMU	On Wood
Multi-Point Hanging Fluorescent Light	4.5	Roof	221		46
Fluorescent Light Fixture	24	Roof	1178		247
Light Fixture	35	Roof	1719		361
Garage light	10	Wall	1392	18	46
Wood cabinet	50	Wall	6960	89	230
Steel Lockers	110	Wall	15312	195	506
Peg board	5	Wall	696	9	23
Ceiling Diffuser	9	Roof	442		93
Bookcase	80	Wall	5568	142	368
Cabinet / Locker	60	Wall	4176	106	276
Metal Lateral File Cabinet (04 drawers)	710	Wall	98832	1257	3266
Whiteboard	53	Wall	7378	94	244
Emergency Respirator Wall Case	14	Wall	1949	25	64
15,000 BTU Window AC	108	Wall	15034	191	497
Metal Conduit, Electric	7	Wall	974	12	32
Fan Wall Mounted, 24 in	50	Wall	6960	89	230
Fan Wall Mounted, 30 in	54	Wall	7517	96	248
Water Dispenser, 5 Gal.	84	Wall	11693	149	386
Fiber Glass Ladder, 6 ft	18	Wall	2506	32	83
Fiber Glass Ladder, 8 ft	26	Wall	3619	46	120
Fiber Glass Ladder, 12 ft	48	Wall	6682	85	221
Hose Reel Retractable	40	Roof	1964		412
Oxygen Tank	120	Wall	16704	212	552
Electrical Box	45	Wall	6264	80	207
Storage Rack	1050	Wall	146160	1859	4830
HVAC Duct	138	Roof	6776		1421
Heater	55	Wall	7656	97	253
Base Cabinet	67	Wall	9326	119	308
Kitchen Cabinet, Wall Mounted	125	Wall	17400	221	575
Book Shelve, Wall Mounted	44	Wall	6125	78	202
TV. 40" Plasma	20	Wall	2784	35	92

In addition to the loading from the blast, gravity loads are imposed on non-structural components affixed to the roof. However, this refinement may be meaningful only for certain components. Using as example the values from Table 1, the gravity effect on wood components may increase the anchorage force by 20%. For the other roof components (i.e. steel purlins, Open Web Steel Joists, (OWSJ)), the gravity effect could be disregarded. Figure 14 and Figure 15 depict the free body diagrams (FBD) of anchorage forces along wall and roof surfaces, respectively.

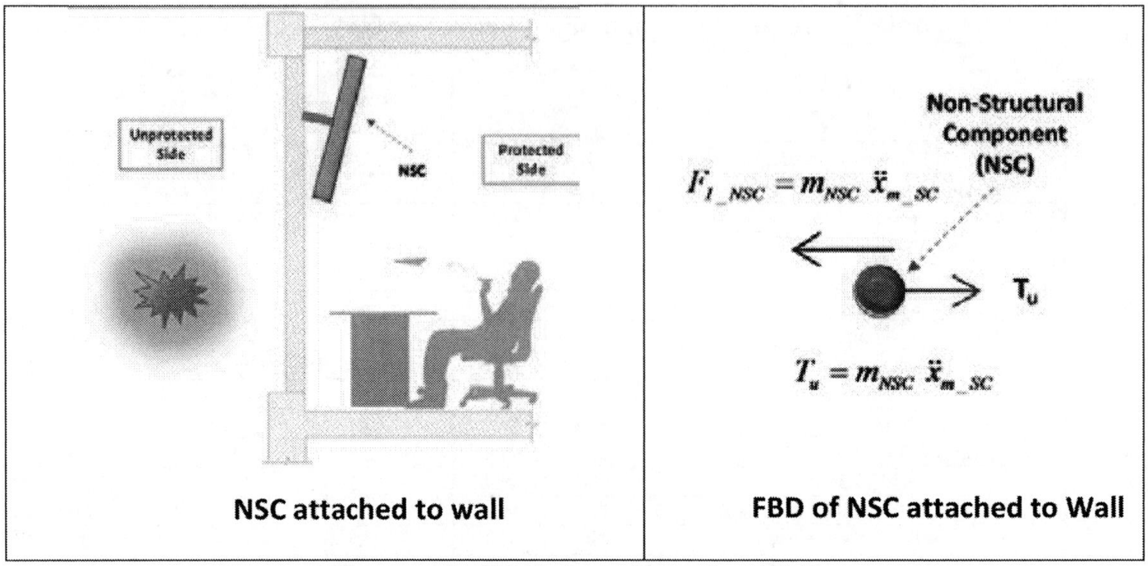

Figure 14. NSC along Exterior Wall

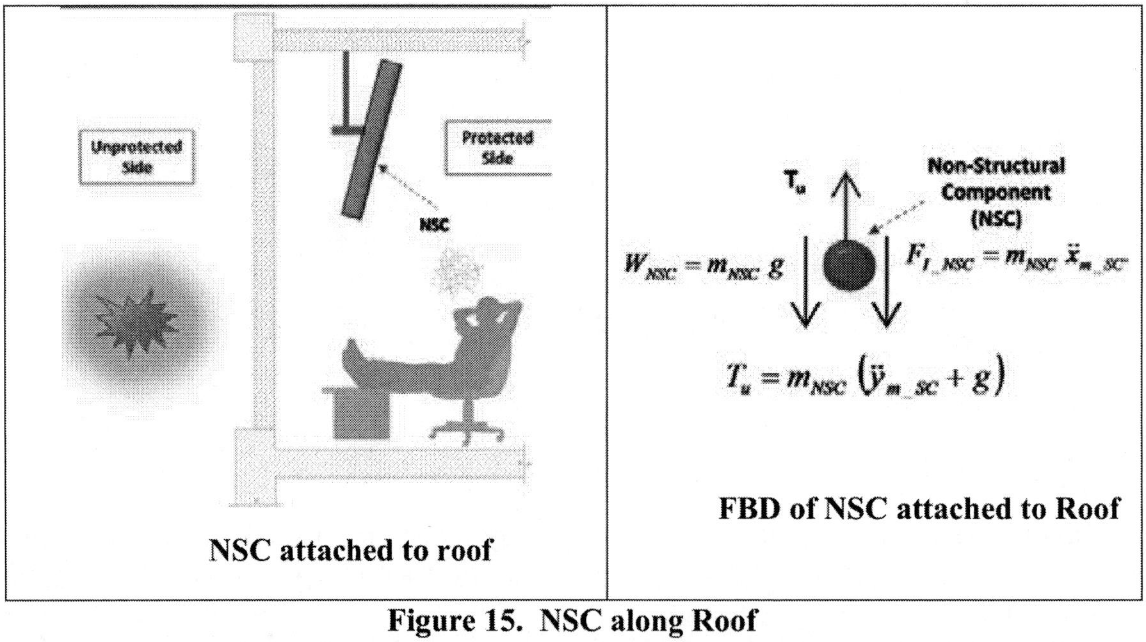

Figure 15. NSC along Roof

Finally, the strength of the provided anchorage used to restrain the NSC must be greater than the required force as shown in Table 2. Also, this retrofit technique may require the connection between the fixture and structural component formed by several pieces of hardware rather than single fasteners (i.e. hooks, chains, straps, etc.). Therefore, all hardware involved in the connection must comply with the required strength through the load path. As an example, Figure 16 depicts the retrofit of wall fans mounted on exterior wall.

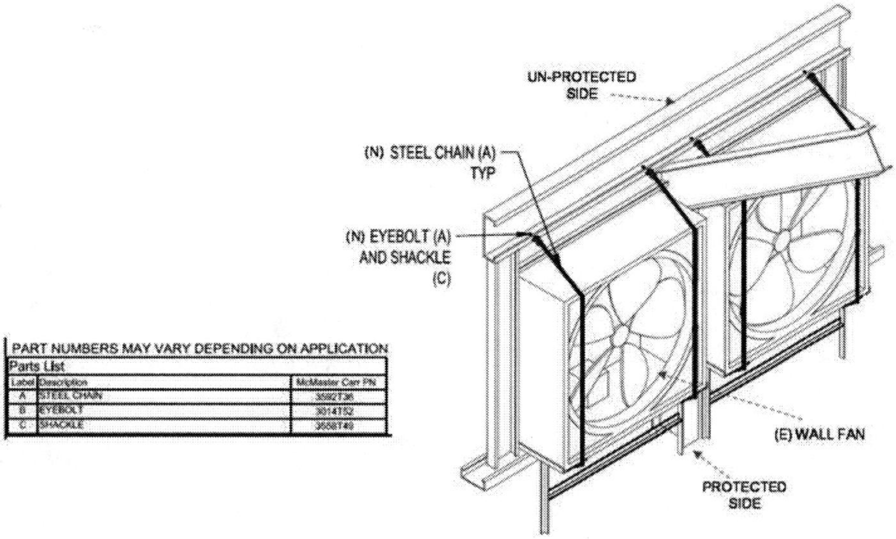

Figure 16. NSC Retrofit Sketch (example)

4. Recommendations

In addition to the anchorage requirements for the coupling retrofit technique described in the Section3.3, some wall/roof conditions may require special attention as follows:

- Placement of anchorage in location that does not induce prying in fasteners; see Figure 17.
- Placing the NSC near the support of the structural component and away from the point of peak deflection may significantly reduce the anchorage requirements; see Figure 18.
- NSCs that are relatively large compared with the structural component length may reduce the anchorage requirements because acceleration of fixture varies along its length; see Figure 19.
- Anchorage for retrofit must be installed into secondary or main frame members capable of supporting the design basis forces.

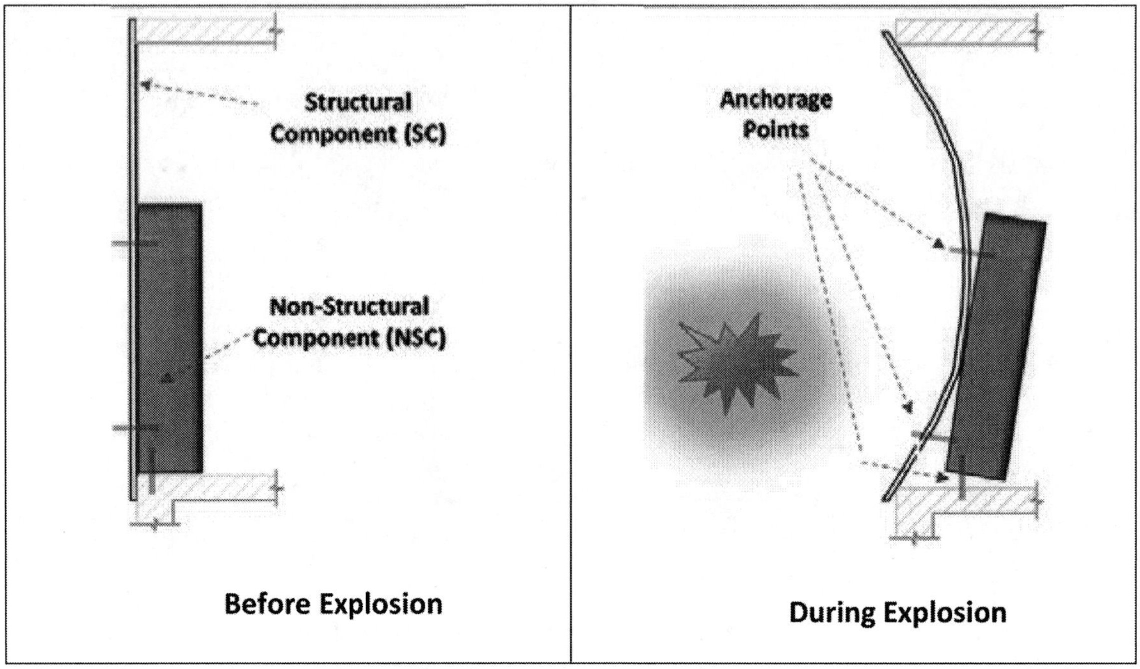

Figure 17. Possible Prying Effect in Fasteners During the Blast Response of SCs & NSCs

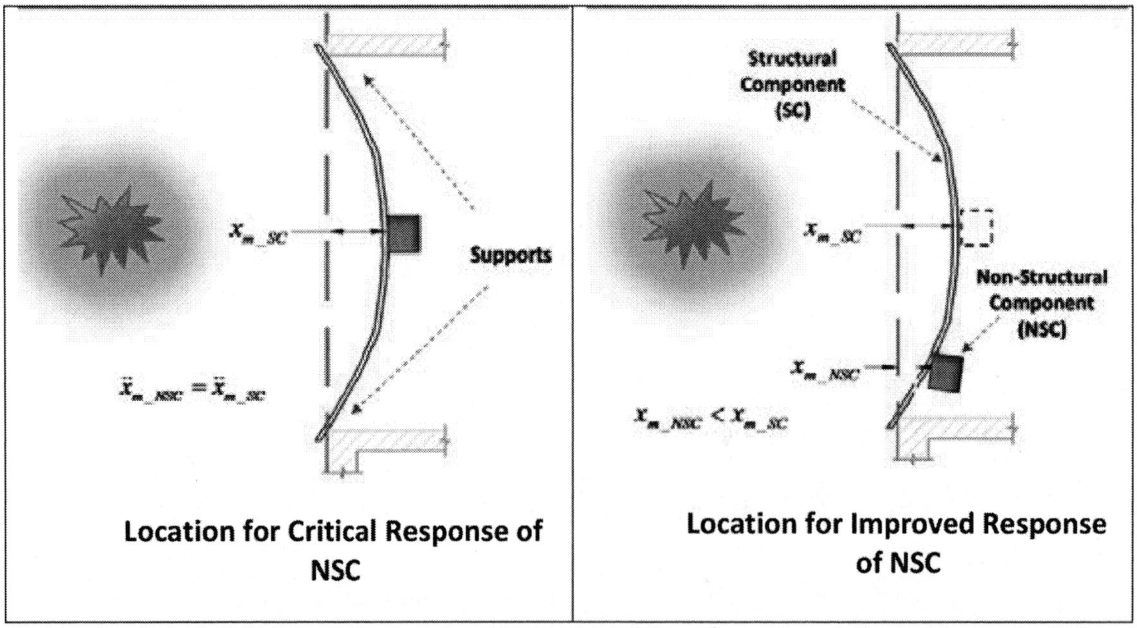

Figure 18. Additional Mitigation Hazard of Non-Structural Component (NSC)

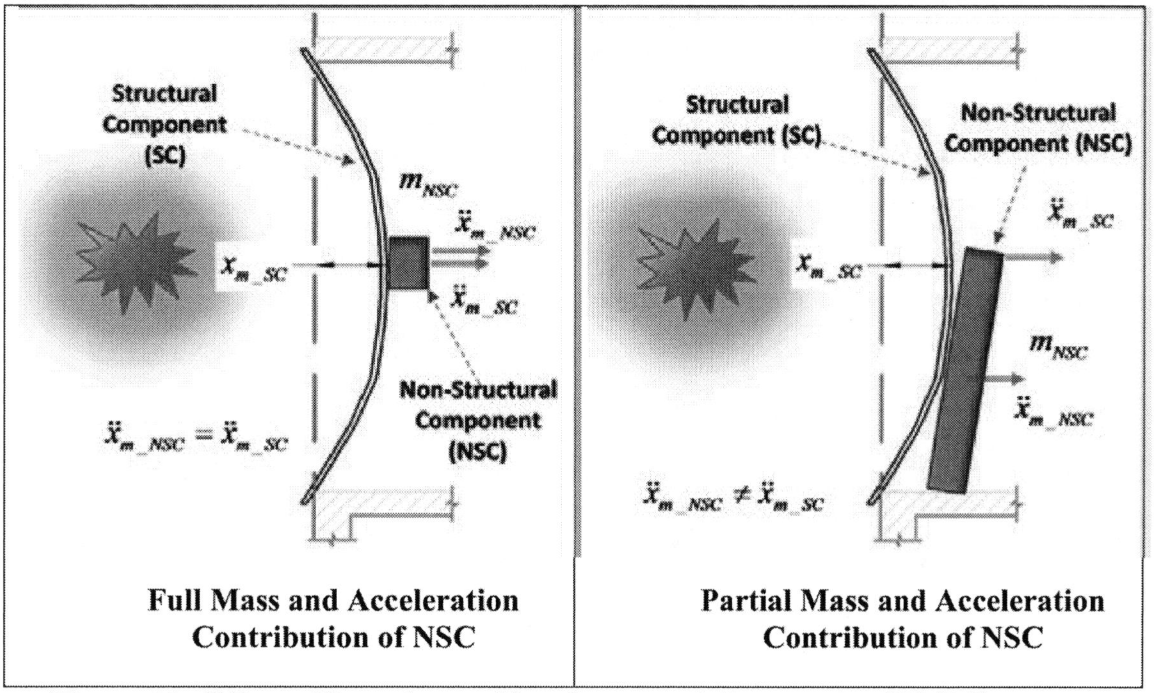

Figure 19. Effect of Location on Non-Structural Component (NSC) Response

5. Conclusions

Evaluation of non-structural building components for the potential to become hazardous debris and mitigation of predicted hazards for buildings subjected to blast load is important to mitigate the hazards to personnel occupying these buildings. A methodology for creating simplified design tool for addressing these hazards was presented and two approaches for retrofitting non-structural components by separating the component from the building structure or by determining the required forces to attach the fixture to the structure were also provided and illustrated.

6. References

[1] Ben Harrison, Sanaa Alaoui, “T2 Laboratories Explosion Damage Assessment”, 34th DDESB Explosives Safety Seminar (2010).

[2] Biggs, J.D., “Introduction to Structural Dynamics,” McGraw-Hill Publishing Company, New York, NY, (1964)

[3] SBEDS, "Single-degree-of-freedom Blast Effects Design Spreadsheets", Ver 5.0, U.S. Army Corps of Engineers Protective Design Center, December 2012

[4] Department of Defense (DoD), Unified Facilities Criteria (UFC) UFC 3-340-02 “Structures to Resist the Effects of Accidental Explosions” (former TM 5-1300), September 2014

[5] ASCE Task Committee on Blast-Resistant Design. "Design of Blast-Resistant Buildings in Petrochemical Facilities" 2nd. Edition (2010)

445437 LOPA Boot Camp – Basic Training with Practical Examples

Abstract:

Layer of Protection Analysis has become a cornerstone risk evaluation tool in the chemical industry since the original CCPS text on the topic was published in 2001. Now, 15 years later, the method is used by many companies as the primary means of determining the adequacy of controls for hazard scenarios as well as the required safety integrity level for recommended safety instrumented systems. This presentation will present the basics of the LOPA methodology including the critical evaluation of the hazard scenario, establishing the initiating event and its frequency, investigation of safeguards and selection of independent protection layers, a brief overview of enabling conditions and conditional modifiers, and comparison of the resulting calculated risk to established targets. Practical examples of the application of LOPA to several scenarios will be used to illustrate the use of the technique.

12TH GLOBAL CONGRESS ON PROCESS SAFETY

Hazards Inherent to Control Systems: Case Studies and Lessons Learned

Brenton L. Cox
Exponent, Inc.
4580 Weaver Parkway, Suite 100, Warrenville, IL 60555
bcox@exponent.com

Stephen W. Garner, Andrew R. Carpenter, and Mark T. Fecke
Exponent, Inc.
4580 Weaver Parkway, Suite 100, Warrenville, IL 60555
sgarner@exponent.com, acarpenter@exponent.com, mfecke@exponent.com

Prepared for Presentation at
American Institute of Chemical Engineers
2016 Spring Meeting
12th Global Congress on Process Safety
Houston, Texas
April 11-23, 2016

Hazards Inherent to Control Systems: Case Studies and Lessons Learned

Brenton L. Cox
Exponent, Inc.
4580 Weaver Parkway, Suite 100, Warrenville, IL 60555
bcox@exponent.com

Stephen W. Garner, Andrew R. Carpenter, and Mark T. Fecke
Exponent, Inc.

Keywords: Process control; Case studies; Lessons learned

Abstract

Control systems are an integral part of almost all chemical processes, regardless of size or complexity. As the demands for performance and safety increase, control systems will only proliferate. With automation comes the ability to develop more complex and interrelated processes, and to respond more rapidly to disturbances. However, increased reliance on control systems, both discrete and continuous, requires increased diligence with respect to their development and testing, as well as the implementation of independent layers of protection.

A control system that provides optimal product for typical feed conditions may behave poorly during abnormal conditions, such as idling or turnarounds. Similarly, independent control systems charged with safety instrumented functions may not be called upon to act until they become necessary to prevent or mitigate an emergency situation. In order to recognize and mitigate hazards associated with control systems, process design and process hazard assessment teams may need to consider these and other potential issues.

Presented here is a brief overview of the hazards associated with control systems. Finally, case studies are used to highlight the need for considering the potential hazards associated with control systems, from the design stage, through startup, operation and turnarounds.

1 Introduction

In 1993, the Center for Chemical Process Safety (CCPS) published *Guidelines for Safe Automation of Chemical Processes*, the seminal text on the subject of process safety with respect to process control systems.[1] However, in the intervening 23 years since its publication, incidents and near-misses related to process control systems continue to occur, and in some cases, a better understanding of the hazards inherent to these control systems by designers, builders, operators and other personnel may have allowed prevention or mitigation of these issues. With the advent of self-driving vehicles and the Internet-of-Things, there can be no doubt that automation and interconnectivity in the chemical process industry will only increase.

Here, we aim to highlight some of the hazards described in reference texts, e.g., those by the CCPS, and through the lessons learned from case studies.

2 Background

In 1993, when the *Guidelines for Safe Automation of Chemical Processes* was published, the microprocessor was about twenty years old, and had already begun to see widespread use within the chemical process industry, through microprocessor-based sensors, operator workstations, and distributed control systems. At the time of its publication, the prevalence and complexity of microprocessors was still in its relative adolescence. For comparison, in 1993, only about 23 percent of American households had computers.[2] In 2016, the average operator has more computational power in their pocket when they arrive at the plant than was available to operate many of the early process control systems.

Similarly, automated process control began with single-loop controllers, primarily subject to single event failures. [3] Today, more advanced control algorithms allow the operator to execute commands directly targeting production specifications, rather than single set-point values such as pressure and flowrate. However, as a result of the increased complexity of the underlying process control system, plants may be more susceptible to equally complex failure modes, including multiple event failures which could include operator error.[4] Nevertheless, the guidance provided within this text remains valid, especially the framework for chemical process safety and the role process control systems and safety instrument systems play within this framework.

Building on *Guidelines of Safe Automation of Chemical Processes*, in 2007, the CCPS published *Guidelines for Safe and Reliable Instrumented Protective Systems*.[5] The latter text guides the reader through the seven core attributes of independent protection layers (IPL) as they apply to the life-cycle of a plant's instrumented protective system (IPS). These attributes are: independence, functionality, integrity, reliability, auditability, access security, and management of change.

Guidelines for Safe and Reliable Instrumented Protective Systems also discusses the application of recognized and generally accepted good engineering practices (RAGAGEPs) that exist within the field of process control and instrumented protective systems. Of particular relevance are the International Electrotechnical Commission's (IEC's) international standard IEC 61511, *Functional Safety: Safety Instrumented Systems for the Process Industry Sector*, and its close counterpart in the United States, ANSI/ISA 84.01, *Application of Safety Instrumented Systems (SIS) for the Process Industry.*

Process control textbooks, e.g., *Process Dynamics and Control* by Seborg, Edgar and Mellichamp, also deal with issues related to process safety.[6, pp. 248-255] However, in the cited textbook, only a cursory discussion of issues is included, and the reader is referred to the other sources for more detail. This background section roughly follows the Seborg, Edgar, and Mellichamp discussion of the components of an automated process control system, with emphasis on some of the specific safety aspects not discussed in the textbook.

2.1 Basic Process Control Systems

The basic process control system (BCPS) is defined as the control equipment installed to support normal production functions. This excludes specialized safety instrumentation and the safety interlock system (SIS). The lifecycle of the BCPS mirrors that of the plant: design, selection, installation, operation, and maintenance. CCPS provides guidance on safety considerations in the selection and design of BCPSs in Chapter 4 of *Guidelines of Safe Automation of Chemical Processes*.[1]

The reliance upon components to fail safely is a common safeguard cited during a process hazard analysis (PHA). However, microprocessor-based process control systems involve complex interactions of software and hardware that may exhibit unanticipated failure modes, and the assurance that all of these failure modes end in a fail-safe state may be challenging. It is important that knowledgeable team members weigh in on the ability of a particular component of the BPCS to fail safely if credit is to be taken during a PHA. For example, some process controllers allow an engineer to specify a fail-safe value for controller outputs.[1, p. 121] In the event of an unexpected controller restart, the outputs will reset to these fail-safe. However, the system should be carefully reviewed for circumstances where this could create an unsafe condition.

An additional consideration is the performance of the BCPS during abnormal conditions. Operators may require additional training or guidance for circumstances such as loss of instrument air, loss of cooling water, a fire, or other out-of-the-ordinary situations.[1, p. 261] Procedures and training may prevent escalation of an incident due to an improper response by the operators.

The BPCS may be used for protective functions, but there are limits to the level of separation that can be achieved between safety and control functions of the same system.[5, p. 275] CCPS guidance suggests that "the risk reduction allocated to a safety function in the BPCS must be less than 10." They also recommend failure modes with a hazard rate greater than 10^{-6}/hr not rely on a BPCS serving both control and protective functions, even if using separate sensors, controllers and elements.

2.2 Sensors and Alarms

Sensors and alarms are typically configured to monitor the process during steady state, and alert an operator when a material deviates from anticipated steady state conditions. However, during shutdowns or idle periods, the sensors and alarms may not function as expected, if at all, and the onset of hazardous conditions may not be recognized by plant personnel. If the sensors and alarms for the system during normal process conditions are to be relied upon to alert operators or maintenance staff of unexpected behavior during a shutdown or turnaround, it is important to consider these non-routine operating conditions.

For example, a recent article in Chemical Engineering Progress discussed an explosion that occurred during a plant shutdown.[6] An EPA investigation into the incident concluded that the incident was at least in part due to a lack of procedures for putting the vessels into a safe state at shutdown, or for monitoring the vessel during shutdown. A key lesson cited by the article was that a hazard assessment of the shutdown step would have revealed that key sensors could not function during shutdown. In order to ensure this is the case, *Guidelines for Safe Automation of Chemical Processes* advises that risk experts consult with process design and instrumentation specialists to "facilitate the understanding of interactions between different portions of the total system and processes."[1, p. 40]

Another concern associated with alarms is the presence of too many alarms with too tight of limits, referred to as "nuisance alarms." In this case, operators may become less responsive to alarms, regardless of their importance.[8] Additionally, during process upsets, many alarms are not critical, and provide the operator with excessive information at a time when quick action is critical.[9] The CCPS guidance suggests considering importance, grouping (priority) and context as criteria in managing alarms.[1, p. 139] It is also important to prove plant operators with sufficient data to take appropriate action when alarms occur, and to log all alarm and event data.

2.3 Safety Interlock Systems

As described by the *Guidelines for Safe Automation of Chemical Processes*, a Safety Interlock System (SIS) "is the backup device that intervenes and takes a system to a safe state when the BPCS fails to do so."[1, p. 140] In order to ensure a failure of the BPCS does not interfere with the SIS, the SIS should be physically distinct from the BPCS, with its own sensors and actuators.[6, p. 252] The alarm panel for the SIS and the BCPS may also need to be physically separated to reduce the possibility that both fail simultaneously.

Components of the SIS are expected to provide accurate process surveillance and timely and effective automated corrective actions. However, many of these systems charged with safety instrumented functions may never be called upon to act until they become necessary to prevent or mitigate an emergency situation. Thus, testing of the SIS is essential to ensure its protective function. The frequency at which SIS components should be tested depends on the safety protection provided, and the safety integrity level availability requirements. *Guidelines for Safe Automation of Chemical Processes* provides detailed guidance on testing procedures and frequency.[1, p. 278-283]

2.4 Interlocks

Interlocks, much like process alarms, should be carefully chosen to not result in unnecessary plant shutdowns and startups.[6, p. 254] The result of an excessive number of interlocks or too sensitive of trip conditions is not only financial. Periods of plant shutdown and startup are often considerably more hazard-prone than steady-state operation. Guidance on the design and selection of safety interlocks is provided in *Guidelines for Safe Automation of Chemical Processes.*

Another aspect to consider is the fail-safe characteristics of the interlock, such as whether the interlock operates by being energized or deenergized. If plant power or other utilities are lost, the safety interlocks should fail to a safe condition.

3 Case Studies

3.1 Furnace Explosion

3.1.1 System Background

The incident occurred at grain and corn processing complex. As part of the corn-to-ethanol process, corn-based material (fiber) was dried in an industrial dryer. The dryer use gas burners which co-currently dried the wet fiber. The wet fiber was fed into the dryer just beyond the combustion zone of the gas fired burners. A process flow diagram (PFD) for the Fiber Dryer #1 process is provided in Figure 1. For reference, the rotating drum component of Fiber Dryer #1 is approximately 12 feet in diameter and 60 feet long.

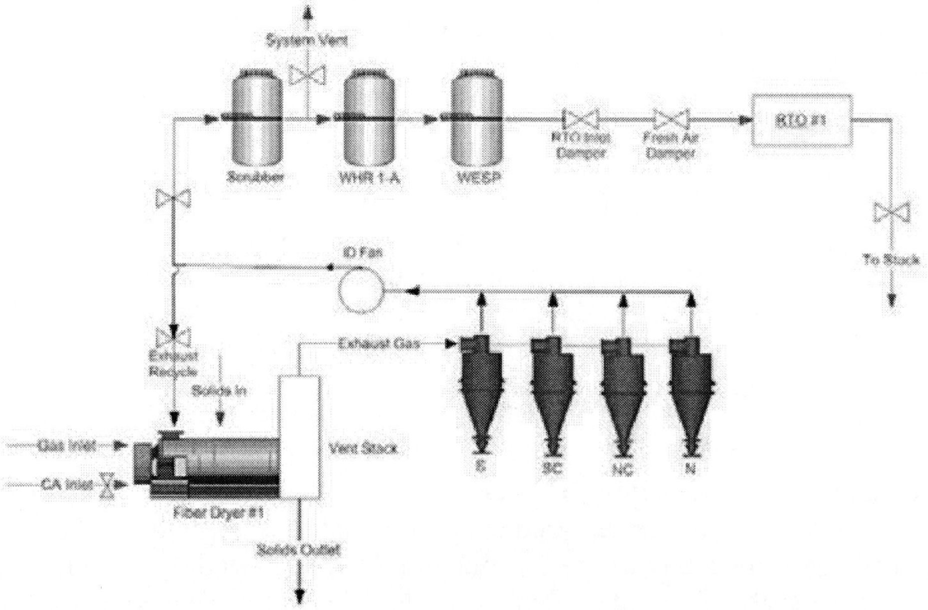

Figure 1. Process flow diagram for the Fiber Dryer #1 process.

The dryer consisted of a burner assembly connected to a large rotating drum which has internal baffles that mix the wet material with the burner exhaust and carried it down the length of the dryer. The dried fiber was discharged from the exit of the dryer via conveyor and then travels to other process equipment within the complex. During normal operation, the exhaust gases from the dryer were fed through several stages of emissions control prior to being vented to the atmosphere. The downstream path from the dryer to the atmosphere included:

1. four cyclone dust collectors (labeled South, South-Center, North-Center, and North),
2. an induction fan (driving force for moving the exhaust gases through the process),
3. a sulfur scrubber (referred to as scrubber),
4. a waste heat recovery unit (WHR 1-A),
5. a wet electro-static precipitator (WESP),
6. a regenerative thermal oxidizer (RTO), and
7. an atmospheric vent stack.

During startups and atypical operations, such as what occurred on the day of the explosion, the exhaust gases may be vented directly to the atmosphere following the scrubber, indicated in Figure 1 as the "System Vent."

Fiber Dryer #1 utilized a 90 million BTU burner that operated using a flammable gas as the fuel source. A portion of the facility complex consists of a biorefinery, which digests organic material and produces a flammable biogas that can be used as a natural gas supplement. This biogas is typically blended with natural gas and fed as the flammable gas fuel to the Fiber Dryer #1 burner.

A simplified process flow diagram for the dryer-only portion of the Fiber Dryer #1 process is shown in Figure 2. The burner portion of the dryer combines combustion air (CA) with flammable gas in a combustion chamber and ignites the flammable mixture. The combustion air and flammable gas flows are modulated via flow control valves (FCVs). The pilot air and pilot gas are only utilized during startup; following successful ignition of the main burners, the pilot lines are deactivated. The main burner consists of two parts: a radial burner assembly and primary burner assembly.

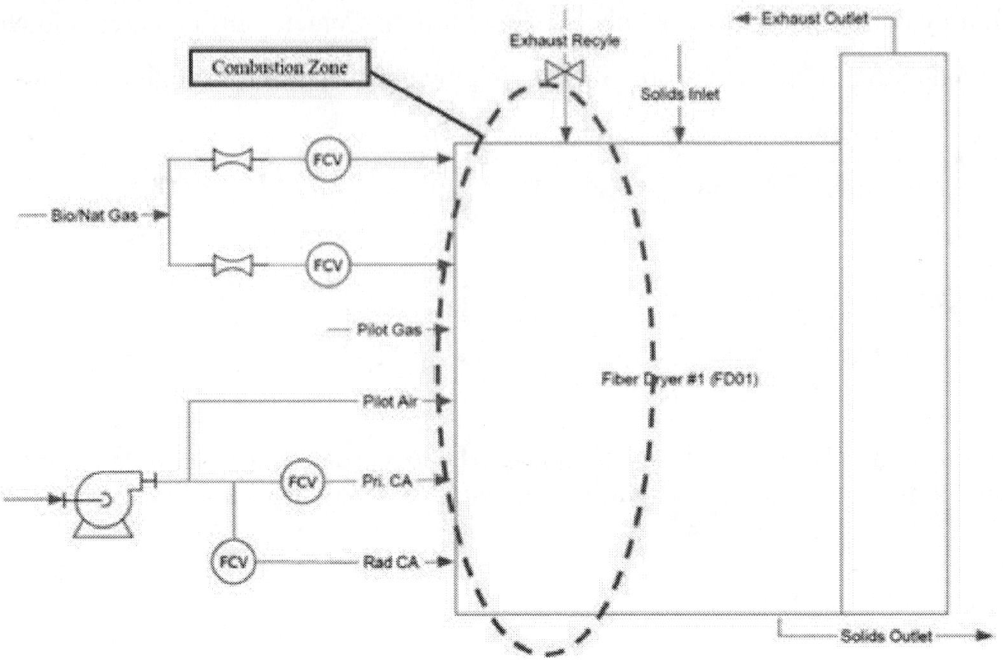

Figure 2. Process flow diagram for Fiber Dryer #1 portion of dryer process.

To control temperatures of the combustion zone of the dryer (indicated by a red circle in Figure 2), a portion of the exhaust gases could be recycled to the burner via louvers at the top of the combustion zone. The Fiber Dryer #1 process was controlled by a distributed control system (DCS). The burner was also controlled by a burner management system (BMS), which provided safety functions and combustion controls and included associated wiring, sensors, controllers, and actuators; the BMS also communicated with the Distributed Control System (DCS). The control system was monitored by plant operators who interfaced with the system from the control room.

3.1.2 Incident Summary

Fiber Dryer #1 was shut down for routine maintenance. The work order describing the maintenance activity for this shutdown was titled "#1 Fiber Dryer Quarterly Inspection." This work order included nine major tasks; one of the major task's description stated "Inspect #1 Fiber Dryer Gas Line Flow Meter and Clean." The primary gas flow rate flow meter was used to measure the gas flow being introduced into Fiber Dryer #1's primary burner. Following completion of the work order, the dryer was restarted.

To restart the Fiber Dryer #1 process, the operators first purged the system with steam. Following the steam purge, the operators initiated a restart on the console and the system performed a pre-programmed restart. The system starts by purging the process using combustion air at 100% flow for 5 minutes. Following the purge, the pilot gas and pilot air are introduced into the combustion chamber and the flammable mixture is ignited. The valves that block the radial and primary gas lines (one set of valves are shared between both) are opened and the radial gas and radial combustion air are introduced into the combustion chamber where it is ignited by the pilot. If flame is confirmed (via a UV flame scanner) within a certain time, the system is considered to have lightoff and the pilot is turned off. If the flame is not confirmed, then the startup process is terminated and must be reinitiated (starting with another steam purge). Immediately after lightoff, the system goes into a standby mode during which the radial flame is at minimum fire position and there is no primary flame. This standby mode will persist until the operator releases the system to "autofire." This is the first operator interaction with the process since initiating a restart. Once in autofire, the primary flame is at minimum fire position and the operator manually then brings the system up to the desired operating temperature. Once the dryer outlet temperature is near the desired condition, the operator then releases the system to modulate, or automatic control. The system remains in automatic until a system trip occurs, an operator switches out of automatic, or a process upset/alarm returns the system to the standby mode (where radial flame is at minimum fire and there is no primary flame).

During the startup process on the day of the incident, three restarts were attempted before lightoff was achieved (referred to as 1st lightoff). Operators and personnel indicated that unsuccessful lightoff during the first several restart attempts is typical for restarts following maintenance. A description of the first three restart attempts follows:

- The unit tripped during the first restart attempt due to a flame safety alarm.
- The unit tripped during the second restart attempt during the purge due to a loss of the purge permissive.
- Successful lightoff was achieved during the third restart attempt.

After radial flame lightoff, the operator initiated lightoff of the primary burner flame. The burner flames were on for approximately 2 minutes before a system alarm brought the burners down to radial burner only standby mode (primary burner off) due to a lid limit switch on the wet fiber feed screw. Approximately 30 seconds later, a BMS permissive caused the burner system to trip due to an interlock.

Following the 1st lightoff and subsequent shutdown, there were three more restart attempts before lightoff was achieved again (referred to as 2nd lightoff). A description of the second set of restart attempts follows:

- The unit tripped during the fourth restart attempt during the purge due to a loss of the run permissive; several other alarms occurred.
- The unit tripped during the fifth restart due to a flame safety alarm.
- Successful lightoff was achieved during the sixth restart attempt.

After the radial flame lit off the second time, the operator initiated lightoff of the primary burner flame. The primary burner was lit and the unit was operating for approximately 5 minutes and 30 seconds before the explosion.

Following a detailed root cause analysis (RCA) it was determined that following the removal and cleaning of the gas flow meter for the burner, the subsequent reinstallation of the gas flow meter was performed incorrectly; specifically, the sensing lines had been reconnected in reverse. As a result of the reversed sensing lines on the primary gas flow rate meter, the primary gas flow rate signal provided to the controller reached its low-saturation setting. When evaluated in the context of the other control signals and system parameters, this indicated an abnormal operating condition, called a process upset. At least three opportunities to identify the existence of a process upset and make timely corrections were possible, including:

1. Detecting a signal under-range (low-saturation) condition on the gas flow rate meter;
2. Detecting a primary gas low flow deviation for the primary gas flow rate control loop (non-responsive control loop);
3. Detecting an air-to-fuel deviation (out-of-range) condition.

The control system was not configured to detect the first of the three conditions, the control system provided no alarms or automatic trips for it, and the operator did not respond to the condition. The control system was configured to detect the last two of these three conditions, and, as the alarm log indicates, did provide alarms for them. However, there were no automatic trips in response to these two alarms, there was no operator response to alarms that were triggered, and there was no operator response to the abnormal operating condition. Neither the operator nor the control system performed any corrective action in response to the abnormal operating conditions (summarized in Table 1). Had the control system or the operator taken timely corrective action, the explosion may have been prevented or its severity reduced.

Table 1. Causal factors 1 through 3: Alarms, Trips, and Operator responses.

Causal Factor	Alarm	Trip	Operator Response
1 - Sensor Under-Range	None	None	None
2 - Non-Responsive Control Loop	Yes	None	None
3 - Air/Fuel Ratio Deviation	Yes	None	None

3.2 Fire at a Specialty Chemical Manufacturer

A specialty chemical manufacturer of a very high purity chemical used in the semi-conductor materials industry suffered a fire which resulted in the total destruction of the facility and an environmental release of chlorine gas. The specialty chemical, which they produce at a purity levels as high as 99.99%, is both pyrophoric and water reactive, and its reaction with the moisture in air can form a solid compound that is both shock and friction sensitive. The chemical is produced and purified in batches, including a multi-day controlled batch distillation step at a very high reflux ratio.

At the time of the fire and explosion, three employees of the facility were in the process of cleaning up a spill of the high purity chemical which had overflowed receiving vessels in the reactor room. Fortunately all three of the employees were able to quickly exit the facility at the onset of the event and were not injured.

3.2.1 System Background

The chemical produced at the facility is first generated in a dedicated reactor vessel and subsequently refined to very high purity via batch distillation with a very high reflux ration. The incident occurred in the distillation room which the chemical was being distilled to a purity level approaching 99.99% and as such, the reactor and other proceeding processes including the first stage of purification will not be discussed further in the paper.

During the batch distillation phase, the crude product is distilled in glass distillation columns operating at a very high reflux ratio. The distillation room included four batch distillation systems which were all completely independent from each other with the exception of sharing a common control system and instrument air supply. Figure 3 below is a schematic of a batch distillation system utilized at the facility:

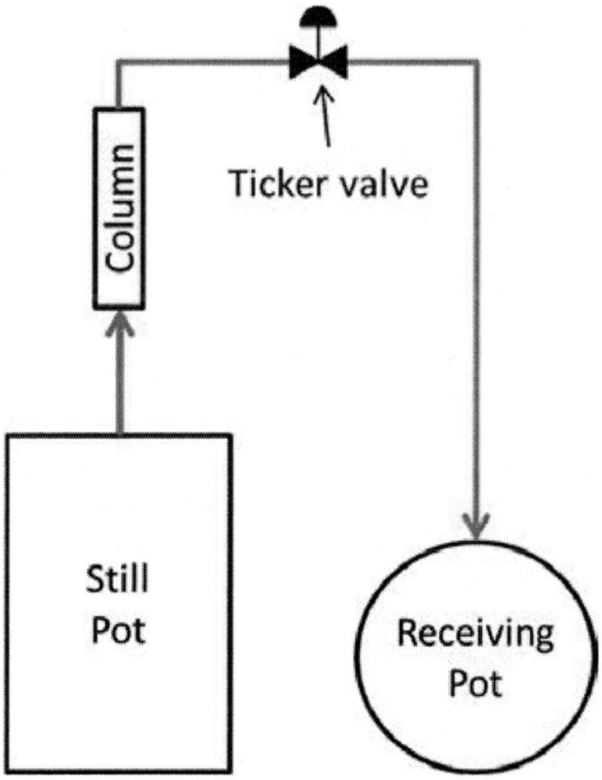

Figure 3. Approximate sketch of batch distillation system.

In simple terms, the crude product is transferred into the still pot and the distillation is accomplished by heating the still pot and pulling off very small amounts of high purity product from the top of the column. The high purity product would be transferred from the column to a receiving vessel by the opening of a solenoid valve. This valve would be opened very briefly allowing the transfer of the distilled chemical. To obtain the desired purity, the reflux ration was set to pull only a few ml of product from the column at a time. The opening of the solenoid valves, as well as the heating of the still pots for all four of the distillation systems was controlled the process control system which had been custom designed and programmed for this application.

3.2.2. Incident Summary

The facility operated the distillation columns in batch mode with each batch requiring several days of processing to complete. The facility was typically only staffed for a single shift during week days, so while the facility was unoccupied, the distillation systems were in operation solely under the control of the control system without any operator oversite. The control system was configured to send alerts and alarms via email and text message to several employees while the facility was unoccupied. During the day of the incident, two of the batch distillation processes were in operation. The incident occurred on the weekend when the facility would not be normally staffed, with the exception of a planned weekend walk through by one of the operators.

The first signs of any problems with the process occurred at approximately 6:00 am on the morning of the incident when the control system sent an email alert that one of the operating still pots had alarmed on a high temperature. The control system was intended to shut down power to the still pot heaters and isolate the distillation system; in essence putting the distillation system into a safe state. It was reported that the emailed alert was first noticed by one of the operators approximately two hours later at approximately 8:00. The responding operator was planning on visiting the facility later in the day for a walk through, but he decided to arrive earlier than he planned to check on the alarm and if possible, return the distillation system to service.

When the responding operator arrived, he observed that there was a white fog (consistent with the presence of spilled high purity product) in the distillation room, and that two of the distillation systems receiving vessels were completely liquid full. One of the overfilled receiving vessels was for the distillation system that was in alarm, while the other overfilled receiving vessel was from the other operating distillation system from which no alarms had been reported. The responding operator called in additional support from the facility to assist in the cleanup and investigation of the alarm. While waiting for additional support, the responding operator noticed that the alarmed distillation system still pot was still in the heating mode, which was a deviation from the intended safe state.

When additional personnel arrived, the responders donned PPE and began the process of cleaning up the spill. During the cleanup activities, a small fire occurred which was attributed to the ignition of an accumulation of dried high purity product from frictional heating while some of the equipment was moved to allow for cleanup of the spill. All three operators safely exited the distillation room. It was quickly determined by the responding operators that this initial, small fire had self-extinguished, and they made the decision to reenter the distillation room and resume cleanup activities, however they made the decision to also stage hand held firefighting equipment in the distillation room as a precaution should another small fire occur. While staging the firefighting equipment inside the reactor room, another fire occurred which very quickly flashed to engulf the entire distillation room. At the time of this ignition, all three responding employees were located just inside the doorway to the distillation room and were able to make an immediate exit from the room. One of the operators was taken by ambulance to a local hospital for medical observation, but was later determined to have not been injured during the incident, and he was subsequently released.

3.2.3 Control system failure

The fire which initiated in the distillation room quickly spread to engulf the rest of the facility, and the responding fire department acted to set up a perimeter around the facility, but due to the concern of the water reactive materials stored inside the facility, allowed the fire to burn itself to competition. As a result, the entire facility was completely destroyed, and no elements of the control system could be recovered for examination. It was determined however that the control system took two unintended actions and failed to take one intended action while responding to the initial high temperature alarm. The control system failed to shut down heat to the still pot as was the intent. The control system also called for the opening of the valve to the receiving vessel (a normally closed valve, and a valve intended to remain closed during the alarm state) as well as call for the opening of the valve to the receiving vessel for the other distillation system in operation which was not reported or found to be in the alarm state. Unfortunately the extent of damage to the facility precluded a thorough investigation into the control system, so the exact nature of the control system failure, including equipment failure or programing failure could not be determined.

4 Conclusions

As the breadth and depth of process control systems increase, the importance of understanding the failure modes and hazards associated with these systems becomes increasingly important. In depth review of guidance from the CCPS and other sources, and sharing lessons learned from process safety incidents and near misses both help foster continued improvement, and help stakeholders grow along with the technology available.

Of particular importance are the instruments and control systems that are relied upon as safeguards. Careful consideration may be necessary to determine circumstances (e.g., specific operating modes, idling, or turnarounds) where instrumentation and controls do not function as intended and thus may not be reliable. Also, for instrumentation or controls not frequently used or monitored during operation, testing may be necessary to demonstrate operability on demand. Lastly, it is important to determine whether instrumented and/or automated safeguards are independent, or may be affected by a common failure mode.

5 References

[1] CCPS. *Guidelines for Safe Automation of Chemical Processes*. Center for Chemical Process Safety, American Institute of Chemical Engineers, New York, NY, 1993.

[2] "Computer Use in the United States." P20-522, United States Census Bureau, October 1997.

[3] "Overcoming the Hurdles in Chemical Process Automation." *Chemical Processing*, November 15, 2002.

[4] R.A. Ogle, D.M Morrison III, A.R. Carpenter. "The relationship between automation complexity and operator error" J. Haz Mat, 159, pp. 135-141, 2008.

[5] CCPS. *Guidelines for Safe and Reliable Instrumented Protective Systems*. Center for Chemical Process Safety, American Institute of Chemical Engineers, New York, NY, 2007.

[6] D.E. Seborg, T.F. Edgar, and D.A. Mellichamp. "Process Dynamics and Control." 2nd Ed., John Wiley & Sons, Hoboken, NJ, 2004.

[7] Ness, A. "Lessons Learned from Recent Process Safety Incidents." *Chemical Engineering Progress*, March 2015, pp. 23-29.

[8] C.S. Connely. "Lack of planning in alarm system configuration is, in essence, planning to fail." *ISA Trans*. Volume 36, Issue 219, 1997.

[9] J.A. Shaw. "Smart Alarm Systems: Where Do We Go From Here?" *InTech*, December 1985.

444116 Impacts of the New National Ozone Standard

Abstract:

Now that EPA has announced a new lower national ozone standard, there will be potential impacts to all sectors of the economy. This includes small industry, large industry, small businesses, large businesses, and potentially every individual consumer in the United States. This presentation will discuss the current state of the ozone air quality and the challenges to meet the new lower ozone standard. Challenges for industry located in ozone non-attainment areas include lower major source thresholds and higher NSR offset ratios for permitting. Challenges for state regulatory agencies include resources needed to address more non-attainment areas and addressing background ozone and international emissions contributions.

Towards a Level Playing Field

Permitting and Regulatory Compliance Around the World

Gordon Combs
Lummus Consultants International
(a CB&I Company)

Environmental Regulations in the USA

- In the beginning ...
  - ✓ 1948 – Federal Water Pollution Control Act ("FWPCA")
    - Established vague and often contradictory framework for national water pollution control policy
    - Established the Water Pollution Control Advisory Board
      - Surgeon General or his designee
      - Department of the Army representative
      - Federal Works Agency representative
      - Department of Agriculture representative
      - Six more Presidential appointees
        - Engineer with expertise in sewage and industrial waste disposal
        - Person with interest in wildlife conservation
        - Representative of municipal government
        - Representative of State government
        - Representative of an affected industry
        - One more person
    - Supported technological research into industrial waste treatment
    - Provided financial support to State and local entities for construction of treatment facilities

LUMMUS CONSULTANTS INTERNATIONAL

Environmental Regulations in the USA

- The Modern Era
  - ✓ 1970 – Environmental Protection Agency ("USEPA")
  - ✓ 1971 – Federal Clean Air Act ("CAA")
  - ✓ 1972 – Federal Clean Water Act ("CWA")
  - ✓ 1974 – Federal Safe Drinking Water Act ("SDWA")
    - Established underground injection control ("UIC") program
    - Does not apply to injection of brine or other fluids associated with the production of oil or natural gas
  - ✓ 1976 – Federal Resource Conservation and Recovery Act ("RCRA")
    - Established definitions of "solid waste" and "hazardous waste"
    - Completed the circle of regulations
      - Take it out of the air and put it into the water
      - Take it out of the water and put it into the ground
      - Take it out of the ground and put it into the air
- Regulatory system based on enforcement ("Command and Control")

A World of **Solutions** April 2016 2

LUMMUS CONSULTANTS INTERNATIONAL

International Regulations

- Europe and other "Developed" Countries
  - ✓ Comparable laws and regulations to the USA
  - ✓ Comparable promulgation schedule to the USA
  - ✓ Comparable enforcement policies to the USA
- Developing Countries
  - ✓ Comparable laws and regulations to the USA (today)
  - ✓ Promulgation schedules
    - Vary widely by country
    - Some nations still developing first regulations in the 1990s
    - Virtually all nations have comparable regulatory system today
  - ✓ Enforcement
    - Varies by country
    - More emphasis on working with industry to reach "win-win" solution

A World of **Solutions** April 2016 3

LUMMUS CONSULTANTS INTERNATIONAL

Project Finance

- Definition
  - ✓ Method of financing in which the lender looks primarily to the revenues generated by a single Project, both as the source of repayment and as security for the exposure, rather than to the parent company; i.e., "non-recourse" financing
  - ✓ Usually for large, complex and expensive Projects such as power plants, petroleum refineries, chemical processing plants, mines, oil and gas production, transportation, telecommunications and environmental treatment facilities
  - ✓ Financing of new Projects and refinancing of existing Projects
  - ✓ Borrower usually a Special Purpose Entity that is not permitted to perform any function other than developing, owning and operating the Project
  - ✓ Loan repayment depends primarily on the Project's cash flow and on the collateral value of the Project's assets
- Reserve-based financing in extractive sectors which is non-recourse and where the proceeds are used to develop one particular reserve (e.g., oil field) is considered to be a Project Finance transaction

A World of Solutions April 2016 4

LUMMUS CONSULTANTS INTERNATIONAL

Financial Institutions

- Multilateral Financial Institutions (Public Sector)
  - ✓ World Bank Group
    - International Bank for Reconstruction and Development ("IBRD") - government Projects
    - International Development Association ("IDA") – government Projects in poor countries
    - International Finance Corporation ("IFC") – loans for private Projects
    - Multilateral Investment Guarantee Agency ("MIGA") – political risk insurance
  - ✓ Regional Institutions
    - InterAmerican Development Bank ("IDB")
    - European Bank for Reconstruction and Development ("EBRD")
    - Asian Development Bank ("ADB")
  - ✓ Export Credit Associations ("ECAs")
    - United States Export-Import Bank ("US Ex-Im")
    - Export Development Canada ("EDC")
    - Japan Bank for International Cooperation ("JBIC")
- Development Banks (Private Sector)
    - Citigroup (USA)
    - Credit Agricole Corporate and Investment Bank (France)
    - Sumitomo Mitsui Banking Corporation (Japan)

A World of Solutions April 2016 5

World Bank Standards

- History
  - ✓ 1987 – Pollution Prevention and Abatement Handbook ("PPAH")
  - ✓ 1990s – Safeguard Policies
  - ✓ 2006 – Performance Standards
  - ✓ 2012 – Performance Standards (current revision)
- Today
  - ✓ Performance Standards – define borrower's responsibilities for managing the Project's environmental and social risks
  - ✓ Environmental, Health and Safety ("EHS") Guidelines – successor to PPAH
  - ✓ Common reference for Projects worldwide
  - ✓ All standards available via the Internet (for free!)
  - ✓ "Does the Project Comply with the World Bank Standards?"

IFC Safeguard Policies – 1990s

OP 4.01 Environmental Assessment

OP 4.04 Natural Habitats

OP 4.09 Pest Management

OP 4.36 Forestry

OP 4.37 Safety of Dams

OP 7.50 International Waterways

OP 4.20 Indigenous Peoples

OD 4.30 Involuntary Resettlement

OPN 11.03 Cultural Property

Child and Forced Labor Policy Statement

LUMMUS CONSULTANTS INTERNATIONAL

IFC Performance Standards - 2006

1. Social and Environmental Assessment and Management Systems
2. Labor and Working Conditions
3. Pollution Prevention and Abatement
4. Community Health, Safety and Security
5. Land Acquisition and Involuntary Resettlement
6. Biodiversity Conservation and Sustainable Natural Resource Management
7. Indigenous Peoples
8. Cultural Heritage

A World of Solutions April 2016 8

LUMMUS CONSULTANTS INTERNATIONAL

IFC Performance Standards – 2012

1. Assessment and Management of Environmental and Social Risks and Impacts
2. Labor and Working Conditions
3. Resource Efficiency and Pollution Prevention (reference the EHS Guidelines)
4. Community Health, Safety and Security
5. Land Acquisition and Involuntary Resettlement
6. Biodiversity Conservation and Sustainable Management of Living Natural Resources
7. Indigenous Peoples
8. Cultural Heritage

A World of Solutions April 2016 9

LUMMUS CONSULTANTS INTERNATIONAL

IFC EHS Guidelines

- Technical reference documents
  - ✓ Performance levels normally acceptable to IFC
  - ✓ Achievable in new facilities at reasonable cost by existing technology
  - ✓ Reference several globally accepted standards (e.g., WHO, USEPA, API, IMO)
- Project required to meet host country or IFC requirements
  - ✓ Whichever is more strict
  - ✓ Site-specific requirements if justified by Environmental Impact Assessment

LUMMUS CONSULTANTS INTERNATIONAL

General EHS Guidelines

- Cross-cutting EHS issues potentially applicable to all industry sectors
- Categories
  - ✓ Environmental
  - ✓ Occupational Health and Safety
  - ✓ Community Health and Safety
  - ✓ Construction and Decommissioning

LUMMUS CONSULTANTS INTERNATIONAL

Sector-Specific EHS Guidelines

- Industry-specific examples of Good International Industry Practice ("GIIP")
- Categories
  - ✓ Agribusiness / Food Production
  - ✓ Chemicals, including
    - Large Volume Petroleum-based Organic Chemicals Manufacturing
    - Natural Gas Processing
    - Petroleum Refining
  - ✓ Forestry
  - ✓ General Manufacturing
  - ✓ Infrastructure, including
    - Crude Oil and Petroleum Product Terminals
    - Ports, Harbors and Terminals
  - ✓ Mining
  - ✓ Oil and Gas
    - LNG liquefaction and regasification
    - Oil and gas exploration, development and production (onshore and offshore)
  - ✓ Power Generation, Transmission and Distribution

LUMMUS CONSULTANTS INTERNATIONAL

Equator Principles

- Purpose
  - ✓ Designed for use by private sector financial institutions – *particularly when one or more multi-lateral institutions is not involved*
  - ✓ Based on policies and guidelines of the World Bank Group
- History
  - ✓ 2003 – 10 banks from 7 countries created the Equator Principles
  - ✓ 2006 – First general revision
  - ✓ 2013 – Current revision (available via the Internet – for free!)
  - ✓ Today
    - 83 banks in 36 countries worldwide
    - 70 percent of international Project Finance debt in emerging markets

LUMMUS CONSULTANTS INTERNATIONAL

Equator Principles Categories and Requirements

1. Review and Categorization
   - ✓ Category A – potentially significant adverse environmental and social impacts
   - ✓ Category B – potentially limited adverse environmental and social impacts
   - ✓ Category C – minimal or no adverse environmental and social impacts
2. Environmental and Social Assessment
   - ✓ ESIA required for Category A and (some) Category B Projects
   - ✓ Alternatives analysis required if GHG emissions exceed 100,000 tonnes CO_2-e per year
3. Applicable Environmental and Social Standards
   - ✓ Designated (developed) countries – host country standards
   - ✓ Non-designated countries – IFC Performance Standards and EHS Guidelines
4. Environmental and Social Management System and Action Plan
5. Stakeholder Engagement
6. Grievance Mechanism
7. Independent Review – pre-financial close
8. Covenants – Borrower agrees to comply (in writing)
9. Independent Monitoring and Reporting – construction and operations
10. Reporting and Transparency

A World of Solutions April 2016 14

LUMMUS CONSULTANTS INTERNATIONAL

Conclusions

- Today, industry cannot "export pollution" by building facilities in developing countries with lax environmental and social requirements
- Some international requirements (e.g., offsite noise control) are stricter for Projects in developing countries than in the USA
- Mitigation of potential social impacts just as important (if not more important) as mitigation of potential environmental impacts
- Financial institutions are a primary driving force behind international environmental and social progress associated with public and private industrial development
- Golden Rule of Business: "Whoever has the gold makes the rules!"

A World of Solutions 15

LUMMUS CONSULTANTS
INTERNATIONAL

CB&I

LUMMUS CONSULTANTS
INTERNATIONAL

A World of **Solutions**

BRACEWELL

AIChE Ethylene Producers' Conference
Ethylene Plant Environmental Panel

Proposed Amendments to EPA's Risk Management Program Regulations

April 12, 2016

Matt Paulson
Bracewell LLP

BRACEWELL

Overview

- Background on EPA's Risk Management Program
- Overview of Existing RMP Regulations
- Review of Proposed Rule
  - Prevention Program Requirements
  - Emergency Response Preparedness Requirements
  - Information Availability Requirements
- Specific Requests for Comment

2

BRACEWELL

Background on Risk Management Program

- 1984, Union Carbide, Bhopal India
- 1985, Waxman (D. CA) introduces first proposed legislation
- 1986 EPCRA
- 1990 CAA Amendments

3

BRACEWELL

Overview of Existing RMP Regulations

- Applicability
  - More than a threshold quantity
  - Of a regulated substance
  - In a process
  - At a stationary source
- Three Main Components
  - Hazard assessment
  - Prevention program
  - Emergency response program

4

BRACEWELL

Overview of Existing RMP Regulations (cont.)

- Three-Tiered Program
  - **Program 1**: No affect to public of worst case release; no accidents with offsite consequence in five years; limited to hazard assessment, coordination, and plan submission
  - **Program 2**: Not eligible for Program 1/not subject to Program 3; streamlined prevention program requirements; additional hazard assessment, management and emergency response requirements
  - **Program 3**: Not eligible for Program 1 and either subject to OSHA PSM standard or listed industry sector; accident prevention requirements (14 PSM elements); Program 2 requirements

5

BRACEWELL

Background on Proposed Rule

- Why is EPA revising the RMP regulations now?
  - 2013 West Fertilizer incident
- The following incidents are also highlighted by EPA in the preamble:
  - BP Texas City
  - Tesoro Anacortes
  - Chevron Richmond
  - Williams Geismar

6

BRACEWELL

Background on Proposed Rule (cont.)

- August 2013 – Executive Order 13650
- December 2013 – OSHA PSM RFI
- May 2014 – Working Group Report
- July 2014 – EPA RMP RFI

7

BRACEWELL

Background on Proposed Rule (cont.)

- July 2014 – EPA RMP RFI (cont.)
  - Areas addressed in RFI but not covered by proposed rule:
    - Regulated Substances List
    - RAGAGEP
    - Safety Critical Equipment
    - Safety Case
    - Facility Siting
    - Automatic Detection and Monitoring
    - Offsite Consequence Data
    - Emergency Shutdown

8

BRACEWELL

Background on Proposed Rule (cont.)

- Dec., 21, 2015 – EPA proposed rule sent to OMB
- February 25, 2016 – Pre-publication version of proposal released
- *Federal Register* publication – March 14, 2016
- Comment deadline – May 13, 2016
  - Unusual, shorter "soft" deadline for information availability-related comments – April 13, 2016
- Final rule issued – Fall 2016?

9

BRACEWELL

Overview of Proposed Rule

- Prevention Program Requirements
  - Incident Investigation
  - Third Party Compliance Audits
  - Safer Technology and Alternatives Analysis
- Emergency Response Preparedness Requirements
  - Coordination with Local Responders
  - Facility Exercises
- Information Availability Requirements
  - LEPC Disclosure Requirements
  - Public Availability of Information
- Specific Requests for Comment

10

BRACEWELL

Prevention Program Requirements

- Incident Investigation
  - Revisions to Definition of "Catastrophic Release"
    - "The owner or operator shall investigate each incident that ~~which~~ (1) resulted in~~, or could reasonably have resulted in~~ a catastrophic release ~~of a regulated substance~~ . . . or (2) Could reasonably have resulted in a catastrophic release (i.e., was a near miss)."
    - "Catastrophic release means a major uncontrolled emission, fire, or explosion, involving one or more regulated substances that ~~presents imminent and substantial endangerment to public health and the environment~~ results in deaths, injuries, or significant damage on-site, or known offsite deaths, injuries, evacuations, sheltering in place, property damage, or environmental damage.

11

BRACEWELL

Prevention Program Requirements (cont.)

- Incident Investigation (cont.)
  - New definition – "root cause"
    - The report shall include . . . "factors that contributed to the incident including the initiating event, direct and indirect contributing factors, and root causes."
    - "Root cause means a fundamental, underlying, system-related reason why an incident occurred that identifies correctable failure(s) in management systems."
  - Use of undefined term – "near miss"
    - Preamble discussion: "any unanticipated release of a regulated substance, and some process upsets could potentially lead to a catastrophic release."

12

BRACEWELL

Prevention Program Requirements (cont.)

- Third Party Compliance Audits
  - When is it required?
    - RMP-reportable accident, or
    - Regulatory authority requires one based on any noncompliance
  - Auditor Qualifications
    - RMP program knowledge, certification, training and PE
    - Source type and process unit experience
    - Experience with applicable RAGAGEP
    - Receive no financial benefit tied to outcome
    - No consulting services for owner/operator in past three years or for three years post-audit

13

BRACEWELL

Prevention Program Requirements (cont.)

- Third Party Compliance Audits (cont.)
  - Audit Report
    - Must provide to EPA
    - Include drafts
    - No legal privileges – becomes publicly accessible
  - Audit Findings
    - 90 days after completion of audit owner/operator must submit response to each finding with a schedule for "promptly" addressing deficiencies
  - Submission to Audit Committee of Board of Directors

14

BRACEWELL

Prevention Program Requirements (cont.)

- Safer Technology and Alternatives Analysis
  - Applies to Program 3 facilities in three industries:
    - paper manufacturing
    - coal and petroleum products manufacturing
    - chemical manufacturing
  - Facilities are required to evaluate the "feasibility" of safer technology and alternatives when conducting PHAs

15

BRACEWELL

Prevention Program Requirements (cont.)

- Safer Technology and Alternatives Analysis (cont.)
  - PHA must include analysis of potential safer technology and alternatives in the following order of preference: IST or ISD, passive measures, active measures, procedural measures
    - IST/ISD: measures that would replace or reduce the use of regulated substances or make operating conditions less hazardous or complex
    - Passive measures: dikes, berms, blast walls
    - Active measures: alarms, safety instrumented systems
    - Procedural measures: policies, operating procedures, training

16

BRACEWELL

Prevention Program Requirements (cont.)

- Safer Technology and Alternatives Analysis (cont.)
  - Definition of "feasibility": "capable of being successfully accomplished within a reasonable time, accounting for economic, environmental, legal, social and technological factors."
  - EPA rejected this same concept in 1996 rulemaking – concluded that the existing PHA requirement was sufficient.
  - What has changed?
    - Safer technologies are "becoming more widely available and being adopted by some companies"

17

BRACEWELL

Prevention Program Requirements (cont.)

- Prevention Program Takeaways:
  - Potential for significant increase in number of formal investigations
  - Mandates identification and correction of management system failure even where no such failures exist
  - Releases with only on-site impacts can trigger formal investigation
  - Third party audits could be required in response to any violation
  - Disclosure of audits could fundamentally undermine audit process
  - Rigid auditor qualifications could exclude many auditors
  - STAA would require burdensome justifications for technology and design choices

18

BRACEWELL

Emergency Response Preparedness

- Coordination with Local Responders
  - Documentation of coordination efforts:
    - Names and contact information of individuals involved
    - Dates of coordination
    - Conclusions
    - Next steps

19

BRACEWELL

Emergency Response Preparedness (cont.)

- Facility Exercises
  - Annual tabletop emergency response exercises
  - Field exercise every five years
  - Annual notification exercises

20

BRACEWELL

Emergency Response Preparedness (cont.)

- Emergency Response Takeaways:
  - Potential to overwhelm local responders
  - Owner/operator would bear enforcement risk – "ultimate burden rests with the owner/operator"
  - Security concerns

21

BRACEWELL

Information Availability Requirements

- LEPC Disclosure Requirements
  - Regulated substances on site
  - Compliance audit summary reports
  - Summaries of incident investigation reports
  - Summary if IST implemented or planned to be implemented
  - Reports on emergency response exercises/drills

22

BRACEWELL

Information Availability Requirements (cont.)

- Public Availability of Information
  - Names of hazardous chemicals
  - MSDSs for all regulated substances at facility
  - Five-year accident history
  - Summary of emergency response program
  - Summary of emergency response exercises
  - LEPC contact information

23

BRACEWELL

Information Availability Requirements (cont.)

- Information Availability Takeaways:
  - Public already has access to much of this information
  - Public release of sensitive security information
  - Unclear how much detail is required in "summaries"

24

BRACEWELL

Specific Requests for Comment

- Increase scope of RCA
- Require incident investigation report completed before restart
- Expand third party audit trigger to include conditions "suggesting a concern"
- Require third party auditor to submit draft report to regulatory authority before provided to owner/operator
- Require implementation of STAA, rather than feasibility analysis
- Create "score card" for third party auditors to grade facilities
- New facility siting requirements (extensive discussion)
- Emergency shutdown (extensive discussion)

25

BRACEWELL

Specific Requests for Comment (cont.)

- Request for Comment Takeaways:
  - Could provide EPA with argument against claim that adding new requirements to final rule not in proposal violates scope of proposal/reasonable outgrowth limitations
  - Facility siting and emergency response most likely to result in new requirements in final rule that were not in proposal

26

BRACEWELL

Matt Paulson
Bracewell LLP

111 Congress Avenue, Suite 2300
Austin, Texas 78701-4061
512.494.3659

711 Louisiana Street, Suite 2300
Houston, Texas 77002-2770
713.221.4959

matthew.paulson@bracewelllaw.com

Texas
New York
Washington, D.C.
Connecticut
Seattle
Dubai
London

bracewelllaw.com

27

Application of Dynamic Simulation in Improving Plant Efficiency

By

Tim Song, Maryn Pols, Prakash Ramdoss

CB&I
2 Riverway, Suite 1300
Houston, TX 77056
Tel: +1 713 375 8711

Prakash.ramdoss@cbi.com

Abstract

In ethylene plants, considerate amount of CO2 and NOx emissions come from gas turbine exhaust gas and furnace flue gas. The hot gas turbine exhaust gas may be reused as oxygen source in the thermal cracking furnaces in place of cold ambient air, thus increasing the overall thermal efficiency, potentially reducing CO2 and NOx emissions from gas turbine combustion chamber and furnace combustion chamber combined. However the integration between gas turbine operation and the furnace operations raises operating flexibility and plant safety concerns, as the combined process needs to be able to operate in both ambient air mode and turbine exhaust gas mode and be able to transit from one mode to another in a safe and robust manner.

This work is part of a revamping project at an ethylene plant, where a new gas turbine is to be installed and integrated with the existing furnaces to improve plant efficiency. The hot gas turbine exhaust gas is routed to the furnaces, instead of the cold ambient air there by reducing fuel need for the furnaces. The work demonstrates that the dynamic process simulation models can be utilized to study the transition between modes of operation, revealing critical process insights. The transient analysis can also be used to determine key operating parameters such as damper opening and closing time parameters to ensure safe and smooth transitions and operations. Consequently, the analysis assures the plant operators with high confidence the robustness of integration between the gas turbine and furnace processes which increases the plant fuel efficiency thereby reducing emissions.

Introduction

The flue gas from the new gas turbine is reused as oxygen source for the thermal cracking furnaces, therefore reducing CO2 and NOx emissions from gas turbine exhaust gas and furnace flue gas combined.

A transient analysis for the flue gas system is required to ensure that sufficient combustion air flow to the cracking heaters is maintained at all times during the system transient conditions to prevent sub-stoichiometric combustion and positive heater firebox pressure leading to personnel safety concerns.

The purpose of the computer modeling is to dynamically represent the combustion air supply system and induced draft flue gas system to allow simulation of the transient response of these systems during upset conditions.

The computer model determines the pressure, temperature, mass flow rate and composition at critical points in the system as a function of time. The interaction of the combustion air, the louvers and dampers and the controllers is very important to an accurate calculation of the dynamic response of the integrated combustion air system. The time delays associated with the transient phenomena in the ducting, the sensor response and the valve motions are modeled to determine if the control systems can respond fast enough to prevent an unstable or dangerous condition. The residual heat in the refractory-lined duct system is taken into account for an accurate representation of the flows. Boundary conditions for the computer analysis are provided by models representing the fans, gas turbine, louvers and dampers, controllers and cracking heaters.

The transients of the upset scenarios described below are analyzed for the new gas turbine duct system.

1. The new gas turbine trip.
2. Trip of trim fan associated with the new gas turbine.
3. Trip of the induced draft fan on a thermal cracking furnace.

The primary purpose of the transient analysis is to ensure that the post-trip interlock logic and control strategy:

- Provides sufficient combustion air flow to the cracking heaters at all times during the system transient conditions avoiding shutting down heaters.
- Prevents sub-stoichiometric combustion.
- Prevents a severe pressure transient which may temporarily lead to positive heater pressures causing concerns regarding personnel safety.
- Avoids potential damage to the combustion air and gas turbine exhaust ducts due to excessive draft pressure.
- Prevents the trip of the GTG as a result of a single heater upset.
- Prevents potentially destabilizing pressure oscillations in the system.

The transients were performed using a computer dynamic simulation model of the new gas turbine combustion air system and induced draft flue gas system which includes all principal components (gas turbine, trim fan, combustion air supply ducting and stack, heater radiant sections, heater convection sections, ID fans and FD fans), control dampers and associated control systems.

The results have been utilized to implement appropriate post-trip interlock logic and control strategies for the revamping project.

System Description

The existing heater island consists of a number of thermal cracking furnaces, with half of them are already integrated with an existing gas turbine, supplemented with ambient air. In order to improve overall plant energy efficiency, a new gas turbine exhaust gas will be reused for the rest of thermal cracking heaters currently using ambient air for combustion.

The existing turbine exhaust gas duct system will be modified and expanded to accommodate the integration of the new machine with the thermal cracking heaters currently operating on 100% ambient air. An overflow connection between the new duct system and the existing duct system will be installed to maximize the turbine exhaust gas utilization in case a heater on the new duct system is shut down. The new duct system will have one operating trim fan without spare.

All thermal cracking furnaces have induced draft fans at the furnace top, while some of them have forced draft fans at air inlet and some do not. Upon a new gas turbine trip, some thermal cracking heaters will trip to forced draft operation while others will trip to natural draft operation. The duct branch to each of these heaters will be equipped with an ambient air intake system.

The ID fans control the pressure at the firebox arch. The furnaces that do not have forced draft fans will each be equipped with a dedicated air intake duct with control damper, which will allow ambient air operation using the ID fan as driving force to pull in combustion air when the gas turbine is down.

The furnaces require additional oxygen for stoichiometric combustion and the trim fan by side of the turbine exhaust duct provides the additional oxygen by mixing ambient air with turbine exhaust gas in the mixer. The turbine exhaust stack provides relief during upset events leading to high pressure in the duct. The main duct is pressure controlled by manipulating trim fan inlet damper, while the trim fan operates at constant speed. All furnaces are lined up along the main exhaust duct. The cross flow duct with a cross damper connects the new duct system to the existing GTG system. Normally the cross flow duct is closed. When one or more furnaces in the new duct system are down, the surplus exhaust gas will be routed to the existing duct for better plant efficiency. The furnaces' FD fans and FD fan dampers during normal turbine exhaust gas operation are closed and the fans are not running. Upon a trip of the new turbine, the furnaces will trip to forced draft operation if there is a forced draft fan associated with it, otherwise natural draft ambient air operation.

Within each heater, the simulation model takes into account the buoyancy effect, burner combustion air side pressure loss, and the convection section hydraulics loss at flue gas side in addition to the induced draft fan and the fan damper characteristics.

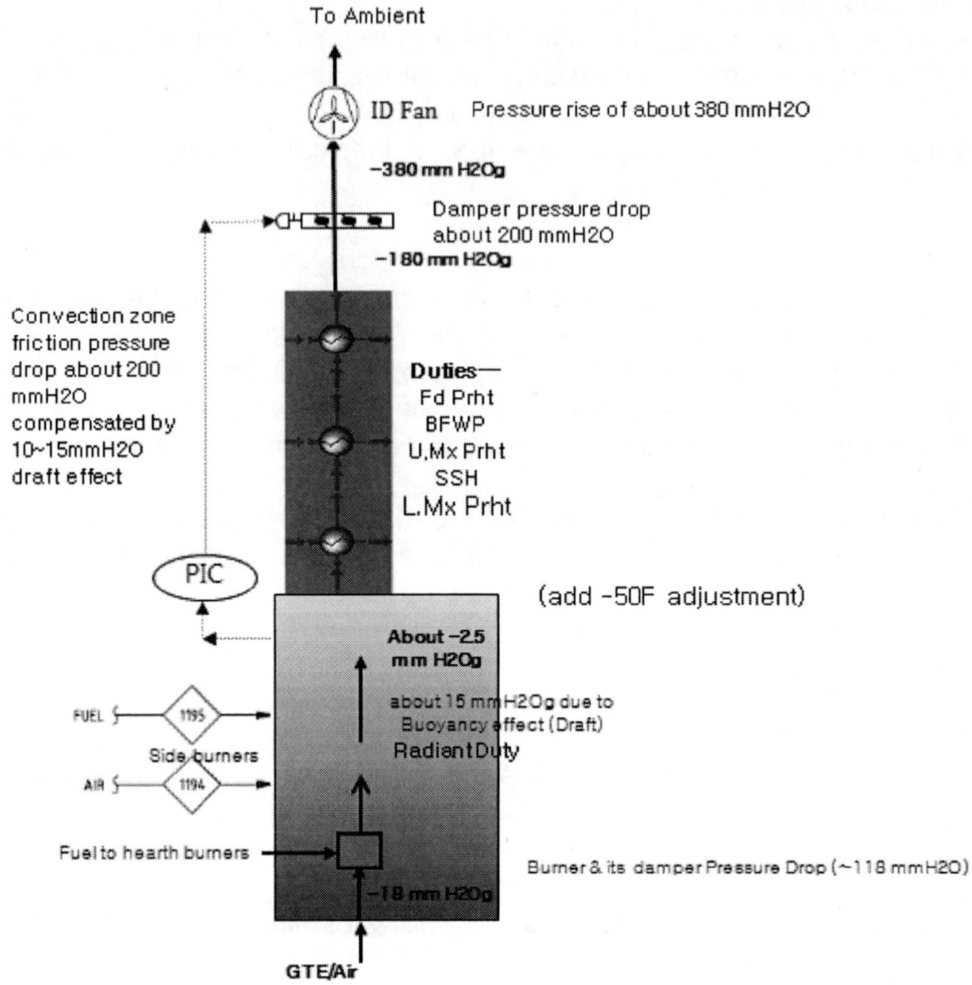

When gas turbine is tripped, both of the flow and temperature decay over time. In order to predict the pressure loss and temperature effect during transition from turbine gas operation to ambient operation and vise versa, duct sizes and duct metal wall thickness, as well as insulation layer are considered in simulation model.

All fans and dampers are modeled with performance curves defined by flow vs pressure head/loss.

Within each furnace, the process side duty demand is correlated to flue gas side by cross limiting control mechanism, such that the coil outlet temperature is properly controlled. The process side feed rate to each furnace is assumed to remain constant during the transients.

On the flue gas side of each furnace, the arch is pressure controlled by the overhead ID fan inlet damper while the ID fans operate at constant speed.

Buoyancy is rigorously considered in the firebox as well as in the flue gas stack according to past engineering practices.

Combustion is taking place in the firebox. The combustion is rigorously calculated thermodynamically, that is, the combustion reaction ensures atomic balance. If there is shortage of oxygen, carbon monoxide will be generated.

The furnace convection section is rigorously modeled at the flue gas side with actual flue gas path volume and measured heat transfer coefficients.

Process side modeling is mostly simplified with duties proportional to the feed rate and inlet temperature. The purpose of modeling feed temperature effect is such that the coil outlet temperature can eventually be adequately predicted and used as input signal to cascade control the firing via cross limiting control mechanism.

Results and Discussion

Gas Turbine Trip

The gas turbine trip is imposed 10 seconds into the simulation. At 10 seconds, the turbine is tripped and the corresponding events are carried out by the simulation model, primarily closing flue gas duct dampers to each furnace while jump open induced draft fan damper opening of each furnace momentarily. The ambient air entry to the furnaces without forced draft fans will be active when duct pressure becomes negative.

During gas turbine trip the concern is the shortage of flue gas/air, therefore oxygen rate, going into each furnace. The simulation predicts the flow rates, as well as oxygen rates, going into a typical furnace are as follows -

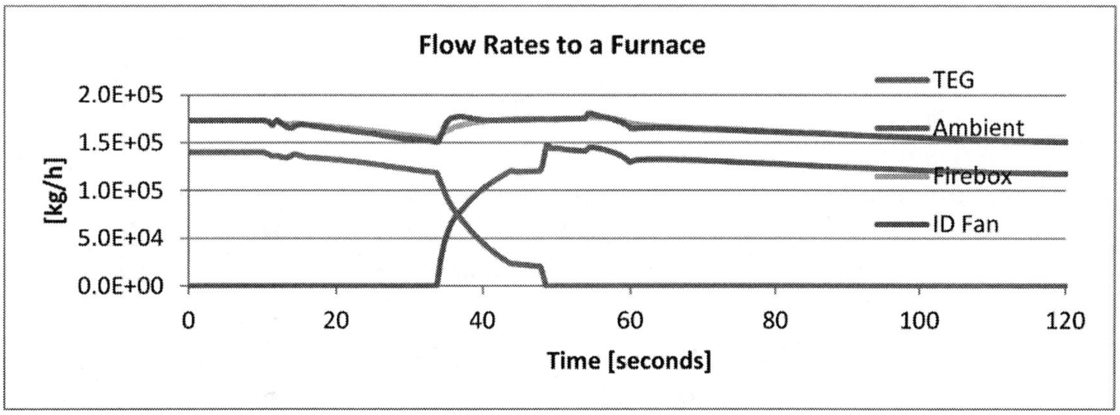

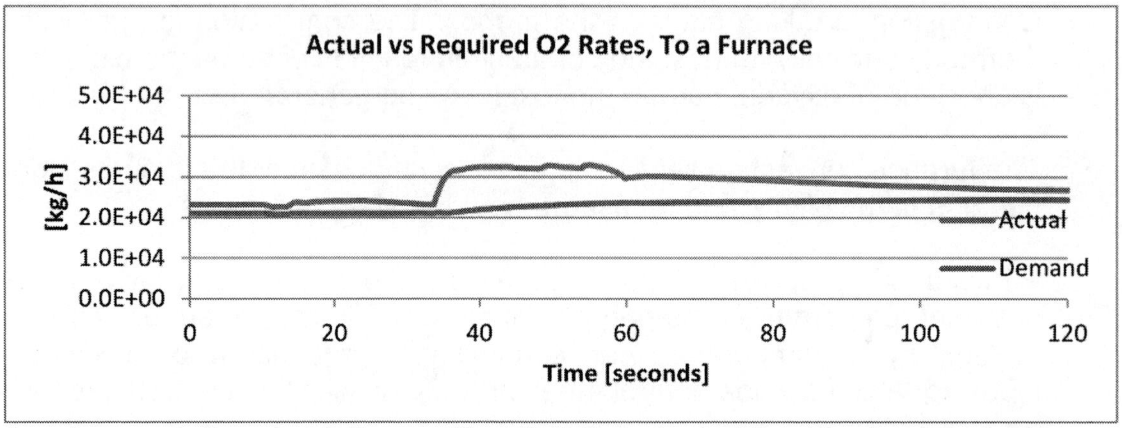

As observed, the simulation verifies that the gas turbine tripping interlocking actions are adequate to ensure sufficient air/oxygen supply to furnaces. As such, sub stoichiometric combustion has been avoided, while production rates remain unchanged.

Trim Fan Trip

The trim fan trip is imposed 10 seconds into the simulation. The sequence of events is implemented according to definition, primarily starting up the forced draft fans for those furnaces with one while leaving the rest of the furnaces intact.

The concern is the potential shortage of oxygen in the furnaces' firebox and positive pressure in the furnaces with forced draft fans.

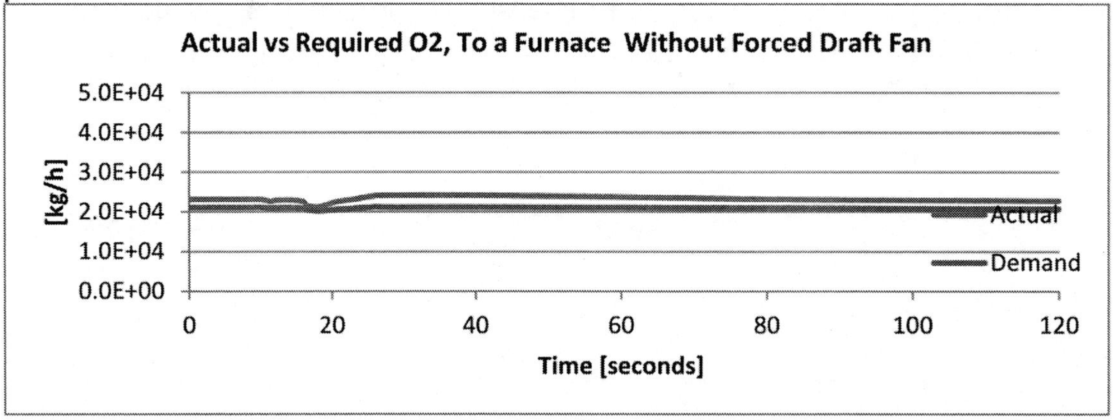

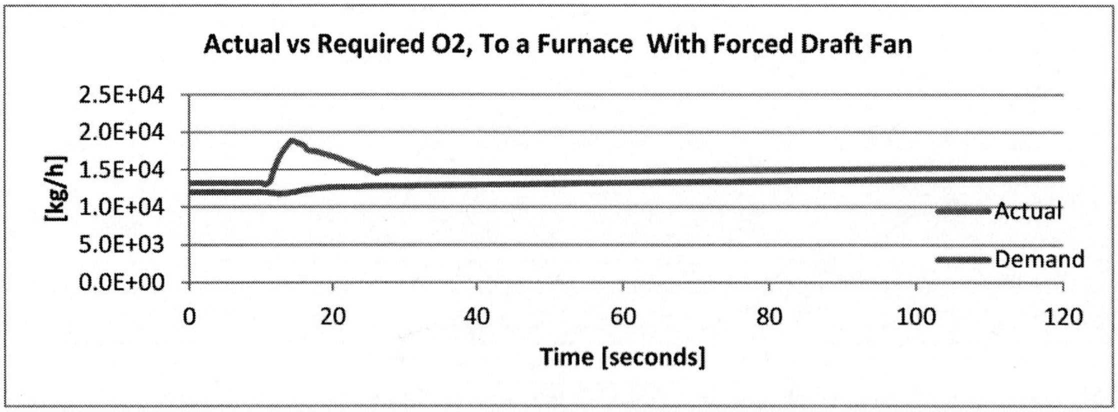

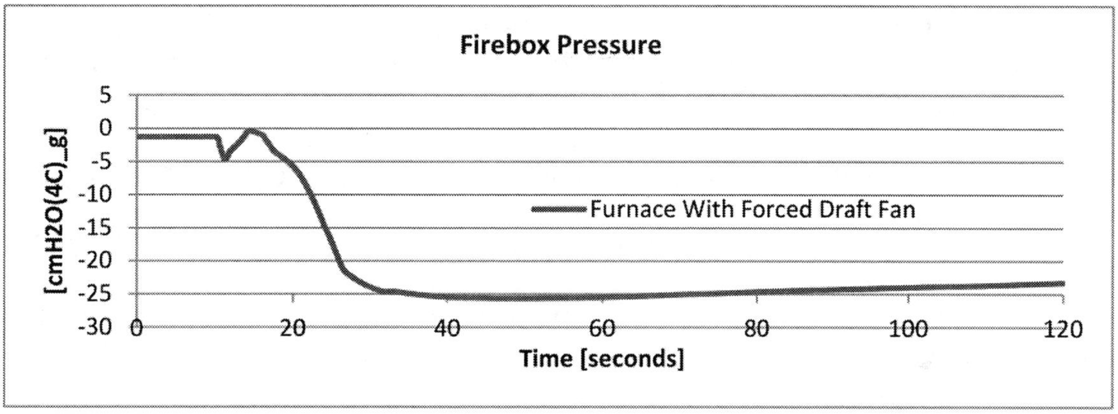

As seen, for a furnace without a forced draft fan, the oxygen supply reaches minimal level 5 seconds into the trip, yet still remains over the oxygen demand at the time. The oxygen supply is ample for furnaces with forced draft fans and the fire box pressure is also below ambient, with the highest pressure at about 5 seconds into the trip.

Induced Fan Trip

The induce fan trip is imposed 10 seconds into the simulation. Some details actions of the trip are listed below:

1. The induced draft fan is tripped and the fan is spinning down in about 3 minutes.
2. Trip close process side feed valve.
3. Trip close all fuel valves.
4. Opening up process side dilution stream valve by 25%.

Since the induced fan takes quite some time to come down, the flue gas damper will stop the flow from furnace inlet side much faster than the fan spinning down process. The net result is that the firebox pressure is predicted to go sub-ambient for as low as -150 mmH2O(g) for the furnace that is tripped, as consequence of its induced fan trip. However the abrupt stop of turbine exhaust gas consumption by the tripped heater does not disrupt the rest of the running heaters. The reduction in flue gas consumption causes the stack damper to open and vent the surplus flue gas to atmosphere. The tripped heater can be shut down in a controlled manner without leading to positive firebox pressure or severe shortage of oxygen in the firebox. Thus, safe operation is anticipated during the induced draft fan trip.

Conclusions

The transient behaviors of the flue gas system bridging turbine exhaust and furnace burners are examined and the results indicate that the system can safely transit from turbine exhaust gas operation to ambient air operation during gas turbine trip, under the condition of a carefully selected series of mitigation measures. That is, both sub-stoichiometric combustion and positive firebox pressure can be avoided.

The results also indicate that the system can safely transit from the induced draft fan trip of a heater, and trim fan trip as long as proper mitigation measures are implemented.

Highlights of the key mitigation measures during gas turbine trip include:

1. For each furnace, switch arch pressure controller and TEG/air intake flow controller to manual mode and open ID fan dampers and TEG dampers on heaters without FD fans to a certain degree while holding TEG dampers opening for heaters with FD fan.

2. When FD fan speed is up, open FD fan damper to a certain degree while closing the corresponding TEG damper, making sure the net flow to the heater remain largely unchanged.

3. When the main duct pressure drops below ambient, open stack damper fully; shutdown trim fan and close its damper; open up ambient air intake dampers to a certain degree. At same time, close TEG dampers such that the net air flow to heaters largely remains constant.

4. Optionally adjust air dampers opening to manage air flows into heaters.

5. Eventually when the process in each heater is sufficiently stable, switch the arch pressure controller back to auto, restore the air damper controls to auto and conclude the transition.

Highlights of trim fan trip are as follows:

This is an event with lesser impacts than the gas turbine trip event. The key measure is to slowly ramp close the TEG damper in about 20 seconds while allowing the combustion air flow controllers to automatically open the ambient damper after a short delay to allow the FD fan discharge pressure to equal with the duct pressure. As a result of operating on FD fans, the surplus TEG will eventually be vented to atmosphere through the stack damper. The firebox pressures can be adequately maintained below ambient while no oxygen shortages are seen.

Highlights of ID fan trip are as follows:

When the fan of a furnace trips, the TEG damper is tripped closed, stopping the combustion air flow into the tripped heater. The remaining running heaters can effectively reject the surplus TEG causing duct pressure to rise to the point that the stack damper pressure controller set point is reached and venting through stack takes place. Positive pressure is not observed in the tripped heater firebox due to residual capacity of the tripped fan during the long period of fan rotor decay time while the faster TEG damper stroke time stops the TEG flow to the heater in just a few seconds, which causes firebox pressure to go down to as low as -150 mmH2O(g).

References

1. Transient Analysis Duty Specification
2. Basis of Design
3. Material Balance
4. Process Description
5. Equipment List
6. Hydraulics Summary
7. Process Control Description
8. PFD and P&ID
9. Interlock Summary
10. Convection section rating tool reports
11. Damper characteristic definition
12. Plot plan
13. Duct Pressure Loss Calculations
14. Reference projects done in the past.
15. Heater burner data sheets
16. Heater data sheets
17. KOM
18. TRIM FAN DAMPER Perf. Curve
19. Pressure Drop Calculation
20. Damper with Actuator Data Sheet
21. Technical Data Table for Process Damper
22. FDF datasheet & curve
23. MOM Study Review Meeting
24. DCS Data for Analysis
25. Actual ID fan damper opening

AIChE Paper 86A

DESIGN, OPERATION AND STARTUP STRATEGY OF A NEW PARALLEL C2 SPLITTER

André Bernard
Process Design Engineer
NOVA Chemicals (Canada) Ltd.

Prepared for Presentation at the 2016 Spring National Meeting
Houston, TX, April 10th – April 14th, 2016

AIChE Paper

DESIGN, OPERATION AND STARTUP STRATEGY OF A NEW PARALLEL C2 SPLITTER

André Bernard
Process Design Engineer
NOVA Chemicals (Canada) Ltd.

Abstract

Feed transition to larger volumes of ethane is known to increase the operating loads in a C2 Splitter that was originally designed for a plant that processes a mix of NGL and Naphtha feedstocks. Revamp options such as changing to high performance trays may sometimes overcome the shortfall in capacity that can be expected due to the additional ethane feed. However, in some cases upsizing the tower or installing a parallel unit become the only viable options. This paper discusses the implications of processing larger volumes of ethane feed on a C2 Splitter along with the revamp options that could be considered. Operation and startup strategies are also discussed.

Introduction

NOVA Chemicals Corunna Site was originally started up in 1977. The olefins unit was designed and operated as a naphtha cracker. Over the years, through revamp and expansion projects; feed flexibility was introduced such that the plant could process a mix of liquid and natural gas liquid (NGL) feedstocks. The last plant expansion was implemented in 2005 and brought the ethylene production to 180% of the original nameplate capacity[1,2]. At that point the plant could process a feed slate with an NGL contribution varying between 30 to 65% of the total feed. The NGL feedstocks were predominantly propane and butane with a limited ethane contribution varying between 5 to 15% of the total feed.

Market conditions for olefins feedstocks around 2010 resulted in a shift by many plants towards lighter feeds. Availability of ethane from shale gas propelled ethane as the feedstock of choice for ethylene plants. This shift in feedstocks market conditions drove a project for converting the Corunna cracker to 100% NGL feed, primarily ethane.

C2 Splitter System

The Corunna site olefins unit process scheme is a front end Demethanizer with back end Acetylene reactors. The feed to the C2 Splitter system is a mixture of mostly ethane/ ethylene coming from the overhead of the Deethanizer after conversion of acetylene in catalytic reactors. The C2 Splitter system prior to the feed conversion project is illustrated on Figure 1. It is a high pressure system, which uses -40 °C propylene refrigerant (C3R) for condensing overhead vapors. Heat to the bottom reboiler is provided by condensing C3R vapor. A side reboiler provides heat integration with charge gas entering the chilling train. Ethylene product is withdrawn at tray 10 and ethane is recovered at the bottom of the tower and recycle to the cracking furnaces. The top 10 trays are used as a pasteurization section, preventing accumulation of methane and hydrogen coming from the Acetylene reactors. A small rectifier tower and a vent condenser using ethylene refrigerant (C2R) allows minimization of the vent stream, which has a relatively high ethylene content. The vent stream is recycled to the charge gas compressor (CGC).

The tower was originally equipped with 122 conventional two pass trays with V1 valves. The tower was first re-trayed in the late eighties with 153 high capacity trays. A second re-tray was done in 2005 with a different type of high capacity trays[3]. The tray count in the 2005 revamp was reduced to 123.

A dilute ethylene product draw was configured in the tower in the mid-nineties in conjunction with a Styrene plant revamp as a mean of increasing the tower capacity.

The side reboiler tube bundle was upgraded with enhanced surface in the late eighties and the original 3 shell and tube condensers were replaced by a single core in kettle exchanger in the 2005 revamp.

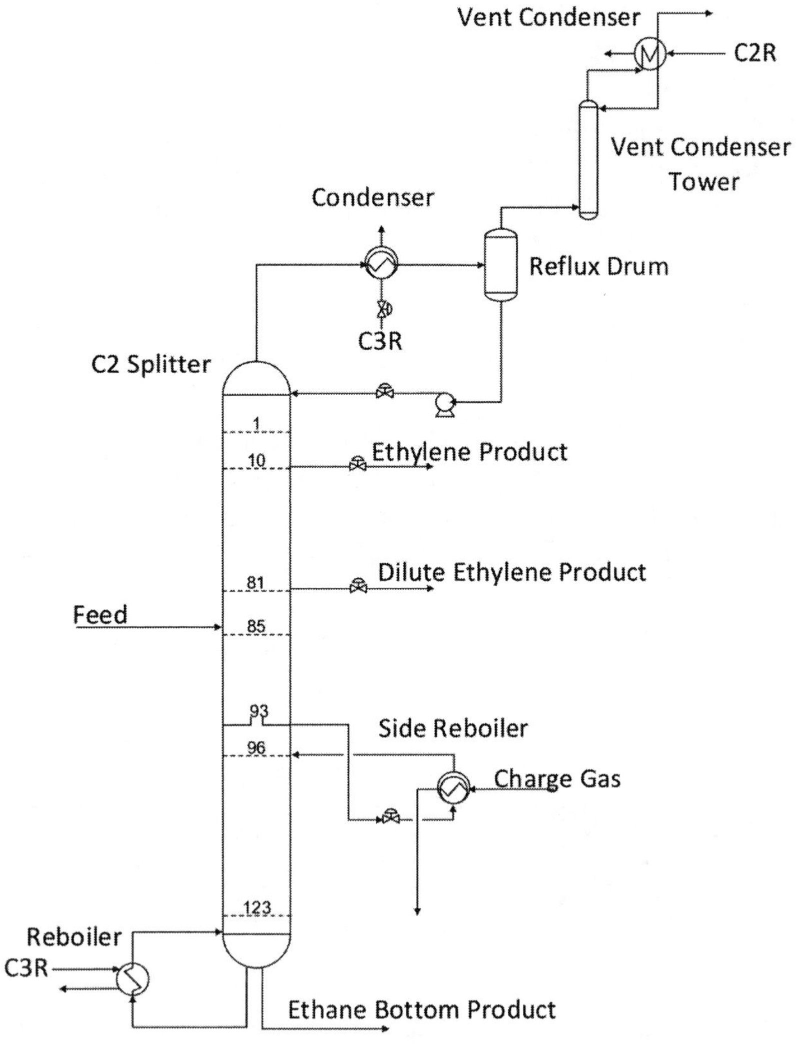

Figure 1: C2 Splitter System

Revamp Project

High crude oil price and availability of ethane from the Marcellus shale in 2010 led to a project for converting the Corunna Ethylene unit to 100% NGL feed. The design feed slate, based on ethane/ butane/ propane; substantially increased the volume of ethane cracking.

The timing of the project was a challenge as the study phase was initiated at the beginning of 2010 with the next plant turnaround schedule for the fall of 2011. Given the short timeline, priorities were given to the identification of tie in's required for executing the project outside the plant turnaround. The 2011 turnaround window was therefore used strictly for providing tie in's required by the project. Identification and design of the tie in's was done as the project was in the basic engineering phase. Detail engineering was started a few months following the turnaround and completed a year later.

Understanding that higher ethane feed would impact the capacity of the C2 Splitter along with a manufacturing desire to eliminate production of dilute ethylene made the existing tower evaluation a high priority. Determination of the tower capacity relative to future operating conditions and modifications required for meeting the project objectives were critical for establishing the list of required tie in's.

Impacts of Feedstocks on C2 Splitter Operation

The plant historical ethane/ ethylene tower feed composition relative to feedstock transition is provided on Table 1. The composition for 100% ethane is also provided as it represent the bookend of the feed spectrum. As shown on Table 1, the ethane fraction in the feed increases as more NGL especially ethane is being cracked.

Table 1: Ethane/ Ethylene Feed Composition Relative to Feedstock

Feedstock	Ethane (mole %)	Ethylene (mole %)
Naphtha	15	84
Mix Naphtha/HAGO/NGL	15 - 17	81 - 85
100% NGL Project (Design)	32	67
100% Ethane Cracking	38	62

Figure 2 shows that although the number of theoretical stages required is not significantly different, the reflux ratio increases by 30 to 40% when comparing ethane to naphtha cracking. It is also observed that slim improvement in reflux ratio are obtained beyond 120 theoretical stages.

Figure 3 shows the implication of the ethane feed composition on the reflux ratio for a tower with 120 theoretical stages and feed at its optimum location. This plot suggest that although the plant had transition from the original 100% naphtha cracking operation to a mix of NGL and liquid feed (Naphtha and HAGO); the ethane feed composition had only increased slightly resulting in a 2.2% increase in required reflux ratio. The ethane feed composition expected by the 100% NGL feed project would result in a reflux ratio increase of 23.6% due to significant increase in ethane cracking. An all-ethane cracking operation would cause the reflux ratio to be 36.6% over what was required on naphtha. For similar ethylene production the internal traffic in the tower would therefore increase with the implementation of the 100% NGL feed project.

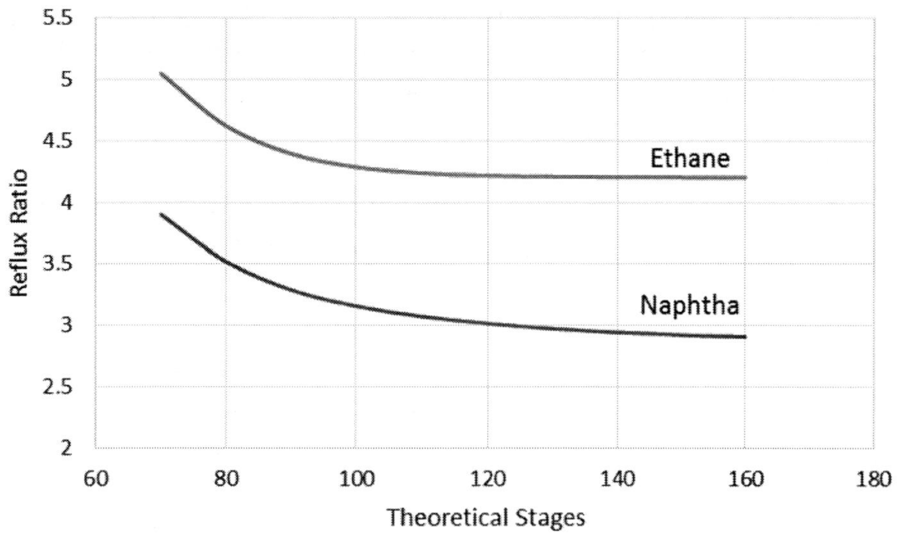

Figure 2: Impact of Theoretical Stages on Reflux Ratio

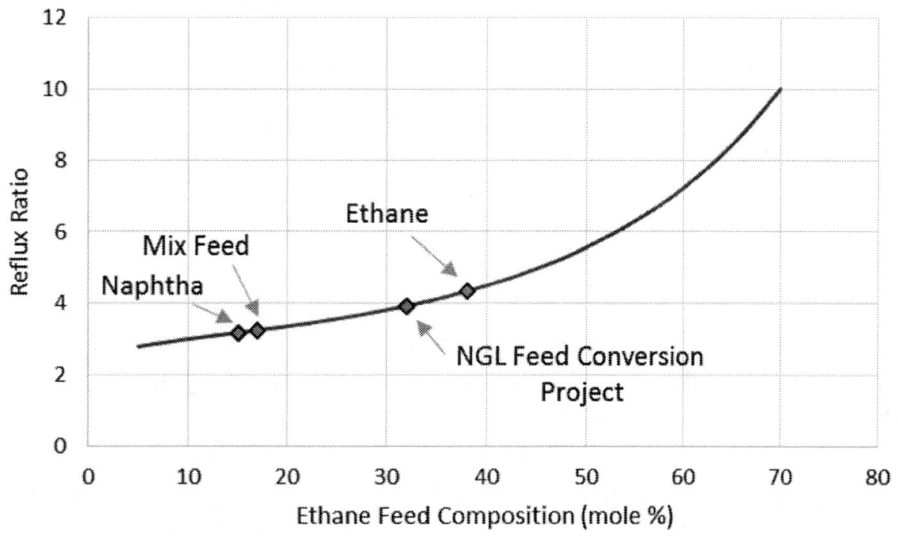

Figure 3: Impact of Ethane Feed Composition on Reflux Ratio

Another interesting observation is the migration of the optimum feed location as the ethane feed composition increases. For the naphtha and mix feed, the optimum feed stage location is located at stage 70 for a tower having 120 stages. The optimum location migrates to stage 65 on a feed equivalent to the 100% NGL feed project and to stage 62 on an all ethane feed. The migration of the optimum feed point is explained by the difference in composition profile, which also causes a change in temperature profile as illustrated on Figure 4a and 4b. For a conventional tower operating with a bottom reboiler and overhead condenser, the change in temperature profile has little significance. However, with side reboiler operation this change may have significant implications as discussed later.

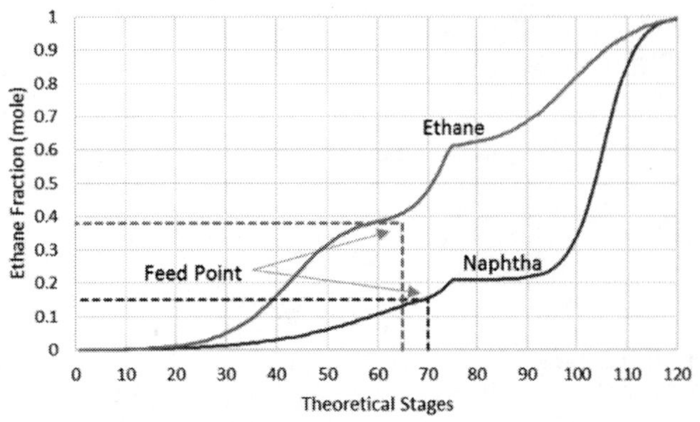

Figure 4a: Ethane Composition Profile

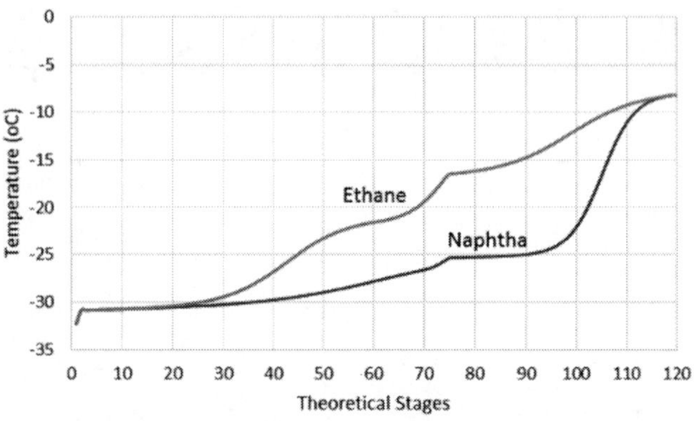

Figure 4b: Temperature Profile

Existing C2 Splitter Evaluation

A performance test was conducted on the C2 Splitter following the 2005 re-tray. The tower was false loaded up to the design load for the re-tray and the overall tray efficiency evaluated at 78%[3]. The tower simulation was adjusted with plant data and provided a relatively accurate prediction of its operation. Using the simulation and tray performance diagram provided by the vendor, the maximum tower capacity without dilute ethylene was evaluated at 110% of the plant required capacity based on the feed slate processed after the 2005 revamp. The maximum capacity was define as the rate where flooding would first occur as predicted by the tray performance diagram.

In the initial stage of the 100% NGL feed project, the tower capacity was re-evaluated based on the new expected feed slate. A first evaluation reveals that the tower would reach the flood limits at 92% of the capacity required by the project; however, to reach this capacity, the heat duty required would exceed the capability of the existing side reboiler by 165%. The warmer temperature profile of the C2 Splitter caused by the higher ethane feed composition as described earlier, would cause the charge gas side of the side reboiler to pinch at a warmer temperature; hence providing less heat duty. Because of the new feed slate, a drop in charge gas flow across the side reboiler is also expected, which would further reduce the duty available for the side reboiler.

Proper balance between the side and bottom reboiler is critical in the tower due to the narrow tray operating window in that section. Tray operating windows for the rectifying section and the tray section located between the side and bottom reboiler are illustrated on Figure 5. Because of the short tray spacing between the side and bottom reboiler (12 in), the trays in that section are limited by downcomer backup. The larger tray spacing in the rectifying section (20 in) causes the jet flood and downcomer backup limits to be more or less equivalent around the expected operating line (left side of Figure 5). The tray section between the side and bottom reboiler (right side of Figure 5) offers a relatively narrow operating range between weeping and downcomer backup, which explain the criticality for proper balance between the side and bottom reboiler. Insufficient side reboiler causes higher duty on the bottom reboiler and would cause the trays to operate above the downcomer backup limit. On the other hand, excessive side reboiler causes lower bottom reboiler duty, which causes the tray to operate below the weep point. Higher ethylene slippage in the tower bottom is typically observed when operating below the weep point.

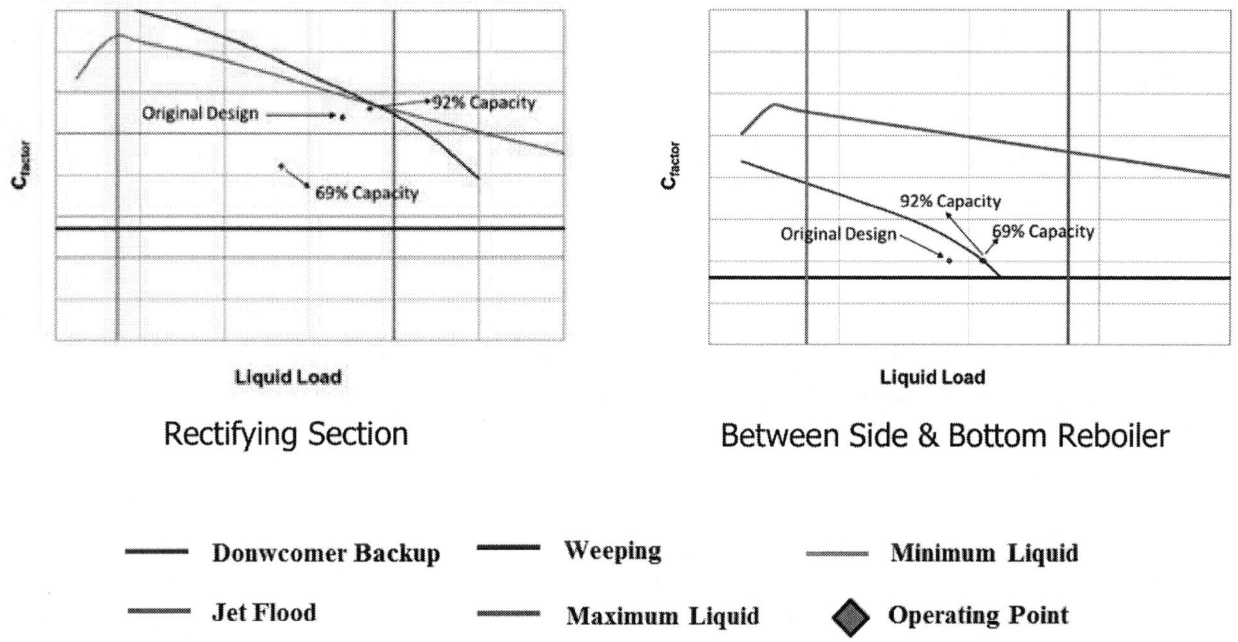

Figure 5: Operating Points vs Tray Operating Limits

Operating the side reboiler with the heat available from the charge gas stream would reduce the capacity of the tower to 69% of the required project capacity. As illustrated on Figure 5, the section between the side and bottom reboiler would operate at the downcomer backup limit, while the trays in the rectifying section would still have significant capacity available. The original design and 92% capacity points show a tower that is more balanced from a hydraulic standpoint when compared to the 69% capacity point.

C2 Splitter Revamp Options

The existing C2 Splitter evaluation concluded that the exiting tower in its current configuration would not support the capacity expected by the project. The following options were considered for meeting the required project capacity.

- Parallel side reboiler with higher cracking severity.
- Tower re-tray
- Parallel side stripper
- Parallel tower

Parallel side reboiler with higher cracking severity

As shown above, without any constraint on the side reboiler the tower capacity increases from 69 to 92% of the project required capacity. Increasing the severity of ethane cracking would reduce the ethane content in the feed thus reducing the tower reflux ratio. A 7% gain in tower capacity was predicted when increasing the ethane conversion by 5% in the furnaces. The combination of a parallel side reboiler using C3R vapor for heating medium along with higher cracking severity on furnaces processing ethane could potentially bring the tower capacity within the target required by the project. Although requiring relatively small capital investments, this option would offer slim capacity upside if any, making the tower sensitive to any transient conditions. Furthermore, the furnaces cycle time would be impacted causing more frequent operation transitions for decoking[4].

Tower Re-Tray

The tower is already equipped with high capacity trays believed to provide the maximum capacity. Change in tray design may provide some upside, for example modifying the trays between the side and bottom reboiler may move the downcomer backup limit, however, any of those changes may provide slim increase in capacity, which would fall well below the capacity required by the project.

Consideration to tray technology operating beyond the system limit by using centrifugal forces[5] or de-entrainment devices[6] has been evaluated. A hybrid design using a combination of high capacity trays along with tray design for operation above the system limits was found capable of increasing the tower capacity to 120% of the project required capacity.

Limited experience with the tray technology brings an element of novelty and risk assessed unacceptable to the project. A tower re-tray would also have to be implemented at a much later plant turnaround which does not align with the project schedule.

Parallel Side Stripper

Building on the idea of a parallel side reboiler, a parallel side stripper may offload the tower and provides the capacity required by the project. The concept as illustrated on Figure 6, involves condensing the existing vapor feed and re-routing it to a new parallel side stripper. The existing side reboiler is maximized with additional material withdrawn to feed to the new side stripper. The overhead vapor from the side stripper feeds the C2 splitter at the current feed point.

This option would increase the capacity to 103% of the project requirements, which is a slim increase in capacity involving significant capital investment. Furthermore, significant energy penalties would be expected due to increases in condenser and reboiler heat duties.

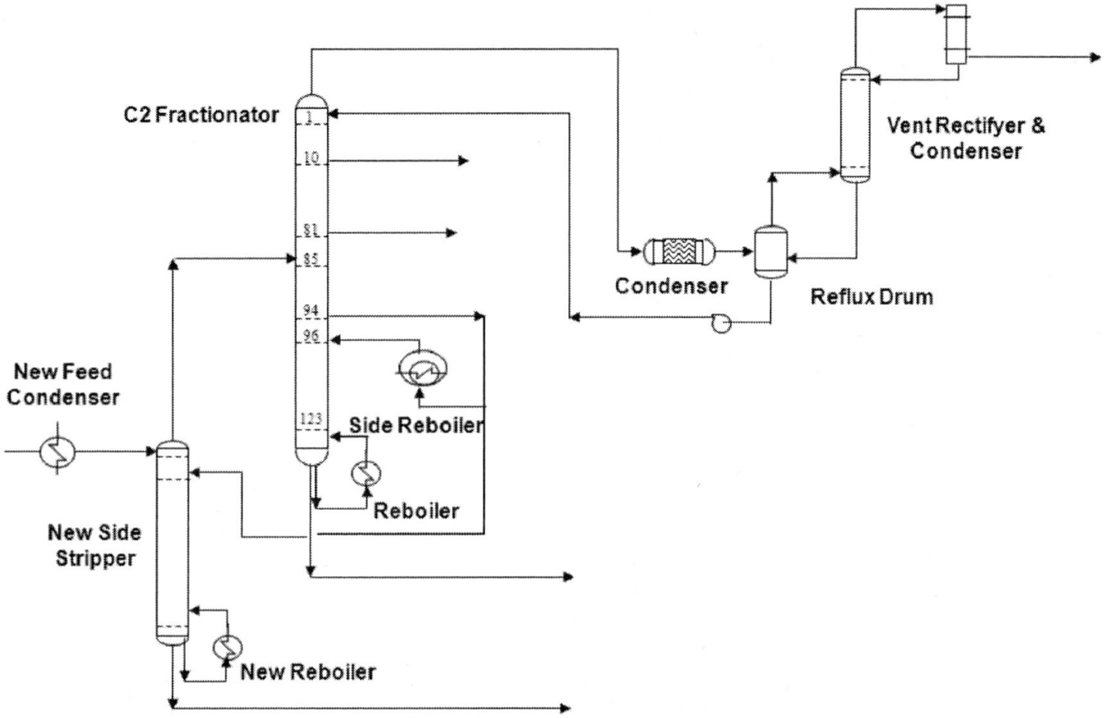

Figure 6: Parallel Side Stripper

Parallel C2 Splitter

A full parallel C2 Splitter system would ensure meeting the production capacity required by the project. With proper tie in's, it can be build and commissioned outside a plant turnaround. Although the most expensive, it was the selected option as it aligns with the project constraints and objectives.

Parallel C2 Splitter Design Considerations

For a given operating pressure, the height of a tower is driven by the number of trays required to achieve the desired fractionation; which is independent of capacity. Capacity will dictate the diameter required for the tower. As described above, the existing C2 Splitter system has a 31% shortfall in capacity in its current configuration. With a parallel side reboiler this gap can be reduced to 8%. Given the magnitude of the investment, designing strictly to meet the project capacity would not make much sense from a long term perspective. Alignment with business plans regarding potential future expansions and feedstocks is imperative for setting up a design strategy for the new tower.

A low pressure C2 Splitter system with a heat pump[7] would reduce the height of the tower while increasing its diameter. However, this option was ruled out as it is more applicable to front end acetylene reactors. Furthermore, it would involve a significant increase in capital given the associated compressor unit and potential operability issues with back end acetylene reactors. The new C2 Splitter was therefore chosen to be a high pressure system similar to the existing tower.

The following criteria were considered for scaling the size of the new tower:

- Future expansion plans.
- Future feedstock strategies.
- Current and future operability.
- Energy efficiency.

Under sizing the tower would limit the ability for expanding the plant capacity while oversizing it would affect energy efficiency. For setting up the size of the new tower a roadmap to expansion was set based on business strategies and other plant constraints. To meet the project capacity, the existing tower would operate at capacity without a parallel side reboiler (69% of the project capacity). The new tower would first be configured with conventional trays. A road map to expansion would be staged by adding a parallel side reboiler on the existing tower followed by a re-tray of the new tower with high capacity trays. Based on those considerations, the new tower diameter was set identical to the existing one. To maximize capacity upside, the initial step was to design conventional trays meeting a turndown rate equivalent to the project capacity (31% of the total required capacity). Further reduction in capacity, if required could be achieved by turning down the old tower.

The capacity of both towers was evaluated for the configuration considered by the project as well as for potential expansion configurations. A capacity range was established as a function of potential feed slate and is summarized on Table 2 relative to the project required capacity.

Further increase in capacity can potentially be accomplished by considering tray technologies that operates beyond the system limit. Although un-acceptable from a schedule and risk standpoint for this project, experience and available data in the future may make this option viable once completing the capacity roadmap described on Table 2.

Table 2: C2 Splitter System Capacity Roadmap

	Old Tower (% of project capacity)	New Tower (% of project capacity)	Total (% project capacity)
Project configuration	60 to 70	70 to 75	130 to 145
Parallel side reboiler on old tower	80 to 92	70 to 75	145 to 165
Parallel side reboiler on old tower & new tower re-tray	80 to 92	85 to 95	165 to 190

Operation Strategy

As illustrated on Figure 7, the effluent from the acetylene reactors splits between the two C2 Splitters. One tower is fed at a fixed rate through a flow controller. Feed to the other tower is regulated by a pressure/ flow cascade control loop with the Deethanizer pressure being the control variable and flow to the C2 Splitter the manipulated variable. The controls are configured such that operation could select which tower is base loaded on flow control with the other tower taking feed variations through the pressure/ flow cascade control loop.

The new tower vent is routed to the existing vent condensing system of the old tower to minimize recycle to the CGC. Provisions were made to vent the new tower directly to the CGC in case of pressure instability. The existing ethylene product drum is shared by both towers.

Reflux, ethylene product draw and reboil rates are control by an advanced control application (APC) as per the existing tower, with the exception that the new tower does not have a side reboiler. The APC application uses analyzers and flows to regulate its control variables for meeting product specifications and keeping the tower in material balance.

For plant startup, a single tower operation will be sustained until feed rates allow a smooth transition to a two tower operation. In a similar fashion for plant shut down, a transition from two to single tower operation will first occur as plant rates are reduced.

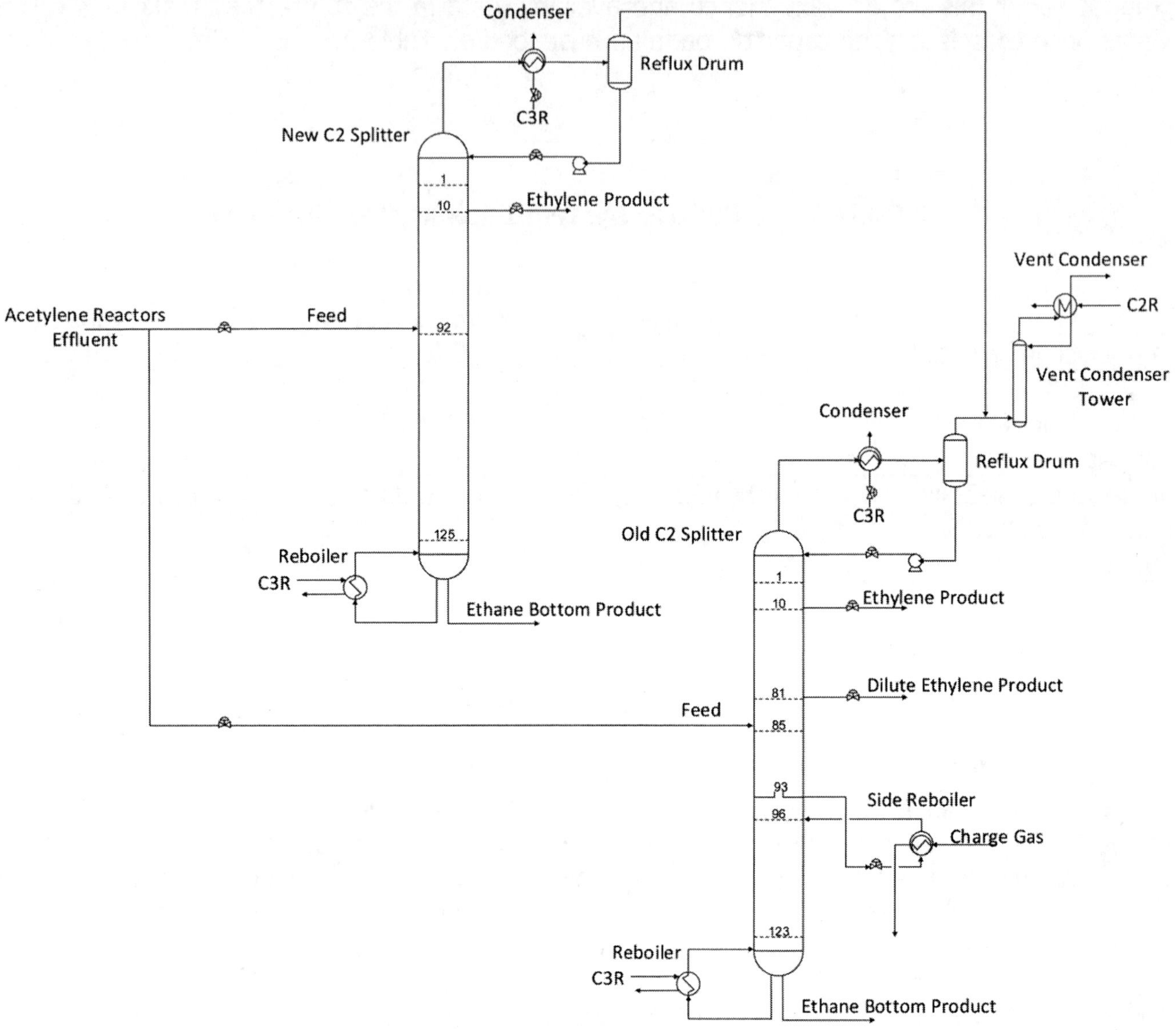

Figure 7: Parallel C2 Splitter System

Fabrication

Process engineering involvement during the release and approval of mechanical drawings of the tower and associated hardware might diminish, however participation remains crucial as other disciplines may not fully understand the implications of certain components in achieving efficient fractionation. Fabrication of a new tower involves different parties such as vessel fabricator, internals (trays, distributors), reboiler and condenser suppliers. Furthermore it is not unusual to see different people involved as work progresses from conceptual engineering to fabrication. Alignment amongst those different players must be scrutinized as drawings are being issued and work progressed. Items such as location of feed nozzles, thermowells, pressure taps, tray rings and manways are to be checked on the vessel drawings. For instance thermowells should be positioned in location exhibiting significant temperature changes and preferably located in the liquid phase near the bottom of downcomers[8]. Tray rings should be consistent with the specified tray spacing. Feed nozzles should be at the expected location with orientation that fits with distributor design. Tray drawings, reboiler and condenser are to be consistent with their associated process datasheets. Any differences should be assessed through revised process datasheets and implications on capacity understood.

It is a good practice to visit the fabrication shop and inspect one tray from each different design. The tray or part of it depending on size, should be laid out on the floor. Downcomer width, flow path length and weir heights are typical measurements that should be done at different places along the length of the tray. On valve trays, a valve count and random measurements of valve lift should be performed. For sieve trays, although tedious, a hole count should be done, which allows verification of the tray hole area. Ensuring that dimensions are within tolerance and consistent with the mechanical drawings will avoid speculation on tray geometry defects in the event of problems once the tower is up and running. When feasible, it is also a good practice to test liquid distributors with water.

Figure 8: Reflux Distributor Water Test

By following the above practices, two design changes were introduced following the inspection at the fabrication shop. Installation of a false downcomer around the reflux distributor was recommended by the design engineer after noticing sideways movement from the liquid jet leaving the nozzles while testing the distributor with water (Figure 8). The false downcomer was to prevent projection of the liquid on the tray deck rather than its intended inlet sump. Projection of the liquid on the tray deck could potentially cause excessive froth on the top tray.

The feed distributor, illustrated on Figure 9; is an H pipe where vapor is projected inwards and outwards from the two laterals. During drawing reviews, it had been pointed out by process engineering that the feed was slightly superheated and that it should not project vapor outwards to the downcomer plates. According to simulation, the vapor feed enters the tower 10 °C above the liquid flowing down the downcomer. During the inspection at the fabrication shop, it was found that holes had been provided on both side of the laterals as shown on Figure 9. Given the risk that the hot vapor may flash the liquid and choke the downcomer, impingement plates were added to the design.

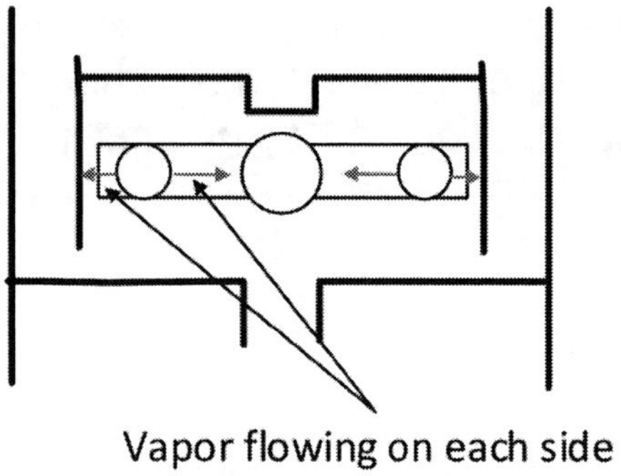

Figure 9: Vapor Feed Distributor

The trays and associated internals were installed at the vessel fabrication shop with the tower on the horizontal position. The vessel fabrication shop was only a few miles from the plant site. However as illustrated on Figure 10 and 11, moving and erecting the tower was a challenging and spectacular event.

Figure 10: New C2 Splitter Transportation from the Fabrication Shop

Figure 11: New C2 splitter Lift

Commissioning

As stated earlier, the existing tower would meet 69% of the project capacity with the design feed slate. This implies that commissioning of the new parallel tower had to occur prior to full feed transition to minimize production impacts. However, a too-soon transition would also involve potential turndown issues.

Commissioning of the new tower was planned with the plant cracking 68% of the expected ethane feed volume. The cracking furnaces feed was broken down into 40% ethane, 10% propane, 20% butane and 30% liquids (naphtha and HAGO). While still cracking liquids, this feed slate lead to a C2 Splitter feed composition relatively close to the one expected by the project but at rates reduced to 78% of target. Dilute ethylene was still produced on the existing tower, providing a 6% boost in capacity. Part of the ethane feed system is integrated with ethylene refrigeration and had not been commissioned at that time. Once fully commissioned, vaporization of the ethane feed would be accomplished by condensing ethylene vapor from the ethylene refrigeration compressor (ERC) discharge. This integration reduces power on the propylene refrigeration compressor (PRC) as condensation of ethylene from the ERC discharge is otherwise accomplished by propylene refrigerant. For commissioning the tower, the ethane feed was vaporized by low pressure steam, compounding the impact of higher PRC power on the steam system.

Until commissioning of the new tower, operation of the existing tower was under the responsibility of the operation group and no significant issues had been reported. Process design got involved in setting up a transition strategy for commissioning the new tower. The plan involved inventorying the tower with ethylene followed by a period of total reflux operation. Beside usual checks for leaks, the total reflux operation also proved the functionality of the overhead condenser and reboiler. It however increased the power required to run the PRC, hence causing a higher steam demand. For feeding the tower, target operating conditions were calculated from simulations with expected outcomes on the towers, the PRC and steam system. The plan was to ramp up feed rates on the new tower to the plan targets while reducing feed on the existing tower. Given a reduced rate operation at 78% of the project target, conditions were calculated to prevent or minimize false load on both towers. Turndown was limited by weeping in the bottom section of each tower. The target operating conditions and the transition plan with expected impacts was presented to operation prior to proceed with the commissioning activities.

As feed was introduced to the new tower and slowly ramped up, operation reported that the steam demand had increased beyond the comfort level of operators and that the propylene refrigeration compressor was operating near its maximum power. Operating conditions were gathered and compared against targets. As shown on Table 3, the feed to the new tower was only at 35% of target and 110% on the old tower. Feed rates to the cracking heaters had also been reduced such that the total feed to both towers was at 88% of target.

Review of the operating conditions listed on Table 3 revealed that the side reboiler duty on the old tower was quite high and causing unnecessary false load. Reducing the side reboiler duty to its target would be associated with a reduction in reflux, hence condenser duty. The PRC power would consequently decrease, which would also help the steam system. Operation was then instructed to slowly ramp down the tower draw rate to the side reboiler. Operation reported early in this transition that the front end of the chilling train was warming up and making the ERC unstable. As discussed in a paper by NOVA Chemicals[9], problems with the C2R system were known prior to commissioning the new tower and were caused by a fouled sub-cooler, which was loading up the ERC 3rd stage; up to that point, those problems should have been independent from the tower operation. Figure 10 illustrates the path of charge gas from the dryers to the front end of the chill train. Charge gas enters the side reboiler at about 15 °C and is first cooled down in the C2 Splitter side reboiler. Heat integration between charge gas and the C2 Splitter reduces the load on the propylene chiller, which uses 1st level propylene refrigeration. Normally reduction of heat duty on the side reboiler should be picked up by the propylene chiller without any consequences on the C2R system. The propylene chiller is a core in kettle exchanger that had been in service for almost 10 years and had always provided relatively tight temperature approach. Simulation work prior to commissioning the tower suggested heat duties within the capability of the exchanger. However, as load was reduced on the side reboiler, the charge gas temperature exiting the propylene chiller warmed up. The additional heat duty was then shifted to the 1st C2R core exchanger in the cold box and contributed to increase the load on the ERC 3rd stage.

Table 3: C2 Splitter Transition Operating Conditions

Old Tower	**% of Recommended Target**
Feed	110
Dilute ethylene draw	58
Reflux	134
Side reboiler draw	267
C3R to reboiler	101
New Tower	**% of Recommended Target**
Feed	35
Reflux	92
C3R to reboiler	90
C2 Splitter System	**% of Recommended Target**
Feed	88

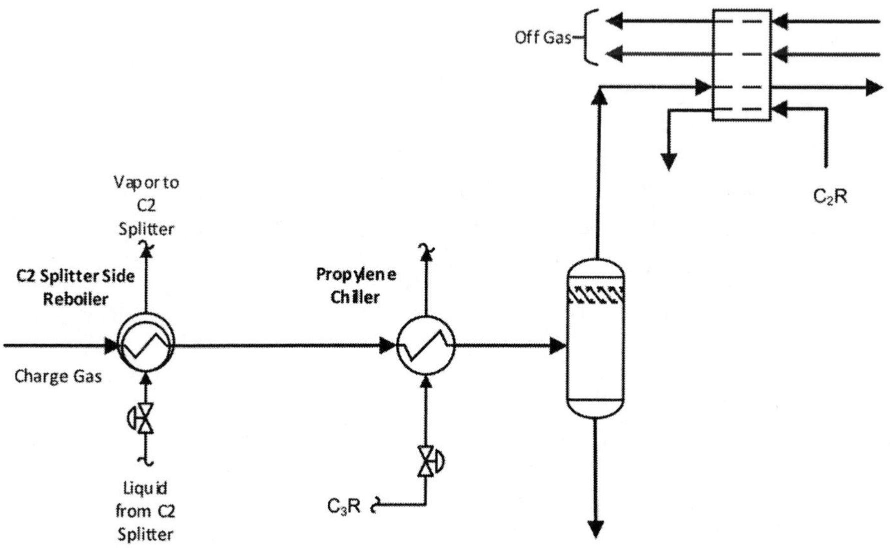

Figure 10: Side Reboiler & Charge Gas Chilling

Accepting that the propylene chiller would not provide the heat duty required, the side reboiler was set back to conditions allowing a stable operation of the ERC. An attempt was made at reducing the load on the bottom reboiler. However, as shown on Figure 11 the operating point on trays located between the bottom and side reboiler were already at the weep point. As a result, ethylene slippage in the bottom of the tower started rising. The bottom reboil rate was therefore brought back to its previous rate.

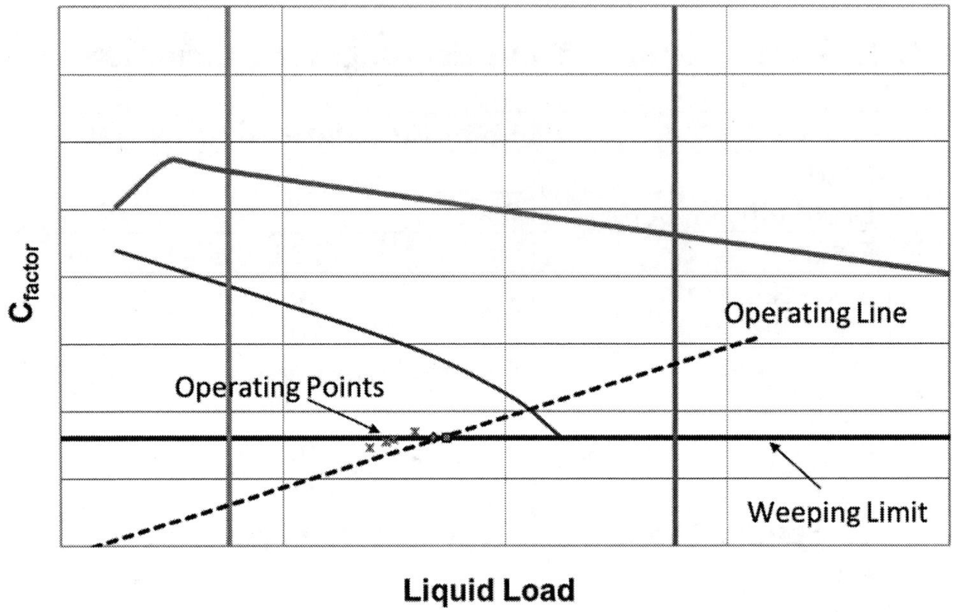

Figure 11: Old Tower Bottom Trays Operating Window

While going over this sequence of events, it became odd that the side reboiler could provide so much heat duty. As previously discussed, limitations in side reboiler duty were expected when cracking larger volume of ethane due to a shift in temperature profile. However, as shown on Figure 12; false loading the tower moved the temperature profile down allowing the tower draw to enter the side reboiler almost 3 °C colder than expected without false load.

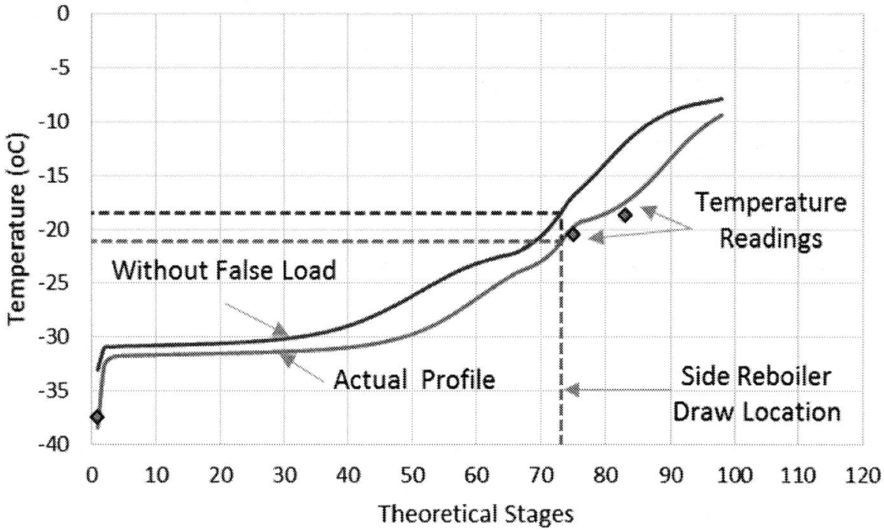

Figure 12: Old Tower Temperature Profile

Recognizing the implication of maximizing the side reboiler heat duty relative to charge gas cooling and stable ERC operation left few options in terms of reducing load on the old tower. Tray weeping limitations in the bottom and the need to satisfy a cooler temperature profile imposed constraints beyond the need of just fractionating ethylene. At that point, it did not make much sense to progress the feed transition to plan targets until gaining a better understanding on constraints encountered. The option of bringing the new tower down was contemplated. However, both towers were fractionating effectively and although high, compressor power and steam loads had dropped to levels acceptable by operation.

While feeding only 35% of the plan target rate, the new tower was stable and provided effective fractionation. The tower was false loaded with the reflux and reboil rates at 90% of targets. As shown on Figure 13, the trays were operated with a good margin from the weeping limit. Venting of non-condensables in the old tower venting system was done smoothly and appeared to be working effectively.

Until resolution of the chill train/ ERC constraints, operation of the towers were left at conditions similar to those reported on Table 3. Increase in plant feed rates would be managed by directing additional feed to the new tower. The old tower would still be operated with conditions ensuring adequate heat exchange with the chill train via the side reboiler.

Operating data on the propylene chiller was gathered for a period covering the previous 600 days of operation. The log mean temperature difference (LMTD) and pressure drop was plotted over time. The results revealed that until about 8 months prior to commissioning the new tower, the propylene chiller LMTD was operating at or below the design value. The pressure drop of the exchanger looked low and relatively steady for the same period. Then a slight increase on the LMTD was observed 6 to 8 months prior to the commissioning activities without any significant increase in pressure drop. The LMTD and pressure drop started showing a constant rise for the following 6 months. Increase LMTD and pressure drop are symptoms that typically correlate with fouling. The timing of the changes appeared to correlate well with the feedstock transition towards larger volume of ethane. The fouling theory was also supported by lab analyses on some of the charge gas streams.

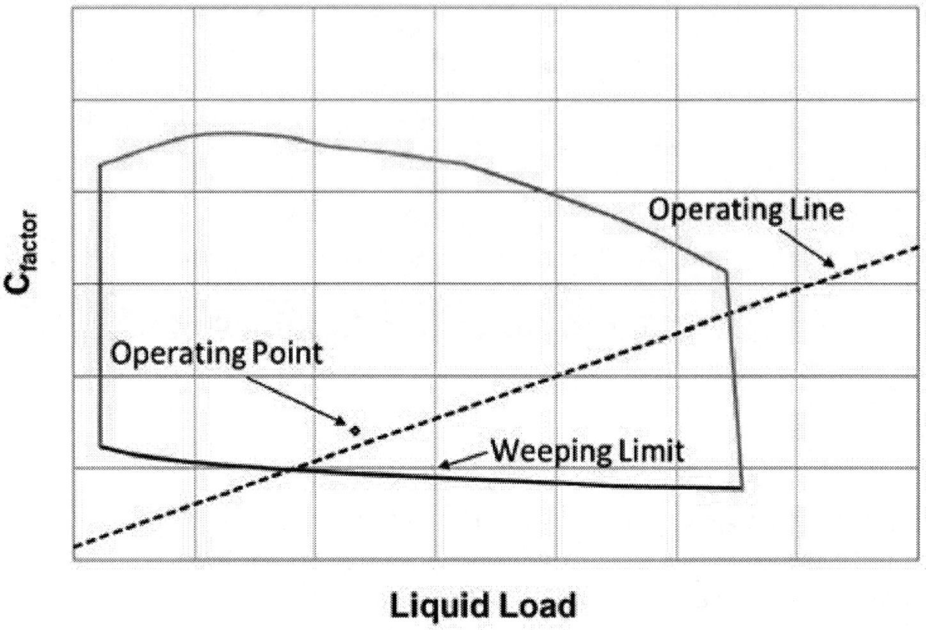

Figure 13: New Tower Stripping Section Tray Window

Implications of the newly discovered propylene chiller constraint were evaluated with simulations relative to the ability of meeting the project feed slate and production targets. Without false loading the old tower, the side reboiler would pinch at warmer temperature and shift the heat load beyond the current capability of the propylene chiller. The residual heat duty would shift to the ERC core exchanger in the cold box. This would not be a problem without limitations on the ERC. However, as experienced during commissioning of the new tower, the ERC was currently experiencing a constraint on the 3rd stage. Although false loading the old C2 Splitter was done during commissioning, false loading for meeting the temperature approach required on the side reboiler would cause a significant increase in condenser duty, which would likely push the PRC power beyond the capability of the machine at the targeted production rate. This evaluation was based on the current state of the propylene chiller; further deterioration in performance would be expected due to the typical nature of exchanger fouling. Given the threat on meeting the project objectives, the project team expedited the design and purchase of a new exchanger. Required tie in's were installed during a short plant outage. The new exchanger was installed and now operates in parallel with the existing units. Since then, the plant has been meeting and exceeding the production targets. The two C2 Splitters have been operating effectively without any problems reported by operation.

Conclusion

Due to the increase in required reflux ratio, higher volume of ethane cracking will reduce the capacity of a C2 Splitter. In this case a reduction from 110 to 92% of the targeted production was evaluated. However, side reboiler limitations and tight tray spacing in the bottom section further reduce the capacity of the tower to 69% of the targeted production. A challenging project schedule imposing construction activities outside normal plant turnaround, limited the number of debottleneck options. Although expensive, a new parallel tower was the only option meeting the project constraints and objectives. To make the investment most profitable, alignment with future business plans and competitive feedstocks was incorporated in the design strategy. Definition of a road map to expansion provided a staged approach for future increase in capacity, ultimately pushing the C2 Splitter system capacity to 163 - 187% of the current project target depending on feedstock.

This work illustrates the importance of thorough process engineering from front end development all the way through commissioning. The use of plant data, simulations and tray operating windows was instrumental in understanding existing equipment limitations, which is imperative for defining debottleneck options. Besides generating process loads for the new tower and tray design, simulation results allowed identifying potential problems with the feed distributor design.

Ethylene plants are highly integrated, and constraints in systems believed non-related at first can shift loads on other systems and limit operation flexibility. During commissioning of the new tower, fouled exchangers increased loads on the ERC and required maximizing heat integration on the old C2 Splitter at the expense of efficient fractionation. Those conditions led to higher power usage on the PRC due to significant false load on both C2 Splitters, which also resulted in higher steam demands. Simulation of the olefins unit ahead of the commissioning activities and definition of process targets along with effective process data gathering tools helped identify the above constraints before they became unmanageable. The use of simulations was also instrumental for establishing new process targets that would aligned with the newly discovered constraints. Although not per original plan, the new tower was commissioned successfully. Further evaluation on the constraints encountered during the commissioning activities provided understanding and early warning, which allowed proper action for removing a constraint that would otherwise be a threat to the success of the project.

Nomenclature

APC:	Advanced Process Control
C2R :	Ethylene Refrigerant
C3R :	Propylene Refrigerant
CGC:	Charge Gas Compressor
ERC:	Ethylene Refrigeration Compressor
HAGO:	Heavy Atmospheric Gas Oil

NGL: Natural Gas Liquid
PRC: Propylene Refrigeration Compressor

References

1. M. de Boer, B. Evans, Ethylene Cracker Capacity Increase And Modernization, Ethylene Producer Conference, 2006 AIChE Spring Meeting, Orlandos, FL
2. A. Bernard, R. Hayden, Planning and Designing the Modernization of the Recovery Area of a Flexible Cracker, Ethylene Producer Conference, 2004 AIChE Spring Meeting, New Orleans, LA
3. A. Bernard, D. Summers, W de Villiers, Improve Product Ethylene Separation, Hydrocarbon Processing, April 2009.
4. S. Katrekar, P. Nair, Technology Leveraged Continuous Process Analysis to Optimize the Performance of Cracking Furnaces, APGC, May 2007, Kuala Lumpur
5. W. de Villiers, P. Wilkinson, D. Summers, Developments in Splitter Revamps, Ethylene Producer Conference, 2004 AIChE Spring Meeting, New Orleans, LA
6. Z.P. Xu, B.J. Nowak, K.J. Richardson, SimulFlow™ Device Capacity Beyond System Limit, Distillation Topical Conference, 2007 AIChE Spring Meeting, Houston, TX
7. D. Leung, J. Hu, W. Lam, Advanced C2 Processing System (ACPS), Ethylene Producer Conference, 1999 AIChE Spring Meeting, Houston, TX
8. H. Z. Kister, "Distillation Design", McGraw-Hill.
9. D. Gent, R. Hayden, Conversion of a Flexible Cracker to a Mostly Ethane Feed, Ethylene Producer Conference, 2016 AIChE Spring Meeting, Houston, TX

AICHE PAPER NUMBER 86B

INTERPRETATION AND APPLICATION OF EN 746-2 REQUIREMENTS TO THE DESIGN AND OPERATION OF BURNER MANAGEMENT SYSTEMS ON CRACKING FURNACES

David J Brown
Senior Engineer – Cracking Furnaces
Total Research & Technology Feluy

Lode De Leye
Senior Engineer – Steam Cracking
Total Research & Technology Feluy

Sandra Gainier
Expert Engineer Furnaces
Total Petrochemicals Gonfreville

Dries De Cooman
Process Engineer Front End Studies
Total Olefins Antwerp

Jim Middleton
Manager of Furnace Technology and Design
Technip Stone & Webster Process Technology

Prepared for Presentation at the 2016 Spring National Meeting
Houston, Texas, April 10-14, 2016

INTERPRETATION AND APPLICATION OF EN 746-2 REQUIREMENTS TO THE DESIGN AND OPERATION OF BURNER MANAGEMENT SYSTEMS ON CRACKING FURNACES

David J Brown
Senior Engineer – Cracking Furnaces
Total Research & Technology Feluy
Lode De Leye
Senior Engineer – Steam Cracking
Total Research & Technology Feluy
Sandra Gainier
Expert Engineer Furnaces
Total Petrochemicals Gonfreville
Dries De Cooman
Process Engineer Front End Studies
Total Olefins Antwerp
Jim Middleton
Manager of Furnace Technology and Design
Technip Stone & Webster Process Technology

Abstract:

In 2010 Total, working closely with its selected licensor and basic engineering contractor, Technip Stone & Webster Process Technology ('TS&W'), developed a philosophy for the burner management system ('BMS') of a pair of new furnaces being designed for Total's Gonfreville site (part of the Normandy Platform.) The key features were operator safety; compliance with the European standard (EN 746-2:1997); and reliability. The idea of having remote ignition (from a local panel) and relying on flame monitoring (no visuals) was rather unusual for a cracking furnace. The operators were especially concerned. In practice the system has proved to be reliable, convenient and, most importantly, safe.

The BMS was adapted by Total and retrofitted to a pair of older furnaces at Gonfreville. Later, after TS&W was awarded a sub-contract for a near-duplicate pair of furnaces for the Antwerp Platform, Total provided detailed feedback from the experience in Gonfreville to allow the design to be fine-tuned. It was also reviewed for compliance with the new edition of the European standard (EN 746-2:2010).

In the paper the authors describe, compare and contrast the design of the BMSs for the three cited references and report on how the systems performed in practice. Finally they consider the implications of the publication of the International Standard, ISO 13577-2:2014, that will eventually supersede EN 746-2 in Europe, and is likely to become the standard adopted by countries which do not currently have a national standard.

Overview

Starting in 2010, TS&W was responsible for licensing and performing the basic engineering of two new pyrolysis furnaces for Total at the Gonfreville site, part of the Normandy Platform. In fall 2012 Total implemented an upgrade of the BMS for two older furnaces, originally started-up in 1988, also at the Gonfreville site. Finally, in mid-2012, TS&W started work on two new furnaces to be installed at the NC-2 Plant, part of the Antwerp Platform. The Antwerp furnaces started up in 2015.

European Standard Requirements

When EN 746-2: 1997 was published it challenged the European petrochemicals industry to a radical rethink of its approach to the design and operation of burner management systems ('BMSs') for cracking furnaces. In 2000 the European Ethylene Producers' Committee ('EEPC') set up a Working Group to consider the implications of the standard (Matthews, 2000). The results were presented at an EEPC HSE ('Health, Safety & Environment') Conference (Overwater, 2001) and then summarized in a final report in 2002. The report confirms that EN 746-2 has the "...status of a national standard in all EU ['European Union'] member state[s] and EFTA ['European Free Trade Association'] countries in September 1997...."; that it "...clearly states its applicability...in petrochemical industrial thermoprocessing...."; and that it covers "...all thermoprocessing equipment installed since September 1997...". Although the standard is not retroactive it could, according to the writers of the report, apply if a furnace undergoes a major revamp. Borealis went further (Enqvist, 2010) and described the development of a strategy to upgrade the BMSs of its old furnaces, some dating back to the early 1970's, and the implementation, starting in 2004.

In late 2009 Total agreed internally some key points on the implementation of EN 746-2 for cracking furnaces as follows:

- It applies to new furnaces and existing furnaces in which burners are replaced.
- It will be applied to floor burners and balcony/terrace-wall burners only and will not be applied to wall burners for the following reasons:
    - Too many small burners
    - Technical issues with pilots and flame detection
    - Has to be covered by procedure
- All floor burners will have continuous pilots.
- A limited amount of floor burners, necessary to reach a minimum flue gas temperature of 750°C, plus a margin for reliability reasons (based upon common sense and operating feedback) will be called 'start-up burners'.
- All start-up burners and their associated pilot burners will have automatic safety shut-off valves ('XVs').
- All start-up burners will be equipped with a remote ignition device, flame detection by means of an ionization rod on their associated pilot burners, and flame detection by means of a UV detector on the main burners.
- Prior to light off, a tightness test to all burners has to be performed. Nitrogen will be used for the test.

- 95% of the burners are either balanced or natural draught: there will be no airflow metering.
- Pilot fuel gas shall be different from the fuel gas to the main burners (natural gas recommended).
- A flue gas temperature measurement at the arch is mandatory.

These general principles were applied in the two projects in Gonfreville. For the Antwerp project we had to consider the changes between the 1997 edition of EN 746-2 and the 2010 edition. The key changes are listed in the table below.

Table 1. EN 746-2 changes.

Clause	EN 746-2:1997	EN 746-2:2010
5.2.2.3 (1997) 5.2.2.3.4 (2010)	Valves controlling capacities greater than 1200 kW shall be equipped with a *leak tightness device.* [And from 3.36 b), definition of leak tightness device, '*....two safety shut-off valves in series, fitted with proof of closure switches to close the fuel line, and a third valve fitted with a switch to prove that it is open, to vent safely the space between them....*' In other words a double-block-and-bleed.]	Automatic shut-off valves controlling capacities greater than 1 200 kW shall be equipped with a *valve proving system.* If this is applicable, then the valve proving system shall comply with EN 1643 or give an equivalent level of safety. The signal of a leaking valve from the *valve proving system* shall stop the current start-up when the test takes place during start up or prevent the next start-up when the test takes place after shut down. This function shall meet the requirements of a protective system according to 5.7.2 and 5.7.3. [3.93 defines a valve proving system as follows, '*...system to check the effective closure of automatic shut-off valves by checking leak tightness NOTE It consists of a programming unit, a measuring device, valves and other functional assemblies....*']
5.2.2.3.1 (2010)	No equivalent	Low cycling applications intended to operate continuously for periods longer than 1 year shall have a redundant safety shut-off system at least comparable with the flow sheet given in Figure C.5 that allows testing the effective closure of the valves at least once a year.
5.2.6.1	For systems where the pilot burner remains in use during main burner operation, separate flame sensors to supervise the pilot and main flames shall be fitted. The main flame sensor shall be so positioned that it cannot in any circumstances detect the pilot flame.	For systems where the pilot burner remains in use during main burner operation, separate flame detector device to supervise the pilot and main flames shall be fitted. The main flame detector device shall be so positioned that it cannot in any circumstances detect the

		pilot flame. *In the case that the pilot burner ignites under all circumstances the main flame, it is sufficient to supervise only the pilot flame providing the flow rate of that pilot burner is checked by a protective System according to 5.7.2 and 5.7.3 (e.g. minimum gas pressure switch).*
Annex C (informative) Typical examples of piping and components	No equivalent	'Figure C.5 — Central pipework for low cycling applications' is the most relevant.

Implementation on new furnace project at Gonfreville

The new furnaces at Gonfreville are back-to-back, single radiant cell USC72SU (symmetrical U) configuration. About 85% of the heat input is provided by 24 x John Zink Ultra-Low NO_x PSMR floor-mounted burners; the remainder of the heat is from 2 x rows of John Zink LPMW wall-mounted burners, 48 in total. The burners are balanced draft. The air dampers on the floor burners are linked with a jackshaft. An end view of one of the furnaces is given in Figure 1.

Figure 1. Furnace overview.

All the floor burners are equipped with continuous pilots. Half of the floor burners are designated as start-up burners. These start-up burners have igniters and ionization flame detectors on the pilots; UV flame detectors to monitor the main flame; and local XVs on the both the pilot fuel supply and the main fuel supply. Typical start-up and non-start-up (a.k.a. 'main') burners are shown in Figure 2.

It is important to note, in passing, that pilots are used as a safe and convenient way to ignite the main flame of the high heat release floor burners, not as a safety device. Total agrees with the philosophy of Platvoet (2013) that "...pilot burners should be considered as ignition devices, not as safety devices..." and follows the requirements of EN 746-2:1997, paragraph 5.2.6.1 to provide "...separate flame sensors to supervise the pilot and main flames..."

The steps up to reaching the prescribed auto-ignition temperature (the so-called "start-up period" defined in paragraph 5.2.6.3

of EN 746-2) are carried out by an operator from a local panel in a safe, remote location. Other operators have to evacuate the area. This is a fundamental part of Total's philosophy. The local panel is shown in Figure 3. Those steps are described in some detail below.

Quite a different approach has to be used with wall-fired furnaces. There are technical reasons why remote ignition and remote flame monitoring are usually not practicable. This is discussed both by Enqvist (2010) and Platvoet (2013). The topic is not covered further in this paper.

Figure 2. Start-up and main burners.

Pre-Ignition Steps

Before igniting the burners a leak test must be performed to confirm the integrity of the fuel gas isolation valves, and the firebox must be purged.

Purge

The furnace purge can be performed concurrently with the leak testing and also after failed or aborted ignition attempts.

The ID fan is running. The arch pressure set point and the floor burner damper positions are set for 25% of the air flow rate at maximum heat input. The purge timer, set for five complete air changes, is started.

After purging, the ID fan continues running to automatically control the arch draft at normal design value.

Figure 3. Local panel.

Leak Test

The system leak tightness test is carried out using nitrogen. It is, in principle, feasible to use the normal fuel gas. There are pros and cons for both approaches. Total's view is that if the leak tightness test fails then there is a risk that fuel gas may leak into the firebox and cause an explosive mixture: in other words a procedure intended to avoid a potentially dangerous situation is responsible for creating a dangerous situation. The disadvantage of residual nitrogen making it difficult to ignite the first burners can be managed – see discussion below.

The fuel gas is blinded. The ID fan is running. The nitrogen supply is connected to the fuel gas header, but the manual valve on the nitrogen supply remains closed. All valves are closed except the manual valves on the fuel gas to pilots and main fuel gas connections on start-up burners.

A hand switch starts the leak test. In a regular, defined sequence the integrity of the various XVs is tested. This procedure, in its slightly improved version, is described in more detail for the Antwerp project.

A fixed and defined time after the successful completion of the leak test, all of the 'valve test ok' indicators will automatically reset. Therefore ignition of the pilots must proceed within the defined time of completion of the leak test.

If the leak test fails then the DCS operator needs to reinitialize the leak test by pressing again the hand switch.

Significant amounts of nitrogen in the fuel gas header could prevent the successful light-off of pilots and burners. The nitrogen is purged from the pilot fuel gas header, and from the main fuel gas header, with a pre-defined sequence of successive pressurize/depressurize cycles using fuel gas. This is efficient: the only trouble encountered is the lighting of the very first pilot, for which the first attempt to light may fail because of small amounts of residual nitrogen.

Ignition Steps

The floor burner dampers are set to about 15% open using the jackshafts. The original intent was to close the dampers for the main, non-start-up floor burners fully (minimum stop) by disconnecting from the jackshafts. In practice this was found to not be necessary.

The first pilot has to be lit within 10 minutes of validation of the atmospheric analysis or else the process has to be done again: atmospheric analysis, validation, launch of the pilot lighting sequence.

The operator starts the automatic pilot ignition sequence by pressing a local panel push button. All pilot spark igniters are energized and after 3 seconds the pilot XVs to the start-up burners open automatically. The ignition timer stops after 5 seconds. The status of the pilot flame detectors, ionization flame type, start-up burners only, is checked. If all 12 pilot flame detectors are positive then the automatic lighting sequence will move on to the next step. If less than 12 positive flame signals are detected then a second, and finally a third, attempt is made.

If some pilots on start-up burners still do not have positive flame signals, the pilot isolation XVs are closed on these burners.

The originally planned success criteria for lighting the pilots were as follows:

- If fewer than 8 pilot flames are shown as positive, or if any 2 adjacent pilot flames fail to be shown as positive, then the furnace start-up will have failed, a trip is invoked and the leak test and purge sequence will have to be repeated.

These were found, in practice, to be too rigorous. The criteria were modified. The revised method is described below in detail for the Antwerp new furnace project below.

After various adjustments to the process system – not discussed in detail in this paper – the fuel gas header is pressurized. The gas firing is gradually increased by opening the XVs to start-up burners, one at a time, following a burner lighting sequence described in the written procedure (no automatic sequence programmed for the burners). If any start-up burner does not have a positive pilot flame signals, the XV cannot be opened. A positive main flame signal from the main flame UV detector is required within the ignition timer limit. If a positive flame signal is not received, the XV on that burner will close. Three tries are allowed to light the main flame on each start-up burners. A further three attempts are allowed after a timed 20 minute delay – this was a modification to the original plan and described in more details for the Antwerp new furnace project.

When an arch temperature of 180°C is reached dilution steam supply is started.

The next phase of the ignition sequence was originally planned as follows:

- When 8 or more main flames are shown as positive for a period of 15 seconds, and no adjacent start-up burners are unlit, the start-up over-ride on the main furnace trip is removed. Subsequently, if less than 8 main flames are shown as positive, the furnace will trip.
- The main fuel gas pressure control is released to DCS control. If the number of positive pilot flame signals falls below 8 before the arch temperature reaches 750°C an alarm warns the operator; if the number of positive main flame signals falls below 8 the furnace trips; if 2 adjacent start-up burners do not show a positive flame signal the furnace trips.

Again, this was modified based on the experience of the first few cold light-offs. The modified procedure, now used at both sites, is described for the new furnace project in Antwerp. See below.

The firing on the start-up burners is increased until the arch temperature is above 750°C (determined by 3oo3 voting of arch temperature). The dampers are properly adjusted.

Once the arch temperature is above 750°C, the XV for the main burner pilots is permitted to be opened, to allow lighting of main burner pilots.

The main burner pilots are manually lit, one at a time, by opening the manual valve to each burner and with either a portable, manual igniter or cross-ignition. If there are any start-up burner pilots not showing positive flame signals the associated XVs are opened and pilots are lit from the built-in igniter or cross-ignition.

Once the arch temperature is above 750°C, the XV for the main burners is permitted to be opened, to allow lighting of main burners. The main burners are then lit from the pilots using the local, manual isolation valves.

The XV for the wall burner header is also permitted to open. The lighting of the wall burners proceeds in the same manner as the main floor burners, except that the wall burners, having a rather low heat release, do not have pilots and are lit directly with a manual igniter or by cross-ignition.

The furnace start up is now complete and all burners are lit.

Figure 4. Older furnaces at Gonfreville

Implementation on existing furnaces at Gonfreville

The older furnaces at Gonfreville are back-to-back, single radiant cell USC48U configuration. They are 100% floor-fired by 24 x John Zink Low NO_x PSFFG burners. The burners are forced draft, air pre-heated with BFW. The furnaces are shown in Figure 4. Pilots were added to the burners as part of the safety upgrade. Half of the burners (12) were designated as start-up burners. The BMS retrofit was basically identical to what was used for the new furnaces.

Implementation on new furnace project at Antwerp

The pair of new furnaces at Antwerp is very similar to the pair of furnaces at Gonfreville. The return of experience ('REX') from Gonfreville enabled some improvements to be made. In addition the new version of EN 746-2 required some rethinking.

After a detailed analysis it was decided that the double-block-and-bleed system could be considered to be equivalent to the valve proving system ('VPS'), newly mandated in the 2010 Edition of the EN. This was accepted by the Notified Body (accredited third-party inspector in EU, also referred to as 'NoBo'). It is worth noting that ISO 13577-2 also does not require a VPS for most cracking furnace applications (see below). De Boer (2013) describes a different viewpoint.

Natural gas was used as the supply for the pilots (see paragraph 5.2.2.5 of EN 746-2). This was not practicable for the Gonfreville projects.

The new edition of the EN introduced a new requirement that ".......a redundant safety shut-off system at least comparable with the flow sheet given in Figure C.5....." shall be provided. Since the purpose of the parallel system is to allow "...testing the effective closure of the valves at least once a year...." the solution proposed for Antwerp, and accepted by the NoBo, was to install a bypass around the 2 x XVs with a removable spool piece that is not connected during normal operation, and with isolation valves that are car-sealed closed (CSC). If and when the by-pass is connected it requires continuous supervision by the field operator.

Purge

This follows the Gonfreville procedure.

Leak Test

The leak test is carried out using nitrogen. Several of the valve tests are performed simultaneously. The first steps are the same as described for Gonfreville.

The fuel gas system venting is tested by opening all XVs up to the first individual burner isolation and opening all atmospheric vents. The PCV's are opened. The header pressure should fall to atmospheric pressure within 180 seconds.

The upstream, primary XVs (main fuel gas and natural gas to pilots) are closed and confirmed to close within 5 seconds.

The nitrogen fill connection manual valve is opened and pressurizes the fuel gas and pilot gas headers upstream of the primary XVs. The field operator depresses the appropriate hand switches on the local panel to confirm that the nitrogen valve is opened and to start the automated valve leak tests.

Test 1: pilot gas primary XV, and Test 5: burner fuel gas primary XV. The downstream and intermediate 'bleed' XVs are confirmed to close within 15 seconds. The PT's between the closed block valves are confirmed to rise by not more than 0.1 bar in 60 seconds.

Test 2: pilot gas secondary isolation and bleed XVs, and Test 6: burner fuel gas secondary isolation and bleed XVs. Pilot gas primary XV is confirmed open within 15 seconds. Downstream pressure is confirmed to rise to above 2.0 barg within 10 seconds. The upstream valve is confirmed closed within 15 seconds. The downstream pressure is confirmed to hold (falls by less than 0.1 bar in 60 seconds). The same procedure is followed for the main

Figure 5. Furnaces and local panel.

fuel gas supply except that a longer time is allowed for pressurizing (larger volume to fill).

Test 3: start-up pilot gas XVs (12 total) and main pilots XV, and Test 7: start-up burner XVs (12 total), main (non-start-up) burners XV and wall burners XV. The pilot header up to the pilot gas XVs and the main pilot XV is pressurized, the main isolation XVs are closed, and then the header is confirmed to hold pressure using the same time-pressure criteria as above. Next the main fuel gas header is pressurized up to the start-up burner XVs, the main burners XV and the wall burners XV. The main isolation (upstream) XVs are closed, then the main header is confirmed to hold pressure using the same time-pressure criteria as above.

Test 4: main pilot gas manual valves test, Test 8: main burner manual valves test, and Test 9: wall burner manual valves test. The pilot header up to the start-up pilot gas XVs and the main pilot gas manual valves is pressurized, the main isolation XVs are closed, and then the header is confirmed to hold pressure using the same time-pressure criteria as above. Similarly the main header is pressurized up to the start-up burner XVs, the main burner manual isolation valves, and the wall burner manual isolation valves. The main isolation (upstream) XVs are closed. Then the main header is confirmed to hold pressure using the same time-pressure criteria as above.

Nitrogen is vented from the fuel gas headers. The fuel gas and pilot gas valve integrity test is complete.

24 hours after the successful completion of the valve integrity test, all of the 'valve test ok' indicators will automatically reset. Therefore ignition of the pilots must proceed within 24 hours upon completion of this test. This is a change from the Gonfreville protocol based on REX: it is not always practicable to go to the ignition steps as quickly as originally specified.

If the leak test has failed then the control room operator will need to reinitialize leak test by re-pressing the appropriate push button.

After the successful completion of the valve integrity test, the nitrogen supply must be isolated from the fuel gas network. Significant amounts of nitrogen in the fuel and pilot gas headers could prevent the successful light-off of pilots and burners. The nitrogen shall be isolated and purged from the fuel gas header and pilot gas header, prior to the lighting of the pilots.

Ignition Steps

The ignition steps have been described in detail for the Gonfreville new furnace project. Here we simply described the differences.

A series of conditions have to be satisfied before the ignition sequence can start:

- It is less than 24 h since the successful completion of the valve integrity test.
- The fuel gas and pilot gas headers have been purged of nitrogen.
- The pilot gas low pressure trip is overridden and the control system is lined up.
- The field operator has confirmed, by pushing a button on the local panel, that the fuel supply is commissioned and that the firebox atmosphere test is complete and passed.
- The 10 min post-purge timer has not timed out (note that the post-purge timer only starts when the damper position and fan speed are adjusted to ignition conditions, not when the minimum purge time of 20 minutes is reached).

For the start-up burner pilot ignition sequence:

- If less than 6 pilot burners are confirmed lit, the pilot ignition sequence is deemed to have failed and the furnace trips.
- If between 6 and 7 start-up pilots are confirmed lit, the ignition sequence is halted for a period of 20 minutes. After this period the ignition attempt cycle counter is reset, and the operator can initiate another 3 pilot lighting attempts by depressing the hand switch on the local panel.
- If 8 or more start-up pilots are confirmed lit, the distribution of the lit pilots is checked to make sure the pattern of light start-up pilots will provide an even firing pattern. If the lit pilots are in an acceptable pattern, a permissive will be provided to open the start-up burner fuel gas XVs.
- If 8 or more start-up pilots are confirmed lit, but the pattern is unacceptable, the ignition sequence is halted for a period of 20 minutes. After this period the ignition attempt cycle counter is reset, and the operator can initiate another 3 pilot lighting attempts by depressing the hand switch on the local panel.

The REX from Gonfreville suggests that 8 start-up burners can heat the furnace to 750°C (auto-ignition, safe temperature) without adjusting the main, non-start-up burner dampers to limit the overall amount of excess air entering the firebox. This means that the main burner dampers do not need to be disconnected from the common jackshafts.

For the burner ignition and furnace warm up phase:

If a positive main flame signal is not received after 3 attempts, the XV on that burner will be locked closed for 20 minutes; but the operator may choose to continue the warm up sequence by attempting to light a different burner. After the 20 minute period a further three attempts to light the main flame of this burner will be allowed.

If the number of positive flame signals from different burner locations (i.e. either positive pilot flame signal or positive burner flame signal; if both flame signals from the same burner location are positive, this only counts as one positive signal) falls below 6 before the arch temperature reaches 750°C the furnace trips. In order to reach an arch temperature of 750°C at least 8 of the 12 start-up burners should be lit.

International Standard Requirements

ISO 13577-2 was published in 2014. It is clear that the origin and source of the new international standard is EN 746-2. There are, however, important differences, most of which have been described by Van der Bij (2015) and by Clauss (2014).

EN 746-2 requires a VPS*, compliant with EN 1643, for automatic XVs controlling capacities greater than 1 200 kW. ISO 13577-2 only requires an automatic [valve proving] system if the industrial furnaces and associated processing equipment (TPE) is "....intended to be started up more than two times [twice] a year...". Hence an automatic VPS should not be required for most cracking furnaces, although each furnace type has to be evaluated individually.

****NOTE:*** *this is a new requirement from 2010. The original version of EN 746-2 only required a "leak tightness device", of which a double-block-and-bleed is one permitted variant. See Table 1.*

EN 746-2 mandates "a redundant safety shut-off system" so that annual in situ testing can be made of the automatic shut-off valves if the furnace operates continuously for more than 1 year (see Table 1 of this paper, ref. paragraph 5.2.2.3.1). The equivalent paragraph of ISO 13577-2 (4.2.2.6) states "...NOTE It is commonly agreed that valves are to be tested annually **unless longer testing intervals can be justified by the risk analysis**....[highlighting by the authors]". Van der Bij (2015) claims that the testing interval can be extended to 3-4 years by carrying out a risk analysis according to the so-called Method D of ISO 13577-4.

EN 746-2 mandates a pre-purge rate at least 25% of the maximum combustion air flow rate. In ISO 13577-2 the rate doubles to 50%.

Conclusions

The standards in Europe for burner management systems have become much rigorous in the last 15 years or so. Although such standards are, strictly speaking, only applicable to new projects they can also apply to revamps in which there are major changes to the combustion system. With the publication of the international, ISO standard, this approach is likely to be adopted in countries outside the EU.

Total and TS&W have developed a stringent but flexible design. It has proved, in practice, to be reliable, convenient and safe.

The requirements are still evolving and certainly the BMS will likely be somewhat different for each furnace type considered.

Acknowledgement

Heurtey Petrochem S.A. was the Engineering, Procurement and Construction ('EPC') contractor for the furnace projects at both Gonfreville and Antwerp.

References:

1. "Industrial thermoprocessing equipment – Part 2: Safety requirements for combustion and fuel handling systems", European Standard EN 746-2: 1997.
2. Matthews, R. (2000), "European Ethylene Producers' Committee [EEPC], EN 746-2 (Industrial Thermo-processing Systems), Working Group Agreed Terms of Reference".
3. Overwater, K. (2001), "EN 746-2 Furnace Protection Work Group - final report –", *EEPC HSE Conference,* The Hague.
4. "European Ethylene Producers Committee, Final Report, Issue Group to consider the application of European Standard EN-746/2 to steam cracking plants", 2002.
5. Enqvist, K. (2010), "State of art BMS requirements - Upgrade of old furnaces and implementation in new furnaces", *AIChE Spring National Meeting*, San Antonio, Texas.
6. "Industrial thermoprocessing equipment – Part 2: Safety requirements for combustion and fuel handling systems", European Standard EN 746-2: 2010.

7. Platvoet, E., Strupp, C., & Theis, G. (2013), "Safety considerations for the design and operation of ultra-low NO_x burners in cracking furnaces", *AIChE Spring National Meeting*, San Antonio, Texas.
8. De Boer, M. (2013), "Burner management system upgrade as a result of cracking furnace firing system replacement", *AIChE Spring National Meeting*, San Antonio, Texas.
9. "Industrial furnaces and associated processing equipment — Safety — Part 2: Combustion and fuel handling systems", International Standard ISO 13577-2:2014(E).
10. Van der Bij, M. (2015), "Safety requirements for furnace fuel handling systems", *EEPC Ethylene Seminar*, Prague.
11. Van der Bij, M. (2015), private communication.
12. Clauss, R. (2014), "IG [Issue Group] Cracking Furnace Safety Final Report", *EEPC Ethylene Seminar*, Warsaw.

449973 Ethylene Cracking Furnace Burner Applications Operating with Significantly Less Than 90 mg/nm^3 NO_X Emissions

Abstract:

This abstract will focus on the field results of eight (8) ethylene cracking furnaces, which use Zeeco GLSF Enhanced-Jet burners at Petrochina Sichuan Petrochemical Company Limited (SPC) that achieve field measured NOx emissions significantly less than 100 mg/Nm^3. Using ultra-low emissions burners in ethylene cracking applications have proven to be an economical solution at Zeeco for achieving lower NO_x levels than 100 mg/Nm^3 from existing fired equipment. Providing a next generation low NO_x burner footprint that fits into ethylene cracking applications has proven to be a challenge in many applications. Ultra low NO_x burner flame pattern(s) or profile(s) that can operate properly within a required burner to burner spacing and heat flux limitations have also proven to be a challenge.

This paper will focus on an application, which uses the GLSF Flat Flame Enhanced-Jet Burner model operating in eight (8) ethylene cracking furnaces that achieved less than 100 mg/Nm^3 in the field. For this application, NOx emissions averaged around 50 mg/Nm^3 in the field. The furnace field measured NOx emissions ranged between 35 mg/Nm^3 (17 ppmv) to 65 mg/Nm^3 (32 ppmv) at design heat release.

For applications requiring less than 100 mg/Nm^3 NOx emissions, Zeeco has developed an ultra-low NOx emissions burner technology, "GLSF Enhanced-Jet Flat Flame Burner", which provides a very small mechanical footprint and produces an excellent heat flux profile without flame to flame interaction. The GLSF Enhanced-Jet floor burners utilize the basic "free-jet" method of mixing fuel gas and inert products of combustion to form a new combustible mixture before entering the combustion air stream. The resulting mixture, burns at a lower adiabatic flame temperature, resulting in lower thermal NO_x production. In addition, the GLSF Enhanced-Jet burner is designed to greatly reduce issues of flame interaction and flame rollover experienced with other burner designs. Flame interaction between burners can cause flame impingement upon the process tubes and increased emissions. Flame impingement can result in shorter run lengths and higher tube metal temperatures which are problematic for prolonged ethylene production.

In conclusion, Zeeco and Petrochina Sichuan Petrochemical Company Limited (SPC) will co-author the paper and provide specific application details and the resulting field emission test results. This paper will focus on an application, which uses the GLSF Flat Flame Enhanced-Jet Burner design in eight (8) ethylene cracking furnace that achieved significantly less than 100 mg/Nm^3 in the field. This application will show that the field measured NO_x emissions were between 35 mg/Nm^3 (17 ppmv) to 65 mg/Nm^3 (32 ppmv) at design heat release for this highlighted project.

444925 Compressor Wash Oil Supply in a Light Feedslate Environment

Abstract:

The shale gas development, primarily in the US, has resulted in lower cost feedstocks for gas cracking and thus the addition of new gas cracking capacity. This increased capacity has resulted in increasing demand for Compressor Wash Oil. Furthermore the transition of some naphtha (liquid) crackers to gas feedstocks has further increased demand for Compressor Wash Oil. With this transition fewer internal streams are available for use as Compressor Wash Oil. Current products used for Compressor Wash Oil are generally C10-C12+ aromatic based. This paper will focus on the desirable properties for a Compressor Wash Oil.

442001 WAO Reliability Improvements

Abstract:

The Wet Air Oxidation (WAO) system used for processing spent caustic at an Ethylene plant was plagued with extensive fouling. It required cleaning of the heat exchangers, piping, and the Wet Air Oxidation reactor at a certain frequency.

An analysis of the foulant was conducted, and improvements implemented accordingly to reduce the fouling. These have led to reduced frequency of cleaning, increased run-time, achieve design temperatures, and increased spent caustic through put to the WAO system. It has also led to a reduction in waste water generated from equipment preparation for cleaning.

AIChE Paper 86f

Simulation-Based Assessment and Mitigation of Fouling in Water Quench and Charge Gas Compressor Sections of Gas Crackers

Mathieu Zellhuber

Process Engineer Petrochemical Plants

Linde AG, Engineering Division

Matthias André

Senior Process Lead Engineer Petrochemical Plants

Linde AG, Engineering Division

Thomas Emmert

Manager Process Engineering Petrochemical Plants

Linde Engineering North America

Gunther Kracker

General Manager Process Engineering Petrochemical Plants

Linde AG, Engineering Division

Prepared for Presentation at the 2016 Spring National Meeting

Houston, Texas, April 10-14, 2016

Simulation-Based Assessment and Mitigation of Fouling in Water Quench and Charge Gas Compressor Sections of Gas Crackers

Mathieu Zellhuber
Process Engineer Petrochemical Plants
Linde AG, Engineering Division
Matthias André
Senior Process Lead Engineer Petrochemical Plants
Linde AG, Engineering Division
Thomas Emmert
Manager Process Engineering Petrochemical Plants
Linde Engineering North America
Gunther Kracker
General Manager Process Engineering Petrochemical Plants
Linde AG, Engineering Division

Abstract: Fouling in the hot section of the separation train is widely recognized as a major challenge for successful, long-time operation of gas crackers. Root causes for fouling can be of very different kinds and are consequently hard to determine at first sight. As for the causes, mitigation measures applied in plant operation are also very diverse, ranging from chemical and wash oil injections to targeted management of recycles and small revamps. It is hence of utmost importance to know when to apply which measure.
Process simulation has become an essential tool for the design of new plants, allowing precise prediction of the separation steps. The present paper shows how Linde uses process simulation for understanding root causes of fouling in operating plants and for developing suitable mitigation measures. Although hot section fouling is of high physical and chemical complexity, we have developed a pragmatic approach to evaluate and compare fouling risks and quantities in the water quench and charge gas compressor sections. The methodology has been successfully applied during a recent case study for an existing plant experiencing fouling in charge gas compressor intercoolers and suction drums.
These improved tools can thus be used to solve actual fouling issues, but also to prevent future problems, for instance before proceeding to a feed change. This permits anticipating potential fouling risks and developing mitigation strategies in advance. The developed prediction methods are similarly used for the design of new plants.

1. Introduction

The potential fouling of equipment items and/or piping sections represents a significant concern for the operation of ethylene plants. In many cases, fouling is observed first in terms of decreased plant efficiency, mostly linked to higher pressure drops and/or thermal resistance. This typically results in higher utility consumption and maintenance effort; in more severe cases, equipment damages, unplanned shutdowns and production losses can occur. The potential impact of fouling issues on the overall plant economics is therefore high; this explains the increased attention brought to this topic by licensors and plant operators.

Fouling cannot be handled in a straightforward, singular manner, because the term stands for the panoply of physical and chemical processes that can affect plant operation all along the gas processing route. This high complexity particularly applies to the hot section of the separation train, i.e. to quench and charge gas compressor sections, where stream compositions, pressure and temperature levels are wide-ranging. As a consequence, different fouling mechanisms may come into play, involving various critical cracked gas components. Moreover, these fouling mechanisms will strongly depend on the specific plant setup.

The resolution of past fouling issues and the avoidance of future ones therefore become challenging tasks. The aim of the present paper is to show how the efficient use of process simulation tools can prove to be very useful in this task.

Given the high physical complexity of the problem, it is clear that the employed simulation tools cannot predict fouling mechanisms in all its details, since this would go well beyond industry-standard computational resources. Instead, the developed approach aims at collecting data from standard process simulations and in-house developed post-processing tools in order to evaluate the fouling tendency in a given case.

This approach was applied in the frame of a study made for an existing plant, which experienced fouling issues in the charge gas compressor section after a change of the feed scenario. The root cause analysis was performed by Linde using process simulation that reproduces real operation data. The analysis methods developed allowed to identify the root causes and to evaluate the effect of potential mitigation measures.

Although the study was performed on a gas cracker with mixed ethane/propane feed, the developed methods can be applied to other plants and feed stocks in a similar way.

2. Fouling in water quench and charge gas compressor sections

2.1 Process description

The main task of the water quench section is to ensure sufficient cooling of the charge gas stream down to cooling water temperature level and consequently to condense significant amounts of dilution steam and heavier hydrocarbons. Its main element is the quench tower; in modern designs, it is divided into a lower water saturation section and, separated by a chimney tray, the upper sections, in which the charge gas is cooled by direct contact with circulating quench water. The heavy end of the hydrocarbon spectrum is partially condensed in the lower and upper column stages and retrieved from the system.

The water quench section is followed by the charge gas compressor, whose main purpose is to elevate the charge gas to a pressure level sufficient for the subsequent cryogenic separation into value products. Moreover, with its repeating sequence of compressor stage, intercooler and suction drum, it acts as a continuation of the water quench section, by performing a multistage condensation of water and heavy hydrocarbon components. The molecular weight of the condensed hydrocarbons is decreasing along the compressor stages, resulting in varying compositions of condensates collected in the suction drums.

The similar functions of water quench and charge gas compressor sections in terms of partial charge gas condensation led us to consider them conjointly regarding fouling risk assessment, as will be described in section 3.

2.2 Fouling mechanisms and influence parameters

Previous works and experience indicate that the solidification of heavy hydrocarbons is a potential fouling threat in the water quench section [1, 2]. Indeed, some of the typical charge gas components like naphthalene have a high melting point (80°C at 1 atm) compared to components with similar molecular weight [3]. As a consequence, part of the hydrocarbon components condensed in the quench tower may solidify, precipitate and cause plugging of column internals. This has for example been addressed by Schmigalle and Tolergard [1], who calculated pour point profiles in the quench tower to determine the fouling risk linked to plugging. They could elaborate appropriate fouling mitigation methods for an existing plant to ensure sufficient margin between pour point and process temperatures and avoid the solidification of heavy hydrocarbons.

In the charge gas compressor section, the formation of polymer material is generally referred to as the most common fouling threat. Previous works describe different types of mechanisms and reactions which lead to the formation of macromolecules that can deposit and cause plugging [4-6]. Very relevant is for instance thermal polymerization, in particular of styrene. These reactions are often reported to occur within the compressor casing where temperatures are highest. Indeed, temperature has an exponential impact on reaction rates. Nevertheless, also other parameters need to be taken into consideration to evaluate the risk of polymerization [6, 7]:

- Amount of monomer components in the system, in particular double-bonded, unsaturated hydrocarbons.
- Residence time of monomer components.
- Amount of solvent material, in particular aromatics.
- Presence of oxygenates promoting polymerization reactions.
- High iron contents due to piping and equipment corrosion.

This forms the theoretical background on which the study was performed for an existing plant. The following section will first describe the actual fouling issue and then describe the root cause analysis made.

3. Investigation of fouling in existing plant

3.1 Plant history and problem description

The considered plant has been operated with 10+ years of service with mixed ethane/propane feed. In recent years, propane furnace load and cracking severity have been increased beyond original design values resulting in altered charge gas compositions. Since last major turnaround in 2013, a continuous pressure drop increase could be observed at the 2nd compressor interstage, i.e. between stages II and III. Plugging material was found in both interstage cooler E2 and suction drum D3, see Figure 1; the latter was causing the major part of the pressure drop. Unplanned shutdown for cleaning was needed in 2015.

The foulant material collected at both locations was analyzed and consisted in both cases mainly of organic material, indicating the presence of heavy pyrolysis components as well as polymer materials.

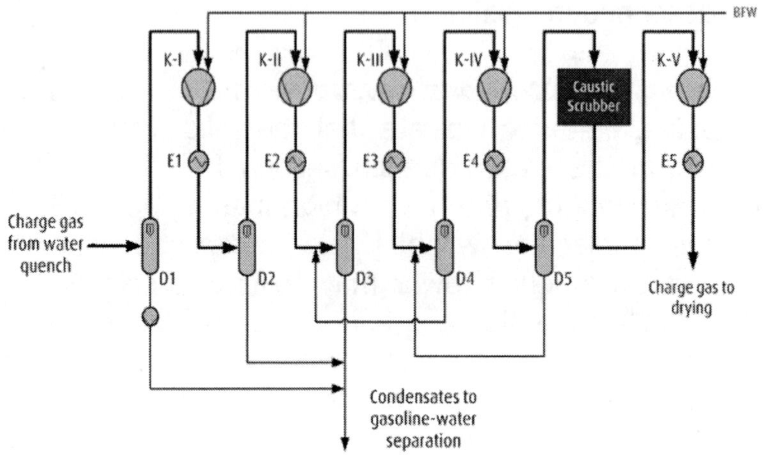

Figure 1: Sketch of considered process topology in charge gas compressor section

3.2 Methods employed for root cause analysis

In order to develop suitable mitigation measures, it is most important to gain a solid understanding of the fouling root causes. Therefore, the investigation was initiated by analyzing operation data provided by the plant owner in order to set up a realistic simulation of current process operation. This included detailed kinetic simulations in ethane and propane furnaces based on actual feed load and conversion data. Furthermore, internal recycle settings and operation set points were adapted in the main process simulation to give a realistic representation of the actual plant operation.

Thereafter, a specific analysis of the process simulation was carried out in order to evaluate which parts of the water quench and charge gas compressor sections are specifically prone to the considered fouling mechanisms.

3.3 Evaluation of fouling risk linked to heavy cracking products

In the present work, the pour point approach described by Schmigalle and Tolergard [1] was employed to evaluate whether solidifying heavy hydrocarbons may contribute to the fouling observed. The pour point calculation is based on proprietary property packages. While in its original application it was applied to the water quench tower only, we applied it now additionally to the charge gas compressor section, more precisely to the intercoolers. Indeed, the foulant material retrieved from the plant was partly composed of heavy components like naphthalene, as mentioned above.

The pour point of the relevant hot section streams is mostly determined by the heavy end of the total cracked gas, which is a very small fraction, in particular for cases of gas cracking. The exact heavy end quantities and compositions are hence difficult to determine with certainty and are moreover sensitive to operating conditions. In order to account for the challenging nature of pour point predictions, an estimation range was determined. This is shown in Figure 2 for intercoolers E1 to E4, in which the process temperature curve is plotted along with the estimated pour point range.

When comparing process and pour point temperatures at the intercoolers, one reckons a potential pour point violation at the second intercooler E2 and, to a slightly minor extent, at the first intercooler E1. It can be followed that, in the first two intercooler stages, fouling can potentially occur due to solidification of heavy cracking products. Indeed, most of the naphthalene is condensed at the second compressor interstage. Further downstream, the hydrocarbons condensed from the cracked gas tend to become lighter, which explains the reduced risk of heavy hydrocarbon solidification at the later compressor stages.

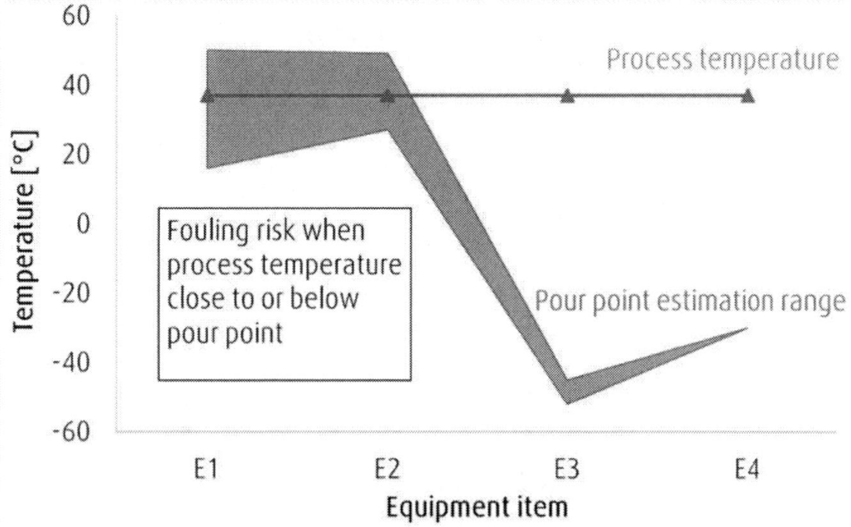

Figure 2: Pour point profile for condensate streams in charge gas compressor intercoolers (base process)

3.4 Evaluation of fouling risk linked to polymerization

As mentioned earlier, polymerization reactions are strongly enhanced by elevated temperatures. This leads to the basic assumption that the polymers found in the foulant material samples were originally formed inside the compressor casing, where temperatures are highest.

The first step in the investigation linked to polymer formation was therefore to look for indications regarding polymerization inside the compressor. Looking at the average outlet temperatures did not support this hypothesis, since they were constantly ranging well below 85°C. In addition, compressor performance data was analyzed in terms of scaled capacity curves. For the given case, no clear indication about polymerization occurring inside the compressor could be found.

As a consequence, other elements of the compressor train were checked more closely. Special interest was brought to the suction drum D3 upstream of the 3rd compressor stage, because this equipment was showing the largest increase in pressure drop. In the investigated plant, the drum D3 receives the cracked gas condensates from the two downstream collection drums D4 and D5, as shown in the flow sketch of Figure 1. We suspected that the continuous recycling and the consequently long residence times of monomer components in the three interconnected drums might promote polymerization.

In order to verify this hypothesis, an average, component-specific residence time was calculated for the three interconnected suction drums D3 to D5. This was done by combining equipment design information and process simulation data in a post-processing step.

The corresponding result is shown in Figure 3. It can be clearly seen that some monomer components are strongly accumulated in the interconnected drums. The residence times, in particular for styrene, are significantly longer than the liquid retention time of a single suction drum due to the constant internal recycling of these species.

These strongly increased residence times lead to high risk of polymer formation in the suction drums, in particular in equipment D3, which is the last element in this recycling cascade.

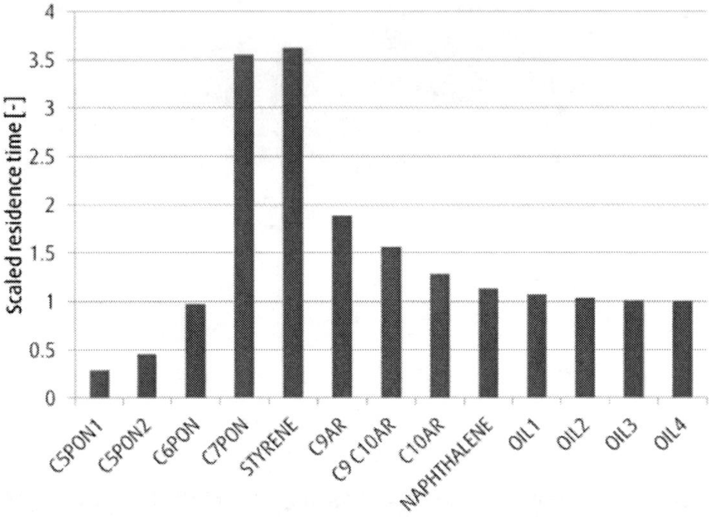

Figure 3: Combined component-specific residence time distributions in suction drums D3, D4 and D5 (base process). Scaled residence time values >1 indicate excessive accumulation of a given component due to plant internal recycles (PON: paraffins, olefins, naphthenes; AR: aromatics).

3.5 Summary of root cause analysis

The investigation led to the following conclusions on fouling root causes for the specific plant:

1. The fouling observed in the second interstage cooler is believed to be originally caused by heavy hydrocarbon components depositing on the exchanger, since the process temperatures are close to or below the estimated pour point temperature range. This is supported by the identification of heavy pyrolysis products in the foulant material samples. The observed polymer formation is supposed to be a consequence of the heavy deposits. The latter will accumulate at the front plate and the tube walls, where they are likely to retain monomers from the gas flow, thereby increasing their residence time and thus facilitating polymerization reactions.

2. In the suction drum upstream of the 3rd compressor stage, polymerization of accumulated monomers is believed to primarily cause the reported fouling. This accumulation is directly linked to the installed routing of cracked gas condensates, which are routed back from other collection drums. Residence times and quantities are significantly increased,

therefore strongly enhancing the risk of polymerization. As for the intercooler, the additional presence of heavy cracking products further increases the probability of solid formation.

Since both root causes are linked to the presence of heavier cracking by-products, it is supposed that the fouling risk was enhanced by the increased propane load and cracking severity conditions at which the plant was operated in recent years, and which were not covered by the original plant design. The impact of furnace feed scenario on condensate quantities and compositions is illustrated exemplarily for intercooler E2 in Figure 4.

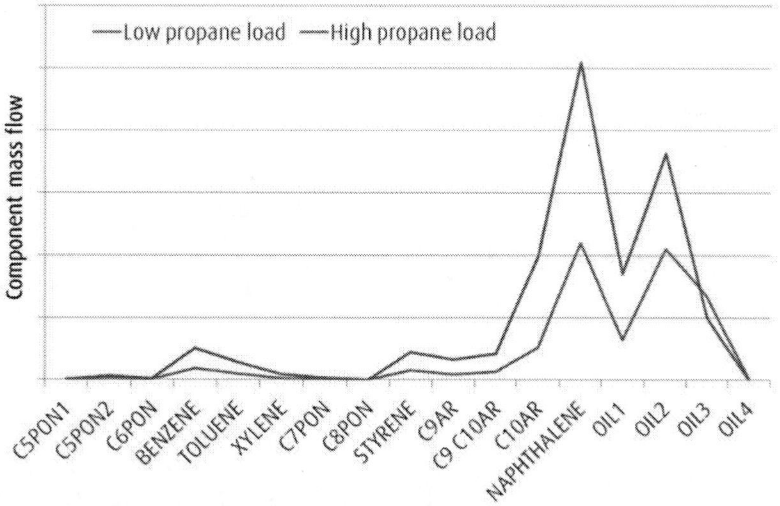

Figure 4: Component mass flow distribution in condensate stream of intercooler E2 for different furnace feed scenarios (PON: paraffins, olefins, naphthenes; AR: aromatics).

4. Developed fouling mitigation measures

Based on the root cause analysis, we defined mitigation measures to reduce the fouling risk linked to both mechanisms. As for the root cause analysis, process simulation and post processing tools were used to evaluate the effect of the proposed process modifications. The following mitigation measures were considered:

1. Recycle pyrolysis gasoline into the water quench system, in order to increase solvent quantities.
2. Optimize previous operation guidelines regarding wash oil quantity and quality for continuous injection into the compressor casing.
3. Modify the process topology by rerouting the condensates from suction drums D5 and D4.

Each measure was evaluated individually regarding its impact on both fouling mechanisms. However, none of these single measures was found to be capable of solving all fouling problems. Instead, a combination of all these measures was tailored which offers an effective fouling mitigation. This follows from Figure 5 and Figure 6, which show pour point profiles and residence time plots, respectively, for this new process setup.

The proposed measures therefore significantly reduce the residence times of potentially polymerizing components in the main process path and provide substantial washing effect at all stages of the cracked gas compressor train.

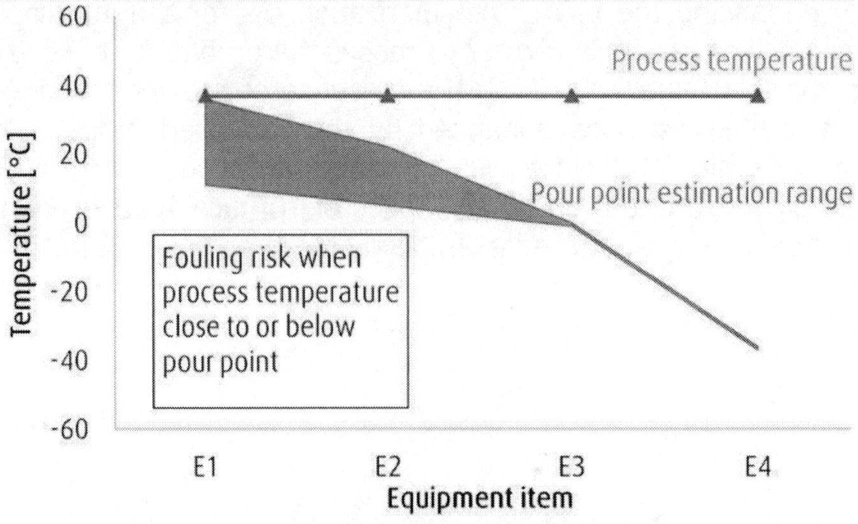

Figure 5: Pour point profile for condensate streams in charge gas compressor intercoolers (optimized process)

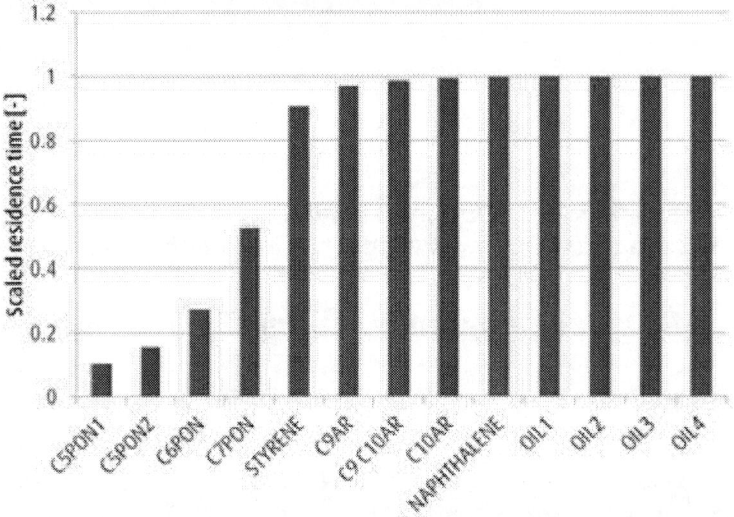

Figure 6: Combined component-specific residence time distributions in suction drums D3, D4 and D5 (optimized process). Scaled residence time values >1 indicate excessive accumulation of a given component due to plant internal recycles (PON: paraffins, olefins, naphthenes; AR: aromatics)

5. Conclusion and general applicability

The paper gives an illustrative example on how process simulation tools can be employed to assist plant operation services in understanding and mitigating fouling problems in water quench and charge gas compressor sections. The numerical tools are building upon standard process simulation methods, combined with an in-house developed post-processing method in order to get quantitative indications on fouling risk. This approach proved to be very effective and useful in a case of an actual plant. The findings described above have been confirmed by an experimental measurement campaign on-site.

The methods developed can readily be applied to other existing plants, independently of the feedstock. They can be used for the troubleshooting of existing fouling issues as well as for the assessment and prevention of future issues. In this manner, before a change in feed or furnace operation, potential fouling problems can be anticipated and taken into account as part of risk evaluation and mitigation strategy. The approach is similarly used for the design of new plants, in particular when increased feed flexibility is requested.

6. References

1. Schmigalle, Holger and Tolergard, Christer (2010), "Fouling Mitigation of Quench Tower System", in *Ethylene Producers Conference, AIChE Spring National Meeting*, San Antonio, Texas.
2. Kolmetz, Karl, Sloley, Andrew W., Zygula, Timothy M., Ng, Wai Kiong and Faessler, Peter W. (2004), "Design Guidelines for Distillation Columns in Fouling Service", in *Ethylene Producers Conference, AIChE Spring National Meeting*, New Orleans, Louisiana.
3. Collin, Gerd, Höke, Hartmut and Greim, Helmut (2012), "Naphthalene and Hydronaphthalenes", in *Ullmann's Encyclopedia of Industrial Chemistry*, Vol. 23, pp. 661-670, Wiley-WCH Verlag GmbH & Co. KGaA, Weinheim
4. Kimball, Russell (2011), "Fouling Mechanisms and Solutions in the Cracked Gas Compressor Train", in *EEPC Ethylene Seminar*, Lisbon, Portugal.
5. Hancock, Jessica M., Rossana, Debby and Arnst, Ted (2013), "Review of Organic Fouling Throughout an Ethylene Train", in *Linde Ethylene Licensor Conference*, Pullach, Germany.
6. Vachon, Jérôme, Hancock, Jessica M. and Robson, Ian (2015), "A chemist's perspective on organic fouling in ethylene operations: update", *Hydrocarbon Processing*, October, pp. 49-53.
7. Emmert, Thomas, Antonas, Sam and Kracker, Gunther (2016), "Full Range Feedstock Flexibility from Naphtha to Ethane", *Ethylene Producers Conference, AIChE Spring National Meeting*, Houston, Texas.

NGL Recovery Plant Technology: Ethane Recovery or Rejection for Changing Market Conditions

Scott Schroeder
Advisian
Houston, TX
scott.schroeder@advisian.com

Prepared for Presentation at
American Institute of Chemical Engineers
2016 Spring Meeting
11th Global Congress on Process Safety
Houston, Texas
April 10-14, 2016

UNPUBLISHED

NGL Recovery Plant Technology: Ethane Recovery or Rejection for Changing Market Conditions

Scott Schroeder
Advisian
Houston, TX
scott.schroeder@advisian.com

Keywords: NGL, Natural Gas, Turbo-expander, cryogenic, LPG

Abstract

In the United States ethane has typically been recovered from natural gas due to its increased value as a petrochemical feedstock. The rapid increase in natural gas production from developing shale plays has resulted in a change from the norm. The increase in ethane supply is exceeding current demand, resulting in depressed ethane pricing; furthermore many of these new sources of natural gas are not well connected to the traditional ethane handling network resulting in high transport costs. In this new environment rejecting the ethane into the gas stream instead of recovering it will be a preferred option for some plants. The rapid development of shale gas production has also led to extensive use of pre-engineered "standard cryo plants" which offer a shortened delivery time compared to a custom engineered plant. These "standard cryo plants" are typically designed to be as flexible as possible, but may not be optimum economic solution for each case. This paper will examine the typical requirements for ethane rejection, the typical performance of "standard cryo plants" and some of the alternate designs that maybe economically attractive for specific cases.

1. Background – Natural Gas Processing

Natural gas from the wellhead is primarily methane but also contains heavier hydrocarbons – ethane, propane, butanes, pentane and c6+ and may also contain H_2S, CO_2 or N_2. Although other compounds can be found, this is typical most of the gas that is commercially produced, either from conventional formations or from the new shale gas formations that have led to a huge increase in natural gas production in the United States. The gas from the wellhead is also quite commonly saturated with water.

The natural gas from the wellhead must be processed to meet sales gas specifications. The exact specifications vary regionally, but can be generalized as follows:

- Water content – the water content is specified to avoid liquid water condensing from the gas in the transmissions network. Typically specified as a water dewpoint or in Lbs H_2O/MMSCF (mg/m^3) of gas.
- H_2S content – the hydrogen sulfide is highly toxic as well as corrosive. Typically specified as grains/100 scf (mg/m^3).
- CO_2 content – Carbon dioxide can be corrosive and reduces the heating value of the gas. Typically specified as a maximum volume percentage.
- N2 content – Nitrogen is inert and reduces the heating value of the gas. Typically specified at a maximum volume percentage, maybe combined value with other combustion inert compounds.
- Hydrocarbon Dewpoint – The hydrocarbon dewpoint is specified so the liquids do not condense in the pipeline transmission system. Typically specified as a maximum temperature at a specific pressure or pressure range.
- Heating Value – The heating value range is specified to allow interchangeability of the gas. Systems that combust natural gas are normally adjusted for a specific heating value and a large fluctuation in the heating value will cause problems with end users. Typically specified in BTU/SCF (MJ/m^3). An additional interchangeability specification such as a Wobbe Number may be used in some cases.

The values for these specifications are regionally dependant and generally have evolved over time due to technological advances, market development and regulations.

The processing of natural gas to separate the hydrocarbons heavier than methane is the area where evolution has occurred due to cycles of technology and market developments. During the operation of the first natural gas pipelines it was discovered that the removal of heavier components was required to prevent liquid slugs from accumulating in low spots in the pipelines. This spurred the development of early dew-point control plants. The development of the automobile increased the demand for gasoline and processes were developed to recover increasing amounts of "natural gasoline" which consists mainly of pentanes and heavier. The processing technology continued to advance resulting in higher recoveries of butanes and then propane. The recovery of propane and butane allowed the development of the LPG (Liquefied Petroleum Gas) market. LPGs are used as a portable fuel since they can be maintained as a liquid under moderate pressures.

Starting in the 1960's, the introduction of the turbo-expander and the development of plate-fin or brazed aluminum heat exchangers allowed for the next step in the technology/market development cycle. This new technology made high levels of ethane recovery economically possible. However, as a pure product ethane is notably different than propane and heavier compounds. Ethane is not easily portable as a liquid and has not developed as a fuel product. Ethane is used almost exclusively as a feedstock to the petrochemical industry, primarily for the production of ethylene by steam cracking.

The growth of the petrochemical industry's use of ethane as a feedstock, especially in North America, has led to deep ethane recovery being standard practice. The typical recovery process is a turbo-expander cryogenic process which recovers an ethane plus mixture. The ethane plus mixture is often referred to as "Y-grade" or raw NGL mix. This ethane plus mixture is transported by pipelines to fractionation plants that separate the ethane, propane, butanes, and natural gasoline into purity products. These fractionation plants typically have petrochemical facilities located nearby to allow the ethane to be delivered by dedicated ethane pipelines. The fractionation facilities also often are located near salt caverns that allow cost effective storage of large volumes of hydrocarbon products.

2. The Shale Gas Revolution

The Shale Gas Revolution, as it has been called, is the rapid increase in production from what are called "unconventional" wells. These "unconventional" wells are drilled in relatively impermeable shale formations. Unlike conventional wells that are drilled vertically and typically flow freely, commercial production from the shale formations requires extended horizontal drilling followed by hydraulic fracturing (fracking). Although horizontal drilling and hydraulic fracturing are not new, it was not until about 2006 that the right conditions of improving technology combined with high natural gas prices and favorable regulation ignited the revolution. The drop in natural gas prices at the end 2008 didn't stop the growth in shale production; with the large improvements made in production efficiency and continued high oil prices, production moved to formations with high liquids content and high productivity to counter the lower gas prices.

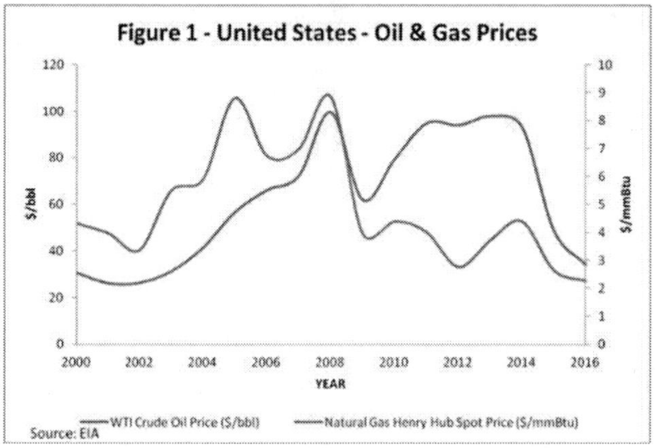

Figure 1 - United States - Oil & Gas Prices

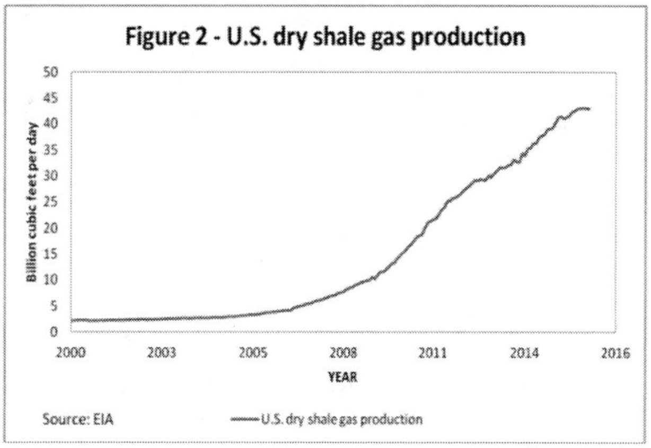

Figure 2 - U.S. dry shale gas production

The locations of the active shale plays are important to understand the drastic shift that has occurred in the North American production. Among the largest active plays, the Barnett and Eagle Ford plays are located close to regions with a history of oil and gas production and enjoy a more developed infrastructure compared to the Bakken, Marcellus and Utica plays. This is especially important with regards to ethane recovery since pipelines or cryogenic transports are needed to transport ethane and ethane is only used as petrochemical feedstock. This infrastructure takes time to develop. The first transportation projects to move ethane from the Bakken to Canada's existing ethylene crackers and from the Marcellus and Utica to the east coast for shipment in refrigerated gas carriers to existing European crackers are just coming online now. The new ethane supply has spurred a wave of proposed ethylene cracker projects positioned to take advantage of the new ethane supplies; however, most of these projects are still in the engineering phase and are not expected to be online until 2017 or later.

Figure 3 – North American Shale Plays

3. NGL Recovery Plants – The Impact of Shale Gas Revolution

As discussed in the background section, cryogenic turbo-expander processes have become the dominate plant type for NGL recovery. Among the many processes that have been developed, the Gas Subcooled Process (GSP) originally patented by the Ortloff Corporation in the late 1970's has become the dominate process for ethane recovery in North America[1]. The advantages of the GSP process are good ethane recoveries of 90+%, good tolerance to inlet CO2 which can freeze at cryogenic temperatures and the flexibility to reject ethane when required. Although there have been many processes that have been developed which have better performance than GSP process, the fact that GSP is now an open art process that provides good overall performance, relatively simple process design and simple operation has led to its continued use as a standard design.

The impact of the rapid growth of shale gas production has been the move towards "standard plants". Although the concept of modularized standard plants is not new to the industry, the high ethane recovery cryogenic plants had remained mostly a custom designed plant where the turbo-expander and plate-fin exchanger were designed specifically for the inlet gas stream and then put on order. The industry was also moving to increasing larger plants to take advantage of economies of scale. The penalty for custom designing these large plants is time. Several vendors realized that for the shale revolution, speed was more important that absolute efficiency. They designed "standard plants" that would work for a range of inlet gas compositions in several "rated" sizes such as 60 MMscfd, 120 MMscfd and 200 MMscfd. The long lead items such as the turbo-expander and plate-fin exchangers were pre-ordered and in some cases the module fabrication was already started before the customer committed to an order. The maximum plant size of 200 MMscfd was based on the ability to modularize the plant to the greatest degree possible while allowing easy transportation of the modules and quick installation. These vendors that had foresight to develop their "standard plants" early captured the majority of the market for new cryogenic plants installed in North America.

Figure 4 – GSP Process

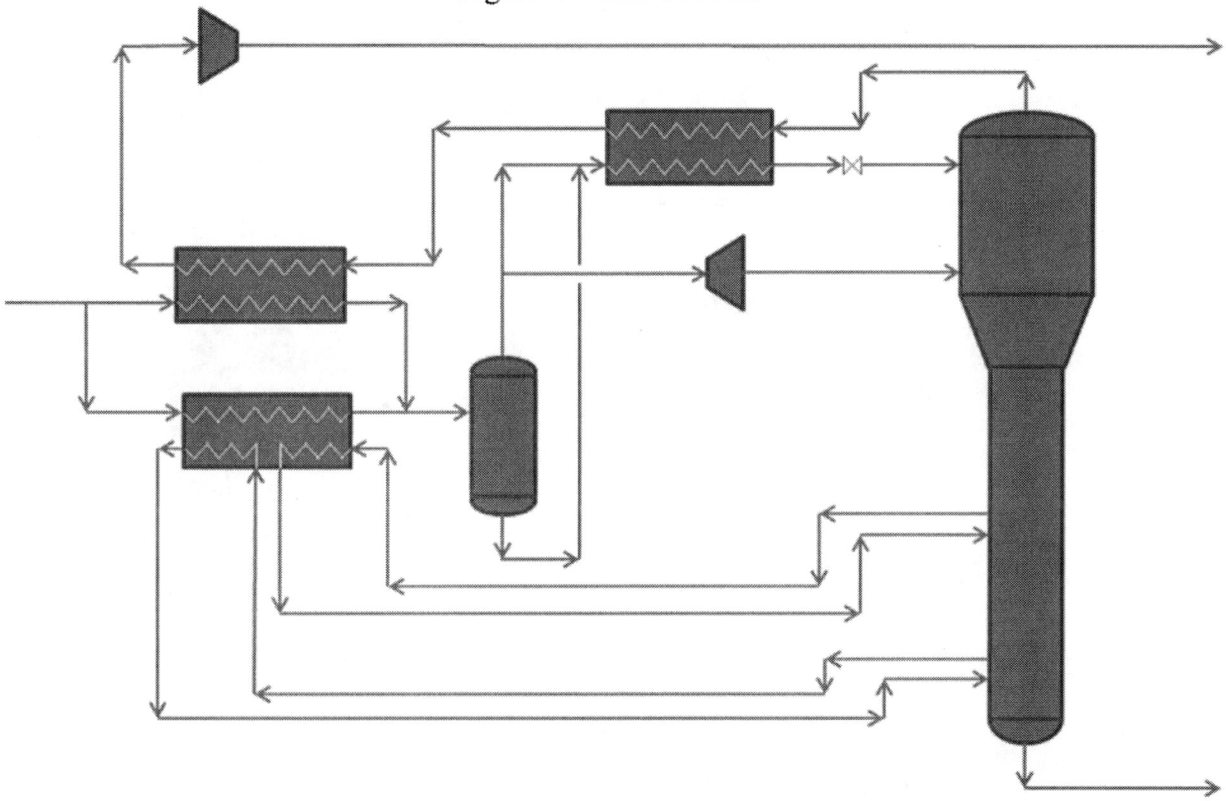

4. NGL Recovery Plants – Post Revolution

Although production from shale gas is not expected to stop, the sharp drop in oil prices at the end of 2014 has tempered the wild growth that was typical during the "Revolution". The industry has now entered into the evolution stage where only the strongest and fittest will survive. In this environment optimization of processing assets, both new and existing, is of prime importance.

The GSP standard plant is good overall, but has it weaknesses. Specifically, although it has the flexibility to reject ethane, high levels of ethane rejection causes the propane recovery to fall as well, possibly dropping into the 80% range in full rejection mode. The lost propane can limit the amount of rejection possible in cases where the high heating value of propane pushes the gas to its specification limit. The lost propane is also lost value as propane typically maintains a positive value for recovery.

There are a number of processes that have improved on the GSP design that allow higher propane recovery during rejection. A typical enhancement is the addition of recycle vapor stream that is used as a top reflux, shown in Figure 5 is Ortloff's Recycle Split Vapor (RSV) process[2]. This addition increases the maximum ethane recovery and maintains higher propane recovery during ethane rejection, but requires a higher initial capital investment. Several other technologies are also available that effectively enhance the reflux compared to the GSP process.

Figure 5 – Recycle Split Vapor Process

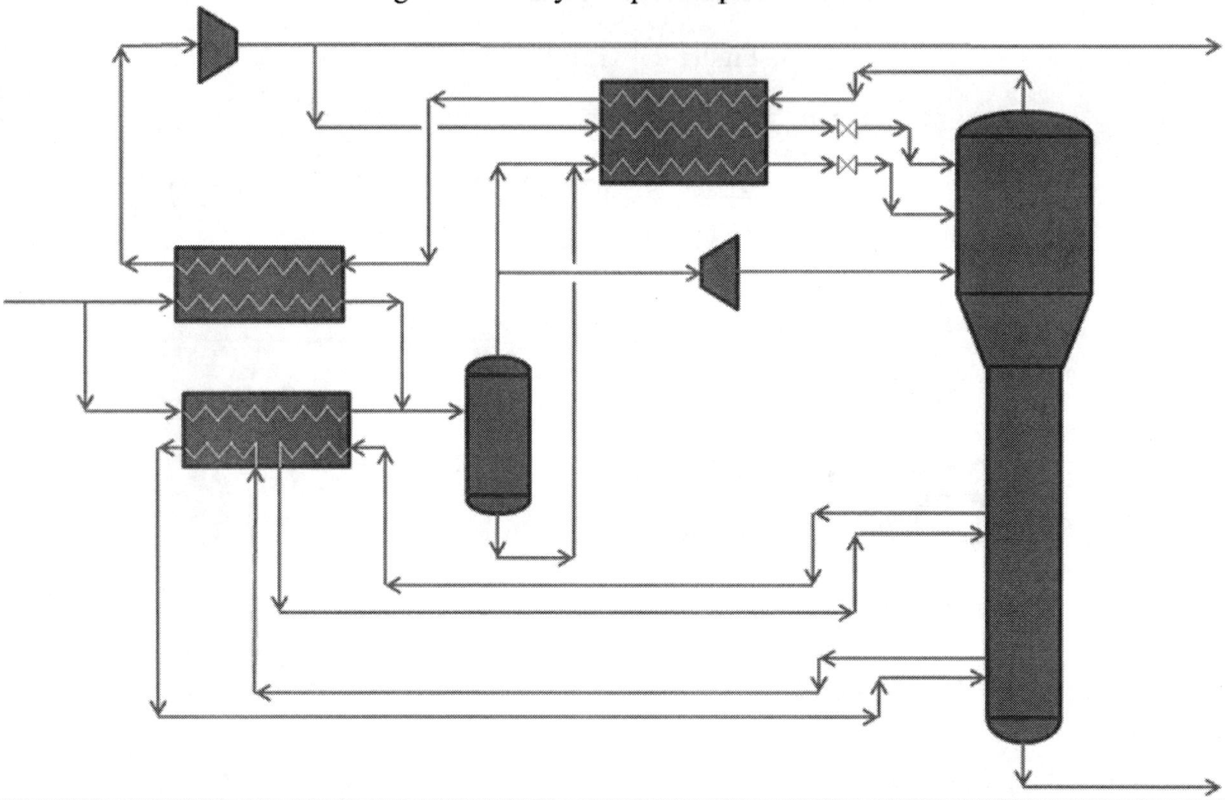

Processes that specifically focus on high propane recovery can typically achieve propane recoveries of 98+%, with nearly complete ethane rejection, with greater efficiency than an ethane recovery process being operated in ethane rejection mode. The designs using turbo-expanders typically use two fractionation towers; however, some designs have combined towers into a single column. Several mixed refrigerant designs that avoid the use of a turbo-expander are also available. If lower levels of propane recovery are acceptable, then a simplified refrigerated plant with a lower capital investment could be considered.

Revamping or retrofitting an existing unit or an uninstalled "standard plant" may be a better option than a new custom plant or simply installing a "standard plant" as-is. As an example Ortfloff Engineers is now specifically marketing their Retro-Flex™ package for this purpose[3]. This package is proposed as a standard add-on to a "standard plant" and bolts into the overhead system of a standard GSP plant to improve the reflux especially under ethane rejection conditions.

Making the decision whether to install a "standard plant" or consider one of the alternatives is currently not as clear. For each location a variety of factors should be considered:

- Constraints – product specifications, regulations, permit requirements
- Market Conditions – expected production rate, expected transportation costs, expected product
- Market Variably – development of sensitivity cases for the market conditions over the expected life of the plant
- Capital Cost – capital cost of the new plant or revamp of a existing unit should consider the possibility of obtaining distressed assets in the current market environment as well as current labor conditions for construction
- Operation Cost – the operating cost of the plant gains more importance as the product margins decrease
- Schedule – the economic advantages of a shorter construction schedule cannot be ignored

By using these factors the economics of each of the options can be compared and the optimum process selected for a specific installation.

5 Conclusion

The shale revolution has brought a wave of change to the North American and world markets, rapid expansion has led to an oversupply condition causing large swings in product pricing. The ethane market has specifically suffered since the ability to transport purity ethane is limited compared to other NGL products and the demand is almost exclusively for ethylene crackers. The low price of ethane has spurred a wave of new ethylene crackers as well as transportation projects to move ethane to existing crackers. However, the drop in crude prices has left many questioning where the market will stabilize.

Purchasing and installing a "standard plant" had become almost a standard practice. The margins were high enough that getting the plant running as soon as possible was the primary concern. As the margins have decreased, spending the time to perform economic screening studies looking at the available options to increase capacity or optimize processing efficiency is increasingly important; the "standard" answer may not be the best answer for all situations.

6. References

[1] Campbell, R.E., Wilkinson, J.D., U.S. Patent 4,157,904

[2] Campbell, R.E., Wilkinson, J.D., and Hudson, H.M., U.S. Patent 5,568,737

[3] Website - http://www.ortloff.com/files/Retro-Flex-01.pdf

Additional references not cited:

[4] "GPSA Engineering Databook" 12th Edition

[5] "Hydrocarbon Gas Liquids (HGL): Recent Market Trends and Issues," U.S. Energy Information Adminstration, November 2014

REACTIVE DISTILLATION FOR PROCESS INTENSIFICATION – A STUDY OF DESIGN PARAMETERS

Gaurav Arora, Mohammad Shafiei
Process Technology, T&I
SABIC
Sugar Land, TX

AIChE Spring Meeting
Houston, TX
12th April, 2016

CHEMISTRY THAT MATTERS™

AGENDA

- Introduction
- Applications
- Case Study – Irreversible Reaction in Series
- Design Variables
- Results and Discussion
- Conclusions

No. 1

INTRODUCTION

- Reactive Distillation: Combination of reaction and distillation in a single equipment.
- Earliest application dates back to 1920s for esterification processes in a series of patents by Backhaus.[1]
- There were over 200 licensed commercial scale reactive distillation processes by the year 2007, many of which are in commercial operation.

Example: A + B $\leftrightarrow$ C + D

$\alpha_C > \alpha_B > \alpha_A > \alpha_D$

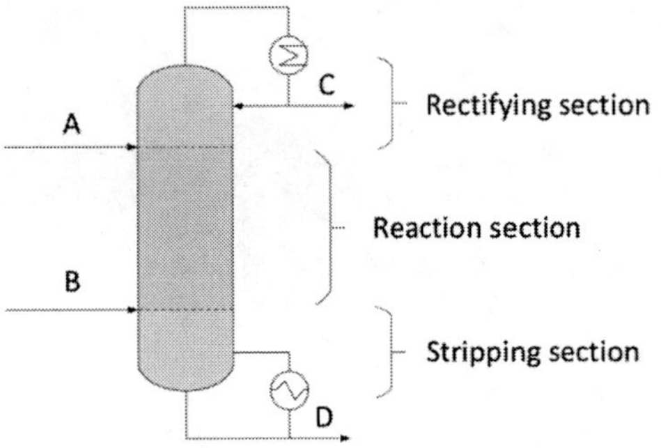

- Advantages
  - Process Simplification
  - Economical Benefits
    - Improved conversion
    - Higher selectivity
    - Improved reliability
    - Capital cost reduction
    - Operating cost savings
  - Environmental, safety benefits
- Constraints
  - Temperature
  - Volatility
  - Reaction Rates
  - Reaction phase
  - Scale-up

[1] US Patents 1400850. 1400851. 1400852

No. 2

APPLICATIONS

- Over 236 types of reaction systems for reactive distillation application.[1]

- Types of reactive distillation systems
  - Reversible or **Irreversible**
  - Catalytic or **Non-catalytic**
  - Catalytic: Homogenous or Heterogeneous
- Types of reaction systems
  - A + B $\leftrightarrow$ C +D
  - A + B $\leftrightarrow$ C
  - A $\leftrightarrow$ B +C
  - A + B $\leftrightarrow$ C + D
B+C $\leftrightarrow$ D + E
  - **A + B $\rightarrow$ C**
C + B $\rightarrow$ D

- Class of problems and Examples
  - Equilibrium limited reactions
- MTBE, ETBE, TAME, ...
  - **Suppressing side reactions**
- Diacetone alcohol, EG, ...
  - Azeotropic system
- Methyl Acetate, ...
  - Temperature control
- Cumene, EG, ...

[1] W.L. Luyben, C.C. Yu, *Reactive Distillation Design and Control*. John Wiley & Sons, Inc.: Hoboken, New Jersey, 2008.

No. 3

CASE STUDY - IRREVERSIBLE REACTIONS IN SERIES

- Problem definition – Minimize rate of formation of [D] thereby maximizing selectivity for [C] by controlling reactant concentration profile using design variables.
- Class of problem – suppressing side reactions
- Reaction system – Irreversible reactions in series

$A + B \xrightarrow{k_1} C$

$C + B \xrightarrow{k_2} D$

$\alpha_B > \alpha_A > \alpha_C > \alpha_D$

$E_{a1} = E_{a2}$

$k_2 > k1$

$\frac{r_2}{r_1} = \frac{k_2\,[C]}{k_1\,[A]}$

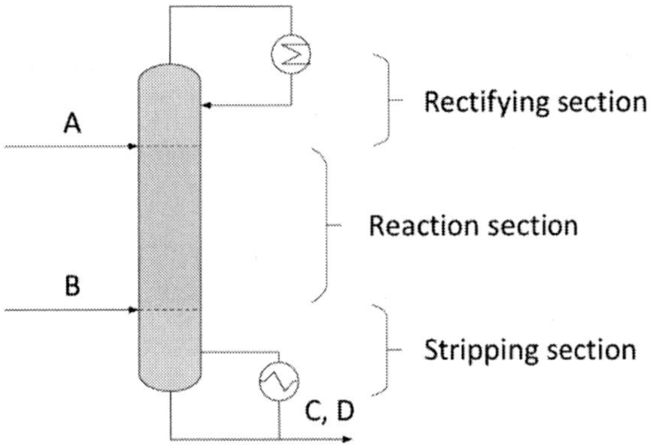

No. 4

DESIGN VARIABLES

- Pressure
  - Conventional distillation: Typically P ↓ ⇨ T↓ ⇨ α ↑
  - Reactive distillation: P ↓ ⇨ T↓ ⇨ α ↑, also *k* ↓ ⇨ hold up volume ↑
- Tray holdup
  - Assume a holdup → design column → calculate liquid height
- Reboiler duty (or distillate rate or distillate/feed ratio etc.)
  - Affects temp profiles, conc. ⇨ phase equilibria along the column ⇨ selectivity
- Number of theoretical stages
  - Higher number of trays does not necessarily reduce the energy requirements.
- Feed location
  - Feed location effects the compositions profiles and therefore the column performance.
- Stoichiometric operation or Excess reactant

No. 5

CONFIGURATIONS

- Feeds fed in exact stoichiometric ratio reduces downstream processing.
- Process configuration with excess feed offers better control of the column.

$A + B \rightarrow C$
$C + B \rightarrow D$

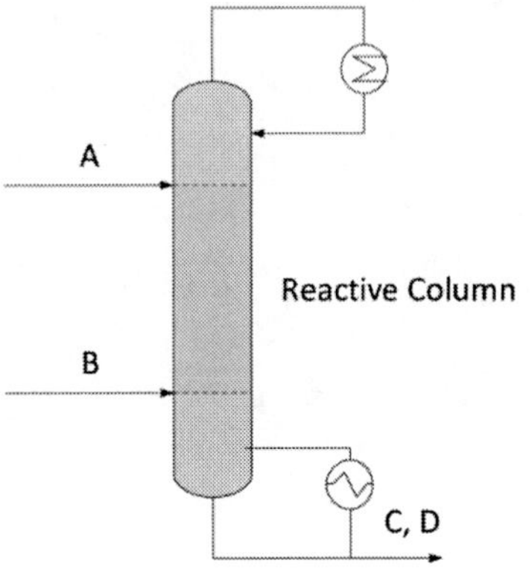

Feed in stoichiometric ratio

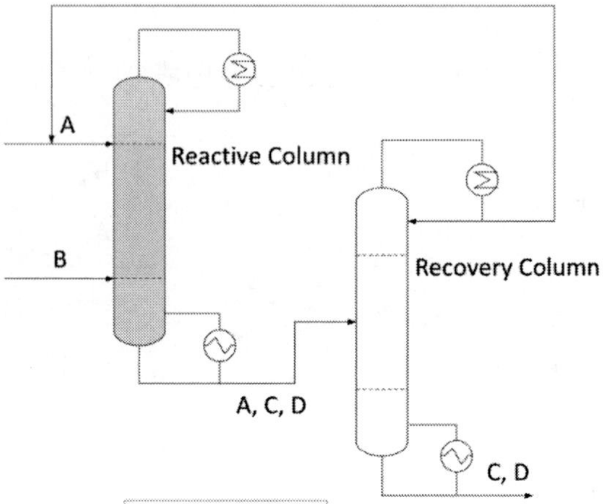

Excess [A]

No. 6

NUMBER OF THEORETICAL STAGES

- Reactant [B] gets consumed moving up the column.
- Unreacted [B] in the vapor phase condenses at the top of the column.
- [B] in the condenser results in formation of [C] followed by [B].

$A + B \rightarrow C$
$C + B \rightarrow D$

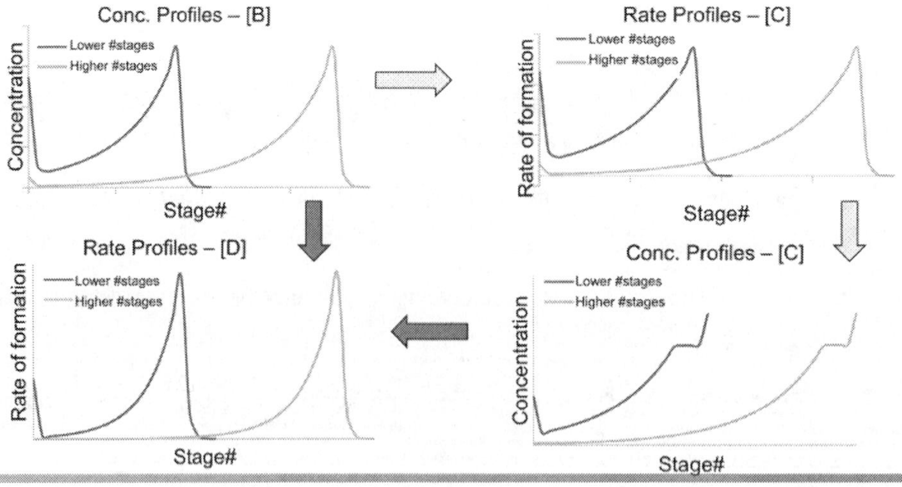

No. 7

NUMBER OF THEORETICAL STAGES

- Higher amount of [B] in the condenser → higher the amount of [D] formed → lower the selectivity for [C]

A + B → C
C + B → D

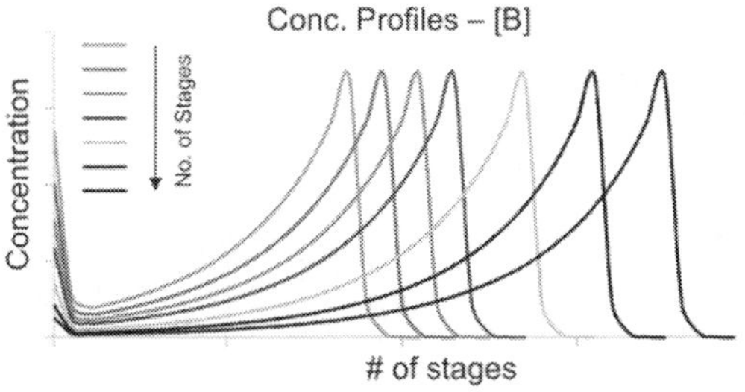

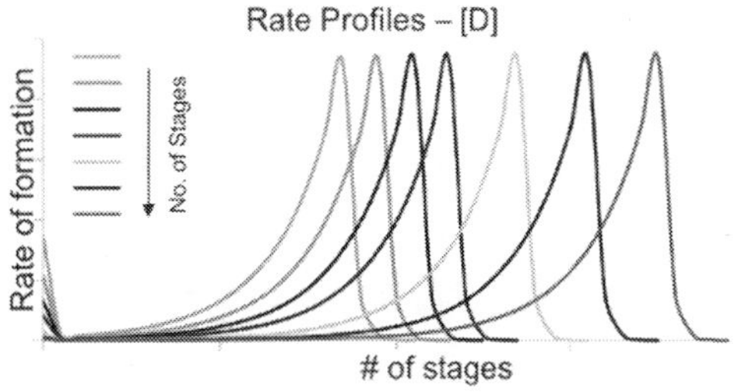

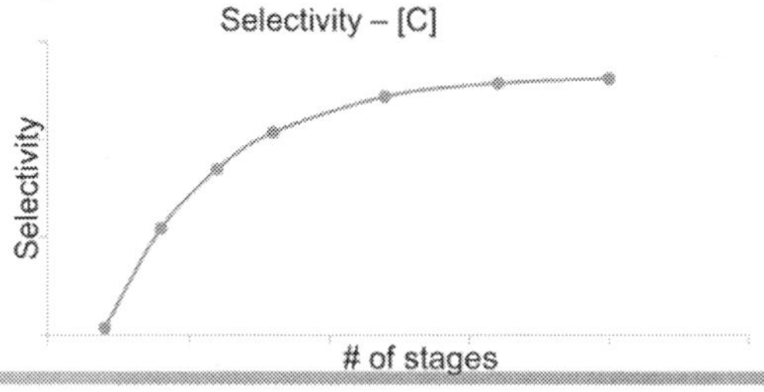

No. 8

FEED LOCATION – [B]

- Reactant [B] gets consumed moving up the column.
- Unreacted [B] in the vapor phase condenses at the top of the column.
- [B] in the condenser results in formation of [C] followed by [B].

A + B → C
C + B → D

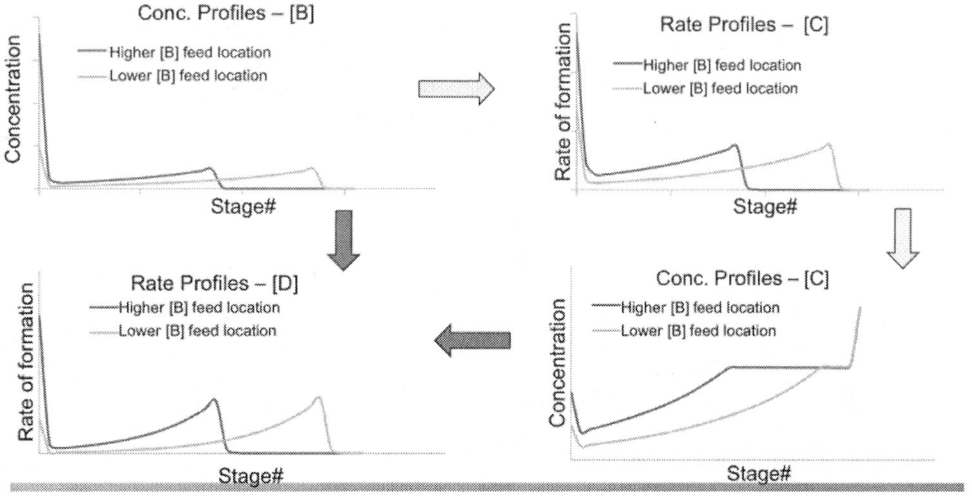

No. 9

FEED LOCATION – [B]

- Higher the amount of [B] in the condenser → higher the amount of [D] formed → lower the selectivity for [C]

A + B → C
C + B → D

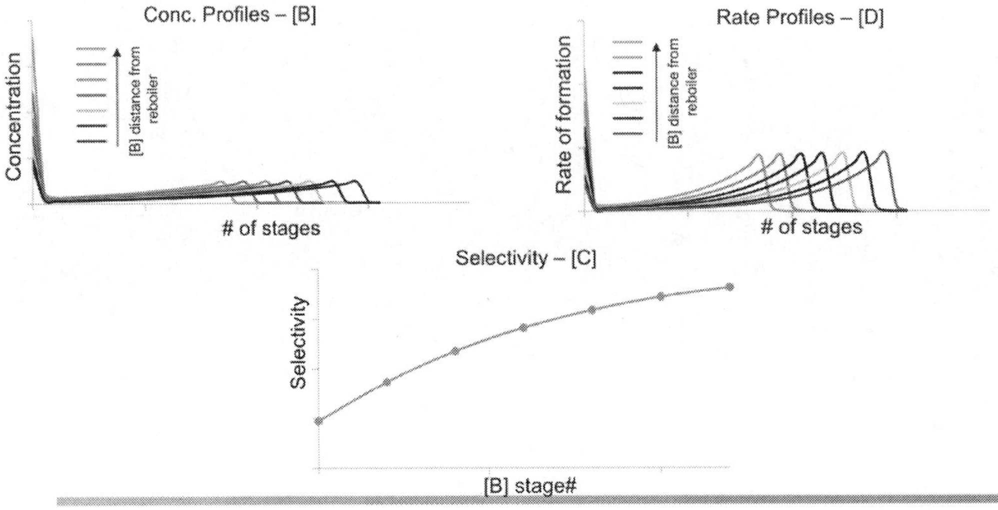

No. 10

FEED LOCATION – [A]

- Higher conc. of [C] on stages above [A] feed location.
- Higher the conc. of [C] → higher the rate of formation of [D] → lower the selectivity for [C]

A + B → C
C + B → D

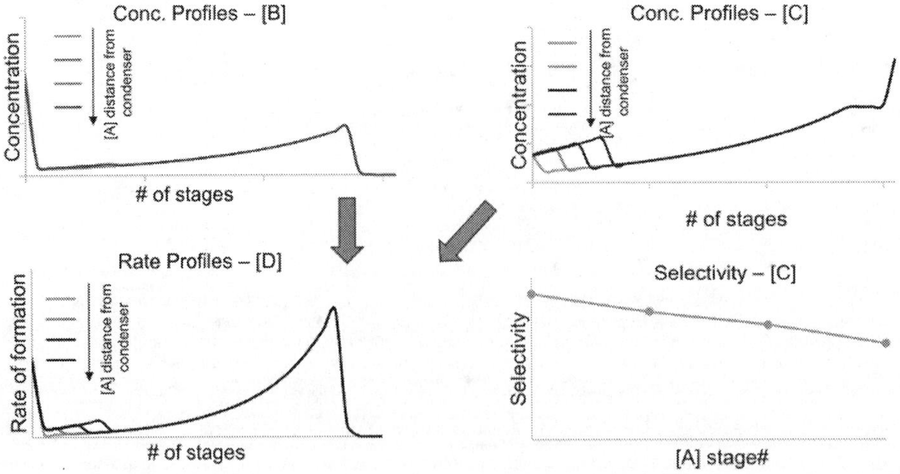

No. 11

REBOILER DUTY

- Lower the duty → lower the vapor traffic in the column
- Larger temperature drop if colder feed is mixed with a lower vapor traffic.
- Colder temp → higher the amount of [B] in the liquid phase → higher the rate of formation of [C] → higher the rate of formation of [D]

A + B → C
C + B → D

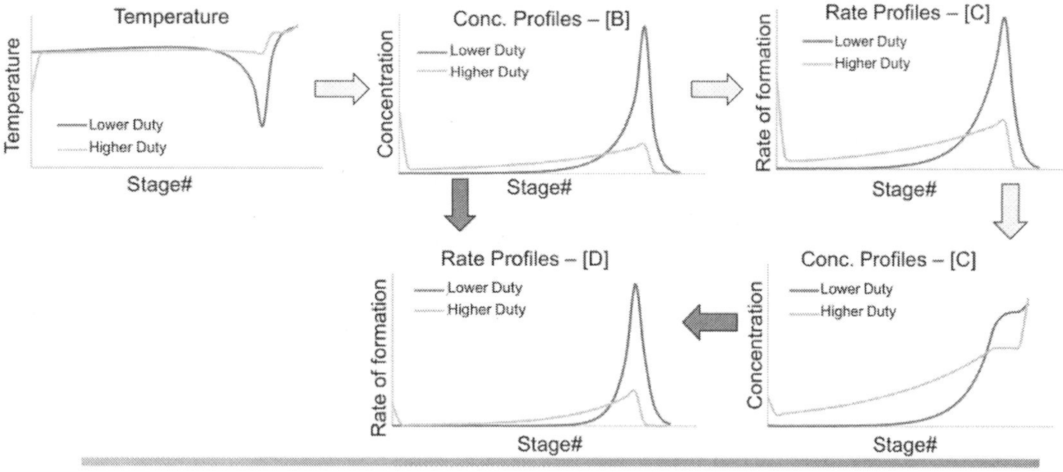

No. 12

REBOILER DUTY

- Higher the amount of [B] at the feed location → higher the rate of formation of [D] → lower the selectivity for [C]

A + B → C
C + B → D

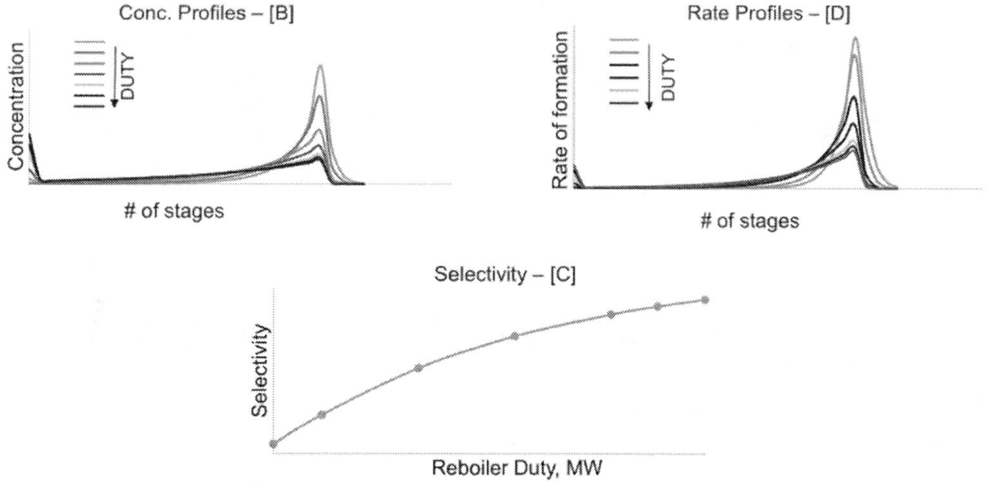

No. 13

FEED FLOW RATIO

- Lower [B] traffic in the column results in higher [C] concentration.
- Higher [C]/[A] → higher r_2/r_1 → lower the selectivity for [C]

A + B → C
C + B → D

$$\frac{r_2}{r_1} = \frac{k_2}{k_1}\frac{[C]}{[A]}$$

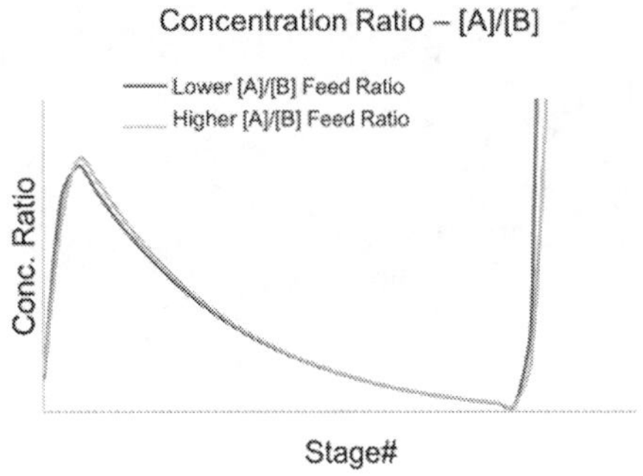

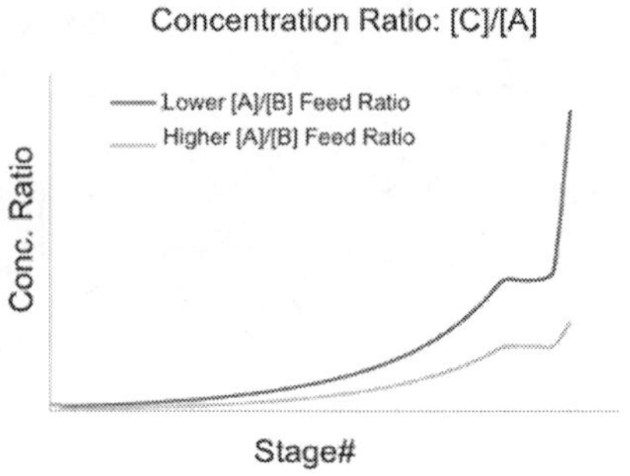

PRESSURE

- Higher pressure → higher temperature → higher rate constants.
- Higher pressure → higher temperature → lower the amount of [B] in liquid.
- Rate of formation of [D] set by competing rate constants and conc. of [B] in liquid.

A + B → C
C + B → D

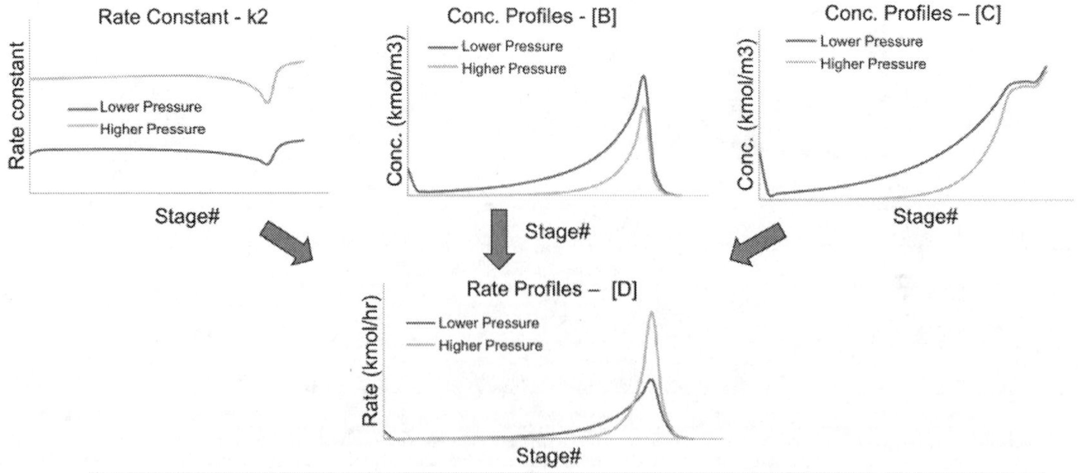

CONCLUSIONS

- Effectivity utilized analysis of concentration and hydraulic profiles in the column to study the effect of design parameters on the reactive distillation performance.
- Analysis allows optimizing the design maximizing economic benefits of the reactive distillation design.
- In general, for a reaction in series system [A + B → C, C + B → D], selectivity for C can be increased by controlling concentrations of the reactants as set by phase equilibria
  - reducing the amount of light component [B] reaching the condenser
  - reducing the ratio of [C]/[A] in the column
  - reducing the concentration of [B] at the feed location
- Similar analysis can be used for obtaining a fundamental understanding of other reactive distillation systems.

No. 16

Design and Operation of B&P Process Podbielniak Liquid-Liquid Separator Centrifuges for AICHE Symposium

B&P Process Equipment and Systems LLC /
B&P Littleford Day LLC
1000 Hess Avenue
Saginaw, MI 48601
sales@bpprocess.com

Presenter: Arthur Bach / B&P Process Engineer

Prepared for Presentation at
American Institute of Chemical Engineers
2016 Spring Meeting
11th Global Congress on Process Safety
Houston, Texas
April 10-14, 2016

UNPUBLISHED

Design and Operation of B&P Process Podbielniak Liquid-Liquid Extractor Separators for AICHE Symposium

B&P Process Equipment and Systems LLC /
B&P Littleford Day LLC
1000 Hess Avenue
Saginaw, MI 48601
sales@bpprocess.com

Presenter: Arthur Bach / B&P Process Engineer

Other Author(s):
Cited B&P Process Equipment Process Engineers involved in research and development of machines.

Keywords: Pod Contactor ®, Podbielniak, Liquid- Liquid contacting centrifuge, continuous process centrifuge, Oil acidulation, pharmaceutical production, Fermentation, separation, extraction, stripping, B&P Process Equipment, Vitamin Extraction, Flavor oils, Essential Oils, Fat, Water-Oil separation

Abstract

The Podbielniak Liquid-Liquid contacting centrifuges are a type of phase separator that uses the enhanced gravitational field for phase separation or contacting for mass transfer. These type of units may be used for a two phase separation of immiscible liquids, distillation applications that could be accomplished by a liquid- liquid solvent extraction (possibly due to temperature concerns for product), tower stripping/ rectification or any gravity settling applications where process purity may benefit due in a higher G field. These may also be considered where tower or tank liquid- liquid phase separation takes up to much plant space or to reduce quantities of a hazardous material held up in process area equipment.

This article will cover some of the concepts for design and examples of operational performance of acetic acid extraction with water from methyl isobutyl ketone and kerosene/ water separation. These are two well known process examples for a Podbielniak Liquid-Liquid contacting centrifuges and other liquid-liquid contacting technologies. Additionally examples of Process implications of centrifuge operation parameters are discussed as are some existing user operation troubleshooting ideas.

Applications include major chemical process steps such as separations, extractions, rectification, ion exchanges, neutralization & washings. Examples are found industry wide in chemical production, Pharmaceutical applications, Petroleum Refining Liquid Ion Exchange, Food industries (Flavor oil extractions) and many other specific examples.

1. Introduction & Background

The Podbielniak Liquid-Liquid contacting centrifuge is a well known example of Liquid-Liquid extraction process equipment which has been in service for many years. It is often referred to as a "Pod Contactor®". It may also be referred to as a Centrifugal Extractor, a Continuous Centrifuge, a Continuous Centrifugal Extractor, or several variations there-in. It is a rotating chamber with 2 phase feeds that takes advantage of the higher gravitational field within the rotating chamber to achieve or improve phase separation, extract a target species or any other chemical process application involving the contact of two immiscible liquid phases.

Initial development originated from, among others, the research and efforts of Dr. Walter J. Podbielniak (1899-1978). Associated patents from him include US Patents 2,004,011 (6/4/1935) & 2,044,996 (6/23/1936) on counter current centrifugal devices consisting of rotating tubes to perform two phase contacting in the 1930s. Additionally, in early 1935 his wife purchased US Patents 1,936,523 & 1,936,524 (11/21/1933) by Adolf Placek for a spiral banded, horizontally rotating chambered device for distillation & absorption. Dr. Todd's patent expanded the application to include contacting of immiscible or semi-immiscible phases in addition to the Gas-Liquid & Vapor-Liquid contacting. Development work was ongoing through the 1930s until 1960 under Dr. Podbielniak's company Podbielniak Centrifugal Super-Contactor Laboratories (and several other iterations) until being sold to Dresser Industries as the Podbielniak Division. Some of the chief processes driving the development came from the use of the Podbielniak separators in penicillin production and Phenol contacting of Lubrication oils, in addition to many others.

Dr. David B. Todd joined the Podbielniak division of Dresser industries in 1963. Under his direction, research and redesign activities were completed to standardize the product lines. The standardized Pod Contactor sizes ultimately became the A-1, B-10, D-18, D-36 & E-48 designations when the division was sold in 1968 to Baker & Perkins, relocating the division and personnel to Saginaw, Michigan; the current B&P Process Equipment site.

Since the initial development and design refinements of these machines, wider chemical industry awareness has decreased. Awareness is kept up by existing users, but it may not be as familiar to outside or newer processes. This paper is to refresh industrial awareness that the Liquid-Liquid contacting centrifuge is a viable and effective technology for any application where two immiscible / partially immiscible liquid phases are to be separated or contacted together for mass transfer and/ or reaction.

2. Concepts for design & operation of Liquid-liquid contacting Centrifuges

2.1 Design

2.1.1 Basics

Figure 1 shows an example of B&P Process Equipment Liquid-Liquid contacting centrifuges:

Figure 1 Example of a B&P D-900 (previous D-36) Liquid- Liquid contactor Centrifuge. Rated Capacity ~68.1 m3/hr (~300 GPM)

Figure 2 illustrates an example of the fluid passage through the unit.

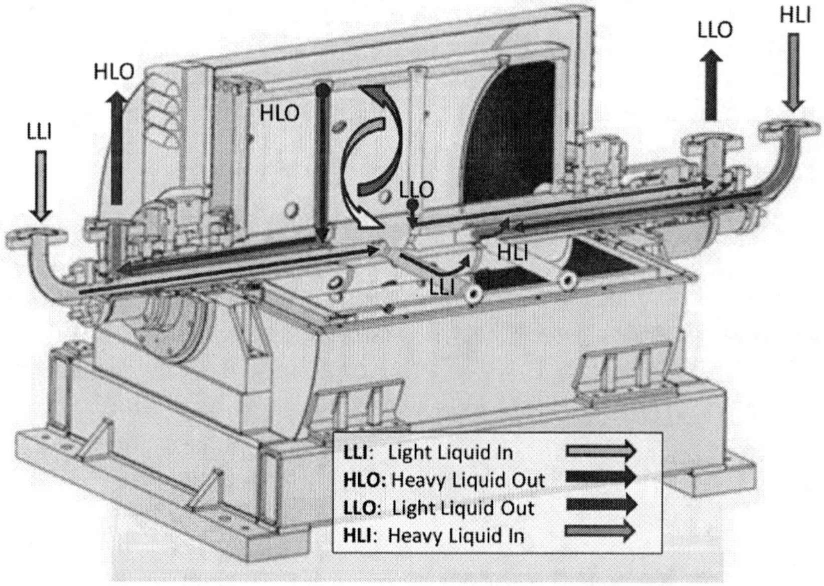

Figure 2 Flow path of feed and discharge streams.
(Note- Flow arrangement of feeds varies depending on process and operational needs)

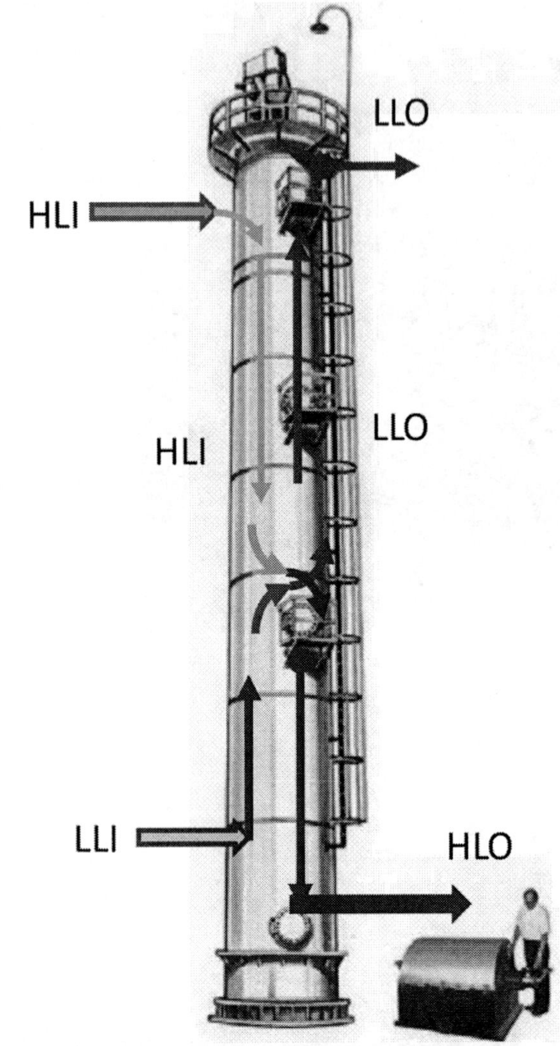

Figure 3 Equivalent stripping Column flow paths

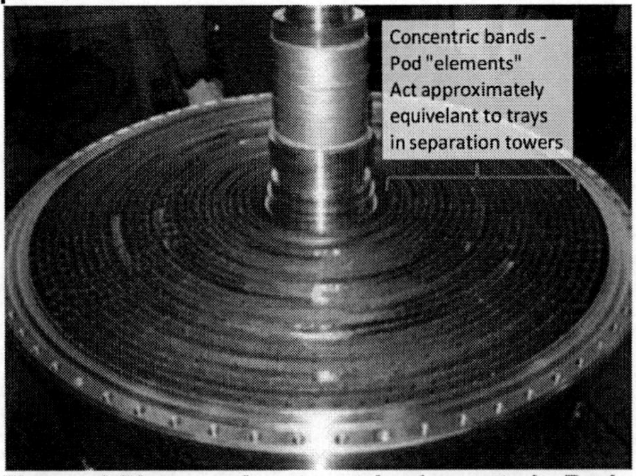

Figure 4 Picture of concentric elements in Pod Contactor Rotor

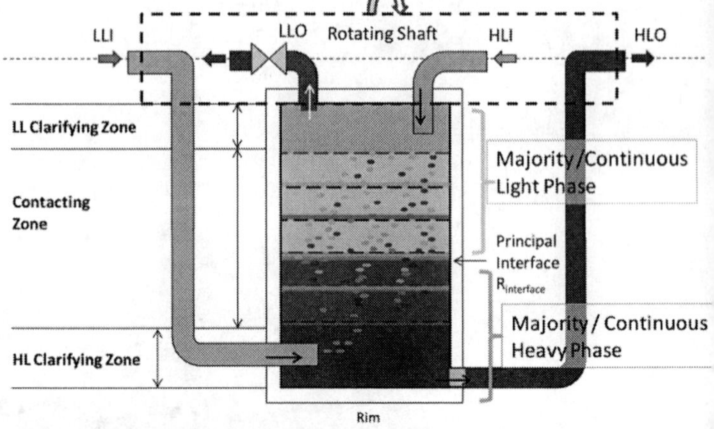

Figure 5 Tower Analogy of Centrifugal Extractor [1, 2]

The radial flow path in a Liquid-Liquid contacting centrifuge is equivalent to a separation tower (or equivalent device) where the Heavy Liquid (HLI) enters near the top of the vessel and works its way past the light phase, through the column trays and to the Heavy Liquid discharge (HLO) on the bottom of the unit. Conversely, the Light Liquid phase will enter (LLI) near the bottom and float is way up to the Light Liquid Out (LLO) discharge.

Physically, the top of the tower corresponds to the rotor shaft discharge path of the LLO; the bottom tower HLO discharge is equivalent to the centrifuge outer rim position; the Pod Contactor HLO discharge path is via discharge tubes which isolates it from the contacting phases and carries it to the shaft to exit the machine.

The Feed streams (LLI & HLI) will enter at feed tubes with discharge holes positioned at radial positions approximately equivalent to the LLI & HLI tower positions

Some contacting centrifuges (such as B&P Centrifugal Extractors) use shaft concentric metal band "elements" (Note Figure 4) to provide multiple individual equilibrium separation stages- parallel to tower trays .

Liquid-Liquid contactors are used in many applications. Table I is a brief example of B&P Process Pod's applications. The full range of processes is in the hundreds or thousands.

Service	Generic case	Some Specific Examples
Petroleum Refinement	Lubrication oil extractions	Color/ impurity removal
	Acid / base/ neutralization treatment of feed stocks for impurity removal	Paraffin Purification, :"White" oil (Hi Purity paraffin), Transformer Oil, Sulfonic Acids
	Acidulation/ caustic treatment for product recovery	Napthenic Acid recovery
Liquid Ion Exchange	Metal recovery	Uranium, Vanadium, Rare Earth Elements
	Acid recovery	Citric Acid recovery
Extraction	Acrylic Ester	
	Aromatic / Flavor oil recovery	Citral (Lemonal), Terpenoids, others
	Pharmaceutical Extractions	Penicillin, Erythromycin, Tylosin, other antibiotics
	Plasticizers	Fumaric Acid /Maleic Acid extraction from DiButyl Phthalate
	Aniline (phenylamine / amino benzene)	
	Butadiene Extraction	Butadiene extraction
	Vitamin recovery	Tocopherol (Vitamin E), Ascorbic Acid (Vitamin C)
Wash	Water Wash	
	Polyols (multi Hydroxyl alcohols, Poly esters)	Precursor for many polymer processes
	Nylon intermediates	
	Polycarbonate wash	

Table I Short Example list of applications [4]

2.1.2 Scaling

Continuous Liquid-Liquid Extraction centrifuge equipment and its understanding remains an under researched area. Historical manufacturers of Liquid-Liquid extractors include Podbielniak Extractors (B&P Process Equipment, formerly: APV Chemical Division, Baker Perkins Inc, Chemical Machinery, Dresser Industries Podbielniak [3]), Alpha-Laval (Sweden), GEA Westfalia Separator GmbH. How scale up is conducted by current manufactures may vary. This presentation covers the current practices of B&P Process Equipment in line with its internal design policies. These practices are from 1st principles, pilot testing and manufacturing experience. Scale up is facilitated by machine designs which accommodate scaled and similar physical concepts such as similar internal liquid velocities, residence times, design available G-fields, pressure operations, etc.

2.1.2.1 Scaling by B&P Process Equipment

Pod Contactor scale up is generally based on testing results from a smaller pilot scale test machine. For best pilot scaling B&P uses the B-250 (~40 GPM) unit. The smaller A-25 model (0.4 GPM) is also used as a proof of concept machine, but scalability is not as well reflected. There is a recent reintroduction of the A-125 model that is expected to scale better to the production size units. Machines may be scaled from one production model to the others by the concept of G-minutes. This is the mathematical product of the multiples of gravity (G's) and the "Total Traffic" residence time (τ_{TT}).

$$\text{G-minutes} = (\text{G's}) * \tau_{TT}$$

Equation 1 G-minutes

G's = Rim Centripetal acceleration/ 1 g

τ_{TT} = Residence time for total feed rate

$$\tau_{TT} = \left(\frac{V_i}{Q_{TT}} \right)$$

Equation 2 Total Traffic Residence Time

V_i = Hold up volume of considered Centrifugal Extractor

Q_{TT} = Total Volumetric Feed rate

$$Q_{TT} = Q_{HL} + Q_{LL}$$

Equation 3 Total Volume Traffic

Q_{LL} = Light Liquid Volumetric Feed rate total

Q_{HL} = Heavy Liquid Volumetric Feed rate total

(Equation 3 may be $Q_{LLI} + Q_{HLI}$ or $Q_{LLO} + Q_{HLO}$)

When the G-minutes from one model to the other are matched, the Extraction stages or separation quality is conservatively expected to be the same. This is supported by many years of observed operating experience. Centrifuge capacity varies from process to process and scale up continues to rely on initial pilot scale trials.

Historical process capacity information may also be used for scaling of similar chemical processes provided it is both available and a close match for process operation conditions. Historical Data does have its limits though. Chemically similar processes can behave differently due to process subtleties, affecting system capacity. Additionally what is defined as acceptable product for one user to may not be acceptable for another user. And certainly, new processes may not have data or operating conditions available.

Machine design also accommodates a "Traffic Velocity" which is set to establish heavy & light liquid fluid velocities at the Pod Contactor elements. Traffic Velocity is set to values known to accomplish effective process performance- IE emulsion prevention, effective mass transfer, and phase separation.

$$\upsilon_{\text{Total Traffic}} = \frac{Q_{TT}}{\Sigma A_{\text{Flow Area}}}$$

Equation 4 Traffic Velocity calculation

υ_{TT} = Total Traffic Velocity (for individual element)
Q_{TT} = Total Volumetric Feed rate
$\Sigma A_{\text{Flow area}}$ = Σ element flow area

Effective process performance judgment is based on observed process characteristics for a process and/ or chemical families. Some of these families include pharmaceutical/ fermentation separations, vitamin extractions from fats or oils, petroleum refinement oil acidulations, clean fluid extractions or other generic families of processes. Other process performance considerations include viscosity or solids content that may or may not be an additional specific process concern.

When B&P refers to Centrifugal Extractor process capacity, the reference is to a specific well known Liquid- Liquid Extraction process: Methyl-Isobutyl Ketone, Acetic Acid and Water. It is a clean service (IE no solids) with a nominal SG difference (ΔSG) of 0.2- which is typical for many water- oil type systems. For a fermentation process separation, capacity is expected to be less than ½ of clean service capacity. Penicillin separation process is the "benchmark" for fermentation services. Different chemical processes can have different capacity ratings. Pod Contactors are able to accommodate down to 0.01 ΔSG, although a capacity reduction occurs as ΔSG gets very small.

2.1.3 Extraction efficiency and stages

The extraction efficiency of centrifugal extractor operation is quantified using the same methods used by other extraction or stripping processes equipment. Theoretical Transfer Stages can be used to quantify process efficiency and be determined by thermodynamic process practices such as these:

- Counter current theoretical stages from the right triangle method from system equilibrium.
- The shortcut McCabe-Thiele stepwise calculations can be used by stepping off the number of stages between the operating line and the equilibrium line.
- The NTU (numbers of transfer units) may also be used to express efficiency.

Figure 6 is an example of the A-25 model centrifugal extractor demonstrated capacity for Water solvent Extraction of Acetic Acid from Methyl Isobutyl Ketone Feed- often used as a bench mark for quantifying centrifugal extractor capacity and comparing with other Liquid-Liquid Extraction behaviors:

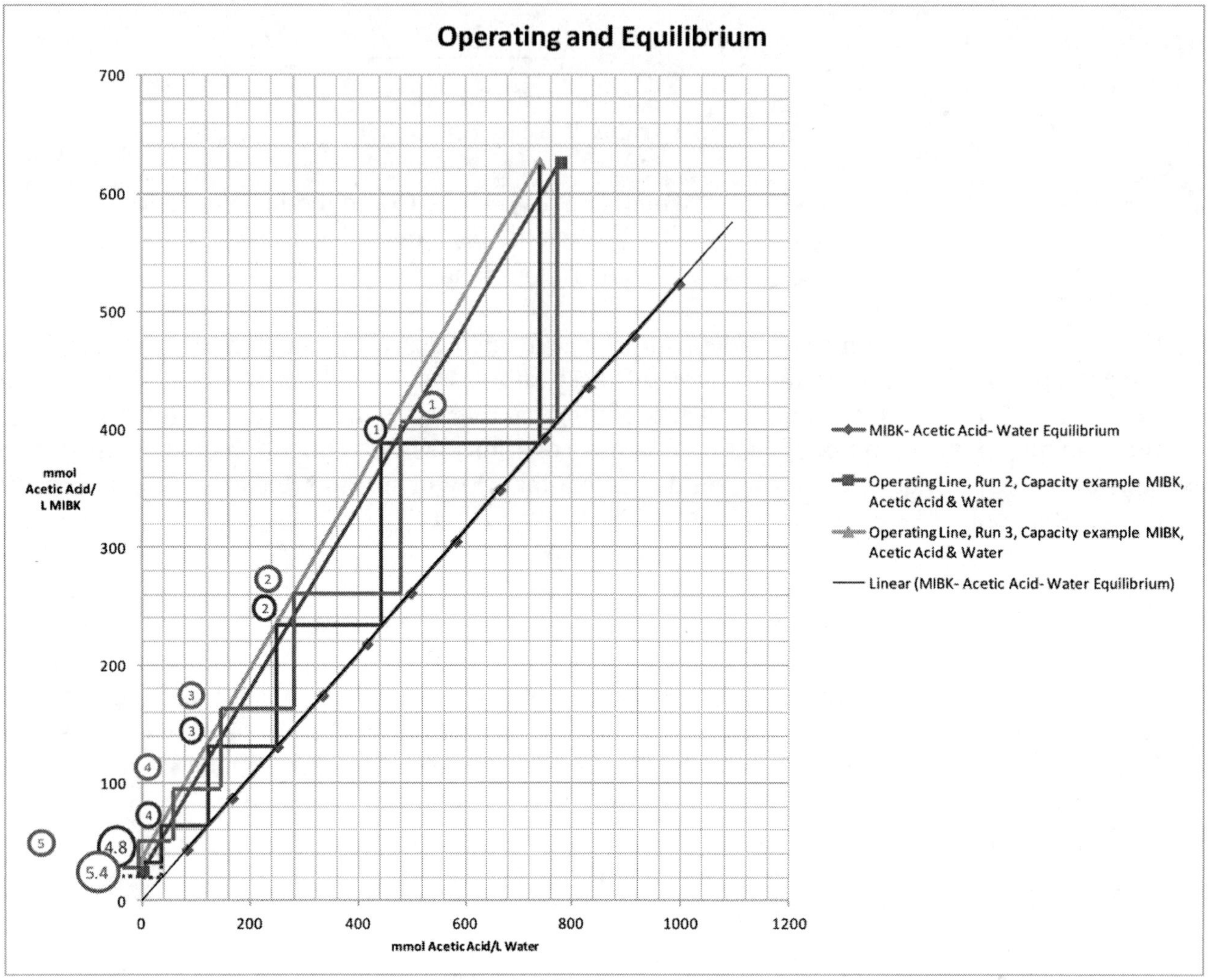

Figure 6 A-1 Pod Contactor Example MIBK- Acetic Acid- Water Extraction stages- Run 2: 5.4 Stages, Run 3: 4.8 Stages.

Both Runs feed rates are MIBK (LLI) = 551 mL/min, H2O (HLI) 755 mL/min (~80% of Flood capacity) at ~38C with standard extraction elements. Run 2 yielded 5.4 Stages and Run 3 yielded 4.8 Stages.

	mmol Acetic Acid / 1 L phase		mmol Acetic Acid / 1 L phase	
	MIBK Light	Water Heavy	MIBK Light	Water Heavy
Graph	Y_1	X_1	Y_2	X_2
Equilibrium Line	43.6	83.0	523.7	996.1

Operating Lines		755 mL/min	551 mL/min	
Graph Axis ID	Y_1	X_1	Y_2	X_2
McCabe Theile Designations	x_N (L_N) Raffinate	y_{N+1} (V_{N+1}) Solvent	x_0 (L_0) Feed	y_1 (V_1) Extract
	LLO	HLI	LLI	HLO
Run 2	25	0	626.5	778.5
Run 3	36.5	0	626.5	738.5

Figure 7 Composition Data for Figure 6

Similar extraction efficiency understanding can be developed for each process under consideration. B&P uses MIBK-Acetone- Water Ternary system for extraction efficiency expression since equilibrium data is readily available. Better understanding can be developed by the amount of testing conducted for an application. End users will often have better equilibrium data available specific to their process and can develop further separation/ extraction performance data as needed.

2.1.4 Effects of Viscosity

Increasing viscosity decreases the Pod Contactor capacity. When a new process without capacity test data is considered for anticipated capacity, it is de-rated by an amount approximately proportional to the inverse of viscosity to 1/3rd power. If pilot scale testing data becomes available, that data is preferred.

This viscosity capacity relationship can be illustrated with this chart and is based on demonstrated historical capacity from various process testing:

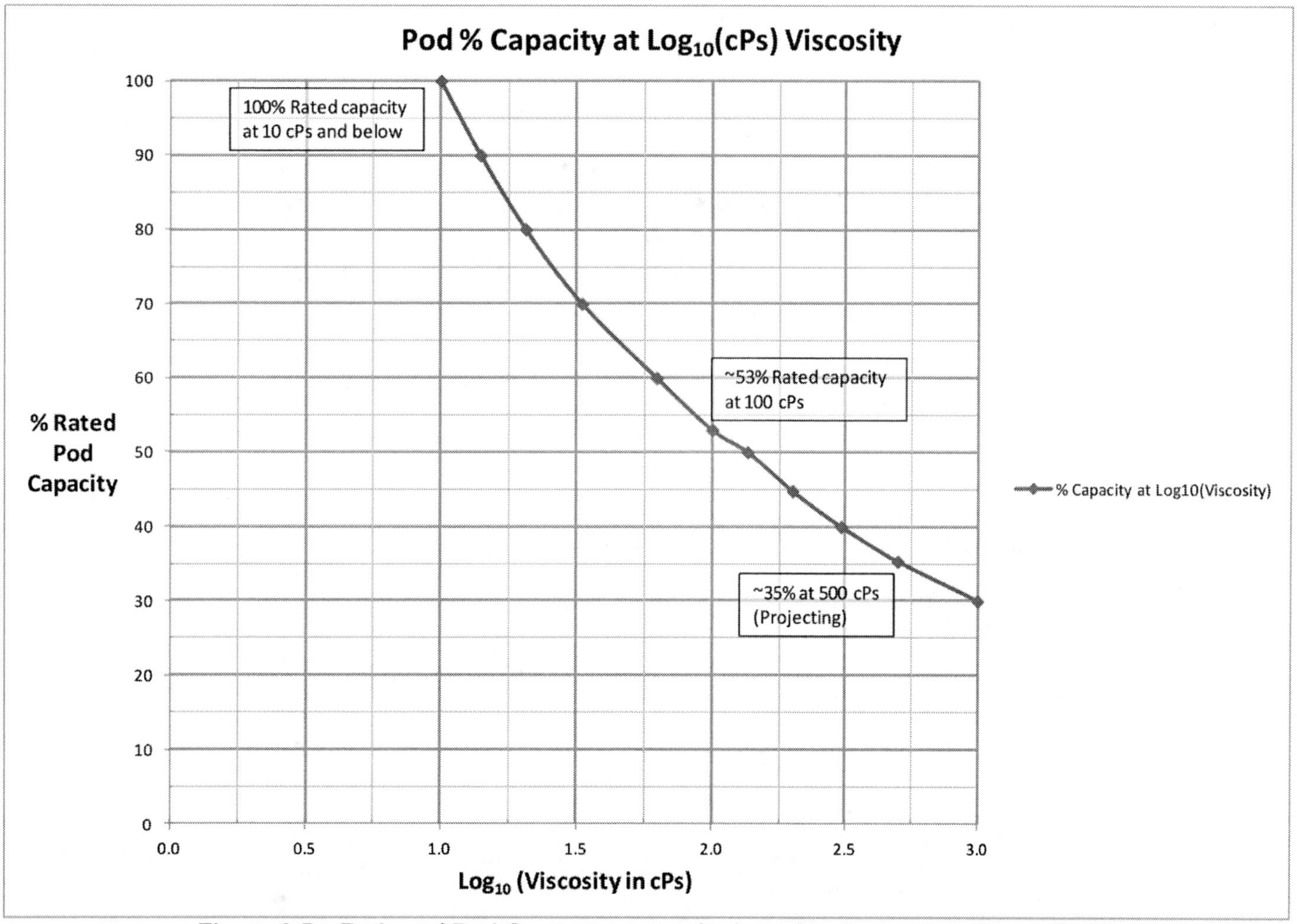

Figure 8 De-Rating of Pod Contactor capacity based on increasing Viscosity [5]

2.2 *Operation of a Pod Contactor:*

The internal pressure for the Centrifugal separator is treated as a Centrifugal manometer [1, 7, 8] - where the pressure is relative to the radial "height" of the column of liquid. The outer Rim equates to the bottom of the tower and the center shaft equates to the top of the tower.

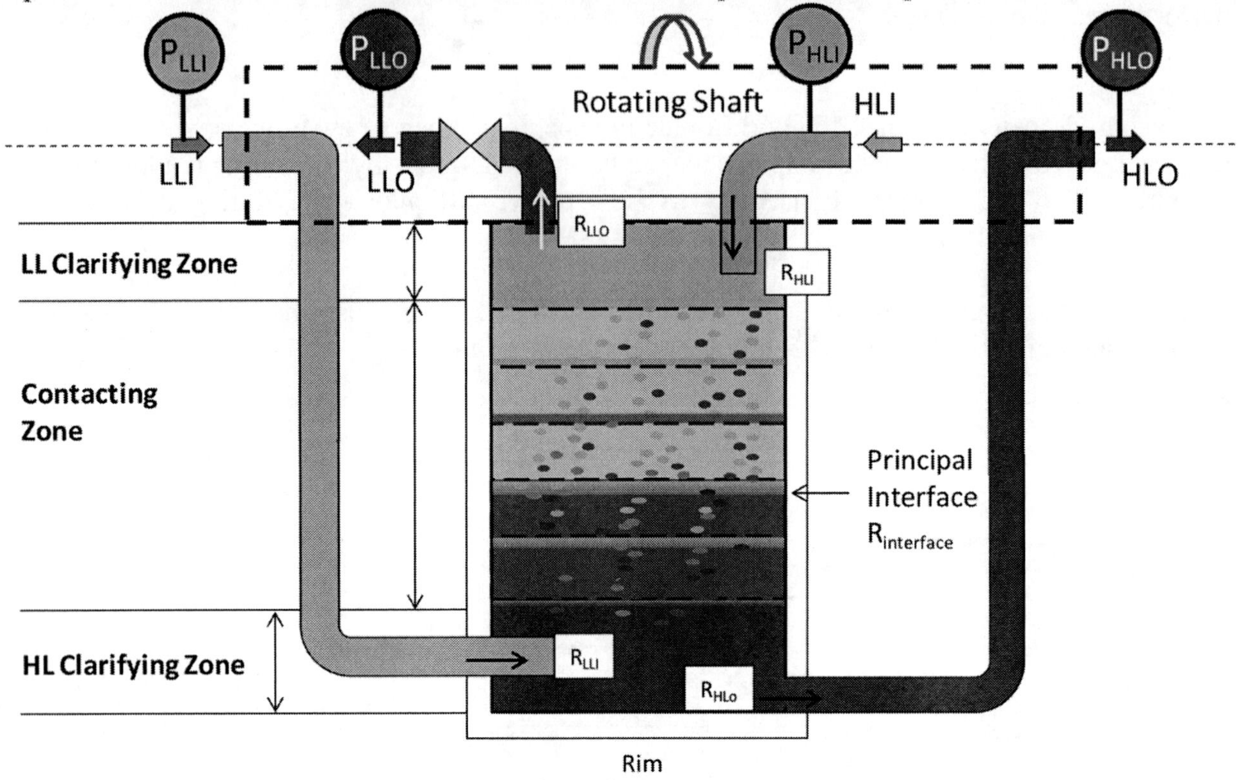

Figure 9 Illustration of Centrifugal Manometer

Streams and process conditions are identified below in Table II by a series of B&P often used abbreviations. Many will be seen in the upcoming operations and equation descriptions.

Variable	Definition
Di	Diameter position of "i" relative to shaft Center line (D_{Rim}, $D_{interface}$, D_{LLI}, D_{HLI}, D_{LLO}, D_{HLO}, etc)
HL	Heavy Liquid
HLI	Heavy Liquid In
HLO	Heavy Liquid Out
LL	Light Liquid
LLI	Light Liquid In
LLO	Light Liquid Out
Pi	Pressure of "i" (P_{LLI}, P_{HLI}, P_{LLO}, P_{HLO}, etc)
Pri	Pressure at Radial position of stream i (P_{RHLI}, P_{RLLI}, P_{RLLO}, P_{RHLO}, P_{RIM}, etc)
Ri	Radial Position of "i" relative to shaft center line
$R_{interface}$	Radial Position of principle interface boundary
SG_i	Specific Gravity of considered species (SG_{HLI}, SG_{LLI}, SG_{LLO}, SG_{HLO}, etc)

Table II Variables around Centrifugal Manometer

Pod Contactor operation is controlled by the differential pressure (P'LLO or ΔP'LLO) of the Light Liquid Out, Heavy Liquid Out. Increased pressure differential ΔP'LLO prevents the Heavy Liquid from overcoming the Light Liquid phase and permits the phases to separate for discharge from the machine. If ΔP'LLO is too high, the light liquid passes out with the heavy liquid. Likewise if it is too low, heavy liquid will pass out with the light liquid stream. Adjusting the ΔP'LLO also sets the radial position of the internal Principle (Major) Interface Boundary and can manage separation quality. The differential pressure (ΔP'LLO) is adjusted by altering the Light Liquid Out back pressure- P_{LLO}.

Back pressure adjustments made at the HLO position-P_{HLO}- do not help the differential pressure to change. Changing the P_{HLO} changes total system pressure; IE P_{LLI}, P_{HLI}, P_{LLO} all change by the same ΔP as the pressure change at the P_{HLO}. There may be process reasons to operate at a higher P_{HLO}, but generally it is not a separation quality control parameter.

The Principle Interface Boundary is a radial boundary where concentric element chambers transition from continuous Light Liquid into a continuous Heavy Liquid. (IE the Light Liquid is continuous for element bands inboard of the interface boundary and the Heavy Liquid is continuous outboard of the Interface):

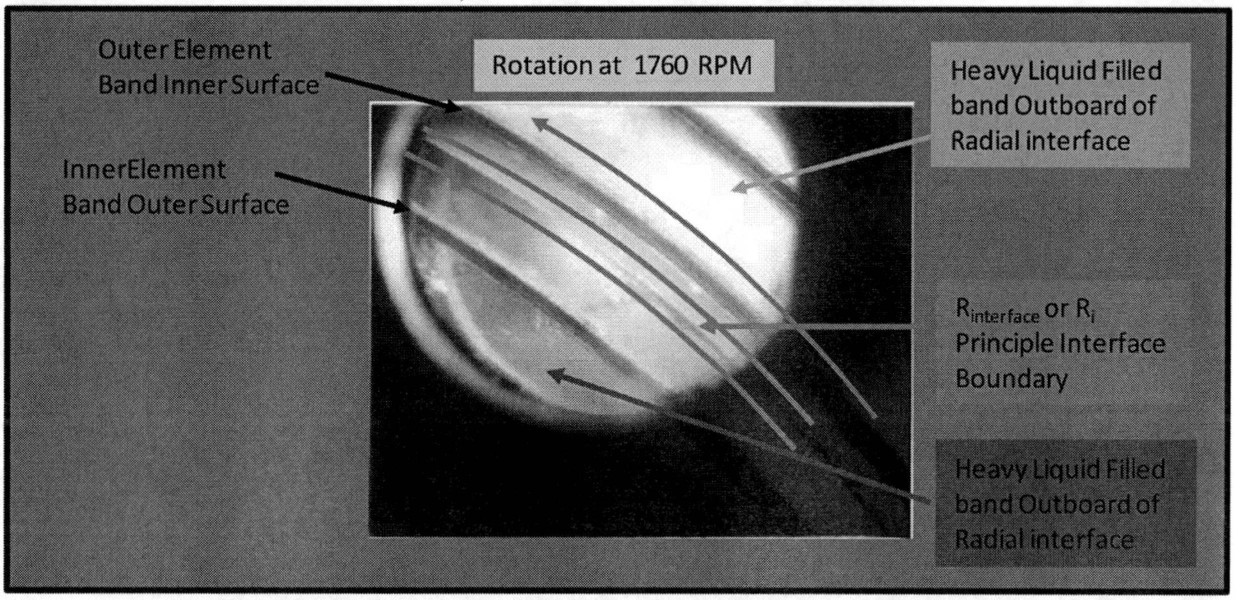

Figure 10 Interface boundary for Pod Contactor operation. [7]

Figure 11 shows the Radial liquid distribution images:

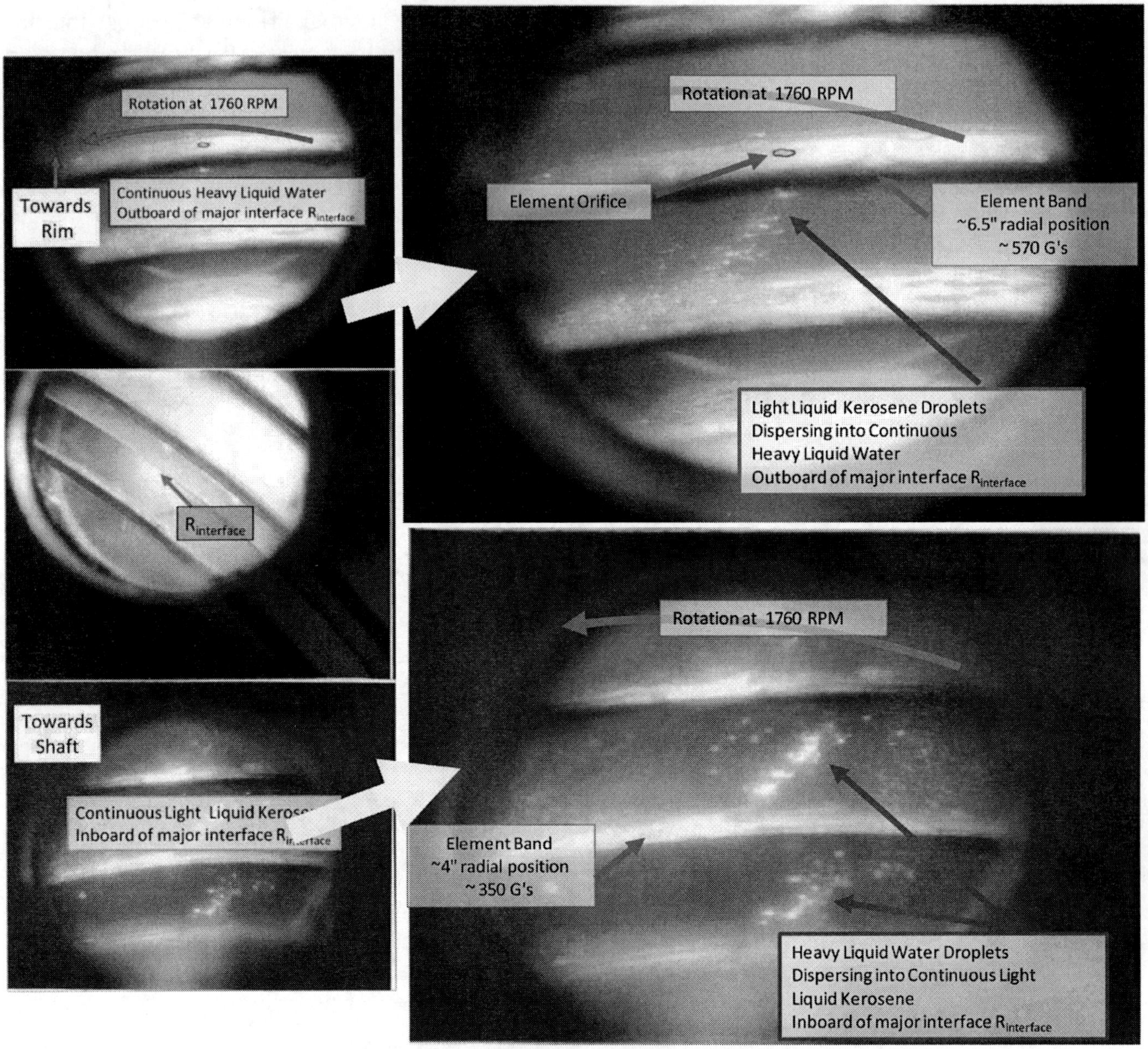

Figure 11 Clear window Pod Contactor images illustrating phase breakup and droplet formation at elements [7]

Pod Contactor rotor speed may also be used to manage Pod Contactor operation. Increase of extractor rotational speed increases the G-field the feed streams are exposed to. It can be used to vary separation quality or capacity- provided the rotor speed remains within the limits of the machine.

Two additional methods for separation / extraction management are the element design and the feed ratios. Element sets can have open area decreased to improve turbulence & extraction efficiency; or the area can be increased to accommodate problem emulsifications, high viscosity or –provided quality remains acceptable to the end user- allow higher feed rates. Feed ratios can be varied so that solvent use is optimized for extraction. Again though, both of these design parameters are limited by the required process quality requirements.

Temperature and operating pressure of the system can be manipulated, but these are generally set by the individual process needs. These can be evaluated during testing based on anticipated or current process conditions. Sometimes temperature increases due to seal friction needs to be managed with higher flow rates, upstream heat exchangers or buffer fluids. Operation adjustments can be made as process experience is gained. Only direct Pod Contactor operation variables are discussed here.

2.2.1 ΔP'LLO- Differential Pressure and Interface boundary ($R_{interface}$)

The ΔP'LLO is used to control where the principle interface boundary ($R_{interface}$) is positioned. The pressure values may also be used to determine that boundaries' radial position. This is relevant due to operating concerns. If the interface boundary is too close to the shaft LLO exit, there can be loss of separation at the LLO position. If it is too close to the HLO exit, there can be loss of separation at the HLO. Operation with $R_{interface}$ too close to one side or the other and the process can be much more susceptible to minor process upsets such as composition, pressure fluctuations, feed rates or other process variables. The best Operating ΔP'LLO (and hence $R_{interface}$ boundary) can be determined from which tests yield the best process quality acceptable to the end user. Additional refinement can be made with gains in process operation experience.

For extraction cases, sometimes there are large gains in extraction efficiency by adjusting the $R_{interface}$. Other times there is little or no gain. Typically the HLI & LLI feed diameters are positioned so the $R_{interface}$ ($D_{interface}$) is midway between the feed positions. Sometimes interface placement is best for just outboard of the LLI or just inboard of the HLI positions. Unfortunately, which operation is better remains an empirical understanding and relies on testing or process experience.

For separations, the best feed may be the inner radial heavy feed position if the feed is mostly heavy phase and low content of the light phase. The outer light feed position may be better if it is a majority of light phase components. But this may not always be the case and a fully intermediate position may be better for the given process. Testing a single stream from either the HLI or the LLI position can yield comparative data for what is an optimal feed location.

The outlet stream differential pressure is calculated as follows:

$$\Delta P'LLO_{TT} = P_{LLO} - P_{HLO}$$

Equation 5 Outlet Stream Differential Pressure

P_{LLO} = Pressure on Light Liquid Discharge
P_{HLO} = Pressure on Heavy Liquid Discharge

To get to pressures inside of the Pod Contactor Rotor, several equations from physics are used. Since pressure is Force/ Area, the first equation will be Force.

$$Force = m * a$$

m = mass
a = acceleration
(g for a manometer pressure calculation)

For uniform circular motion, the magnitude of centripetal acceleration a_r, is:

$$a_r = \frac{|v|^2}{r} = r\omega^2$$

a_r = magnitude of centripetal acceleration
v = speed in distance/sec
r = radial position under consideration
ω = angular speed (s^{-1}) = (N rpm)*(2π/60)

As in a conventional manometer, the pressure (force/ area) is a product of the (mass * acceleration) over area. Extending this definition to find pressure P at any point R on a rotating cylinder of liquid and neglecting effects of gravity due to orientation,

$$P = \frac{Force}{Area} = \frac{m * a_r}{A} = \frac{(\rho * Vol) * a_r}{A} ==> \frac{(\rho * (\pi r^2 w)) * (r\omega^2)}{2\pi r w} ==> \frac{\rho r^2 \omega^2}{2}$$

$$\frac{dP}{dr} = \rho \omega^2 r ==> dP = \rho \omega^2 r \, dr$$

$$\int_0^P dP = \int_0^R \rho \omega^2 r \, dr ==> P = \frac{1}{2} \rho \omega^2 R^2$$

To express this equation in Specific Gravity, Rotational speed in RPMs and generic units of radial length, this becomes:

$$P_i = \left(\frac{2\pi^2 \rho_{water}}{3600}\right) SG_i N^2 R^2 ==> \left(\frac{\pi^2 \rho_{water}}{1800}\right) SG_i N^2 R^2$$

From this, the following expressions can be used for English customary units:

$$P_i = \left(0.513 * 10^{-6}\right) N^2 R_i^2 SG_i$$

Equation 6 Single Phase Liquid Leg Pressure at radial position

P_i	=	Liquid Leg Pressure in PSI
N	=	Rotor speed in Revolutions/ Minute (RPMs)
R_i	=	Liquid leg radial position (Inches)
SG_i	=	Specific Gravity of liquid leg under consideration (Dimensionless)

This expression can be used for SI units:

$$P_{i\,(N/m^2)} ==> \left(5.483\right) SG_i N^2 R_i^2$$

Equation 7 Single Phase Liquid Leg Pressure (SI)

P_i	=	Liquid Leg Pressure in N/m^2
N	=	Rotor speed in Revolutions/ Minute (RPMs)
R_i	=	Liquid leg radial position (meters)
SG_i	=	Specific Gravity of liquid leg under consideration (Dimensionless)

The interface boundary can also be obtained from the pressure balance of the heavy and light liquid. This equation can be used with English units:

$$\Delta P'LLO = \left(0.513 * 10^{-6}\right) N^2 R_{interface}^2 \Delta SG_{HL-LL}$$

Equation 8 ΔP'LLO at given major interface boundary $R_{interface}$

Rearrangement to solve for $R_{interface}$ becomes

$$R_{interface} = \sqrt{\frac{\Delta P'LLO}{\left(0.513 * 10^{-6}\right) N^2 \Delta SG_{HL-LL}}}$$

Equation 9 Major Interface boundary position calculation

$R_{interface}$	=	Interface position from shaft center in inches
N	=	Steady State rotational speed in RPM
ΔP'LLO	=	Outlet Differential Pressure in PSI
ΔSH_{HL-LL}	=	SG Difference between Heavy and Light Liquid

Other equations appropriate to available units can be derived.

2.2.2 Effects of Rotor Speed on Capacity

Centrifugal extractors may be operated at different speeds depending on process needs. B&P design limits rim pressure are approximately at ~625 PSI differential pressure. Since it is a pressure limitation, the Specific Gravity of the heaviest phase will set the rotor speed limits for a given process. The rotation speed limit becomes reduced by the following relationship:

$$N_{Max} = N_{model\ RPM\ Limits} * \sqrt{\frac{1.0}{SG_{HL}}}$$

Equation 10 Rotational Speed De-rating due to SG>1.0

N_{max}	=	Model RPM limit for process (SG_{HL} >1)
$N_{Model\ RPM\ limits}$	=	RPM limits for machine under consideration
SG_{HL}	=	SG of heaviest condition of Heavy Liquid

(See Table VI In Appendix for values suitable for B&P machines)

The Model RPM limit is not increased if the involved process fluids have an SG < 1.0. This is due to various design limits and not specifically associated with the pressure limits.

Gains may be made for the process by running slower or faster. With respect to machine process capacity, the increase in RPMs expands the feed rate capacity as illustrated in Figure 12 and Figure 13.

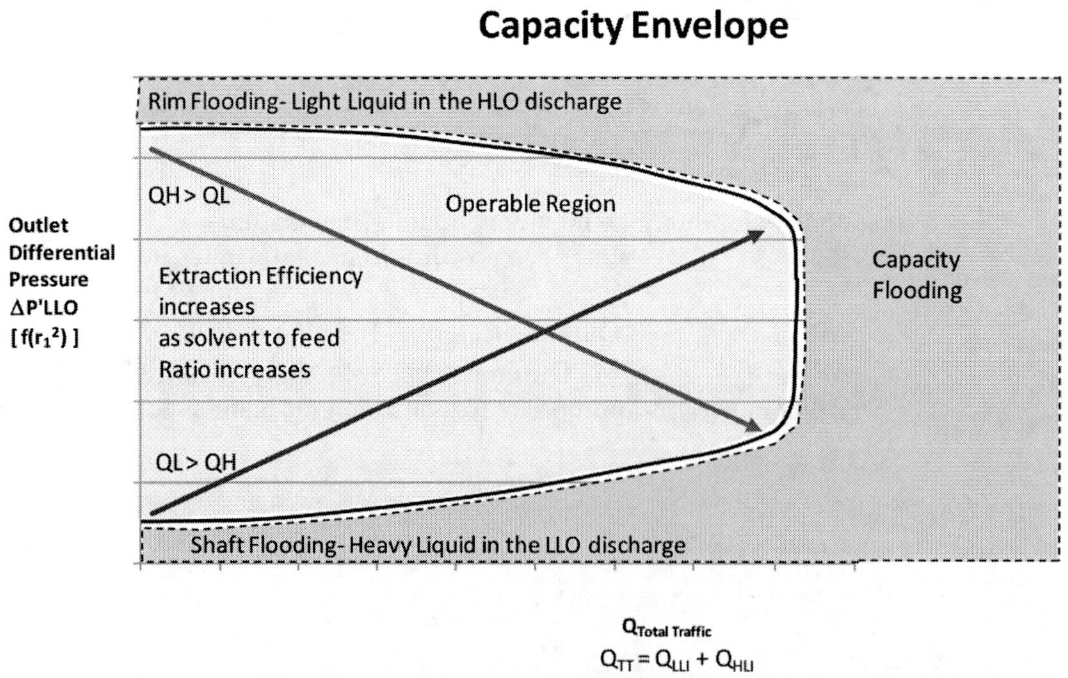

Figure 12 Qualitative Operable Capacity Envelope for Centrifugal Extractor [6]

When the RPM is increased or decreased, the operable region expands or contracts as illustrated in Figure 13

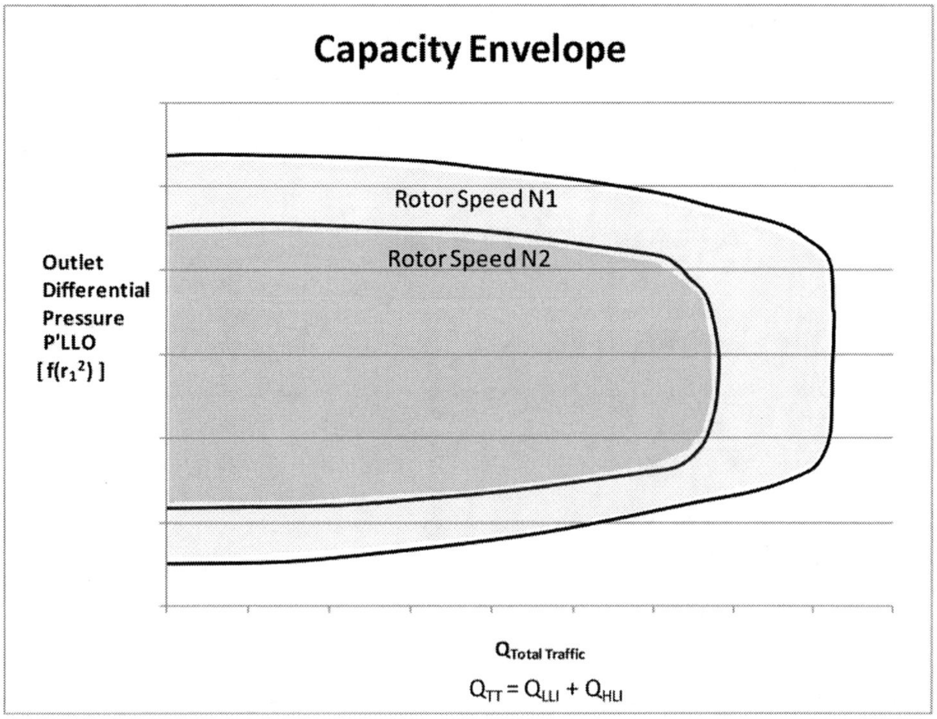

:

Figure 13 Effect of Rotor Speed Variation- N1> N2. [1, 6]
(Speed Variation must be within rotor speed limits for design and process)

2.2.3 Effects of Internal element design

Specific demonstrations of the above capacity envelop have been documented with the Kerosene & Water system. Similar behavior has been noted with other processes. Here is an illustration of the Kerosene & water capacity envelope:

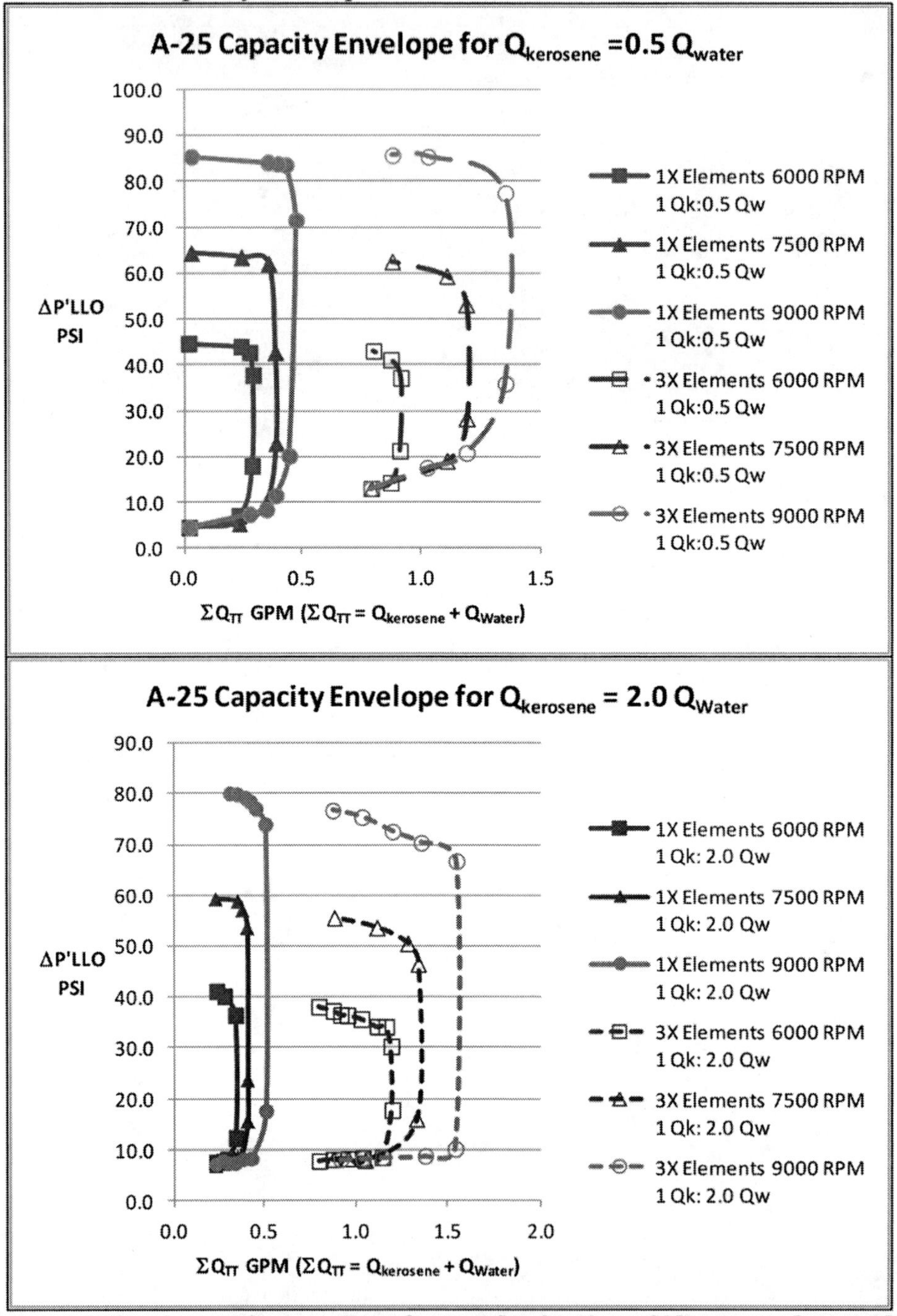

Figure 14 Flood capacity envelope for two feed ratios of Kerosene & water[1]

Two styles of elements were tested for the Flood capacity charts. The 1X element set is a "standard" A-25 extraction element set and is the set most often used with the A-25 demonstration unit. It features 4 to 6 holes per sheet metal contacting element band and there are ~10 of 14 bands in the contacting zone. The 2 inner most and outer most elements have more open area and allow coalescing of the light and heavy phases before they pass out of the machine. The 1X set is also used with the example MIBK- Acetic Acid- Water extraction and is selected due to its observed better extraction performance.

The "3X" set carries 3 times the number of holes/ single element and was a set used to explore possible capacity increases when increasing the Percent open area in the machine's element set. This increase in open area results in lower traffic velocities at the process streams. This can be good for easily emulsifying systems or if high viscosity is present. The process cost however can be a reduction in extraction efficiency; likely attributable to lower velocity & turbulence. Lower velocity & turbulence allow larger droplets, reducing exposed phase surface area. Conversely, higher velocity results in higher turbulence and more & smaller droplet formation, tending to promote mass transfer.

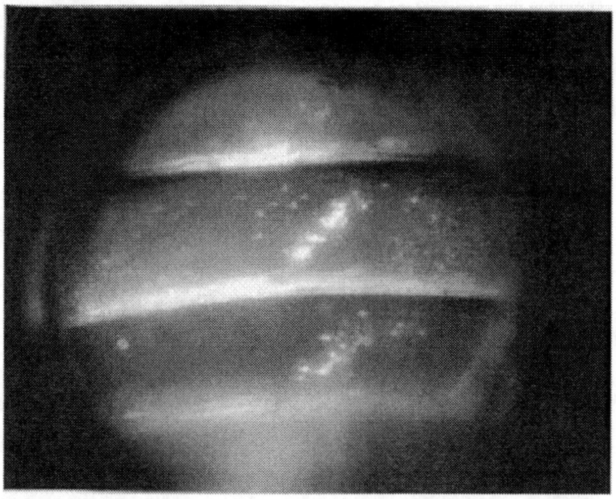

Figure 15 Example droplet dispersion Heavy Water droplet into Continuous Light phase

2.2.4 Effects of varying feed ratios

Figure 16 is a comparison of the 1X element with respect to variation in the Light ($Q_{Kerosene}$) & Heavy (Q_{water}) liquid feed ratios (Excerpt of same data in previous charts):

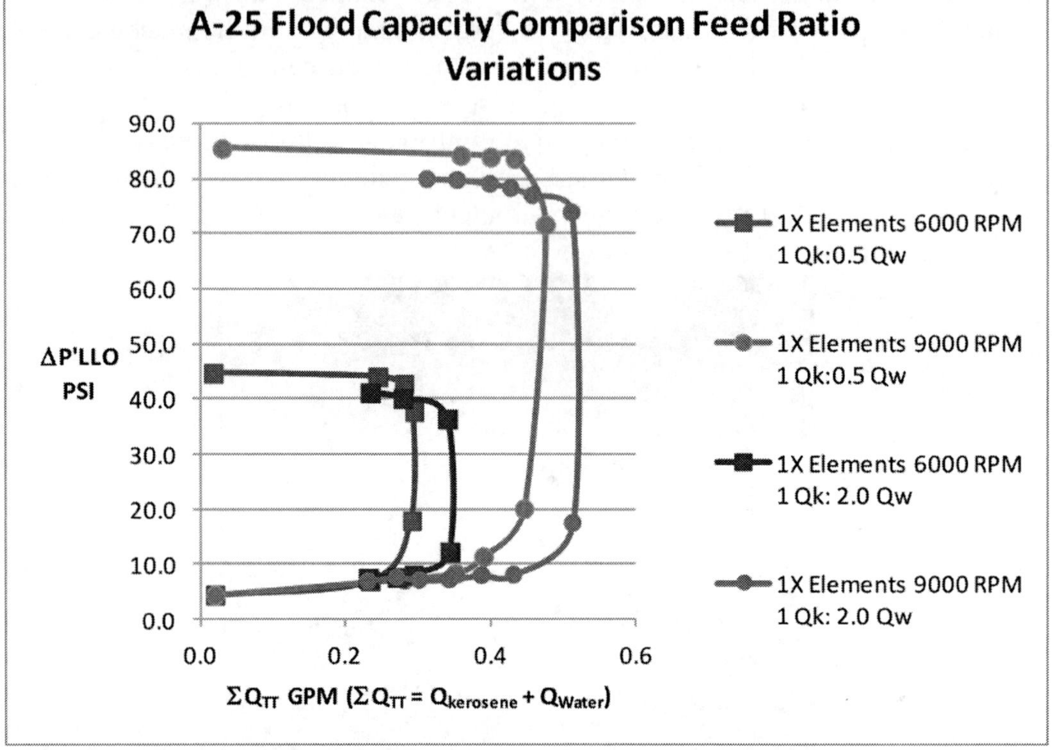

Figure 16 Comparison of Kerosene/ Water system Flood capacity when Feed ratio is changed

The ΔP'LLO envelope contracted somewhat, but for this particular system the flood the P'LLO values remain relatively close (at 9000 RPMS, ~85 PSI ΔP for 1 $Q_{kerosene}$: 0.5 Q_{water} versus ~79 PSI ΔP for 1 $Q_{kerosene}$: 2.0 Q_{water}) on the Rim Flooding (LLO in the HLO).

B&P observes and expects this same behavior in other Liquid- Liquid extractions. Full mapping of the capacity envelope can be plotted, but requires a large number of tests. Often for limited time and resource initial pilot tests, feed ratios are matched to existing process gravity separation equipment and full capacity is not fully charted. Some feed ratio variation can be made based on test performance and the amount of material available, but overall, the interest level of the process owner contributes to how well a capacity envelope is defined.

Some processes get an initial operation and the party is satisfied with quality without further process optimization. Other times, operational or historical experience offers a wider view of the capacity envelop and may offer better or alternative feed ratios to consider for the pilot scale testing. Because extensive material may be required, conducting a pilot plant Pod Contactor test at the customer's site can be more advantageous than shipping material for a remote site trial.

2.2.5 Existing process troubleshooting

As a final note concerning operation for those who may already be using these machines- When doing process troubleshooting for centrifugal contactors consider these process troubleshooting tips:

- If there is Heavy Liquid in the Light Liquid Out (LLO) position, the Light Liquid Out pressure (P_{LLO}) may be too low
- If there is Light Liquid in the Heavy Liquid Out (HLO) position, the Light Liquid Out pressure (P_{LLO}) may too high
- If there is entrainment at both positions at previously demonstrated Differential pressure ($\Delta P'LLO$ / P_{LLO} / P_{HLO}), the feed rate may be too high.
- A seal leak may be suspect if entrainment is on one machine side or the other at a previously demonstrated successful feed rate (Q_{TT}) and operation differential pressure ($\Delta P'LLO$).

For further and more in-depth analysis, the cited sources carry more details and further background than can be covered in this discussion. Copies of the Podbielniak specific articles may also be obtained from B&P process Equipment upon request.

3. Conclusions:

Liquid-Liquid Contacting centrifuges such as Pod Contactors used for extractions and/ or separations remain a valuable tool in many commercial chemical processes. These may be used as a simple two phase separation, some distillation applications that could be preferentially accomplished by solvents (possibly due to temperature concerns for product), tower stripping/ rectification of immiscible phases or any gravity settling application that may benefit in process purity due to the higher G field. Their smaller size can also be advantageous where large process equipment hold-up can be a concern or where better product separation quality for less space is desirable.

References remain vague when discussing what is considered for scaling of continuous extraction centrifuges and this paper covered what B&P Process Equipment considers when evaluating and scaling an application. Other methods may be devised, but B&P has been using these practices with success for many years.

Factors that optimize process operation include

- Adjusting thermodynamic equilibriums available to the process during passage through the unit
    - Flows rates to adjust residence time τ
    - Operation pressures
    - Differential pressure
    - Temperatures.
- Operational speed adjustments
    - Adjust to optimize mass transfer and/ or separation quality
    - Bypassing emulsify regimes of operations

- Fluid velocity profile management on internal centrifuge design-
  - Internal design adjustment to accommodate viscosity or emulsification
  - Adjusted internal turbulence/ mixing to increase phase mass transfer surface area

The method B&P uses to scale from one process rate to another relies on the mathematical product of the maximum multiple of gravity and the residence time. Ideally, when a process is to be scaled, a pilot plant test is conducted. This provides operation data such as best feed ratios and outlet differential pressures that will be applicable to the full scale production units. Scale up may then be accomplished with the outlined equations. See Table VI in the Appendix for examples of design values suitable for B&P Centrifugal Extractors.

Historical test data can also provide reliable scale up for operation and capacity provided operating conditions are closely matched. For many existing Liquid-Liquid contacting & separation processes this data may be available. In the absence of historical data or pilot testing, chemically similar applications may have close operation behavior and conditions, providing a basis for initial test planning and or potential scale up rates.

Applications include major chemical process steps such as separations, extractions, rectification, ion exchanges, neutralization & washings. Examples are found industry wide:

- Large bulk and general chemical production
  - Fumaric Acid extraction, Polycarbonate & poly-alcohol washing & others
- Pharmaceutical applications
  - Penicillin, Erythromycin, Vitamin extractions & others
- Petroleum Refining
  - Acid treatment, “White” oil purifications (High purity refined mineral oils), contaminate stripping, neutralization & washing steps & others
- Liquid ion exchange
  - Citric acid, rare earth metal ion recovery & others
- Food industries
  - Flavor and essential oil extractions & others

4. Appendix Data

	1X perforation 6000 RPM 1 $Q_{Kerosene}$: 0.5 Q_{Water}		1X perforation 7500 RPM 1 $Q_{Kerosene}$: 0.5 Q_{Water}		1X perforation 9000 RPM 1 $Q_{Kerosene}$: 0.5 Q_{Water}		3X perforation 6000 RPM 1 $Q_{Kerosene}$: 0.5 Q_{Water}		3X perforation 7500 RPM 1 $Q_{Kerosene}$: 0.5 Q_{Water}		3X perforation 9000 RPM 1 $Q_{Kerosene}$: 0.5 Q_{Water}	
	□		□		□		□		□		□	
Data Point	□P'LLO	GPM	□P'LLO	GPM	□P'LLO	GPM	□P'LLO	GPM	□P'LLO	GPM	□P'LLO	GPM
1	4.5	0.02	4.5	0.02	4.5	0.02	13.0	0.79	13.0	0.79	13.0	0.79
2	7.0	0.23	5.3	0.24	7.3	0.28	14.3	0.87	19.1	1.11	17.6	1.03
3	18	0.29	9.9	0.35	8.4	0.35	21	0.91	28	1.19	21	1.19
4	38	0.29	23	0.39	11.5	0.39	37	0.92	53	1.19	36	1.35
5	43	0.28	43	0.39	20	0.45	41	0.87	59	1.11	78	1.35
6	44	0.24	62	0.36	72	0.47	43	0.80	63	0.88	85	1.03
7	45	0.02	63	0.25	84	0.43					86	0.88
8			64	0.03	84	0.40						
9					84	0.36						
10					85	0.03						

Table III 1 Flood Rate & Differential pressure for $Q_{Kerosene}$: 0.5 Q_{Water} Table [1]

	1X perforation 6000 RPM 1 $Q_{Kerosene}$: 2.0 Q_{Water}		1X perforation 7500 RPM 1 $Q_{Kerosene}$: 2.0 Q_{Water}		1X perforation 9000 RPM 1 $Q_{Kerosene}$: 2.0 Q_{Water}		3X perforation 6000 RPM 1 $Q_{Kerosene}$: 2.0 Q_{Water}		3X perforation 7500 RPM 1 $Q_{Kerosene}$: 2.0 Q_{Water}		3X perforation 9000 RPM 1 $Q_{Kerosene}$: 2.0 Q_{Water}	
	□		□		□		□		□		□	
Data Point	□P'LLO	GPM	□P'LLO	GPM	□P'LLO	GPM	□P'LLO	GPM	□P'LLO	GPM	□P'LLO	GPM
1	7.6	0.23	6.8	0.23	7.0	0.23	7.8	0.80	8.1	0.87	8.1	0.92
2	7.7	0.27	7.3	0.27	7.7	0.27	8.1	0.87	8.2	0.96	8.3	1.03
3	8.1	0.29	7.4	0.30	7.2	0.30	8.2	0.96	7.9	1.05	8.5	1.14
4	12	0.34	8.5	0.35	7.3	0.34	8.3	1.03	16	1.33	8.8	1.38
5	36	0.34	9.2	0.38	8.0	0.39	8.5	1.14	46	1.34	10.2	1.54
6	40	0.28	16	0.41	8.1 PSI	0.43	18	1.20	51	1.28	67	1.55
7	41	0.23	24	0.41	17	0.51	30	1.19	54	1.11	70	1.35
8			54	0.41	74	0.51	34	1.16	56	0.88	73	1.20
9			57	0.38	77	0.46	34	1.11			75	1.03
10			59	0.36	78	0.43	36	1.03			77	0.87
11			59	0.23	79	0.40	36	0.95				
12					80	0.35	36	0.92				
13					80	0.31	37	0.88				
14							38	0.79				

Table IV 1 Flood Rate & Differential pressure for $Q_{Kerosene}$: 2.0 Q_{Water} Table [1]

Run Number	P (psig) Q (ml/min) T (°C/°F)	Light Liquid Out	Heavy Liquid Out	Heavy Liquid In	Light Liquid In	Remarks
	P					
	Q					
	T					
	P					
	Q					
	T					
	P					
	Q					
	T					

Table V Example of typical Contacting Centrifuge data collection during tests

B&P Process Centrifugal Contactor design dimensions							
Model	Inside Rim Diameter	Camber width	Hold up Volume	RPM limits at SG = 1.0 ($N_{Model\ RPM\ limits}$)	G's at RPM Limit	Clean Process rated	Fermented Process
	Inches	Inches	US Gal	RPM	Multiples of g	US GPM	US GPM
A-25	7.0"	1.0"	0.16	10,000	10,030 G's	0.40	0.20
B-250	23.0"	10.0"	15.0	3,200	3,340 G's	40	25
D-450	34.0"	18.0"	58.0	2,100	2,130 G's	150	65
D-900	34.0"	36.0"	116.0	2,100	2,130 G's	300	130
E-1200	44.0"	48.0"	260.0	1,600	1,630 G's	600	260

Table VI Generally accepted design parameters for B&P Contactor centrifuges [1]

3. References

[1] Dr. Todd, David, "Performance of Centrifugal Extractors", Proceedings of the International Solvent Extraction Conference", Lyon, Vol 3, 2380-2398, September, 1974, Print

[2] Bach, A, Chan, T, "Pod Presentation-Jan2014.PPTX", B&P Process internal documentation, Tech Center\Library\Presentation, 2014, electronic

[3] Aweibel, L, "History Over View", *Introduction & History.doc* B&P Process Pod Contactor internal documentation, Marketing\Tech Papers\PODS\1999 Pod Sales Manual, 1998, electronic

[4] Bush, Lisa, "3- Pod Contactor Applications List.doc", B&P Marketing documentation, Marketing\Sales Resource Manual\b - II Pod Contactor\2 - Overview, 2009, electronic

[5] B&P Process internal process documentation sources, "De-rating Pods due to Viscocity.pdf", ENGINEER\Calculations\Pods\Pod Process Design Manual\, (Various Dates), electronic

[6] "Capacity Envelope", *Pod Contactor Process*, 22, B&P Process internal process documentation sources- Pod Process Design Manual\01 Podbielniak Centrifugal Contactor Process Design Rev 1.pdf, (Various Dates), electronic

[7] Todd, D.B, Davies, G.R, Lange, H.A, "Behavior of drops in a centrifugal field", B&P Process internal process documentation sources- Pod Process Design Manual\Pod clear window rotor flow pictures & presentation\Behavior of Drops in a Centrifugal Field.pdf, (1999), electronic

[8] Davies, G.R. "The History of the Podbielniak Contactor", Tech Center\Library\POD\The History of the Podbielniak Contactor- Gordon R. Davies.PDF

Additional references not cited:

[9] Todd, D.B & Podbielniak, W.J., "Advances in Centrifugal Extraction", Chemical Engineering Progress, 61, No. 5, 69, 1965

[10] Todd, D.B, "Improving Performance of Centrifugal Extractors", Chemical Engineering, 79, No. 16, 152, 1972

[11] *Pod Contactor Process*, B&P Process internal process documentation sources- ENGINEER\Calculations\Pods\Pod Process Design Manual\01 Podbielniak Centrifugal Contactor Process Design Rev 1.pdf, (Various Dates), electronic

443334 Expanding Applicability of Integrated Separation Systems

Abstract:

Methathesis of mixed butenes to propylene and hexene will be used as a case study. An integrated separation system will be explored as a potential process intensification. This approach expands the applicability of such systems to processes involving a combination of reaction and multicomponent separations

444320 Process Development for Synthesis of Bio-Based Plasticizers

Abstract:

E3Tec and MSU are pursuing the commercialization of an integrated, energy-efficient process technology for the production of bio-based plasticizers. Organic acids, typically produced by fermentation of carbohydrate feedstocks, constitute an important class of bio-renewable platform chemicals that can be further converted to useful products. Bio-based plasticizers are currently gaining significant attention because they can be used in place of petroleum-based phthalate plasticizers that have shown to have health concerns in certain consumer products. E3Tec received and recently completed US DOE SBIR Phase I and II projects. The objective of the project was to design and evaluate processes using heat integrated reactive distillation (HIRD) with side reactors and PerVap membrane for esterification of organic acids. The model chemical system studied in the SBIR project was esterification of citric acid with ethanol to produce tri-ethyl citrate (TEC). ASPENPlus™ design method was developed and validated with the MSU pilot-scale test unit. The process analysis is based on ASPENPlus™ integrated with validated rate-based models of side reactors. The primary focus has been to develop a design methodology for rapid scale-up of bio-based processes. The validated ASPENPlus™ design model is applied for synthesis of two leading bio-based plasticizer candidates, dioctyl adipate (DOA) and dioctyl succinate (DOSX). Techno-economic merit comparisons based on C-Footprint, CAPEX, and OPEX for both the new and current commercial process technologies will be presented.

444077 Energy Efficient Subcritical Oxidation of High COD Wastewaters

Abstract:

Supercritical wet oxidation is the holy grail for destruction of tough to biodegrade species in industrial wastewaters - ones with high COD and not necessarily high BOD. Subcritical approaches are nearly as effective but equally restrained by high energy costs. A process intensification approach to equipment design and energy integration can find a sweet spot for implementing an economic solution.

445134 ProcessMD: Capturing the Heartbeat of our Plants through Advanced Analytics

Abstract:

The Industrial Gases market is characterized by production plants that balance both capital and operating costs along with the important need for the continual supply of product. It is therefore critical to maximize asset productivity while concurrently maintaining baseline reliability requirements. In response to these operational objectives, companies like Air Products and Chemicals, Inc. have sought to acquire the requisite process insight through advanced data modeling and analytics.

With the advent of digital control systems and data historians, manufacturing companies now have a preponderance of data; however, these high-dimensional datasets are often characterized by low signal-to-noise ratio along with a high degree of correlation. Advanced analytics in the form of state-of-the-art statistical multivariate techniques can denoise process signals, as well as capture the correlation amongst these signals, transforming huge plant datasets into actionable information.

ProcessMD, a patented web-based, predictive-monitoring and fault-diagnostic platform, has been developed to provide foresight and insight about asset performance. At the foundation of this novel technology platform are predictive models built uniquely for each plant or piece of equipment. These data-driven, auto-adaptive models provide advanced alerts of subtle abnormalities and deviations from expected behavior across a whole plant or its specific components. The predictive analytics focuses on diagnostics to enable proactive intervention – unlocking value from data.

444202 Big Data Analytics in Oil & Gas

Abstract:

Big data, and big data analytics, have existed for some time in oil and gas, particularly in the form of seismic data analyzed using high performance computing clusters. However, these days, data from other sources is increasing, and tools specifically designed to perform analysis on large diverse data sets have emerged. In oil and gas, high volume and high velocity data occur routinely, while data of great variety are the norm, and significant issues around data veracity exist as well. The emergence of big data analytics in oil and gas is permitting the industry to address challenges that could not be satisfactorily handled previously. These techniques can be applied across the full value chain, from exploration to drilling to production and processing, thru shipping and handling of supply chains and trading. This talk will examine applications of big data analytics in oil and gas as well as areas where big data analytics has not been applied but could add value.

445141 Moving the Needle through Transaction Level Analytics

Abstract:

Dow's Advanced Analytics team provides actionable insights to key business and corporate leaders using a wide range of computational methods. We are continuously striving to integrate any sources of information that can make our models more timely and accurate to improve decision making. Most attempts at understanding and predicting customer demand have either been non-analytical surveys of customer needs or have used time series methods of forecasting aggregate demand over customer segments. At Dow we have developed a predictive analysis of customer demand which brings the world of analytics to the understanding of individual customers. A crucial aspect of this modeling is the understanding of the high volume and velocity of order dynamics, such as changes to existing order prices, quantities, and order delivery dates. The value of this work extends beyond better forecasting to allow prescriptive analytics for prioritizing marketing activity and earlier detection of problems in the supply chain before they become evident.

442429 Adequate Pressure Relief

Abstract:

Pressure relief valves are reliable and effective means of overpressure protection that protect personnel and equipment from the dangers of overpressure. Relief systems design validation and ensuring appropriate relief protection should take place not only during initial design but on a regular interval as changes are made to equipment and/or processes. Changes and tie-ins can drastically influence the capability of existing systems to safely handle overpressure situations. This presenstation reviews the requirements and illustrates how seemingly minor process modifications result in severly underrated pressure relief scenarios.

Additionally a case study is presented for a boiler plant which shows distribution pattern of equipment not meeting code requirements as well as distribution pattern of various types of deficiencies/concerns.

The presentation also describes benefits of rigorously simulating column systems while estimating relief requirements, steady state set up for simple systems, simulation set up for complex systems with convergence issues and benefits of dynamic simulations.

PREVENTION OF COMBUSTIBLE DUST EXPLOSIONS

12TH GLOBAL CONGRESS ON PROCESS SAFETY

Steven D. Emerson
Emerson Technical Analysis, LLC

Tutorials in Process Safety – LPS
2:30 PM Tuesday 12 April 2016

Speaker:
Steven D. Emerson

President – ***Emerson Technical Analysis, LLC***
Corpus Christi, TX & Las Vegas, NV

The firm specializes in the safe management of hazardous, explosive, energetic or flammable materials

Active in GCPS (especially PPSS, PSM^2)

PE (ChE), BSc, MEA, PhD

2

PREVENTION OF COMBUSTIBLE DUST EXPLOSIONS

Steven D. Emerson
Emerson Technical Analysis, LLC

ETA 3

OUTLINE OF TALK

1. Fundamentals
2. Unique Dust Explosion Features
3. Types of Combustible Dusts
4. Equipment Susceptible to Dust Explosions
5. Prevention & Mitigation
6. Standards & Regulation

ETA 4

FUNDAMENTALS

Imperial Sugar Company - Port Wentworth, GA
February 7, 2008
13 Fatalities and 42 injuries

ETA 5

What Is a Dust Explosion ?

Many solid materials become highly flammable when reduced to fine powders.

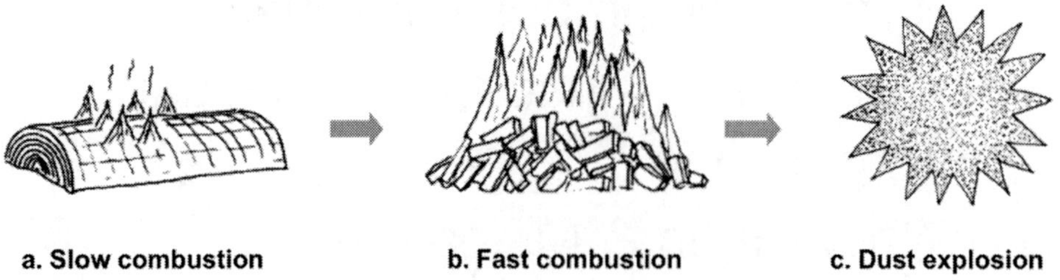

ETA 6

Fire Triangle

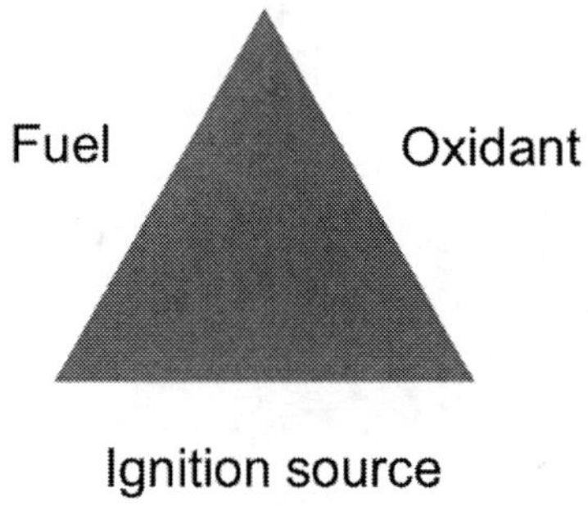

The fire triangle

Like all fires, a **combustible dust fire** occurs when:

fuel (the combustible dust) is exposed to
heat (an ignition source)
in the presence of
oxygen (air).

Removing *any one* of these elements eliminates the possibility of a fire --- or an explosion.

7

Dust Explosion Pentagon

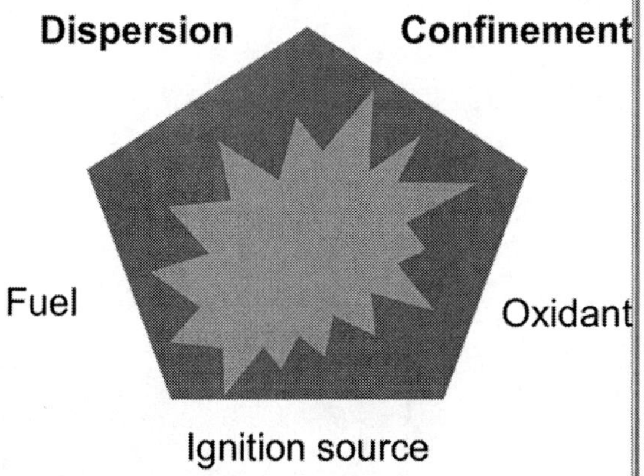

The dust explosion pentagon

A **combustible dust explosion** requires two additional elements:

Dust suspension and
Confinement.

Removal of *either* the suspension or the confinement *prevents* an explosion.

However, fire may still occur.

8

UNIQUE EXPLOSION FEATURES

West Pharmaceutical Services – Kinston, North Carolina
January 29, 2003
Polyethylene powder
6 Fatalities and 38 injuries

ETA 9

Primary and Secondary Explosions

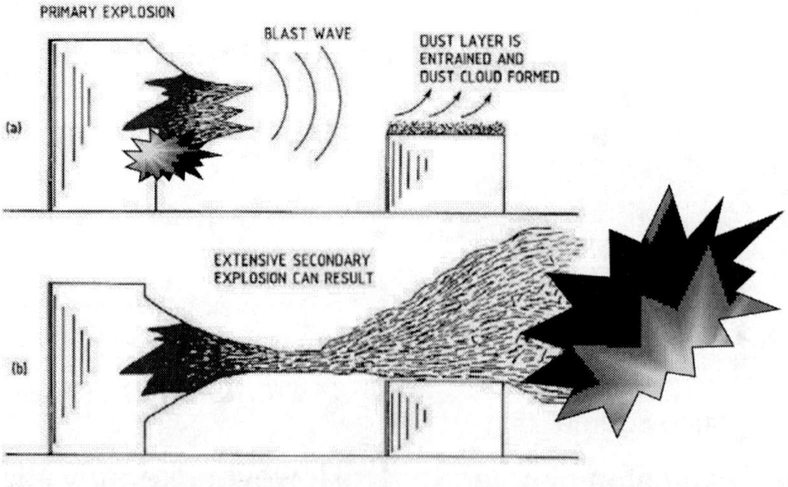

Figure 1.7 *Illustration of how the blast wave from a primary explosion entrains and disperses a dust layer, which is subsequently ignited by the primary dust flame*

ETA 10

Important Distinction Combustible Dust Explosion vs. Vapor Cloud Explosion

- Primary Explosions of Combustible Dust Clouds usually ***inside*** Process Equipment
  - (e.g., mills, baghouses, cyclones, etc.)
- VCEs usually ***outside*** Process Equipment, because equipment is supposed to contain *only* fuel.

ETA 11

Explosible Range

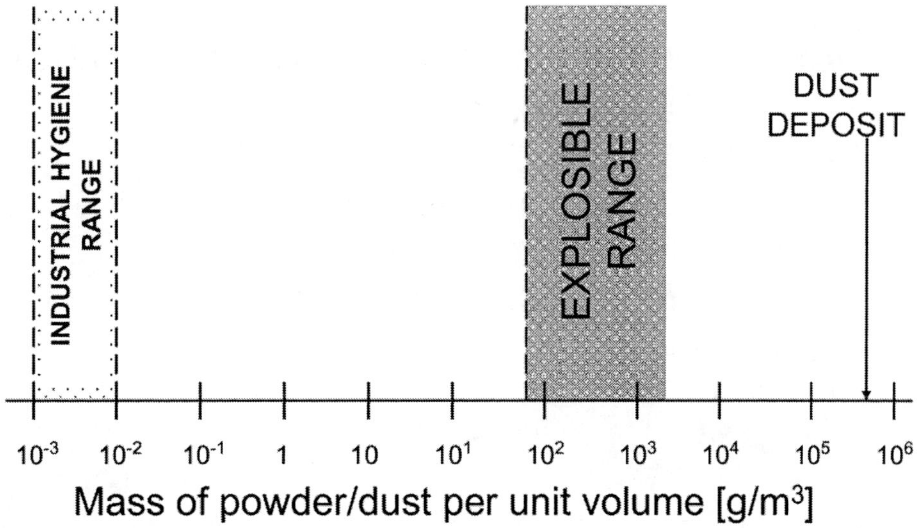

Eckhoff, Rolf, Dust Explosions in the Process Industries, 3rd ed, Gulf Publishing (2003)

ETA 12

TYPES OF COMBUSTIBLE DUSTS

Hayes Lemmerz International – Huntington, IN
October 29, 2003
aluminum dust
1 Fatality and 1 severely burned

ETA 13

Combustible Dusts

Dusts that can cause explosions include:

- **Metals** (e.g., Al, Mg, Ti)
- **Chemicals** (e.g., drugs, dyestuff)
- **Plastics** (e.g., urea formaldehyde resin, polyethylene, polystyrene)
- **Food** (e.g., starch, flour, sugar, cocoa, feedstuffs)
- **Wood**
- **Coal**

ETA 14

Dust Chemistry

Heats of combustion per mole O_2 consumed

SUBSTANCE	OXIDATION PRODUCT	KJ/mole O_2
Mg	MgO	1240
Al	Al_2O_3	1100
Fe	Fe_2O_3	530
Cu	CuO	300
Sucrose	CO_2+H_2O	470
Starch	CO_2+H_2O	470
Polyethylene	CO_2+H_2O	390
Carbon	CO_2	400
Coal	CO_2+H_2O	400
Sulphur	SO_2	300

ETA 15

Equipment for explosibility measurements

1m³ Vessel

1000 L

ETA 16

Equipment for explosibility measurements

Useful to measure

P_{max},
$(dP/dt)_{max}$, K_{st},
LOC,
MIE,
MEC

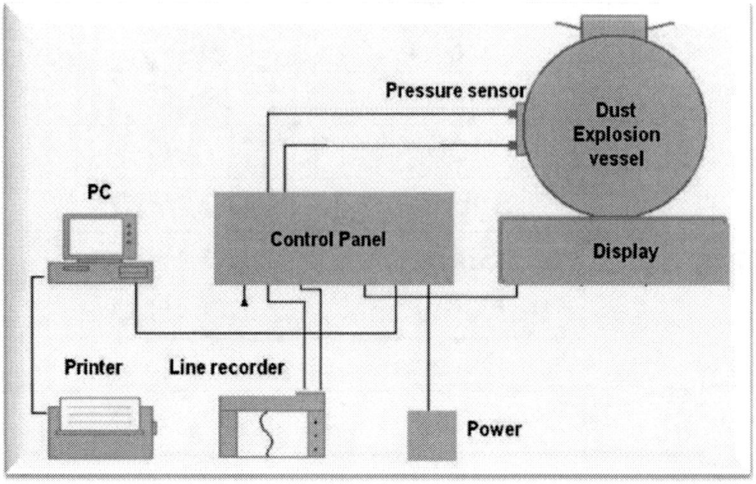

ETA 17

Pressure/time-diagram of a dust explosion

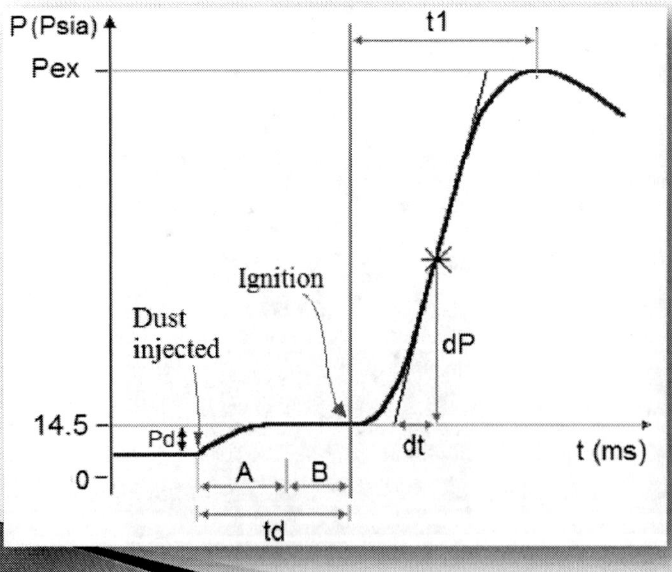

t_d: total delay time

P_d: Expansion pressure

t1: Duration of combustion

P_{ex}: Explosion overpressure

dP/dt: Rate of pressure rise with time

ETA 18

Explosibility characteristics

Combustible dusts are classified by their K_{st} values in increasing order of **explosion violence** as follows:

$\mathbf{K_{st}} > 300$	Group St3: very strong	
$200<\mathbf{K_{st}}=< 300$	Group St2: strong	*Increasing*
$0<\mathbf{K_{st}} =<200$	Group St1: weak	*Explosibility*
$\mathbf{K_{st}} = 0$	Group St0: non-explosive	

Dust Classification

Kst	**St3**	**e.g. fine metal dusts**
	St2	**e.g. organic pigments**
	St1	**e.g. grain dust**
	St0	**non-explosive**

EQUIPMENT SUSCEPTIBLE TO PRIMARY EXPLOSIONS

Port Grain Silo – Corpus Christi, TX
1981
9 Fatalities and 30 Injuries

Eckhoff, Rolf, Dust Explosions in the Process Industries, 3rd ed, Gulf Publishing (2003)

ETA 21

Equipment Involved in Dust Explosions

EQUIPMENT	US (1985 – 1995)		UK (1979 – 1988)		Germany (1965 – 1980)	
	No.	%	No.	%	No.	%
Dust Collectors	156	42%	55	18%	73	17%
Grinders	35	9%	51	17%	56	13%
Silos/ Bunkers	27	7%	19	6%	86	13%
Conveyors	32	9%	33	11%	43	10%
Dryers/ Ovens	22	6%	43	14%	34	8%
Mixer/ Blender	>12	>3%	7	2%	20	5%
Other/Unknown	84	23%	95	31%	114	27%
Total	372	100%	303	100%	426	100%

Guidelines for Safe Handling of Powders and Bulk Solids
CCPS, AIChE.

ETA 22

Silos and Hoppers

A smoldering nest in a dust deposit can be discharged into an combustible dust cloud and initiate a dust explosion.

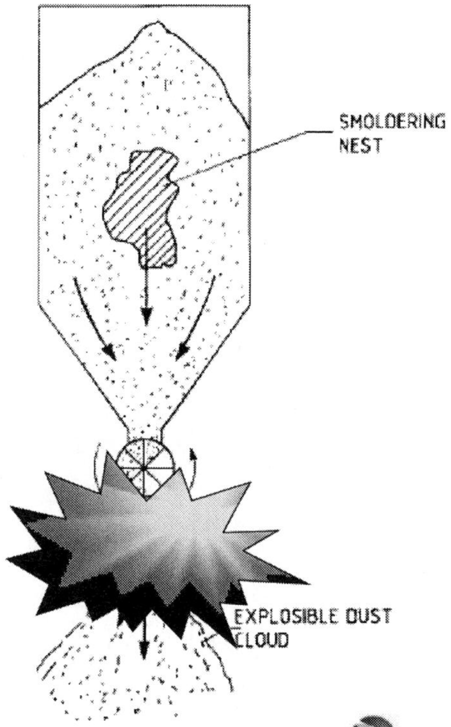

Eckhoff, Rolf, Dust Explosions in the Process Industries, 3rd ed, Gulf Publishing (2003)

23

PREVENTION and MITIGATION

24

How to Prevent

PREVENTION		MITIGATION
Preventing ignition	**Preventing dust clouds**	
Smoldering	Inert gases	Isolation
Open flames	Inert dust	Venting
Hot surfaces	Intrinsic inerting	Pressure-resistant containment
Electric sparks. Electrostatics	Dilution of dust concentration	Automatic suppression
Mechanical strikes		Housekeeping

Prevention - Ignition

- Electrical equipment
  - Class II, Division 1 and 2
    - Combustible Dusts
  - Class I and Class III may also be an issue depending on other materials on site
  - Consider equipment both inside dust handling equipment and outside

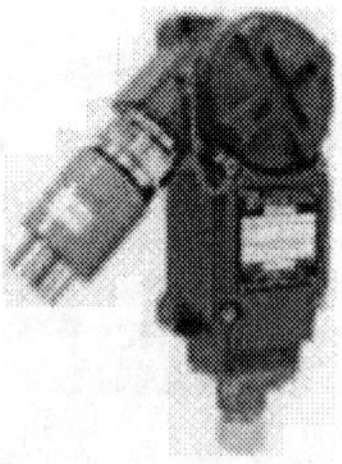

Mitigation - Isolation

- **Screw conveyors**

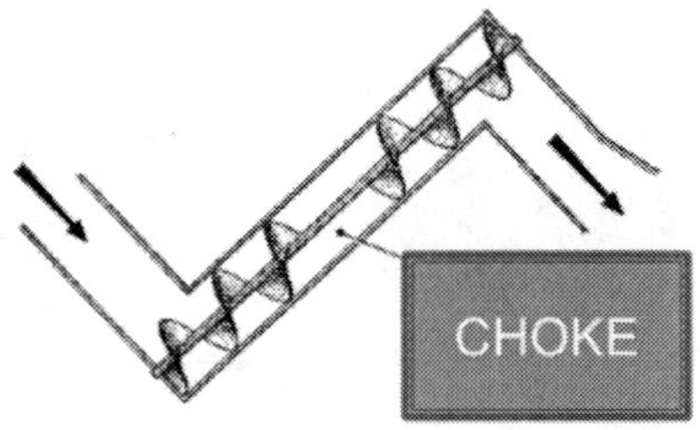

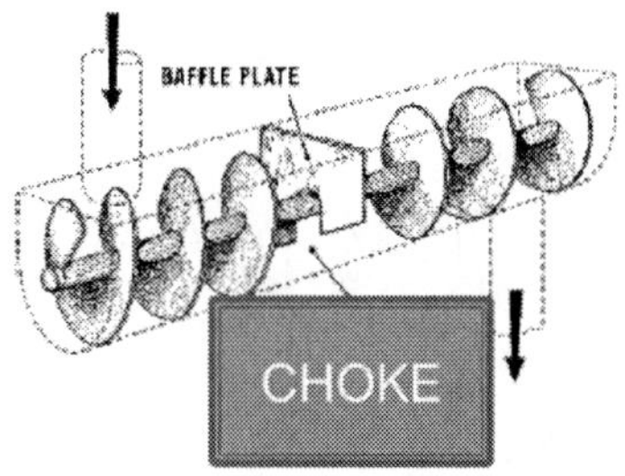

Screw conveyors - prevent transmission of dust explosion

ETA 27

Mitigation - Isolation

- **Rotary locks**

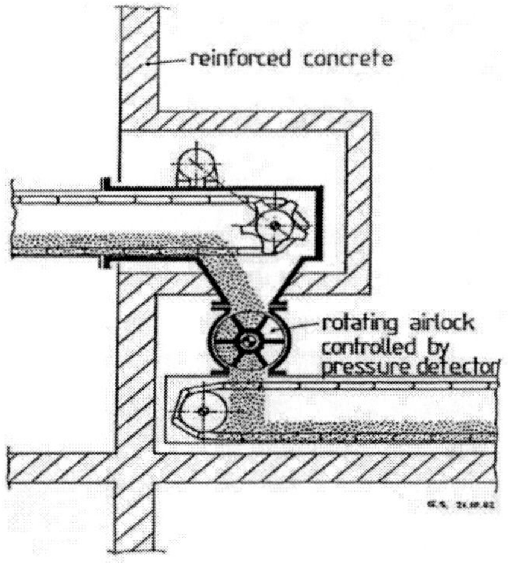

ETA 28

Mitigation – Isolation & Venting

- **Passive devices in ducts**

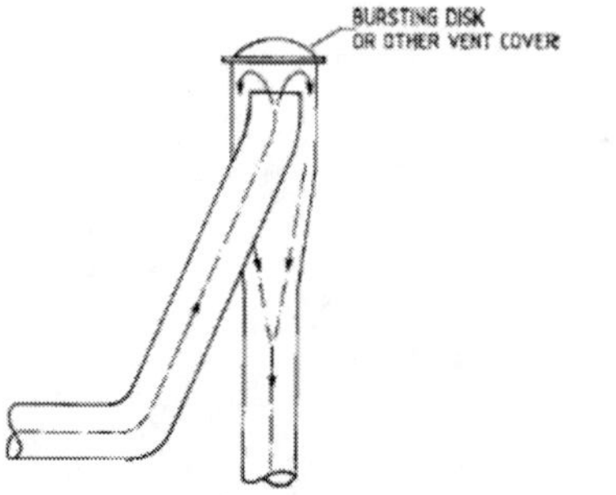

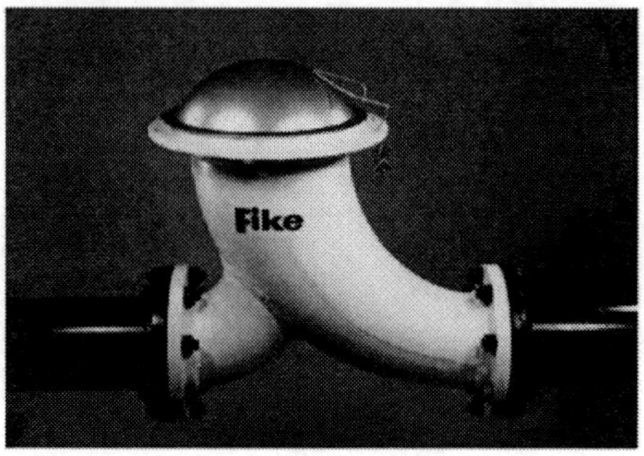

To interrupt dust explosions in ducts
Both vents and changes flow direction

ETA 29

Mitigation - Isolation

- **Active devices in ducts**

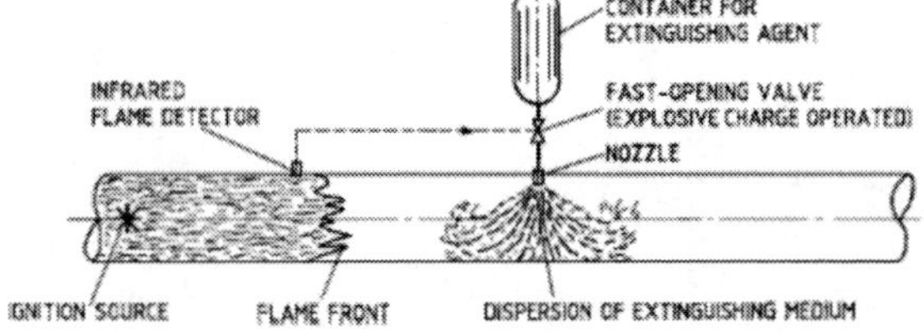

Extinguishing agent ahead of flame

Automatic valve closing ahead of flame

ETA 30

Prevention & Mitigation HOUSEKEEPING

1/32"	3/64"	1/4"	1/2"
0.031"	0.047"	Typical accumulations above ceiling at West Pharmaceuticals	
NFPA 654	U.S. Dime Coin		

ETA 31

HOUSEKEEPING Dust Layer Thickness

- Assume a 10' x 10' x 10' room
- Theoretical thickness to yield Optimum Concentration (OC)

Dust Material	OC oz/ft^3	Depth
Cornstarch	0.5	0.15"
Cork	0.2	0.125"
Sugar	0.5	0.075"
Wood	1.0	0.47"
Polyethylene	0.5	0.180"

NFPA 499 Classification of Combustible Dusts and Hazardous Locations for Electrical Installations, Appendix A

ETA 32

STANDARDS & REGULATIONS

NFPA Standards for Prevention of Fires and Explosions in:

61 - Ag Products and Food
484 - Combustible Metals
652 - Fundamentals – NEW
654 - Combustible Particulate Solids
655 - Sulfur
664 - Wood processing and Woodworking

---

68 – Deflagration Venting
69 – Explosion Prevention Systems

ETA 33

OSHA Timeline

•1999-2003	High-Profile Catastrophic Dust Explosions (e.g., Ford, West)
•November 2006	CSB issues report on "*Combustible Dust Hazards*"
•October 2007	OSHA NEP on Combustible Dust (reissued March 2008)
•February 2008	Imperial Sugar Savannah, GA
•October 2009	OSHA Status Report on NEP
•October 2009	OSHA ANPR issued
•January 2010	Comments due on ANPR
•Present	Internal OSHA review.

ETA 34

OSHA DUST NEP

4900 violations at 1100 inspected facilities

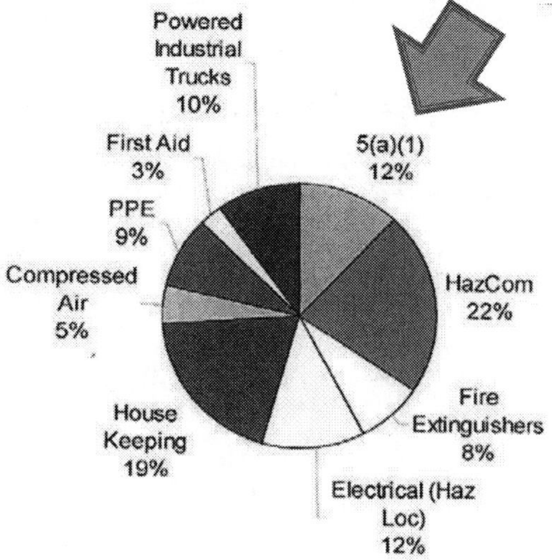

ETA 35

THANK YOU

QUESTIONS?

ETA 36

444734 Enterprise Manufacturing Intelligence (EMI) for Batch Plant Data Systems

Abstract:

Dow collects and stores millions of data across its global manufacturing sites every day using a wide variety of data historians to monitor product and process variables during production runs. Because of the complexity of these database systems, extracting and analyzing this vast amount of data into meaningful information for the plant can be complex and time consuming, thus reducing opportunities to improve certain areas of the process. In order to extract the most significant and immediate value from that data, process design knowledge and operational experience can be used to summarize and aggregate plant data to expose the most relevant information in the right context. The goal of Enterprise Manufacturing Intelligence (EMI) is to make these high value relevant data available in real time to all levels of production management (Operations to Plant and Business Management). Therefore, plants can take advantage of this information to make improvements in raw material usage, asset utilization by reducing unnecessary process wait times and customer satisfaction through improved product quality by reducing product and process variation. This applies to both batch and continuous processes. A Dow case study will be discussed in the presentation to show the benefits of EMI in a batch process.

Through the development and advancement of new commercially available systems[1], key batch process components can be monitored and tracked for a single production run through a unique batch identifier. This software provides easy and fast segmentation of production data into batches, campaigns or other logical groupings for easier analysis and production reporting to optimize process conditions. Because of the defined start and stop time throughout the batch steps, the data associated with each batch can be evaluated either individually or from one batch to another. Historically, this has not been easy to track data for a particular parameter from one batch to another. The isolation of this data provides plants key information about the production process that can help drive continuous improvement across the batch process.

Now that the batch data has been compiled into uniquely identified production runs, plants can take advantage of EMI dashboard technology to quickly understand the performance of their process. EMI allows for continuous monitoring of batches and "flags" when an abnormal process condition has occurred; based on historical batch trends and process targets. This technology also has the capability to merge process data with product analysis to ensure all pertinent information is readily available to plant personnel. This has been extremely difficult to do in the past. This can significantly improve the analysis time and prevent any reoccurrence of incidents prior to the next batch being produced. The implementation of EMI allows for continuous monitoring of long-term batch trends to drive productivity improvements proactively and utilize Dow's assets more efficiently and effectively.

[1] Marty Moran, "Extending the Value of MES Technology into New Applications – An Industry White Paper," www.AspenTech.com, 2013

443534 Using Data Mining Techniques for Process Improvement in Polypropylene Plants

Abstract:

Determination of the most significant manufacturing process parameters can be very valuable in improving product quality and plant reliability. Given the huge amount of recorded data and complexity of chemical plants, data mining is crucial to uncover information, knowledge, and patterns from the historical data.

This presentation describes how to facilitate decision making process in a polypropylene plant by using data mining techniques such as variables screening and modeling. High liquid hydrocarbon level within the polymer product is a chronical safety issue in polypropylene plants. Since the high hydrocarbon level can often be the cause of either plant shutdowns or reducing the production rate, understanding the root cause of this problem has a significant impact of the plant reliability and economics.

Data visualization and multivariate analysis techniques are used to identify relative significances of process variables and evaluate their effects on the hydrocarbon level. As a result of this study, new operational trials are proposed and tested in the plant. The result of trials is used to study the impact of changing in significant variables on reducing hydrocarbon level.

440552 Big Data Doesnt Really Mean Big Intelligence

Abstract:

Big data is the IT phrase of the moment in almost every industry. The concept of collecting huge amounts of data to learn, manage and drive your business forwards is attractive to all managers, especially as without it you may feel that you're being left behind. But without knowing how to collect data effectively, knowing what data has value, and how to effectively and efficiently use the volume of information, Big Data can be a hindrance, not a help.

In Process Safety, we rely on data coming from the field in many instances. Collecting this data can be time consuming and costly, so making the most of the data you have is paramount, as is ensuring the data you collect is seen as a valuable asset by everyone in your organisation. But turning that data into intelligence to guide you on your path to Process Excellence is what it's really all about. Data is as valuable an asset as any other. Look after it. Nutrue it. Value it. And make sure you use it.

In this paper we examine the core principles of making data work for you to give you the best usable, actionable intelligence possible.

445906 React Now to Reactive Hazards: Early identification is crucial for your safety concept

Abstract:

I've been a Process Engineer with Evonik for 3.5 years and this is my first job since graduating college in 2012. For 2 of my years here I have been working on a capital investment project that has posed quite a challenge in the area of process safety. The process begins with an extremely energetic exothermic reaction that produces an unstable intermediate. The temperature, pH, and ratios of raw materials and catalyst all have to be closely controlled in order to keep the reaction under control. The effects of various process upsets had to be studied to determine the severity of the outcome and what safety measures would be effective. In order to do this over 25 adiabatic calorimeter tests were performed as part of the basic engineering for the project. In several of the cases, venting with a pressure relief device was not feasible which resulted in many additional high integrity interlocks. This highlighted how important it is to understand the safety risks involved in any given project, and it is imperative to dedicate resources to resolving safety risks early since they can result in major changes. The significant take away from this project for me is that in this industry no matter the financial or schedule impact, there will be no compromise when it comes to safety!

445907 An early lesson and a late night: how I learned the real design expert is the user

Abstract:

During my first few weeks as a full-time engineer, safety truly hit home. My co-worker and I had been tasked with working on a fermentation filtration experiment skid. The skid held about 50 gallons of fermentation broth, and was testing the ability of a filter to separate solids from the liquid. We set the system up, tightened the bolts, and secured the lines, then started up the pump. Liquid quickly started shooting out at high pressure, as we realized a line had disconnected. We fished through the slippery fermentation liquid and quickly hit the emergency stop button, preventing further loss of contamination. At least 20 gallons had been released into the lab, making a long day of clean up.

The situation occurred because there were no procedures, because of inexperienced operators, and because of failure to do a full system review or walk through prior to startup. The system designer, however, was successful in his thoughtful placement of the emergency stop buttons. Process engineers, product designers, and researchers all have an important responsibility in preventing accidents similar to this. It is vital to design with the operator at the forefront of every decision in order to create success. It is equally important to ensure the operators understand how to operate the equipment, and what to do if it fails.

445909 Personal Growth and Mission Born from Tragedy

Abstract:

Matt McGinnis, an engineer for The Dow Chemical Company, discusses the day PSM hit home when a co-worker passed away 4-weeks after exposure to a hot chemical solution causing 2nd and 3rd degree burns over 70% of his body. Matt reflects on the department moral, the intense investigation, media attention and subsequent legal involvement that has opened his eyes to the importance of process safety in the chemical industry.

Transition Path Sampling for Process Safety Events

Ian H. Moskowitz, Warren D. Seider, and Amish J. Patel
Department of Chemical and Biomolecular Engineering
University of Pennsylvania
Philadelphia, PA 19104-6393

Jeffrey E. Arbogast
American Air Liquide
Newark, DE 19702

Prepared for Presentation at
American Institute of Chemical Engineers
2016 Spring Meeting
12th Global Congress on Process Safety
Houston, Texas
April 11-23, 2016

Abstract:

Chemical manufacturing processes have the potential for catastrophic accidents – those involving loss of human life or major environmental impacts. The numerous ways through which a complex chemical process can fail is limited neither by our experience, nor by our imagination; consequently, so-called *un-postulated* safety-event scenarios pose a major concern. By combining extensive chemical process data, process modeling and state-of-the-art path sampling algorithms, this work aims to uncover rare and un-postulated process safety events, to identify secondary process variables that can portend such events (so they can be incorporated in alarm systems), and to characterize their rates of occurrence.

Transition path sampling (TPS) has been a widely used by the nano-scale modeling community. Its goal is to identify rare-event trajectories that are worthy of investigation; such as the dissociation of a weak acid in aqueous solution or the flipping of phospholipids in a lipid bilayer. The types of rare-events that are best handled by transition path sampling are those in which the time-scale of the event is far shorter than the time-scale over which the event may take place. This ratio of time-scales is similar to a safety-event in a chemical manufacturing process; the event may occur once every five years of operation, while the event takes place over just a five-hour period. To simulate hundreds of such events, a typical process simulation with expected disturbances would have to be simulated for 500 years! It would be computationally beneficial to use a sampling technique that focuses on the safety trajectories of interest, and excludes the vast number of safe operating trajectories.

The details of this Monte-Carlo approach are outlined, and the ability to discover un-postulated rare events demonstrated. The safety trajectories investigated with TPS are used to build event-specific alarms, whose simulated false-positive and false-negative rates are explored. Extensive process, control, and alarm data are essential and are highlighted throughout the presentation. A dynamic model of an air separation unit (ASU) is the test bed for the TPS technique.

Introduction

In response to typical disturbances in manufacturing processes, the basic process control system normally maintains operation within safe thresholds. Occasionally, large disturbances violate operation and safety thresholds (as special-cause events) and alarm and safety interlock systems are required to return the process to its typical operating conditions. The latter events occur rarely, but can be expensive (due to unplanned shutdowns and/or lost product) and even rarer they can be catastrophic (with human health losses, major environmental impacts, …). There is clear motivation to study these events and to identify trends amongst them, but the available process and alarm data are statistically insufficient.

To gain a fuller understanding of how a process moves from typical operation into a safety-event, the process could be simulated over long time periods, during which occasional safety-event trajectories would be encountered, assuming appropriate disturbances and fluctuations are included. To obtain a statistically sufficient set of trajectories (typically on the order of just a few hours), the simulation time would have to be on the order of decades! Clearly this would be computationally infeasible, and a technique is needed to target safety-event trajectories while sharply reducing the time spent simulating typical operations.

A similar challenge, identified in the molecular dynamics community, involves the molecular dynamics of reacting molecules. The reaction events occur very quickly – on the order of nanoseconds, wheras the time between reactions is on the order of milliseconds. Said differently, a simulation millisecond would be needed to capture a nanosecond long reaction trajectory. While the time (and length) scales of the rare events in molecular dynamics are far smaller than those studied in process simulations, the *ratio of rare event times and the waiting period between them* is similar.

This work uses a dynamic simulation model of a simplified, non-commercial air separation unit (ASU) as a test bed for demonstrate application for the identification of rare event trajectories. The rare event that this presentation will focus on is the impurity of the LAR product due to excess nitrogen. While this particular rare event has limited likely safety consequences, it provides a test bed example on which to propose, develop, and refine a technique for generating many trajectories of a rare event. For this particular hypothetical example, the rare-event of concern occurs when the mole fraction of nitrogen rises above 0.01. As the application of the method expands, safety-events with significant potential safety consequences can be efficiently and effectively studied. The methods presented in this case study are generally applicable for many manufacturing processes and rare safety-events.

Overview of Transition-Path Sampling (TPS)

In TPS, which has evolved over the past two decades, the probabilities of locating rare reactive paths are increased significantly [1,2]. First, just one rare trajectory leading to the event of interest is identified. Then, a new "trial" trajectory is generated by perturbing the initial trajectory, and when the trial trajectory is also reactive, it is accepted with a finite probability. A new trial trajectory is then generated, and the procedure is repeated. By focusing only on the reactive paths, the waiting time between the trajectories of interest is eliminated, and consequently, computation times are sharply reduced.

To begin a TPS algorithm, using a dynamic (molecular or process) model, it is necessary to locate one rare trajectory from an initial state to a desired rare target state. One is illustrated schematically in Figure 1a. Here, it proceeds from an initial state (e.g., acceptable nitrogen in the LAR – green) and ends at a rare state (e.g., impure LAR product – red). Then, any time is randomly selected (t') and the independent variables (e.g. the liquid molar holdup and composition of each tray, see the *TPS Demonstration* section) are perturbed (Δx). The model is integrated forward from t' to final time t_f, and in reverse from t' to 0. When another desired rare trajectory is found with its end points lying within the red and green zones, as in Figure 2b, the new trajectory is accepted with a finite probability that depends on the relative likelihood of observing the two trajectories. The trajectories that are a result of more typical process initial conditions and fluctuations are selected at a higher rate. The TPS algorithm is presented in Table 1.

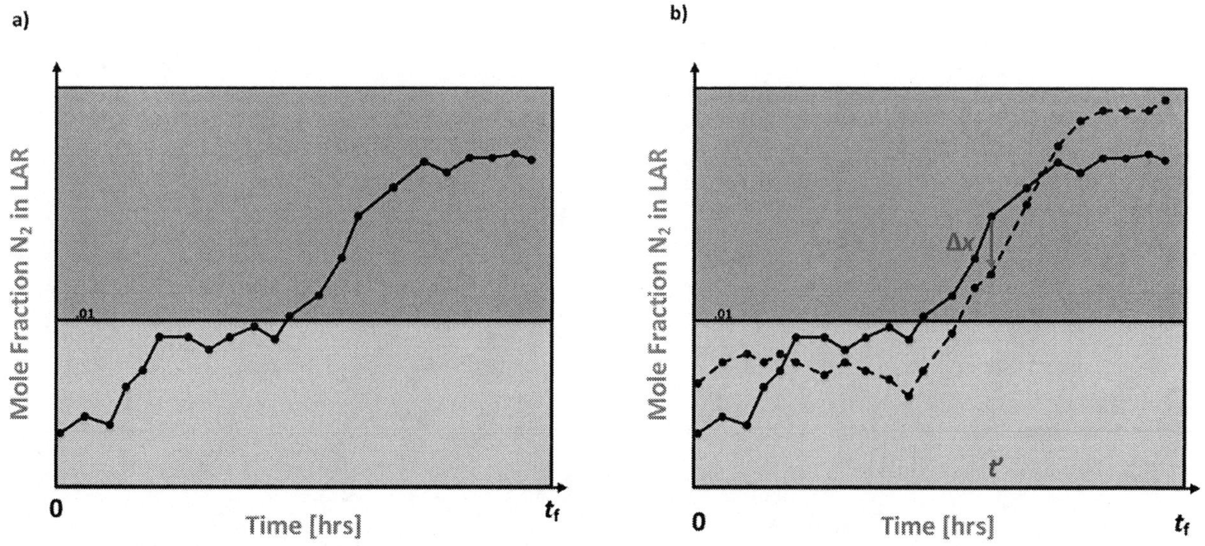

Figure 2. Typical trajectories between initial desired states

Table 1. TPS Algorithm

1.	Identify an initial safety event trajectory
2.	Choose a random time, t', along the trajectory
3.	At t', perturb state variables, x_i, to x_i'
4.	Integrate forward from t' to t_f, and backwards from t' to 0
5.	Determine if this trial trajectory identifies a safety-event
6.	If yes, accept the trial safety event trajectory as the new trajectory with a probability, min $(1,p'/p)$. Here, p'/p is the relative likelihood of the occurrence of the trial perturbed trajectory with respect to the unperturbed trajectory. If the trial trajectory is not accepted, set the previous trajectory to be the new trajectory.
7.	Return to step 2.

Example ASU Model

The proposed approach is demonstrated using a dynamic model of a simplified, non-commercial ASU process, motivated by the process flow diagram shown by NETL[3], and further modeling considerations were motivated by Huang et al. [4]. Saturated vapor air is separated into a nitrogen,oxygen, and argon product in this process, shown in Figure 2. Three distillation columns are used – a high pressure column (HPC) at 5.5 bar beneath a low pressure column (LPC) at 1.25 bar, plus a crude argon column (CAR) for separating argon. As connecting streams expand from 5.5 bar to 1.25 bar, saturated liquids are maintained at a constant temperature. Three components are included in this model – nitrogen, argon, and oxygen, listed in order of their volatilities.

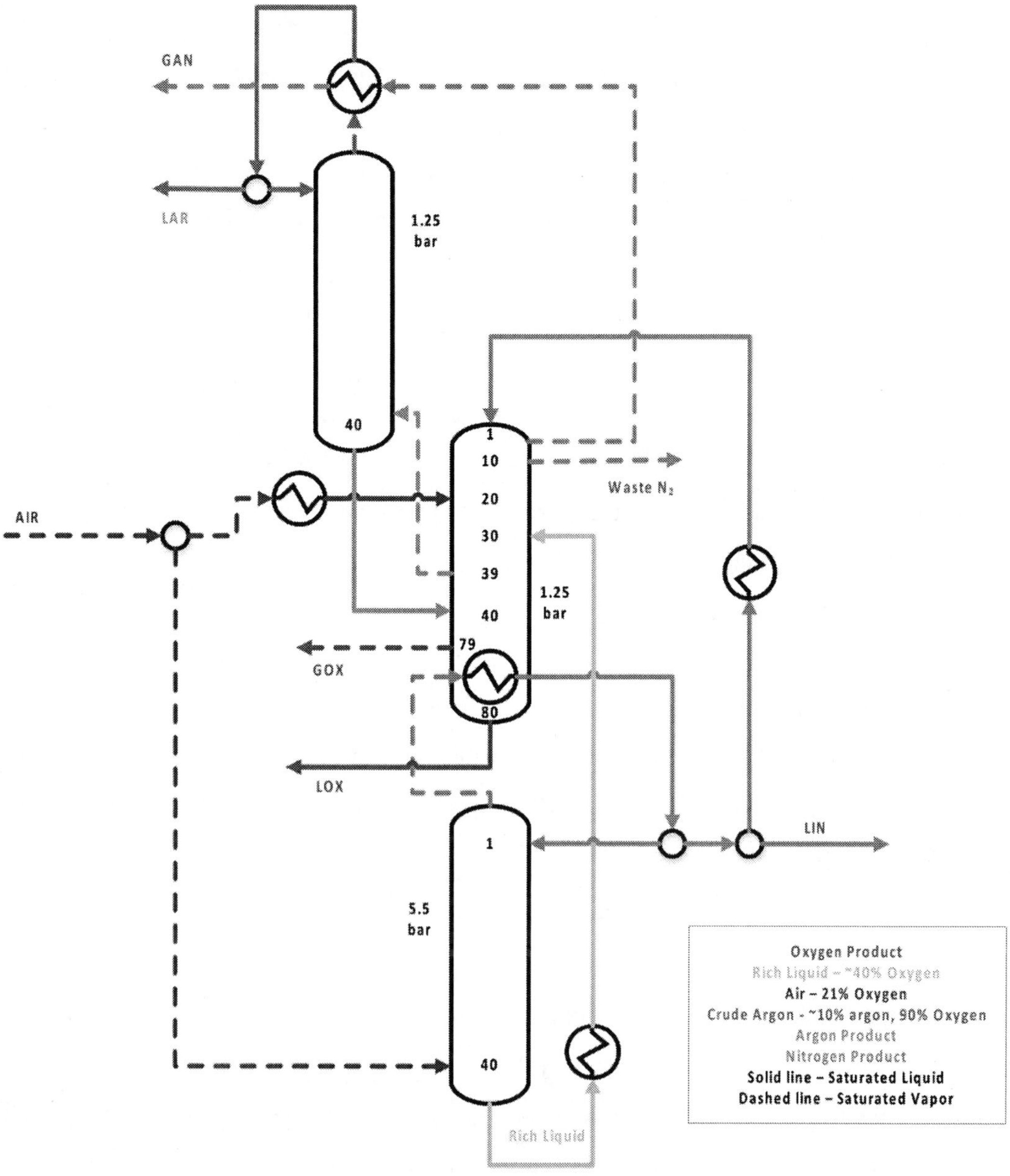

Figure 2. Process flow diagram of the ASU.

Saturated vapor air is fed to the sump of the HPC. Off the top of this column, nitrogen product is condensed while boiling liquid in the sump of the low-pressure column (LPC), with a fraction of the liquid nitrogen (LIN) drawn as product, a fraction returned to the HPC as reflux, and the remainder sent to the top of the LPC as reflux. The oxygen-rich liquid from the sump of the HPC is fed to the middle of the LPC, along with saturated liquid air. Gaseous nitrogen product (GAN) is taken off the top of the LPC, with liquid oxygen (LOX) and gaseous oxygen (GOX) product is removed from the bottom of this column. A vapor side-stream of approximately 10% argon, 90% oxygen, and trace nitrogen is sent to the sump of the CAR. Here, liquid argon (LAR) is produced using the GAN stream for cooling in the condenser. Some of this LAR is taken as product, with the remainder refluxed to the CAR column.

In the simulated model, nearly all of the nitrogen entering the CAR escapes in the LAR product. Because nitrogen is the lightest component, it rises up the column – making the LAR difficult to condense if nitrogen content is too high. While nitrogen is removed in the LAR product stream, some is returned in the column reflux. When the LAR purity cannot be met, the CAR column undergoes abnormal operation – not ideal for maintaining safety. For this reason, the excursion of nitrogen content in the LAR product is considered a safety-event for the context of this presentation. In a similar commercial ASU process, the impact upon plant operation would be especially pronounced due to the significant heat integration of the plant, the use of rich liquid from the HPC (instead of GAN) in the condenser of the CAR column, and the return of vaporized rich liquid to the LPC [5].

The trays of each column are modeled at equilibrium using the MESH (mass balance, phase equilibrium, summation of mole fractions, and heat balance) equations, an empirical equation relating the liquid holdup to the liquid flow rate, and the Peng-Robinson equation of state. This model is modified from that presented by Huang et al. [4] for model-predictive control. The overall material balance of tray i is represented by Eqn. (1):

$$\frac{d}{dt}(M_i) = L_{i-1} + V_{i+1} - L_i - V_i + F_i \tag{1}$$

where M_i is the liquid molar holdup, L is the liquid molar flow rate, V is the vapor molar flow rate, and F_i is the molar feed flow rate (zero for most trays). It should be noted that only the liquid holdup is considered here – the vapor, being far less dense than the liquid, has a negligible holdup. The component material balances describe the composition of each tray, and is shown in Eqn. (2):

$$\frac{d}{dt}(x_{i,j}M_i) = x_{i-1,j}L_{i-1} + y_{i+1,j}V_{i+1} - x_{i,j}L_i - y_{i,j}V_i + z_{i,j}F_i \;\forall\, j \in \{N_2, Ar, O_2\} \tag{2}$$

where x is the liquid molar fraction, y is the vapor molar fraction, and z is the feed molar fraction (phase unspecified). The heat balance is:

$$\frac{d}{dt}(M_i h_i^L) = h_{i-1}^L L_{i-1} + h_{i+1}^V V_{i+1} - h_i^L L_i - h_i^V V_i + h_i^F F_i \tag{3}$$

where h^L is the liquid molar enthalpy, h^V is the vapor molar enthalpy, and h^F is the feed molar enthalpy (phase unspecified).

An empirical relationship for the liquid flow rate as a function of the liquid molar holdup is assumed:

$$L_i = k_d M_i \tag{4}$$

where k_d was determined by Huang et al. [4] to be 0.5 min^{-1} .

The vapor-liquid equilibrium in each tray is calculated by equating the mixture fugacities of the species in the vapor and liquid phases:

$$\phi^L_{i,j} x_{i,j} = \phi^V_{i,j} y_{i,j} \tag{5}$$

where $\phi^L_{i,j}$ is the liquid fugacity coefficient of species j on tray i and $\phi^V_{i,j}$ is the vapor fugacity coefficient of species j on tray i, each calculated using the Peng-Robinson equation of state, at the temperature of each tray, T_i, the column pressure, and the mole fractions of the associated phases. The enthalpies are also calculated using the equation of state as described by Peng et al. [6]. Lastly, the mole fractions sum to unity in the vapor on each tray:

$$1 = \sum_j y_{i,j} \tag{6}$$

TPS Demonstration

The first step of the TPS algorithm, shown in Table 1, is the identification of a rare safety-event trajectory. This trajectory begins with little nitrogen in the CAR column – thus, there is little nitrogen entering the column from the LPC. The vapor mole fraction profile along the LPC is shown in Figure 3a at the start of this trajectory – notice that the argon composition peaks close to tray 39, the tray whose vapor sidedraw is fed to the CAR column, while the nitrogen fraction is sufficiently low. At the beginning of this trajectory (which spans the total of 24 hours) overly aggressive setpoints for the production rate of GOX and LAR are simulated. These setpoints are probably unlikely in typical operation, as discussed shortly. With large amounts of vapor from the bottom of the LPC column, liquid higher in the column vaporizes to replace it, and the argon bubble begins dropping along the column. The mole fraction profiles at 1-, 4-, and 12-hr are shown in Figures 3b, 3c, and 3d, respectively. As the nitrogen enters the CAR, it rises into the LAR product, whose nitrogen mole fraction along the trajectory is shown in Figure 3e. This trajectory, with the mole fraction of nitrogen exceeding 0.01, predicts a safety-event – it begins with less than 0.001 mole fraction nitrogen in the LAR product and finishes with almost 0.02.

a)

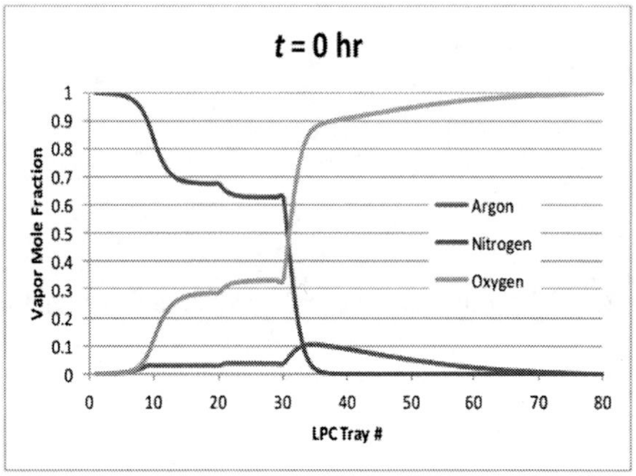

b)

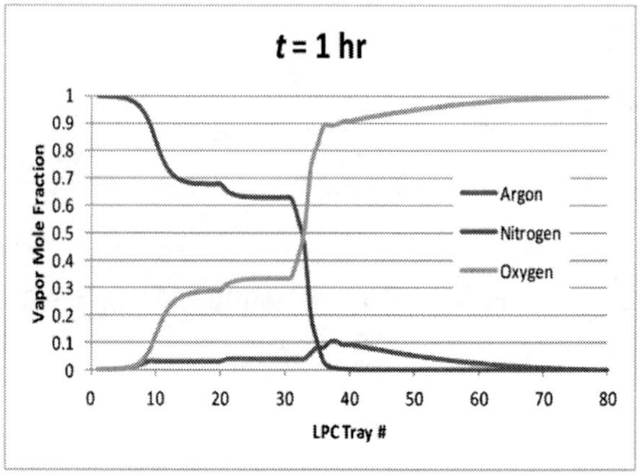

c)

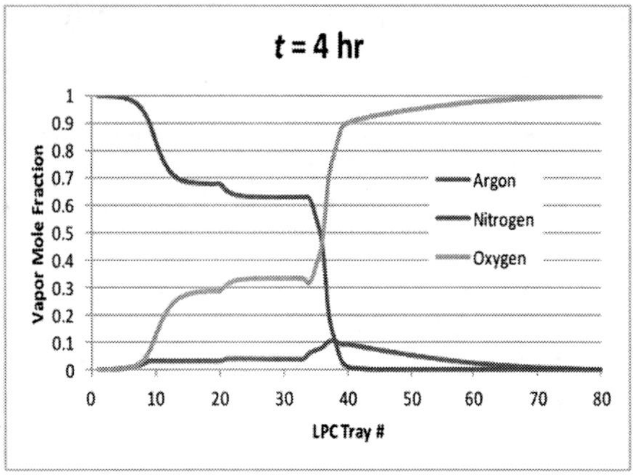

d)

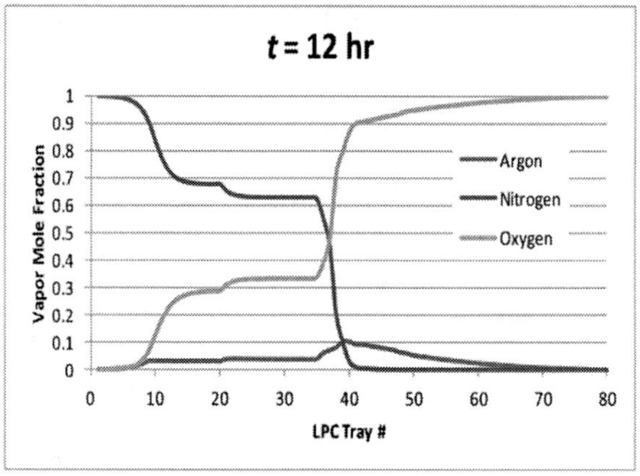

e)

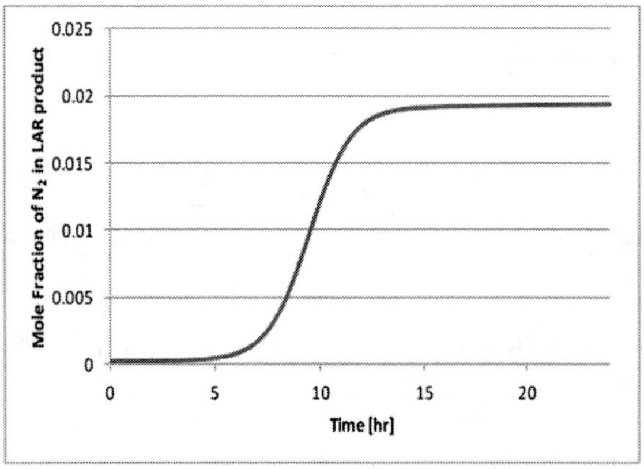

Figure 3. Mole fraction profiles after GOX and LAR setpoints in LPC are increased: (a) initial profiles, (b) 1-hr, (c) 4- hr, (d) 12-hr. (e) The initial safety-event trajectory is displayed, with nearly 2% nitrogen in the final LAR product at t_f.

The second and third steps involve choosing a random time t' (t' = 18 hours) and perturbing all the independent variables, respectively. The independent variables on each tray are $\left\{\frac{dM_i}{dt}, \frac{dx_{i,j}}{dt}, P_{SP}\right\}$, where P_{SP} is the product flow rate setpoint of each of the three liquid products. Each independent variable is sampled from a normal distribution centered at $\left\{\frac{dM_i}{dt}|_{t'}, \frac{dx_{i,j}}{dt}|_{t'}, P_{SP}|_{t'}\right\}$, with a user-defined variance. These distributions play a critical role in determining the types of trajectories uncovered using TPS, to be discussed in the oral presentation. The fourth step of the TPS algorithm involves the forward integration of the perturbed system to $t = t_f$, and the backward integration to $t = 0$. The trial trajectory is shown in Figure 4 – although there is less nitrogen in the LAR at t_f, it is also a rare event of concern. Step six involves determining the relative probability of the trial trajectories compared to the initial trajectory. Likelihood distributions are built using long simulations, generating simulated data by which the model initial conditions and P_{SP} perturbations are evaluated. The 'unrealistic' P_{SP} values used to generate the initial trajectory are relaxed into more realistic values through iterations of TPS. Said differently, as more realistic demand profiles are chosen using random perturbations, the trajectories become more realistic, while continuing to generate safety events. In this way, new and un-postulated safety events are generated, and postulated safety events are better understood. The likelihood distributions play an important role in determining the safety event trajectories generated by TPS – and will be discussed in the oral presentation. As this algorithm is repeated, new safety event trajectories are generated, one of which is displayed in Figure 4.

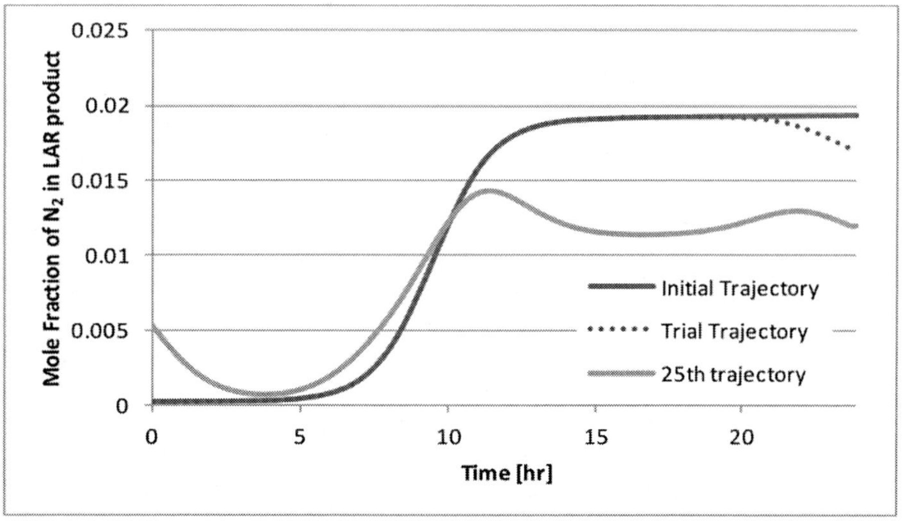

Figure 4: Various rare safety-event trajectories generated using TPS.

Conclusions and Future Work

TPS is an efficient method for generating rare safety-event trajectories. As highly unlikely initial trajectories are relaxed into more realistic trajectories, un-postulated safety events may be discovered. The approaches for generating and accepting these trajectories leave room for user flexibility – these must be tuned appropriately. As trajectories are generated, common trends among them are identified. The long term goal is to influence operation and design enabling previously un-postulated, not adequately appreciated, and/or poorly understood, safety-events to be anticipated and handled appropriately. This novel technique may provide an efficient, stochastic approach to studying rare safety-events in manufacturing processes and should be applicable across all manufacturing industries.

Work Cited

1. P.G. Bolhuis, D. Chandler, and P.L. Geissler. "Transition path sampling: Throwing ropes over rough mountain passes, in the dark," Annual Review of Physical Chemistry, pp. 291-318, 2002.
2. C. Dellago, P. Bolhuis, and P. L. Geissler. "Transition path sampling," Advances in Chemical Physics, pp. 1-84, 2002.
3. Commercial Technologies for Oxygen Production. National Energy Technology Laboratory, available at http://www.netl.doe.gov/research/coal/energy-systems/gasification/gasifipedia/commercial-oxygen, Accessed on March 15, 2016.
4. R. Huang, V. M. Zavala, and L. T. Biegler. " Advanced step nonlinear model predictive control for air separation units," Journal of Process Control, pp. 678-685, 2009.
5. History and Technological Process – Cryogenic Air Separation, The Linde Group, available at http://www.linde-engineering.com/internet.global.lindeengineering.global/en/images/AS.B1EN%201113%20-%20%26AA_History_.layout19_4353.pdf, accessed on March 17, 2016.
6. D. Y. Peng and D. B. Robinson. "A new two-constant equation of state," Industrial & Engineering Chemistry Fundamentals, pp. 59-64, 1976.

Analysis of Overflow Incidents: Searching for the Demons of Situation Awareness

Nicholas P. Sands, CAP, PE
Manufacturing Technology Fellow, DuPont Protection Technology

Palaniappan Chidambaram
Global Solutions Architect, DuPont Sustainable Solutions
DuPont Company (Singapore) Pte Ltd
1 Maritime Square, #10-54, HarbourFront Centre, Singapore 099253

Abstract

Alarms are intended to provide indications to operators to take the necessary actions to prevent undesired consequences. Thus, alarms are often considered as a layer of protection during Process Hazard Analysis (PHA) or Safety/Hazard Review. Without due consideration for human factors that are necessary to ensure its effectiveness, alarms only give a false sense of security.

In several process industry incidents, the designed layers of protection using the operator response to alarm failed. There are several mechanisms for this failure relating to systems or human factors. Among the human factors, there are several situation awareness demons such as attentional tunnelling, data overload and misplaced salience as identified by Endsley (2011).

This paper will examine several overflow incidents for the causes specifically related to the failure of alarms, the human factors component and situation awareness demon, and how it can be addressed during the facility life cycle. The incidents presented in this paper are from a major chemical manufacturing company and from the chemical industry.

Summary

Alarms are commonly used as layers of protection to prevent vessel overflow events in the process industries. A review of industry incidents shows that human error is often a factor in overflow incidents. However in several incidents, the designed layers of protection using the operator response to alarm failed to prevent the consequence. There are several mechanisms for this failure, relating to systems or human factors. Among the human factors, there are several situation awareness demons such as attentional tunnelling, data overload, and misplaced salience as identified by Endsley (2011).

Background

A review of articles on incidents shows that overflow incidents continue to occur, and that human error is a significant factor in overflow events. Further, human error as a factor in overflow incidents may be increasing. The studies indicate that an increased focus on human factors in design may reduce the overflow incidents.

Overfilling of vessels can occur in any process that handles liquid and is of primary concern when the liquids are hazardous in nature, with the potential to cause personnel, environmental, asset and business impact in case of loss of containment. A study of 242 storage tank accidents in North America, Asia and Europe that took place between 1960 – 2003, shows that 30 percent of the incidents were due to operational and maintenance error. Overfilling was identified as the most frequent type of incident in operational error category. Among the 15 overfilling incidents reviewed, 13 cases led to fire and explosion (Chang, 2006).

A further survey of 330 accidents shows human error as the cause in 72 incidents and among them, 12 were overfilling incidents. A review of incidents data set in 21st century showed that the human factor contribution increased to 35 percent for the period 2000-2013 as compared to their contribution share in 1961-2013 (Hemmatian, 2014).

Waite (2013) highlighted 10 tank overfilling incidents that happened between 1972-2009; with eight out of 10 incidents resulting in fatalities and one incident resulting in 43 injuries. These incidents are similar in nature to that of refinery incidents in which columns have been flooded, resulting in loss of containment through the relief system. Waite highlights the need to challenge and review the operator mental models such as reliance on pressure or liquid head without understanding their relation with density, and on out-flow information from a vessel to confirm liquid is not accumulating and to address them adequately during Process Hazard Analysis with special focus on human factors (Waite, 2013).

While these analyses point toward human error, a more focused review of the human and organizational factors that influence such errors will enable prevention of such incidents. A detailed review of factors that led to failure of the operator response to alarms, including failures to alarm, may lead to changes in the design of protection layers. A framework was developed in a recent paper, using concepts from the standard ANSI/ISA-18.2-2009 Management of Alarm Systems for the Process Industries and human factors concepts from the airline industry.

Alarms and Alarm Management

The definition of an alarm, as stated in standard IEC62682 Management of Alarm Systems for the Process Industries, is an audible and/or visible indication to the operator of an equipment malfunction, process deviation or abnormal condition that requires a timely response. The operator response to alarms can correct conditions such as product quality problems, equipment malfunctions, and operating issues that could lead to potentially catastrophic process hazards. Because of the range of consequences related to the alarms directed to a single operator, the alarm system has to be managed for them to be effective. In an unmanaged alarm system, lower priority alarms can obscure the operator's response to critical alarms. The identification of alarms as a protection for process hazards such as a vessel overflow, the design of the alarms and the design of the Human Machine Interface (HMI) are important steps in management of the alarm system as shown in the alarm management lifecycle, described in ISA-18.2 and IEC62682 (Figure 1).

The lifecycle puts activities into stages, and the stages can be mapped to other lifecycles or processes. The identification stage of the lifecycle would include Hazard Review and Layer of Protection of Analysis (LOPA) in a project process or the Hazard and Risk Assessment and Allocation of Protection Layers stages of the Safety Instrumented System lifecycle. Identification activities create potential alarms for the rationalization stage. If the identified potential alarm satisfies the criteria for rationalization, it becomes an alarm. Often alarms identified as safeguards in a PHA, do not pass rationalization for reasons such as the operator not having an action to prevent the consequence or the time to take that action.

The detailed design stage of the lifecycle includes the activities of basic alarm design, advanced alarm design, and HMI design. Alarm dead bands and time delays are part of the basic alarm design. Suppressing the alarm based on process or other conditions is part of the advance alarm design. The way alarms are presented, including the alarm priorities, is part of the HMI design. Mistakes in the basic design of an alarm can cause an alarm to become a nuisance.

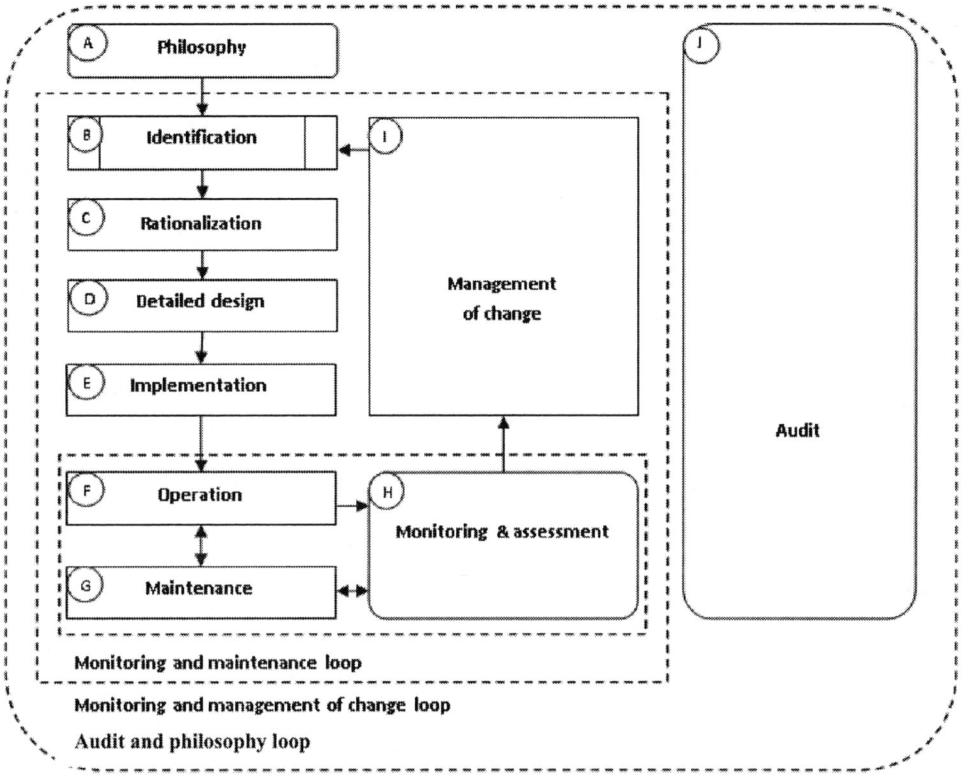

Figure 1 The Alarm Management Lifecycle

Identification of Alarms – Hazard Review

Operator responses to indications and alarms are often considered as effective safeguards without giving much consideration as to their effectiveness and independence in HAZard and OPerability Study (HAZOP) or "What If" analysis. Often it is due to insufficient understanding of what is an alarm and the requirements of an alarm to be an effective safeguard.

It is essential to educate the team that alarms should not be identified as safeguards if the operator:

1) does not have a corrective action than can break the chain of events and bring the process to a safe state;
2) does not have time to take the corrective action between the time the alarm occurs and the time the consequence occurs (process safety time); and
3) training is not provided for operators and documentation of corrective action is not available.

By doing so, a number of configured alarms can be reduced, hence minimizing alarm flood potential and encouraging a culture of responding to alarms if it is activated, instead of ignoring or suppressing them. It is essential for the HAZOP/"What If" team to resist the temptation to add alarms that are ineffective and management systems to ensure those added are managed effectively.

Operator Response to Alarms

ISA-18.2 provides a model for the operator response to an alarm, as shown in figure 2. The model is useful as a starting point for evaluating human error. If the alarm is annunciated, but the operator does not see the alarm, the failure is in detection. If the operator detects the alarm, but does not choose the correct response to the alarm, the failure is in diagnosis. If the operator determines the correct action, but cannot complete the action or complete the action in time, the failure is in the response to the alarm (Dunn, 2013).

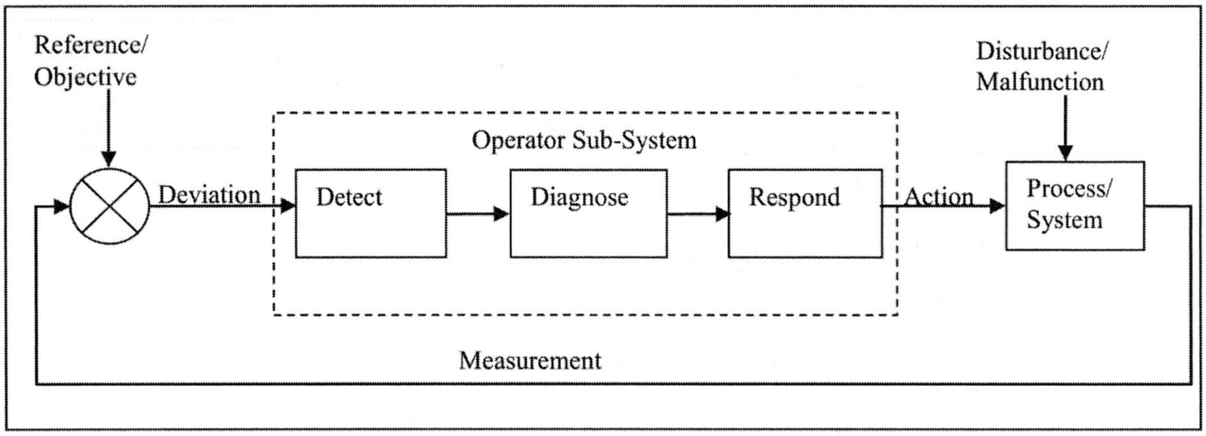

Figure 2. Feedback model of operator process interaction

Alarm Failure Mechanisms

Dunn et al combined a set of alarm failures in a fishbone diagram, shown in figure 3. The top half of the figure are failure mechanisms related to operator response and the bottom half of the figure are failure mechanisms related to the design and maintenance activities of the alarm management lifecycle (Dunn, 2013).

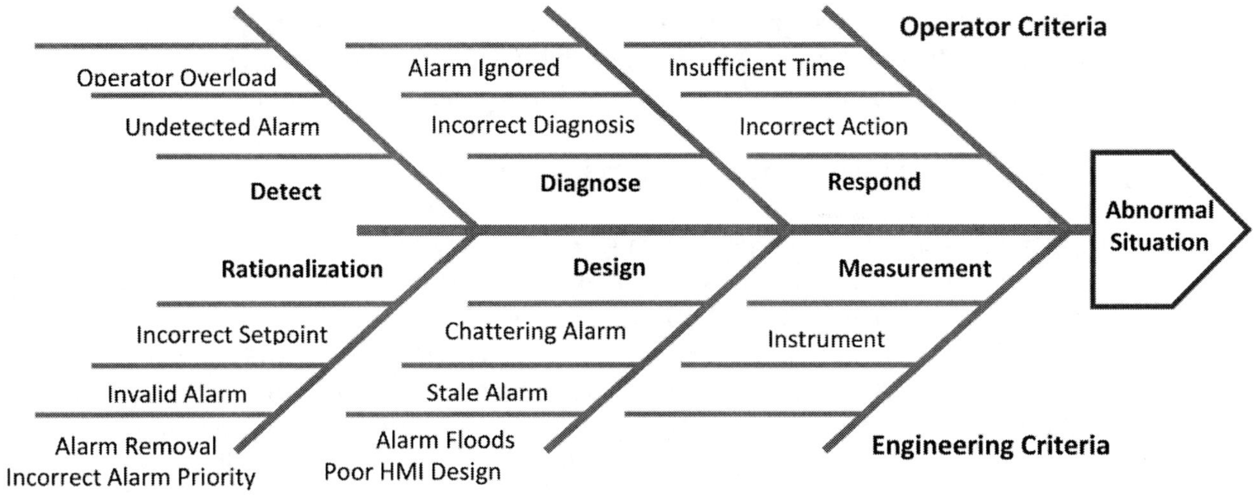

Figure 3. Failure Mechanisms (Fishbone)

Situation Awareness Demons

To further evaluate the human factors that contribute to human error in response to an alarm, the concepts of situation awareness and detrimental factors or situation awareness demons, can be borrowed from the work of Endsley (2011). "Situation awareness is being aware of what is happening around you and understanding what that information means to you now and in the future" (Endsley, 2011).

The situation awareness demons were restated specific to alarm management (Dunn, 2013):

- Attentional Tunnelling – Focusing on one area or issue to the extent that alarms from another area or issue are excluded;
- Requisite Memory Trap – Over dependence on operator memory for effective response to alarms, perhaps under different operating conditions;
- Workload, Anxiety, Fatigue and Other Stressors – (WAFOS) physical or psychological stress that can impact performance;
- Data Overload – Over dependence on the operator to make sense of excessive amounts of alarm information;
- Misplaced Salience – Incorrect alarm priority or HMI representation of alarm importance and other status information;
- Complexity Creep – complex response or complex HMI features that increase response complexity;
- Errant Mental Models – Thought process that incorrectly interprets alarms or mistakenly discounts relevant alarms; and
- Out-of-the-Loop Syndrome – Process or response automation that causes the operator to not know what to do when automation fails.

Neither the failure mechanisms nor the situation awareness demons are mutually exclusive, therefore more than one can be related to a specific incident. It is helpful to identify only the minimum number of failures or demons related to an incident.

Alarms – Are we crediting appropriately in LOPA?

Operator response to alarms is often given more credit without a detailed evaluation and such practices result in a higher residual risk. In 2007, as part of HSE initiative, Bellamy conducted a review of seven LOPA studies from different companies related to overfill scenarios in tank farm. He concluded that the demonstration of independence of protection layers and justification for credit assigned for operator response to high-level alarms was lacking in general (Bellamy, 2007). It is critical to ensure that operator response to alarms are effective, especially for batch processing facilities, unloading/loading operations and in non-routine operations, where reliance on human intervention is high. Myers (2013) emphasizes the need to consider the site specific culture while estimating human error probabilities. Myers also highlighted typical issues in the LOPA when considering human error as initiating events and human intervention as independent protection layers.

Kletz (2002) highlights that incident reports are superficial, and often stop at the first evident cause or immediate causes without further investigation into ways to avoid hazards, organizational factors that contributed to the incidents and identification of the factors that influenced the human error instead of just listing human error as cause. This results in increased potential for incident recurrence. Many incident investigations do not contain the details necessary to identify the failure mechanisms and the situation awareness factors.

Questions that capture some of the information during an investigation specific to alarms include:

- Was there another process upset occurring at the same time?
- What was the alarm rate at the time of the incident?
- Was the alarm or alarms of interest frequent alarms or nuisance alarms?
- Was the alarm or alarms of interest easy to distinguish in the HMI?
- Was consequence of deviation and associated response documented and training provided?

Overflow Incidents Analysis

The incidents described below are reduced to the facts needed for the failure analysis:

Incident 10 Tank Overflow

During a start-up of a process unit, a tank of organic solvent, above its flash point, overflowed causing a vapor cloud, which did not ignite. The level rise generated at least two different alarms prior to the overflow. Two other units in the span of control of the same control room operator were conducting maintenance tests that generated a significant number of alarms. The operator did not respond to the tank-level alarms. A field operator reported the vapor cloud.

The facts related to the alarm system failures can be stated as follows:

- The operator was distracted by the maintenance testing in the two units.
- The operator did not respond to the tank high-level alarms.
- The tank high-level alarms were not nuisance alarms.

The failures can be mapped to the following failure mechanisms:

- Diagnose Failure
    - Operator did not respond to the tank high-level alarms.
        - Operator was distracted by the maintenance testing.
        - Operator may have believed the alarms were triggered by maintenance testing.

The failures can be mapped to the following SA conditions:

- Attention Tunnelling
    - Operator was distracted by the maintenance testing.
- Errant Mental Model
    - Operator did not connect tank high level alarm to unit that was in start-up.

Incident 11 Sump Overflow

Hazardous waste sump overflowed when redundant sump pumps lost prime and cavitated due to failure of common seal flush system. High level and high-high level alarms sounded before overflow. A process upset was occurring simultaneously and the control room operator was addressing the upset and did not respond to the alarms. A field operator identified the overflow. ~500 gallons of material was spilled.

The facts related to the alarm system failures can be stated as follows:

- The operator was focused on the upset in the operating area.
- The operator did not respond to the sump high-level alarms.
- The sump high-level alarms were not nuisance alarms.

The failures can be mapped to the following failure mechanisms:

- Diagnose Failure
    - Operator did not respond to the sump high-level alarms.

        - Operator was focused on the upset in the operating area.

The failures can be mapped to the following SA conditions:

- Attention Tunnelling
    - Operator was focused on the upset in the operating area.

Incident 12 Sump Overflow

During a plant shutdown, a check valve failed, allowing backflow of a hazardous material from a tank to a sump. An alarm indicated the high sump level condition, and the control room operator contacted the field operator to empty the sump. Later a second high sump level alarm occurred and the control operator did not contact the field operator. The sump overflowed, spilling ~4,500 lbs of material until the spill was discovered.

The facts related to the alarm system failures can be stated as follows:

- The operator was less attentive to alarms during the plant shutdown.
- The operator did not respond to the second high level sump alarm.
- The sump high level alarms were nuisance alarms due to rain.

The failures can be mapped to the following failure mechanisms:

- Diagnose Failure
    - Operator did not respond to the sump high level alarm.
        - Operator was not as vigilant due to the plant shutdown.

The failures can be mapped to the following SA conditions:

- Errant Mental Model
    - Operator was not as vigilant due to the plant shutdown.

Incident Analysis summary

The results of above discussed overflow incidents 10-12 have been included in the previous set of analysis by Dunn (2013) to demonstrate the importance of understanding situational awareness and alarm failure mechanisms. As evident from table 1 below – errant mental models and inability to diagnose are key contributors to incidents where alarms failures were identified as contributors:

	Failure Mechanisms (FM)							Situation Awareness (SA)								
Incident	Rationalization	Design	Measurement	Detect	Diagnose	Respond	FM Totals	Attention Tunneling	Requisite Memory Trap	Workload, Anxiety, Fatigue...	Data Overload	Misplaced Salience	Complexity Creep	Errant Mental Models	Out-of-the-Loop	SA Totals
1		X		X	X		3			X	X			X		3
2		X	X	X	X		4							X		1
3		X	X	X			3									0
4	X				X		2	X						X		1
5	X				X		2							X		1
6					X		1							X	X	2
7					X		1	X			X					1
8					X		1							X		1
9					X		1				X		X	X		3
10					X		1	X						X		1
11					X		1	X								1
12					X		1							X		1
	2	3	2	3	11	0	18	4	0	1	3	0	1	9	1	14
% of FM/SA Total	11%	17%	11%	17%	44%	0%		14%	0%	7%	21%	0%	7%	57%	7%	
FM/SA % of Incidents	17%	25%	17%	25%	92%	0%		33%	0%	8%	25%	0%	8%	75%	8%	

Table 1. Overflow Incidents – Alarm Failure Analysis

Conclusions

Overflow incidents are common, and the contribution of human error is a significant factor. In this paper, we have highlighted the need to ask the right questions during incident investigations in order to learn and understand the human factors related to alarm failures. We have demonstrated the importance of understanding the human factors related to alarms and addressing them adequately during the alarm life cycle using alarm failure analysis framework. Improvements can be made by ensuring situation awareness demons and potential failure mechanisms are addressed adequately in the project life cycle when alarms are identified as safeguards and credited as protection layers.

References

Bellamy, L.J., Geyer, T.A.W., 2007, *Development of a working model of how human factors, safety management systems and wider organizational issues fit together,* Health and Safety Executive 2007

Chang, J. I., Lin, C-C., 2006, *A study of storage tank accidents*, Journal of Loss Prevention in the Process Industries 19 (2006) 51–59

Dunn, D. G., Sands, N. P., Stauffer. T., 2013, *When Good Alarms Go Bad: Learnings from Incidents*, 70th Annual Instrumentation and Automation Symposium – Texas A&M University

Endsley, M. 2011, *Designing for Situation Awareness: An Approach to User-Centered,* 2nd Design, Second Edition

Hemmatian, B., Abdolhamidzadeh, B., Darbra, R.M., Casal, J., 2014, *The significance of domino effect in chemical accidents,* Journal of Loss Prevention in the Process Industries 29 (2014) 30-38

Kletz, T.A, 2002 *Accident Investigations - Missed Opportunities,* Institution of Chemical Engineers Trans IChemE, Vol 80, Part B, January 2002

Myers, M.P, 2013, *Layer of Protection Analysis e Quantifying human performance in initiating events and independent protection layers*, Journal of Loss Prevention in the Process Industries 26 (2013) 534-546

Waite, P., 2013, *The Chemical Engineer Magazine,* iChemE (March 2013)40-44

INHERENTLY SAFER PROCESS DESIGN FOR VERTICAL KNOCK-OUT DRUMS AND VAPOR/LIQUID SEPARATORS IN GAS PROCESSING

By
George Maliakkal
Bechtel Corporation
Oil, Gas and Chemicals
Houston, Texas
gjmaliak@bechtel.com

David Rodriguez
Bechtel Corporation
Oil, Gas and Chemicals
Houston, Texas
drodrigu@bechtel.com

Prepared for Presentation at
American Institute of Chemical Engineers
2016 Spring Meeting
12th Global Congress on Process Safety
Houston, Texas
April 11-13, 2016

INHERENTLY SAFER PROCESS DESIGN FOR VERTICAL KNOCK-OUT DRUMS AND VAPOR/LIQUID SEPARATORS IN GAS PROCESSING

George Maliakkal
Bechtel Corporation
Oil, Gas and Chemicals
Houston, Texas
gjmaliak@bechtel.com

David Rodriguez
Bechtel Corporation
Oil, Gas and Chemicals

Keywords: Inherently Safer, Vapor Liquid Separators, Pool Fire, Depressuring, Fireproofing

ABSTRACT

In gas processing and LNG production, large diameter (> 2 meter) vertical vapor/liquid separator vessels and compressor suction knock-out drums are common. These separate small (<0.1wt%) fraction of flammable liquid from large volumes of gas. The liquid level in these vessels is typically controlled above the bottom tangent line of the vessel permitting storage of large quantity of flammable liquid at the bottom of the vessel, often in excess of the threshold volume for pool fire. To safeguard against hazards of pool fire, such vessels must be designed for fire-depressuring. Piping, adjacent structures and equipment associated with the vessels must also be fireproofed. Inclusion of a small diameter (< 0.5 meter) boot below the bottom of the vessel enables relocation of the level control from the main vessel to the boot. The large liquid inventory in the bottom of the main vessel is removed. The liquid inventory in the boot will be reduced, at least by one order of magnitude to well below the threshold for pool fire, eliminating pool fire potential. This makes the system inherently safer while eliminating costs associated with fire-depressuring and fireproofing.

1.0 INTRODUCTION

In gas processing plants and LNG plants, high pressure-large diameter vertical vessels handling hydrocarbon vapor and liquid are common. The hydrocarbon liquids in these drums are generally highly volatile and flammable. These vessels may be compressor suction drums or vapor/liquid separators. These vessels separate small fractions of liquid from large volumes of gas. The liquid level in these vessels are normally controlled in the bottom section of the vessel, above the bottom tangent line. For large diameter vessels, this type of level control will result in large inventory of flammable liquid to be present in the vessel all the time. The inventory can be in excess of the threshold volume for pool fire, which may occur in case of a leakage or spill. In that case, to safeguard against pool fire, equipment, structures and piping in the vicinity of the vessel must be fireproofed. Also, the process system associated with the vessel must be designed for fire-depressuring. These requirements make the system more expensive to design and build, even as the hazard of fire associated with carrying large inventory of flammable liquid continues to be present. The method proposed here, of controlling the liquid level in a boot at the bottom of the vessel with inventory less than the threshold volume for pool fire, will make the vessel and associated system inherently safer while reducing capital expenditure to design and build.

2.0 TRADITIONAL LEVEL CONTROL IN VERTICAL DRUMS

2.1 VESSELS WITH HEMISPHERICAL HEADS

Figure 2.1 shows typical level control for a vertical vapor/liquid separator (in this case a compressor suction drum) in a gas processing plant, where the feed gas to the drum generally has only a small fraction of entrained liquid. The volume below the high liquid level is considered as inventory for pool fire assessment. This includes the volume in the cylindrical section as well as the hemispherical section. The generally accepted industry threshold for pool fire is that the flammable liquid volume be greater than 5 cubic meters or ≈180 cubic feet. Generally, to facilitate fabrication, large diameter vessels with high design pressures are built with hemispherical heads. The wall thickness for hemi-head is one half of that for ellipsoidal head for vessel with the same diameter and design pressure.

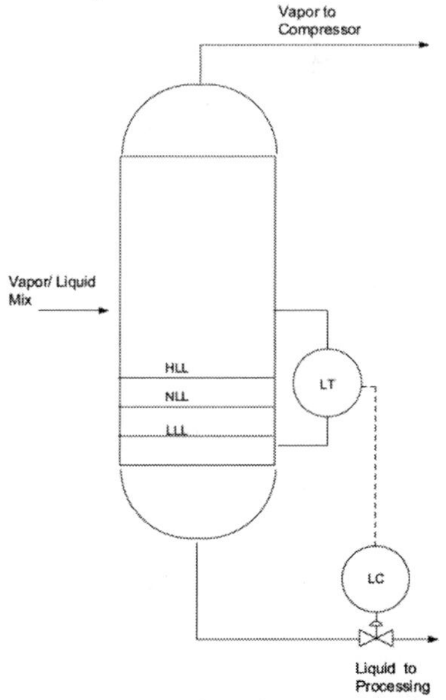

Fig. 2.1: Vessel with Hemispherical Head

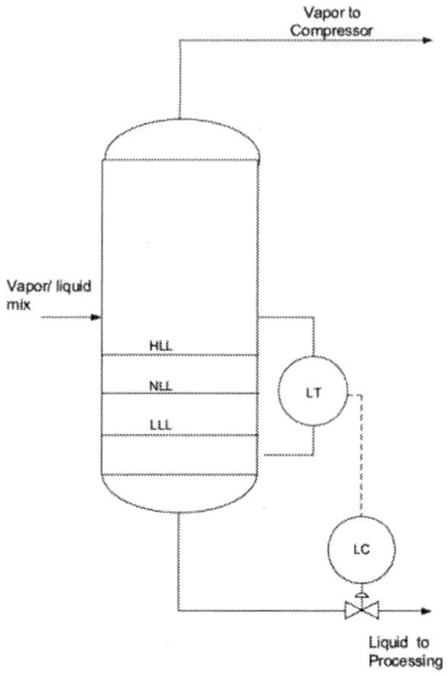

Fig. 2.2: Vessel with 2:1 Ellipsoidal Head

2.2 *VESSELS WITH ELLIPSOIDAL HEADS*

Vessels with lower design pressures will have 2:1 ellipsoidal heads. Clearly, the vessels with ellipsoidal heads carry less inventory in the head. Figure 2.2 illustrates vertical vessel with ellipsoidal head. Whether hemispherical or ellipsoidal head, the total liquid inventory below high operating liquid level should not exceed the guideline (5 m^3) if one has to avoid pool fire.

3.0 APPLICATIONS OF LEVEL CONTROL IN A BOOT

3.1 *HORIZONTAL FILTER/COALESCERS IN GAS PROCESSING*

It is common practice in hydrocarbon industry to collect the separated liquids in a "boot" and remove the liquids under level control from the boot, when the vapor/liquid separator is horizontal. Such is the case with filter/coalescers. Typical illustration is shown in Figure 3.1.

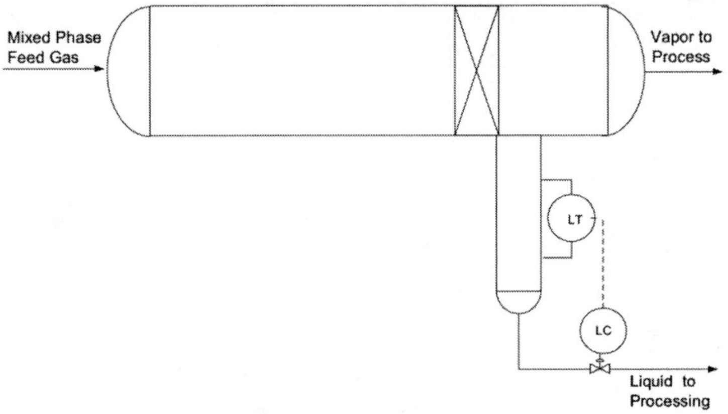

Fig. 3.1: Filter/Coalescer in Gas Processing

3.2 *VACUUM COLUMN IN CRUDE OIL PROCESSING*

In petroleum refining industry level control in a boot is used at the bottom of vacuum tower in crude processing, with the intent to minimize residence time of vacuum resid in order to prevent coking at the prevailing high temperature. This is illustrated in Figure 3.2.

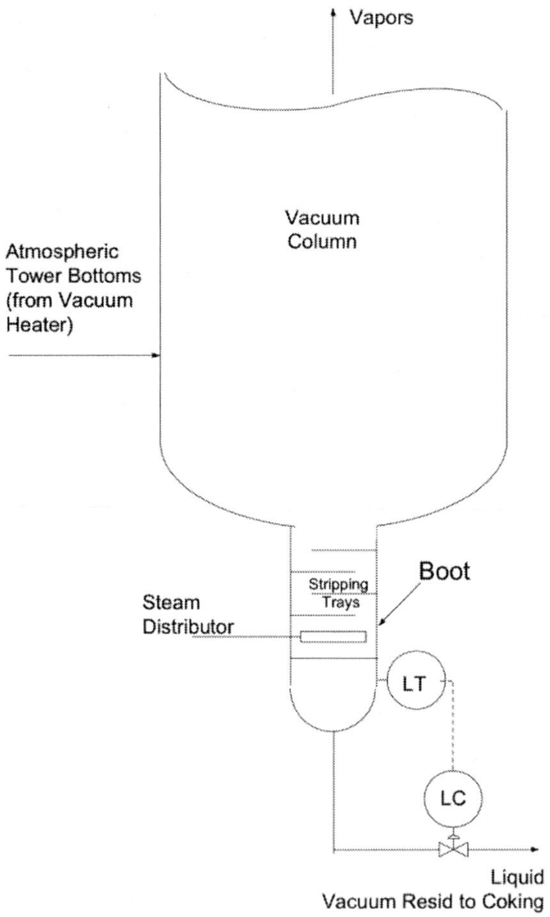

Fig. 3.2: Vacuum Column with Boot in Crude Oil Processing

4.0 BASIS FOR LIQUIDS REMOVAL BY LEVEL CONTROL IN A BOOT

As described in Section 3.0 a "boot" design is used for applications of liquids removal both in horizontal and vertical vessels. The liquid removal operation is independent of the vapor/liquid separation process that occurs in the body of the main vessel. Therefore, liquid removal by level control in a boot can be incorporated into any vessel for vapor/liquid separation, regardless of vessel geometry. Only question is: are there tangible benefits for doing so and under what circumstances?

4.1 APPLICABILITY OF LEVEL CONTROL IN A BOOT FOR VERTICAL VAPOR/LIQUID SEPARATORS

In LNG and gas processing, large volume flow rates of gas, sometimes exceeding a billion standard cubic feet of gas per day, are flowed through vapor/liquid separators and compressor suction drums in order to remove traces of liquid. The prevailing pressures range from low to high, from 30 barg to 120 barg. The presence of liquid can be because of entrainment from previous processing step, condensation due to temperature drop or due to retrograde condensation due to pressure drop. These liquids are volatile hydrocarbons, typically molecular weight that of heptane. The proportion of liquid present in these gas streams is rather small, generally less than 0.1 weight %. Large diameter vessels are needed to process the large volumes of gas. High design pressures leads to high wall thickness for the vessel. Vertical vessels 5 meters in diameter with 125 millimeter wall thickness are not uncommon.

The traditional method of removing liquid from these vertical vessels by controlling the level of liquid in the main diameter results in carrying large inventory of liquid in the bottom of the vessel. For a 5 meter diameter vessel with hemispherical bottom, this inventory below high operating liquid level is estimated to be 60 cubic meters, which is far in excess of the 5 cubic meter limit for pool fire. With the addition of a small diameter "boot" at the bottom of the vessel, the level control for the small amount of liquid being removed in the separator can be shifted to the boot. Because only very small amounts of liquid is collected in the vessel, the boot can be as small as a tenth of the diameter of the main vessel. The boot can be designed to carry less than 1 cubic meter of liquid inventory, with no liquid inventory in the main vessel. The inventory is much lower than the threshold 5 m^3 for pool fire. The avoidance of pool fire makes the design inherently safer with significant benefits related to safety and cost. The traditional level control and level control in a boot are compared for a typical vapor/liquid separator in Figures 4.1a and 4.1b.

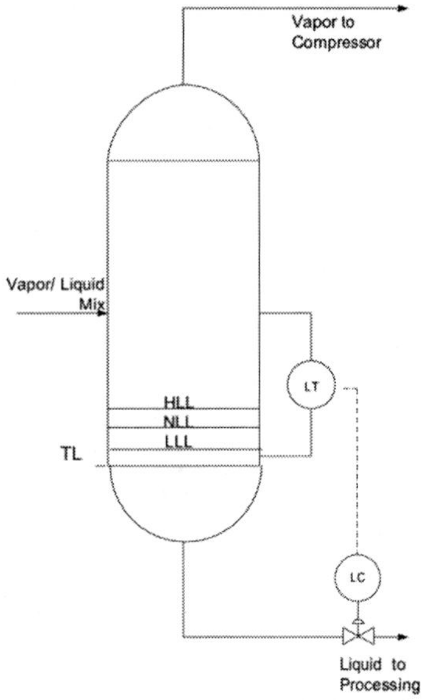

Fig. 4.1a Traditional Level Control

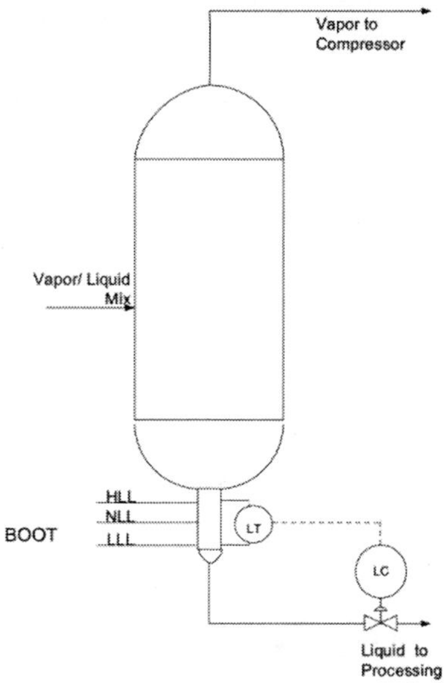

Fig. 4.1b Level Control in a Boot

5.0 SAFETY BENEFITS OF LEVEL CONTROL IN A BOOT

5.1 AVOIDANCE OF POOL FIRE

The liquid level control in a boot design is applicable to practically all vapor/liquid separators and compressor suction drums encountered in gas processing, being that these gas streams have only very small liquid content. The inventory of flammable hydrocarbons in the boot will be well below the threshold for pool fire, thus avoiding hazards associated with pool fire in the event of a spill. In comparison, control of level in the body of the vessel comes with large inventory of flammable hydrocarbons with the associated hazard of pool fire.

5.2 AVOIDANCE OF FIRE DEPRESSURING

API 521 requires that vessels subject to pool fire be designed for fire depressuring. For vessels thicker than 1", the mandatory depressuring criterion is from design pressure to 50% of design pressure in 15 minutes in the case of a fire emergency. If there is no potential for a pool fire, as is the case with boot level control, mandatory depressuring rate is not applicable. Depressuring can be spread over longer time to suit the convenience of plant operations and maintenance. Accordingly, the depressuring load and depressuring header size can be much smaller with the boot option. The relief load to the flare is also reduced.

6.0 COST SAVINGS OF LEVEL CONTROL IN THE BOOT

6.1 AVOIDANCE OF FIRE PROOFING

With pool fire comes the requirement for fireproofing. The fireproofing is required for all equipment, structures and piping within several meters radius around the vessel and several meters height above grade. Material such as *pyrocrete 241* or equal is used for fireproofing. The cost associated with such fire proofing can run into the thousands of dollars, depending on the size of the vessel. With no fire proofing needed with a boot design, the cost savings can be substantial.

6.2 SAVINGS RELATED TO VESSEL SIZE/COST RELATED TO BOOT

The boot allows the high liquid level for the vessel to be set in the boot, whereas with level control on the main vessel diameter, the high liquid level will be set well above vessel tangent line. Once the high liquid level is moved to the boot, the full depth of the head is free of liquid. The chance of liquid re-entrainment due to inlet vapor distribution is eliminated. The T/T height of the vessel can be at least less by length equal to one radius in the case of hemi-head vessel and half radius for vessel with ellipsoidal head. In comparison, the design with level control in the boot will require the addition of a small diameter cylindrical section, below the bottom of the vessel.

7.0 EXAMPLE VERTICAL VESSEL DESIGN WITH BOOT

Figure 7.0a and 7.0b shows process details of a vertical vessel (compressor suction drum) design with and without boot in gas processing application.

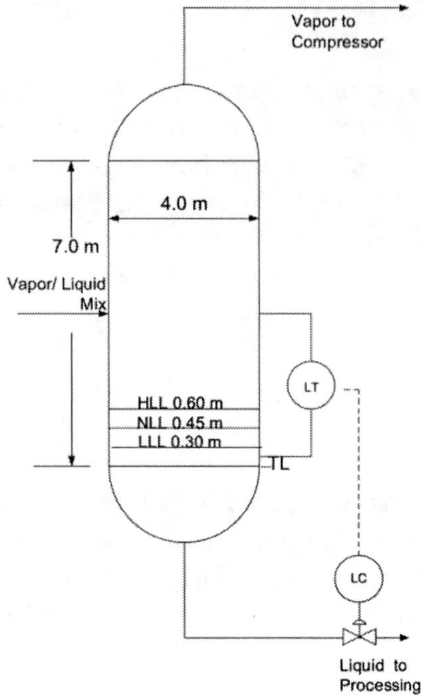

Fig. 7.0a Traditional Level Control

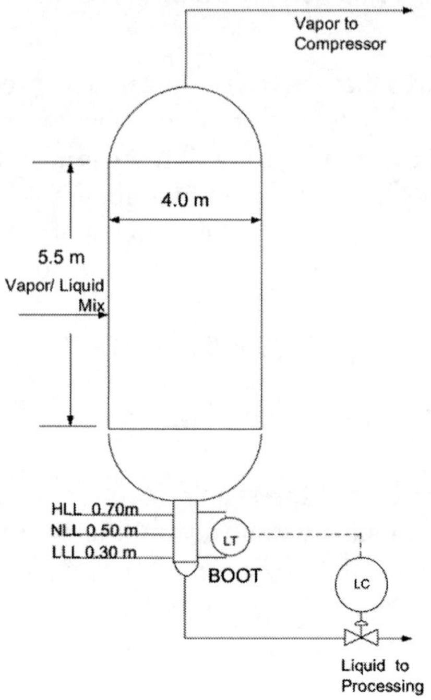

Fig. 7.0b Level Control in a Boot

Feed gas is natural gas with maximum of 0.1 wt% liquid dispersed in the gas. For residence time purposes, the assumption is that the liquid is completely separated in the vessel. The two designs are compared in Table 7.0. With the boot, the inventory of flammable liquid is reduced from 24.5 m3 to 0.2 m3, a hundred-fold reduction, which is far below the threshold for pool fire of 5 m3. Accordingly no fireproofing or no fire depressuring are needed for design with a boot.

TABLE 7.0

EXAMPLE VERTICAL VESSEL DESIGN W/ AND W/O BOOT

COMPRESSOR SUCTION DRUM

Row No.	CONFIGURATION		VERTICAL VESSEL	VERTICAL VESSEL WITH BOOT
1	LEVEL CONTROL LOCATION		ON MAIN DIAMETER OF VESSEL	ON THE BOOT
2	GAS FLOW RATE	kg/hr	880,000	880,000
3	GAS FLOW RATE	mmscfd	1,050	1,050
4	LIQUID CONTENT	wt%	0.1	0.1
5	LIQUID (TYPICAL)		HEPTANE	HEPTANE
6	VOLUME RATE OF LIQUID	m^3/hr	1.35	1.35
7	VESSEL DIAMETER	m	4	4
8	VESSEL T/T	m	7	5.5
9	BOOT DIAMETER	m	-	0.5
10	BOOT HEIGHT	m	-	1
11	HLL (above bot tan)	m	0.6	-
12	LLL (above bot tan)	m	0.3	-
13	HLL (above BOOT bot tan)		-	0.7
14	LLL (above BOOT bot tan)		-	0.3
15	VOL BETWEEN HLL AND LLL	m^3	3.8	0.08
16	LIQUID RESIDENCE TIME (LLL TO HLL)	minutes	165	4
17	VESSEL FLAMMABLE LIQUID INVENTORY (BELOW HLL)	m^3	25	0.2
18	DESIGN FOR POOL FIRE		YES	NO
19	FIREPROOFING REQUIRED		YES	NO
20	FIRE DEPRESSURING REQUIRED?		YES	NO
21	FIRE DEPRESSURING RATE	kg/hr	310,000	84,000
22	DEPRESSURING HEADER SIZE	in.	16	10

Note on Residence Time:

While controlling the liquid level in the main diameter gives a residence time of 165 minutes between high and low liquid levels, the level control on the boot gives a residence time of 4 minutes. A residence time of 4 minutes is consistent with process engineering practice for liquids pressured out of vessels, as is the case with most of these separators that operate at relatively high pressures in gas processing applications. The potential for vapor breakthrough is likewise no different than other similar process systems. With 165 minutes of residence time for the main diameter level control, the level control valve is exercised very infrequently, whereas with the boot design, it is exercised more frequently, which may be in favor of boot design.

8.0 CRITERIA FOR SELECTING A BOOT FOR LEVEL CONTROL

Vessels of various diameters with ellipsoidal head and hemispherical head were evaluated to determine the breakpoint diameter at which boot design is suggested to avoid pool fire. Since the subject analysis applies only to separators with small fraction of liquid compared to the vapor flow, the operating levels would be set by minimum gaps between various level settings for purposes of level measurement rather than by residence time considerations. This led to a minimum high liquid level (HLL) of 0.625 meter above bottom tangent line. The volume of liquid inventory below this level was calculated for various diameters for vessels with hemispherical and ellipsoidal heads. The table below indicates that below diameter of 2.48 meters for ellipsoidal heads and below diameter of 2.18 meters for hemispherical heads, the flammable inventory is less than the threshold of 5 cubic meter for pool fire. Boot level control is suggested for vessels above these diameters to avoid pool fire by the "5 m^3 inventory" criterion and the consequent mandatory mitigation requirements.

TABLE 8.0

Vertical Vessel with Ellipsoidal (2:1 SE) Head					
Diameter	meter	2.00	2.25	2.48	2.50
	feet			8	>8
Volme, head	m3	1.05	1.49	2.00	2.05
HLL (above BTL)	m	0.625	0.625	0.625	0.625
Volume, cylinder (to HLL)	m3	1.96	2.49	3.02	3.07
Liquid Inventory	m3	3.01	3.98	5.02	5.12
Meets Pool Fire Criterion OF 5 m3?		No	No	Yes	Yes
Candidate for Vessel with Boot?				Yes	Yes

Vertical Vessel Hemi Head						
Diameter	meter	2.00	2.18	2.20	2.25	2.50
	feet		7		>7	
Volme, head	m3	2.09	2.71	2.79	2.98	4.09
HLL (above BTL)	m	0.625	0.625	0.625	0.625	0.625
Volume, cylinder (to HLL)	m3	1.96	2.33	2.38	2.49	3.07
Liquid Inventory	m3	4.06	5.05	5.16	5.47	7.16
Meets Pool Fire Criterion OF 5 m3?		No	Yes	Yes	Yes	Yes
Candidate for Vessel with Boot?			Yes	Yes	Yes	Yes

As was indicated in the example in Section 7.0, level control in the boot reduces flammable inventory to very low levels, to a fraction of a cubic meter. The hazard associated with carrying flammable inventory is nominally proportional to the inventory, and any means by which this inventory could be reduced should be considered for safer design. Thus, even vessels with diameter as low as 1 meter (with inventory ≈1 cubic meter) are candidates for level control in a boot.

9.0 CONCLUSION

Exercising liquid level control in a boot below the bottom rather than in the main vessel section is a viable and preferred method for removing separated liquid from vertical vapor/liquid separators and compressor suction drums that are encountered in gas processing and LNG plants, where the liquid fraction in the mixed phase feed stream is small. For vessels over 2 meter diameter, the inventory of flammable liquid below high liquid level is likely to be greater than 5 cubic meters, the generally accepted threshold spill volume for pool fire. Fire-depressuring and fireproofing of equipment, structures and piping in the vicinity of the vessel become mandatory. For such vessels, having liquid level control in a boot reduces the inventory to a fraction of a cubic meter, making the vessel inherently safer without the need for fire depressuring and fireproofing. Even for smaller diameter (<2 meters) vertical separator vessels, the significant reduction in flammable inventory when considering level control in a boot as compared to level control in the main vessel may warrant a boot design because of being inherently safer.

10.0 REFERENCES

[1] Pressure-relieving and Depressuring Systems, API Standard 521, Sixth Edition, January 2014

[2] Fireproofing Practices in Petroleum and Petrochemical Processing Plants, API Recommended Practice 2218, Third Edition, July 2013

Abstract ID# 445580 – Accepted by *12th GLOBAL CONGRESS ON PROCESS SAFETY*

Title: Basic Concepts in Risk Analysis

Author: Vic Edwards, VHE Technical Analysis, Houston, TX
vhe@alumni.rice.edu

The chemical process industries often work with large inventories of materials that are hazardous due to flammability, explosivity, reactivity, corrosivity, and/or toxicity. The resulting hazards of these process systems must be identified, eliminated when possible, minimized, and/or mitigated.

Risk from a hazardous process event is a function of both the consequences of the event (what could happen) and the frequency with which it could occur (how often). It is essential to minimize risks in the chemical process industries. This is done by analysis of the components of the risk, coupled with management of the risks by reducing consequences and/or frequencies of potentially hazardous events.

Minimization of risk is the primary goal of process safety management (PSM). Presented here are the basic concepts of risk analysis that are at the core of PSM.

VICTOR H. (VIC) EDWARDS, PHD, PE

CHEMICAL ENGINEERING CONSULTANT

- **Director Process Safety (retired), IHI E&C, 30 years**
- **2015 Norton H. Walton/Russell L. Miller Award for Process Safety Engineering**
- **Chair, 2013 Global Congress on Process Safety**
- **Chair, 1992 Process Plant Safety Symposium**
- **Five DuPont Awards for HSE Excellence**
- **2007 HSE Excellence in Design Award**
- **2003 Service Award-Texas A&M University Process Safety Center**
- **1998 Employee of the Year**
- **Projects include biochemical and chemical processes, biomass energy, coal gasification, methanol-to-gasoline, LNG, chemical weapons demilitarization, hazardous waste disposal, offshore hydrocarbons facilities**
- **>60 publications, including editor-Section 10 of Perry's Chemical Engineers' Handbook**
- **AIChE Fellow & Life Member; Member ACS, AAAS, NFPA, NSPE, and New York Academy of Sciences**

BASIC CONCEPTS IN RISK - OUTLINE

- Many Process Chemicals are Hazardous, Particularly in Large Quantities
- The Chemical Industry is One of the Safest Places to Work
- What is Risk?
- Hazard Identification and Risk Management
- Example - What-If Analysis
- ALARP
- Risk Matrix
- Layers of Protection and LOPA Example
- Recommended Reading
- Questions?

PUBLIC PERCEPTION – CHEMICALS AND THE INDUSTRY ARE *RISKY!*

- Everything is made of chemicals
- The chemical process industries give us many valuable products-from agriculture to potable water
- Yet the public perceives all things *"Chemical"* as *a Risk* to health and safety
- This perception is partly due to the public's limited knowledge about chemicals and the industry
- This perception is also partly due to the media coverage whenever there is a significant industrial incident
- This perception is also due partly to the occasional large process incidents (Bhopal-thousands killed; tens of thousands injured; BP Gulf of Mexico)

THE CHEMICAL PROCESS INDUSTRIES - AMONG THE SAFEST OCCUPATIONS

US Fatal Occupational Injury Rates for 2012 - for a Few Occupations	
Fishers and Related Workers	120.8
Logging Workers	129.9
Structural Iron and Steel Construction	37
Roofers	42.2
Agriculture - Crop Production	32.5
Taxi and Chauffeurs	16.2
Protective Services (police, fire fighters)	15
Average Fatality Rate	3.4
Grocery Store	2.5
Chemical Manufacturing	1.8
Finance and Insurance	0.3

Fatal Injuries per 100,000 full-time workers per year

- Roy Sanders, Chemical Process Safety, 4th Edition, 2015

WHAT IS RISK?

- *RISK is the combination of three attributes:*

1. ***Hazard*** - What can go wrong?
2. ***Consequence*** **-** How bad can it be?
3. ***Likelihood*** **-** How often might it happen?

- Guidelines for Risk-Based Process Safety, CCPS, 2007

RISK IN THE PROCESS INDUSTRIES

- The chemical process industries often have large inventories of hazardous materials, often at high temperatures and pressures.
- Loss of containment can lead to major fires, explosions, and/or toxic gas releases.
- Because of careful and skillful design, construction, operation, maintenance and management of process facilities, the frequency of major process incidents is low.
- The risk of large consequence process incidents is comparatively low (but still too high) because the frequency is kept low.

HAZARD IDENTIFICATION AND RISK ANALYSIS (HIRA)

- ***Hazard*** *– Caged Tiger:* **Safe Condition**

  A Hazard is a Source of Danger
- ***Initiating Event*** *– Gate Left Open - Escaped Tiger:* **Risk Condition**
- ***Consequences*** *– Dangerous Situation!!! – About to Suffer the Consequences of Hazard's Damaging Capability*

- Roberto Blanco, Process Safety Progress, 2014

RISK MANAGEMENT IS.....

- *WHAT*.....the systematic application of management policy, procedures, and practices to the tasks of analyzing, assessing, and controlling risk.....
- *WHY*.....in order to protect employees, the general public, and the environment as well as company assets while avoiding business interruption.

- Guidelines for Chemical Process Quantitative Risk Analysis, CCPS, 2nd Edition, 2000.

HAZARDS IDENTIFICATION AND RISK ANALYSIS (HIRA): NON-SCENARIO BASED

- Preliminary Hazard Analysis
- Safety Review
- Relative Ranking
- Checklist Analysis

- Guidelines for Hazard Evaluation Procedures, 3rd Edition, CCPS, 2008.

HAZARDS IDENTIFICATION AND RISK ANALYSIS (HIRA): SCENARIO-BASED

- What-If/Checklist Analysis
- Hazard and Operability Review (HAZOP)
- Failure Modes, Effects and Criticality Analysis (FMECA)
- Event Tree Analysis (ETA)
- Fault Tree Analysis (FTA) – *Detailed - Quantitative Risk Analysis*
- Cause-Consequence Analysis (CCA) and Bow-Tie Analysis
- Layer of Protection Analysis (LOPA)
- *- Guidelines for Hazard Evaluation Procedures, 3rd Edition, CCPS, 2008.*

HIRA TEAMS

- Most hazard identification and risk analysis activities are conducted by a team of experts
- Team members typically represent process technology, process engineering, process safety, operations, maintenance, mechanical, instrumentation and control
- Some team members attend all meetings, but specialists may be brought in only when appropriate
- For small facilities, meetings may last a day or two, but large facilities may require weeks or months to complete
- Some HIRA studies, such as Fault Tree Analysis, may be conducted by one analyst.

TYPICAL WHAT-IF ANALYSIS WORKSHEET

Area:	Meeting Date:
Drawing Number:	Team Members:

What If	*Hazard*	*Consequence*	*Safeguards*	*Recommendations*

WHAT-IF EXAMPLE: DAP REACTOR TYPICAL WHAT-IF QUESTIONS

1. The wrong feed material is delivered instead of phosphoric acid?
2. The phosphoric acid concentration is too low?
3. The phosphoric acid is contaminated?
4. The valve controlling phosphoric acid feed is closed or plugged?
5. Too much ammonia is supplied to the reactor?
6. Agitation stops in DAP Reactor?
7. Valve on outlet of DAP Reactor is closed or plugged?

- CCPS, 2008

EXAMPLE WHAT-IF ANALYSIS

- **What If** – *Wrong feed material is delivered instead of phosphoric acid?*
- **Hazard** – *Contaminant incompatibility*
- **Consequences** – *Potentially hazardous phosphoric acid or ammonia reactions with contaminants, or production of off-specification product*
- **Safeguards** – *Reliable vendor; plant material handling procedures*
- **Recommendation** – *Ensure adequate material handling and receiving procedures and labeling exist*

WHAT RISK IS ACCEPTABLE?

- ***ALARP*** – As Low As Reasonably Practicable
- The concept that efforts to reduce risk should be continued until the incremental sacrifice (in terms of cost, time, effort, or other expenditure of resources) is grossly disproportional to the incremental risk reduction achieved.

  \- *Guidelines for Risk-Based Process Safety*, CCPS, 2007

RISK MATRIX - EXAMPLE

Risk	Serious Danger in Immediate Area	Serious Danger Inside Battery Limits	Serious Danger Site Wide	Serious Danger Offsite
More Than Once Per Year	Action Required Unless Risk ALARP	Action Required at First Opportunity	Immediate Action Required	Immediate Action Required
Once Every Few Years	Action Required Unless Risk ALARP	Action Required Unless Risk ALARP	Action Required at First Opportunity	Immediate Action Required
Once in the Facility's Lifetime	No Action Required	Action Required Unless Risk ALARP	Action Required Unless Risk ALARP	Action Required at First Opportunity
Not Expected in the Facility's Lifetime	No Action Required	No Action Required	Action Required Unless Risk ALARP	Action Required Unless Risk ALARP

PROTECTIVE LAYERS IN PROCESS SYSTEMS

- At the core – Basic Process Design
- BPCS and Operator Response
- Safety-Instrumented Systems
- Pressure Relief Systems
- Procedures and Training
- Emergency Response
- Community Response

LAYER OF PROTECTION ANALYSIS (LOPA)

- Semi-Quantitative Tool for Analyzing and Assessing Risk
- Uses Order-of-Magnitude Categories for
  - Initiating Event Frequency
  - Consequence Severity
  - Probability of Failure-Independent Protective Layers(IPLs)

LOPA EXAMPLE-COOLING WATER FAILURE-RUNAWAY REACTION WITH LOSS OF CONTAINMENT

- ***Risk Tolerance–One in 1,000,000 years_Frequency = 10^{-6}***
- Frequency of Loss of Cooling Water-Once in 10 years = 10^{-1}
- Probability of Reactor Runaway on Cooling Water Loss = 0.5
- ***Frequency of Unmitigated Consequence = 5×10^{-2} per year***

Thus Unmitigated Risk is Unacceptably High

LOPA EXAMPLE – MITIGATIONS & RISK

- BCPS Alarm and Human Action – 10^{-1}
- Pressure Relief Valves – 10^{-2}
- Safety Instrumented System (SIS) – 10^{-3}
- Total of All Independent Protection Layers – 10^{-6}

- *Frequency of Mitigated Consequences – 5 x 10^{-8} per year*
- *Risk Tolerance – 10^{-6} per year > 5 x 10^{-8} per year*

- *Therefore, Risk Tolerance Met with SIS*

RECOMMENDED READING

- Guidelines for Risk Based Process Safety, AIChE CCPS and John Wiley, Hoboken, NJ (2007).
- Guidelines for Hazard Evaluation Procedures, 3rd Edition, AIChE CCPS and John Wiley, Hoboken, NJ (2008).
- Guidelines for Engineering Design for Process Safety, 2nd Edition, AIChE CCPS and John Wiley, Hoboken, NJ (2012).
- Roy E. Sanders, Chemical Process Safety – Learning from Case Histories, 4th Edition, Elsevier, Inc., Oxford UK (2015).

OUR NEXT SPEAKER, DR. VINICIUS SIMOES, WILL ADDRESS QUANTITATIVE RISK ANALYSIS

THANKS FOR YOUR ATTENTION!

QUESTIONS?

SPARE SLIDES

RISK ANALYSIS IS…..

- *WHAT…..*The development of a quantitative estimate of risk…..
- *HOW…..based on engineering evaluations and mathematical techniques for combining estimates of incident consequences and frequencies.*

- Guidelines for Chemical Process Quantitative Risk Analysis, CCPS, 2nd Edition, 2000.

RISK ASSESSMENT IS.....

- ***WHAT.....*The process by which the results of a risk analysis are used to make decisions.....**
- ***HOW.....*<u>either</u> through relative ranking of risk reduction strategies.....**
- **.....<u>or</u> through comparison with risk targets (risk criteria).**

  - Guidelines for Chemical Process Quantitative Risk Analysis, CCPS, 2nd Edition, 2000.

INHA UNIVERSITY

Optimization and Economic Evaluation of Industrial Wastewater Treatment Using Electro-oxidation Process

Hongmin Lee, Sangsun Lee, Yoonkyung Kim, Dohyung Lee, Sunghyeon Baeck, Sungwon Hwang*

Graduate School of Chemistry and Chemical Engineering, Inha University

Corresponding author* : sungwon.hwang@inha.ac.kr

2016. 04. 13

Contents

1. Introduction

1. Introduction

- SK Energy wastewater treatment process

  Wastewater collection pond & tanks → Oil separator → Flocculation/Coagulation tank → DAF→ Bio-reactor → Clarifier → Sand filter → Activated carbon filter → Effluent pond
- Physical/Chemical + Biological + High class Treatment
- COD_{mn} removal 200 mg/L → 10 mg/L

1. Introduction (Con't)

- Wastewater discharged from refinery or petrochemical plant contains high concentration of organic pollutants and toxic compounds during turnaround period.
- Conventional wastewater treatment system is generally ineffective in removing non-degradable organic. It requires additional pre-treatment system.

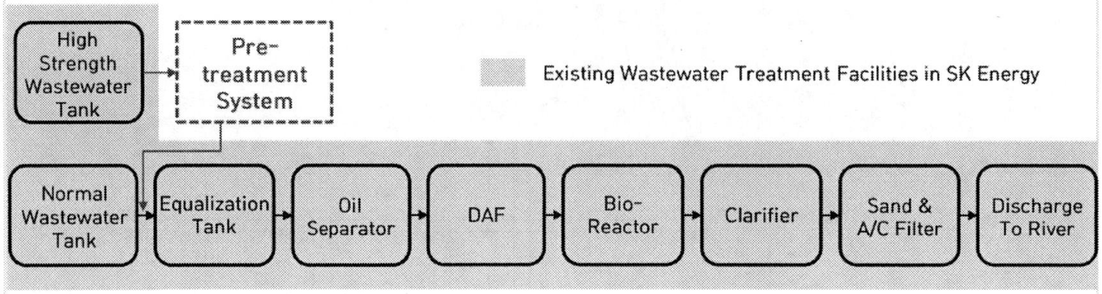

1. Introduction (Con't)

- Available Technologies

① **Incineration**

- High temperature (820-1100℃)
- Direct combustion with excess oxygen from air in flame
- Applied mainly for concentrated wastewater with COD > 100 g/L

② **Wet Air Oxidation (WAO) Process**

- Elevated temperature (250-300℃) and high pressure (100-150bar)
- Usually Cu^{2+} is used as a catalyst.
- The optimal COD is in the domain : 50 g/L > COD > 15 g/L

③ **Oxidation with strong oxidants**

- The oxidation of organic pollutants with strong oxidants (H_2O_2, O_3) at room temperature
- The oxidation takes place in the presence of catalyst and UV radiation
- Treatment of dilute wastewater with COD < 5 g/L

2. Fundamental of Electro-oxidation

2. Fundamental of Electro-oxidation

1. Electro-oxidation reaction mechanism

$$S + H_2O \rightarrow S(\cdot OH) + H^+ + e^-$$

$$S(\cdot OH) + R \rightarrow S + Mineralization\ products + H^+ + e^-$$

$$S(\cdot OH) \rightarrow S + \frac{1}{2}O_2 + H^+ + e^-$$

- Items to be considered for electro-oxidation reaction mechanism :
  - Organic oxidation reaction, OER (Oxygen Evolution Reaction)
  - Adsorption enthalpy of S-OH

2. Fundamental of Electro-oxidation (Con't)

2. Oxidation power of the anode material

Electrode	Oxidation potential(V)	Overpotential of O_2 evolution(V)	Adsorption enthalpy of S-OH	Oxidation power of the anode
RuO_2-TiO_2	1.4-1.7	0.18		
IrO_2-Ta_2O_5	1.5-1.8	0.25		
Ti/Pt	1.7-1.9	0.3		
Ti/PbO_2	1.8-2.0	0.5		
Ti/SnO_2-Sb_2O_5	1.9-2.2	0.7		
Ti/BDD	2.2-2.6	1.3		

2. Fundamental of Electro-oxidation (Con't)

3. Fabrication of Ti/SnO_2-Sb_2O_5 electrode

① Spray pyrolysis

② Sol-gel process

③ Sputtering

④ Chemical vapor deposition (CVD)

⑤ Electron beam evaporation

⑥ Thermal oxidation

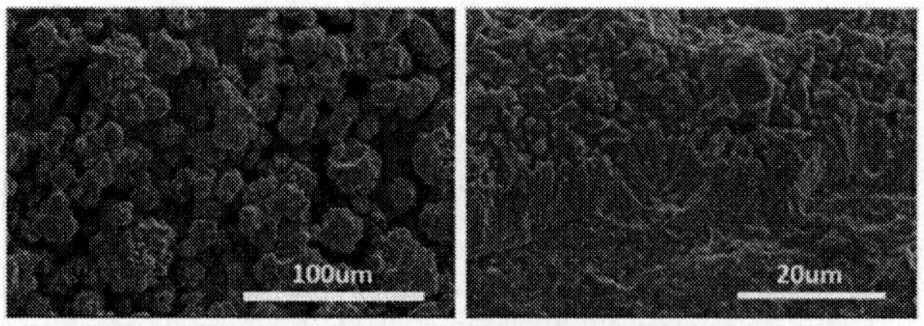

SEM images of Ti/SnO_2-Sb_2O_5 electrode

3. Results of the Experiment

3. Results of the Experiment

- **Experimental condition**

- Wastewater composition (ppm)

COD	pH	H_2S	T-N
70,000	8~10	3,000	3,000

Electrode area	6 x 7.5 = 45cm^2
Anode	Ti/SnO_2-Sb_2O_5
Cathode	Hastelloy
Current density	80mA/cm^2
Electrolyte volume	450mL
Temperature	50℃
Time	9h

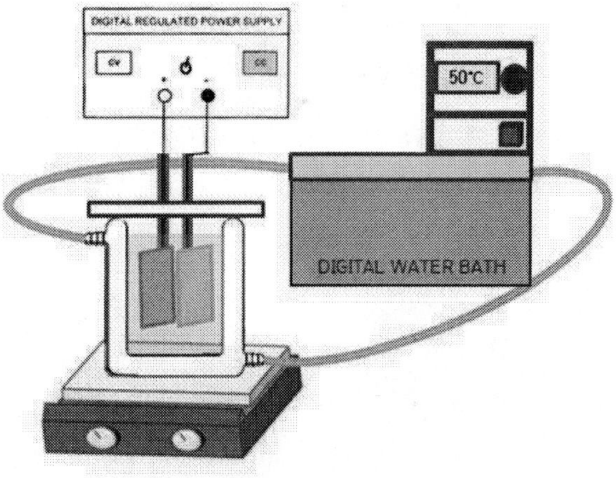

<Electro-oxidation Reactor >

3. Results of the Experiment (Con't)

- **Electrode preparation : Ti/SnO_2-Sb_2O_5**

1) Ti plate pre-treatment

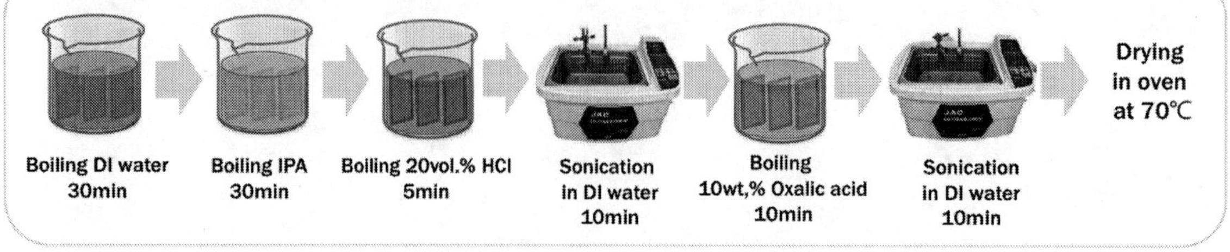

2) Spray coating

(SnO_2 : Sb_2O_5 = 7:3, 8:2, 9:1)

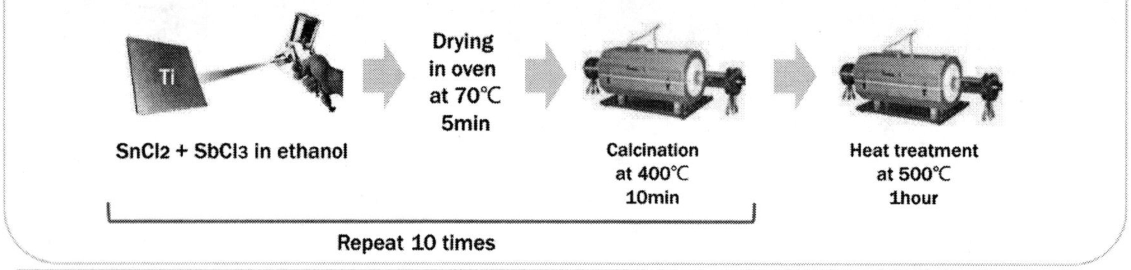

3. Results of the Experiment (Con't)

- Design and operating parameters

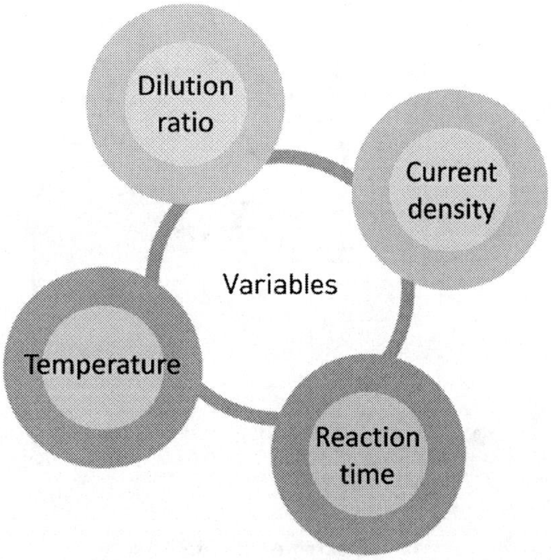

- Design the experiments using Central Composite Design (CCD)
- The experiment under various operating conditions
- Optimization of the variables
- Impact on economic evaluation such as payback period (PBP)

3. Results of the Experiment (Con't)

- Electro-oxidation experiments results

Anode	$Ti/SnO_2\text{-}Sb_2O_5$
Cathode	Hastelloy
Electrode area	6x7.5=45 cm^2
Electrolyte volume	450 mL
Time	9h
Initial COD concentration	70,000~80,000 ppm
Dilution ratio	0, 5, 10, 50
Current density	16, 48, 80 mA/cm^2
Temperature	40, 50, 60 ℃

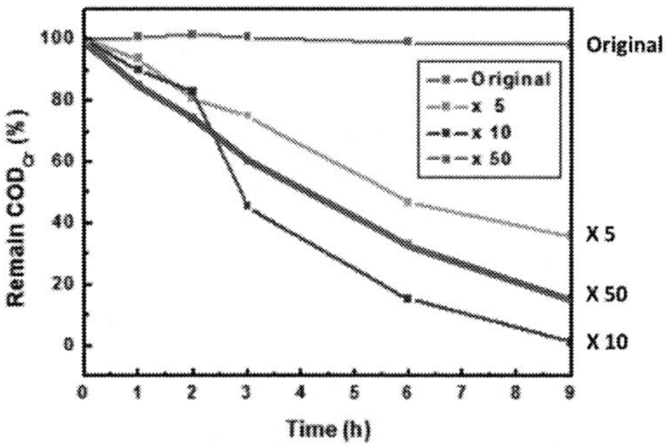

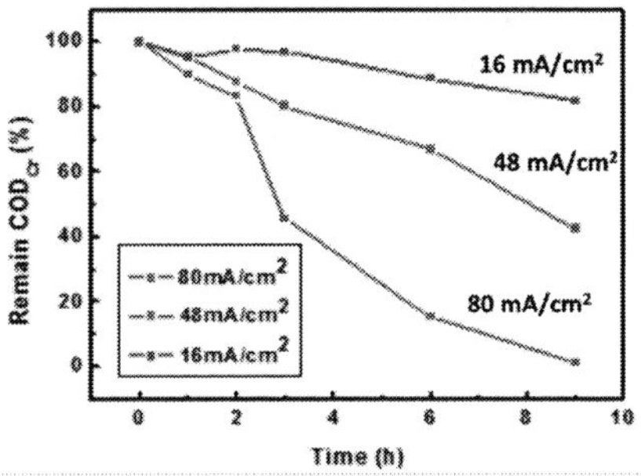

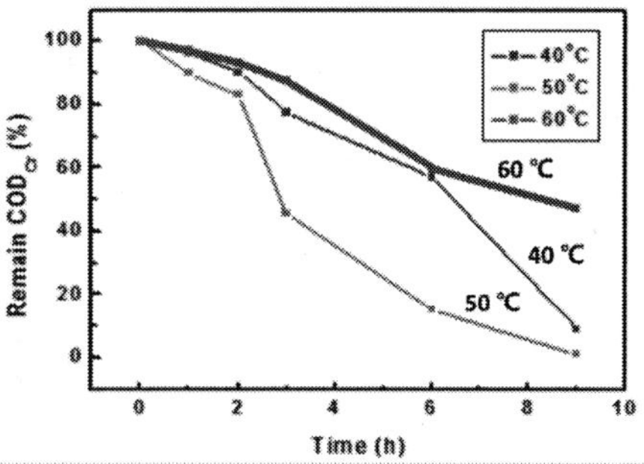

INHA UNIVERSITY

4. Results of the Modeling and Optimization

4. Results of the Modeling and Optimization

- Central Composite Design(CCD)

  ① Box and Wilson, 1951

  ② $y = b_0 + \sum b_i x_i + \sum b_i x_i^2 + \sum b_{ij} x_i x_j$

  - Linear terms in each of the variables$(x_1, x_2, \ldots, x_n)$
  - Squared terms in each of the variables$(x_1^2, x_2^2, \ldots, x_n^2)$
  - First order interaction terms for each paired combination

  ③ Relationship between codes and values

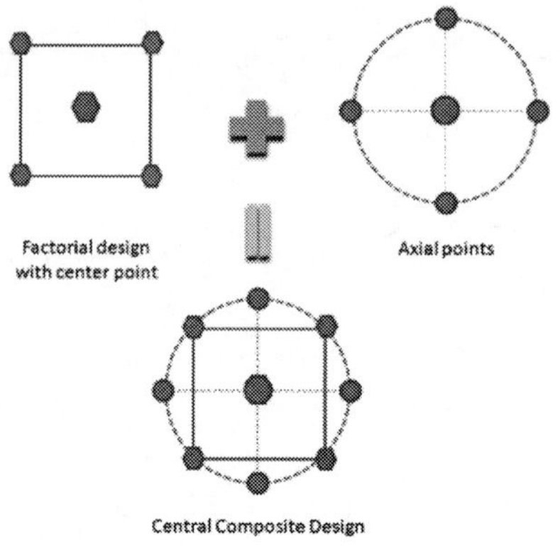

Codes	Values
-β	x_{min}
-1	$[(x_{max} + x_{min})/2] - [(x_{max} - x_{min})/2\alpha]$
0	$(x_{max} + x_{min})/2$
+1	$[(x_{max} + x_{min})/2] + [(x_{max} - x_{min})/2\alpha]$
+β	x_{max}

$\alpha = 1^{k/4}$

k = number of independent variables

4. Results of the Modeling and Optimization (Con't)

No.	Coded Level				COD removal(%)	
	x_1	x_2	x_3	x_4	Observed	Predicted
1	-1	1	0	-1	0	0
2	-1	1	0	0	1	5
3	-1	1	0	1	2	27
4	0	1	1	-1	6	6
5	0	1	0	0	43	43
6	0	1	0	1	64	69
7	1	1	0	-1	15	18
8	1	1	0	0	59	61
9	1	1	0	1	80	93
10	1	-1	0	-1	4	0
11	1	-1	0	0	9	14
12	1	-1	0	1	17	17
13	1	0	0	-1	4	6
14	1	0	0	0	29	36
15	1	0	0	1	59	53

No.	Coded Level				COD removal(%)	
	x_1	x_2	x_3	x_4	Observed	Predicted
16	1	1	0	-1	10	18
17	1	1	0	0	75	61
18	1	1	0	1	100	93
19	1	1	-1	-1	6	2
20	1	1	-1	0	38	54
21	1	1	-1	1	90	95
22	1	1	0	-1	10	18
23	1	1	0	0	75	61
24	1	1	0	1	100	93
25	1	1	1	-1	4	0
26	1	1	1	0	32	27
27	1	1	1	1	52	50
28	-1	-1	-1	-1	0	0
29	0	0	0	0	20	15

4. Results of the Modeling and Optimization (Con't)

- Central Composite Design(CCD)

$$CODremoval(\%) = 15.73 + 29.98x_1 + 25.48x_2 - 46.26x_3 + 18.79x_4$$
$$-9.63x_1^2 + 1.78x_2^2 - 20.74x_3^2 - 5.98x_4^2 - 1.61x_1\,x_2$$
$$-21.27x_1x_3 + 4.79x_1\,x_4 + 59.87x_2\,x_3 + 14.07x_2\,x_4 - 9x_3\,x_4$$

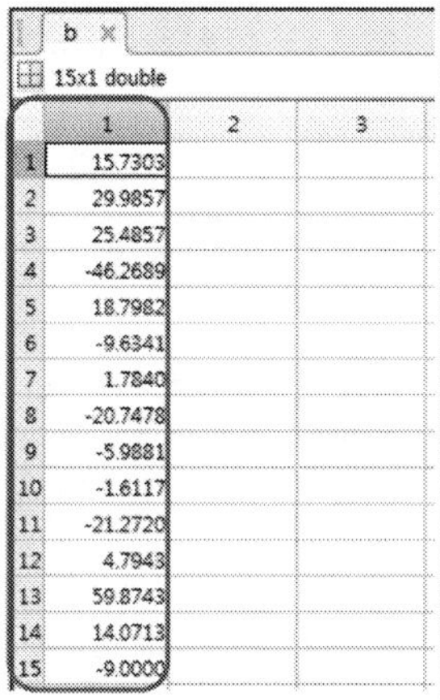
b ×
15x1 double

	1	2	3
1	15.7303		
2	29.9857		
3	25.4857		
4	-46.2689		
5	18.7982		
6	-9.6341		
7	1.7840		
8	-20.7478		
9	-5.9881		
10	-1.6117		
11	-21.2720		
12	4.7943		
13	59.8743		
14	14.0713		
15	-9.0000		

- Coefficient of determination

$$R^2 = \frac{SSR}{SST} = \frac{Sum\ of\ squares\ explained\ by\ reegression}{Total\ sum\ of\ squares}$$

$$= \mathbf{0.92}$$

4. Results of the Modeling and Optimization (Con't)

- Optimization of experimental results

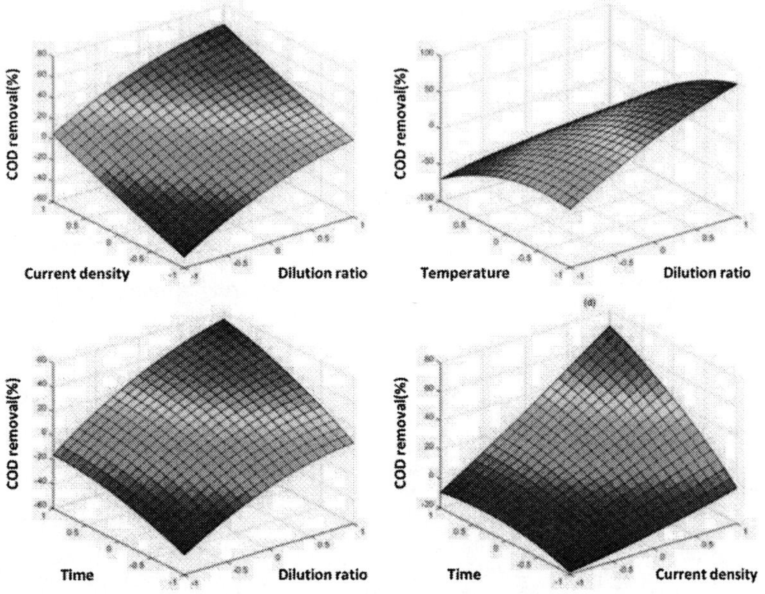

4. Results of the Modeling and Optimization (Con't)

- Optimization of experimental results
  - MATALB Optimization (Levenberg-Marquardt)

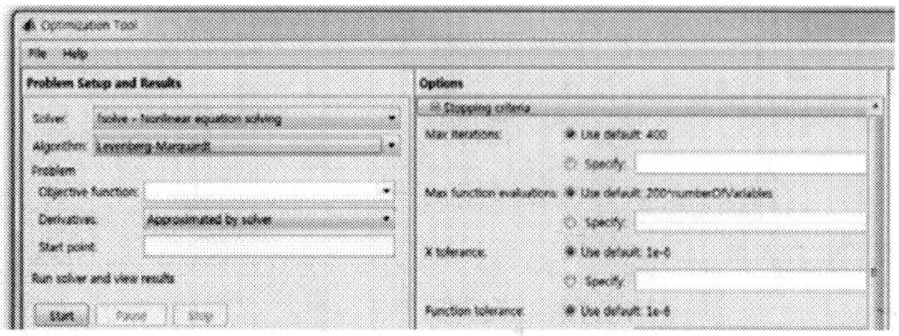

Objective : Y = 70 (COD removal efficiency)

Optimization variables : x_1, x_2, x_3, x_4

Subject to : $0 < x_1 < 16, 16 < x_2 < 80, 40 < x_3 < 50, 1 < x_4 < 9$

- Optimization results

① Dilution ratio(x_1) = 7.24 : 1

② Current density(x_2) = 57 mA/cm^2

③ Temperature(x_3) = 43.1 ℃

④ Reaction time(x_4) = 6.7 h

COD removal efficiency = 69.8%

4. Results of the Modeling and Optimization (Con't)

- Cost evaluation of electro-oxidation process
  - Main equipment : Electrochemical cell, the electrodes and the power supply
  - For industry applications, filter-press flow cell is widely recommended

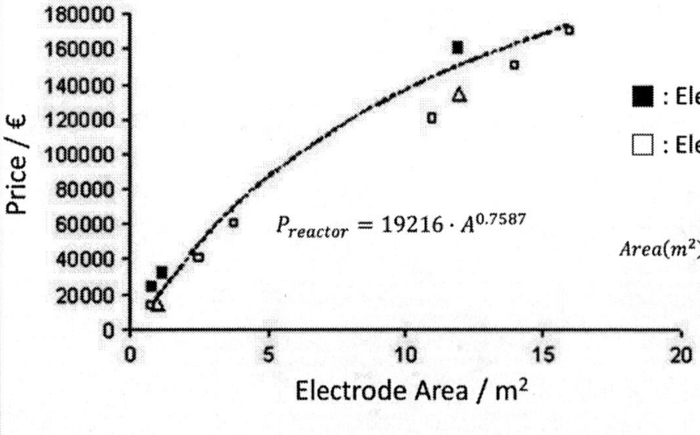

■ : Electrocell, Skjem, Denmark

□ : Electrochemistry Encyclopedia data

$$Area(m^2) = \frac{Q(A\,h\,dm^{-3}) \times 1000 \times flow\,rate(m^3\,d^{-1})}{j_{applied}(A\,m^{-2}) \times 24}$$

Canizares, Pablo, et al. *Journal of Environmental Management* 90.1 (2009): 410-420.

4. Results of the Modeling and Optimization (Con't)

- Cost evaluation of electro-oxidation process
  - Filter-press flow cell, electrode, electrical energy cost
  - Filter-press flow cell : $ 1.2 million for a 100m^3 per day process
  - The price of electrode : 6,500 – 10,800 $/m^2
  - Electrical energy cost : 0.08 $/kWh
  - Operating cost and maintenance cost : $ 540,000 per years

Estimated Payback Period (PBP) = 3.3 years

INHA UNIVERSITY

5. Conclusions and Future Works

5. Conclusions and future works

① Experiments

- CCD with 4 parameters and 3 level was used to design the experiments.
- The electrochemical oxidation of industrial wastewater with Ti/SnO_2-Sb_2O_5 electrodes
- Various experimental conditions : dilution ratio, current density, temperature and reaction time

② Modeling and Optimization

- The coefficients of CCD were estimated from the experimental results using MATLAB.
- MATLAB optimization was used to maximize the removal efficiency of COD under constrained condition.

(Dilution ratio = 7.24 : 1, Current density = 57 mA/cm^2, Temperature = 43.1 ℃, Reaction time = 6.7 hours)

③ Cost Evaluation

- Filter-press flow cell was used to estimate the expense of the process (i.e. $1.2million for a 100m^3 per day).
- The estimated operating cost and maintenance cost : $540,000 per year
- Estimated PBP: 3.3 years

5. Conclusions and future works (Con't)

Future works

1 Detail Design of Electro-oxidation process

2 Scale-up study for field application

3 Extended application to other industries

INHA UNIVERSITY

Q & A

This research was respectfully supported by Engineering Development Research Center (EDRC) funded by the Ministry of Trade, Industry & Energy (MOTIE). (No. N0000990)

"Development of Integrated Interactive Model for Subsea and Topside System to Evaluate the Process Design of Offshore Platform", funded by Ministry of Trade (project No. 10060099).

444092 Using Steel Slag for CO_2 Capture in Saturated Packed Bed

Abstract:

Carbon dioxide capture/sequestrate is the most promising options to reduce the CO_2 emissions into the atmosphere. Industrial mineral carbonation of alkaline wastes, an increasingly promising component of CO_2 capture/storage, may play an important role. Particularly the steelmaking slags are of interest because of their high calcium contents. In addition to CO_2 fixation, the carbonation of these waste materials also reduces the cost associated with handling the waste and minimizes the release of some harmful substances to the environment.

Considering the after mentioned advantages, a significant number of research initiatives have been undertaken at different organization in the world. These findings indicates that the prospect of carbonation of steel slag to become an economically-feasible CO_2 mitigation option. However, the successful implementation of this technology is likely to rely on the (i) availability of affordable and recyclable solvents/additives, (ii) accelerating and maximizing the CO_2 fixation by engineering the reactions involve of in order to achieve appreciable sequestration rate and (iii) detail economic analysis.

The proposed research mainly focused on enchantment of the CO_2 fixation using steel making slags by accelerating the reaction between CO_2 and the minerals in the slags. It is expected that the development and implementation of the proposed technology will create new opportunities both in the area of research and industrial applications to address both the CO_2 emission and steel management issues.

443922 New Approach to Using 3-D Dispersion Modeling for Accidental Toxic Gas Releases in Downtown Seoul

Abstract:

Chlorine is one of the most widely used chemicals in the world, especially for water purification plants near populated areas. So, if an accidental chlorine release occurs in a water purification plant, the damage would fatally spread to the residents and widely affect people.

In this paper, 3D model simulations are performed for a possible Cl2 leak scenario in the urban area of Seoul, South Korea where the population density is very high as well as numerous buildings exist near the plant in operation. FLACS is used to predict the dispersion results for the area whose size is around 5 by 3 km. In fact how precisely implementing 3D geometries is the key for successful 3D simulation. For that a methodology is suggested to rapidly implement geometry by importing CAD files provided by the South Korea National Spatial Information Clearinghouse, and processing it with another software. Finally, each FLACS simulation result was compared and analyzed with other dispersion modeling programs that do not reflect the real geometry.

NAVIGATING THROUGH REFINERY WASTEWATER TREATMENT PROCESSES AND PLANT CONFIGURATIONS.

Carlos D. Claros, P.E.,
CDM Smith, Houston, TX

Abstract

Petroleum refineries deal with a wide range of environmental compliance concerns encompassing wastewater discharges, air pollutant emissions and solids waste. Refineries must comply with stringent effluent discharge limitations which require extensive level of treatment and monitoring. Effluent discharges to surface water must pass whole effluent toxicity (WET) testing and meet very low levels of heavy metals and organics constituents. Some refineries discharge to Publicly-Owned Treatment Works (POTW), which impose less stringent limits, requiring varying levels of pretreatment and monitoring. Treatment processes used in refineries include oil water separators, dissolve air or gas floatation, activated sludge, nitrification-denitrification, selenium removal, filtration, adsorption and membrane processes. These processes generate residual sludge that needs to be dewatered and disposed. Control of hazardous air pollutants from treatment units is also an important consideration due to Volatile Organic Compound (VOC) emissions limitations. This paper presents an overview of typical wastewater treatment processes and configurations utilized in refineries.

Keywords: refinery, wastewater, compliance.

Refinery wastewater characteristics

Refinery wastewater consists of process wastewater, utility wastewater, and contaminated stormwater. Non-contaminated storm water runoff can be discharged directly to surface water provided it meets the regulatory limits. Utility wastewater consist of cooling tower blowdown, boiler blowdown, and residuals from water treatment. Sanitary wastewater is typically segregated and conveyed in a separate collection system.

Process wastewater originates from various process units including desalting, light oils, distillation, fractionation, alkylation, and reforming processes. Process discharges contain free and emulsified oil from leaks and spills, bottom tank draws; emulsions incident to chemical treatment; and condensates from distillate separators. Alkaline waste streams are discharged from the treatment of oil product with alkaline reagents and the sweetening of oil to remove mercaptans. Major wastewater streams are sour water, scrubber water from reformer catalyst regeneration, spent potassium hydroxide from alkylation, desalting wastewater, and caustic wash from isomerization. US EPA[1,2] has conducted surveys and studies for the various sources of wastewater generation and their characteristics.

In a typical refinery, process wastewater and utility wastewater are discharged to the process sewer collection system connected to refinery wastewater treatment plant (WWTP). During rain events, contaminated runoff also drains to the process sewer. The combined wastewater presents high variability in both flow and quality, so the WWTP must be designed to attenuate this variability to protect downstream treatment processes susceptible to upset and shock loads.

Table 1. Characteristics of Selected Process Discharges*.

Waste Stream Type	Condensates (Sour Waters)		Spent Caustic	Water Draws	Stripped Sour Water	Emulsion	Refinery Separator Inlet
Source of waste	*Low Sulfur Gas Separator*	*Light Oil Recovery*	*Naphtha Wash FCC*	*Slop Oil Treatment*	*Sour Water Stripper*	*Desalting Unit*	
Alkalinity, ppm	1518	69-13100	247900	--	--	739	0 - 92
Ammonia, ppm	500	3- 8500	-	--	30 - 100	2 – 100	--
Chemical Oxygen Demand (COD), ppm	1204	214-16,255	901200	22000 – 56000	600 – 1200	865-3031	--
Oil, ppm	3	6-230	1	4,900 – 10300	< 10	32-713	3900 – 6500
pH	8.5	5.0-9.2	13+	10 – 10.2	8	9.26	2.4 – 6.2
Phenol, ppm	0.06	0-213	309,300	--	< 200	4.1 - 100	0.5 – 7.9
Sulfide, ppm	600	10-5000	0-3380	--	1 - 10	1 – 100	--
Total Suspended Solids (TSS), ppm	-	14	678	60 - 940	< 10	< 500	131 – 678

* Adapted from Nemerow (1987) and IPIECA (2010)

Table 1 shows a list of conventional wastewater parameters for selected process discharges. Tank water draws, spent caustic, and desalter discharges have high organic loading and pH that could upset the wastewater treatment system if not properly managed within the treatment scheme. Condensates contain high concentrations of ammonia (NH_3) and sulfides, and are collected in the sour water tank for segregated treatment and reuse.

Other constituents in process discharges are chlorides, fluorides, mercaptans, heavy metals, selenium, volatile and semi-volatile organic compounds (VOCs and SVOCs), dioxins, and phenolic compounds. Toxic dioxin congeners (2,3,7,8 TCDD and 2,3,7,8 TCDF) have been found in treated effluent of activated sludge plants. The dioxins emanate from the scrubber water from reformer catalyst regenerator.

Effluent Limitations

Refineries effluent discharges are regulated by federal Effluent Limitation Guidelines (ELGs) set forth in 40 CFR Part 419 – Petroleum Refining Point Source Category. The U.S. EPA delegates the authority for issuance of facility discharge permits to the state and local agencies. The permit effluent limits are determined based on the following factors: refining operations category, crude processing capacity, age of plant, and whether the facility is a direct or indirect discharger. The wastewater treatment overview presented here-in focuses on the topping and cracking subcategories.

Table 2 shows the range of effluent concentration limits observed in recent permits issued to six U.S. refineries. Permitted parameters given in mass loading (e.g. metals) were converted to concentrations using the average annual flow. The direct dischargers included refineries with discharge of treated wastewater to an estuarine bay, a river, and a creek. The indirect dischargers included refineries with treatment plants including aerobic biological process that currently discharge to a POTW.

Table 2 – Range of Effluent Limits Observed in Recently Renewed Permits

Parameter	Direct Discharge (Surface Water)		Indirect Discharge (POTWs)
	Monthly Average	Daily Maximum	Daily Maximum
pH	6 - 9	6 - 9	5.0 – 12.5
Oil and Grease	10	10 -21	100
TSS	25 - 35	40 - 50	250 – 300
BOD	30 - 40	60 - 70	250 – 300
COD	200 - 300	450 – 550	--
Ammonia	3 - 20	35 – 45	100
Sulfides	0.15 – 0.2	0.005* – 0.45	3600
TPH	--	--	50
Chloride	--	300	--
Phenols	0.2	0.5	--
Benzene	--	--	0.05 - 0.5
Chromium	0.2 – 0.5	1.0	0.7 - 6.0
Hexavalent Chromiun	10 – 30	30 - 70	40
Mercury	--	0.001	0.005 - 0.2
Selenium	--	0.006*	0.05 – 7.5
Zinc	--	0.30*	2.0
Copper	0.01	0.025*	0.01
Lead	--	0.008*	0.05 – 0.15
Arsenic	--	0.20	0.2 – 1.6

* Limit based on dissolved concentrations

Federal effluent limitations and guidelines include metal limits only for total and hexavalent chromium; however, state and local permits have included other metal compounds based on surface water quality concerns. **Table 2** shows daily maximum concentrations for various metals including selenium, mercury, vanadium, and chlorides. Achieving such low limits of chlorides and selenium for surface water disposal may prove capital and O&M expensive.

Federal pretreatment standards are used by POTWs to establish effluent limits for categorical pollutants. In addition, the regulations allow POTWs to add pollutants or impose more stringent limits of regulated constituents when such limitations are necessary to implement the provisions of applicable Local or State regulations.

BWON Regulations

Benzene Waste Operations NESHAPs (BWON) regulations require emissions controls for waste streams with high concentrations of benzene (>10 ppmw). Waste streams from the crude unit desalter, sour water overflows, diesel coaslecers, and tank water draws commonly fall under this category. These are segregated and designated as BWON streams.

Refineries have been upgrading collection sumps and front-end treatment equipment such as American Petroleum Institute (API) separators, equalization tanks, and dissolved-air flotation (DAF) units with air tight covers and emissions controls to comply with BWON.

Treatment processes

The refinery wastewater treatment plant (WWTP) consist of the following general treatment process sequence:

- Oil and Solid Removal
  - Gravity Separation
  - Floatation
- Equalization
- Physico-chemical Treatment
- Biological Treatment
- Tertiary Treatment (Polishing)

Figure 1 shows the general treatment plant scheme used at refineries. Variations from this simplified process flow diagram are discussed in sections below. Segregated treatment is conducted for sour water and BWON discharges before entering the wastewater plant. IPIECA[3] describes these processes and technology options for each in more detail.

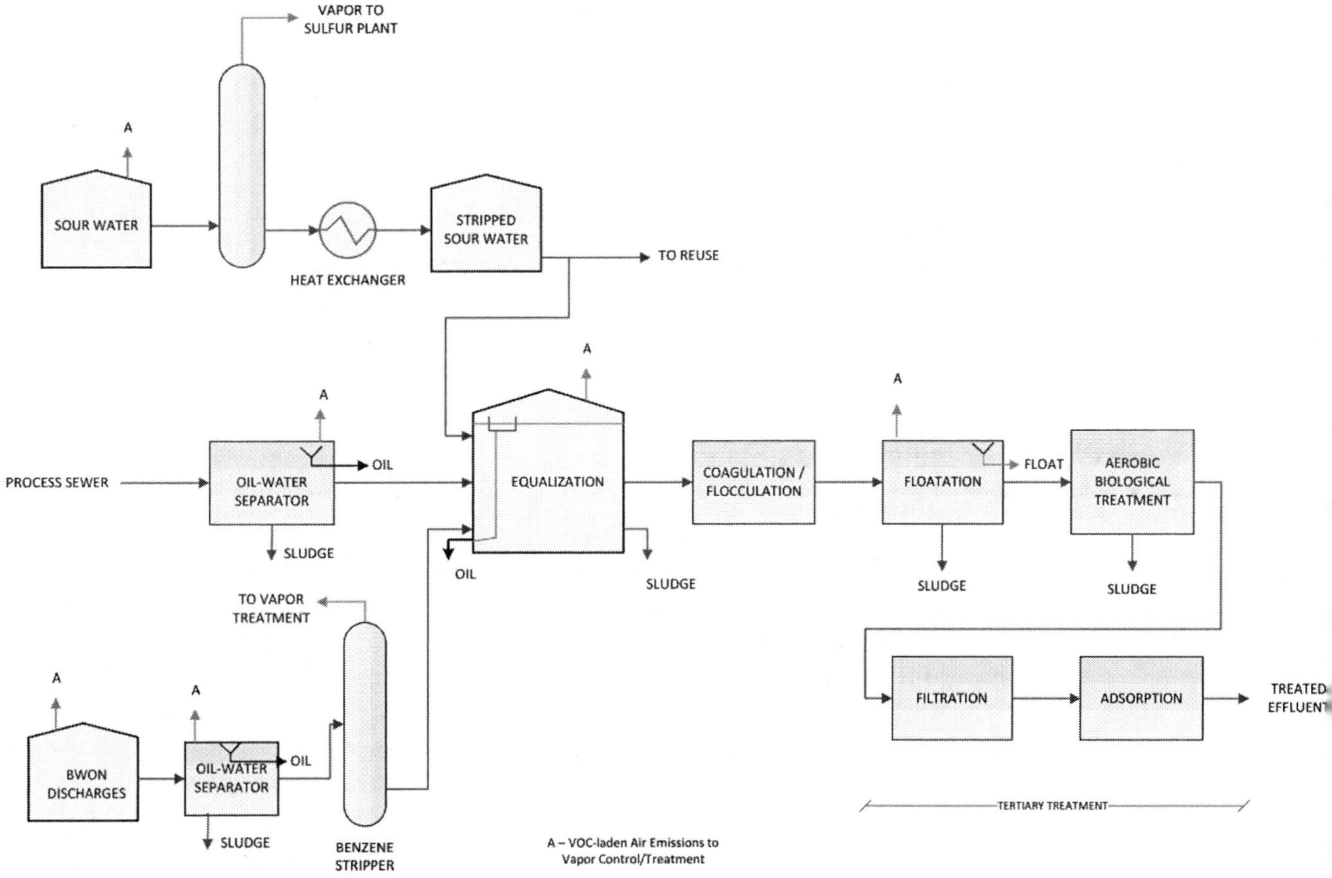

Figure 1 – General Refinery Wastewater Treatment Plant Process Flow

Oil and Solids Removal

The first step is to remove free-phase oil and suspended solids before the water is sent to equalization. API separators are the most commonly used gravity separation systems at the front end of the plant.

Some refineries use Coaslecing Plate Separation (CPS) units such as Corrugated Plate Interceptor (CPI), which is smaller in footprint than APIs. A common problem with CPI is that if not properly maintained they tend to foul and plug with light solids present in refinery wastewater, impairing separation efficiency. Demulsifiers may be used to facilitate oil separation in CPS units.

Secondary oil separation can be provided by a floatation unit following the API or CPI units, however this results in larger floatation units than if installed downstream of equalization.

Equalization

Equalization tanks attenuate wastewater flow and quality variability to protect downstream processes from fluctuations caused by high-strength waste discharges. These tanks are operated at a constant water level in flow through mode and keep available capacity for storm events.

Equalization tanks deprived of API separators are not common but are in place at some refineries where the oil layer is removed by skimming and/or lateral drain port withdrawals. Equalization tanks will accumulate sludge over time, reducing its effective volume, even more so for tanks not protected by API separators.

Physicochemical Treatment

Following equalization, the next step is the removal of emulsified oil by physicochemical treatment consisting of:

- pH adjustment
- Coagulation/Flocculation
- Dissolved or Induced Gas Floatation
- Chemical Oxidation

Coagulation and flocculation assists with oil de-emulsification and removal of heavy metals. Coagulation and flocculation uses ferric chloride and polymer to agglomerate metal precipitates into flocs that can be readily floated. In refineries, due to the presence of oil and low density solids, floatation is a more effective process than settling.

There are two general types of floatation processes used in refineries: Induced Air Floatation (IAF) and Dissolved Air Floatation (DAF) units. IAF removal efficiencies are typically less than for DAF units.

Some facilities use two stages of floatation, an IAF followed by DAF, where the IAF separates emulsified oil while the second stage removes metal by coagulation, flocculation and floatation. This configuration is robust and can produce a good effluent quality, however is more capital intensive.

Stringent BWON air regulations are an important driver for upgrades from DAF to DGF (G: nitrogen gas) units, because DGF units do not require a continual air sweep flow that DAF units need to keep VOC concentrations below lower explosive limits. The use of an air sweep in DAF creates a large volume of benzene-laden air that requires treatment by air pollution control devices.

Chemical oxidation reduces the effect of chelating compounds that may interfere with metal precipitation (e.g. cyanides or ammonia), oxidizes sulfides, and breaks oil emulsions. Chemical oxidation is an enhancement to physicochemical train that improves the efficiency of metal precipitation and helps with treatment of spent caustic discharges.

Segregated Treatment

Process discharges that pose significant pollutant loading are segregated in a separate equalization tank and/or pretreated before they are sent to the refinery WWTP.

BWON discharges

They contain high concentrations of benzene, which bring about emission concerns related to BWON regulations. These include desalter effluent, diesel coalescers, coker knock-out drum, and oil storage tank draws.

These waste streams are sent to a separate equalization tank equipped with oil skimmers for can be subsequently treated by steam or gas stripping to remove benzene and other VOCs. The VOC vapor stream is sent to an air pollution control device. The treated effluent is cooled and sent to the refinery WWTP.

Sour Water

Sour water consists of steam condensates generated in various refinery processes. Sour water carries low hydrocarbon content, but relatively high concentrations of hydrogen sulfide (H_2S) and NH_3, as shown in **Table 1**. This stream was also found to be the major source of selenium in refinery wastewater. Sour water stripping by either direct steam injection or reboiler steam removes H_2S and NH_3 from water but selenium remains in the water phase. The vapors go to the sulfur plant and the Sour Water Stripper (SWS) effluent is then reused in the desalter, flue gas scrubber, and hydrotreater make-up. The water that is not reused is sent to the refinery WWTP.

Sour water from delayed cokers and catalytic crackers also contain phenols and cyanides. These waters may be treated in a dedicated stripper to remove H_2S and NH_3, and then SWS effluent reused in the desalter where phenolic compounds can removed, reducing its loading to the refinery WWTP.

Selenium Treatment

Some refineries have been given very low selenium limits in the range of 5- 10 ppb during recent permit renewals. This has led refineries to assess selenium sources and implement upgrades for selenium treatment either at-source or end-of-pipe treatment, or as a combination of both. Feasible technologies are:

- SWS effluent treatment
    - o Ion Exchange
    - o Activated Sludge followed by Anoxic-Anaerobic Selenium Reduction
- End-of-pipe treatment
    - o Iron co-precipitation
    - o Specialty clay adsorption
    - o Biological Anoxic-Anaerobic Selenium Reduction

Ion exchange has been successfully implemented as an at-source treatment of the SWS effluent but it has a high O&M and capital costs.

Anaerobic-anoxic biological treatment is an attractive alternative for selenium removal with low production of sludge, and is more cost competitive to ion exchange. This has been successfully piloted at a refinery as an end-of-pipe polishing step for indirect discharge, but full-scale implementation has been limited. This process causes an increase in WET toxicity due to sulfide formation, hence subsequent air oxidation would be required if discharging to surface water.

Biological Treatment

After free-phase and emulsified oil and metals have been removed, biological treatment removes dissolved organics and soluble oil. There are various biological treatment systems used at refineries.

Activated sludge operated in extended aeration mode has been reliably used for many years. Some systems have been retrofitted with support media for nitrification. Aerated lagoons have also been utilized over the years but refineries face regulatory pressures to phase them out due to more stringent VOC emissions regulations.

Aerobic Membrane Bioreactors (MBRs) are in the inception phase of implementation with several pilot tests showing promising results, and a few being taken to full-scale implementation. A variation of MBR uses suspended granular activated carbon (GAC) to help the reactor to better cope with shock loads of recalcitrant organics. Membrane materials available are PVDF and ceramic-type. The benefits of this technology are high organic removal efficiencies and nitrification/denitrification, which can help refineries meeting stringent toxic organics limits (e.g. benzene of less than 5 ppb) and nitrates associated with surface water discharge.

Tertiary treatment

Tertiary treatment (effluent polishing) consists of process units that remove TSS, recalcitrant trace organics and trace metals not removed in biological treatment. The typical tertiary treatment consists of filtration and media adsorption. Sand-anthracite media filtration are commonly used at refineries. Granular activated carbon (GAC) vessels are namely used for removal of trace organics although trace metal concentrations are also removed.

New pollutants of concern and/or more stringent effluent limitations are triggering WWTP upgrades at refineries with some occurring in this final treatment stage. For instance, new selenium and mercury limits are very low, and feasible treatment options include specialty adsorption media (modified clays), ion exchange, and anoxic biotreatment. Specialty GAC media can also be used for sulfides removal.

Some facilities discharging to surface water are being imposed low chlorides and TDS limits, which can be removed by Reverse Osmosis (RO). Alternatively, these constituents can be removed by at-source treatment of high TDS streams by desalination technologies (e.g. humidification-dehumidification).

Conclusions

New compounds of concern and more stringent effluent limitations have been affecting refineries at permit renewal for both direct and indirect discharge. Several treatment technologies and process configurations are available to upgrade and/or retrofit the existing plants to meet updated permit limits. There are new emerging technologies that do not have a track record in refineries but are worth considering during the feasibility analysis. Ultimately, the selected option for implementation needs to be pilot tested to confirm feasibility and demonstrate performance in achieving desired removal efficiencies.

Indirect discharge (POTWs) relieve facilities from more stringent surface water discharge limitations. However, potential upsets in the refinery WWTP may cause upsets in the POTW processes that could lead to disruptions of the refinery operation due to interruption of service while POTWs restore their plant operations. The refineries must protect against these events by identifying pollutant sources that could adversely affect both the onsite and POTW plants, and implement controls to minimize or manage these discharges.

Some refineries discharging to POTWs have treatment plants that provide extensive treatment, with some even including tertiary treatment. When considering treatment modifications and/or upgrades, treatment options that reuse treated effluent can prove beneficial to reduce pollutant discharge load, and in turn better meet effluent permit limitations.

As some refineries use more, or completely convert to, shale crude oil as feedstock, more studies are needed to understand whether there are changes in wastewater composition that could impact existing WWTP plant performance and/or discharge permit limitations.

References

1. US EPA (1996) - "Preliminary Data Summary for the Petroleum Refining Category", Office of Water, Engineering and Analysis Division.
2. US EPA (2004) – "Technical Support Document for the 2004 Effluent Guidelines Program Plan, Section 7 – Petroleum Refining" – EPA-821-R-04-014, Office of Water, Engineering and Analysis Division.
3. IPIECA (2010) – "Petroleum Refining Water/Wastewater Use and Management", Operations Best Practice Series.
4. Nemerow, L.N. (1987) – "Industrial Water Pollution: Origins, Characteristics, and Treatment; R.E. Krieger Pub.

445435 Effective Operational Discipline Programs

Abstract:

To achieve successful process safety performance and consistent management of process risk, effective process safety programs must include emphasis on safety culture and leadership, process safety systems, and operational discipline (OD). Effective OD programs help ensure that well-designed process safety systems are successfully implemented and help achieve consistent, desired, day-to-day performance. Failure to follow process safety system requirements can result in unsafe behaviors, unsafe acts, and ultimately, serious incidents and injuries. A focus on OD requires continued commitment by company leadership to ensure appropriate systems, procedures, and safe work practices have been established and by workers at all levels in the company to consistently follow and meet these requirements. This presentation reviews the benefits, key concepts, and practical applications of effective OD programs for evaluating and improving OD in order to help improve process safety performance.

445434 Best-in-Class MOC Implementation – Case Studies

Abstract:

So your facility has an MOC procedure, right? You hear complaints about the MOC process being burdensome, and you worry that people aren't doing MOCs when they ought to. What do companies do that are successful at MOC?

Successful MOC practitioners have several key characteristics. First, they treat Management of Change just like a project--not as a form that needs to be filled in. The project orientation engenders the formality and respect that MOCs deserve. Second, successful companies think broadly in terms of the business benefits that the MOC procedure/process can bring them. While MOC is required for PSM covered processes, the methodology can also be leveraged for replacements-in-kind, as well as smaller capital projects. And finally, successful MOC practitioners simply "get on with it". They realize that, after 24 years since the PSM regulations were promulgated, most companies are into a "sustaining" mode, complacency may have set in, and it is time to take the next steps in MOC evolution. They take action and avoid waiting for perfect solutions and universal commonality.

This presentation will present a number of case studies, from companies that have recently updated their MOC processes, and use those case studies to illustrate the key learnings.

Facility Integrity:
Learnings from the Equipment Lifecycle

Robert F. Wasileski, III
Baker Engineering and Risk Consultants, Inc.
11011 Richmond Avenue, Suite 700, Houston, TX 77042
RWasileski@BakerRisk.com

Prepared for Presentation at
American Institute of Chemical Engineers
2016 Spring Meeting
11th Global Congress on Process Safety
Houston, Texas
April 10-14, 2016

UNPUBLISHED

Facility Integrity: Learnings from the Equipment Lifecycle

Robert F. Wasileski, III
Baker Engineering and Risk Consultants, Inc.
11011 Richmond Avenue, Suite 700, Houston, TX 77042
RWasileski@BakerRisk.com

Keywords: Mechanical Integrity, Reliability, Lifecycle, Near Miss

Abstract

The Chemical Processing Industry (CPI) has witnessed growth in mechanical integrity (MI) programs, which have evolved from standards-based compliance, to continuous improvement programs, and on to risk-based programs. For instance, operators (i.e., manufacturers) have redesigned corporate standards, plant-level procedures and field practices to keep pace with incident learnings and published best practices, such as those produced by the American Institute of Chemical Engineers' (AIChE) Center for Chemical Process Safety (CCPS). Furthermore, the property insurance industry has duly taken note of this MI evolution, giving rise to a significantly greater focus on MI programs during insurance surveys and inspections. Meanwhile, incident investigators remain in demand as relatively large incidents with MI-related causes continue to occur in the CPI. This apparent disconnect naturally raises questions in regards to subjects such as "risk based inspection (RBI)", "reliability centered maintenance (RCM)," "inspection frequency", "industry best practice" and "inspection, testing and preventive maintenance (ITPM)".

MI programs play a considerably important role in the process safety lifecycle of equipment. The engineering design phase of a project may be on the order of just days, up to several years for a complete plant. Procurement and construction typically follow a similar timeline proportionate with the engineering design phase. However, once a facility becomes fully operational, the time devoted towards operation and maintenance will normally far out-weigh the engineering, procurement and construction (EPC) period, and may last for many decades. As such, the process safety equipment lifecycle (PSEL) is typically dominated by the MI program, relative to other lifecycle phases such as the EPC period or decommissioning and retirement.

The PSEL is explained in this article, with an in-depth examination of the central elements which should be embraced by a comprehensive MI program. While the article demonstrates the breadth and depth of MI, it aptly proposes an innovative approach towards the management system that should form the foundation of a robust MI program. The proposed framework leverages the common "Onion Skin" diagram in the context of the equipment lifecycle to create an intuitive approach to MI management. This unique management system framework is comprehensive, sound, and yet practical for implementation at facilities of most any size.

1. Glossary (Terminology, Definitions and Acronyms)

1.1 The author uses or infers to a variety of acronyms and terminology throughout this article. Where a definition does not contain a reference, the definition is that of the author. Terminology, definitions and acronyms are listed in Appendix A of this article.

2. Introduction and MI Background

> "*Operators sometimes regard tests and inspections as a nuisance, interfering with the smooth operation of the plant. Operator training should emphasize that protective equipment is there for their protection and they should 'own' it.*" – Trevor Kletz [1].

According to the Marsh Energy Practice, the world's leading property insurance broker, major accidents due to mechanical integrity related failures continue to make a significant contribution to large losses in the CPI. Marsh comments, "...loss totals in the refining industry have continued to *trend upwards over the last few years. Piping failures or leaks* (corrosion or incorrect metallurgy) and start-up and shut-down events *continue to be significant causes.*" [2] This quote may not be particularly surprising to some, given the fact that it directly references the refining industries. However, given the relative age of most refineries to other facilities such as those in petrochemicals and oil & gas, this upward trend in piping failures should be a concern to all readers from every sector of the CPI. Logically, if the age of the facility is a significant correlating factor towards piping (and other mechanical-related) failures, the future holds cause for concern.

MI programs began to receive serious attention perchance in the early 1990s, with the promulgation of the United States *Process Safety Management (PSM) Standard for Highly Hazardous Chemicals*, OSHA 29CFR1910.119. This U.S. Standard contains a relatively lengthy discourse on "Mechanical Integrity", and the performance-based requirements for those facilities covered by the Standard. Once the Standard was promulgated into law, most facilities adopted a compliance-based maintenance program to meet the requirements of the Standard. However, MI programs have evolved and matured since then, with the CPI adopting integrated approaches and philosophies to MI management such as run-to-failure maintenance, reliability-centered maintenance and, most recently, risk-based maintenance.

This article will briefly explain basic concepts such as the PSEL, the familiar "onion skin diagram", and then go on to propose a new framework for MI management programs which leverages the onion skin diagram in the context of leading-edge practices promoted by respected organizations such as the AIChE/CCPS.

3. Facility Lifecycle Explained

With respect to preventing Process Related Incidents (PRIs), perhaps one of the greatest challenges an organization may face is changing the mindset of employees from an "equipment-based" to "facility-based" mechanical integrity program. In retrospect, an arguable failure of the U.S. Process Safety Management (PSM) Standard was the over-stated emphasis on flammables and what OSHA refers to as "Highly Hazardous Chemicals". Since the early 1990s this Standard has largely shaped the CPI into a mindset that MI programs are *limited to* what is known as a "covered process" under the Standard. In practice, some companies have intentionally created artificial boundaries around "covered" equipment, in order to limit the scope of coverage of the plant's MI program. In some cases, the "covered process" literally is separated or "bounded" by a single flange on a line containing a flammable or toxic process material. The concern with this approach is that MI programs, in some cases quite robust programs, are not being applied to hazardous areas where significant process risks may be present. These artificial boundaries have potentially limited the scope of MI programs to equipment which is only required by law to have an MI program. This approach may create vulnerabilities in the plant and amplify the opportunity for PRIs beyond the user-defined "covered process" (See Table 1).

The risk of a PRI outside of the "covered process" is a very real possibility. These are process risks which involve a mindset change from "equipment-based" MI programs to a "facility-based" MI program. The scope of a facility-based MI program is clearly broader in nature. However, with practical approaches to MI such as Run-to-Failure and Risk-Based Maintenance, this increase in scope can be managed with available resources, while still controlling the risks posed by equipment outside of the "covered process" areas. Thus, the PSEL needs to be viewed as being applicable and necessary to the entire facility – not just "covered process" equipment – using a risk-based decision making process (See Figure 1).

The PSEL begins at the conceptual design phase (Phase 1). In practice, this phase may occur in a Research & Development (R&D) facility, pilot plant, or with process engineering (e.g., with an EPC). During Phases 1 through 4 it is *essential* for the Process Design Basis (PDB narrative) and Basic Design Data (BDD) to be developed and documented for process equipment, critical buildings, infrastructure and so forth. The PDB and BDD forms the foundation upon which the final plant or facility will be operated. BDD often includes chemical and physical material properties, materials of construction, reliability data, and information on operating history. However, the scope of BDD is not limited to traditional process safety information; in certain cases it may be necessary to complement the scope of data with information from other disciplines (e.g., Occupational Health, Toxicology, Product Stewardship,etc.) to ensure the data is complete and thoroughly documented [6].

Table 1: Examples of CPI Incidents *Outside* of the PSM "Covered Process" Boundaries [3], 1

EVENT NO.	EVENT DATE	EVENT TYPE	LOCATION	PRIMARY PROCESS MATERIAL INVOLVED[2]	FAILURE TYPE OR DAMAGE MECHANISM	DAMAGE (2013 U.S. Dollars)[3]
1	07/04/2012	Fire and Explosion	Bangkok, Thailand	Kerosene	Internal vessel (stripper) fire	$140,000,000
2	09/28/2011	Fire	Palau Bakom, Singapore	Blended refined products	During maintenance inside a pump house building	$150,000,000
3	04/20/2010	Fire and Explosion	Deepwater Horizon Rig (Macondo Well), Mississippi Canyon Block #252, Gulf of Mexico	Crude Oil	Well blow-out during maintenance / abandonment activities	$4,500,000,000[4]
4	11/22/2002	Fire and Explosion	Port of Mohammedia, Morocco	Waste oil	Hot surface ignition following a loss of primary containment	$200,000,000
5	09/21/2001	Explosion	Toulouse, France	Ammonium Nitrate	Internal vessel (tower) explosion	$680,000,000
6	08/14/2001	Fire	Lemont, Illinois, U.S.	Pyrophoric material	Ruptured pipework leading to air ingress	$370,000,000
7	04/09/2001	Fire	Wickland, Aruba	High flash point oil and hot surface ignition	Block valve failure (leak) during maintenance activities	$250,000,000
8	12/13/1994	Explosion	Port Neal, Iowa, U.S.	Ammonium Nitrate	Unknown	$370,000,000
9	08/07/1994	Fire	Ryazan, Russia	Crude oil	Procedural: Inadequate equipment preparation for hot work	$180,000,000
10	06/20/1991	Explosion	Dhaka, Bangladesh	Anhydrous Ammonia	Failure of a welded joint on a process vessel	$140,000,000

[1]Incidents listed in Table 1 occurred outside of the U.S. PSM "covered process" area, (a) if the incident occurred in the United States or (2) where the incident occurred outside of the United States, the incident would have been outside of the "covered process" area and/or exempt from the U.S. PSM Standard for other reasons, such as the primary process material involved.

[2] "Primary Process Material" listed is the initiating material that led to the event. Other process materials within the facility may have contributed to the total damage; however, those process materials were secondary in nature due to "domino" or knock-on effects.

[3]Damage values inflated to December 2013 U.S. Dollars using the Nelson-Farrar Petroleum Plant Cost Index. Further, estimated damage values in Table 1 are "ground up" values, limited to *property damage alone*. Other costs, such as business interruption, injuries and fatalities, and liability claims are <u>not</u> included in the estimated damage.

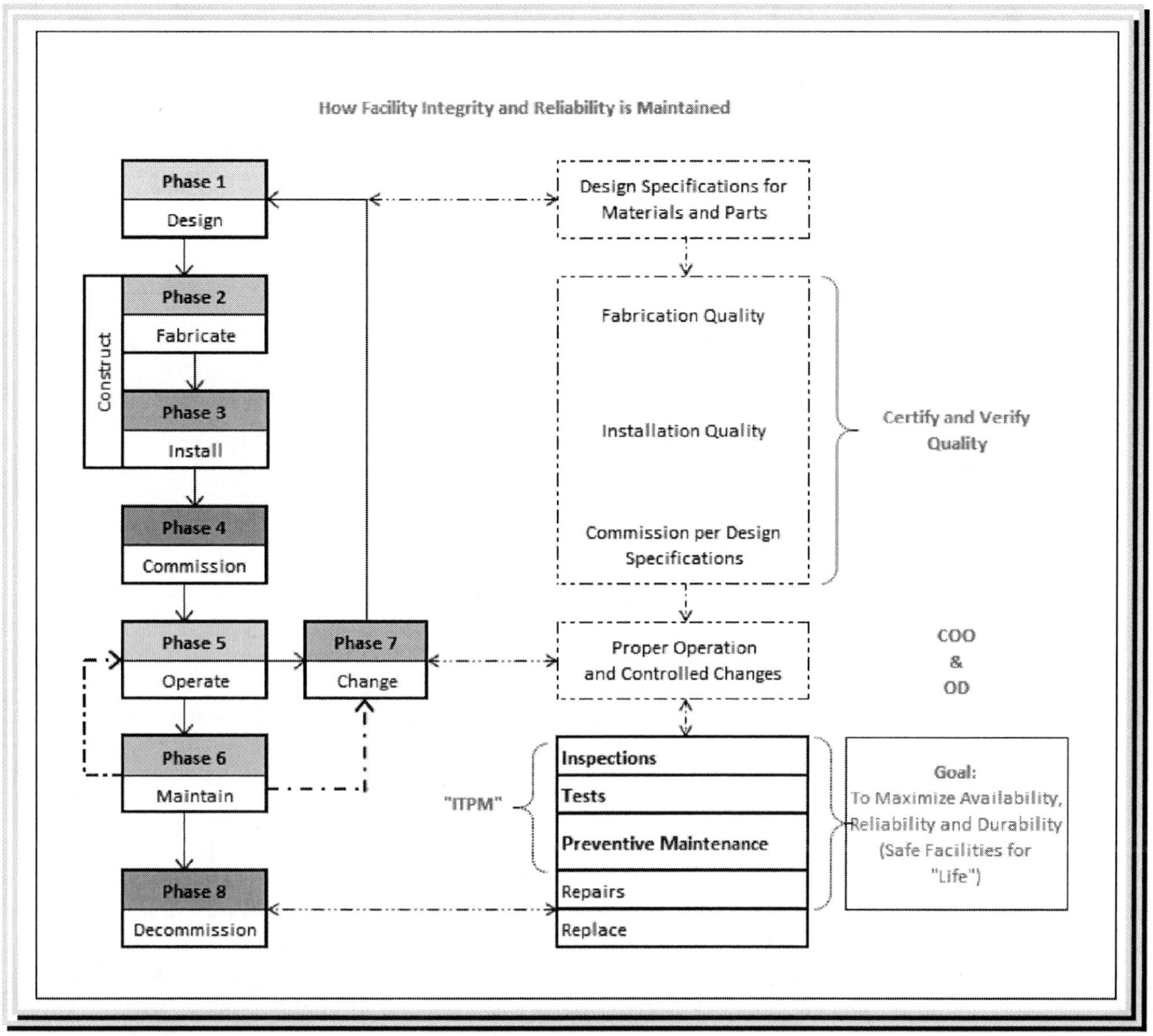

Figure 1: PSEL Diagram [5]

Phases 5 through 7 represent the *operational period* of the facility, which may last for decades and far out-weigh Phases 1 through 4 (in terms of length of time). With the operational period dominating the time element of the PSEL, it is particularly important to operate the facility in a manner which preserves the original process design basis. Preservation of the process design basis is achieved through attentive execution of the following four elements:

1. *Conduct of Operations (COO)*
2. *Operational Discipline (OD)*
3. *Management of Change (MOC)*
4. *Inspection, Testing and Preventive Maintenance (ITPM)*

Phases 5 through 8 of Figure 1 illustrate the relationship between operations, maintenance and engineering (via "Changes" to the facility) which act to preserve the original process design basis of the facility, with the support of robust management systems.

Figure 2 illustrates the PSEL with respect to time, demonstrating how the failure rate of (critical) equipment changes over time. In Figure 2, the previously explained Phases 1 through 4 are contained in the "Early Life" period. Phases 5 through 7 are within the "Intrinsic" period, and Phase 8 occurs in the "Wear Out" period of the distribution.

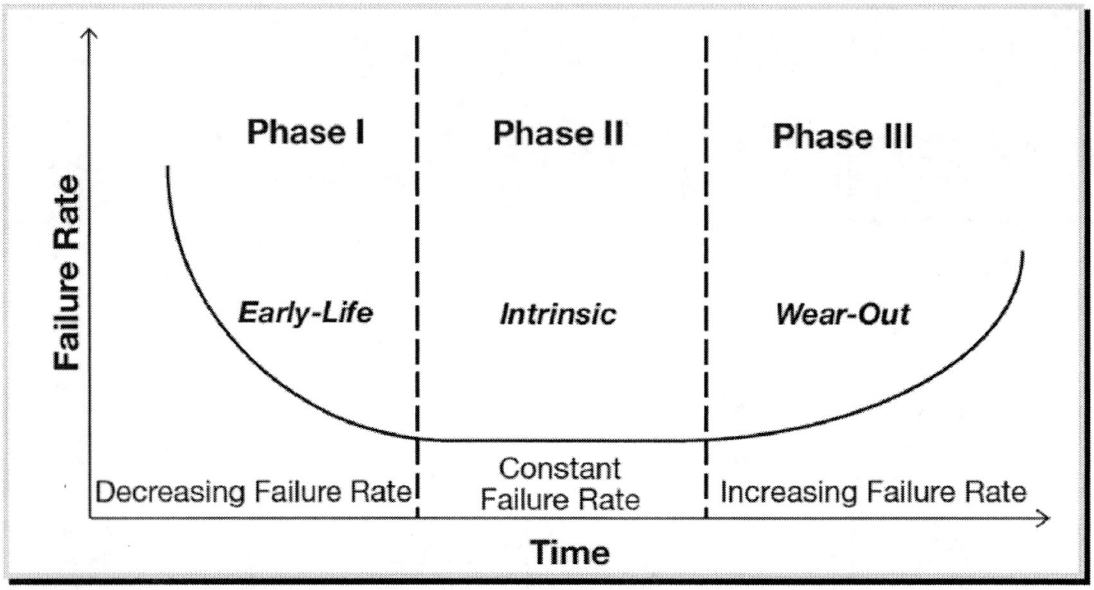

Figure 2: PSEL Expressed Using the Weibull Distribution [7]

"The Weibull distribution can be applied to mechanical component reliability, biology, weather, finance, and a broad range of other systems that obey the characteristics of the bathtub curve. Early in the lifecycle, component failures rates decrease (Phase I), eventually leveling off and remaining relatively constant (Phase II). As the system approaches the end of its life (Phase III), components begin to wear out and the failure rate increases again."

4. "Facility Integrity Program" Scope

The "onion skin Diagram" is perhaps the most useful and practical illustration of the scope of a "Facility Integrity Program." Figure 3 illustrates this scope, and further identifies the core elements and process safety barriers (i.e., layers of protection) that comprise the Facility.

Legend:

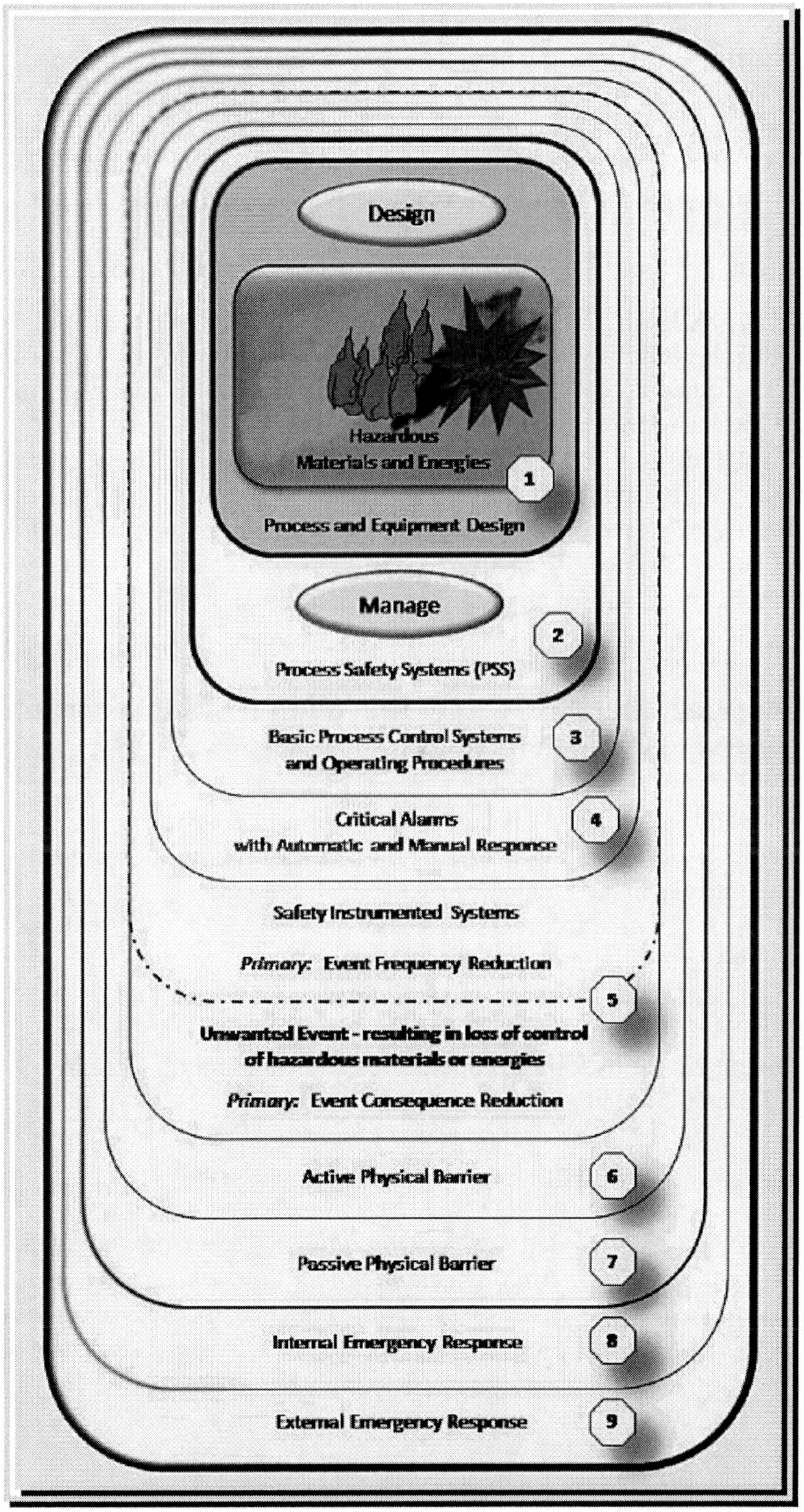

Figure 3: Onion Skin Diagram: The Scope of a "Facility Integrity Program" [8]

The barriers identified in Figure 3 are detailed below. The relevance of the barriers is that every barrier identified in the figure needs to be included in the scope of the facility integrity program. Therefore, it is *strongly recommended* for practitioners, operations and maintenance to:

a) Identify the critical equipment in their plant which is included in these barriers;
b) Fully understand the PDB, BDD and purpose of each equipment within the barrier (i.e., protection layer); and,
c) Operate and maintain that equipment in accordance with the original design basis, until which time the equipment is properly decommissioned [9].

The barriers in Figure 3 are as follows:

1. Process Design, including Inherently Safer Design (ISD) applications.
2. Process Safety Management (PSM) Systems.
3. The Basic Process Control System (BPCS) and Standard Operating Procedures (SOPs).
4. Critical Alarms **with** Operator Response.
5. Instrumented Protective Systems (IPS) and Safety Instrumented Systems (SIS) (automated systems).
6. Active Physical Protective Systems.
7. Passive Physical Protective Systems.
8. Plant (Internal) Emergency Response Teams.
9. External Emergency Responders (e.g., Local Emergency Planning Committees).

5. Coupling the PSEL and the Onion Skin Diagram: The *New* "Facility Integrity Management System"

The new Facility Integrity Management System is created by coupling the previously described PSEL with the Onion Skin Diagram. When this union is created, five critical elements of the management system are created:

1. Element 1: Documenting the PDB and BDD.
2. Element 2: Establishing and Reinforcing COO and OD.
3. Element 3: Establishing ITPM on critical equipment.
4. Element 4: Ensuring Competencies and Conducting Training.
5. Element 5: Managing Deviations.

5.1 The Five Management System Elements Explained

"A major issue for organizations is being able to learn from incidents and capture those lessons into design, procedures, training, maintenance and other programs". "But one factor that is often overlooked is the types of incidents that are tracked or investigated. [3]" -Dr. M. Sam Mannan, PE, CSP, DHC, Regents Professor and Director, Mary Kay O' Connor Process Safety Center

5.1.1 Documenting the PDB and BDD

Documentation of the PDB and BDD may include a broad range of documents which serve to support both hazard identification and risk assessment (HIRA) studies (e.g., Hazard and Operability (HazOp) studies) and Management of Change (MOC) programs. These documents must be kept *evergreen* and used as *"baseline"* for HIRA studies and MOCs, in order to thoroughly evaluate the risk of proposed changes against the original baseline design of the plant. Examples of information which is typically included in this element include:

a) Process chemistry narratives;
b) Process flow diagrams (PFDs);
c) Piping and instrumentation diagrams (P&IDs);
d) Nameplate capacities;
e) Area electrical classification diagrams;
f) Safe operating limits, and
g) Critical operating parameters.

MOCs inherently act to make changes to the facility. The PDB and BDD, when kept evergreen, allow engineering and other practitioners to evaluate the risk of changes in a meaningful manner. Conversely, a proposed facility change cannot be thoroughly evaluated for changes in process risk if the necessary baseline documents are not available, because the PDB and BDD is not an accurate reflection of the current facility.

The author has witnessed several cases where either incorrect or incomplete BDD documentation has resulted in accidents or cost increases on large capital expense projects to correct the deficiencies. For example, sizing and design data for pressure safety devices (PSDs) is normally developed during the EPC stage of a new facility. If, through MOC practices during the operational phase, the PSD data is not kept evergreen this documentation becomes increasingly unreliable over time. This slow "creeping" action of leaving PSD design basis documentation in a static, unchanged state will often become a significant issue during plant expansions, where the expansion project affects a relatively large number of PSDs in the facility. In several actual cases, the author has seen the cost of a plant expansion project dramatically increase due to outdated PSD design documentation. In order for the expansion project to proceed, all of the PSD design documentation within the project scope had to be individually evaluated, re-engineered and updated at a significant cost to the project. These evaluations, in each case, were unforeseen at the start of the project and, as such resulted in substantial scope change and cost increases to the projects.

Another real example of *incorrect* BDD resulted in an internal deflagration within a newly installed catalytic thermal oxidizer (CatOx). In an article written by the author titled "*The Role of Basic Design Data in Preventing Explosions within Fired Equipment: A Case Study* [6]", the purpose of the BDD is explained in great detail. However, in summary, a project was executed to collect various waste streams from points throughout a process in a common vent header, and effectively destroy the hydrocarbons with a CatOx in order to comply with environmental permits. During the EPC stage of the project, the method by which the BDD was developed steered the project team to incorrectly assess and document the design basis for operating the CatOx. Upon initial start-up of the CatOx unit, a mechanical failure occurred, resulting in a blow-out of a pressure relief panel – a safety critical device. A second attempt to re-start the unit allowed air ingress into the vent header, creating a flammable atmosphere that entered the combustion chamber of the CatOx. When this flammable mixture contacted the open ignition source (i.e., the flame) in the combustion chamber, a damaging internal deflagration resulted, setting the project back several months.

These examples, while not necessarily catastrophic in nature, are real events and they do in fact emphasize the importance of developing and maintaining evergreen PDB documentation and BDD.

5.1.2 Establishing and Reinforcing COO and OD

Element 2 addresses the day-to-day operations of the facility. This element goes beyond the "process equipment" itself, and includes provisions such as staffing (e.g., minimum operating complements), building use and occupancy practices (e.g., changes in building use which could impact vulnerabilities and/or design and construction requirements), spare part stocking philosophies and temporary safety system impairments. **Operations owns the risk** pertaining to day-to-day operations and these various provisions, and therefore must adhere to COO and OD practices in order to safely manage the risks in the facility on an on-going basis. Referring back to Table 1, incidents number 1, 5, 6 and 9 arguably have components of failed COO and OD which contributed to these particular losses when examined with greater detail.

MOC practices must consider not only process risks, but also a spectrum of other risks that may impact personnel, operability of the plant, environmental concerns and so forth. For example, changes in occupancy within a critical building can increase the risk profile of the facility by concentrating the risk to employees via an increase in building occupancy. While this is not an equipment change per se, it is a *change* that MOC practices should evaluate with a facility siting study in order to understand the change to personnel risk. In similar fashion, changes in staffing and workforce may adversely impact the operability of the plant. Again, this is not an equipment change but one that does warrant a MOC review. This type of change underscores the importance of establishing a minimum operating complement for the facility, particularly with the operations staff. The minimum operating complement should be considered as PDB documentation, and maintained in an evergreen status with physical changes to the plant, such as plant expansions and decommissioned units.

Wolf and Wasileski have written on the importance of establishing a robust COO and OD program in their article "*THE ILLUSION OF ATTENTION: Are There Gorillas in Your Plant?*"[7]. The failure to notice an unexpected stimulus in your field of vision while performing other attention-demanding tasks is a cognitive phenomenon known as inattentional blindness. The article presents experimental evidence which suggests that our ability to recognize hazards (in the field) is more impaired than we think. The thrust of the article goes on to present a proactive model for Organizational MOC which seeks an optimum level of awareness for operating staff. The relevance of this article with respect to

COO and OD is that MOC plays an important and integral role to successfully managing Element 2.

5.1.3 Establishing ITPM on critical equipment

Element 3 addresses maintenance and reliability, and their role in managing facility risks. Of particular importance in this element is the need for written ITPM plans for each equipment and facility included in the onion skin diagram, *especially for barriers 3 through 7* (the main protection layers used to prevent or mitigate PRIs). Written ITPM plans must include clear instructions for maintenance personnel pursuant to ITPM methods, frequencies/intervals, and **contingency measures** that must be in place to manage process risk during temporary safety system impairments which are required to carry out ITMP activities. This requires clear, close communication with operations during ITPM activities.

Perhaps no other incident exemplifies this need for close communication between operations and maintenance than the fire and explosion on July 7th, 1988 on the Piper Alpha drilling rig in the North Sea. Inflated 2003 U.S. dollars puts the property damage cost of this loss at over $1.8 Billion [3], in addition to 167 fatalities. Much has been written about this accident, in articles, books and other learning material. Possibly one of the most important lessons learned from the Piper Alpha accident has been described in an article by M. Elisabeth Pate-Cornell [10]. Pate-Cornell writes, "*...a piece of equipment (a critical pump with one redundancy) had been turned off for repair and the night crew that operated the platform had not been informed of it. This problem, in turn, was mostly a failure of the "permit-to-work system" that did not ensure proper communications. Things would have not taken catastrophic proportions...*" if not for the lack of communications between operations and maintenance. Pate-Cornell goes on in her article to describe numerous additional shortcomings in communication between operations and maintenance personnel, which are categorically described as "Human Errors". As she goes on to summarize, "*The human errors, questionable decisions, and bad judgments that have been identified above and contributed to the accident can therefore be divided into four categories: (1) design decisions; (2) production and expansion decisions; (3) personnel management; and (4) inspection, maintenance, and correction of detected problems*". This statement alone reinforces not only the importance of COO and OD, but also once again points to the importance of Element 1 which addresses process design.

5.1.4 Ensuring Competencies and Conducting Training

Engineering staff must receive training and instruction towards their responsibilities under Element 1. Element 1 is essentially the foundation of the facility integrity management system. As such, any errors or deficiencies within Element 1 can have adverse impacts towards the other four elements in the management system. For example, if BDD is not fully documented for a particular process, HIRA studies and MOCs become overly cumbersome, difficult endeavors to properly execute and evaluate risks. In similar fashion, operations and maintenance both must be properly trained such that personnel understand their responsibilities under Elements 2 and 3. Operations needs to understand they own the risk of the facility that they operate. Therefore, it is the responsibility of operations to ensure equipment has been restored to a "safe to operate" condition following ITPM activities. Furthermore, maintenance personnel must be in close communication with operations during ITPM activities such that operations is aware of maintenance activities being carried out, equipment that has been temporarily impaired for inspections, and any changes made to process equipment.

In light of these conditions, other management systems such as Permit-To-Work (PTW) and Safety System Impairments [11] become critical systems. These management systems need to clearly describe the nature of the work taking place, the location in the plant where the work will be performed, the anticipated duration of the work, and approval signatures from those with accountability for the equipment, with an escalation process to address work that lasts longer than originally anticipated.

5.1.5 Managing Deviations

The fifth element of the facility integrity management system captures deviations – the unplanned acts and undesired events that occur in a process. Unplanned acts are often undesirable acts or unsafe acts, with no readily observable negative consequences to the process (See Figure 4).

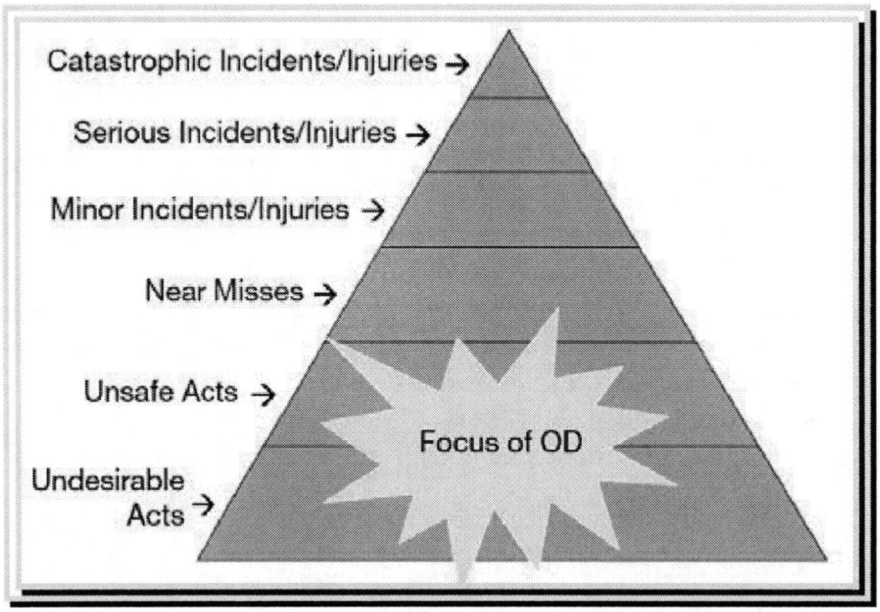

Figure 4: Process-Related Incident Pyramid [8]
"*A program to reduce the OD-related problems at the bottom of the triangle helps to reduce the possibility of more serious undesired events at the top*"

Conversely, undesired events are often revealed through the ITPM program, and frequently fall into the category of a "near miss". Equipment failures, both "revealed" and "unrevealed", demands placed upon safety systems (i.e., protection layers), spurious activation of safety systems, and unintentionally impaired equipment fall into this fifth element. Each of these "unplanned acts" and "undesired events" must be treated as an incident, and subjected to an incident investigation commensurate with the actual and/or potential consequences of the finding or observation. In investigating these acts and events as an incident, the facility will understand the basic causes, learn from the investigation and be prepared to put measures in place to prevent a recurrence from escalating higher into the PRI Pyramid (e.g., to a serious or catastrophic incident). This fifth element of managing deviations through incident investigation completes the PSEL by providing a circular path back to Elements 1 through 4, such that deficiencies in any element may be corrected, and thus reducing the probability of a more severe incident.

6. Conclusions

MI programs are clearly necessary for managing process risk in a manufacturing facility. While relatively short, the engineering design phase of a project is the starting point for a robust MI program. Procurement and construction typically follow a similar, relatively short timeline which is commensurate with the engineering design phase. However, once a facility becomes fully operational, the length of time devoted towards operation and maintenance normally far outweighs the EPC period, and in many cases lasts for decades. As a consequence, the PSEL is dominated by the operation of the plant and the ITPM program.

When the U.S. PSM Standard was promulgated into law, most facilities adopted a compliance-based maintenance program to meet the requirements of the Standard. However, MI programs have evolved and matured since then, with the CPI adopting blended approaches and philosophies to MI management such as run-to-failure maintenance, reliability-centered maintenance and, most recently, risk-based maintenance. While these approaches each have their own merit, the CPI needs to continue evolving and maturing by adopting a "**facility integrity**" management system approach to curtail equipment-related failures, which remain a leading cause of large-scale accidents.

7. References

[1] **Kletz, Trevor,** "*Lessons From Disaster: How Organizations Have No Memory and Accidents Recur*", Gulf Publishing Company, 1993 Institution of Chemical Engineers, ISBN: 0-88415-154-9.

[2] **Marsh Energy Practice**, "*The 100 Largest Losses: 1972-2009, 21st Edition*", Marsh Ltd, 2010.

[3] **Marsh Energy Practice**, "*The 100 Largest Losses: 1974-2013, 23rd Edition*", Marsh Ltd, 2014.

[4] **Dittrick, Paula**, "BP reaches deal with DOJ over Macondo well blowout in gulf", *Oil and Gas Journal*, November 15, 2012, Houston, TX.

[5] **J. A. Klein, and B. K. Vaughen**, *Process Safety: Key Concepts and Practical Applications*, CRCPress, Taylor & Francis, Boca Raton (to be published in 2016).

[6] **Wasileski, Robert F.**, "The Role of Basic Design Data in Preventing Explosions within Fired Equipment: A Case Study", *Process Safety Progress*, Vol. 26, No. 3, September 2007, DOI 10.1002/prs.10195.
[7] **Elliot M. Wolf, and Robert Wasileski**, "THE ILLUSION OF ATTENTION: Are there Gorillas in Your Plant?", *Chemical Engineering Progress,* July 2014, pp. 28-33.
[8] **J. A. Klein, and B. K. Vaughen, "**Implement an Operational Discipline Program to Improve Plant Process Safety"*, Chemical Engineering Progress,* June 2011, pp. 48-52.
[9] **Wasileski, Robert F.**, "Retired and Dangerous: Out of Service Equipment Hazards", *Professional Safety, Journal of The American Society of Safety Engineers (ASSE)*, July 2014, pp. 33-40.
[10] **M. Elisabeth Pate,** "Learning from the Piper Alpha Accident: A Postmortem Analysis of Technical and Organizational Factors", *Risk Analysis*, Vol. 13, No. 2, 1993, pp. 215-232.
[11] **David K. Lewis and Fred Henselwood**, "Safety System Impairments and the Need to Manage Peak Risk", *Process Safety Progress*, Vol. 31, No. 4, December 2012, DOI: 10.1002/prs.11535.
[12] **AIChE/CCPS**, Process Safety Glossary, http://www.aiche.org/ccps/resources/glossary
[13] **AIChE/CCPS**, "*Guidelines for Mechanical Integrity Systems*", CCPS / John Wiley & Sons, Inc., 2006, ISBN: 13-978-0-8169-0952-0.
[14] **Benac, Daniel J.**, "Using Failure Analysis of Near-Misses to Avoid "Two-Peats" and "Three-Peats"", *Journal of Failure Analysis and Prevention*, Vol. 15, Issue 6, DOI 10.1007/s11668-015-0048-z.

Appendix A: Glossary

AIChE – American Institute of Chemical Engineers.

Availability – The fraction of time that the system is actually capable of performing its mission. [12]

CCPS – Center for Chemical Process Safety (an AIChE Technological Community).

Compliance-Based Maintenance (CBM) – A maintenance strategy whereby the principal goal of the maintenance program is to comply with the minimum regulatory requirements set forth by the governmental authority which has jurisdiction in the region of interest (e.g., *U.S. 29CFR1910.119 "Process Safety Management of Highly Hazardous Chemicals"*). [13]

Conduct of Operations (COO) – The embodiment of an organization's values and principles in management systems that are developed, implemented, and maintained to (1) structure operational tasks in a manner consistent with the organization's risk tolerance, (2) ensure that every task is performed deliberately and correctly, and (3) minimize variations in performance. [12]

Critical Equipment – Equipment, control rooms, occupied buildings and other facilities, which upon failure, may lead to a PRI with a consequence severity exceeding that of the organization's tolerable risk criteria. While this definition does not explicitly exclude knock-on and domino effects, it does assume that the damage is primarily limited to the plant and/or unit from where the PRI originates.

Deviation – An unplanned event or condition which does not comply with the design intent of the equipment and/or facility under consideration.

Equipment (and Process Equipment) – A piece of hardware which can be defined in terms of mechanical, electrical or instrumentation components contained within its boundaries. [12]

Facility – The physical location where the management system activity is performed. In early life-cycle stages, a facility may be the company's central research laboratory or the engineering offices of a technology vendor. In later stages, the facility may be a typical chemical plant, storage terminal, distribution center, or corporate office. [12] A "Facility" includes process equipment, infrastructure, buildings, process control hardware and software, and other materials used for activities such as storage, processing, manufacturing, etc. [9]

Fitness for Service (FFS) – A systematic approach for evaluating the current condition of a piece of equipment in order to determine if the equipment item is capable of operating at defined operating conditions (e.g., temperature, pressure). [12]

Hazard Identification and Risk Assessment (HIRA) - A collective term that encompasses all activities involved in identifying hazards and evaluating risk at facilities, throughout their life cycle, to make certain that risks to employees, the public, or the environment are consistently controlled within the organization's risk tolerance. [12]

Human Reliability Analysis (HRA) – A method used to evaluate whether system-required human-actions, tasks, or jobs will be completed successfully within a required time period. Also used to determine the probability that no extraneous human actions detrimental to the system will be performed. [12]

Inspection, Testing and Preventive Maintenance (ITPM) - Scheduled proactive maintenance activities intended to (1) assess the current condition and/or rate of degradation of equipment, (2) test the operation/functionality of equipment, and/or (3) prevent equipment failure by restoring equipment condition. [12]

Management of Change (MOC) - A management system to identify, review, and approve all modifications to equipment, procedures, raw materials, and processing conditions, other than replacement in kind, prior to implementation to help ensure that changes to processes are properly analyzed (for example, for potential adverse impacts), documented, and communicated to employees affected. [12]

Mechanical Integrity (MI) Program – A management system for ensuring the ongoing durability and functionality of equipment and/or facilities. [13]

Near Miss – A precursor event of negligible consequence, which may precede a deviation or process-related incident that has significant, undesirable consequences. If near-misses are identified, investigated, and the reasons for occurring are resolved, communicated, and documented, then major failures can be reduced. [14]

Onion Skin Diagram – A graphical illustration of a Facility which shows inherent process hazards at the center of the diagram, followed by successive layers of protection that are normally in place to stop the progression of a PRI. Note: The successive protection layers are intentionally ordered ("layered") in the sequence upon which a demand would be placed on that specific layer of protection.

Operational Discipline (OD) – The performance of all tasks correctly every time; OD is the execution of the COO system by individuals within the organization. [12]

OSHA – United States Occupational Safety and Health Administration

Preventive Maintenance – Maintenance that seeks to reduce the frequency and severity of unplanned shutdowns by establishing a fixed schedule of routine inspection and repairs. [12]

Process-Related Incident (PRI) – The materialization of a process hazard into an adverse event such as fire, explosion, undesirable chemical reaction, personnel exposure to hazardous materials, unnecessary complexity in the operation, and the unintended transfer of hazardous material from one operating area to another. [9]

PSM Standard – United States Standard OSHA 29CFR1910.119 "*Process Safety Management of Highly Hazardous Chemicals*"

Reliability – The probability that, when operating under stated environment conditions, process equipment will perform its intended function adequately for a specified exposure period. [12]

Reliability Centered Maintenance (RCM) – A systematic analysis approach for evaluating equipment failure impacts on system performance and determining specific strategies for managing the identified equipment failures. The failure management strategies may include preventive maintenance, predictive maintenance, inspections, testing, and/or one-time changes (e.g., design improvements, operational changes). [12]

Risk Based Inspection (RBI) – A risk assessment and management process that is focused on loss of containment of pressurized equipment in processing facilities, due to material deterioration. These risks are managed primarily through equipment inspection. [12]

Run-to-Failure Maintenance (RTFM) – A maintenance strategy whereby facility management proactively commits to equipment failures by ensuring that: (1) spare parts are obtained, (2) OEM manuals and/or repair procedures are available, and (3) applicable training is conducted to perform the repair. [13]

Safety Instrumented Function (SIF) – A system composed of sensors, logic solvers, and final control elements for the purpose of taking the process to a safe state when a process deviation occurs at a predetermined condition or state.

Safety Integrity Level (SIL) – Criterion defining the acceptable probability of failure on demand for a safety instrumented function. [13]

Safety Instrumented System (SIS) – One or more safety instrumented functions combined to protect a process or a key process component. [13]

Tuning PID Controllers on Integrating Process with Complex Dynamics

James Beall
Principal Process Control Consultant
Emerson Process Management

16 AIChE
Spring Meeting, Houston, TX

EMERSON.
Process Management

Agenda

- Background
- Tuning Method
  - Identify the process type as Integrating
    - Integrator Plus Lag
    - Integrator Plus Lead
  - Measure Process Dynamics
  - Choose "Speed of Response", Lambda
  - Calculate Tuning
- Case Studies
  - Integrator Plus Lag
  - Integrator Plus Lead
- Questions

16 AIChE
Spring Meeting, Houston, TX

EMERSON.
Process Management

"PID = Proportional, Integral, Derivative"

- **P = proportional** - output contribution is "proportional" to error (Set Point – PV)
- **I = Integral** - output contribution is from the "integration" of the error
- **D = Derivative** - output contribution is based on the rate of change of the error. Actually, often based on the rate of change of the PV.

16AIChE Spring Meeting, Houston, TX

EMERSON. Process Management

Caution!!! – Units for Tuning Constants

- Controllers have different units for their PID tuning constants. For this presentation:
  - **Proportional** action will be expressed in **Gain**
  - **Integral** action will be expressed as **Reset Time**, **Tr,** in **seconds**
  - **Derivative** expressed as **Rate** in units of **seconds (not used in this presentation)**

16AIChE Spring Meeting, Houston, TX

EMERSON. Process Management

Caution!!! – PID "Form"

- Controllers can have a different "**Form**" of the PID algorithm. In some cases this can significantly impact the response of the same tuning parameters
  - Forms include
    - Standard (aka Ideal, Non-interactive)
    - Series (aka Classical, interactive)
    - Parallel
  - The tuning calculations in this presentation are based on the "Series" Form. Conversion to "Standard" Form tuning is provided in Appendix A

16AIChE
Spring Meeting, Houston, TX

EMERSON.
Process Management

Tuning Procedure

- Step 1: Identify the basic process type
  - Self-regulating
  - Integrating
    - Integrator plus lag
    - Integrator plus lead
- Step 2: Measure process dynamics
  - Must use %Process Variable (%PV) and %OUT
  - %ΔPV = ΔPV_Eng_Units*(100%/Span_PV_Eng_Units)
  - %ΔOUT= ΔOUT_Eng_Units*(100%/Span_OUT_Units)
- Step 3: Choose desired closed loop response time, "Lambda"
- Step 4: Calculate tuning constants

16AIChE
Spring Meeting, Houston, TX

EMERSON.
Process Management

Step 1 – Identify Self Regulating or Integrating

Self Regulating Process

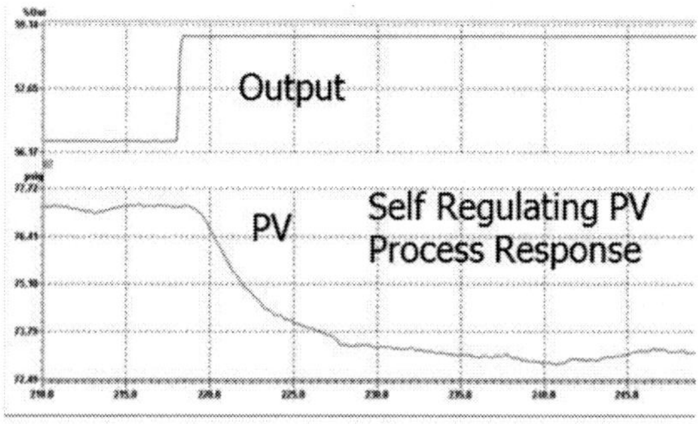

Integrating Process

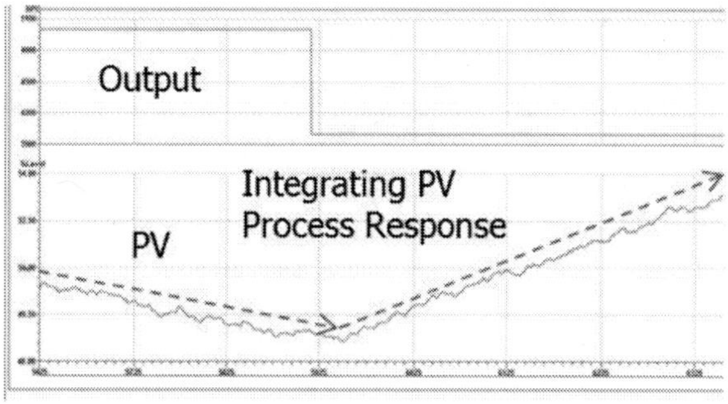

16AIChE
Spring Meeting, Houston, TX

EMERSON.
Process Management

Step 1 – Integrating, Complex

Integrator

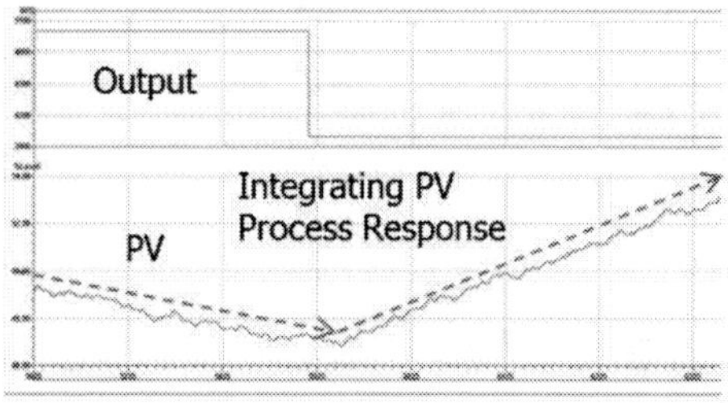

Integrator + Lag

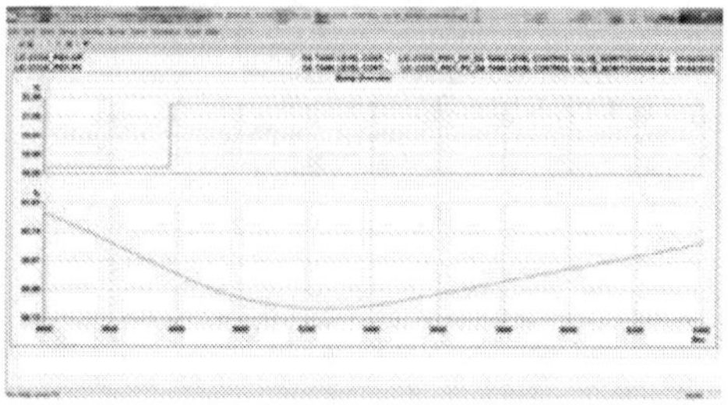

Integrator + Lead

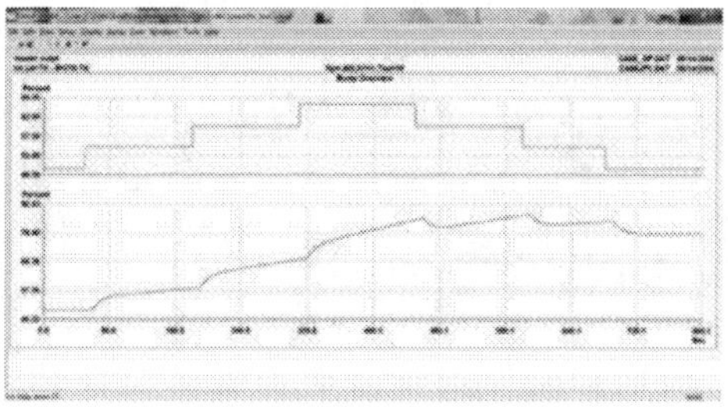

16AIChE
Spring Meeting, Houston, TX

EMERSON.
Process Management

Integrating Process - Open Loop Response

- Most levels
- Some temperatures
- Some Pressures (e.g. large pressure drop across valve)

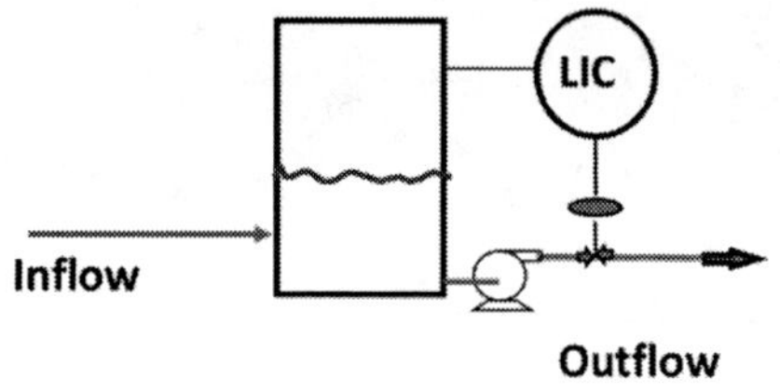

16AIChE Spring Meeting, Houston, TX

EMERSON. Process Management

Step 2 - Measure the Process Dynamics

- There is no "final value of the PV" so you can't measure $\frac{\Delta\%\mathrm{PV}}{\Delta\%\mathrm{OUT}}$ for Process Gain!
- So, you use the "change in slope of %PV"
- Integrating Process Dynamics are:
  - Process Gain, Kp = $\frac{\%\mathrm{PV\ Slope\ Final} - \%\mathrm{PV\ Slope\ Initial}}{\Delta\%\mathrm{OUT}}$
  - Dead Time, Td in seconds

16AIChE Spring Meeting, Houston, TX

EMERSON. Process Management

Step 2a - Measure the Process Dynamics-Kp

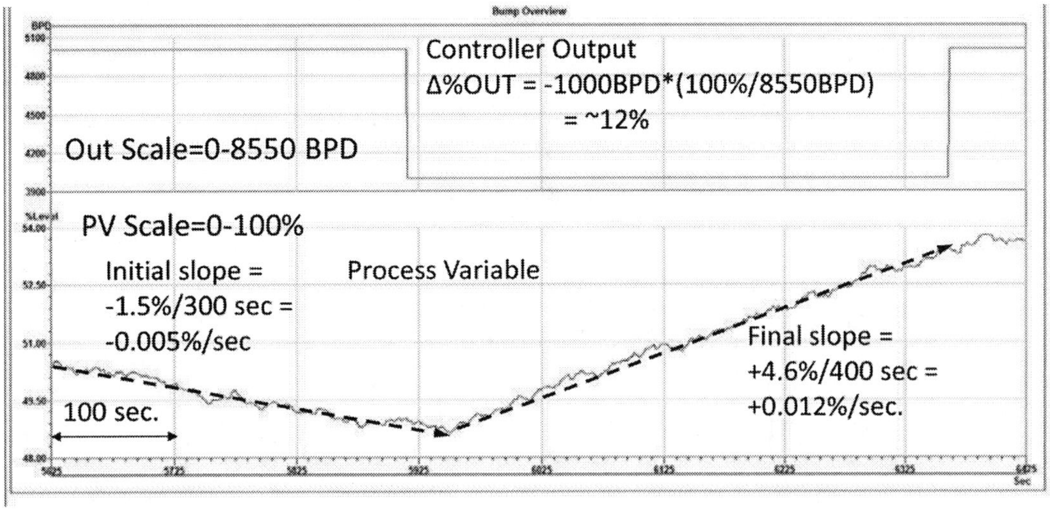

Kp = (final slope - initial slope)/ Δ%OUT

$$Kp = \frac{0.012\%PV/sec - (-0.005\%PV/sec)}{-12\%OUT} = -0.0014\%PV/sec/\%OUT$$

16AIChE
Spring Meeting, Houston, TX

EMERSON.
Process Management

Step 2b - Measure the Process Dynamics-Td

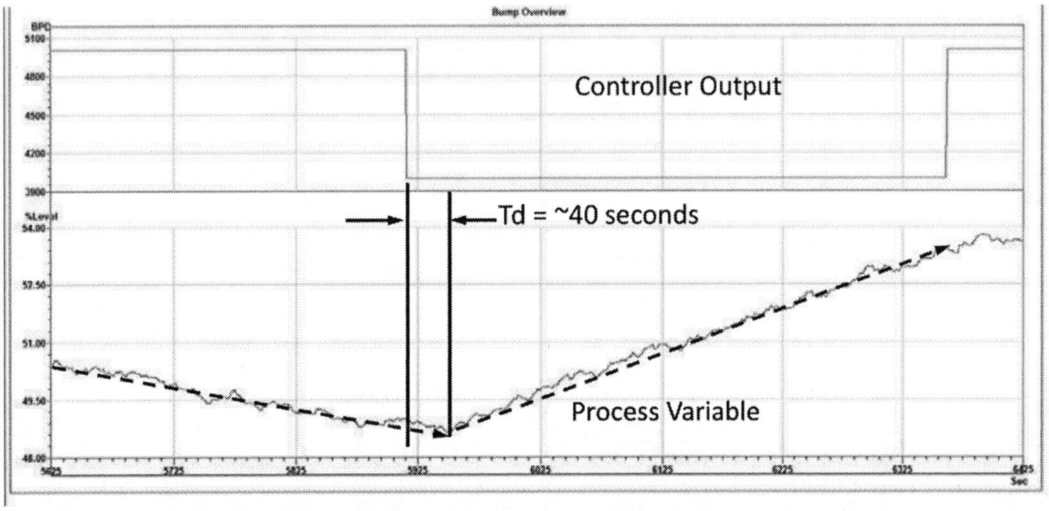

Process Model for this Integrating Process is:

- Process Gain = Kp = -0.0014%PV/sec/%OUT
- Dead Time = Td = 40 seconds

16AIChE
Spring Meeting, Houston, TX

EMERSON.
Process Management

Step 2c - Measure the Lag

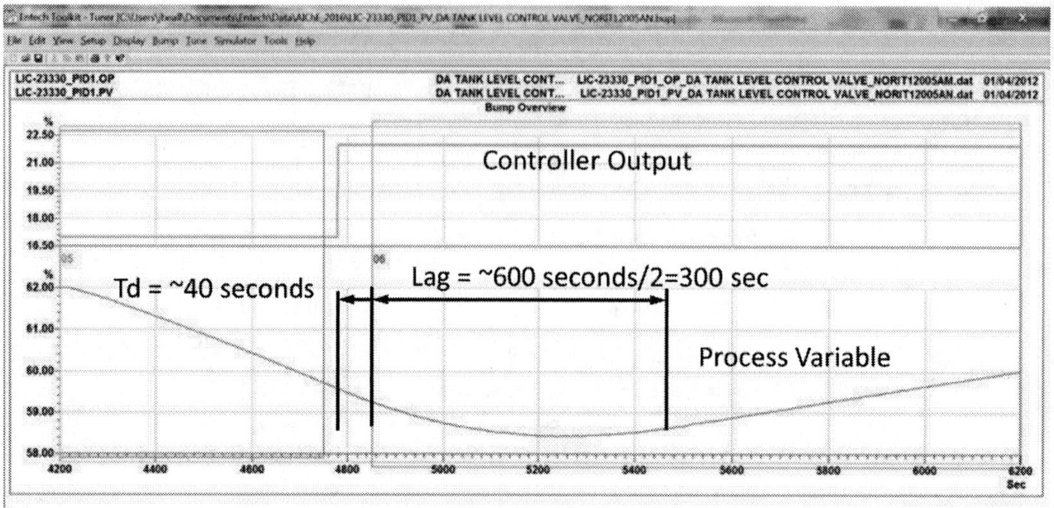

Process Model for this Integrating Process is:

- Process Gain = Kp = -0.0014%PV/sec/%OUT
- Dead Time = Td = 40 seconds
- Lag = 289 seconds

16AIChE Spring Meeting, Houston, TX

EMERSON. Process Management

Step 2c - Measure the Lead

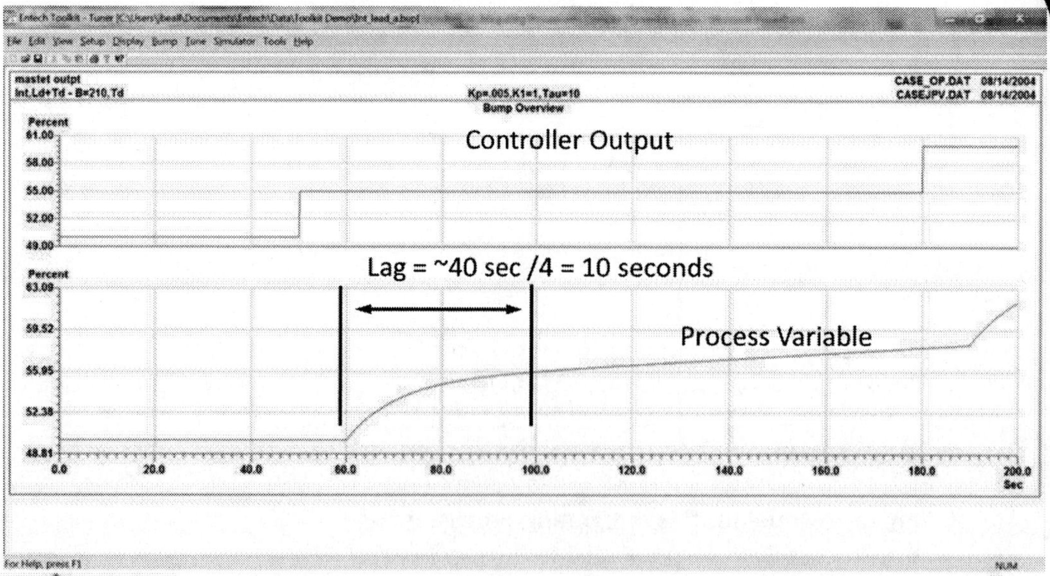

16AIChE Spring Meeting, Houston, TX

EMERSON. Process Management

Step 3 – Choose the Lambda

Step Set Point Change Response

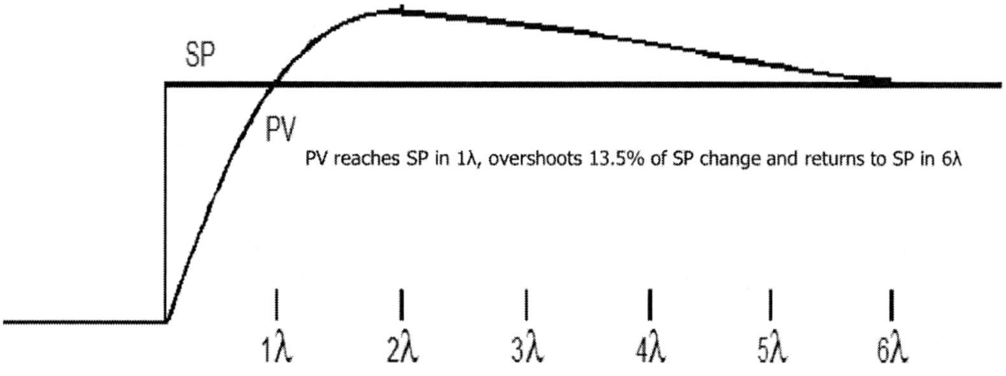

16AIChE
Spring Meeting, Houston, TX

EMERSON.
Process Management

Step 3 – Choose the Lambda

Step Load Change Response

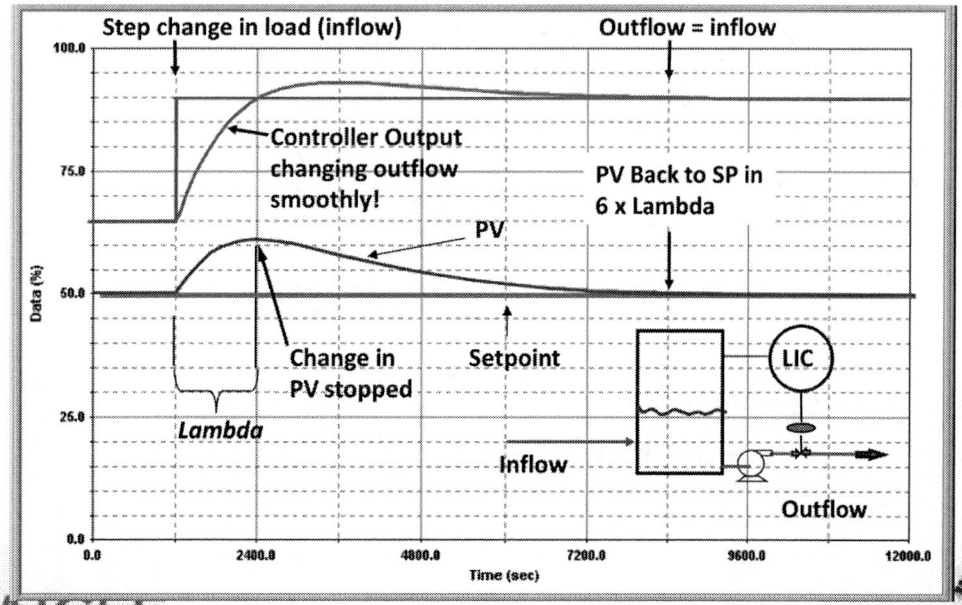

16AIChE
Spring Meeting, Houston, TX

EMERSON.
Process Management

Step 4a – Integrator: Calculate the Tuning Constants

- For an Integrator plus dead time process:
  - Reset Time, Tr = $2\lambda + Td$
  - Controller Gain, Kc = $\frac{\mathrm{Tr}}{|\mathrm{Kp}|\,(\lambda+\mathrm{Td})^2}$
  - Rate = 0 seconds
- Note, Reset Time is not constant. It varies with Lambda as does the Controller Gain

16AIChE
Spring Meeting, Houston, TX

EMERSON.
Process Management

Step 4a – Integrator: Calculate the Tuning Constants

- Using Lambda = 600 seconds (more info later!)
- From example, Kp=-0.0014/sec., Td=40 sec
- Reset Time, Tr = $2\lambda + Td$ = 2(600)+40= 1240 sec
- Controller Gain, Kc = $\frac{\mathrm{Tr}}{|\mathrm{Kp}|\,(\lambda+\mathrm{Td})^2}$

$$= \frac{1240 \text{ seconds}}{0.0014/\text{sec}^*(600 + 40 \text{ sec})^2}$$

$$= 2.2$$

- Rate = 0 seconds
- Be sure to use the same time units for Kp, λ & Td

16AIChE
Spring Meeting, Houston, TX

EMERSON.
Process Management

Step 4 a– Integrator: Simulated SP Response

λ = 600 sec

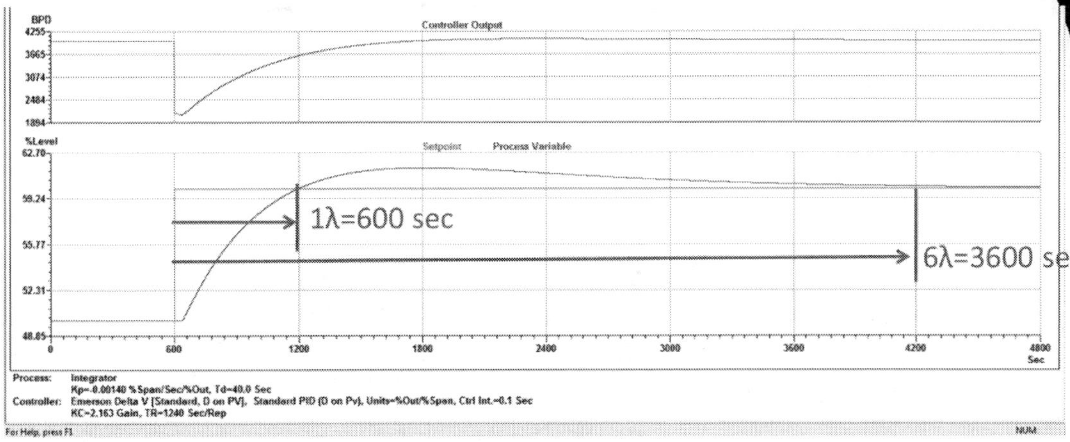

λ=600 sec., Kc = 2.2, Tr=1240 sec, Rate=0 sec

16AIChE
Spring Meeting, Houston, TX

EMERSON.
Process Management

Step 4a – Integrator: Simulated Load Response

λ = 600 sec

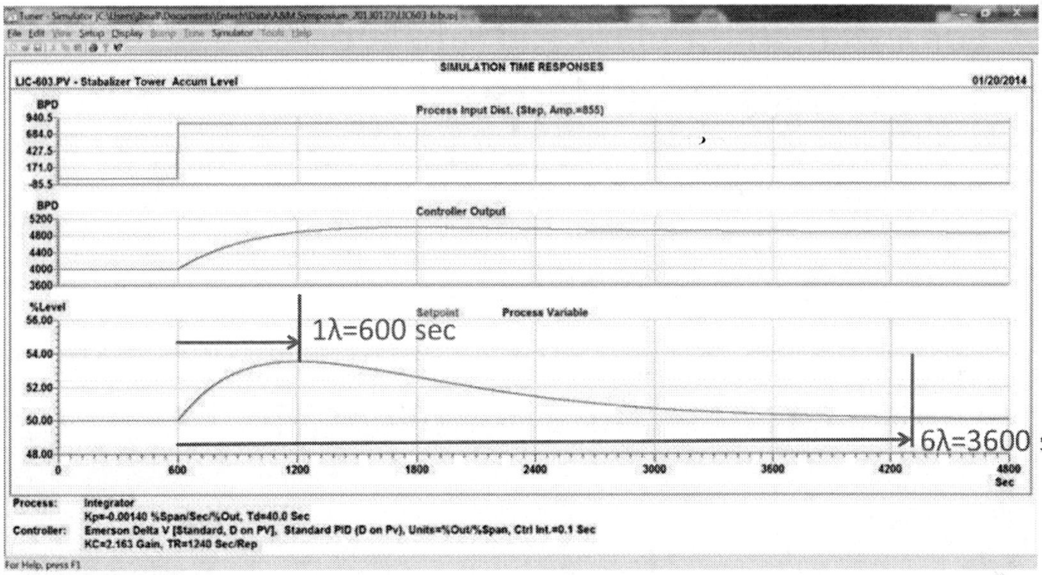

λ=600 sec., Kc = 2.2, Tr=1240 sec, Rate=0 sec

16AIChE
Spring Meeting, Houston, TX

EMERSON.
Process Management

Lambda for Integrating Processes

For the previous example of Kp = -0.0014%PV/sec/%OUT and Td = 40 seconds:

Lambda	Kc	Tr
600 sec	2.2	1240 sec
300 sec	4	640 sec
200 sec	5.5	440 sec
120 sec	7.8	280 sec

Note that for a lower Controller Gain, Kc, a "slower" or longer reset time is calculated. "Tune by feel" often results in too "fast" (short) reset time which will cause oscillation of an integrating loop.

16 AIChE
Spring Meeting, Houston, TX

EMERSON.
Process Management

Step 4b– Integrator + Lag: Calculate the Tuning Constants

- For an Integrator plus dead time process:
  - Reset Time, Tr = $2\lambda + Td$
  - Controller Gain, Kc = $\dfrac{\mathrm{Tr}}{|\mathrm{Kp}|\,(\lambda+\mathrm{Td})^2}$
  - Rate = Lag (seconds)
- Note, Reset Time is not constant. It varies with Lambda as does the Controller Gain

16 AIChE
Spring Meeting, Houston, TX

EMERSON.
Process Management

Step 4b – Integrator: Calculate the Tuning Constants

- Using Lambda = 600 seconds (more info later!)
- From example, Kp=-0.0014/sec., Td=40 sec., Lag = 289 sec.
- Reset Time, Tr = $2\lambda + Td$ = 2(600)+40= 1240 sec
- Controller Gain, Kc = $\dfrac{\text{Tr}}{|\text{Kp}|\ (\lambda+\text{Td})^2}$

$$= \frac{1240 \text{ seconds}}{0.0014/\text{sec}*(600 + 40 \text{ sec})^2}$$

$$= 2.2$$

- Rate = 289 seconds
- Be sure to use the same time units for Kp, λ & Td

16AIChE
Spring Meeting, Houston, TX

EMERSON.
Process Management

Step 4 a– Integrator + Lag: Simulated SP Response λ = 600 sec

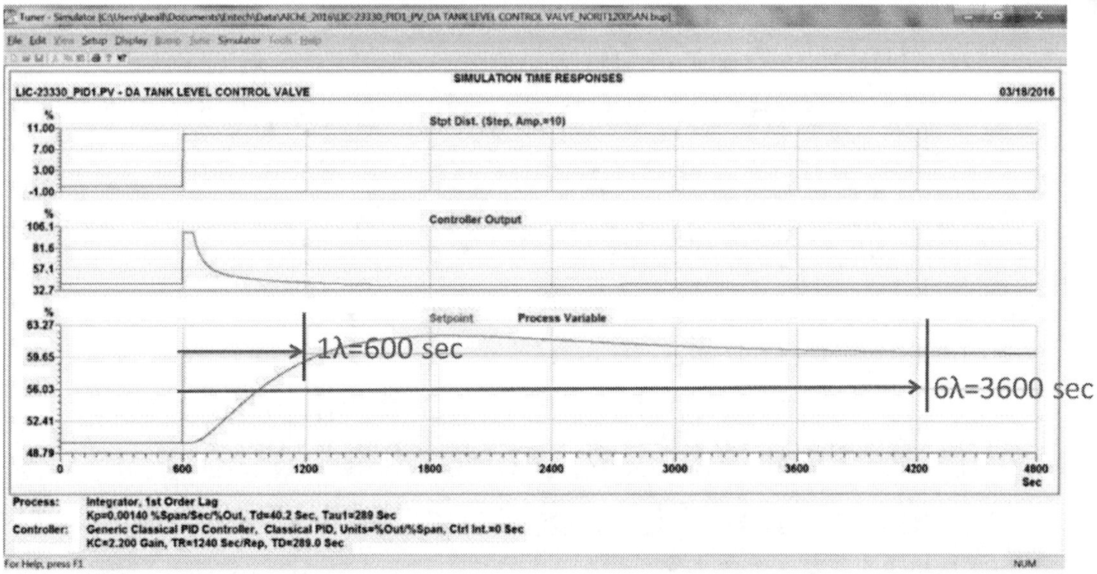

λ=600 sec., Kc = 2.2, Tr=1240 sec, Rate=289 sec

16AIChE
Spring Meeting, Houston, TX

EMERSON.
Process Management

Step 4a – Integrator + Lag: Simulated Load Response
λ = 600 sec

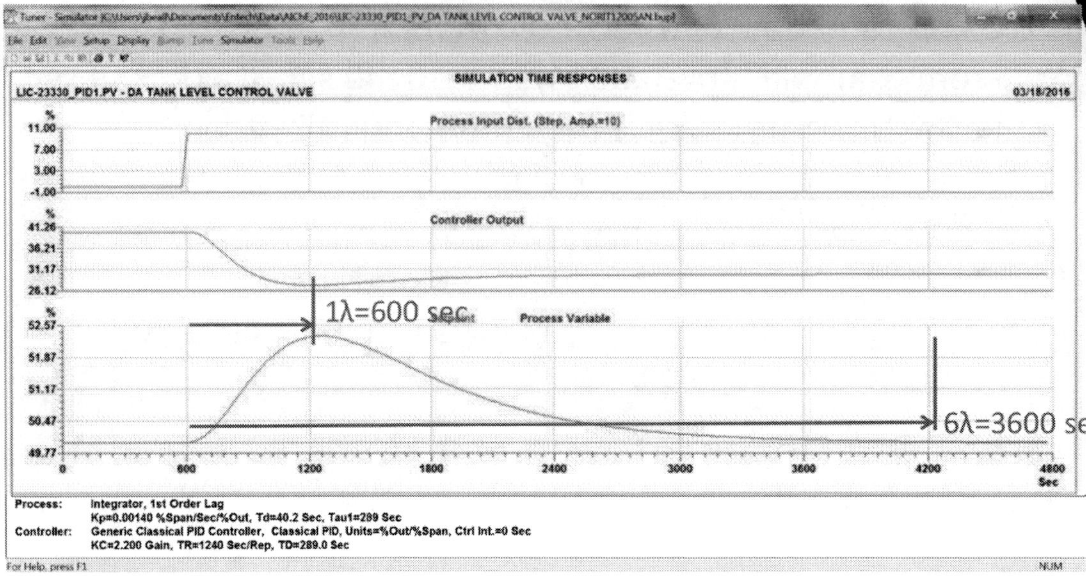

λ=600 sec., Kc = 2.2, Tr=1240 sec, Rate=289 sec

16AIChE
Spring Meeting, Houston, TX

EMERSON.
Process Management

Choosing Lambda

- For "tight" load regulation, choose a smaller Lambda – use caution if Lambda < 3*Dead time. Watch out for excessively high controller gain (though it might be appropriate)
- For cases you want the PV to absorb variability and reduce the variability of the manipulated variable (controller output) – choose a "slow" Lambda for "averaging" control

16AIChE
Spring Meeting, Houston, TX

EMERSON.
Process Management

Choosing Lambda

- Required Lambda $\sim \frac{2*APD}{Kp*MLD}$ where
  - APD = Allowable %PV Deviation from SP
  - MLD = Maximum Load Disturbance in units of % of controller output span
- Can be used for "tight" or "loose" control

16AIChE Spring Meeting, Houston, TX

EMERSON. Process Management

Case 1 – "Tight" Level Control

- Need tight level control
- Level was held close to SP but product outflow was varying dramatically
- Non-uniform distribution of additive in polymer product threating produc quality/sales
- The level controller had been re-tuned but it did not seem to help

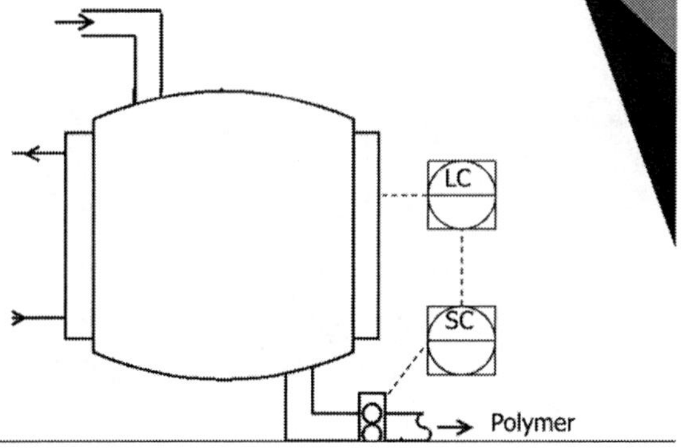

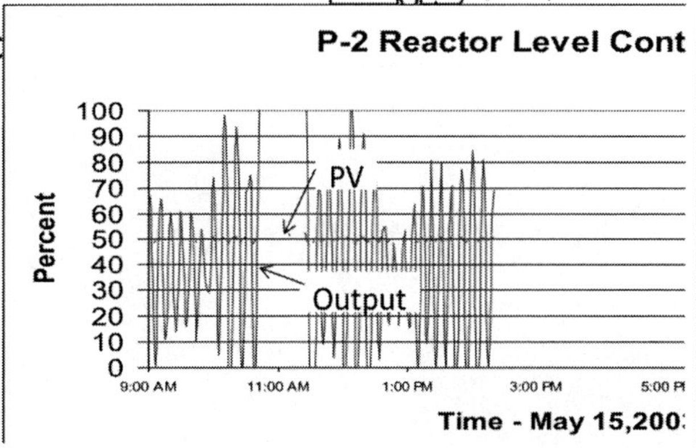

16AIChE Spring Meeting, Houston, TX

Case 1 – "Tight" Level Control

- Measure process dynamics
- ADP = 5%
- MLD = 25%
- Lambda = ~$\frac{2*APD}{Kp*MLD}$

=2*5%PV
0.00029%PV/sec%out * 25%
= ~1400 seconds

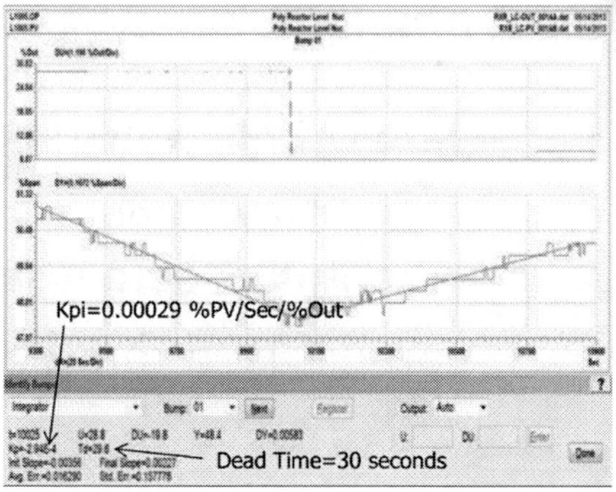

16AIChE
Spring Meeting, Houston, TX

EMERSON.
Process Management

Case 1 – "Tight" Level Control

Calculate Tuning for Lambda = 1400 seconds

- Reset Time, Tr = $2\lambda + Td$ = 2(1400)+30= **2830 seconds**
- Controller Gain, Kc = $\frac{\text{Tr}}{|\text{Kp}|\ (\lambda+\text{Td})^2}$

$$= \frac{2830 \text{ seconds}}{0.00029/\text{sec}*(1400 + 30 \text{ sec})^2}$$

= **4.8**

- Rate = **0 seconds**

16AIChE
Spring Meeting, Houston, TX

EMERSON.
Process Management

Case 1 – "Tight" Level Control

Test with simulation, ADP=5%, MLD=25%

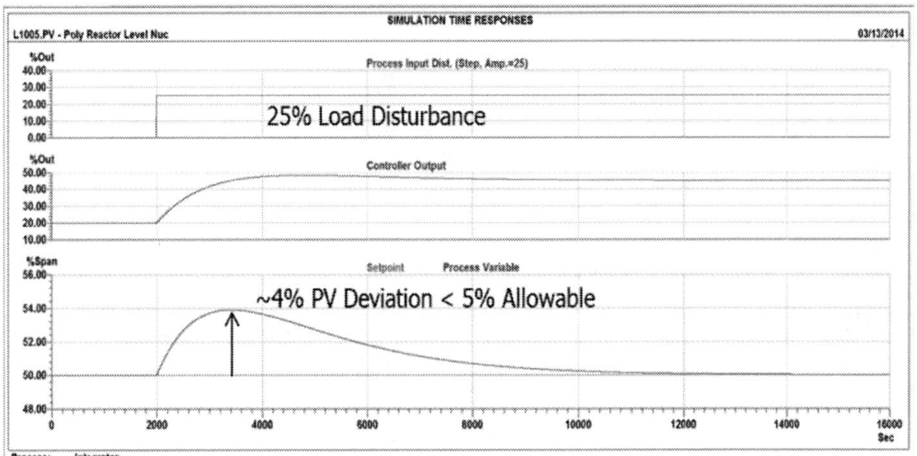

Process: Integrator
Kp=-0.000290 %Span/Sec/%Out, Td=29.6 Sec
Controller: Generic Standard PID Controller, Standard PID, Units=%Out/%Span, Ctrl Int.=1 Sec
KC=4.800 Gain, TR=2930 Sec/Rep

16AIChE
Spring Meeting, Houston, TX

EMERSON.
Process Management

Case 1 – "Tight" Level Control

Results on actual project

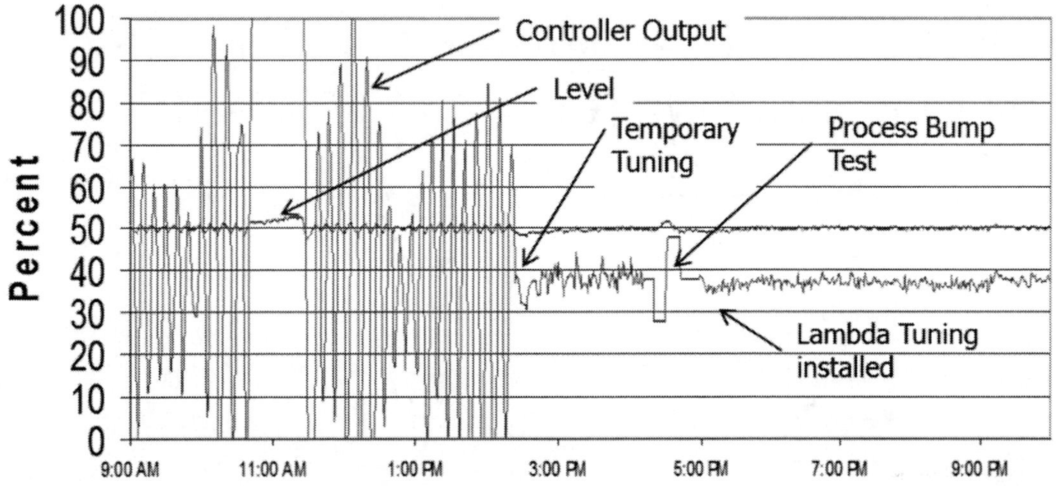

16AIChE
Spring Meeting, Houston, TX

EMERSON.
Process Management

Attenuate Variability with Control/Equipment

- "Capacity" in the process can be used to attenuate or absorb variability
- Primary source of process capacity is level control
- To utilize level control as a capacity tune the controller only "fast" enough to hold the PV within the allowable PV deviation (APD) for a maximum load disturbance (MLD)
- Called "Averaging" or "Surge" control

16AIChE
Spring Meeting, Houston, TX

EMERSON.
Process Management

Case 2 – "Averaging" Level Control

- Need to impact of inflow fluctuations
- Need to keep level within 30% of SP if 2 of 4 reactors shutdown (1 reactor = 20% load)

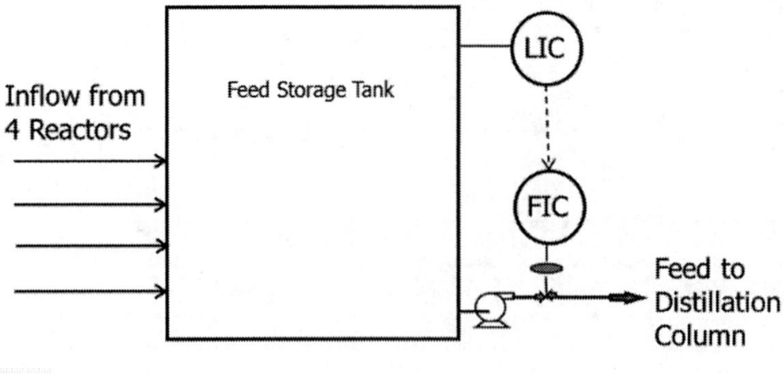

16AIChE
Spring Meeting, Houston, TX

EMERSON.
Process Management

Case 2 – "Averaging" Level Control

- Measure process dynamics
- ADP = 30%
- MLD = 40%
- Lambda = ~$\frac{2*APD}{Kp*MLD}$

  =2*30%PV

  0.000216%PV/sec%out * 40%

  = ~6900 seconds

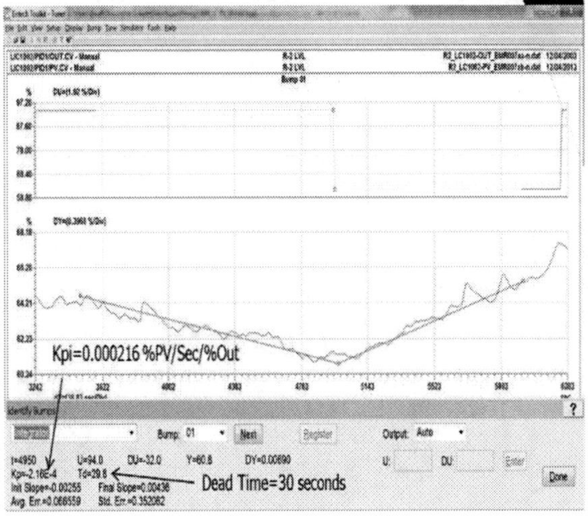

16 AIChE Spring Meeting, Houston, TX

EMERSON. Process Management

Case 2 – "Averaging" Level Control

Calculate Tuning for Lambda = 6900 seconds

- Reset Time, Tr = $2\lambda + Td$ = 2(6900)+30= **13,830 seconds**
- Controller Gain, Kc = $\frac{\text{Tr}}{|\text{Kp}|\ (\lambda+\text{Td})^2}$

  $= \frac{13{,}830 \text{ seconds}}{0.000216/\text{sec}*(6900 + 30 \text{ sec})^2}$

  = 1.33

- Rate = **0 seconds**

16 AIChE Spring Meeting, Houston, TX

EMERSON. Process Management

Case 2 – “Averaging” Level Control

Test with simulation, ADP=30%, MLD=40%

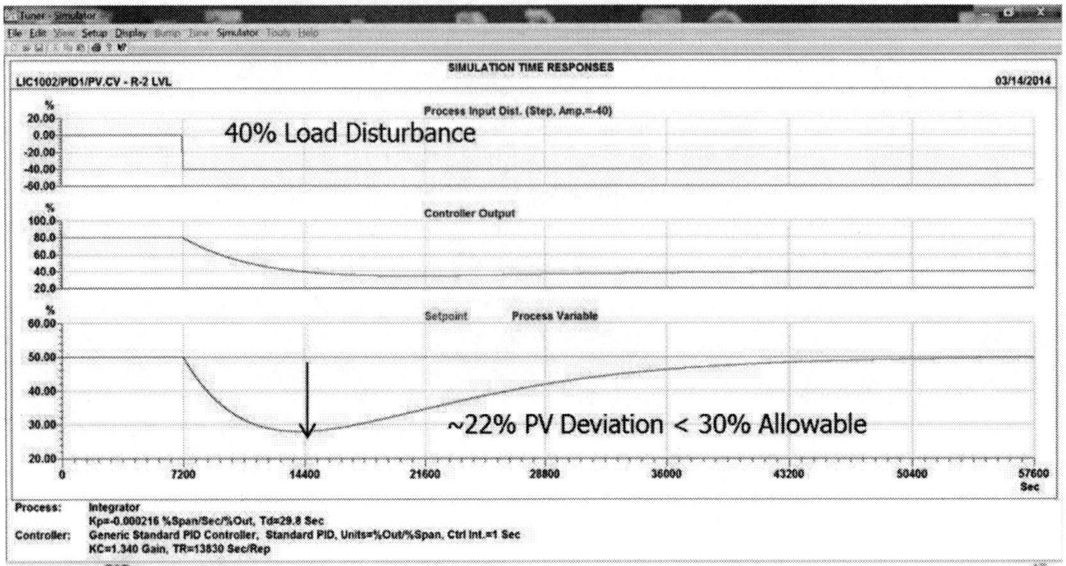

16 AIChE Spring Meeting, Houston, TX

EMERSON. Process Management

Case 2 – “Tight” Level Control

Attenuation of Lambda vs. Ziegler-Nichols Tuning

+-5%, 120 sec. period sinusoidal load disturbance

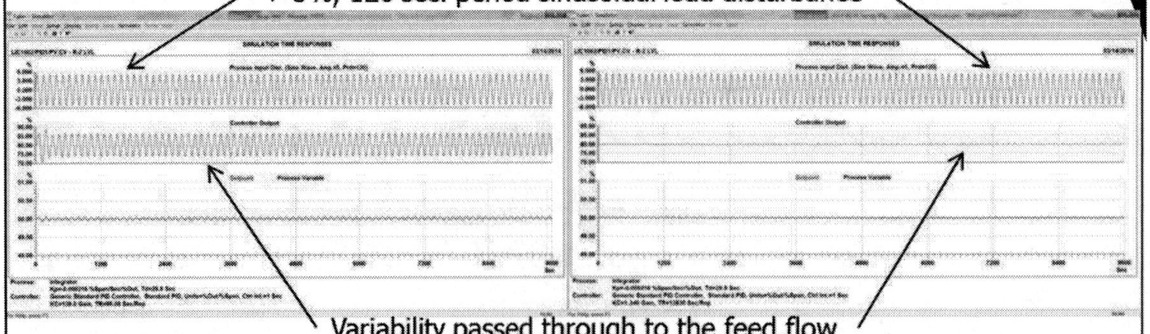

Variability passed through to the feed flow

Zeigler-Nichols Reaction Rate Tuning | Lambda Tuning, Lambda = 6900 seconds

16 AIChE Spring Meeting, Houston, TX

EMERSON. Process Management

Complex Integrators

- These "complex" integrator responses that can be handled with Lambda Tuning Methods (all include dead time)

- Integrator + lag
- Integrator + lead
- Integrator non-minimum phase

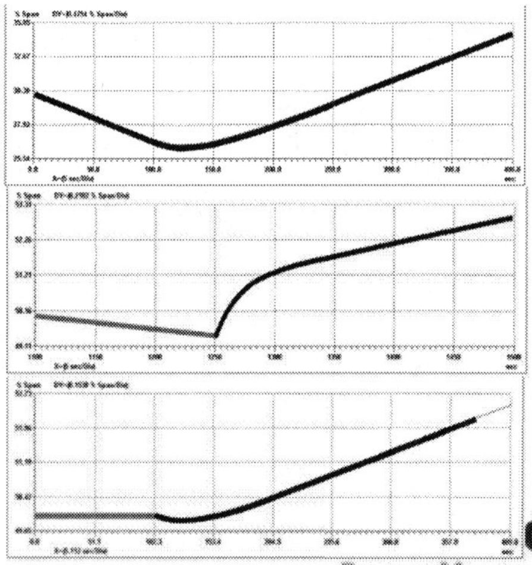

16AIChE
Spring Meeting, Houston, TX

ON
Process Management

Questions?

James.Beall@Emerson.com

903-235-7935

16AIChE
Spring Meeting, Houston, TX

EMERSON
Process Management

Modernizing Control Functionality

Easy Steps to Start Process Optimization

Laurie R. Ben
James F. Beall
Emerson Process Management

16AIChE
Spring Meeting, Houston, TX

EMERSON.
Process Management

Agenda

- Introduction – Taking advantage of New Technology
- Examples – Best Practices & Lessons Learned
  - Split Range Duty
  - Smith Predictor
  - Gain Scheduling/Adaptive Control
  - Signal Characterization
- Summary

16AIChE
Spring Meeting, Houston, TX

EMERSON.
Process Management

Introduction

- Modernizing Control System – Decision Done Check!
- 'Do I' or 'Don't I' take advantage of new technology
  - Implemented older cumbersome techniques?
  - Complex custom strategies that handle automation challenges?
  - Hard to implement & maintain?
- Standard features that are field proven, practical, and easy to implement
  - Easier to implement & maintain
  - More efficient and effective control strategies
  - Achieve improvements in process performance
  - Lays the foundation for optimization

16AIChE
Spring Meeting, Houston, TX

EMERSON.
Process Management

Agenda

- Introduction
- Examples – Best Practices & Lessons Learned
  - Split Range Duty
    - Control Challenge
    - Legacy Solution
    - Newer Technology Solution
    - Summary
  - Smith Predictor
  - Gain Scheduling/Adaptive Control
  - Signal Characterization
- Summary

16AIChE
Spring Meeting, Houston, TX

EMERSON.
Process Management

Split Range Duty – Control Challenge

- Use steam and hot oil as energy source for distillation column for temperature controller.
- Hot oil is “free energy” and should be maximized
- Control Scenarios
  - Maximize hot oil and supplement with steam
  - Base load steam (if bad meter) and supplement with hot oil
  - Base load hot oil flow (if bad meter) and supplement with steam
- Constraints
  - Minimum and maximum hot oil valve position and flow rate
  - Minimum steam flow rate

16AIChE
Spring Meeting, Houston, TX

EMERSON.
Process Management

Split Range Duty – Legacy Solution

– Old legacy system implemented this feature using a velocity PID with incremental output and a “micro expert rules system”

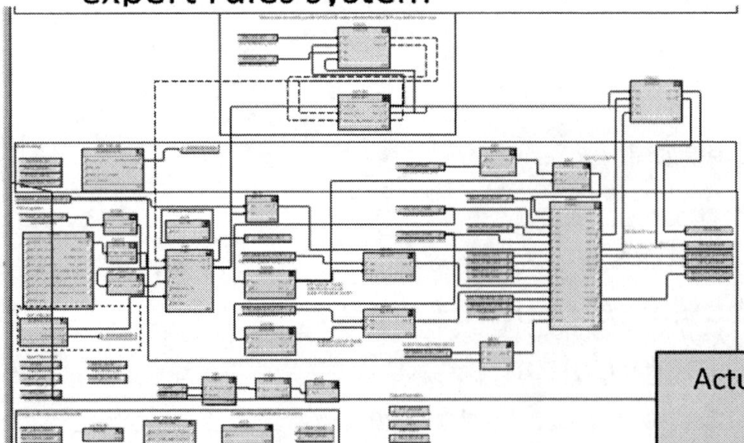

Actually Duplicated in New system, but still…

- Hard to maintain and modify
- Difficult to troubleshoot
- Suboptimal performance
- Cumbersome/Kludge/Ugly

16AIChE
Spring Meeting, Houston, TX

Split Range Duty – New Technology Solution

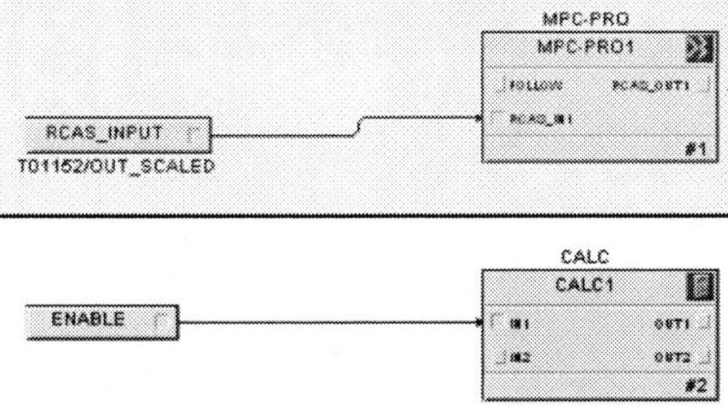

Embedded MPC Function Block provides all control functionality! EASY!!!

Simple calculations to calculate limits

16AIChE Spring Meeting, Houston, TX

EMERSON. Process Management

Split Range Duty – New Technology Solution

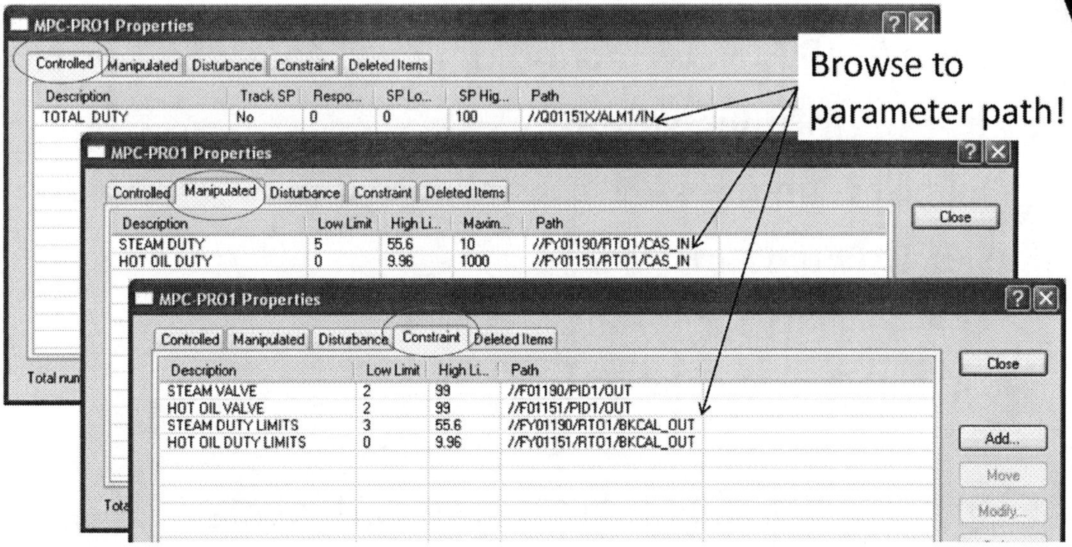

16AIChE Spring Meeting, Houston, TX

EMERSON. Process Management

Split Range Duty – New Technology Solution

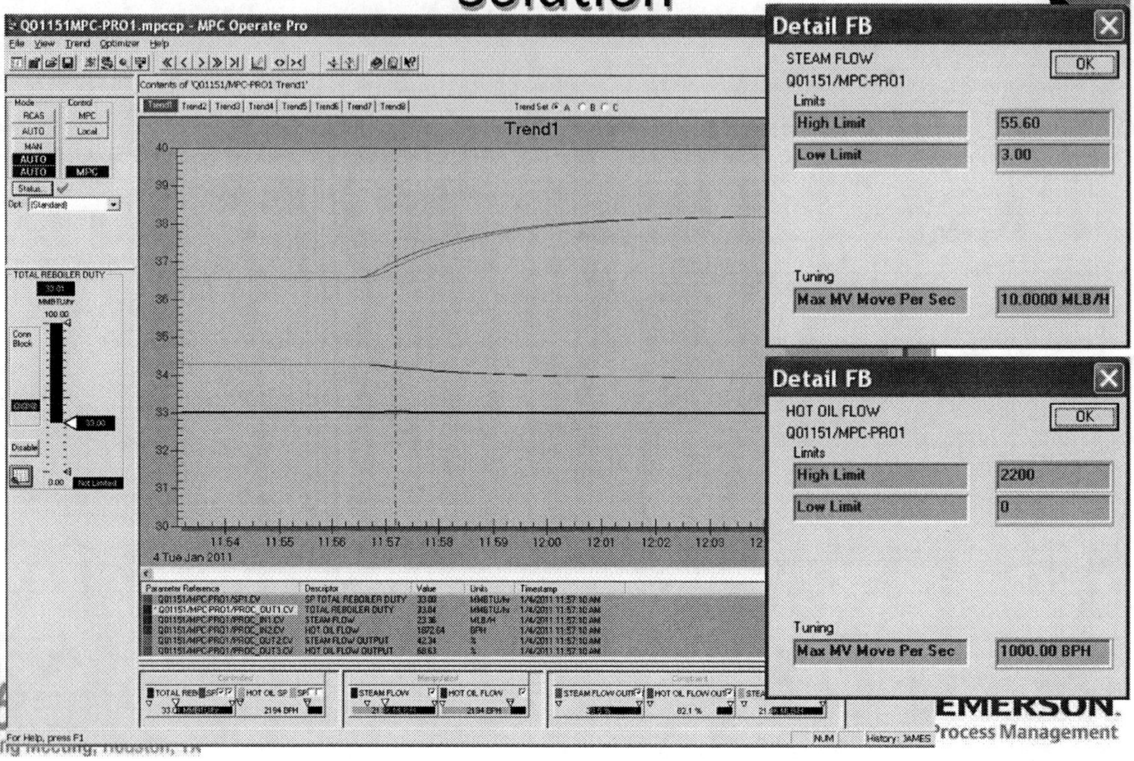

EMERSON. Process Management

Summary Split Range Duty Control

Embedded MPC (New Technology) vs Legacy Solution

- Better performing solution
  - Bumpless transfers and constraint control
- Easier to Maintain
  - Change limits and settings
  - Simpler implementation
  - Easy to understand, train others, document
- Easily expands to handle future constraints
- Reduce energy costs!

16 AIChE Spring Meeting, Houston, TX

EMERSON. Process Management

Agenda

- Introduction
- Examples – Best Practices & Lessons Learned
  - Split Range Duty
  - Smith Predictor
    - Control Challenge
    - Legacy Solution
    - Newer Technology Solution
    - Summary
  - Gain Scheduling/Adaptive Control
  - Signal Characterization
- Summary

16 AIChE
Spring Meeting, Houston, TX

EMERSON.
Process Management

Smith-Predictor PID – Control Challenge

- Control of processes that are "dead time dominant" (dead time > time constant)
- Typically custom configured
  - some legacy control systems have preconfigured options
- Algorithm uses a process model
  - gain, dead time, and time constant
- Challenge: as process dynamics change from model used, performance degrades rapidly

•High maintenance when process dynamics change
•Performance degrades rapidly as model mismatch increases
•Operations view distorted by Pseudo PV

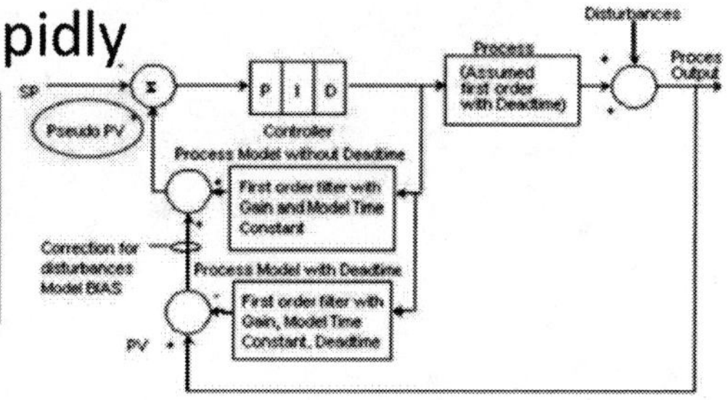

16 AIChE
Spring Meeting, Houston, TX

Smith-Predictor PID

Easy to implement the Smith-Predictor if standard function block exists, but still ...

- High Maintenance when process dynamics change
- Performance degrades rapidly as process dynamics change from configured model

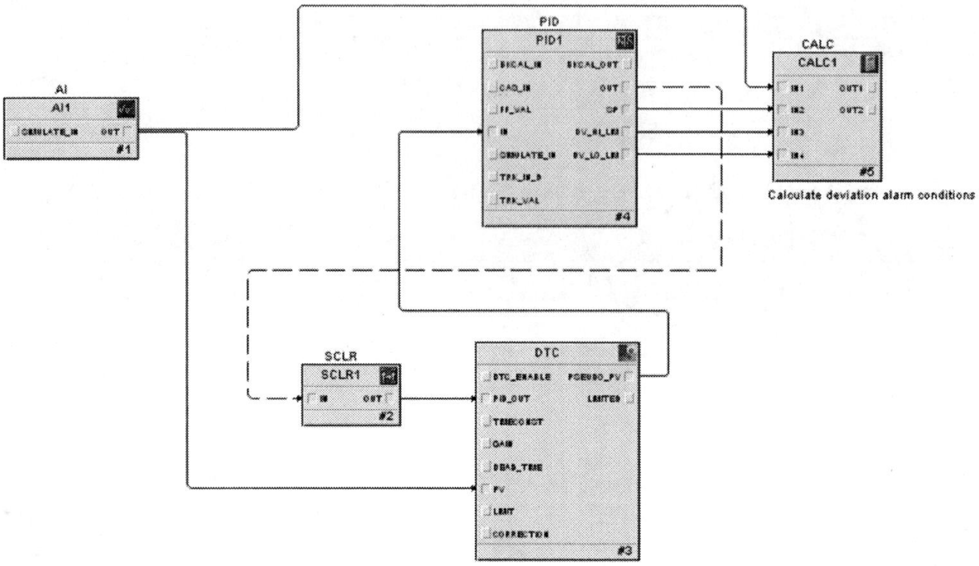

16 Spring Meeting, Houston, TX

EMERSON. Process Management

Smith-Predictor PID – New Technology Solution

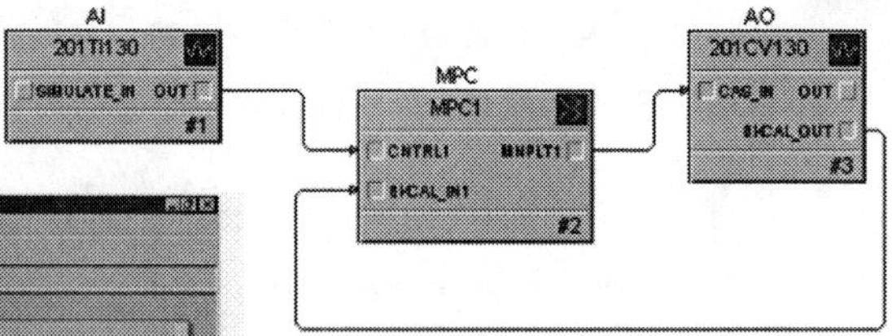

- Embedded MPC Soluti

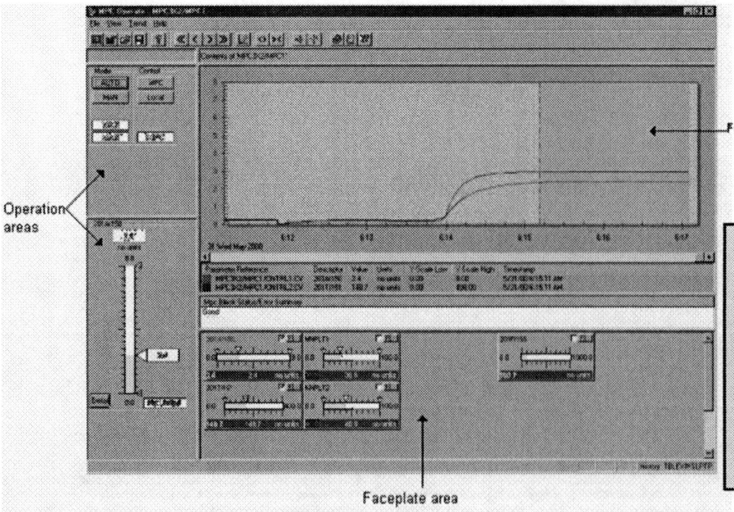

•Provides Future Prediction
•Provides MPC Faceplate – no configuration needed
•More robust under changing process dynamics

16AIChE Spring Meeting, Houston, TX

EMERSON. Process Management

Summary Smith-Predictor PID

Embedded MPC (New Technology) vs Legacy Solution

- More robust solution under a changing process model
- Provides visual display of the "future" PV
- Save money through better control performance

16AIChE Spring Meeting, Houston, TX

EMERSON. Process Management

Agenda

- Introduction
- Examples – Best Practices & Lessons Learned
  - Split Range Duty
  - Smith Predictor
  - Gain Scheduling/Adaptive Control
    - Control Challenge
    - Legacy Solution
    - New Technology Solution
    - Summary
  - Signal Characterization
- Summary

16AIChE Spring Meeting, Houston, TX

EMERSON. Process Management

Gain Scheduling – Control Challenge

- Legacy systems have limited capability to modify tuning based on process conditions
- Some Legacy systems had add-on application software
  - e.g. PROVOX "Gain Scheduler" system provided 3 sets of Gain, Reset, and Rate applied in 3 separate regions of a "state variable"
    - State variable = PV, OUT or "other"
    - Transition between tuning sets interpolated over "dead band"
- Beware: some suspend loop execution when tuning parameters change, so don't change the tuning every scan

16AIChE Spring Meeting, Houston, TX

EMERSON. Process Management

Gain Scheduling – Legacy Solution

- Custom code
- Multiple points, registers, displays
- Restricted (slower) update rate

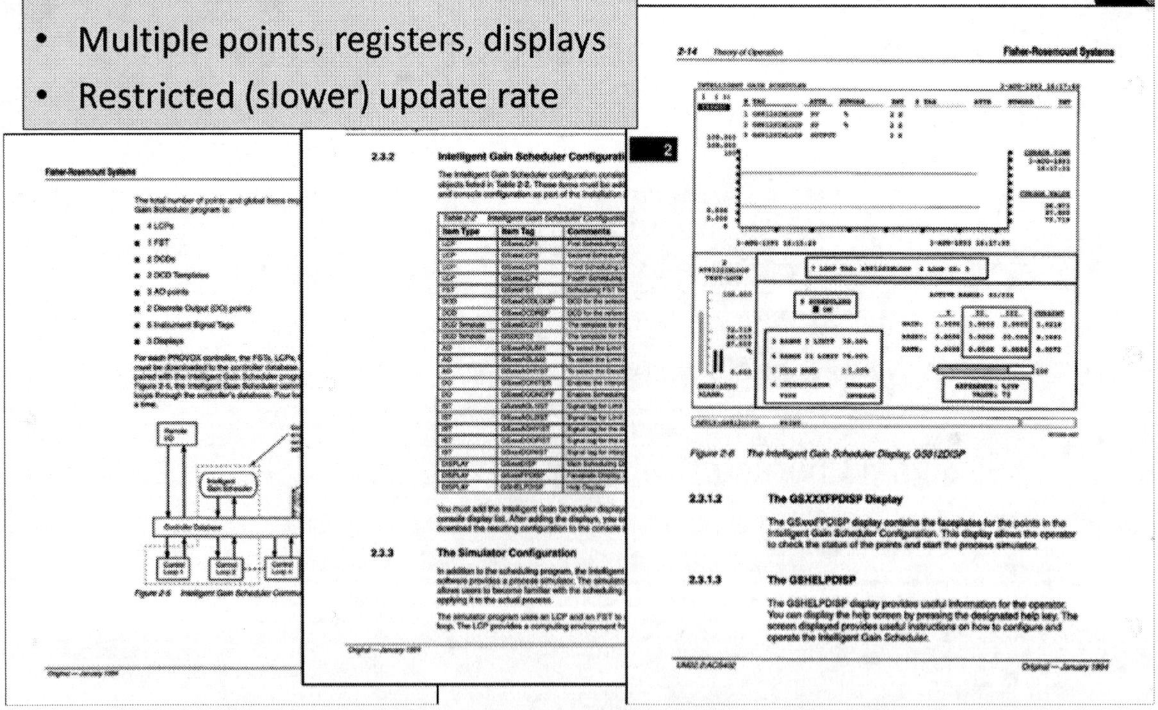

Gain Scheduling – New Technology Solution #1

Standard Modules

- Standard configuration
- Preconfigured faceplate & detailed display
- Easy to implement
- Tuning updates at loop execution

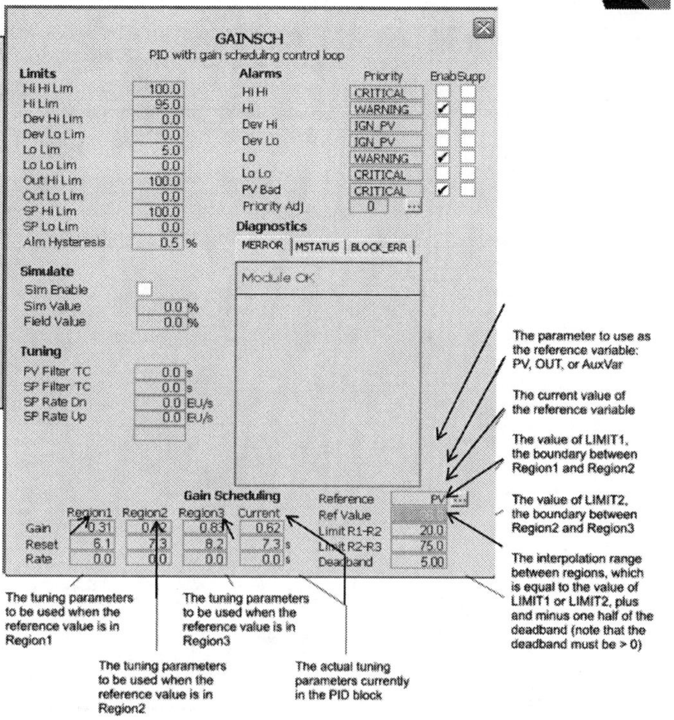

16 AIChE
Spring Meeting, Houston, TX

Gain Scheduling – New Technology Solution #2

New Technology "Adaptive Control"

- "Learns" process models from normal operating moves
- Allows up to 5 tuning regions of the "state variable"
- Partial Adaptive Control
  - Use "approved model" for each region per loop selected tuning rule
- Optional Full Adaptive Control
  - Continuous "Learning" of process model in each region
  - Updates tuning based on 'learned' model while in each region per loop selected tuning rule
  - Set allowable deviation (%) from approved model in each region

16 AIChE
Spring Meeting, Houston, TX

EMERSON.
Process Management

New Technology – Adaptive Control

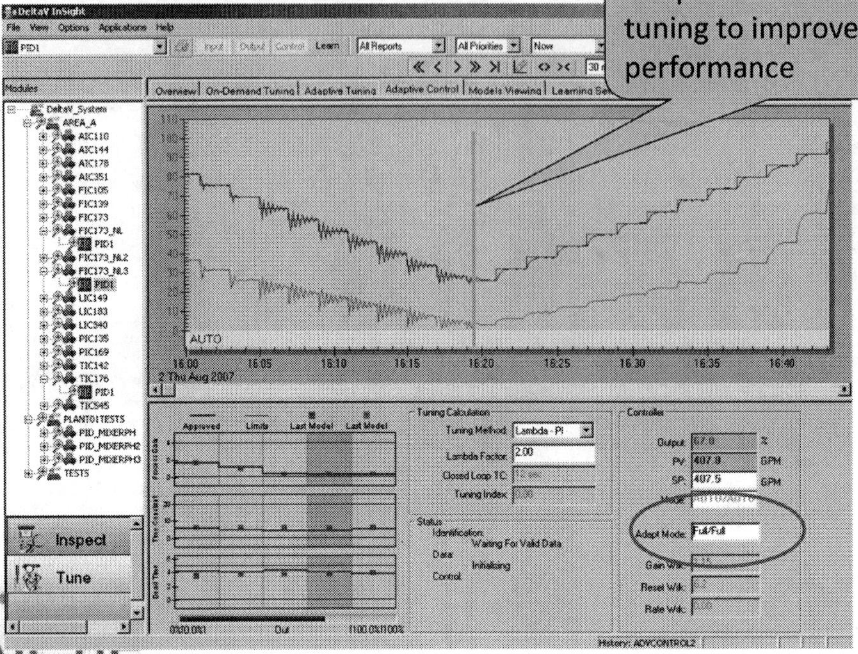

Adaptive control adjusts tuning to improve control performance

16AIChE
Spring Meeting, Houston, TX

EMERSON.
Process Management

Summary Gain Scheduling

New Technology Solution(s) vs. Legacy Solution

- Gain Scheduling easy to implement and understand!
  - Standard configuration
  - Preconfigured faceplate & detailed display
  - Tuning updates at loop execution
- Adaptive Control provides additional benefit and ease of implementation
  - Learns models during normal operation – NO STEP TESTs!
  - More Regions – Up to 5
  - Automatic tuning updates when process model changes
- Improved control performance saves money!

16AIChE
Spring Meeting, Houston, TX

EMERSON.
Process Management

Agenda

- Introduction
- Examples – Best Practices & Lessons Learned
  - Split Range Duty
  - Smith Predictor
  - Gain Scheduling/Adaptive Control
  - Controller Characterization
- Summary

16AIChE
Spring Meeting, Houston, TX

EMERSON.
Process Management

Summary

- Practical application of control features that deal with common challenges
- Highlighted best practices & lessons learned
  - Split Range Duty
  - Smith Predictor
  - Gain Scheduling/Adaptive Control
  - Signal Characterization
- Do look at new functionality available when modernizing
  - Allows jump start to process optimization
  - Leads to improved process performance
  - Easier to implement and maintain over the life investment

16AIChE
Spring Meeting, Houston, TX

EMERSON.
Process Management

Contacts & Questions?

Laurie.Ben@Emerson.com
512-834-7237

James.Beall@Emerson.com
903-235-7935

16 AIChE Spring Meeting, Houston, TX

EMERSON. Process Management

Specifying Control Valve Performance to Achieve Process Performance Goals

James Beall

Principal Process Control Consultant
Emerson Process Management

16AIChE
Spring Meeting, Houston, TX

EMERSON.
Process Management

Introduction

- Importance of Valve Performance
- Standards and References
- Dead Band and Resolution
- Response Time
- Loop Gain
- Comments on Valve Troubleshooting
- Summary
- Speaker Info

16AIChE
Spring Meeting, Houston, TX

EMERSON.
Process Management

Control Loops with Excessive Variability

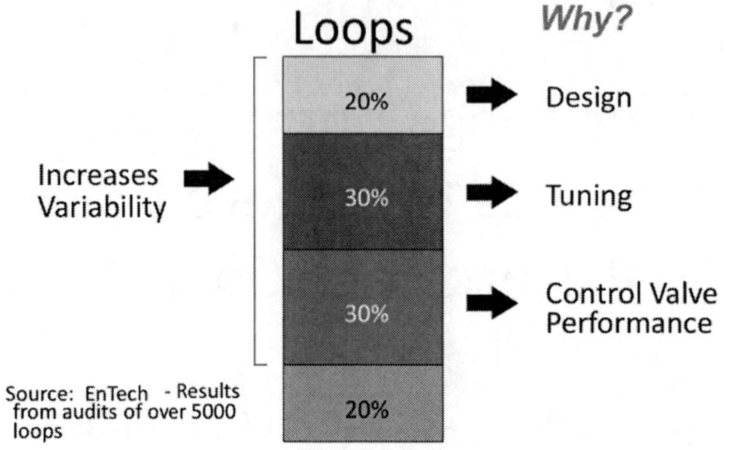

"The undesirable behavior of control valves is the biggest contributor to poor loop performance and the destabilization of product uniformity".

W. L. Bialkowski, President
EnTech Control Engineering

16 AIChE
Spring Meeting, Houston, TX

EMERSON.
Process Management

LNG Plant, Main Gas Feed Flow Loop

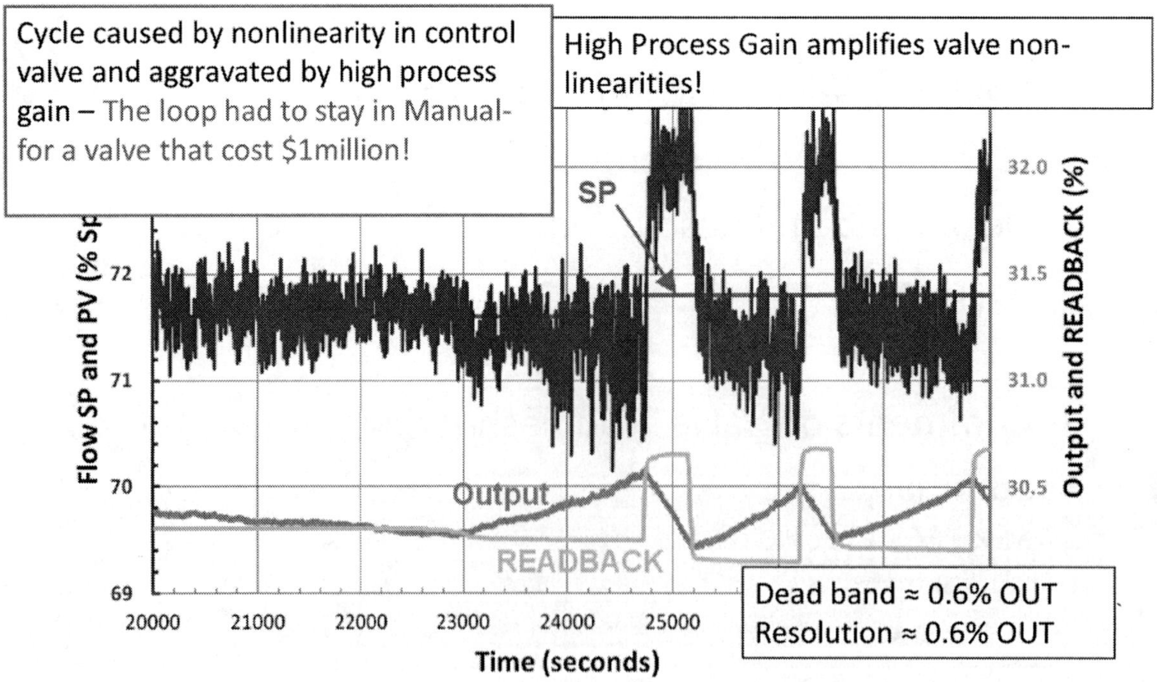

Vitamin plant—Steam Valve

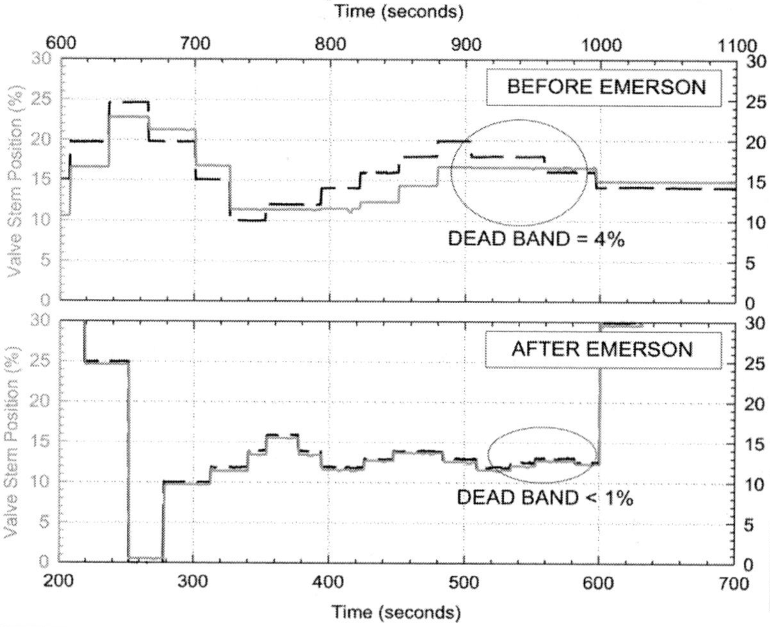

As-found positioner

Digital Valve Controller (DVC) installed same day

Reduced trim installed the following week

Batch time reduced 27% and steam consumption reduced

16AIChE Spring Meeting, Houston, TX

EMERSON. Process Management

Chemical Plant – Steam Valve

Flow cycles in automatic

Flow

PID Output

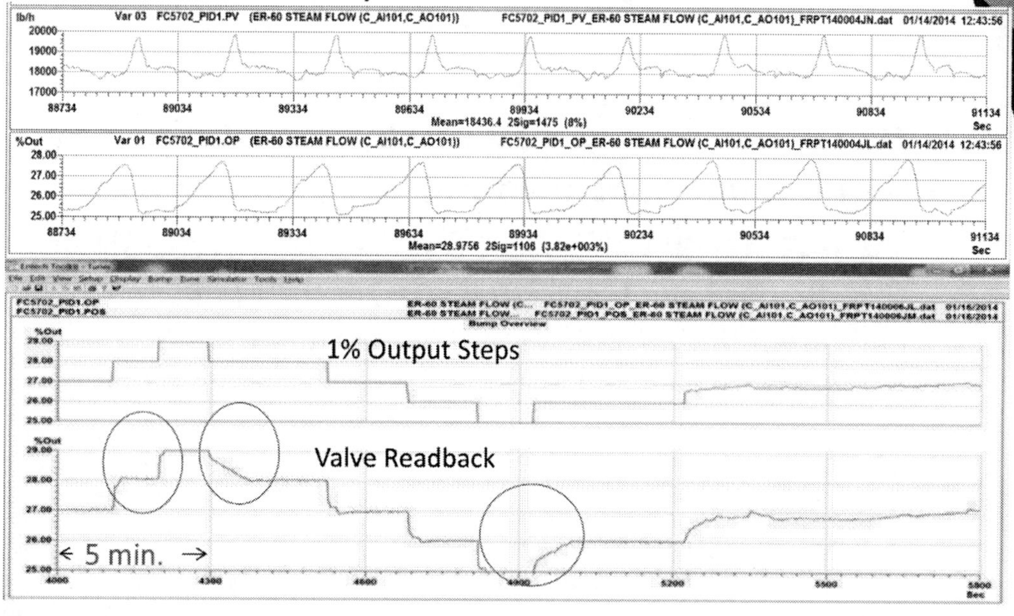

Step test: Valve had a 30 – 90 second Response Time! – Root cause of cycle!

16AIChE Spring Meeting, Houston, TX

EMERSON. Process Management

Control Valve Performance

- ANSI/ISA–S75.25 (Standard) - *Test Procedure for Control Valve Response Measurement from Step Inputs*
  - Defines metrics and test procedure
  - Does NOT provide guidance to specify valve performance requirements
- ANSI/ISA–TR75.25 (Technical Report) - *Control Valve Response Measurement from Step Inputs*
- *EnTech Control Valve Dynamic Specification V3.0*
  - V2.0 update to V3.0 to align with ANSI/ISA–75.25
  - DOES provide guidance to specify valve performance requirements

16AIChE
Spring Meeting, Houston, TX

EMERSON.
Process Management

Definitions from ANSI/ISA-S75.25

- **Dead band: the range through which an input signal may be varied, with reversal of direction, without initiating an observable change in output signal** [ANSI/ISA-S51.1-1979 (R1993)]. In this standard and ISA-TR75.25.02-2000, it is defined in percent of input span. Note that in some other literature this definition is used for dead zone.

16AIChE
Spring Meeting, Houston, TX

EMERSON.
Process Management

Definitions from ANSI/ISA-S75.25

- **Resolution: smallest step increment of input signal in one direction for which movement of the output is observed.** Resolution is expressed as percentage of input span. The term in this document means: the tendency of a control valve to move in finite steps in responding to step changes in input signal applied in the same direction. An example of this is when the valve is driven by a stepper motor or an AC motor or when the pneumatically actuated control valve sticks in place, having stopped moving after the previous step change.

16AIChE Spring Meeting, Houston, TX

EMERSON. Process Management

Resolution

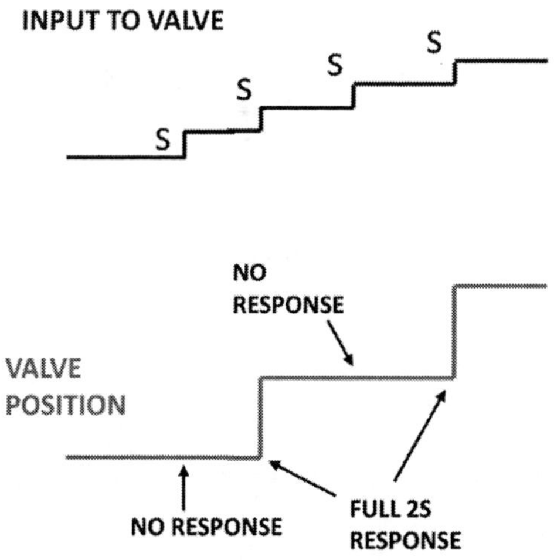

SMALLEST
INPUT CHANGE IN THE SAME
DIRECTION TO
GET A RESPONSE

S<RESOLUTION<2S

16AIChE Spring Meeting, Houston, TX

EMERSON. Process Management

Resolution

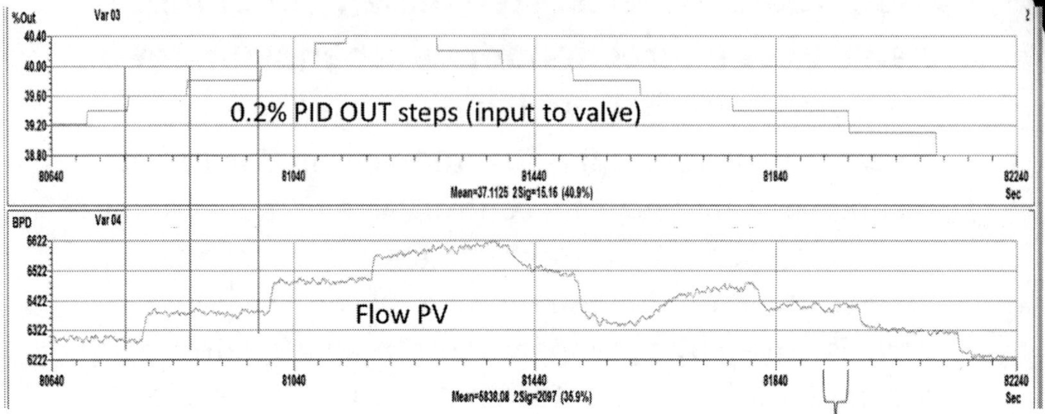

0.2%< Resolution < 0.4%

40 seconds

16AIChE Spring Meeting, Houston, TX

EMERSON. Process Management

Dead Band

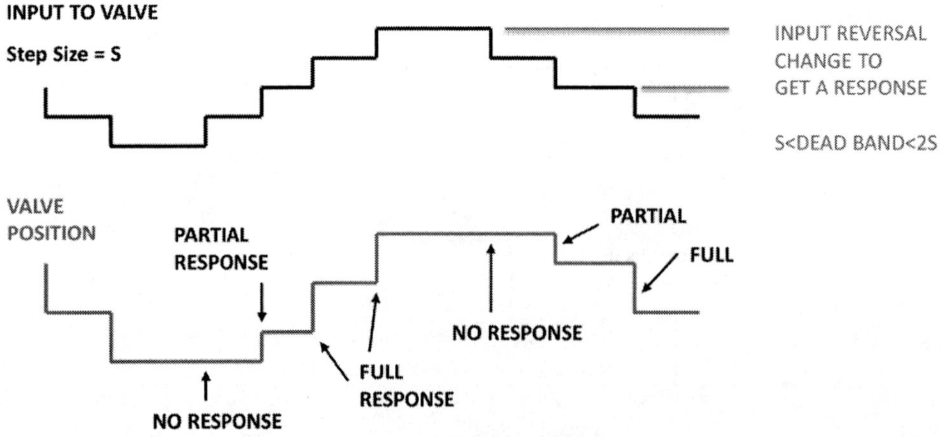

16AIChE Spring Meeting, Houston, TX

EMERSON. Process Management

Dead Band

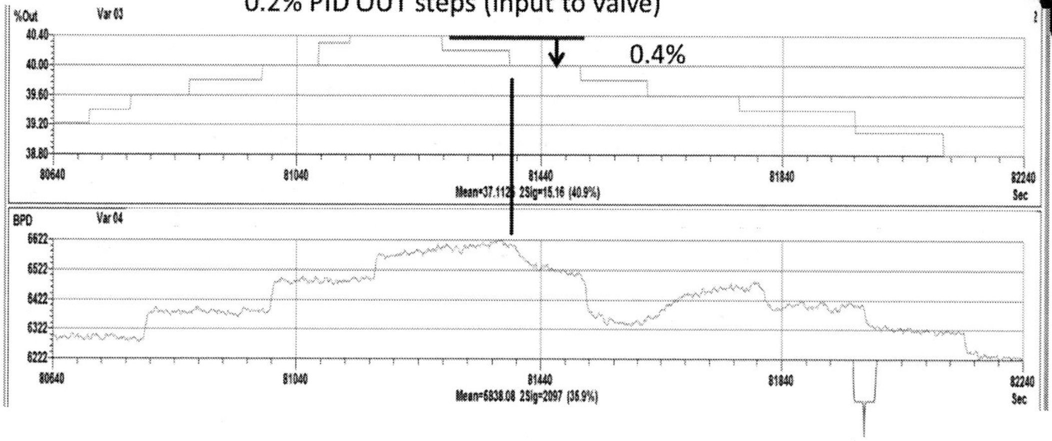

0.2% < Dead Band < 0.4%

40 seconds

16AIChE
Spring Meeting, Houston, TX

EMERSON.
Process Management

Dead Band and Resolution Cause Limit Cycles

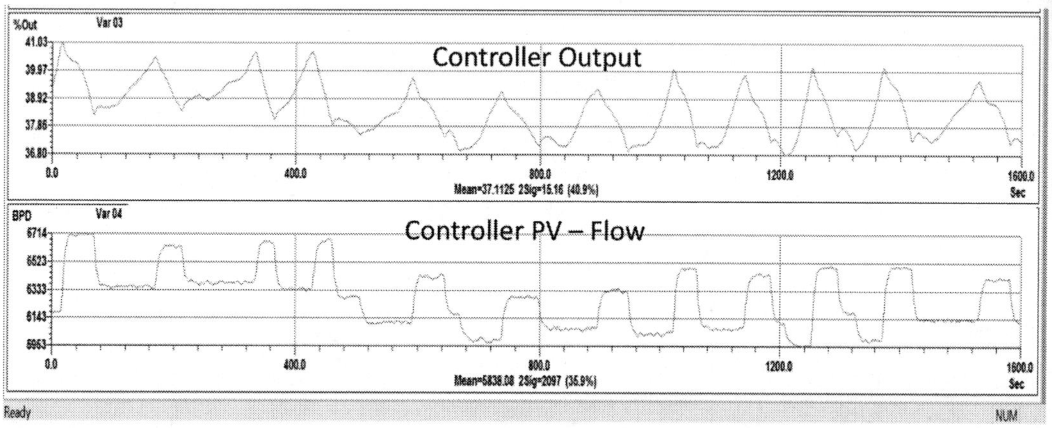

16AIChE
Spring Meeting, Houston, TX

EMERSON.
Process Management

Definitions from ANSI/ISA-S75.25

- **step response time (T86):** the interval of time between initiation of an input signal step change and the moment that the response of a dynamic reaches 86.5% of its full steady state value. The step response time includes the dead time before the dynamic response.

16AIChE Spring Meeting, Houston, TX

EMERSON. Process Management

Step Response Time (T_{86})

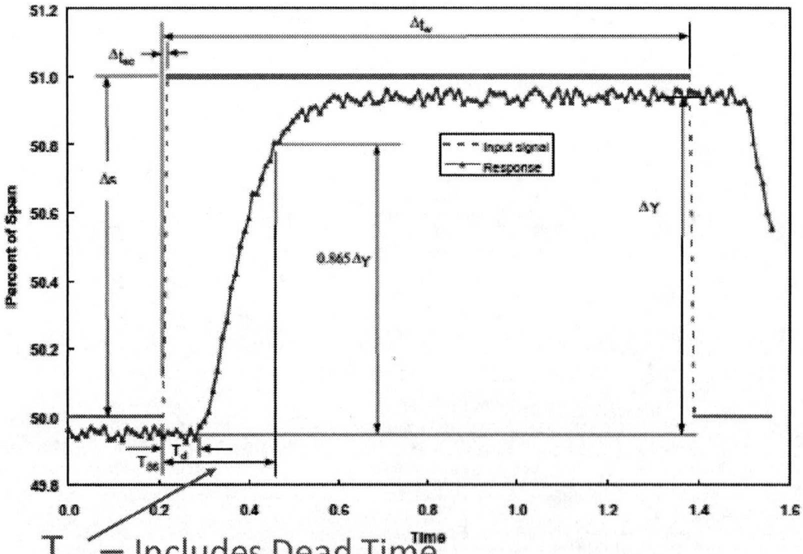

Figure 2 — Typical step change and response without overshoot

16AIChE Spring Meeting, Houston, TX

EMERSON. Process Management

Step Response Time VS. Step Size

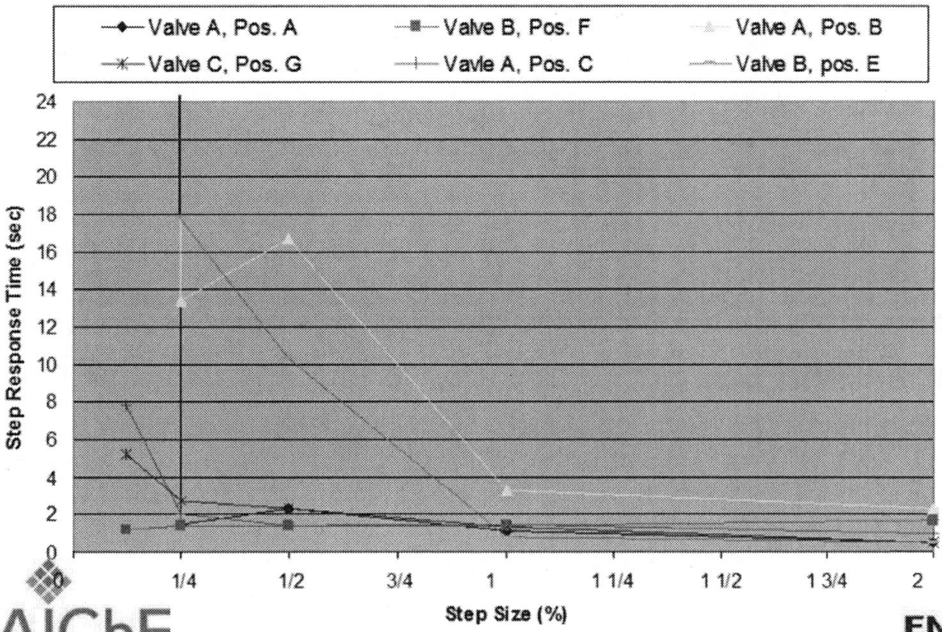

16AIChE Spring Meeting, Houston, TX

EMERSON. Process Management

From ANSI/ISA–TR75.25

- **Region 1** is defined as small input steps which result in no measurable movement of the closure member within the specified wait time.
  - Comment: This is in the region of dead band and resolution.

- **Region 2** is defined as input step changes which are large enough to result in some control valve response with each input signal change, but the response does not satisfy the requirements of the specified time and linearity.

16AIChE Spring Meeting, Houston, TX

EMERSON. Process Management

From ANSI/ISA–TR75.25

- **Region 3** is defined as step changes which are large enough to result in flow coefficient changes which satisfy both the specified maximum response time and the specified maximum linearity.
  - Comment: This is the range of “typical” control moves
- **Region 4** is defined as input steps larger than in region 3 where the specified magnitude response linearity is satisfied but the specified response time is exceeded.
  - Comment: This is typically caused by velocity limited operation.
- *EnTech Control Valve Dynamic Specification V3.0* is similar, using A, B, C, D instead of 1, 2, 3, 4.

Step Response Time VS. Step Size

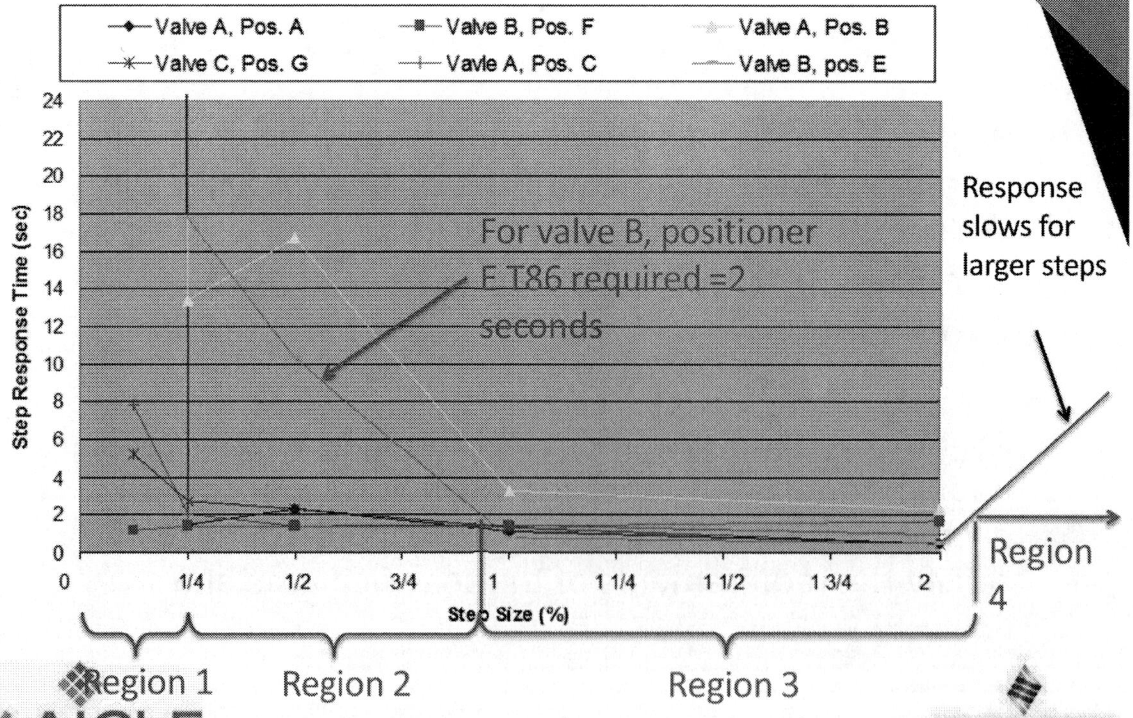

16AIChE Spring Meeting, Houston, TX
EMERSON. Process Management

Region 2 – Causes Limit Cycle

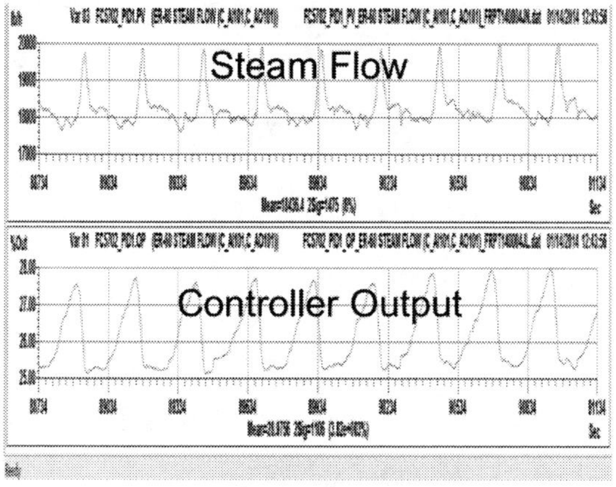

Limit cycle in Auto

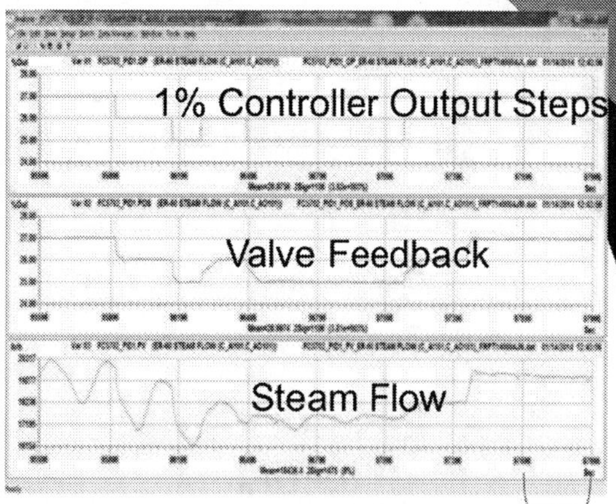

Step test in Manual

300 sec.

Slow response of control valve, due to slow response of valve positioner, causes cycling of PV. Solution – retune the digital positioner!

16AIChE Spring Meeting, Houston, TX

EMERSON. Process Management

Region 2 – Causes Limit Cycle

Region 2: 0.2 - 0.6%; T_{86} = 43 seconds; Region 3 T_{86} = 1 second

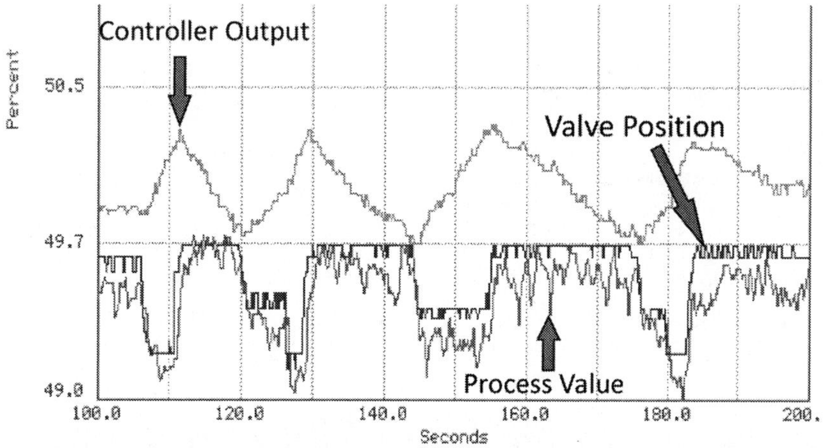

Slow response of control valve, due to slow response of valve positioner, causes cycling of PV. Solution – replace the pneumatic positioner with a high performance digital positioner!

16AIChE Spring Meeting, Houston, TX

EMERSON. Process Management

Summary: Step Response Time

- **Region 1 and 2** are bad for control performance
- **Region 3** performance provides the most linear operation of the valve and hence the best control performance
- **Region 4** performance (large step, slow or velocity limited) is typically inconsequential to control performance except in special cases like anti-surge control
- So, **minimize** the size of **Region 1 and 2**, **maximize** the size of **Region 3**

16AIChE
Spring Meeting, Houston, TX

EMERSON.
Process Management

"Gain"

- **General:** change in *output* divided by the change in *input*

$$Gain = \frac{\Delta Output}{\Delta Input}$$

- **Process control:**

$$Process\ Gain = \frac{\Delta Controller\ PV\ (\%\ of\ span)}{\Delta Controller\ Output\ (\%\ of\ span)}$$

16AIChE
Spring Meeting, Houston, TX

EMERSON.
Process Management

Loop Gain

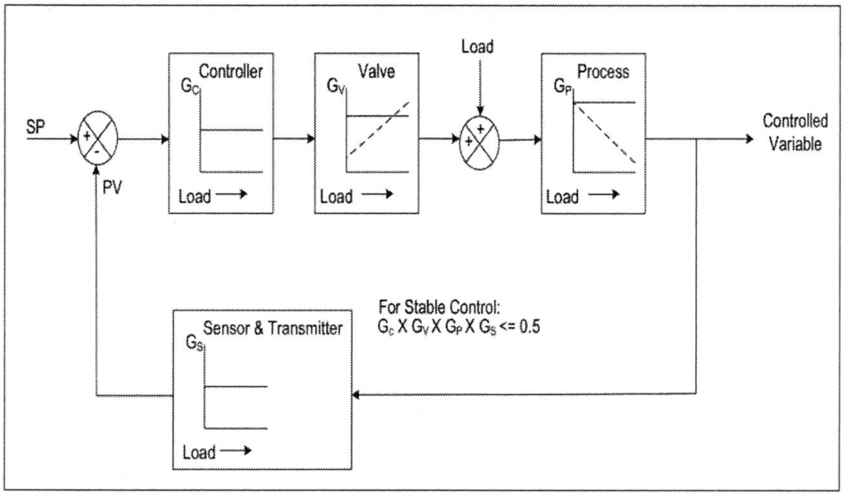

Total Loop Gain = (Controller Gain) x (Valve Flow Gain) x (Process Gain) x (Transmitter Gain)

Valve flow gain magnitude and consistency are important!

Control Valve Flow Characteristics

Control Valve Inherent Flow Characteristics :

The relationship of flow through the valve VS. valve position at constant process conditions and pressure drop

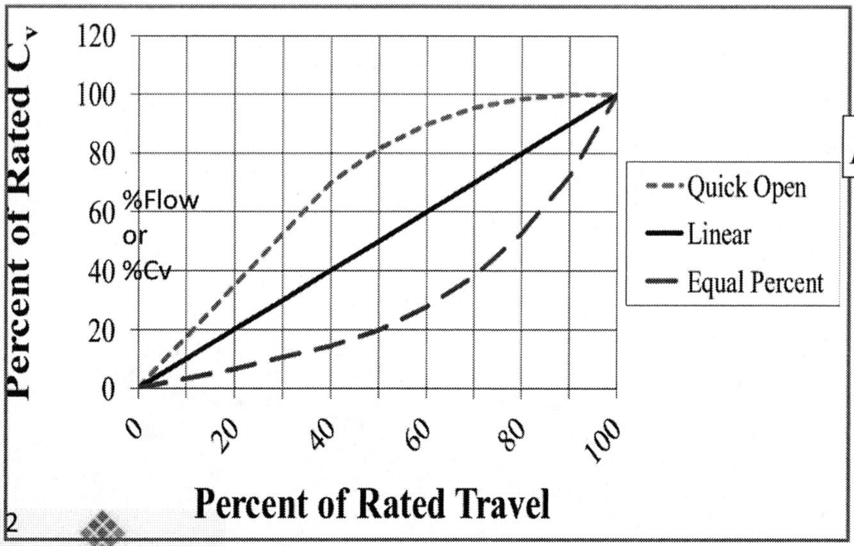

$$Flow = Cv\sqrt{\Delta P / SG}$$

C_V = valve flow coefficient (a function of valve position)
*Thus, at constant ΔP and SG, %Flow = % Rated Cv

Total Loop Gain & Installed Valve Gain

- **Recall: Valve gain magnitude and consistency are important!**
- **Valve Flow Gain as Installed**

$$ValveFlowGain = \frac{\Delta Flowrate(\%)}{\Delta StemPosition(\%)}$$

$$Flowrate = Cv\sqrt{\Delta P / SG}$$

- **Thus, the Valve Flow Gain as installed, depends on inherent valve flow characteristics AND the actual ΔP across the valve**

16AIChE
Spring Meeting, Houston, TX

EMERSON.
Process Management

Valve Trim Selection Example:

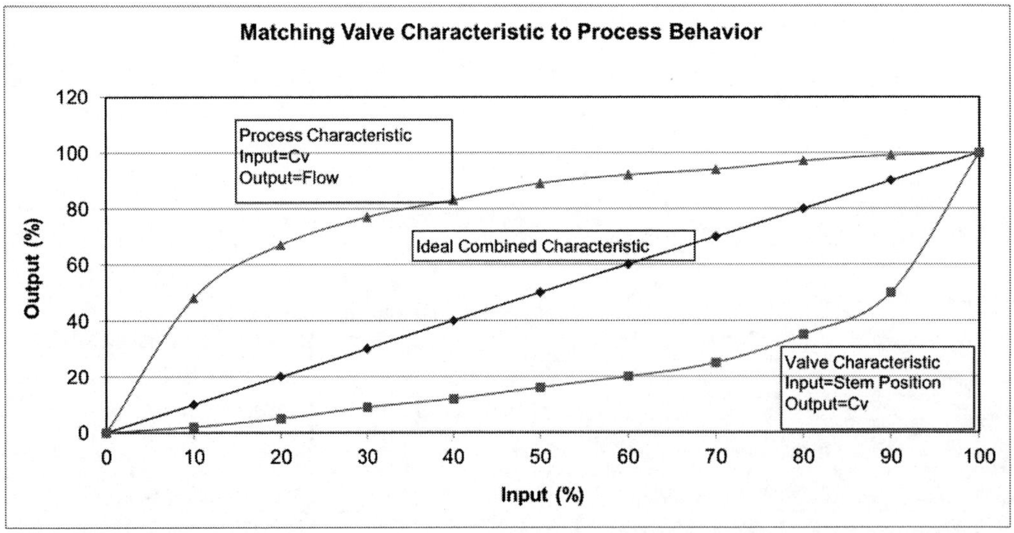

16AIChE
Spring Meeting, Houston, TX

EMERSON.
Process Management

Installed Gain – Goals and Selection

From a "% of flow span* perspective"...

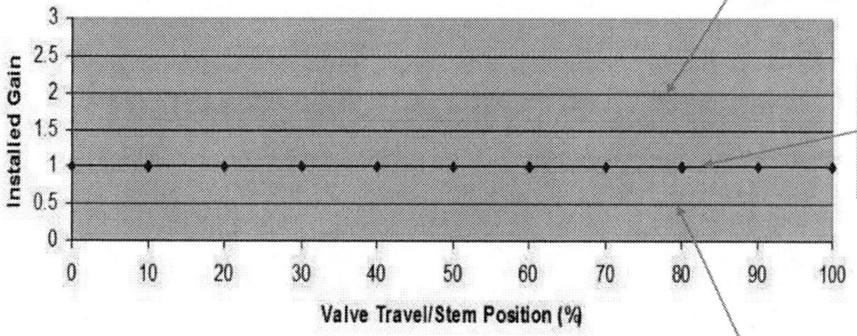

Gain ≤ 2.0 (to avoid amplification of valve non-linearities)

Gain as close to 1 as possible and as constant as possible

Gain ≥ 0.5 (to reach 100% PV)

* Could be span of flow controller, or maximum valve flow or maximum desired flow

16 AIChE
Spring Meeting, Houston, TX

EMERSON.
Process Management

Installed Characteristics and Process Gain

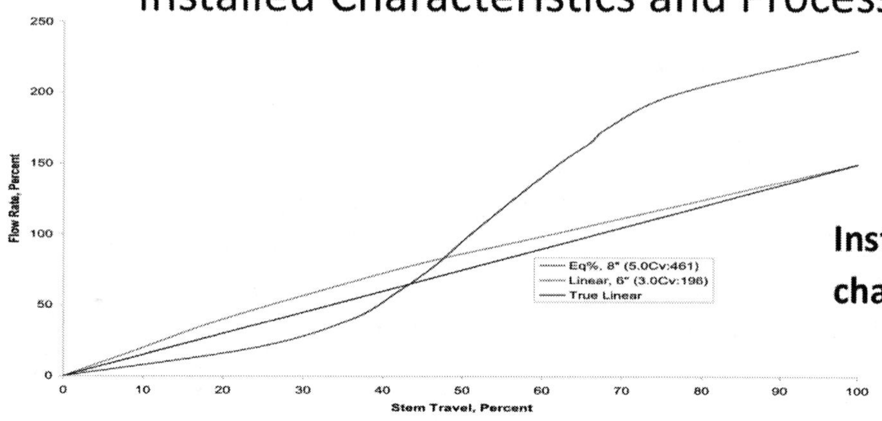

Installed characteristics

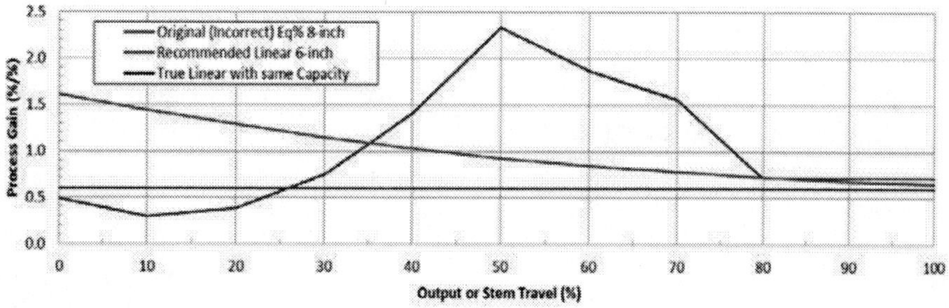

Installed gains

16
Spring Meeting, Houston, TX

EMERSON.
Process Management

Impact of Change in Valve Gain

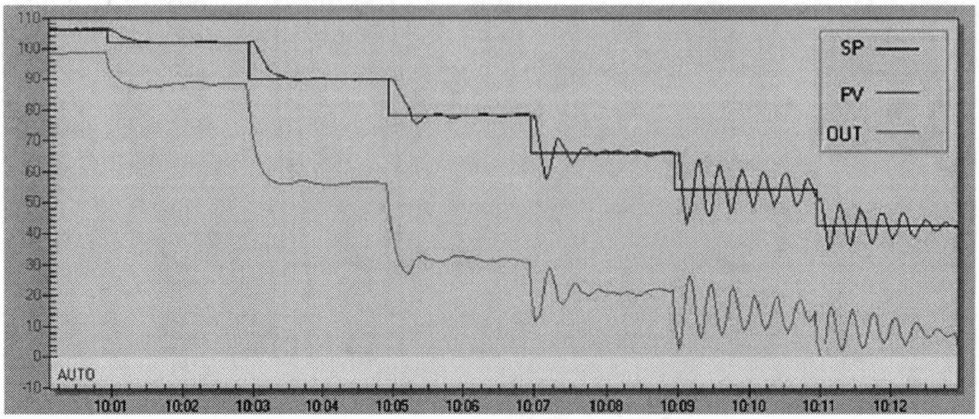

Process Gain increases as controller output decreases, causing loop instability.

16AIChE
Spring Meeting, Houston, TX

EMERSON.
Process Management

Gain Characterizer

- A signal (gain) characterizer on the output of the controller, or in a smart valve positioner, can help linearize the total process and/or valve flow gain
  - Note that the use of this technique can cause confusion. The actual output to the valve, should be displayed to the operator.
  - An alternate method is to adjust the controller tuning based on the valve position (controller output)
- However, note that the characterizer DOES NOT reduce the PROCESS non-linearities due to valve non-linearities
  - If the characterizer REDUCES the process gain, it correspondingly INCREASES the effective valve non-linearity (e.g. dead band) resulting in the same effective PROCESS non-linearity!

16AIChE
Spring Meeting, Houston, TX

EMERSON.
Process Management

Specifying Valve Performance

- Process or Valve Gain – close to 1 %PV/%OUTPUT and "constant as possible"
  - "Span" could be span of flow controller if applicable, or maximum valve flow or maximum specified flow (latter two choices assume 0 min. flow)
- Key Valve Tracking Performance Parameters
  - Valve Dead Band
    - Note Process Dead Band=Process Gain*(Valve Dead Band)
    - Thus, required Valve Dead Band = Allowable Process Dead Band / Process Gain
  - Resolution
    - Note Process Resolution = Process Gain*(Valve Resolution)
    - Thus, required Valve Resolution = Allowable Process Resolution / Process Gain

16AIChE
Spring Meeting, Houston, TX

EMERSON.
Process Management

Specifying Valve Performance

- Response Time Parameters
  - Response Time (T_{86}) - < 1/3 of process time constant
  - Size of Region 1- (~=dead band/resolution) - bad, minimize
  - Size of Region 2 – bad, minimize
  - Size of Region 3 – wide as possible but over 10% typically not required
  - Size of Region 4 – not of great consequence to control
- *EnTech Control Valve Dynamic Specification V3.0* helps *SPECIFY* control valve performance

16AIChE
Spring Meeting, Houston, TX

EMERSON.
Process Management

Summary

- Many operational problems can be prevented, and process control significantly improved, by proper selection of the control valve
  - Consider control loop stability and performance during valve procurement!
  - Valve gain, dead band, resolution, response time regions are clearly defined parameters
  - Severe control valve nonlinearities **cannot** be corrected by the process controller
- Applies across all process industries

16AIChE
Spring Meeting, Houston, TX

EMERSON.
Process Management

Presenter – James Beall

- Principal Process Control Consultant, Emerson Process Management
- 34 Years experience in process control
  - 15 Years Emerson Process Management
  - 19 Years Eastman Chemical Company
- Chairman ISA 75.25 Committee - Control Valve Performance Testing
- James.Beall@emerson.com 903-235-7935

EMERSON.
Process Management

16AIChE
Spring Meeting, Houston, TX

EMERSON.
Process Management

444988 MKS Data Analytics Solutions Umetrics Suite - A Platform for BIG Data Analytics, Multivariate Process Modeling, Monitoring and Prediction

Abstract:

MKS Data Analytics Solutions contributes in the area of BIG Data Analytics as a world leading provider of user friendly software packages for multivariate data analysis (MVDA) and design of experiments (DOE) through the Umetrics Suite of products. Software solutions are graphically driven and complete with surrounding services ranging from training to consulting, implementation and support. With 25 years of experience MKS Data Analytics Solutions is proud to offer easy-to-use robust and mature solutions as well as a highly praised customer support program. The Umetrics Suite includes the products MODDE Go, MODDE Pro, SIMCA, SIMCA-Q and SIMCA-online and application implementations span a range of industries including biopharmaceutical, petrochemical, chemical, food, semiconductor and pulp and paper.

The presentation will highlight a number of success stories of off-line and on-line applications of Umetrics Suite while emphasizing graphical power and capability of rapidly extracting information from and making sense of volume datasets originating from complex arrangements of multiple data sources. Visualization and effective utilization of software tools to fulfill objectives, create return on investment and increase the bottom line will be discussed.

Examples of applications include real-time multivariate process monitoring, open and closed loop multivariate model predictive control, prediction of critical process quality attributes, increasing process understanding, process optimization and process trouble shooting. MKS Data Analytics Solutions Umetrics Suite is utilized throughout research, development and manufacturing, both for continuous and batch processes.

444832 Modeling and Monitoring of Batch Processes in Relative Time

Abstract:

Batch processes are widely used in many industries. Typically, raw materials are combined in a suitable batch vessel before a chemical, physical or biological transformation take place, resulting in the end product. In many cases the control of the batch process is recipe driven and not adjusted to accommodate raw material variation, changes in uncontrollable factors, etc. to ensure the best possible end product quality. Real time control of a batch process saves the industry money and resource due to less rework and rejects, as well as the opportunity for real time quality assurance of the end product.

There are some solutions available for batch monitoring and control but typically they are assuming equal lengths of batches, i.e. the batch starts at the same chemical or biological time t_0 and has the same number of time points for all batches. Obviously this is resulting in problems if the batches do not meet these criteria. Alternative approaches to handle uneven batch lengths include replacing time with a maturity index or using dynamic time warping. In both these approaches complications can occur if the first measurement does not coincide with the true t_0 and the batch evolution is non-linear, which is often the case.

In this presentation, a better approach accommodating both uneven batch lengths and unknown true t_0 is proposed. The approach is based on projections in the score space so all control options used in MSPC are valid and available.

443289 Leveraging Big Data with Modeling & Self-Service Analytics for Rapid Decision Making / Capstone Technology

Abstract:

Businesses today have more data than ever to interpret. Data is arriving from multiple sources at an alarming velocity, volume and variety. The speed and ability of businesses to collect, present, analyze, interpret and decide separates the elite from the masses. How do you exploit this volume of data from so many sources without crippling decision making? It would be great to throw all of this data into a Super Computer and output the optimal solution. However, optimal manufacturing decisions require not only processing the data, but also delivering it to the right people with a tool that keeps up with the speed of thought. This presentation will focus on available commercial software that does just this. 6 keys include 1) Provide access to all manufacturing and enterprise data sources 2) Use data aggregation for speed & filtering 3) Apply calculations, statistics & logic for development of KPI's 4) Layering of asset structure for ease of analysis and segmenting of data by assets & asset types. 5) Visualization, Advanced Analytics and Modeling of all underlying data 6) Use of industry standard protocols to provide easy integration of domain specific applications.

CSB INVESTIGATION OF THE WEST (TEXAS) FERTILIZER FIRE AND EXPLOSION

Presenter:
Johnnie A. Banks, CFEI, Supervisory Investigator,
U.S. Chemical Safety and Hazard Investigation Board, Washington, DC
johnnie.banks@csb.gov

Prepared for Presentation at
American Institute of Chemical Engineers
2016 Spring Meeting
12th Global Congress on Process Safety
Houston, Texas
April 10-14, 2016

CSB INVESTIGATION OF THE WEST (TEXAS) FERTILIZER FIRE AND EXPLOSION

Presenter:
Johnnie A. Banks, CFEI, Supervisory Investigator,
U.S. Chemical Safety and Hazard Investigation Board, Washington, DC
johnnie.banks@csb.gov

Keywords: *Fertilizer Grade Ammonium Nitrate (FGAN), Hazard Awareness, Emergency Planning and Response, FGAN Storage Practices, Land Use Planning*

Abstract

On April 17, 2013, a fire and explosion occurred at the West Fertilizer Company (WFC), a fertilizer blending, retail, and distribution facility in West, Texas. The violent detonation fatally injured 12 emergency responders and three members of the public. Local hospitals treated more than 260 injured victims, many of whom required hospital admission. The blast completely destroyed the WFC facility and caused widespread damage to more than 150 offsite buildings. The WFC explosion is one of the most destructive incidents ever investigated by the U.S. Chemical Safety and Hazard Investigation Board (CSB) as measured by the loss of life among emergency responders and civilians; the many injuries sustained by people both inside and outside the facility fenceline; and the extensive damage to residences, schools, and other structures.

The CSB conducted an exhaustive investigation that lasted more than two years and which yielded 19 recommendations directed to ten recipients. The tragic loss of life to the civilians and emergency responders, in addition to issues related to community-wide property damage were central in the development of these recommendations, and cover a range of elements which include:

- Regulatory oversight
- Insurance
- Emergency response
- Emergency planning
- Land use planning
- Technical conditions

This presentation will provide an overview of the incident, consider the key findings from the investigation, and illustrate the relationship of these findings to the recommendations. Click here for the final West Fertilizer Investigation Report and related videos and information.

Process Safety Incidents: Big Picture Revelations

Paul Baybutt
Primatech Inc.
Columbus, Ohio, USA
paulb@primatech.com

Prepared for Presentation at
American Institute of Chemical Engineers
2016 Spring Meeting
12th Global Congress on Process Safety
Houston, Texas
April 11-23, 2016

Process Safety Incidents: Big Picture Revelations

Paul Baybutt
Primatech Inc.
Columbus, Ohio, USA
paulb@primatech.com

Keywords: process safety instruments; process safety management; risk management programs

Abstract

Many incidents have helped to define and develop process safety. Each has provided valuable learning opportunities. However, it is even more important to identify insights that can be obtained from an analysis of a large set of incidents that represents those that typically occur. This larger picture illuminates trends and commonalities and provides learning opportunities that are even more important than the causes of any one individual incident.

The Chemical Safety Board has published the results of numerous investigations of process safety incidents and the Center for Chemical Process Safety has published information on incidents that have helped to define process safety. These data have been analyzed to identify commonalities and trends so that measures to help protect against future incidents can be developed. Recommendations are made to address key issues identified.

1.0 Introduction

Process safety regulations have been promulgated in various parts of the world to address protecting process plants against catastrophic incidents in response to the occurrence of a number of major incidents such as Flixborough, Seveso, and Bhopal [1]. However, despite the widespread adoption of process safety practices, major incidents continue to occur. The continued occurrence of such incidents is unacceptable and the issue of what can be done to protect against them must be addressed. This paper addresses the issue of why such incidents continue to occur.

Process safety regulations contain requirements for periodic audits for programs to identify any omissions or deficiencies and some process safety programs require periodic management reviews. However, while omissions and deficiencies must be identified and corrected, knowledge of what omissions and deficiencies have actually resulted in process safety incidents is essential to help prevent further incidents.

Compilations of incident descriptions have been published by various authors, for example, [1 - 7]. Usually, incidents are organized in various ways, such as by root cause, release type, equipment involved, operation being conducted, and process safety element involved. Descriptions of individual incidents also can be found, for example, in reports from the U.S. Chemical Safety Board (CSB) [8], and in journals such as the Loss Prevention Bulletin [9]. Such publications usually focus on providing a description of the incident, identifying its causes, and providing recommendations for corrective actions. Certainly, lessons can be learned from each incident. However, it is also important to take an overall view to identify commonalities and trends across incidents so that measures can be identified that will help to protect against future incidents. In particular, the identification of issues that have contributed to multiple incidents is especially important so that future incidents may be prevented. For example, the CSB has flagged reactive hazards [10], combustible dusts [11], and hot work [12] as significant issues across multiple incidents. This paper identifies additional issues that were important contributors in numerous incidents.

CSB investigation reports have been analyzed concerning management system failures [13], process hazard analysis (PHA) [14], and the hierarchy of controls [15]. The evolution of process safety standards and legislation following nine key process safety incidents has been addressed in a recent article [16].

Section 2 identifies the incidents reviewed and the analysis approach followed. Section 3 identifies and discusses the common contributors that were found for the incidents studied and provides recommendations to address them. Section 4 provides a discussion of the results. Section 5 provides some additional noteworthy observations that were made during the review of incident descriptions. Section 6 identifies three issues that underlie all of the other issues. Each underlying issue is discussed and recommendations are made for addressing them. Conclusions are presented in section 7.

2.0 Incidents Addressed and Analysis Approach

Incidents investigated by the Chemical Safety Board (CSB) [8] and incidents described by the Center For Chemical Process Safety (CCPS) [1] were reviewed. CCPS provides descriptions of over 100 incidents. Some date back to the early 20th century but most occurred within the past 40 years. They cover a variety of industries including chemicals, oil and gas production and refining, aviation, shipping, and nuclear power generation. The CSB has issued over 60 incident investigation reports, case studies, and safety bulletins dating back to 1998 when the CSB began investigating incidents. The CSB incidents cover various sectors of the process industries. The CSB currently is investigating one incident from 2010, three from 2013, three from 2014 and one from 2015. Sixty-eight CSB incidents and 12 CCPS incidents making a total of 80 incidents were studied (see Tables 1 and 2, respectively). About 40% of the CSB incidents occurred in processes covered by the Occupational Safety and Health Administration's (OSHA's) process safety management (PSM) regulations. The descriptions of incidents and their causes were taken at face value.

Table 1. CSB Incidents Reviewed.

No.	Incident	Incident Date	Report Date	PSM[1]	PHA[2]
1	Sonat Exploration Co. Catastrophic Vessel Overpressurization	03/04/1998	09/21/2000	N	N
2	Morton International Inc. Runaway Chemical Reaction, Explosion and Fire	04/08/1998	08/16/2000	N	Y
3	Concept Sciences Hydroxylamine Explosion	02/19/1999	02/01/2002	Y	Y
4	Tosco Avon Refinery Petroleum Naphtha Fire	02/23/1999	03/21/2001	Y	N
5	Bethlehem Steel Corporation Gas Condensate Fire	02/02/2001	12/06/2001	N	N
6	BP Amoco Thermal Decomposition Incident	03/13/2001	05/20/2002	N	Y
7	Motiva Enterprises Sulfuric Acid Tank Explosion	07/17/2001	08/28/2002	N	Y
8	Georgia-Pacific Corp. Hydrogen Sulfide Leak	01/16/2002	11/20/2002	N	N
9	Third Coast Industries Petroleum Products Facility Fire	05/01/2002	03/06/2003	N	U
10	DPC Enterprises Festus Chlorine Release	08/14/2002	05/01/2003	Y	U
11	First Chemical Corp. Reactive Chemical Explosion and Fire	10/13/2002	10/15/2003	N	N
12	Environmental Enterprises Hydrogen Sulfide Release	12/11/2002	09/17/2003	N	U
13	Catalyst Systems Inc. Reactive Chemical Explosion and Fire	01/02/2003	10/29/2003	N	N
14	BLSR Operating Ltd. Vapor Cloud Deflagration and Fire	01/13/2003	09/17/2003	N	U
15	West Pharmaceutical Services Dust Explosion and Fire	01/29/2003	09/23/2004	N	N
16	Technic Inc. Ventilation System Explosion and Fire	02/07/2003	08/20/2004	N	N
17	CTA Acoustics Dust Explosion and Fire	02/20/2003	02/15/2005	N	N
18	D.D. Williamson & Co. Catastrophic Vessel Failure	04/11/2003	03/12/2004	N	N
19	Honeywell Chemical Releases (3 incidents)	a. Chlorine 07/20/2003 b. Antimony pentachloride 07/29/2003 c. Hydrogen fluoride 08/13/2003	08/08/2005	Y N Y	Y N N
20	Isotec / Sigma Aldrich Nitric Oxide Explosion	09/21/2003	08/24/2004	Y	Y
21	Hayes Lemmerz Dust Explosions and Fire	10/29/2003	09/27/2005	N	N
22	DPC Enterprises Glendale Chlorine Release	11/17/2003	02/28/2007	Y	Y
23	Giant Industries Oil Refinery Explosion and Fire	04/8/2004	10/26/2005	Y	Y
24	MFG Chemical Inc. Toxic Gas Release	04/12/2004	04/11/2006	Y	N
25	Formosa Plastics Vinyl Chloride Explosion	04/23/2004	03/06/2007	Y	Y
26	Sterigenics Ethylene Oxide Explosion	08/19/2004	03/30/2006	Y	Y
27	Marcus Oil and Chemical Tank Explosion and Fire	12/03/2004	06/06/2006	N	N

No.	Incident	Incident Date	Report Date	PSM[1]	PHA[2]
28	Acetylene Service Company Gas Explosion	01/25/2005	01/26/2006	Y	Y
29	BP Texas City Refinery Explosion and Fire	03/23/2005	03/20/2007	Y	Y
30	Formosa Plastics Propylene Fire and Explosions	10/06/2005	07/20/2006	Y	Y
31	Bethune Point Wastewater Plant Explosion and Fire	01/11/2006	03/13/2007	N	N
32	Synthron Chemical Reaction and Vapor Cloud Explosion	01/31/2006	07/31/2007	Y	N
33	Partridge Raleigh Oilfield Explosion and Fire	06/05/2006	06/12/2007	N	N
34	Universal Form Clamp Co. Explosion and Fire	06/14/2006	04/10/2007	Y	N
35	EQ Hazardous Waste Plant Explosions and Fire	10/05/2006	04/16/2008	N	U
36	CAI / Arnel Chemical Plant Explosion	11/22/2006	05/13/2008	Y	N
37	Valero Refinery Propane Fire	02/16/2007	07/09/2008	Y	Y
38	Barton Solvents Explosions and Fire	07/17/2007	06/26/2008	N	U
39	Xcel Energy Company Hydroelectric Tunnel Fire	10/02/2007	08/25/2010	N	Y
40	Barton Solvents Flammable Liquid Explosion and Fire	10/29/2007	09/18/2008	N	U
41	T2 Laboratories Inc. Reactive Chemical Explosion	12/19/2007	09/15/2009	N	N
42	Imperial Sugar Company Dust Explosion and Fire	02/07/2008	09/24/2009	N	U
43	Goodyear Heat Exchanger Rupture and Ammonia Release	06/11/2008	01/27/2011	Y	U
44	Packaging Corporation Storage Tank Explosion	07/29/2008	03/04/2010	N	N
45	Bayer CropScience Pesticide Waste Tank Explosion	08/28/2008	01/20/2011	Y	Y
46	INDSPEC Chemical Corporation Oleum Release	10/11/2008	09/30/2009	Y	Y
47	Allied Terminals Fertilizer Tank Collapse	11/11/2008	05/26/2009	N	N
48	Veolia Environmental Services Flammable Vapor Explosion and Fire	05/04/2009	07/21/2010	Y	N
49	ConAgra Natural Gas Explosion	06/09/2009	02/04/2010	N	U
50	CITGO Refinery Fire and Hydrofluoric Acid Release	07/19/2009	12/09/2009	Y	Y
51	Caribbean Petroleum Refining Tank Explosion and Fire	10/23/2009	10/21/2015	N	N
52	Silver Eagle Refinery Explosion	11/04/2009	04/14/2014	Y	U
53	NDK Crystal Inc. Explosion	12/07/2009	11/14/2013	N	U
54	DuPont Corporation Toxic Chemical Releases (3 incidents)	a. Methyl chloride 01/22/2010 b. Oleum 01/23/2010 c. Phosgene 01/23/2010	09/20/2011	U U Y	U U Y
55	Kleen Energy Natural Gas Explosion	02/07/2010	06/28/2010	N	U
56	Tesoro Refinery Explosion and Fire	04/02/2010	05/01/2014	Y	Y
57	Horsehead Holding Company Explosion and Fire	07/22/2010	03/11/2015	N	U

No.	Incident	Incident Date	Report Date	PSM[1]	PHA[2]
58	Millard Refrigerated Services Ammonia Release	08/23/2010	01/15/2015	U	U
59	E. I. DuPont De Nemours Co. Hotwork Explosion	11/09/2010	04/19/2012	Y	Y
60	AL Solutions Dust Explosion	12/09/2010	07/16/2014	N	U
61	Carbide Industries Fire and Explosion	03/21/2011	02/07/2013	N	U
62	Hoeganaes Corporation Flash Fires and Explosion	05/27/2011	01/05/2012	U	U
63	Chevron Refinery Fire	08/06/2012	01/28/2015	Y	Y
64	US Ink Fire	10/09/2012	01/15/2015	U	U

[1] Covered by OSHA's PSM standard.
[2] PHA was performed
Y – Yes
N – No
U – Unknown

Table 2. CCPS Incidents Reviewed.

No.	Incident	Incident Date
1	ELF refinery, LPG tank farm, Feyzin, France	01/04/1966
2	DSM Nypro plant, Flixborough, UK	06/01/1974
3	ICMESA, chemical plant, Seveso, Italy	07/10/1976
4	BP hydrocracker unit, Grangemouth, UK	03/21/1987
5	Shell refinery, FCC unit, Norco, LA	05/05/1988
6	Occidental Petroleum, Piper Alpha Platform, North Sea, UK	07/06/1988
7	Phillips Chemical Company, polyethylene reactor, Pasadena, TX	10/23/1989
8	Total refinery, FCC unit, La Mede, France	11/09/1992
9	Texaco Oil Refinery, FCC Unit, Milford Haven	07/24/1994
10	Napp Technologies, Lodi, NJ	04/21/1995
11	Tosco Avon Refinery, hydrocracker unit, Martinez, CA	07/21/1997
12	Sonatrach LNG plant, Skikda, Algeria	01/19/2004

All incident descriptions were first reviewed to identify commonalities. A second reading of the incident descriptions was performed to confirm the initial findings and document characteristics of the incidents. A third reading of the incident descriptions was conducted to corroborate the findings and provide further detail.

3.0 Incidents Addressed and Analysis Approach

This section identifies common contributors to the incidents. Of course, each incident has its own individual characteristics but there is a remarkable degree of similarity in some deficiencies and omissions across the incidents. The focus in this paper is on issues common to many of the incidents.

Incidents where each issue was a contributor are identified, the percentage of CSB incidents where the issue was present is given, aspects of each issue are discussed, and recommendations are made to address the issues.

3.1 Process Design

Many incidents occurred in processes that arguably had deficient designs {1, 3, 6, 15, 16, 17, 18, 21, 25, 29, 31, 34, 37, 40, 41, 42, 57, 58, 64} and {*10, 11*}[3]. This issue was present in 28% of CSB incidents and 18% of incidents covered by OSHA's PSM standard. Deficiencies arose in the initial design and as a result of modifications.

Hazards were not recognized, designs created hazards, safety systems were inadequate, process instrumentation was insufficient, human factors issues were not addressed, hazards of scale-up were not recognized, and throughput was increased without review. In some cases deficient equipment design contributed to incidents {54a, 54c}. Possibly designers were unaware of the problems and/or design reviews were inadequate but these reasons are not acceptable.

Design deficiencies can be addressed by making process design and engineering an element in process safety programs [17]. Examples of topics that should be addressed by requirements for process design and engineering include design practices, design reviews, inherent safety, hazardous area classification, pressure relief, alarm management, damage mechanisms, human factors, facility siting, competency, and training for designers. Also, process design and engineering should be subject to the same management system requirements as other elements in a process safety program such as written policies and procedures, document management, assigning roles and responsibilities, auditing, management review, and continuous improvement. Process design and engineering should be an element of process safety management systems in order to provide a specific focus on the activities involved and to ensure the application of management system practices to it. Also, the competency of process designers must be addressed.

For new designs and process modifications, proof of safety should be required rather than proof of danger. This will require a change in culture.

[3] Braces or curly brackets, that is, { }, are used to enclose the number designations of incidents from Table 1, using normal font, and Table 2, using italic font.

3.2 *Safeguards*

Many incidents involved processes that were not adequately protected by safeguards {2, 7, 8, 9, 11, 12, 13, 17,18, 20, 21, 22, 24, 25, 26, 28, 29, 30, 32, 34, 35, 36, 37, 38, 40, 41, 42, 46, 48, 49, 50, 51, 54a, 54b, 54c, 60, 62, 64} and *{1, 3, 4, 5, 6, 9, 11}*. This issue was present in 56% of CSB incidents and 57% of incidents covered by OSHA's PSM standard. Needed safeguards were not present or those that were present were insufficient, unreliable, disabled, or lacked independence. Commonly, the hierarchy of controls was not applied (see Figure 1). Often reliance was placed on procedural safeguards which are at the lower end of the hierarchy (19a, 22, 25, 29, 32, 46, 54c}. Process risks should be matched with safeguards in the hierarchy of controls. Higher consequence scenarios, such as a chlorine release in a densely populated area, require safeguards higher in the hierarchy and likely multiple, independent safeguards. Procedural safeguards typically are viewed as less reliable, and, therefore, inadequate for protection against high-consequence events.

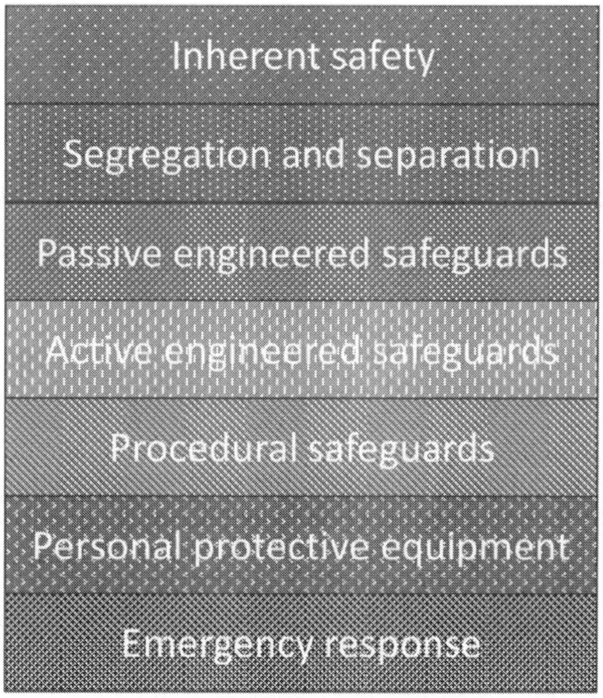

Figure 1. Hierarchy of Controls.

Inherently safer designs or practices could have prevented or mitigated some incidents {29, 39, 55, 56, 54b, 63}. Currently, there are no requirements to address inherently safer systems or the hierarchy of controls in Federal process safety regulations. Process safety regulations in New Jersey require an inherently safer technology review [18] and in Contra Costa County, California require that inherently safer systems be addressed [19]. Proposed amendments to California's process safety regulations for refineries contain a requirement to perform a hierarchy of controls analysis [20].

The operability of safeguards was an issue during several incidents. Explosions and fires disabled safeguards {15, 30, 42} and {*7*}, and toxic releases and fires prevented access to controls {19a, 30, 37, 50} and {*6*}.

Currently, there is no requirement in US process safety regulations to address the design basis for safeguards, evaluate or document the effectiveness of process safeguards, or show that safeguards adequately reduce risks. Proposed amendments to California's process safety regulations for refineries contain a requirement to perform a hierarchy of controls analysis and a safeguards protection analysis (SPA) [20]. The SPA should be performed before a PHA study and updated after the PHA study has been completed. The SPA should include the preparation of a Safety Requirements Specification (SRS) for each individual safeguard [21]. An SRS documents the functional and integrity requirements for a safeguard. Functional requirements describe what the safeguard does to prevent a hazardous state of the process and integrity requirements specify how well the safeguard should perform and what is needed to support that performance level. The SPA should validate that safeguards adequately protect against the hazards of the process, have a specified integrity level, have been installed correctly, are functional, are being maintained appropriately, etc.

Once the SPA has been performed, the PHA team can focus on the qualification of safeguards for specific hazard scenarios [21, 22]. Qualification of safeguards addresses whether safeguards are appropriate for each hazard scenario in a PHA study. Information developed in a PHA study may necessitate revisions to the SPA and SRSs. For example, it may be determined that a redundant power supply is needed to support the integrity required for a scrubber system. SPAs should be updated during management of change (MOC) reviews and when developing recommendations from incident investigations.

3.3 Operations and Maintenance

Problems with procedures were common {1, 2, 3, 5, 6, 11, 12, 13, 14, 18, 19b, 19c, 21, 22, 25, 28, 29, 32, 33, 36, 41, 42, 45, 46, 47, 51, 54c, 59, 61, 62} and {*4, 9*}. . This issue was present in 44% of CSB incidents and 43% of incidents covered by OSHA's PSM standard. In numerous incidents, procedures had not been developed, were not followed, were incomplete, or were otherwise deficient. Such situations are unacceptable. Properly written procedures must be developed following best practices for all process operations and activities before they occur and for all stages and modes of operation. Procedures are needed to deal with any situation where there is the potential for a process safety incident. Procedures must be workable in all circumstances where they apply and personnel need to be trained in the procedures and practice them regularly in drills. Facility personnel must be competent.

In six CSB investigations that identified problems with standard operating procedures (SOPs), the CSB noted that each incident involved SOP deviations that became "necessary violations" to get the job done {45}. Such situations cannot be tolerated.

Existing process safety regulations require written operating procedures. However, the regulations should be clarified to encompass all process tasks and activities and a mechanism is needed to ensure the regulatory requirements are followed.

Deficiencies in maintenance were present in a number of incidents {5, 12, 21, 29, 31, 32, 51, 54c}. Issues regarding the adequacy of inspections were common {4, 7, 47, 53, 56, 62} and {2, 8}. For some incidents, there was no or an inadequate preventive maintenance (PM) program {11, 13, 16, 41} and no or inadequate MI program for some incidents {10, 23, 29, 52, 61, 63}. Such issues were present in 34% of CSB incidents and 32% of incidents covered by OSHA's PSM standard. Existing process safety regulations address the integrity of equipment but a mechanism is needed to ensure the requirements are followed for all safety-critical equipment. Current US process safety regulations require that covered equipment comply with recognized and generally accepted good engineering practices (RAGAGEP) but the term is not defined. OSHA issued a Request for Information (RFI) in advance of rulemaking to revise the PSM standard [23]. Amongst other topics, the RFI sought comment on whether an explicit definition of RAGAGEP should be provided, and, if so, what definition should be used. The current OSHA PSM regulation provides a limited list of covered equipment [24] and the RFI sought comments on whether mechanical integrity requirements should apply to all equipment designated as safety-critical equipment.

The CCPS Risk Based Process Safety (RBPS) element, Conduct of Operations, provides requirements for operational discipline that tolerate no deviation from approved procedures, even if the outcome is inconsequential or desirable [26, 27]. Appropriate work procedures must be determined, documented, and enforced. The importance of following procedures must be emphasized to facility personnel. The adoption of an element on operational discipline would help in this regard coupled with an element that addresses process safety culture. The CCPS RBPS program contains both these elements but they are not included in existing US regulations. Additionally, deviations from approved procedures should be identified and investigated as process safety incidents.

An operational discipline element would encourage and enforce strict adherence to established procedures and imbue facility personnel with the belief that no operation or activity should be performed without a governing procedure. When people's lives are at risk it should not be acceptable to perform activities without proper preparation or by improvising.

Another deficiency present in several incidents was the acceptance of practices as normal and acceptable by facility personnel that outside observers would view as deviant and unacceptable {17, 19c, 21, 29, 42, 57, 62, 63}. This phenomenon has been called "normalization of deviance" [28]. It involves allowing operation of a process that deviates from established limits by relaxing performance standards rather than identifying and correcting the underlying causes of deviations so the deviations become the new, normal operating conditions. Hazards have been normalized in some cases {12, 27, 46, 56}. Normalization of deviance was present in 18% of CSB incidents and 18% of incidents covered by OSHA's PSM standard.

While the acceptance of a lesser performance standard in such cases may not immediately result in an incident, continued successful operation instills a sense of false security. It will be only a matter of time before uncorrected hazardous conditions eventually result in an incident.

Unfortunately, facility personnel can become so accustomed to exposure to hazardous situations that the risks are not recognized even when their lives are at stake. Just because a facility has not experienced a catastrophic incident, perhaps ever in its lifetime, does not mean one could not happen tomorrow.

Normalization of deviance can be addressed by ensuring that any deviations from established performance standards are identified and resolved as soon as they occur. Operational discipline will help in this regard.

3.4 Abnormal and Non-Routine Operations

Many incidents involved non-routine operations and situations {1, 3, 4, 5, 6, 10, 11, 16, 17, 19c, 20, 21, 23, 24, 26, 27, 29, 32, 33, 39, 41, 43, 45, 46, 47, 49, 52, 54a, 54c, 55, 56, 58, 59, 60, 61, 62, 63, 64} and {*1, 7, 9*}. In some cases, they were not recognized as non-routine operations and were not reviewed to ensure that work could proceed safely. Flaws in hot work {7, 31, 33, 44, 59} and other safe work practices {4, 5, 6, 11, 14, 23, 39, 43} and {*6*} also contributed to incidents. Lack of equipment isolation was a contributor to some incidents {11, 33, 59}. Abnormal and non-routine operations were involved in 62% of CSB incidents and 64% of incidents covered by OSHA's PSM standard.

People play an intimate role in non-routine operations. This both increases the likelihood of human error being a factor and places people closer to process hazards. Furthermore, a number of incidents occurred in processes that involved manual operations or were manually controlled {2, 12, 13, 16, 18}.

Current US process safety regulations contain requirements to develop and implement safe work practices and utilize hot work permits. However, the regulations do not specifically address non-routine operations. There are requirements for written procedures under the Operating Procedures element for specific phases of operation listed in the regulations (initial startup, normal operations, temporary operations, emergency shutdown, emergency operations, normal shutdown, and startup following a turnaround or after an emergency shutdown). However, other phases of operation are not specified including commissioning, standby, bypass, unattended operation, maintenance on live equipment, regeneration, troubleshooting, operation with disabled safety systems, cleaning, decommissioning, etc. Process safety programs must address all phases of operation that are or may be experienced by a process. Emphasis should be placed on non-steady state operations including abnormal operations because they are the process states in which most incidents have occurred.

Other elements of process safety programs besides operating procedures should include requirements to address non-routine operations, for example, mechanical integrity and PHA [29]. A separate element on non-routine operations may be merited to provide requirements in a single place. Absent specific requirements to address non-routine operations, many non-routine operations are likely not to be viewed as such and their hazards not recognized or addressed.

Procedures are needed for identifying and planning for non-routine work and abnormal situations, e.g. restart after emergency shutdown or an extended unplanned shutdown. Processes should be operated defensively and conservatively. Procedures should specify alternative courses of action if safety preconditions cannot be met. Furthermore, job hazard analysis [29] should be performed immediately prior to any non-routine activity.

3.5 Human and Organizational Factors

Human and organizational factors contributed to many incidents, for example, valve positions were not clear {23, 51}, valve operation was difficult {56}, lighting was insufficient {51}, operators were distracted {18}, operators were fatigued {29, 45}, operators had insufficient information {57}, procedures were complex and lengthy {45}, communications were problematic {25, 29, 43, 45}, layout was error prone {25}, staffing was not sufficient {25, 29, 45, 51}, operators became desensitized to an alarm due to its history of unreliability {54a}, the work environment encouraged operations personnel to deviate from procedures {29}, a computerized control system was poorly designed {29}, a new control system posed challenges {45}, and there were difficulties with organizational structure {25, 29}. Undoubtedly, problems with safety culture played an important role in many incidents, with BP Texas City being the best known {29}. Most CSB investigations have not explicitly identified contributions to incidents from human and organizational factors so it is not possible to provide meaningful statistics for their contributions to incidents. Undoubtedly, human and organizational factors played a role in most accidents.

Fundamentally, all of the incidents reviewed resulted from human failures ranging from design to operation to maintenance. Even when equipment failures such as corrosion were the basic cause of an incident, the underlying cause usually was human failure {63}.

Some incidents involved conflicts and tradeoffs between process safety and other imperatives {4, 7, 29, 39, 45, 51, 54c, 58, 63}, often financial but sometimes other issues such as noise attenuation {36}.

OSHA's PSM standard requires that "the process hazard analysis shall address human factors" [24]. However, it is not sufficient simply to require PHA studies to address human factors. Certainly, PHA teams must address the role of human factors in influencing human failures but this approach leads to insufficient consideration of human factors issues for processes. Commonly, human factors are addressed by using a checklist with questions that are asked about the entire process so that analysts can try and think of any problems. A much more detailed approach is needed. Otherwise, many aspects of human factors will be missed because of practical limits on the level of detail that can be addressed in PHA studies, limitations on the ability of PHA methods to address human factors, and time pressures on PHA teams. Also, human factors issues affect each of the elements of a process safety program including PHA itself [31]. It is not reasonable to expect a PHA team to address human factors issues that impact its own performance. The Contra Costa industrial safety ordinance requires the development and implementation of a written human factors program [19]. Proposed amendments to California's process safety regulations for refineries contain a requirement to develop, implement and maintain an effective written human factors program that goes beyond the requirements of Contra Costa's industrial safety ordinance [20].

All process safety management programs should have a distinct element that addresses human factors for processes with specific requirements to be addressed. Many aspects of human factors for process safety have been addressed by CCPS [32] and they should be covered by process safety programs.

For many incidents, it is difficult to believe that the conditions that resulted in the incident were allowed to occur. An independent observer easily can conclude that surely someone could have intervened at some stage in the sequence of events that occurred and taken action to prevent the incident. However, such a view does not take account of the psychological factors that influence decision making by people, particularly in groups and when under pressure to achieve competing goals. Factors such as group think, peer pressure, and other cognitive biases interfere with objective decision making [33]. Questionable decision making played a role in numerous incidents, for example {58, 63}. Facility personnel need to apply critical thinking in their daily work activities [34]. Critical thinking can be taught and key facility personnel should be so trained.

3.6 Process Hazard Analysis

Failures in PHA are a common finding in CSB incident investigation reports. Indeed, issues with PHA were present for all CSB incidents where a PHA study was performed. Of particular note is that CSB found that incidents were not identified by the PHA studies performed. However, many of the incidents involved deliberate operation of processes in ways that a PHA team would not consider credible. PHA teams look for ways in which a process may deviate from its design intent but usually the specified design intent identifies what should be done, not what should *not* be done. It is the latter that occurs in many incidents and, while PHA teams try to identify ways that a process may be misoperated, it is unreasonable to expect a PHA team to identify the ways in which a process may be maloperated. Moreover, there is essentially an infinite number of ways in which a process may be maloperated and it would be impractical to ask a PHA team to try and identify all of them. Maloperation must be addressed through operating discipline, proper change reviews, and designing robust processes.

In some cases, the CSB found that PHA studies did not adequately address safeguards because safeguards listed in PHA studies were non-specific or general {19a, 56, 63} and, in many cases, relied too heavily on administrative procedures. For example, safeguards were recorded as "design", "standard operating procedures", "testing and inspection", and "proper emergency response procedures". PHA studies must identify specific safety-critical equipment, for example, a specific relief valve, and specific safety-critical actions, for example, activation of an emergency quench system by an operator. PHA teams should consider the hierarchy of controls when evaluating existing safeguards and deciding on recommendations for further risk reduction. PHA team leaders and team members must be competent [35].

Some CSB incident investigation reports noted the failure to validate critical assumptions used in PHAs {45, 56, 29}. However, such validations should not be part of PHA. They should be performed before a PHA study is performed [21]. Such validations require a different skill set than that required for PHA team members. Moreover, if they were included in PHA studies they would add complexity to the studies and detract from their primary goal of scenario identification. Of course, the adequacy of safeguards for individual hazard scenarios must be addressed by the PHA team.

Some incidents have occurred when more than one activity was being performed simultaneously or sequentially and the activities conflicted {4, 7, 8, 25, 33, 43, 47, 48, 59}. Simultaneous and sequential operations (SIMOPS) reviews [36] should be performed as adjuncts to PHA studies to address such issues. However, they are not yet common and there are no regulatory requirements to perform them.

PHA studies were known to be performed for 22 of the 68 CSB incidents that were reviewed, although incident descriptions did not indicate whether or not a PHA study was performed for 21 of the incidents. However, it is likely that a PHA study was not performed for most of these incidents. Of the 22 PHA studies that were performed, only 4 studies were conducted for processes not covered by OSHA's PSM standard (see Table 1). This finding suggests that PHA studies are only performed when required by regulations.

3.7 Process Changes

Many incidents involved changes that occurred but were not reviewed for their impact on process safety {2, 4, 6, 7, 8, 16, 17, 18, 21, 25, 27, 29, 37, 45, 46, 54c, 59, 63} and {*2, 9, 11*}, and inadequate reviews were performed in some cases {7, 15, 19a, 29, 32, 45, 54a, 56, 63}. Issues with process change reviews were present in 34% of CSB incidents and 39% of incidents covered by OSHA's PSM standard. Omissions and deficiencies can arise because changes are not recognized as requiring MOC reviews, personnel lack awareness of MOC requirements, or MOC reviews are bypassed for expediency. These problems relate to the competency of facility personnel and operational discipline. If personnel are involved in unauthorized and unreviewed changes, they cannot be viewed as competent to work in the facility in the same way as if they maloperated a process or performed maintenance work incorrectly. The establishment of a competency management program should help in this regard. It should be reinforced by operational discipline.

Also, MOC reviews required by the OSHA PSM standard have a narrow definition of covered changes. They cover only changes to PSI, procedures, and facilities that affect a covered process [24]. These requirements are neither global nor sufficiently specific and they are open to interpretation. Organizational changes have played a role in several incidents {29, 45, 56, 64} but they are not currently covered by the regulations, although OSHA's RFI addresses this omission [23]. However, incidents have involved other types of changes not explicitly defined in US process safety regulations including changes in feedstock {4, 63}, addition or removal of equipment {16, 22, 37, 42, 59}, degraded equipment {59}, scale-up of recipes {32}, and prolonged operation {6}. Furthermore, MOC programs should encompass operational deviations and variances, and abnormal situations as well as preplanned changes [37]. Regulations and best practices should specify that any change that may impact on or affect a release of a highly hazardous chemical should be covered by MOC requirements.

Additionally, activities involved in making process changes can be hazardous and are not covered by current US process safety regulations. MOC reviews of changes should include the performance of job hazard analysis for the activities involved.

3.8 Previous Incidents

Information from previous incidents that could have helped to prevent incidents was not used in many cases {2, 5, 6, 11, 17, 19a, 20, 21, 25, 26, 29, 41, 42, 45, 51, 53, 54a, 54c, 60, 61, 63, 64} and {*4, 11*}. Issues with previous incidents were present for 32% of CSB incidents and 29% of incidents covered by OSHA's PSM standard. Previous incidents were not investigated properly, or at all, and near misses were ignored. The issue of retaining information from previous incidents in the corporate memory of companies is well known [38]. However, information cannot be retained if it has not first been developed.

Facility personnel may not have the time or inclination to study incident descriptions and they may not see or understand their relevance to their own facilities. Reliance cannot be placed on people to address lessons learned from previous incidents unless process safety programs contain specific requirements to do so. OSHA's PSM standard requires that previous incidents be addressed [24]. However, the regulatory requirement is part of the PHA element which simply states that the PHA shall address previous incidents. This weak requirement is likely to result in the inadequate consideration of how information from previous incidents can be used to prevent repeat or similar incidents. Process safety programs should have a separate element on previous incidents, including near misses, that requires companies to identify relevant previous incidents from within the company and from other companies in the process industries, compile pertinent information on them, analyze the information in the context of reducing risks in existing processes, and develop recommendations to do so and track them to implementation. The previous incidents review should be performed periodically, e.g. annually, and documented. It should address both the issue of ensuring that incidents are identified, reviewed, and lessons learned are implemented, and that information is retained.

3.9 Facility and Stationary Source Siting

The proximity of facilities to members of the public was an issue in multiple incidents {2, 3, 9, 10, 15, 16, 17, 18, 19a, 20, 22, 23, 24, 25, 26, 27, 29, 30, 32, 35, 36, 41, 45, 46, 48, 51, 52, 53, 63}. This issue was present for 43% of CSB incidents and 64% of incidents covered by OSHA's PSM standard. It raises concerns regarding land use planning in the US. Proximity of facility personnel to hazards in plants also was an issue in numerous incidents {11, 19a, 20, 21, 25, 26, 29, 32, 41, 48, 61} and {*2, 5, 8, 9, 12*]. This issue was present for 16% of CSB incidents and 25% of incidents covered by OSHA's PSM standard. The consequences of some incidents have been exacerbated by the presence of non-essential personnel {29, 49, 56}.

OSHA's PSM standard requires that "the process hazard analysis shall address facility siting" [24] and EPA's RMP rule requires that "the process hazard analysis shall address stationary source siting [39]. However, it is not sufficient simply to require PHA studies to address facility and stationary source siting. Certainly, PHA teams must address the impact of facility and stationary source siting on hazard scenarios but this approach likely leads to insufficient consideration of facility and stationary source siting issues for processes. Aspects of facility and stationary source siting will be missed because of practical limits on the level of detail that can be addressed in PHA studies, limitations on the ability of PHA methods to address facility and stationary source siting, and time pressures on PHA teams.

Process safety management programs should have a distinct element that addresses facility and stationary source siting for processes with specific requirements to be addressed. These requirements should include topics that often are overlooked including using designs that facilitate minimal operator presence; limiting the number of people exposed to hazards, particularly at more hazardous times, by moving unnecessary people away from non-routine operations; training and requiring people to leave when hazardous conditions or situations arise; and assessing the potential for domino events.

The problem of the proximity of people to hazards is compounded by poor emergency planning and response which was present for about 40% of these incidents (see next section).

3.10 Emergency Preparedness and Response

Deficiencies in emergency preparedness and response existed in many cases {5, 8, 10, 11, 16, 19a, 20, 22, 24, 25, 32, 34, 35, 39, 42, 43, 45, 51, 54a, 54b, 63, 64} and {*1, 3, 6, 10, 12*}. This issue was present for 32% of CSB incidents and 39% of incidents covered by OSHA's PSM standard. Deficiencies occurred in hazard communication, evacuation procedures and emergency egress, community notification, emergency preparedness, knowledge of responders, and coordination between responders. OSHA's RFI addresses requiring facilities to coordinate emergency planning with local emergency response authorities [23].

An effective way to help ensure that emergency response plans are in place and will work is to hold regular, periodic drills to test them. It is important to ensure that emergency preparedness and response correlates emergency response with actual hazard scenarios and addresses access to fire water monitors and HAZMAT equipment during incidents and provisions for containment of fire fighting water to prevent environmental pollution.

A hazard analysis should be performed for emergency response activities to identify issues before emergency response is needed. Such a hazard analysis should incorporate task analysis, job safety analysis, and a SIMOP study.

3.11 Audits and Inspections

CSB investigations found that audits by companies, regulators, insurance providers, and others often did not identify process safety omissions and deficiencies and that audits and inspections missed issues that contributed to incidents in several cases {17, 18, 22, 42, 56, 59, 64}. Inadequate audits were identified in several cases {4, 22, 29}. Issues with audits and inspections were present for 13% of CSB incidents and 18% of incidents covered by OSHA's PSM standard.

The CSB noted that in the case of a refinery incident {56} a refinery National Emphasis Program (NEP) inspection was carried out about 18 months before the incident occurred but the mechanical integrity issues that were at the heart of the incident were not identified. NEP audits are the most comprehensive ones that have been performed by regulators. This audit was conducted by a team of people that spent several weeks at the refinery and it focused on the unit that subsequently experienced the catastrophic incident investigated by the CSB.

OSHA's PSM standard requires that a company conduct a compliance audit every three years to "verify that the procedures and practices developed under the standard are adequate and are being followed" [24]. This terse requirement can make for perfunctory audits. Furthermore, process safety compliance audits do not focus on fundamental inadequacies in the design of a process safety program; rather they evaluate the design against regulatory requirements and their implementation. Audits may not properly address non-routine operations because usually they do not occur when audits are performed. Additionally, audits are conducted under time limitations and use sampling techniques.

More comprehensive audits should be performed. They should cover all aspects of a process safety program including safety culture, roles and responsibilities, Auditors must be competent and should be certified [40]. OSHA's RFI addresses requiring third-party audits, increasing the required frequency of compliance audits, and requiring specific time frames for responding to deficiencies found by audits. There is an international standard for auditing management systems [41]. Process safety regulations should require audits be conducted in compliance with the standard.

3.12 Equipment Damage Mechanisms

Equipment damage mechanisms played an important role in several incidents {4, 10, 19a, 20, 53, 56, 63}. Interestingly, all but one of these incidents occurred in a facility covered by OSHA's PSM standard. The CSB has noted the need to perform damage mechanism reviews (DMRs) {63}. Proposed amendments to California process safety regulations include requirements for conducting DMRs [20].

3.13 Process Safety Regulations

Incidents occurred in facilities both covered and not covered by US process safety regulations. Many facilities that are not covered by process safety regulations in the US pose serious process safety risks, for example, facilities handling reactive chemicals and dusts. These issues have been recognized by the CSB [10, 11].

In some cases, facilities that are covered by process safety regulations had not implemented a process safety program {24, 32, 34, 36, 48}.

3.14 Compliance with Industry and Company Standards and Practices and Regulations

Non-compliance with industry and company standards and practices were common in numerous incidents {1, 4, 8, 9, 13, 15, 16, 17, 21, 24, 27, 28, 29, 31, 33, 40, 42, 43, 47, 50, 51, 56, 54b, 54c, 62, 64} and {*2, 7*}. This issue was present for 38% of CSB incidents and 29% of incidents covered by OSHA's PSM standard. Also, there were failures to follow regulations for multiple incidents {8, 12, 16, 24, 25, 33, 34, 36, 39, 40, 41, 64}. This issue was present for 18% of CSB incidents and 14% of incidents covered by OSHA's PSM standard.

Non-compliance with standards, practices and regulations is a fundamental issue for process safety management. If standards, practices and regulations are ignored, the comprehensiveness of their content does not matter and incidents will continue to occur. Certainly, management review and continuous improvement may go some way to address this problem and the inclusion of this CCPS RBPS element [25] was addressed in OSHA's RFI [23]. The RFI also addressed the inclusion of the CCPS RBPS element on measurement and metrics [25] to track the effectiveness of process safety management programs. Regulators in California have proposed amendments to the state's process safety regulations for refineries that will require the use of process safety performance indicators [20]. Public disclosure of process safety performance would encourage management to pay close attention to ensuring those measures reflect favorably on the company. Thus, for example, it can be expected that there would be fewer action items that remain unclosed past their due dates. Of course, if such an approach were to be applied, it may still be ignored by some companies. Changes in regulatory enforcement may be needed.

A recent case in the UK contrasts markedly with US practices. Phosphine was released from a plant requiring 4,500 people to shelter in place for 2-3 hours and causing disruption to a motorway [42]. The U.K. Health and Safety Executive (HSE) investigated the incident and found that an equipment failure caused loss of containment. The company was fined a total of £333,000 and ordered to pay costs of £110,000 after pleading guilty to an offence under the U.K. Health and Safety at Work Act of 1974 having failed to properly assess and act upon the risk of the equipment failing. Such a penalty for an incident of relatively low consequences contrasts with smaller penalties for greater consequence incidents in the United States.

After the prosecution, an HSE inspector said, "This case should serve as a warning to other companies dealing with harmful substances that they need to get their processes absolutely right, in order to ensure the safety of the public, if they don't they will face the consequences." In virtually, every incident analyzed in this paper, the company did not get it right. There are very few, if any process safety accidents; they are preventable incidents.

3.15 Industry Standards and Guidelines

Numerous CSB incident investigation reports identified flaws and omissions in industry standards and recommended practices {9, 14, 16, 25, 26, 29, 36, 37, 48, 51, 53, 54c, 55, 56, 63} from organizations such as the American Petroleum Institute (API), the American Society of Mechanical Engineers (ASME), the Compressed Gas Association (CGA), and the National Fire Protection Association (NFPA). Issues with industry standards and guidelines were present for 22% of CSB incidents and 32% of incidents covered by OSHA's PSM standard.

Such standards and practices usually are subject to periodic review and updating as part of a process of continuous improvement so such omissions and deficiencies ultimately can be addressed. However, the CSB has criticized the permissive nature of some of these standards and recommended practices {56} which leave options open to companies not to follow some of the guidance. Also, the CSB has criticized the process by which standards are set by some organizations, for example API {56}.

This finding suggests that CSB should investigate a diversity of incidents to identify as many deficiencies and omissions as possible in industry standards and guidelines.

3.16 Product Stewardship

Some incidents involved the use of highly hazardous chemicals obtained by one company from another which did not recognize their hazards {14, 17}. In another case, a company was not informed by a supplier that process safety regulations would need to be implemented before receiving shipments {24}. While chemical supplier companies will not necessarily know the details of how purchasers intend to use the chemicals, they should at least communicate their known hazards to the purchaser. Process safety considerations should be a key part of product stewardship initiatives such as that in the ACC's Responsible Care Code of Management Practices [43]. Suppliers of highly hazardous chemicals have responsibilities in supplying such chemicals to other companies.

Some additional issues were identified for a few CSB incidents (see Table 3).

Table 3. Additional Issues Noted for CSB Incidents.

Issue	Examples
Process boundaries defined too narrowly	59
Inadequate workforce participation	56 / *24*
Inadequate process safety information	3, 6
Inaccurate or incomplete safety data sheets	2, 5, 15, 38
Loss of process design and safety information due to poor management of records	11
Lack of hazard recognition and understanding	14, 15, 21, 26, 31
Inadequate communication of plant hazards	5, 17
Need for positive materials identification during commissioning, modifications, and maintenance	10
Inadequate or no pre-startup safety review	3, 13, 29, 45
No or inadequate near miss reporting	54c
Inadequate or no incident investigation	12, 20
Inadequate contractor management	39, 59, 64
Outstanding action items	2, 4, 6, 7, 16, 25, 45, 54b, 54c, 63 / *27*
Inadequate oversight by management	4, 7, 10, 12, 29, 32

4.0 Discussion of the Analysis

A summary of the results of the analysis is provided in Table 4. Clearly the top issues are abnormal / non-routine work and safeguards which are about equally important for all incidents and for PSM incidents and stationary source siting for PSM incidents. These are followed by procedures which are about equally important for all incidents and for PSM incidents and stationary source for all incidents. Compliance with standards and practices, process changes, emergency preparedness and response, maintenance, and previous incidents come next.

Table 4. Summary of Analysis for CSB Incidents.

Issue	% Incidents	% PSM Incidents
Design	28	18
Safeguards	56	57
Procedures	44	43
Maintenance	34	32
Normalization of deviance	18	18
Abnormal and non-routine operations	62	64
Process changes	34	39
Previous incidents	32	29
Stationary source siting	43	64
Facility siting	16	25
Emergency preparedness and response	32	39
Audits and inspections	13	18
Compliance with industry and company standards and practices	38	29
Failure to follow regulations	18	14
Industry standards and guidelines	22	32

Notable differences between the issues for all incidents and PSM incidents occur in some cases. There are fewer issues for both design and compliance with standards and practices for PSM incidents than for all incidents, but there are more issues for PSM incidents than for all incidents for each of stationary source siting, facility siting, and industry standards and guidelines. The difference for stationary source siting is particularly significant.

The lowest percentage of incidents with any of the issues is 13% and the highest is 64%. This is poor process safety performance.

Time trends for the CSB incidents are provided in Table 5. The numbers of incidents displayed per year in the table largely reflects resources available and decisions made by the CSB. However, it appears that the trend for virtually all issues is for them to continue.

Table 5. Time Trends for CSB Incidents.

Issue	1998		1999		2000		2001		2002		2003		2004		2005		2006		2007		2008		2009		2010		2011		2012	
	N	P	N	P	N	P	N	P	N	P	N	P	N	P	N	P	N	P	N	P	N	P	N	P	N	P	N	P	N	P
Design	1	-	-	1	-	-	1	-	-	-	5	-	-	1	-	1	1	1	2	1	1	-	-	-	2	-	-	-	1	-
Safeguards	1	-	-	-	-	-	1	-	4	-	4	2	-	3	-	3	1	3	2	1	1	1	2	2	3	1	1	-	1	-
Procedures	2	-	-	1	-	-	1	-	3	-	5	2	-	1	-	2	1	1	1	-	2	2	1	-	-	2	2	-	-	-
Maintenance	-	-	-	1	-	-	2	-	2	1	3	-	-	1	-	1	1	1	1	-	1	-	2	1	-	2	2	-	-	1
Normalization of deviance	-	-	-	-	-	-	-	-	1	-	2	1	1	-	-	1	-	-	-	-	1	1	-	-	1	-	1	-	-	1
Abnormal and non-routine operations	1	-	-	2	-	-	3	-	1	1	4	2	1	3	-	1	2	1	2	-	2	3	1	1	4	3	2	-	1	1
Process changes	1	-	-	1	-	-	2	-	1	-	5	-	1	1	-	2	-	1	-	1	-	2	-	-	1	3	-	-	-	1
Previous incidents	1	-	-	-	-	-	2	-	1	-	2	2	-	2	-	1	-	-	1	-	1	1	2	-	2	1	1	-	1	1
Stationary source siting	1	-	-	1	-	-	-	-	1	1	4	3	1	4	-	2	1	2	1	-	-	2	2	2	-	-	-	-	-	1
Facility siting	-	-	-	-	-	-	-	-	1	-	1	2	-	2	-	1	-	1	1	-	-	-	-	1	-	-	1	-	-	-
Emergency preparedness and response	-	-	-	-	-	-	1	-	2	1	1	3	-	2	-	-	1	2	1	-	1	2	1	-	2	-	-	-	1	1
Audits and inspections	-	-	-	1	-	-	-	-	-	-	2	1	-	-	-	1	-	-	-	-	1	-	-	-	-	2	-	-	1	-
Compliance with industry and company standards and practices	1	-	-	1	-	-	-	-	2	-	5	-	1	1	-	2	2	-	1	-	2	1	1	1	1	2	1	-	1	-
Failure to follow regulations	-	-	-	-	-	-	-	-	2	-	1	-	-	2	-	-	1	2	3	-	-	-	-	-	-	-	-	-	1	-
Industry standards and guidelines	-	-	-	-	-	-	-	-	1	-	2	-	-	2	-	1	-	1	-	1	-	-	2	1	1	2	-	-	-	1

N – Non PSM Covered
P – Covered by PSM

There is an average of 5 issues per incident and as many as 13 issues. Some CSB investigations appear to be more thorough than others so the number of issues identified in the investigation reports does not necessarily reflect all of the issues that were present. Consequently, these numbers may be low. However, it is clear that most incidents involved multiple issues and this conclusion would not change. It is likely that the underlying reason for multiple issues being present for the majority of incidents is a systemic problem with the process safety management system and process safety culture.

5.0 Further Observations

This section provides some additional noteworthy observations that were made during the review of incident descriptions.

5.1 Age of Plants

Many process facilities in the US were constructed some time ago. Of course the longer a process operates the more exposure there is to risk. Plant aging factors need to be addressed by process safety programs.

Equipment deteriorates over time and this has played a role in some incidents {63}. Damage mechanism reviews can help to address this issue. Old equipment may be used, for example, a tank constructed in 1929 failed catastrophically in 2008 {47}. Use of old equipment must be subject to close review.

As time passes, industry standards and best practices evolve but the designs of processes are frozen in time. Thus, gaps between current industry standards and best practices and existing company standards and practices may exist. Use of dated standards and practices has played a role in several incidents {30}. Outdated technology also has played a role in some incidents {29}. Process safety programs should require periodic formal reviews to identify and evaluate the feasibility of adopting current standards, best practices, and technology retrospectively. Current standards, best practices, and technology should be addressed when developing recommendations to reduce risk, for example, in PHA studies, MOC reviews, and incident investigations. OSHA's PSM standard does not require employers to evaluate updates to applicable RAGAGEP or to examine new RAGAGEP after evaluating and documenting compliance with mechanical integrity requirements. OSHA's RFI sought comments on the best approach for requiring the evaluation of updates to applicable RAGAGEP. However, RAGAGEP is defined narrowly by CCPS which is the definition being advanced by OSHA. It covers generally approved ways to perform specific engineering, inspection, or mechanical integrity activities, such as fabricating a vessel, inspecting a storage tank, or servicing a relief valve. There are many more types of standards and practices that could be included in the definition of RAGAGEP.

Informal procedures may become institutionalized with the passage of time with operators passing them on verbally without others being aware {46}. Key tribal knowledge may be lost through departures of personnel including retirements {11, 21, 54}. An aging workforce may display diminishing physical and mental faculties. These are human factors issues that should be addressed by a human factors program.

Often process facilities have been built in rural areas away from the public but over time housing developments are built near the facility. This issue was present for several incidents. It must be addressed through land use planning.

5.2 Magnitude of Consequences

The consequences of some incidents could have been much worse. For example, releases occurred under favorable weather conditions {22}, with the wind blowing away from populated areas {10}, when projectiles created by explosions missed hitting adjacent processes {11}, and at a time when the number of people potentially exposed was at a minimum {3, 36}. In some cases, people exposed to serious hazards miraculously escaped harm {63} and avoided being caught in the collapse of structures {49}.

These situations were serendipitous. However, it does suggest a hazard control strategy in which riskier activities, such as non-routine work, should be timed for when the risk to receptors is at a minimum. Also, facility personnel in the vicinity of a riskier activity who are not part of the activity should be relocated for the duration of the activity. Such considerations should be an integral part of planning for non-routine work.

Some injuries occurred when people investigated process abnormalities {32, 45, 64} or tried to fix leaks {62, 63}. Such situations must be taken seriously and not considered part of normal operations. Appropriate precautions must be taken.

6.0 Underlying Issues

This section identifies three issues that underlie all of the other issues. Each underlying issue is discussed and recommendations are made for addressing it.

6.1 Competency of Facility Personnel

The CSB has not explicitly addressed the issue of competency in many incident investigations. However, it was mentioned as a contributor in at least one CSB incident {29} and two CCPS incidents {*2, 4*}. In reading incident descriptions it is hard not to reach the conclusion that the people responsible for the processes were not competent in one or more ways in many cases. This issue likely underlies most of the process safety incidents that have occurred.

The decisions made by personnel ranging from managers to operators can be questioned for many incidents. While facility personnel were not obviously incompetent, in numerous cases clearly they were not competent to perform the work required of them safely. In particular, facility personnel often lacked an awareness and/or understanding of one or more critical aspects of a process, such as its hazards {6, 8, 11, 24, 32, 36, 39, 41, 44, 54a, 54c, 62}. Inadequate operator and mechanic training was an issue in a number of incidents {1, 2, 3, 10, 11, 12, 13, 14, 18, 21, 22, 25, 28, 29, 32, 39, 42, 45, 64}.

OSHA's RFI sought comments on the addition of an element on process safety competency [23] as described by CCPS [25]. Certainly, facilities should implement a competency management program for process safety but the CCPS RBPS competency approach should be modified [44]. Critical personnel should be certified and periodically re-certified. For example, operators and mechanics should not be allowed to operate or maintain a process without demonstrating their competence, including such topics as knowledge and understanding of process hazards, the higher risks of non-steady state operations such as startup, previous incidents, the safe operating envelope for the process, handling abnormal and emergency situations, conditions under which the process must be shut down, use of permit-to-work systems, and the identification and management of changes. Responsibility for supervision of the operation of a process should be formally assigned to a single person and the responsibility formally transferred to another individual as necessary. There must be supervisory oversight of processes at all times.

In some incidents, the availability of technical expertise has been an issue {29, 32, 45}. When critical decisions are faced regarding process safety, someone must be present who understands mass and energy balances, physical and chemical phenomena that may occur, process controls and safeguards, hazards present, and the safe operating envelope, that is tolerable ranges of values of safety-critical parameters. Individuals should be designated as engineering and technical authorities who ensure the proper application of engineering standards and technical practices and adjudicate when deviations are proposed. They must be available for consultation at all times and they must have the authority to shut down processes without the need for consultation with managers and without recrimination. They must be free of production or commercial pressures and should report directly to upper management. A stop work authority that authorizes personnel to stop work in the face of imminent risk or dangerous activity and an ultimate work authority that requires designation of who has the authority for operational safety and decision making are addressed by OSHA's RFI [23].

6.2 Standards for Process Safety Elements

Process safety programs comprise multiple elements such as employee participation, PSI, PHA, etc. For example, OSHA's PSM standard has 14 elements [24] and CCPS's RBPS program has 20 elements [25]. However, it is possible for companies to address each of these elements in widely varying ways because their requirements are general in nature and do not address the effectiveness of their contribution to risk reduction. For example, PHA studies can be conducted with widely varying levels of quality. A solution to this problem is to establish minimum performance standards for each element. These standards should be established by an international organization such as ISO so they apply around the world. There is precedent for such standards in the IEC 61511 standard for safety instrumented systems [45 - 47] and the Hazard and Operability Study (HAZOP) standard [48], although the latter is outdated. These standards are still performance-based in nature, as should be new standards for other process safety elements.

One benefit of standards for process safety elements is that they can be updated more easily than regulations. Current US process safety regulations are static and unable to adapt to innovation and advances in the management of major hazard risks.

This problem could be overcome if the US process safety regulations simply established overall requirements for each process safety element and required companies to follow RAGAGEP for each element. There is already a precedent for doing so with the PSI and mechanical integrity PSM elements although RAGAGEP is interpreted more narrowly in that context. The proposal is to broaden the meaning of RAGAGEP, establish RAGAGEPs for all process safety elements, and require that RAGAGEPs be followed. The process safety regulations would effectively update automatically as RAGAGEPs evolve. Continuous improvement is an essential element of a management system for any surety such as quality, environmental protection, and occupational health and safety, and is part of CCPS's RBPS program [25].

6.3 Safety Goals and Risk Tolerance Criteria

In an least one incident, the CSB concluded that one company had knowledge of and acceptance of high risk {29}. Most companies do not really know what risk their process facilities pose because they do not evaluate risk as there are no requirements to do so under the current US process safety regulatory regime. Indeed, it is possible that similar facilities may pose quite different risks because the regulations do not contain any safety goals or risk tolerance criteria.

The goal of process safety programs is to ensure safety and control risks. Consequently, some form of safety goals or risk tolerance criteria are needed [49]. Absent some metric to judge achievement of process safety goals, resources may be allocated inappropriately. There is a current precedent for the use of numerical risk tolerance criteria provided by the New Jersey Toxic Catastrophe Prevention Act (TCPA) process safety regulations [18]. Although the particular approach used is erroneous [50], companies in the state have accepted the use of numerical criteria.

Many jurisdictions around the world employ the tolerability of risk framework developed by the UK HSE including the As Low As Reasonably Practicable (ALARP) principle [51]. Its use should be incorporated into US regulations. Although the ALARP principle usually incorporates the use of numerical criteria, it can be used without specifying such criteria but using its essential element that risk should be reduced until the incremental expenditures of resources is grossly disproportionate to the value of the incremental risk reduction achieved. This is a performance-based requirement that is consistent with the current philosophy of US process safety regulations. Use of the ALARP principle would help to accommodate improved process safety practices without the need for periodic modification of regulations.

One of the difficulties faced by PHA teams is the use of risk ranking in studies to determine if recommendations for risk reduction, that is, additional safeguards, are needed. The CSB has been critical of the subjectivity involved and the criticism is warranted [31]. The CSB has suggested that Layers of Protection Analysis (LOPA) be used for this purpose. Of course, LOPA requires the use of risk tolerance criteria that are allocated from facility criteria [52]. Proposed amendments to the California process safety regulations include requirements for such an analysis [20]. Simply put, the adequacy of safeguards cannot be determined without risk tolerance criteria.

Use of risk tolerance criteria with LOPA raises the issue of what type of risk is being addressed and should be addressed. Certainly, no loss of life is acceptable but in US society single fatalities are more tolerable than multiple fatalities as evidenced by the relative attention paid by the media to single versus multiple fatality incidents. Perhaps the focus in process safety should be on protecting groups of people, particularly members of the public.

All process safety elements would benefit from the use of appropriate performance indicators [25, 53, 54]. Taken together, their use helps to control overall facility and process risks.

7.0 Conclusions

This survey of incidents covered a range of company and facility sizes, and types of processes. Also, it included facilities covered and not covered by process safety regulations. The study results can be used by companies to focus on improving areas where other companies have fallen short.

Of the 80 incidents studied, only the explosion resulting from corrosion failure of piping at the Shell refinery in Norco, LA, that occurred in 1988 {5} was truly unexpected owing to a unique corrosion phenomenon that occurred under conditions that were expected to be less corrosively onerous and with pH within the tolerable range. However, the failure occurred as a result of process changes and it is possible that the incident could have been prevented. All other incidents were predictable and therefore preventable.

Process safety programs aim to identify possible incidents before they occur and implement practices to prevent them. Remarkably, many of the incidents reviewed were attributable to deficiencies or omissions in the design, operation, or maintenance of processes. Therefore, it should have come as no surprise that these processes experienced incidents.

The quote, "It is said that there is nothing new in safety - it has all happened before" was published by CCPS [1]. Virtually every incident or something like it had happened somewhere before. All the incidents involved some type of deficiency or omission in adhering to established process safety practices. In many cases there were multiple deficiencies and omissions. All of these incidents were predictable and preventable. Just because an event may be described as unthinkable does not necessarily mean it is unpredictable.

Several recommendations can be made to strengthen process safety programs and regulations.

7.1 Recommendations for Operating Companies

- Add process safety elements on:
    - Design and engineering

    - Human factors review
    - Facility and stationary source siting review
    - Safeguards protection analysis
    - Operational discipline
    - Competency management
    - Previous incidents review
    - Best practices review
- Address non-routine operations throughout all process safety elements.
- Require the use of JSA for non-routine work.
- Expand the types of changes addressed under the MOC element to include any change that may impact on or affect a release of a highly hazardous chemical.
- Make compliance audits more comprehensive.
- Require annual emergency response drills for facilities.
- Require that process risks be managed using the ALARP principle.
- Require facilities to set safety goals and/or risk tolerance criteria and demonstrate compliance with them.

7.2 Recommendations for Standards Organizations

- Develop international standards for key process safety elements that can be referenced by process safety regulations.

7.3 Recommendations for the Chemical Safety Board

- Expand the incident attributes that are addressed in investigations to gain more insights into process safety performance, for example, explicitly address competency, human and organizational factors, safety culture, plant aging factors, and operational discipline.
- Provide an appendix to each investigation report that characterizes each attribute for the incident. Not only is information on deficiencies and omissions valuable but also knowledge of what was done correctly or that there were no problems in certain areas is valuable.
- Compile a data base of incidents and their attributes that is searchable.

- Provide periodic analyses of the full data base of incidents looking for commonalities and trends and make recommendations for improvements in process safety practices.
- Publish the criteria used for the selection of incidents that are investigated by the CSB and update as appropriate.
- Expand on the current documentation in incident investigation reports to explain how the criteria were applied in selecting the incident for investigation.

The CSB adopted a motto in its early days. It still applies:

Salus populi suprema lex esto.

(People's safety should be the highest law.)

8.0 References

[1] J. Atherton and F. Gil, Incidents That Define Process Safety, Wiley-AIChE/CCPS, New York, NY, 2008.

[2] K. Gugan, Unconfined Vapor Cloud Explosions, Gulf Publishing, Houston, TX, 1979.

[3] T. A. Kletz, Learning from Accidents, 3rd edition, Routledge, Abingdon, UK, 2001.

[4] T. A. Kletz, What Went Wrong?: Case Histories of Process Plant Disasters and How They Could Have Been Avoided, 5th edition, Elsevier, Burlington, MA, 2009.

[5] T. A. Kletz, Still Going Wrong!: Case Histories of Process Plant Disasters and How They Could Have Been Avoided, Butterworth-Heinemann, Burlington, MA, 2003.

[6] Roy E. Sanders, *Chemical Process Safety: Learning from Case Histories*, Butterworth-Heinemann; 4th edition, Butterworth-Heinemann, 5th edition, 2015.

[7] The 100 Largest Losses 1974-2013, Large Property Damage Losses in the Hydrocarbon Industry, 23rd Edition, Marsh and McLennan.

[8] http://www.csb.gov.

[9] http://www.icheme.org/lpb.aspx

[10] Hazard Investigation: Improving Reactive Hazard Management, U.S. Chemical Safety Board, Washington, DC, 2002.

[11] Combustible Dust Hazard Study, U.S. Chemical Safety Board, Washington, DC, 2006.

[12] Seven Key Lessons to Prevent Worker Deaths During Hot Work In and Around Tanks, Safety Bulletin, U.S. Chemical Safety Board, Washington, DC, 2010.

[13] A. S. Blair, Management system failures identified in incidents investigated by the U.S. Chemical Safety and Hazard Investigation Board, Process Safety Progress, Volume 23, Issue 4, pages 232–236, December 2004.

[14] M. Kaszniak, Oversights and omissions in process hazard analyses: Lessons learned from CSB investigations, Process Safety Progress, Volume 29, Issue 3, pages 264–269, September 2010.

[15] Amyotte, P. R., MacDonald, D. K. and Khan, F. I. (2011), An Analysis of CSB Investigation Reports Concerning the Hierarchy of Controls. Process Safety Progress, Volume 30, Issue 3, pages 261–265, September 2011.

[16] T. Kerin, The evolution of process safety standards and legislation following landmark events—what have we learnt?, Process Safety Progress, DOI: 10.1002/prs.11762.

[17] P. Baybutt, Process Design - A Missing Process Safety Element (letter to the editor), Process Safety Progress, Volume 35, Issue 1, March, 2016.

[18] New Jersey Department of Environmental Protection, Title 7, Chapter 31, Toxic Catastrophe Prevention Act Program, Consolidated Rule Document.

[19] Contra Costa County Ordinance, Chapter 450-8-Risk Management.

[20] Proposed General Industry Safety Order (GISO) §5189.1, Process Safety Management for Petroleum Refineries, State of California, Department of Industrial Relations, Division of Occupational Safety and Health, Version 5.0 - September 14, 2015.

[21] P. Baybutt, On the Validation of Safeguards for Process Hazard Analysis, Process Safety Progress, Volume 32, Issue 2, pps. 165–169, June 2013.

[22] P. Baybutt, "What risk reduction measures should be credited in process hazard analysis?", Process Safety Progress, Vol. 31, No. 4, pps 359–362, December, 2012.

[23] Process Safety Management and Prevention of Major Chemical Accidents, Request for Information, Federal Register Volume 78, Number 236, Monday, December 9, 2013.

[24] Process Safety Management (PSM) of Highly Hazardous Chemicals, 29 CFR 1910.119.

[25] Guidelines for Risk Based Process Safety, Wiley-AIChE/CCPS, New York, NY, 2007.

[26] Risk Management Program for Chemical Accident Prevention, 40 CFR Part 68.

[27] Conduct of Operations and Operational Discipline, Wiley-AIChE, New York, NY, 2011

[28] D. Vaughn, Risky Technology, Culture, and Deviance at NASA, 1st edition, University of Chicago Press, Chicago, IL, 1996.

[29] P. Baybutt, Process hazard analysis for phases of operation in the process life cycle, Process Safety Progress, Vol. 31, No. 3, pps 279–281, September, 2012.

[30] G. Swartz, Job Hazard Analysis: A Guide to Identifying Risks in the Workplace, Government Institutes, Rockville, MD, 2001.

[31] P. Baybutt, The Role Of People And Human Factors In Performing Process Hazard Analysis And Layers Of Protection Analysis, J. of Loss Prevention in the Process Industries, Vol. 26 pps.1352-1365, 2013.

[32] Human Factors Methods for Improving Performance in the Process Industries, Wiley-AIChE/CCPS, New York, NY, 2007.

[33] P. Baybutt, The Psychology of Decision Making in Process Hazard Analysis, 11th Global Congress on Process Safety, Austin, TX, April 26 - 30, 2015.

[34] B. N. Moore and R. Parker, Critical Thinking, Eleventh Edition, McGraw- Hill Education, New York, NY, 2015.

[35] P. Baybutt, Competency requirements for process hazard analysis (PHA) teams, Journal of Loss Prevention in the Process Industries, Volume 33, January 2015, Pages 151-158.

[36] Guidance on SIMOPS, IMCA M 203, International Marine Contractors Association, March, 2010.

[37] Management of Change, Safety Bulletin No. 2001-04-SB, U.S. Chemical Safety Board, 2001.

[38] T. A. Kletz, Lessons from Disaster: How organisations have no memory and accidents recur, Institute of Chemical Engineers, Rugby, UK, 1993.

[39] Accidental Release Prevention Requirements: Risk Management Programs Under Clean Air Act Section 112(r)(7), Federal Register / Vol. 61, No. 120 / Thursday, June 20, 1996.

[40] P. Baybutt, Competency requirements for process safety auditors, Process Safety Progress, Volume 34, Issue 2, pages 179–182, June 2015.

[41] Guidelines for auditing management systems, ISO 19011:2011, International Organization for Standardization, Geneva, Switzerland, 2011.

[42] http://press.hse.gov.uk/2016/chemical-company-fined-for-dangerous-gas-release/?ebul=hsegen&cr=8/08-feb-16, last accessed 2/8/16.

[43] Responsible Care® Process Safety Code, American Chemistry Council.

[44] P. Baybutt, The meaning and importance of process safety competency, Process Safety Progress, DOI: 10.1002/prs.11767.

[45] ANSI/ISA—84.00.01—2004 Part 1 (IEC 61511-1 Mod), Functional Safety: Safety Instrumented Systems for the Process Industry Sector — Part 1: Framework, Definitions, System, Hardware and Software Requirements.

[46] ANSI/ISA—84.00.01—2004 Part 2 (IEC 61511-2 Mod), Functional Safety: Safety Instrumented Systems for the Process Industry Sector — Part 2: Guidelines for the Application of ANSI/ISA—84.00.01—2004 Part 1 (IEC 61511-1 Mod).

[47] ANSI/ISA—84.00.01—2004 Part 3 (IEC 61511-3 Mod), Functional Safety: Safety Instrumented Systems for the Process Industry Sector — Part 3: Guidance for the Determination of the Required Safety Integrity Levels — Informative.

[48] Hazard and Operability Studies (HAZOP Studies) – Application Guide, IEC 61882:2001, International Electrotechnical Commission, Geneva, Switzerland, 2001.

[49] P. Baybutt, Letter to the Editor, Managing process risks, Process Safety Progress, Volume 34, Issue 2, page 106, June 2015.

[50] P. Baybutt, Numerical Risk Tolerance Criteria in the United States: a Critique of the Risk Criterion Used for the New Jersey Toxic Catastrophe Prevention Act, Journal of Loss Prevention in the Process Industries 32 (2014), Volume 32, November 2014, Pages 428-435.

[51] P. Baybutt, The ALARP Principle in Process Safety, Process Safety Progress, Vol. 33, Issue 1, Pages: 36–40, March 2014.

[52] P. Baybutt, Allocation of risk tolerance criteria, Process Safety Progress, Volume 33, Issue 3, pages 227–230, September 2014.

[53] Guidelines for Process Safety Metrics, Wiley-AIChE, New York, NY, 2009.

[54] Guidelines for Integrating Management Systems and Metrics to Improve Process Safety Performance, Wiley-AIChE, New York, NY, 2016.

Accidents during Turnarounds, Cleanings, and other Infrequent Operations

Michael C. Stern
Exponent, Inc.
Natick, Massachusetts
mstern@exponent.com

Vijay Somandepalli,[1] Noah Budiansky,[1] Ali Reza,[2] Timothy J. Myers[1]
Exponent, Inc.
[1]Natick, Massachusetts
[2]Los Angeles, California

Prepared for Presentation at
American Institute of Chemical Engineers
2016 Spring Meeting
12th Global Congress on Process Safety
Houston, Texas
April 11-13, 2016

Accidents during Turnarounds, Cleanings, and other Infrequent Operations

Michael C. Stern
Exponent, Inc.
Natick, Massachusetts
mstern@exponent.com

Vijay Somandepalli, Noah Budiansky, Ali Reza, Timothy J. Myers
Exponent, Inc.

Keywords: Process Safety, Turnarounds, Startup, Shutdown, Cleaning

Abstract

Infrequent operations, such as turnaround services and cleaning operations, pose unique hazards due to their lesser familiarity to operators and plant personnel. For example, cleaning chemicals and methods can have unforeseen reactions with scale material, the chemistry of which may not be well-understood. Turnaround services may create a false sense of security because the plant is not in operation, which eliminates many of the commonly encountered hazards. Turnaround services, however, can be hazardous, especially if proper precautions have not been taken when shutting down the plant. Avoidance of accidents during these infrequent operations requires similar hazard analyses to those for standard operations, but these may not be thoroughly performed if the operation is assumed safe. Multiple case studies will be presented on accidents that have occurred during infrequent operations and the lessons learned from those incidents will be discussed.

1 Introduction

Operations including startups, shutdowns, turnarounds, and cleanings can pose additional risk for facilities due to their infrequency and uniqueness. The infrequency of these types of operations provides less opportunities for operators to gain experience and understanding of the procedures. The uniqueness, such as replacing or modifying a part that has been relatively untouched for years, can lead to unintended changes in the process when operation resumes.

Variability is a common risk factor for cleaning and turnaround operations. The same cleaning procedure or chemicals that might have worked for a certain type of scale could be hazardous if used for to clean out a different contaminant. Turnarounds and cleanings may also involve the use of contractors who are less familiar with the facility and may require site specific training to safely operate.

In this paper, the heighted risks of different infrequent operations are discussed and relevant case studies are presented. The case studies, some previously published and some presented for the first time, span a wide range of industries from manufacturing facilities to chemical plants, but all experienced or observed failures during an infrequent and transient operations.

2 Cleaning Operations

Both scheduled and unscheduled cleaning operations are often required at chemical and manufacturing facilities to maintain the equipment functionality and minimize contaminants in the products. Cleaning with chemicals can pose hazards due to undesired chemical reactions between the scale material and the cleaner. In certain cases when multiple cleaners are used, the cleaners could react to form hazardous substances or gases. Table 1, shown below, provides a few examples of chemical mixtures that could form explosive, toxic, or flammable substances leading to hazardous conditions. As will be described in the Condea Vista case study below, even steam cleaning can lead to an explosion in a process vessel when contacted with an aluminum powder containing sludge.

Table 1. Potential reaction products and hazards for certain chemical mixtures

Chemical Type One	Chemical Type Two	Reaction Products	Hazards
Acetone	Peroxides	Acetone-Peroxides	Explosives
Salts with (Cl^-, F^-, CN^-) ions	Acids	Acid gases	Corrosive and toxic
Metal Powders	Water or Steam	Hydrogen	Flammable

When cleaning areas that are already hazardous locations, such as areas near equipment containing flammable liquids or combustible dusts, the cleaning process can be a potential ignition source if proper precautions are not taken. Routine maintenance or cleaning actions can also release combustible dust accumulations creating a hazardous environment in an area where ignition controls are not present, leading to a fire or explosion. The case studies of the Jahn Foundry and CTA Acoustics combustible dust explosions provide examples of how ignition of disturbed accumulations can lead to catastrophic explosions.

2.1 Condea Vista Chemical Plant Explosion[1]

On October 13, 1998, an explosion and fire occurred inside a 3000-gallon Hastelloy reactor (R-2) that was part of the linear alkyl benzene (LAB) process at the Condea Vista plant in Baltimore, Maryland. Process materials released from the various vessels and process piping by the explosion fueled a fire that took approximately 2 hours to extinguish.

The investigation revealed that prior to the explosion, the bottom of R-2 was plugged with approximately 180 gallons of a sludge-like mixture that likely consisted of a mixture of coarse aluminum powder, aluminum chloride and various hydrocarbons. The explosion occurred approximately 10 minutes after steam was introduced at the bottom of the reactor in an effort to break up the sludge accumulation.

Figure 1. Photographs showing the overview of the explosion (left) and top of the exploded R-2 reactor (right).

A series of small-scale chemical reactivity tests were performed to investigate the potential reactions between steam and manufactured sludge that was similar to the material accumulated inside the R-2 reactor. Rapid runaway chemical reactions were observed after steam was injected into this mixture, while pressures in excess of 700 psig and temperatures in excess of 400 F were recorded before the test vessels ruptured.

A review of possible chemical reactions between steam, aluminum and aluminum chloride confirms that the predominant reaction mechanism is the exothermic hydrolysis of aluminum chloride to produce HCl vapor. Significant amounts of hydrogen gas can also be generated as the aluminum reacts with the acid/steam mixture after it is stripped of its protective oxide coating. Exponent concluded that the root cause of this accident was direct and sustained contact between the steam and accumulated sludge at the bottom of the reactor.

2.2 Dust Explosion at CTA Acoustics[2,3]

At approximately 7:30 am on February 20, 2003, a fire in at the CTA Acoustics facility in Corbin, Kentucky, lead to multiple secondary explosions and fires resulting in the death of seven workers and 37 injuries.

CTA Acoustics manufactured thermal acoustic insulation. The insulation material consisted of fiberglass bound together by phenolic resin powder. The product was used in automotive applications including the insulation on the hood or floor board of a car. CTA had four fiberglass forming lines, 401, 402, 403 and 405. The CSB report on the incident stated that the explosion originated near line 405 where dust generated during cleaning was ignited by the oven on line 405.

The CSB investigation concluded that workers started a routine cleaning of the dust collection system on line 405 thirty minutes before the incident occurred. The cleaning was expected to take approximately one hour. The night shift had reported significant clogging on the line. During the cleaning, the operators observed that the transition leading to the baghouse was the source of the clog and proceeded to use compressed air to break up the clog. When the baghouse was restarted at approximately 7:30 am, an initial explosion occurred that spread to the rest of the facility through secondary explosions fueled by clouds of combustible dust that were dislodged during the initial explosion.

2.3 Dust Explosion at the Jahn Foundry[3]

Figure 2. Photographs showing dust deposits on exhaust hoods in Shell Mold department (left) and the structural damage to the foundry (right).

At approximately 3:04 PM, on February 25, 1999, a series of explosions and a subsequent fire caused extensive damage within the Shell Mold building at the Jahn Foundry Corp. in Springfield, MA. Three employees suffered fatal injuries and nine others were injured. The foundry produced gray iron cast products for various customers.

The casting process included the production of shell molds, into which the molten material was ultimately poured. The molds were produced from a 95/5 percent mixture of sand and phenolic resin, added as a thermosetting bonding agent to hold the mold together. Even though the resin was supplied in the form of a fine powder, a dust collection system was not installed in the Shell Mold building. However, an odor control system designed to remove volatile organic compounds emitted from the resin during heating of the shell molds was sufficiently powerful to entrain fine particulates. The odor control system consisted of hoods or canopies above each shell mold station that vented into duct manifolds that connected to a scrubber tower where volatile organic compounds were removed.

An engineering investigation of the incident concluded the explosions were initiated due to the ignition of resin dust dislodged from an overhead exhaust duct or inadvertently dispersed during routine maintenance. A subsequent deflagration (dust explosion) within the odor control system was fueled by excessive, resin rich deposits of dust within the exhaust ducts. A second, larger dust explosion in the upper half of the Shell Mold building caused the majority of the damage to the structure. This second explosion was fueled by resin rich dust deposited on various horizontal surfaces on the upper level of the Shell Mold building.

Personnel at Jahn Foundry were generally unaware of the specific dust explosion hazard associated with this product and the foundry did not comply with the requirements of NFPA 654, *Standard for the Prevention of Fire and Dust Explosions from the Manufacturing, Processing and Handling of Combustible Particulate Solids*

3 Startup and Shutdown Operations

Operator inexperience and unidentified operational differences can lead to additional hazards when starting up or shutting down a plant. For plants that only shut down for maintenance annually or less, the operators will have only limited experience with the tasks required for startup and shutdown due to the infrequency of the operation. As a result, lessons learned from previous experiences may be less likely to be remembered, recorded, and passed on to future operators.

Differences that occur in the process during startup can be difficult to predict and assess. In systems with dust generation, for example, the properties of the collected dust during startup can vary from the dust generated during steady-state operations. It can be difficult to quantify these differences because process sampling during startup is complicated by the inherent transience of the operation.

In processes that include recycle loops or process material recycled downstream, the feedstock during startup can vary in considerable but unidentified ways. Small process changes, equipment maintenance and replacement, and other alterations to the system can introduce small but significant changes into the performance of the system that could lead to equipment failures if the changes are not anticipated. The case of the ammoxidation reactor incident described below provides an example of a chemical reactor that was damaged during startup due to the changes that occurred while the reactor was offline.

3.1 Ammoxidation reactor failure during startup

Several startup attempts of an ammoxidation reactor led to unplanned shutdowns within hours of starting due to over-temperature alarms. The ammoxidation reactor was a packed-bed counter-current tube-in-shell reactor where the tubes were packed with catalyst and an organic heat transfer fluid was pumped through the shell to cool the tubes and remove heat from the exothermic reaction. After a two-day period involving several attempted startups that were followed with subsequent shutdowns due to over-temperature conditions, the reactor was taken offline. Inspection of the reactor revealed a significant leak of heat transfer fluid and severe localized damage to the top tubesheet. The reactor was repaired by replacement of tubesheet sections and a number of tubes.

The reactor was brought back online and was used to produce a different isomer and performed without issue. After two months, the reactor was switched back to producing the original isomer. After two unsuccessful startup attempts that led to shutdowns due to over-temperature conditions, the reactor was inspected and thermal damage was observed.

Investigation of the reactor incidents revealed that prior to the startup, new catalyst was installed to replace the old catalyst that had been in the reactor. The catalyst packing gradient had also been adjusted such that higher concentrations of catalyst were present near the beginning of the reactor tubes. The new catalyst packing gradient and the increased activity of the fresh catalyst both contributed to increased reaction rates within the reactor during operation. It was also discovered that the heat transfer fluid reacted exothermically when exposed to conditions inside the reactor tubes.

It was determined that the failure of the reactor was a direct result of thermal expansion of tubes that occurred as feed rates were increased in the reactor and over-temperature conditions occurred. The thermal expansion led to the mechanical failure of a tube-to-tubesheet connection, which allowed heat transfer fluid to enter the tube side of the reactor. Exothermic reaction of the heat transfer fluid in the tube side of the reactor during repeated startups of the reactor led to the melting and heat damage of the tubesheet.

4 Plant Turnarounds

Plant turnarounds that involve significant maintenance, cleaning, replacement of equipment, and plant modifications can pose additional hazards due to the uniqueness and unfamiliarity of the activities, the involvement of contractors who are not as intimately aware of the hazards at the plant, and the difficulty of entirely purging the process lines and equipment of hazardous chemicals. Corrosion or contaminant buildup that occurred during the operation of the process can alter the condition of the process equipment in ways unbeknownst to the operators until it is inspected during the turnaround. The case study presented below provides an example of equipment that became corroded during operation but was detected before it created hazardous conditions.

4.1 Corrosion in a Flue Gas Desulfurization Absorber

All equipment is subject to both periodic planned shutdowns to conduct routine maintenance and inspections and unplanned shutdowns to replace components or make repairs. During these shutdowns careful inspections of the equipment are important to prevent accumulating degradation of a system or fix other unforeseen issues.

During a routine shutdown to perform periodic cleanout of a flue gas desulfurization (FGD) absorber associated with an incineration plant, corrosion damage was observed in the corrosion resistant liner of the absorber. FGD systems are designed to remove sulfur dioxide and other aggressive gaseous species from power plant and incineration emissions by spraying the flue gas with a limestone slurry, producing a calcium sulfite (gypsum) or other reaction species solid that falls to the bottom of the tower. The FGD slurry environments can have pH values as low as 1 and chloride concentrations as high as 200,000 ppm. The resulting slurry can be very corrosive and highly alloyed metals (or non-metallic coatings) are required to prevent corrosion of the liner to prevent exposure of the underlying carbon steel support structure.

In this case, 2205 duplex stainless was selected as the absorber liner was selected to prevent corrosion based on the nominal operating conditions, but corrosion of the duplex stainless steel liner was observed around a cleanout door. The corrosion damage had penetrated through the stainless steel liner in the time frame since the last inspection and caused minor damage to the underlying structure. A corrosion failure analysis was conducted and it was determined that the isolated environment in the gap between the cleanout access door and the door sill was more aggressive than the nominal conditions in the absorber. Laboratory corrosion testing indicated that corrosion resistance of the stainless steel liner, that was highly resistant to corrosion under the nominal design conditions, was significantly reduced in this area due to accumulated ions resulting in a reduced pH. In this case, periodic inspection of the equipment resulting in detection of corrosion degradation that was ultimately repaired prevented more serious failures from occurring.

5 Discussion and Conclusions

A review many of the industrial failures that have been investigated by Exponent shows a significant number that occurred during infrequent and transient operations including startups, shutdowns, cleanings (scheduled and unscheduled), and chemical transfers. While the specific incidents vary substantially in their magnitude and cause, an unknown change is the system can be traced to the cause. In three cases it was accumulation (combustible dust or aluminum sludge) and in another cases it was a change in a process variable (catalyst packing) that occurred during a turnaround that caused a failure upon startup.

Preparation for these upsets require thorough process hazard analyses (PHAs) and an effective management of change (MoC) program to identify these key variables before they become an issue and develop strategies for operators to employ when these events occur. The case of the FGD unit highlights the importance of careful inspection during turnarounds to identify corrosion or other changes in equipment. In some cases, such as combustible dust incidents, significant industry and governmental focus has raised awareness of the hazards of combustible dust. An OSHA National Emphasis Program and updated NFPA standards have been developed to address these hazards. In other cases, the lessons learned from the incidents should be used in future PHAs to mitigate the likelihood of similar incidents in the future.

6 References

[1] A. Reza, A. Kemal and P. Markey. "Runaway Reactions in the Aluminum, Aluminum Chloride, HCl and Steam System. An Investigation of the 1998 CONDEA Vista Explosion in Maryland," 36th Annual Loss Prevention Symposium. AICHE Spring National Meeting, March 2002.

[2] CSB. *Combustible Dust Fire and Explosions: CTA Acoustics*. United States Chemical Safety and Hazard Investigation Board, Report No. 2003-09-I-KY, February, 2005.

[3] T. Myers and A. Ibarreta. "Investigation of the Jahn Foundry and CTA Acoustics dust explosions: similarities and differences," *Journal of Loss Prevention*, pp. 740-745, 2009.

Biomass Industry Explosion Case Study

Amy E. Theis, P.E.
Fauske & Associates, LLC
16W070 83rd Street, Burr Ridge, IL USA 60524-5802
Theis@Fauske.com

Timothy L. Cullina, M.S., P.E. and Kenneth N. Kurko
Fauske & Associates, LLC
16W070 83rd Street, Burr Ridge, IL USA 60524-5802
Cullina@Fauske.com and Kurko@Fauske.com

Prepared for Presentation at
American Institute of Chemical Engineers
2016 Spring Meeting
12th Global Congress on Process Safety
Houston, Texas
April 11-23, 2016

Biomass Industry Explosion Case Study

Amy E. Theis, P.E.
Fauske & Associates, LLC
16W070 83rd Street, Burr Ridge, IL USA 60524-5802
Theis@Fauske.com

Timothy L. Cullina, M.S., P.E. and Kenneth N. Kurko
Fauske & Associates, LLC

Keywords: Explosion, thermal stability, combustible dust, biomass, incident, dust hazard analysis (DHA), management of change (MOC)

Abstract

The focus of this case study is an explosion that occurred in a biomass processing facility. There were three pieces of equipment involved including a dryer, conveyor and a cyclone. Thankfully, there were no injuries or fatalities due to this incident. This event occurred at a biomass company, resulting in damaged equipment with one piece needing to be completely replaced. Other equipment was able to be repaired and returned to service.

Material handled in the process had been observed to smolder on prior occasions, but the cause of these near miss incidents was not investigated or identified. The material was a potential combustible dust and there was also the possibility for flammable gas to be generated due to decomposition. Through material characterization, several hazards were identified. Combustible dust and calorimetry testing were employed to evaluate the thermal stability, reactivity and combustibility of the material. In addition, off-gases were evaluated to determine if flammable gas was generated. This case study will outline the test design strategy to obtain key process safety information, identify the multiple hazards based on the test results, review existing explosion safeguards and determine a postulated series of events that could have led to the incident. Lessons learned from the incident will be reviewed including recommended risk reduction strategies that could be used to prevent future events. One example is the importance of obtaining appropriate process safety information to facilitate data driven decisions to be made regarding safe operating conditions.

1 Introduction

An explosion occurred at a biomass processing facility. Fortunately no one was injured or killed in this incident. It is believed that a fire, smoldering or other ignition source began in the conveyor. Due to a lack of explosion protection, this led to a subsequent explosion in a connected piece of equipment. Findings indicate that the source of the explosion was consistent with a combustible dust deflagration that appeared to have ignited in or near the transfer equipment and accelerated through the duct work to a cyclone.

Lessons learned presented in this paper are based on experimental test results provided by the customer as well as testing performed by Fauske & Associates, LLC after the incident. Deficiencies that yield opportunities for improvement at this facility include a lack of hazards identification for the process, lack of combustible dust management systems and lack of a risk assessment.

2 Background

2.1 Process Description

In the process, the biomass is fed to a dryer that heats the material to elevated temperatures. The biomass has some residence time in the dryer before it is transferred through the bottom to a conveyor. Forced air is fed to the dryer and then goes to a cyclone where the entrained biomass is collected downstream. See Figure 1 for a process flow diagram.

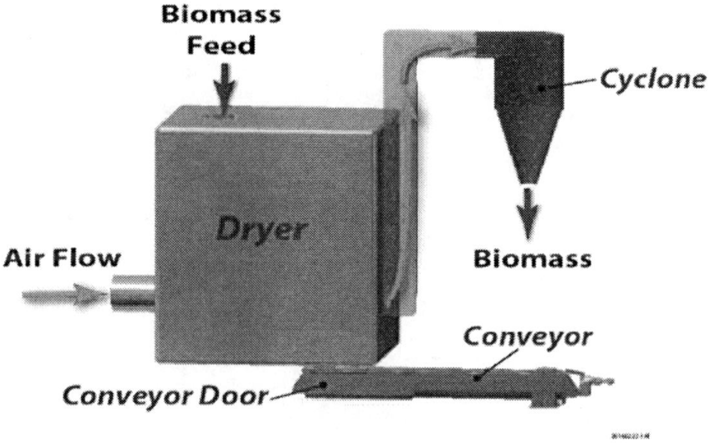

Figure 1 Biomass Process Flow Diagram

2.2 Explosion Event Description

On the day of the explosion event, a fire occurred earlier during the process. It is postulated that heated material accumulated in the conveyor, which led to self-decomposition at elevated temperatures, ultimately resulting in a fire. Employees used a water spray to suppress the flames, but perhaps this did not fully extinguish embers.

When sufficient oxygen was available, hot embers from the transfer equipment could have been drawn into the duct work and provided an ignition source for airborne biomass particulate solids. After ignition of dust, gases, or both, the rapidly expanding pressure wave and flame front moved into all of the available space in the first few milliseconds of the event.

The pressure wave blew the conveyor access door open and ejected burnt material. The flame front propagated to the dryer, fueled by airborne biomass then proceeded to the cyclone, ultimately leading to a rupture along the top of the cyclone. No personnel observed either explosive discharge. All witnesses reported hearing only one explosion. The access door opening and the cyclone failure could have occurred within less than 100 milliseconds of each other, producing a noise identifiable to witnesses as a single explosion. Pressure effects were observed downstream of the cyclone as well.

2.3 Previous Incidents

Based on history from processing similar biomass, personnel believed that smoking and fires, while not desirable, were typical and expected during the process. Fires had not only occurred during processing, but also during transportation of a by-product generated during the process.

In addition, biomass color and odor were not consistent during process runs. Operators believed there was a correlation between initial biomass color, intermediate color/odor, smoking events and fires. A lack of familiarity with combustible dust and gas hazards, especially the potential for deflagrations and explosions by both the management and operators contributed to the incident.

3 Hazard Evaluation

3.1 Material Properties

3.1.1 Biomass #1

Before the explosion, limited combustible dust and calorimetric data were available for this type of biomass. Previous combustible dust test results of a related biomass (Biomass #1) include a minimum explosive concentration (MEC) well above concentrations expected during the normal process conditions and a minimum ignition energy (MIE) value greater than 500 mJ. These results indicate that Biomass #1 was a combustible dust, but was not sensitive to ignition by electrostatic discharge.

Differential Scanning Calorimeter (DSC) testing of Biomass #1 identified one significant exotherm. This exotherm was detected above the normal operating temperature of the process and was confirmed by Carius tube testing. The Carius tube testing also indicated that non-condensable gas was formed during the event.

3.1.2 Biomass #2

After the incident, testing was performed on a sample obtained from biomass being processed at the time of the event (Biomass #2). First, DSC testing was conducted on Biomass #2 per ASTM E 537-12, *Standard Test Method for the Thermal Stability of Chemicals by Differential Scanning Calorimetry* [1] to determine the endothermic and exothermic activity of the material. A Differential Scanning Calorimeter was utilized for this test in conjunction with a 20 μl gold-plated stainless steel high pressure crucible. The DSC cell was heated at a preset rate while the differential heat flow, as a function of temperature, was measured and recorded. Five (5) DSC experiments were run on the sample at different heating rates: 1.0, 2.0, 4.0, 8.0, and 16.0°C/min.

The lowest heating rate DSC test showed that exothermic activity was detected at temperatures just less than the maximum temperature of the normal operating range for the process. Test results indicated that Biomass #2 yielded exothermic activity at lower temperatures compared to Biomass #1. The DSC results are summarized in Figure 2.

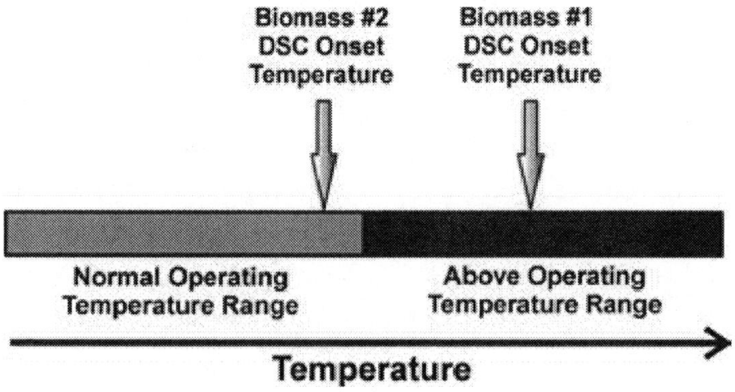

Figure 2 Thermal stability summary for biomass materials

The DSC data were evaluated to determine kinetic parameters for Biomass #2. These results showed that the material exhibited a self-accelerating decomposition temperature (SADT) value according to the United Nations Department of Transportation (UN DOT) H-1 test definition [2] that was within the normal operating temperature range. Similar evaluations were performed for smaller quantities of material (10 kg compared to 50 kg) and were also within the normal operating temperature range of the process.

Kinetic parameters were also used to determine adiabatic times to maximum rate for Biomass #2 as a function of temperature. These results indicated that the temperature that would yield a 24 hour adiabatic time to maximum rate would be in the middle of the normal operating temperature range of the process.

The next test performed on Biomass #2 was conducted using an Accelerating Rate Calorimeter (ARC), which is a high thermal inertia adiabatic calorimeter that is used to obtain data regarding the relationships between time, temperature, and pressure for exothermic reactions. The ARC utilized a 10 ml spherical test cell constructed of stainless steel with ¼" neck. In this experiment, the ARC was run in isothermal mode in order to monitor pressure generation. The instrument was programmed to track a reaction if exothermic activity was detected or maintain the desired temperature. ARC testing was conducted per ASTM E 1981-98, *"Standard Guide for Assessing Thermal Stability of Materials by Methods of Accelerating Rate Calorimetry"* [3].

The ARC isothermal test was performed on Biomass #2 at the maximum operating temperature range of the process over a time period of approximately 6 days. The temperature of the ARC test remained constant for the duration of the experiment. The evidence of non-condensable gas generated indicates that a reaction did occur. After the sample cooled, the non-condensable gas was subjected to a flame. No changes in the flame were observed, indicating that flammable gases were not present.

The last test performed on Biomass #2 was to determine the Minimum Explosible Concentration (MEC) using a 20-L Siwek chamber. The MEC is the minimum concentration of a combustible dust that is capable of propagating a deflagration through a uniform mixture of the dust in air, under the specified conditions of the test. Data obtained from this MEC test method can provide a relative measure of the concentration of a dust in the form of a uniform cloud necessary for an explosion. The test was performed per ASTM E1515, *Standard Test Method for Minimum Explosible Concentration of Combustible Dusts* [4].

Results from the MEC test on Biomass #2 showed that the MEC measured was within the range of what could be expected to occur within the process. If this concentration was present in the process, the material could be readily ignited by a burning/smoldering ember. A comparison of MEC values are shown in Figure 3.

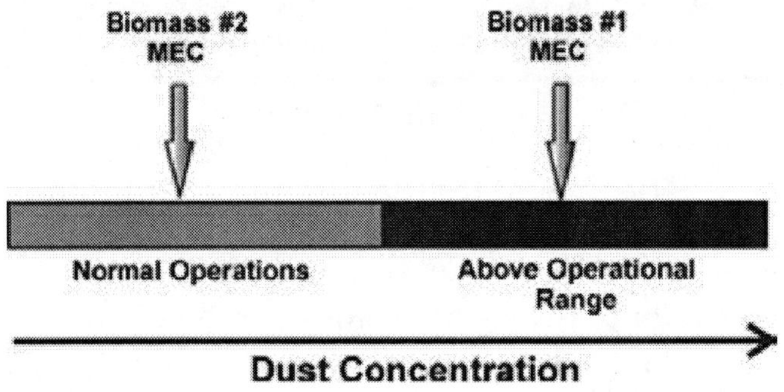

Figure 3 Minimum explosible concentrations summary for biomass materials

3.2 Incident Evaluation

It is postulated that heated material accumulated in the conveyor, which led to self-decomposition at elevated temperatures, ultimately resulting in a fire. Though fire-fighting measured were taken, it is possible that remaining material in the conveyor continued to heat or smolder. This could provide sufficient heat or an ember to serve as an ignition source for airborne particulate solids.

The combustible dust hazards of Biomass #2 handled in an existing process were not adequately identified or understood. It followed that safeguards were not in place to manage the risks associated with each piece of equipment in the process because the hazards and consequences were not characterized or evaluated. Since the hazards were not characterized at the time of the incident, employees were not made aware of the hazards associated with processing the material.

There was a concern whether flammable gas was present that could have created a hybrid mixture in addition to the combustible dust in the system. The presence of a small quantity of volatiles well below the lower flammability limit (LFL) of the gas can create a hybrid mixture with combustible dust that effectively reduces the minimum ignition energy and the minimum explosive concentration required to initiate a deflagration. Testing indicated that non-condensable gas formed during the isothermal experiment was not flammable. The gas tested may not be representative of gases formed during the deflagration event or even after the material was burned, so it is not possible to conclude whether flammable gases were present during the deflagration event. Additional testing is recommended to further evaluate this concern.

The lack of explosion prevention or mitigation devices on the equipment and absence of appropriate training are both a result of this inadequate hazard recognition. Installation of a properly designed explosion vent and installation of isolation technology could have mitigated damage to the cyclone. Figure 4 illustrates the pressure increase in a vessel both with and without the application of explosion mitigation by venting.

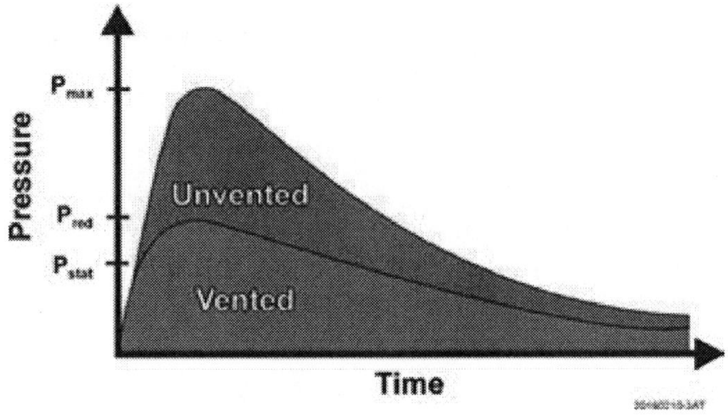

Figure 4 Progress of Explosion With and Without Explosion Protection

There was no record of a dust hazard analysis being performed for this process. Many of the required elements of a comprehensive combustible dust management program as outlined in National Fire Protection Association (NFPA) standards were not utilized to reduce the risk of dust fires and explosions.

3.3 Lessons Learned

This process included equipment covered by NFPA 654, *Standard for the Prevention of Fire and Dust Explosions from the Manufacturing, Processing, and Handling of Combustible Particulate Solids*. While processing Biomass #1, testing was performed to characterize combustible dust hazards and thermal stability for the material. Results indicated that the material was thermally stable in the normal operating temperature range and MEC values were above what would be expected during normal operations. At that time, no further protective measures were implemented. Data was gathered, but a risk basis of safety was not formally established.

When the raw material for the process changed from Biomass #1 to Biomass #2, this modification did not trigger additional testing to be performed. A management of change process, as required in NFPA 654, would have triggered the need to characterize Biomass #2 since it was different from Biomass #1. Even though the process equipment and technology remained the same, the facility changed raw materials which should have triggered Biomass #2 to be evaluated for combustible dust explosibility characteristics. This evaluation should have included thermal stability testing which were also performed previously for Biomass #1.

Fires were not uncommon in this process when processing Biomass #2 and the cause of these fires was not evaluated. It did not appear that an incident investigation process was in place at this facility, as required per NFPA 654 or the most recent NFPA 652, *Standard on the Fundamentals of Combustible Dust*. It appeared that management had allowed an acceptance of deviation to become the norm since fires were expected in the process. The response was for the operator to extinguish the fire and continue processing.

It was not evident if an inspection, cleaning and maintenance schedule was identified for pieces of equipment covered by NFPA 654 and 652. Material was allowed to accumulate in the conveyor which could have self-heated and led to fires observed in this equipment. It is not clear that inspection and access panels were in place to allow adequate cleaning of ductwork and the conveyor. It would be appropriate to establish and review cleaning and maintenance schedules per NFPA 654 and 652 requirements.

A dust hazard analysis (DHA) was not performed on this process. Given the operating temperatures for the dryer, the material should additionally be evaluated for thermal stability. This process should have triggered the need for testing to characterize the material for combustible dust hazards as well as for thermal stability. Once the hazards were identified, the equipment could then be evaluated to determine if additional protective strategies such as explosion protection and isolation were necessary.

After hazards have been identified, employees should be trained and provided guidance on required work practices in order to ensure their safety. Insufficient training contributed to the near miss of exposing employees to the deflagration. Without knowledge of the deflagration potential and the likely effects, employees were preparing to position themselves in high risk areas during the unintended explosion vent.

Data reviews after the incident revealed that employees did not understand the full intent of information recorded during the process. Design of the data forms likely contributed to the uncertainty regarding data collected. Employees did not have an awareness regarding the critical safety parameters and were not adequately trained on how to respond to deviations in the process.

Lessons learned from this case are summarized as follows:

- There was a lack of formal risk based safety documentation for the process.
- A management of change (MOC) process did not trigger data to be obtained when a change in the raw materials for the process were made.
- Sufficient data for the material/process were not obtained to establish safe operating conditions and identify potential upset scenarios.
- Performing a DHA could have identified the hazards, potential ignition sources, potential consequences and provided a framework to evaluate the adequacy of safeguards or mitigation strategies to determine if additional protection was needed.
- Proper hazard identification for the process is critical to ensure that adequate safeguards are in place and that employees receive adequate and appropriate training regarding normal operations and emergency response procedures.

- Routine fires during processing and transportation should serve as warning signs that trigger an investigation to identify preventive and corrective actions.
- The inspection, maintenance and cleaning procedures for dust handling equipment should be consistent with NFPA 654 and 652 requirements.

4 Conclusion

Testing performed after the incident indicated that the new material (Biomass #2) handled in the process was thermally unstable during normal operating conditions as well as being a combustible dust hazard.

In summarizing the lessons learned, it appears that several combustible dust management systems required by NFPA 654 and 652 did not exist, or if they existed, were not effective to prevent this incident. Key process safety information was not obtained to identify the hazards of the process. The company did not understand the warning signs for a potential combustible dust explosion in this process. Performing a DHA on this process could have identified the need to obtain relevant combustible dust and thermal stability data and understand that normal operating conditions could lead to an upset scenario. This information could then be used to propose changes to the process and determine if existing systems and equipment were adequately designed to protect against the consequences of the identified upset scenarios.

The following steps are recommended for identifying hazards associated with a process that contains equipment that could potentially handle combustible dust. These steps are consistent with the practices recommended in NFPA 652.

- Identify equipment in the process that is intended to handle particulate solids and the desired processing conditions (temperature and pressure operating range, equipment design requirements)
- Identify material to be handled in the particulate handling equipment
- Identify the hazards (safety data sheets, combustible dust hazard data, thermal stability data as warranted by process conditions, identify potential consequences due to upset conditions, perform a DHA)
- Understand the consequences in terms of severity and likelihood (determine if flash fire and/or explosion scenarios exist, determine if normal operating conditions or process deviations could lead to thermal decomposition, fire and/or explosion scenarios, identify if flammable gas could be generated)
- Determine process safety controls, safeguards and mitigation strategies (protect against overpressure situations, control and alarm set points, change in process design or procedures, explosion protection systems)
- Implement and maintain the safety measures (document safe operating ranges, perform preventive maintenance and inspections, train appropriate personnel, implement emergency response plans, employ other procedures as identified in NFPA 652)

5 References

[1] ASTM Standard E 537-12. *Standard Test Method for the Thermal Stability of Chemicals by Differential Scanning Calorimetry*, ASTM International, West Conshohocken, PA, 2012.

[2] United Nations. Recommendations on the Transport of Dangerous Goods: Manual of Tests and Criteria (6th ed.), New York and Geneva, 2015.

[3] ASTM Standard E 1981-98 e2. *Standard Guide for Assessing Thermal Stability of Materials by Methods of Accelerating Rate Calorimetry*, ASTM International, West Conshohocken, PA, 2012.

[4] ASTM E1515-14. *Standard Test Method for Minimum Explosible Concentration of Combustible Dusts*, ASTM International, West Conshohocken, PA, 2014.

[5] NFPA. NFPA 652, *Standard on the Fundamentals of Combustible Dust, 2016 Edition.* NFPA, Quincy, MA, September 2015.

[6] NFPA. NFPA 654, *Standard for the Prevention of Fire and Dust Explosions from the Manufacturing, Processing, and Handling of Combustible Particulate Solids, 2013 Edition*, Quincy, MA, June 2012.

[7] Ebadat, V. Testing to Assess Explosion Characteristics of Dust Clouds, NFPA Symposium on Dust Explosion Hazard Recognition and Control, 2009.

[8] U.S. Chemical Safety and Hazard Investigation Board. Investigation Report: Combustible Dust Hazard Study. Rep. No. 2006-H-1, 2009.

[9] Grossel, S.S. and R.G. Zalosh. Guidelines for Safe Handling of Powders and Bulk Solids. Center for Chemical Process Safety, AIChE, New York, 2005.

Atmospheric Tank Explosions: Mechanisms and an Unexpected Case Study

Michael S. Schmidt, P.E.
Principal, Bluefield Process Safety LLC
3101 Cherokee Street
Saint Louis, Missouri 63118, USA
bluefieldsafety@gmail.com

Prepared for Presentation at
American Institute of Chemical Engineers
2016 Spring Meeting
12th Global Congress on Process Safety
Houston, Texas
April 11-13, 2016

Atmospheric Tank Explosions: Mechanisms and an Unexpected Case Study

Michael S. Schmidt, P.E.
Principal, Bluefield Process Safety LLC
3101 Cherokee Street
Saint Louis, Missouri 63118, USA
bluefieldsafety@gmail.com

Keywords: storage tank, atmospheric tank, tank explosion, caustic

Abstract

Most of the literature on catastrophic failures of low pressure tanks focuses on the ignition of explosive mixtures in the vapor space of those tanks, and rightly so, because most catastrophic failures of atmospheric tanks involve explosions of such mixtures. Moreover, there is a general sense in the industry that other than explosions of flammable mixtures, the only other hazard associated with low pressure tanks is the occasional dramatic implosion of a vessel that is inadequately protected against vacuum.

In fact, there are a number of mechanisms that can lead to the catastrophic failure of a low pressure tank that have nothing to do with combustion or unprotected vacuum. Under certain circumstances, even a tank equipped with an atmospheric vent and containing nothing other than salt water can explode, with disastrous impacts.

This paper reviews the mechanisms for catastrophic failure of low pressure tanks, both implosion and explosion, and serves as a reminder for experienced process safety practitioners and as a tutorial for new process safety practitioners of what to look for during a hazard review. It also includes a case study of an atmospheric caustic tank explosion that resulted from a previously unreported mechanism that involved neither combustion nor vacuum.

1 Introduction

It is probably fair to say that atmospheric tanks are not very high on most people's list of hazards to worry about. Usually, they just sit there, storing liquids, doing what they are designed to do. Issues are typically the result of overfilling the tank and, on occasion, leaking components. Nothing sudden and catastrophic. When the words "sudden and catastrophic" are associated with atmospheric tanks, it is usually in relation to the flammable hazards of the liquids being stored. Chang and Lin [1] conducted a study of 242 storage tank incidents and found that 145 of them involved fire and another 61 involved explosion, mostly of flammable vapors from the liquids being stored. Outside of flammability issues, sudden and catastrophic incidents associated with atmospheric tanks are not what we typically worry about. Setting flammability issues aside, which have been addressed thoroughly elsewhere [2], [3], [4], [5] there are three types of atmospheric tank failures that can be sudden and catastrophic: collapses, implosions, and explosions.

Before continuing, however, it is important to first establish what is meant by the term, "atmospheric tank." The ASME boiler and pressure vessel code [6] divides vessels into two categories: pressure vessels, which operate at pressures greater than 15 psig and low pressure vessels, which operate at 15 psig or less. The BPVC applies only to pressure vessels. There are two API standards that apply to low pressure vessels. API 620 [7] applies to tanks that operate between 2½ psig and 15 psig, and at temperatures not greater than 250°F. API 650 [8] applies to tanks that operate at pressures up to 2½ psig, and at temperatures not greater than 200°F. OSHA defines an atmospheric tank as "a storage tank which has been designed to operate at pressures from atmospheric through 0.5 psig" [9].

All atmospheric tanks are low pressure tanks, but not all low pressure tanks are atmospheric tanks. There is not a consensus as to the limit—API 620 uses 2 ½ psig, OSHA uses 0.5 psig, others [10], [11] use a limit somewhere in between. For the purposes of this paper, an atmospheric tank is not intended to operate under vacuum and is not intended to operate at greater than 1 psig, or 30 inch water column. Human lungs have a greater operating pressure range.

2 Catastrophic Atmospheric Tank Failure

2.1 Collapse

One of the most famous catastrophic tank failures of the industrial era is the Great Molasses Flood of 1919 in Boston, Massachusetts [12]. The 50-foot tall, 90-foot diameter cast iron tank, built in a rush during World War I, ruptured and the wave of molasses killed 21 people. There was no fire. There was no explosion. The tank collapsed because it was under-designed for its application.

A more recent example is the failure of a liquid fertilizer storage tank at the Allied Terminals facility in Chesapeake, Virginia in 2008 [13]. The tank, originally designed for service with petroleum products, ruptured while being filled with a 32% solution of urea-ammonium nitrate (UAN). Two welders who were working on the tank as it was being filled for the first time after weld repairs were made were seriously injured when the tank collapsed.

2.1.1 Mechanisms for catastrophic failure

Atmospheric tanks fail for three reasons:

- The design or construction of the tank was inadequate.
- The tank is converted to a service that exceeds the capabilities of the initial design.
- The tank deteriorates over time.

Inadequate design or construction. While an atmospheric tank does not need to withstand high pressures, it does need to withstand the hydrostatic head of the liquid being stored. Atmospheric tanks collapse because the material of construction is not strong enough for the service, as was the case in the Great Molasses Flood, where the tank walls were 50% too thin, even for the relaxed standards of the 1910s [12]. Atmospheric tanks also collapse because the welds holding the tank together are inadequate. The Chemical Safety Board's report on their investigation into the Allied Terminals tank collapse includes an appendix of similar incidents, in which the listed cause of the incidents is "faulty welds" in more than 90% of the incidents where a cause is known [13]. Because of these issues, the CSB report notes that "tanks are more likely to fail when being filled to the maximum level for the first time."

Converted to a new service. Atmospheric storage tanks are designed for a specific service. Many tanks are designed to store petroleum products, for which a typical specific gravity is around 0.8. Simply converting the use from hydrocarbons to water increases the hydrostatic pressure on a tank by 25% when filled to the same level. Converting the use from hydrocarbons to 32% UAN, which has a specific gravity of 1.33, increases the hydrostatic pressure on a tank by two thirds. Converting the use from hydrocarbons to 50% caustic, which has a specific gravity of more than 1.5, almost doubles the hydrostatic pressure on a tank. Converting the service of atmospheric storage tanks to higher density liquids can lead to catastrophic tank failure by exceeding the design strength of the tank. Likewise, increasing the "safe fill level" has a similar impact on hydrostatic pressure, which can also lead to catastrophic tank failure.

Deterioration with time. Lastly, atmospheric storage tanks, like all equipment, deteriorate with time. Walls thin due to corrosion and erosion. Ideally, the thinning is uniform, steady, and slow. Catching this kind of deterioration during a tank inspection that includes thickness testing is straightforward. A tank with an adequate corrosion allowance will give lots of warning before the condition of the tank warrants a change in the operating conditions.

But not all deterioration is uniform. Tanks where nozzles, dip pipes, or eductors are aimed at the wall or floor of the tank are subject to localized thinning. Even a stream running down the interior wall of a tank can cause localized thinning. This weak spot then become the nucleus of a catastrophic failure. Stresses concentrate at a weak spot, the weak spot fails, and the stresses are immediately propagated through the rest of the material. The tank unzips and can collapse suddenly and catastrophically, often leading bystanders to believe there has been an explosion.

2.1.2 Protections against catastrophic failure

The causes of catastrophic tank failures also give a strong hint as to the protections that might be used against them.

Inadequate design or construction. This is best addressed with good equipment specifications according to recognized and generally accepted good engineer practice (RAGAGEP) and appropriate pre-startup safety reviews (PSSRs) to assure that the tank, as designed, constructed, and installed, meets those specifications. During commissioning, when many vulnerabilities in the tank are revealed, personnel should be kept out of harm's way, based on the assumption that the tank will fail.

Converted to a new service. Converting to a new service is a clear case for management of change, and the MOC review should consider changes in level, density, and corrosion rates, both general and local.

Deterioration with time. All equipment deteriorates, which is why a mechanical integrity program that adheres to RAGAGEP is essential. Tank inspections should test for thickness in enough spots to assure that general thinning is detected. Interior inspections should also look for specific evidence of localized thinning, especially at points where nozzles, dip pipes, or eductors are directed at the walls or floor of the tank.

2.2 Implosion

Atmospheric tanks are not designed to withstand vacuum, and they do not have to be exposed to very much vacuum to implode suddenly and catastrophically. The emptier the tank is when exposed to vacuum, the more dramatic the implosion. Tank implosions make such dramatic film that there are countless YouTube videos of tank implosions.

The lore that has been told and retold is of the maintenance team that was steaming out an atmospheric tank to prepare it for vessel entry, but did not finish before the end of the day. Rather than conduct the vessel entry at night, they closed up the tank and left. During the night, a storm front moved in, showering the tank with cold rain. The steam condensed, creating a vacuum, and when the crew returned in the morning they found the tank crumpled like an empty beer can. The crew from *Mythbusters* set out to prove or disprove this scenario using a DOT 111 tank car and decided that this story was "busted" [14]. *Mythbusters* notwithstanding, atmospheric storage tanks are not DOT 111 tank cars and the scenario should be considered credible.

Unlike explosions, implosions do not produce shock waves or flying shrapnel. Breaches may occur at the creases that form, but generally any loss of containment is not significant. The primary hazard associated with implosions are the potential injuries to personnel in the immediate area. The Center for Disease Control and Prevention reports of a worker killed while repairing a tank when it imploded [15]. Similarly, a worker in Spokane, Washington was killed in 2011 when the sewage digester he has checking imploded [16].

<u>2.2.1 Mechanisms for catastrophic failure</u>

Atmospheric tanks implode when they experience an internal pressure lower than they are capable of withstanding, i.e. a vacuum. Source of vacuum include the deliberate vacuum created by vacuum pumps and steam vacuum jets, as well as the incidental vacuum created by the pump suction of other pumps during pump out. Even in the absence of a pump, a gravity drain to empty a tank and other forms of emptying a tank of its liquid contents will create a vacuum. Finally, as described above, phase changes, particularly condensing vapors will create a vacuum.

<u>2.2.2 Protections against catastrophic failure</u>

In the absence of a source of vacuum, protections are unnecessary. However, all atmospheric tanks have a source of vacuum, if for no other reason than they are emptied. Protection include vacuum breakers, pressure/vacuum conservation vents, nitrogen pads, and most commonly, atmospheric vents. As long as these are properly sized and designed and remain fully operational, they are sufficient to protect against catastrophic implosion.

2.3 Explosion

Most atmospheric tank explosions involve the storage of flammable liquids, where it is the explosion of the flammable material that is the concern and which is outside the scope of this paper. There are also physical explosions that can occur, resulting from overpressure in the tank for reasons other than sudden combustion.

2.3.1 Mechanisms for catastrophic failure

Atmospheric tanks explode when they experience an internal pressure higher than they are capable of withstanding, which is not much. Sources of this high pressure include the pressure of inlets to the tank and heating within the tank.

Inlet line pressure can be of either gases or liquids. When the tank is open to atmosphere, the pressure drop at the inlet is complete, and the pressure in the inlet is not transferred to the interior of the tank. However, when the tank is blocked off from the atmosphere, the pressure in the tank will approach the pressure of the inlet line. Gases will flow into the tank until the pressure in the tank matches the supply pressure. Liquids will flow into the tank, compressing the headspace, until the headspace pressure matches the deadhead pressure of the liquid line. A tank does not need to be liquid full to reach the deadhead pressure of the liquid line.

Heating a vessel can result from either external heating or the heat of reaction, planned or unplanned, within the vessel. In either case, the liquid will expand and the vapor pressure will increase. Liquid expansion is only a concern when the tank is blocked off from atmosphere and the tank is completely full of liquid. In this case, since liquids are incompressible fluids, any additional expansion will sharply increase the pressure. However, atmospheric tanks are not absolutely rigid, so the effect will be to cause the tank to bulge, rather than to cause it to burst.

Increased vapor pressure is the real concern. Vapor pressure increases as temperature increases, and is independent of head space. An atmospheric tank blocked in while containing any liquid will experience an increase in pressure when heated, and atmospheric tanks are not designed to contain much increase. The vapor pressure of water, for instance, will increase by 1 psi when heated from 72°F to 113°F; it will increase by 5 psi when heated from 80°F to 165°F.

2.3.2 Protections against catastrophic failure

The key to protecting against physical explosion of atmospheric tanks is to assure that any source of pressure is released from the vessel before the pressure becomes too high. The obvious means is an atmospheric vent. Other protections include pressure/vacuum conservation vents and venting entryways or "trashcan lid."

To be useful, however, it is imperative that these protections be properly sized and designed and then remain fully operational. A vent that is too small or blocked is no protection. An atmospheric vent may be directly to atmosphere, or pass through a seal pot. Many creatures—

birds, wasps, rodents—find the outlet of an atmospheric vent a perfect place to build a nest. Bird screens, when in place, will keep out birds and rodents, but not nest-building insects. Likewise, a seal pot that has frozen solid will not vent. Vent lines that contain monomer vapor, whether directly to atmosphere or through a device can become plugged with polymer or simply glued shut.

Pressure vessels can be protected with rupture discs and pressure relief valves, but these require set pressures that are too high to be of use in atmospheric tanks. Likewise, pressure vessels are frequently protected from inlet pressure with regulators on the inlet. Unfortunately, atmospheric tanks operate at such low pressures that to regulate inlet pressures down to the operating pressure of the tank would reduce flow to unacceptably low rates.

Weak-seam roof tanks are frequently used to protect against the failure of atmospheric tank walls in the event of overpressure. They are designed so that the roof-to-shell joint will fail before any of the seams in the shell or the floor of the tank fail, thus preventing a release of the contents. However, this approach must be used with care. Accessories, such as railing or stairs, can make the roof-to-shell joint stronger than intended, as can the workmanship of an enthusiastic welder. (Welders are rarely scolded for making their welds too strong.) In general, no credit should be given for a weak-seam roof on a tank smaller than 10 meters (30 feet) in diameter because of the concern that the joint will be too strong [17]. When a weak-seam roof is used, the roof itself should be restrained with some kind of tether or hinge. Otherwise, the roof, ripped away from its weak seam, will be launched as a projectile.

3 A Case Study

While process air was blowing through the transfer line from a caustic storage tank to a caustic day tank, the caustic day tank failed catastrophically, launching the top head of the tank approximately 125 feet, where it came down through the roof of a nearby process building and triggered an impromptu evacuation in which one employee fell and had to receive medical treatment.

Figure 1. Caustic Day Tank head, after coming down through the roof of a nearby process building.

3.1 Process Background

The process consisted of a 3-inch transfer line constructed of 304 stainless steel that connected the discharge of a caustic transfer pump at the caustic storage tank to the top of a caustic day tank. A 50% solution of NaOH in water entered the caustic day tank through a dip pipe to the bottom of the tank. The transfer line was purged after each transfer of 50% caustic solution from the caustic storage tank to the caustic day tank by blowing out the line from the caustic storage tank to the caustic day tank for about 10 to 15 minutes with process air.

When the transfer line was being prepared for maintenance, there was a more extensive purge procedure. The normal procedure for purging the transfer line was to blow out the line with process air, blowing from the caustic storage tank to the caustic day tank for about 10 to 15 minutes. The next step was to flush the line with city water for 5 to 10 minutes, again flowing from the caustic storage tank to the caustic day tank. After the water flush, any residual water in the transfer line was blown out with process air, again blowing from the caustic storage tank to the caustic day tank for about 10 to 15 minutes.

The transfer line was routinely purged following transfers and was purged successfully for maintenance a month before the incident.

3.2 Incident Description

On a balmy, spring Friday morning, preparation began for maintenance on a leaking caustic transfer line from the caustic storage tank to the caustic day tank at about 9:30 am. The first step – purging the caustic from the transfer line with process air – was completed without incident.

The second step – purging residual caustic from the transfer line with city water – started at around 9:50 am. This step was also completed without incident, and at around 10 am the operator disconnected the water hose and left it to drain while he turned his attention to other matters in the unit.

At around 10:30 am, the operator returned to the caustic storage tank, disconnected the water hose, and reconnected the air line. He then opened the air supply valve in order to blow the water out of the transfer line.

At 11:10 am (the time shown on the time-stamp from the video monitoring the area), while the process air was still blowing through the transfer line from the caustic storage tank to the caustic day tank, the caustic day tank failed catastrophically. The pressure was sufficient to separate the conical head from the top of the tank and launch it approximately 125 feet, where it came down through the roof of a nearby process building to the east.

While the sides and bottom of the caustic day tank remained intact after the incident, the damage was enough to consider the caustic day tank a total loss. The piping attached to the top head of the caustic day tank was also damaged beyond repair. There were no fatalities or serious injuries associated with the incident, but one employee did require medical treatment for a sprained ankle following a fall that occurred during his evacuation from the immediate area of the incident.

3.3 Possible Issues

An explosion of a caustic tank can result from:

- Sources of pressure considerably higher than the design of the tank was capable of handling,
- Blockage in the equipment that lowered the pressure the tank was capable of handling, or
- Material failures in the tank that lowered the pressure the tank was capable of handling.

Guidance documents from suppliers of 50% caustic solution suggest the following as sources of high pressure, causes of blockage in caustic equipment, and causes of equipment material failure at pressures lower than design pressure [18], [19].

3.3.1 Possible sources of high pressure in a generic caustic tank

- Reaction of caustic solution with reactive metals (aluminum, magnesium, zinc, tin, chromium, brass, and bronzes made with zinc or tin), releasing hydrogen that forms a flammable mixture with air and subsequently ignites.
- Plant air or nitrogen used to transfer caustic solution by pressure.
- Steam used to clean out transfer lines.
- Direct steam heating of frozen caustic, exposing equipment to steam pressure.
- Pump deadhead pressure in unvented tank or line.
- Diluting strong caustic with water, rather than diluting strong caustic by adding it to water.
- Heating caustic solution while a crust of frozen caustic is formed over the top of the liquid

3.3.2 Possible causes of blockage in generic caustic equipment

- ❖ Block valves inadvertently closed.
- ❖ Lines plugged with frozen caustic solution.

3.3.3 Possible causes of generic equipment failure at lower than design pressure

- ❖ "Caustic embrittlement" – stress-corrosion failure which occurs at areas of mechanical stress or where fabricating operations produce high residual stress such as at bends in piping and at welds.
- ❖ Using iron and steel as materials of construction for equipment handling 50% caustic solutions at temperatures above 140°F. While iron and steel are the usual materials for equipment handling 50% caustic solutions below 140°F, nickel, nickel alloys, and stainless steel are required at higher temperatures and for higher concentrations.
- ❖ Using unsupported plastic pipe, which can fail because of the high density of 50% caustic solution, and using fiberglass reinforced plastic pipe except for specific conditions for which it is suited.
- ❖ Continued heating after caustic soda is melted, which may damage the special lining on the inside of railcars. The temperature of 50% caustic solution should never exceed 150°F (66°C) in lined cars.
- ❖ Wet or saturated steam above a maximum of 15 psig and 250°F. Steam at higher temperatures will damage rail car linings.

3.4 Conditions at Incident Scene

3.4.1 Noted from video

There were several dozen closed circuit video cameras positioned around the plant, three of which captured evidence of the catastrophic failure of the caustic day tank:

- ❖ The truck entrance camera, located about 600 feet southeast of the caustic day tank and aimed northwest.
- ❖ The tram gate camera, located about 275 feet southwest of the caustic day tank and aimed northeast.
- ❖ The main crosswalk camera, located about 175 feet west of the caustic day tank and aimed east.

The time stamps on the truck entrance camera and the tram gate camera coincide closely, and both showed the initial cloud release from the tank rupture at 11:10:19. The main crosswalk camera did not show the cloud release from the tank rupture, but shuddered at the time the cloud was released. Both the truck entrance camera and the main crosswalk camera captured an employee falling down and then getting up and walking away, hardhat in hand, at 14 seconds after the tank rupture.

Figure 2. Detail of screen capture from truck entrance camera at 11:10:19.647, showing launch of the caustic day tank top and associated cloud.

The main crosswalk camera, which was closest to the caustic day tank, shuddered for 0.4 sec, consistent with the passing of a shockwave. Neither of the more distant cameras showed the shudder. The more distant cameras, however, showed the initial cloud released. The cloud formed and was completely dispersed within 2 seconds. There was no evidence at all in those videos of combustion or of a secondary BLEVE. The shockwave confirms that there was an explosion. The absence of combustion or BLEVE and the quick dispersion of the cloud that was released confirm that the explosion resulted from simple overpressure, not from the ignition of a flammable mixture of hydrogen and air in the headspace or from the boil-up of the contents of the vessel.

The truck entrance camera was the only view that showed the launch of the top head. In the first 0.2 seconds, a cloud appeared above the west side of the tank. Then, 0.2 seconds after the initial release of the cloud, the conical head fully separated from the tank and began its 125-foot trajectory eastward, where it came down through the roof of a nearby process building. The video suggests that the separation began on the west side of the conical head and the break proceeded almost instantly around the tank head, propelling the conical head upward before the pressure inside the tank could be relieved.

3.4.2 Noted from field observations

The most relevant observation was that of the condition of the vent from the caustic day tank. A 3-inch gooseneck constructed from glued heavy-wall PVC pipe and elbows, the vent was attached to the top of the caustic day tank at a flanged nozzle 40 inches from the center peak of the conical head. The vent sheared off during the incident just above the flanged connection.

The internal surface of the vent (the surface that would have been visible from the outside while it was still attached to the functioning tank) was open and free of any obstruction. However, from the point the nozzle was attached to the top head, through the flange and up to the first elbow, the vent was encrusted with a white solid that tested positive for NaOH. The morphology of the solid was consistent with a slow accumulation over an extended period of time, rather than the result of an incident of overfilling the tank and then freezing in the vent. Moreover, there was no evidence of accumulation or discoloration on the head under the discharge of the vent, which would have been consistent with discharge through the vent onto the surface of the conical head.

The accumulation of solids within the vent choked it down to a diameter of about a half inch, and the rough, convoluted surface of the choked vent would have rendered it less useful than a clean half-inch pipe as a vent.

Figure 3. Vent nozzle, showing build-up of caustic.

Likewise, there were no solids built up on the interior walls of the vessel or on the interior surface of the head. This suggests that the buildup inside the vent line was the result of an entrained mist of 50% caustic solution being carried to the vent, where it collected and evaporated, leaving behind solid NaOH. The monthly vent inspections conducted on the tank since it was put into service in late 2003 would not have detected the gradual accumulation of solids.

Figure 4. View of interior of the caustic day tank after the incident. The thin white solid is caustic that dried there at the time of the incident. The interior shows some staining, but no build-up of the NaOH encrustation seen in the vent nozzle.

A thorough inspection of the vents of the two caustic tanks in the area showed no similar build-up of solids. However, one of the caustic storage tanks was put into service about 5 years before the incident and the other caustic storage tank was put into service about a year before the incident, so there would not have been the same opportunity for buildup.

The caustic day tank was specified to be constructed of 304 stainless steel. A field inspection of the vessel was consistent with that specification. The metal looked like stainless steel and was not magnetic; short of testing the metallurgy, this was considered enough to conclude that the tank was built as specified. Suppliers of 50% caustic recommend that carbon steel is a suitable material of construction for up to 140°F; over 140°F, they recommend 304 stainless steel to provide good corrosion resistance. This tank was operated at ambient temperatures. There was no reason, then, to believe that temperature or corrosion played any role in this incident.

The conical head separated completely. An inspection of the head showed that the separation was at the weld of the head to the 3"x3" angle that formed the top flange of the vertical cylinder of the tank. While the failure of an atmospheric tank is typically expected to be in the form of a split, where pressure is released without accompanying shockwave or shrapnel, an explosive separation of the head from the body of the vessel is not unheard of, e.g., Barton Solvents, Valley Center, Kansas, on July 17, 2007 [20].

Figure 5. The separation of the head from the body of the caustic day tank was at the weld of the head to the 3"x3" angle that formed the top flange of the vertical cylinder of the tank.

3.4.3 Noted from interviews

The pressure gauge on the process air was not read at the time of the incident. The production supervisor noted that the pressure should be set at around 35 to 40 psig, while the maintenance engineer stated that the process air pressure is normally around 80 psig, and that the regulator was normally wide open, with little motivation to crank it back closed. This is consistent with the pressure reading observed in the field, which indicated that the process air pressure was at 90 psig. The caustic day tank itself has a design pressure of atmospheric (0 psig) and was pneumatically tested at 5 psig at the time it was manufactured in 2003.

The supply pressure for the city water used to flush the transfer line is around 40 to 60 psig. This means that the line could have been successfully flushed into the caustic day tank, even if the pressure in the tank had risen to 35 psig during the initial purge. During the second process air purge, which lasted for around 45 minutes before the incident, there would have been more time for the pressure to build.

3.5 Contributing Factors

3.5.1 Non-contributing factors

The investigation was able to eliminate many of the generic issues identified by caustic solution suppliers.

In regard to possible sources of high pressure, the materials of construction – 304 stainless steel and PVC – make hydrogen formation as a result of reaction with reactive metals an extremely unlikely factor. This is further supported by the absence of any evidence of combustion during the incident. Likewise, neither steam nor a pump were in use, so steam pressure and pump deadhead pressure did not factor in. Temperatures were well above 65°F, the temperature at which frozen 50% caustic solution become a concern.

The purging procedure did include a step where water was added to strong caustic, a step that is normally discouraged. However, this did not create the high pressure that resulted in the incident, and overall, is a reasonable way to deal with the flush water.

In regard to possible causes of blockage in caustic equipment, there were no block valves to be inadvertently closed, and ambient temperatures were high enough to prevent lines from being plugged with frozen caustic solution.

In regard to possible causes of equipment failure at lower than design pressure, the materials of construction – 304 stainless steel and PVC – also eliminate the concerns with caustic embrittlement and with temperatures above 140°F. There was no unsupported plastic pipe to fail, and steam was not in use.

3.5.2 Direct causes

A "direct cause" is the uncontrolled release of energy or hazardous material that cannot be absorbed safely. In the case of this incident, the direct cause was high pressure in the caustic day tank, the source of which was the approximately 80 to 90 psig process air used to blow down the transfer line from the caustic storage tank to the caustic day tank.

3.5.3 Indirect causes

An "indirect cause" is the action or condition that results in direct causes. In the case of this incident, the indirect cause was the plugged tank vent on the caustic day tank. A contributing factor was probably the extended time that process air was allowed to blow into the tank. However, the vent was gradually becoming clogged and eventually, even shorter periods would have been sufficient to overpressure the vessel.

3.5.4 Basic causes

A "basic cause" is the policy, procedure, or decision that leads to indirect or direct causes. There were three basic causes. The first basic cause was the procedure that called for blowing fluids into the caustic day tank at pressures higher than the rated pressure of the tank. While pressures higher than atmospheric pressure, either as a result of pressurized fluids or pumping, are absolutely necessary to effect transfer into the tank, the higher pressure also made a functioning atmospheric vent an imperative element of normal operation.

The second basic cause was a tank design and operation that resulted in carrying entrained 50% caustic solution mist to the vent, where it could deposit and plug the vent.

The third basic cause was the vent inspection procedure. In hindsight, it is now clear that monthly visual inspection did not detect that the vent was becoming plugged. In the absence of this hindsight, there is nothing to fault the vent inspection procedure. Caustic solution is not volatile, and there is no general expectation in the industry that vents on caustic solution tanks are subject to plugging. However, now that the problem has been identified, it should be addressed.

3.6 Recommendations

3.6.1 Vent design to assure pressure relief is adequate

Dow Chemical has the following recommendation in regard to vents on caustic solution storage tanks: "In addition to a solution high-level alarm, the tank should have a liquid-level gauge and a vent. The vent should have a cross-sectional area at least four times that of the fill line (which is normally 2"), extending from the top of the tank to within about 3' of ground level. The vent, which also acts as an overflow, should be visible from the tank truck or tank car loading area so that personnel will see and/or hear caustic discharging from this overflow pipe in case the tank is overfilled. Valves should never be installed in the vent line. It must be open to the atmosphere at all times" [18].

In this particular case, the vent and fill are both 3-inch lines. The vent pipe is a gooseneck, rather than a pipe that extends to within 3 feet of ground level. Modifications to the vent piping may be helpful.

3.6.2 Vent inspection to catch any buildup

It was unexpected to discover that the vent was plugged with caustic. The likely cause of the plugging was the deposit of entrained mist in the vent. The mist, composed as it was of 50% caustic solution, would be the source of solids. The practice of blowing process air into the tank through the dip pipe assures that there for mist to be entrained after every transfer into the tank. The evidence is that process air carries the mist all the way to the vent. As a result, this is an issue that should be expected in any tank containing solutions of dissolved solids that can be entrained as mist.

In this particular case, the purpose of the vent inspection was to look for evidence of external blockage, primarily as the result of natural activity, e.g., bird nests, wasp nests, etc. In the case of tanks storing solutions of dissolved solids, a more thorough inspection of the vents is in order. This may involve the use of boroscopes and cameras, or as an alternative, the installation of inspection and clean-out ports in the vent line. The frequency of inspection should be adjusted as experience dictates, beginning with quarterly inspection, but probably settling on annual inspection.

3.6.3 Limit use of dip pipes for inlet piping

The practice of blowing process air or nitrogen into a tank through a sub-surface dip pipe assures that mist will be entrained. While the density of aqueous solutions such as 50% caustic is high enough to lead to the expectation that mist droplets will settle out before reaching a vent, especially if the level is low enough, droplets that are small enough will be carried to the vent, regardless of the level. Operating at lower liquid levels will slow down the rate of blockage, but not stop it, and the improvement would come at the cost of lost capacity.

A more effective solution is to eliminate the dip pipe so that process air is not blown through the caustic solution to create a mist in the first place. Neither Dow Chemical nor PPG Industries show dip pipes in their example configurations of storage tanks [18], [19]. Because aqueous solutions of solids are neither flammable nor non-conducting, the usual concerns regarding static discharge from free-falling liquid are simply not an issue.

In this particular case, the dip pipe also serves to avoid erosion of the head and vessel walls that result from feeding directly through a tank nozzle. Cutting the dip pipe off a few inches below the entrance to the head and fitting it with a slight elbow to direct the free-falling liquid away from the wall still avoids erosion problems but eliminates the mist that forms by sparging process air through the caustic solution.

3.6.4 Pressure regulation to reduce potential for overpressure

Atmospheric tanks are frequently connected to high pressure sources. Examples include fill lines, where the pressure can be as high as the deadhead pressure of the transfer pump; water supplies, which typically run around 40 to 60 psig; and plant air, which is often in the range of 80 to 120 psig. Any one of these can be a source of overpressure, but to regulate them down to atmospheric pressure would render them useless. While the source pressure must be above atmospheric pressure in order for these connections to work, they should not be any higher than necessary.

In this particular case, the plant air supply used to blow out the lines would have been just as effective at 20 psig as it was at 90 psig. Regulating the air supply pressure down creates no burden. The water pressure and the transfer pump pressure need to be where they are.

4 Conclusions

Atmospheric tanks are relatively fragile, especially when compared to vacuum vessels, low pressure vessels, and high pressure vessels. They are also widely used and considered relatively benign. However, because they are considered relatively benign, they are often overlooked as a hazard and neglected as a maintenance priority.

When atmospheric tanks contain flammable materials, the consequence of concern is most often a tank fire. Beyond that, the typical consequence of concern associated with atmospheric tanks is a leak. The impacts of these events are usually limited to injuries and low-consequence impacts to the environment, although there have been some notable exceptions [21], [22]. Rarely is the concern one of sudden, catastrophic failure. However, an atmospheric tank can fail suddenly and catastrophically, with life-threatening consequences. The catastrophic failure can result in a collapse, an implosion, or an explosion. Collapses result when a tank is badly installed or when a tank is misused. The best way to avoid collapses is to first make sure the tank is properly designed and installed, and then to use it within the limits of that design.

Explosions can result from the ignition of flammable vapors, but surprising to many, can also occur for purely physical reasons. Implosions result primarily when sources of vacuum are overlooked or neglected. Atmospheric tanks are called atmospheric tanks because they vent to atmosphere. When that vent becomes inoperable, it becomes all too easy for the pressure in an atmospheric tank to become lower or higher than the thin walls of the tank can withstand. So it is imperative for operators to understand what causes a vent to become inoperable, and to prevent that from happening. The response in a HazOp to "pressure-too high" or to "pressure-too low" cannot be, "That can't happen, we're open to atmosphere."

5 Recommendations

5.1 Regarding catastrophic tank collapse

To avoid catastrophic vessel collapse, apply rigorous management of change and pre-startup safety review when putting an atmospheric tank into new or different service. In particular, pay attention to the impact of using a tank for higher density liquids. When a tank is first put into service is the time when the tank is most vulnerable. Resist the temptation to fast-track atmospheric tank commissioning when it is being filled for the first time; act as though you know it will fail and deploy personnel accordingly.

Once an atmospheric tank has been successfully put into service, it should provide good service. The greatest concern will be wall thinning that compromises the strength of the vessel walls. Ideally, the thinning will be uniform and caught with a minimum of testing. However, uniform thinning is rarely the actual experience, so special attention needs to be paid to localized thinning due to erosion, which can only be detected by looking for it. Pay attention to where nozzles, dip pipes, and eductors are aimed and confirm that weak spots are not being created.

5.2 Regarding catastrophic tank implosion

Avoiding catastrophic tank implosion requires identifying and controlling sources of vacuum and assuring that measures to protect tanks from implosion remain operable. Obvious vacuum sources include vacuum pumps, pump out and other forms of emptying a tank of its liquid contents, and condensing vapors (typically, but not limited to, steam). When the presence of any of these is unavoidable, they should only be present with full awareness of their hazard and the total reliance on the protection device.

All of the devices to protect atmospheric tanks against implosion have mechanisms by which they can fail. Facilities that take credit for these devices – atmospheric vents, conservation vents, inerting pads, and vacuum breakers – need to periodically validate that they are working as designed. The means to validate the functionality of protective devices needs to be designed with the same care with which the device itself is designed.

5.3 Regarding catastrophic tank explosion

In addition to the obvious concerns of flammable vapor combustion, avoiding catastrophic tank explosion requires identifying and controlling sources of pressure. Obvious pressure sources include high inlet gas pressure, high inlet liquid pressure, and heating. Again, the presence of any of these may be unavoidable, so they should only be present with full awareness of their hazard and the importance of the protective devices.

As with any device, the devices to protect atmospheric tanks against overpressure and explosion have mechanisms by which they can fail. The term "fail-safe" does not mean they cannot fail; it means that in the absence of energy sources, they are supposed to go to the safe state. It does not mean that they cannot fail to a dangerous state. Facilities that take credit for these devices – typically atmospheric vents and conservation vents – need to periodically validate that they are working as designed. As with all protective devices, the means to validate the functionality of protective devices needs to be designed with the same care with which the device itself is designed.

For atmospheric vents in particular, the design needs to allow for easy inspection of the entire vent. While this may involve the use of boroscopes and cameras, the installation of inspection and clean-out ports in the vent line may be more cost effective and easier to implement. It is not enough to have inspection measures, however. They must be used and at a sufficient frequency to detect a developing problem in time to address it. The frequency of inspection may be adjusted as experience dictates, but two failed inspections in a row indicates that the test interval is too long.

5.4 Regarding use of dip pipes in atmospheric storage tanks

Dip pipes have two important features when used in atmospheric storage tanks. The most important is that when the liquid being stored is both non-polar and flammable, dip pipes prevent free-falling liquid, which is known to create static sparks, a source of ignition of the flammable vapors. The other feature is that dip pipes direct liquid away from areas of the tank vulnerable to erosion. When these features are not required, dip pipes should not be used.

This is of particular concern when the liquid being stored is a solution of dissolved solids. When the line and its dip pipe are blown clear, which is a fairly common practice, the gas used to blow the line clear will sparge through the liquid and carry some quantity of mist to the vent. At the vent, the droplets will collect, evaporate, and leave a deposit of solids. Unless inspections explicitly look for this deposit, it is likely to go unnoticed. When dip pipes are installed and their use with solutions of dissolved solids involves blowing them out, vent inspections for blockage should not be limited to looking for nests or other biological activity, but should also include inspection for solids build-up.

6 References

[1] Chang, J.I. and C-C Lin. "A study of storage tank accidents," *Journal of Loss Prevention*, 19 (2006) 51-59.

[2] Atherton, W. and J.W.Ash. *Review of Failures, Causes & Consequences in the Bulk Storage Industry*, National Lightning Safety Institute. 2006. http://lightningsafety.com/nlsi_lls/Causes-of-Failures-in-Bulk-Storage.pdf (accessed 02-Sep-2015).

[3] EPA. "Catastrophic Failure of Storage Tanks," *Chemical Safety Alert*, EPA 550-F-97-002b, Chemical Emergency Preparedness and Prevention Office, USEPA, Washington, DC. May 1997. www.nrt.org/production/nrt/nrtweb.nsf/AllAttachmentsByTitle/A-107tanks/$File/tanks.pdf?OpenElement (accessed 02-Sep-2015).

[4] Marsh. "Atmospheric Storage Tanks," *Risk Engineering Position Paper-01*, Rev 1.0, Marsh & McLennan Companies, February 2011. uk.marsh.com/Portals/18/Documents/Atmospheric%20Storage%20Tanks_lowres.pdf (accessed 02-Sep-2015).

[5] OGP. "Storage incident frequencies," Risk Assessment Data Directory, Report No. 434-3, International Association of Oil & Gas Producers, March 2010. http://www.ogp.org.uk/pubs/434-03.pdf (accessed 02-Sep-2015).

[6] ASME. *Boiler and Pressure Vessel Code, Section VIII-Rules for Construction of Pressure Vessels*, 2015 Ed. American Society of Mechanical Engineers, New York, NY. 2015.

[7] API. *API 620: Design and Construction of Large, Welded, Low-pressure Storage Tanks*, 12th Edition. American Petroleum Institute, Washington, DC. 2013.

[8] API. *API 650: Welded Tanks for Oil Storage*, 12th Edition. American Petroleum Institute, Washington, DC. 2013.

[9] OSHA. *29 CFR 1910.119, Process safety management of highly hazardous chemicals.*, Occupational Safety and Health Administration, U.S. Department of Labor, Washington, DC. Last revised 08-Feb-2013.

[10] API. *API 2000: Venting Atmospheric and Low-Pressure Storage Tanks*, 7th Edition. American Petroleum Institute, Washington, DC. 2014.

[11] Ennis, T. "Pressure Relief Considerations for Low-Pressure (Atmospheric) Storage Tanks," *IChemE Symposium Series No. 151*, Institution of Chemical Engineers, Rugby, UK. https://www.icheme.org/communities/subject_groups/safety%20and%20loss%20prevention/resources/hazards%20archive/~/media/Documents/Subject%20Groups/Safety_Loss_Prevention/Hazards%20Archive/XIX/XIX-Paper-63.pdf (accessed 02-Sep-2015).

[12] Schworm, P. "Nearly a century later, structural flaw in molasses tank revealed," The Boston Globe, January 14, 2015. www.bostonglobe.com/metro/2015/01/14/nearly-century-later-new-insight-into-cause-great-molasses-flood/CNqLYc0T58kNo3MxP872iM/story.html (accessed 15-Feb-2016).

[13] CSB. *Investigation Report: Allied Terminals, Inc.-Catastrophic Tank Collapse*. No. 2009-03-I-VA, U.S. Chemical Safety and Hazard Investigation Board, Washington, DC, May 2009. 49 pp.

[14] Savage, A. "On MythBusters' FIRST-EVER Implosion," *TESTED.* January 22, 2016. www.tested.com/art/makers/561346-mythbusters-first-ever-implosion/ (accessed 16-Feb-2016).

[15] FACE. "Worker Dies While Repairing a Vacuum Evaporator Tank in Virginia," FACE 8725. Fatal Accident Circumstances and Epidemiology Project, NIOSH, CDC, Atlanta, GA. October 15, 2014. www.cdc.gov/niosh/face/In-house/full8725.html (accessed 16-Feb-2016).

[16] Clouse, T. "Ruling upheld in sewage death: Supreme Court says CH2M Hill liable for $6.5 million," The Spokesman-Review, Spokane, WA, May 27, 2011. www.spokesman.com/stories/2011/may/27/ruling-upheld-in-sewage-death/ (accessed 16-Feb-2016).

[17] CCPS. *Guidelines for Engineering Design for Process Safety*, Center for Chemical Process Safety, American Institute of Chemical Engineers, New York, NY. 2010. p. 110.

[18] Dow. *Caustic Soda Solution Handbook*, Form Number 102-00011-0810XBBI, The Dow Chemical Company, Midland, Michigan. August 2010. 23 pp.

[19] PPG. *NaOH-Caustic Soda*, #0100 Rev 0406 pdf, PPG Industries, Inc., Monroeville, Pennsylvania. 2008. 63 pp.

[20] CSB. *Case Study: Barton Solvents.* No. 2007-06-I-KS, U.S. Chemical Safety and Hazard Investigation Board, Washington, D.C., June 26, 2008. 14 pp.

[21] DoJ. *Former Freedom Industries official sentenced for role in chemical spill.* U.S. Attorney's Office, Southern District of West Virginia, February 1, 2016. www.justice.gov/usao-sdwv/pr/former-freedom-industries-official-sentenced-role-chemical-spill (accessed 15-Feb-2016).

[22] Asahi. Police push for charges against TEPCO, execs over pollution problems after nuclear disaster. The Asahi Shimbun, Asia & Japan Watch. October 3, 2015. http://ajw.asahi.com/article/0311disaster/fukushima/AJ201510030052 (accessed 15-Feb-2016).

Explosions in SMPO Hydrogenation Unit

Presenter/Principal Author: Nick Gonzales
Affiliation: Shell Global Solutions
Address: 3333 Hwy 6 South, Houston TX 77082
Email address:nick.gonzales@shell.com

Prepared for Presentation at
American Institute of Chemical Engineers
2016 Spring Meeting
12th Global Congress on Process Safety
Houston, Texas
April 11-23, 2016

LEARNING FROM INCIDENTS
AWARENESS ALERT

For Industry-Wide Sharing September 2015

Explosions in SMPO Hydrogenation Unit

Figure 1 – Hydrogenator structure after the explosion.

Target audience for this alert

- Process Safety and Process Engineers
- Production staff in Refining/Chemical and hydrogenation processes

What happened

On June 3rd 2014, a Styrene Monomer and Propylene Oxide (SMPO) unit was in the pre start-up phase after the scheduled shutdown and replacement of catalyst in the hydrogenation reactors. The reaction system contained two fixed-bed adiabatic, trickle-flow reactors. Operations were heating up the hydrogenation unit in preparation for the catalyst reduction step. The heat-up was performed by the circulation of Ethyl Benzene (EB) through the reaction system under a nitrogen blanket and continuous nitrogen purge. During this heat up, one hydrogenation reactor and an additional upstream separation vessel suddenly exploded. Two contractor workers, who were in the immediate process area, were injured with burns from hot catalyst particles. The explosion caused extensive damage requiring over a year to repair.

Why it happened

The incident investigation concluded that both the reactor and vessel had failed because of overpressure, resulting from unexpected exothermic reactions between Ethyl Benzene (EB) and copper chromite oxide catalyst which produced an excessive amount of gas.

Prior to the incident, it was generally believed that under inert nitrogen conditions EB would not react with the passivated catalyst. However, it was subsequently determined and proven in the laboratory that two exothermic reactions actually occurred during the heat up:

- EB reacted with active oxygen present in the catalyst (mainly but not limited to copper chromate $CuCr^{VI}O_4$). The catalyst contained more copper chromate than the previous catalyst used at the site. The redox reaction started when the temperature exceeded approximately 90°C. The target temperature at the end of the heat up phase was 130°C, but due to the exothermic reaction the temperature increased further in local dry zones of the trickle bed reactors.
- Copper oxide ($Cu^{II}O$) was reduced by EB when localized temperatures reached 180°C. This redox reaction generated large amounts of heat and gaseous molecules (such as CO_2 and H_2O) leading to the overpressure.

The early heat generation by the EB-chromate reaction, in combination with reduced cooling capacity in localized dry zones created localized temperatures in the reactor bed, high enough to initiate the EB-copper oxide reaction. Contributing factors include:

Low nitrogen flow: In a trickle flow reactor, sufficient nitrogen flow is required to achieve a proper liquid distribution from the distribution trays to wet the catalyst in the catalyst bed. The actual nitrogen flow was lower than needed, which led to non optimum liquid distribution in both trickle bed reactors causing dry zones in the top part of the catalyst beds.

Low EB flow: The EB flow to the hydrogenator was, at times, lower than needed for proper wetting of the catalyst and the flow variations led to preferential or rivulet flow and stagnant wet zones in the trickle flow reactor catalyst bed. These conditions led to non optimal heat removal and as a consequence high localized temperatures

Increasing Pressure: About thirty minutes before the explosion, the vent gas valve on the second separator tripped closed on high level and stayed closed. The system pressure increased up to the nitrogen supply pressure and caused the nitrogen flow to the reactor to decrease. The increasing pressure reduced the evaporative cooling of the EB from the catalyst.

Reactive hazard understanding: A Reactive Hazard Assessment (RHA) had been conducted in 2011 for this section of the SMPO unit. The RHA was based on process development work, design knowledge, literature search and application of a tool that was developed and used for identifying unintended reactions. The RHA only identified the reactions with hydrogen during the reduction phase. The reactions of EB with the catalyst in the absence of hydrogen were not identified in the RHA. This, together with the previous operational reduction experiences, reinforced the belief that EB was an inert medium in combination with the catalyst.

Lessons learned

- The passivated copper chromite hydrogenation catalyst contains bonded oxygen in the form of copper chromate and copper oxide prior to reduction. Where a hydrocarbon and oxidizing substance (i.e. both a "fuel" and reactive oxygen) are combined and in sufficient quantities, an exothermic reaction may occur and the only mitigation available may be the total amount of cooling provided.
- Heating up in EB, prior to the reduction step with hydrogen, was seen as normal and safe from a chemical reaction point of view. This influenced the choices made during the process design, process development work, the design of the trickle bed reactor, qualification of the catalyst and operational procedures.
- The purpose of a Reactive Hazard Assessment (RHA) is to identify consequences of the potential reactions, intended and unintended, between the components of the reaction streams. At the time of the Assessment, The composition of the reactants and the potential initiation temperatures were not understood adequately. Post incident investigation suggests that suitable analytical tools such as Thermogravimetric Analysis (TGA) may help to identify such exothermic reactions.
- A series of incremental and seemingly innocuous changes can allow unknown reactions to proceed and run out of control, resulting in a severe incident. Introduction of a new catalyst or catalyst supplier can introduce new components or changes even in small quantities of components that could lead to unidentified and unintended chemical reactions. The catalyst selection and subsequent change processes should address these with the focus on reactive hazards in both transient and steady state operations. Note that this may also apply to changes in process chemicals in other services (e.g. polymerization inhibitors, cleaning solutions, etc.).

444197 Distillation Tray Operation

Abstract:

Distillation trays come in different types. Examples include sieve trays, valve trays, bubble cap trays, baffle trays, and dual flow trays. Of these, sieve and valve trays are the most common.

The understanding of the hydraulic and mass transfer characteristics of trays is important for successful design, operation, and troubleshooting of trayed distillation columns.

This paper provides an overview of distillation tray design and operation with particular emphasis on the hydraulic phenomena observed in trayed columns. The influence of tray geometry on the performance of trays will be discussed.

Photos and videos from various FRI tray tests will be shown.

444200 Distillation Packing Operation

Abstract:

Distillation packings are widely used in industrial distillation columns, particularly for low-pressure-drop applications. Distillation packings are generally classified as random packings, structured packings, and grids.

This paper presents an overview of the operation of packed distillation columns, including the hydraulic and mass transfer characteristics of packed columns. The influence of packing geometry on the performance of packed distillation columns will be discussed.

Photos and videos from various FRI tests will be shown.

443684 Acid Gas Absorption Fundamentals

Abstract:

Absorption is a standard unit operation that has been studied by chemical engineering students for decades. One of its primary uses is for the removal of acid gases, like CO_2 and H_2S, from natural gas streams. Historically, aqueous amines and hot carbonate solutions have been used to perform these absorptions. Many different amines, solution strengths, activators and passivators have been tested and employed industrially. Approximately half of the industrial towers are trayed; half, packed. Foaming adversely affects the hydraulic performance in these systems.

Acid gas absorption is receiving appreciable new attention as concerns rise regarding CO_2 in the atmosphere. In addition, the United States has recently experienced significant growth opportunities in shale gas production which require improvements in gas treating technologies. This presentation reviews acid gas absorption and regeneration fundamentals. Several PFD's will be shown to illustrate important concepts. New absorbent solutions will be discussed, including ionic liquids. Flue gas applications will be discussed, including the difficulties associated with column pressure drops and very large column diameters. Rate-based modeling approaches will also be reviewed.

443442 Introduction to Rate Based Mass Transfer Models

Abstract:

This presentation is aimed at providing insights into rate based mass transfer simulations, and how they are different from an equilibrium (ideal stage) model simulations.

Rate based mass transfer models are built on five key elements –

- Mass and energy balances around individual phases on a tray or in a packed segment
- Thermodynamic phase equilibrium

- Equilibrium across interfaces
- Kinetic effects on mass transfer rates
- Mass and heat transfer rate models for transport across interfaces.

In addition, the five key elements have interwoven relationships with chemical, physical, phase and transport properties based on the chemical species being used and the actual mass transfer hardware. In sharp contrast, an equilibrium stage model uses only mass and energy balances around an entire ideal stage plus thermodynamic phase equilibrium and use empirical correction factors to account for efficiencies and mass transfer hardware. For applications such as absorption, where the mass transfer efficiencies are poor (5-25%) equilibrium stage based models prove to be grossly non-predictive, while rate based models are mechanistic, and predictive. This presentation will focus on the characteristics of true mass transfer rate based models, contrast them with the ideal stages and introduce briefly the ProTreat ® process simulator.

443535 Process Investigations Using Radioisotope Techniques

Abstract:

In all chemical plants and refineries there generally is not enough instrumentation in the right places in our processes. During the design phase, decisions were made between knowing more about the process, which means buying more instruments, and saving capital funds. When a problem appears with the operation of a vessel, there are always gaps in the knowledge of what is happening due to a lack of instruments in the places where knowledge is not routinely required. Process engineers and operations personnel need to know what is happening so they can take corrective action. When knowledge is missing, they often take actions that make matters worse.

For distillation and separation processes Chemical Engineers have historically relied upon plant process measurements such as flow rates, temperatures, and pressures and model or simulation results for data with which to solve troubleshooting or re-design problems. An array of on-line diagnostic services are available that can provide additional data offering real-time information on how pieces of process equipment are actually operating.

The most common application is Gamma Scanning. This test is primarily applied, but not necessarily limited to, distillation or separation columns. Gamma scans provide a density profile of the internal process of operating distillation columns and other process vessels. The density profile can be used to diagnose the hydraulic operating conditions of mass transfer devices such as damage to internals, flooding, degree of entrainment or weeping, liquid levels on trays and distributors, liquid distribution through packed beds, etc. The presenter will show case studies where scanning revealed vital process information that helped solve an operating problem, or helped make a revamp/re-design successful.

Additional derivatives from gamma scanning of detecting flow distribution patterns will be demonstrated. The ThruVision Scan technique is a specialized horizontal gamma scan used to generate a topographic profile of the internal cross-sectional density of process equipment. This profile is useful for the detailed study of liquid flow distribution through packed columns as well as having applications beyond distillation columns. A hybrid application involving tracers and scanning on fixed bed catalytic reactors will also be discussed.

Another technique involving radioactive materials is the Tracer Test, in which a radioactive tracer is introduced into the process for investigating the process flows. External radiation detectors are placed at strategically chosen locations outside vessels and piping to monitor the tracer flow through the process. Either liquid, gas or a solid tracer can be utilized depending on process conditions and the fluid to be studied. This tracer technique can identify liquid or gas maldistribution in fixed or fluidized bed reactors. The residence time through a tank or reactor can be measured and compared with the calculated residence time. Observed differences can relate directly to the degree of mixing or channeling inside the reactor. These tracer techniques have also been used in identifying leakage of heat exchangers and valves.

Paper 192 A: DMTO and Beyond: DICP's Sustainable Innovations in Technologies for on-Purpose Production of Light Olefins from Methanol

Zhongmin Liu
Dalian Institute of Chemical Physics, Chinese Academy of Sciences
Zhongshan Road 457, Dalian, 116023, China
Email: liuzm@dicp.ac.cn

Mao Ye
Dalian Institute of Chemical Physics, Chinese Academy of Sciences
Zhongshan Road 457, Dalian, 116023, China
Email: maoye@dicp.ac.cn

Prepared for Presentation at
American Institute of Chemical Engineers
2016 Spring Meeting
11th Global Congress on Process Safety
Houston, Texas
April 10-14, 2016

UNPUBLISHED

DMTO and Beyond: DICP's Sustainable Innovations in Technologies for on-Purpose Production of Light Olefins from Methanol

Zhongmin Liu
Dalian Institute of Chemical Physics, Chinese Academy of Sciences
Zhongshan Road 457, Dalian, 116023, China
Email: liuzm@dicp.ac.cn

Mao Ye
Dalian Institute of Chemical Physics, Chinese Academy of Sciences
Zhongshan Road 457, Dalian, 116023, China

Keywords: methanol-to-olefins, catalyst, process development

Abstract

The rapid development in past thirty years has triggered a significant increase of the light olefins (including ethylene, propylene, and butylene) production capacity in China, and it is expected the demand will continue to expand in the coming years. In view of China's tremendous coal reserves, the Dalian Institute of Chemical Physics (DICP) developed a fluidized bed methanol-to-olefins technology (DMTO). The DMTO technology was successfully commercialized in 2010 in Shenhua Baotou's coal-to-olefins (CTO) plant (with an ethylene and propylene production capacity of 600 kt/a). So far seven commercial units using DMTO technology have been put into stream in China. As part of the DICP's ambition of sustainable innovations for on-purpose production of light olefins from methanol, a dual functional catalyst was developed for converting methanol to light olefins and cracking C4+ to ethylene and propylene simultaneously. By use of this catalyst, a second generation MTO technology (DMTO-II) has been developed and successfully implemented into a recent started up CTO plant with an ethylene and propylene production capacity of 670 kt/a. This paper will introduce the research and development (R&D) activities at DICP on DMTO and DMTO-II technologies, which covers the zeolite and catalyst synthesis, process development, reactor scale-up, and commercial operations. Beyond that, DICP's strategy of innovative catalyst and process research for on-purpose production of light olefins from methanol will also be highlighted.

1. Introduction

The rapid development of China's economics in the past three decades has triggered a significant increase of the demand of light olefins (including ethylene, propylene, and butylene) and thus the production capacity in China. From 2005 to 2010, China's production capacity of ethylene was increased from 7.58 to 15.19 Mt/a while the production capacity of propylene was increased from 8.86 to 15.83 Mt/a. It is expected the demand of light olefins will continue to expand in the coming years. It was estimated that at the end of 2015 China's production capacity of ethylene

Note: Do not add page numbers. Do not refer to page numbers when referencing different portions of the paper

could be further increased to 27.00 Mt/a and that of propylene increased to 24.00 Mt/a. Ethylene and propylene are conventionally produced from petrochemical feedstock via naphtha thermal cracking and fluid catalytic cracking (FCC) processes. It is since the oil crisis in 1970s that there have been continuously growing efforts in finding alternative resources other than oil to produce light olefins (Chang, 1984; Chang and Silvestri, 1977; Liang et al., 1990). Researchers in Mobil discovered that a class of synthetic shape-selective zeolites, namely ZSM-5, could be used to convert methanol to light olefins. Methanol-to-olefins (MTO) opens an important way to produce light olefins from non-oil resources since methanol can be readily produced from coal, natural gas, and biomass via synthesis gas by existing and proven technologies (Chang, 1984; Chang and Silvestri, 1977; Keil, 1999; Tian et al., 2015). Later on, scientists from the Dalian Institute of Chemical Physics (DICP), Chinese Academy of Sciences found that a silicoaluminophosphate catalyst (SAPO-34), which was first synthesized by Union Carbide in 1980s, could greatly improve the selectivity to light olefins in MTO process (Liang et al., 1990). Since then, MTO process has attracted considerable interests and the researches span catalyst synthesis, reaction mechanism, reaction kinetics, process development, and reactor scale-up.

In August 2010, the world's first commercial MTO unit (600 kt/a of ethylene and propylene production) was successfully started up in Shenhua's Baotou coal-to-olefins plant in north China (Liu et al., 2014). This unit is based on the MTO process developed by DICP, and its success represents an important milestone and breakthrough of MTO process development. In this paper, the research and development (R&D) activities at DICP on MTO technologies will be introduced, which covers the zeolite and catalyst synthesis, process development, reactor scale-up, and commercial operations, and focuses on the DMTO and DMTO-II process. Beyond that, DICP's strategy of innovative catalyst and process research for on-purpose production of light olefins from methanol will also be highlighted by two examples.

2. Catalyst development

2.1 Synthesis of Silicoaluminophosphate (SAPO) molecular sieves

SAPO molecular sieves were first synthesized by Union Carbide Corporation in 1984. Among them, SAPO-34 which has the CHA structure shows excellent catalytic performance in MTO reaction. The CHA framework topology is characterized by the cylinder-like cages with eightmembered ring openings. SAPO-34 thus has small pore, medium acidity, and high thermal/hydrothermal stability, and is the ideal breeding ground for hydrocarbon pool intermediates in the MTO reaction.

Scientists at DICP had made great efforts to the synthesis and crystallization mechanism of SAPO-34. In the late 1980s, scientists at DICP first reported the excellent catalytic performance of SAPO-34 in the MTO reaction and its high stability during regeneration. The SAPO-34 synthesized at DICP could keep more than 80% relative crystallinity after steam treatment at 800 °C for 45 h. These efforts confirmed that SAPO-34 is a promising MTO catalyst for industrial application.

In order to further make the SAPO-34 catalyst available for industrial applications, researches on the suitable templates for SAPO-34 synthesis had also been conducted since templates may influence the morphologies, microstructures, and acidic properties of the catalyst. It has been found that triethylamine (TEA), diethylamine (DEA), and their mixtures would be good choice. On one hand, the amines were much cheaper than the traditional tetraethylammonium hydroxide (TEAOH). On the other hand, the template has a lower boiling point (below 100 °C) than water, and the amine could be easily separated from the mother liquid through flash distillation and reused for the synthesis. This could significantly reduce the cost for catalyst synthesis and waste treatment.

In addition, DICP also conducted researches to control the catalytic performance of SAPO-34 via altering the Si coordination environments. They studied the mechanisms governing the nucleation, crystallization, and Si distribution for a better control of composition and Si environmental of the synthetic product, and it has been found that SAPO-34 with low Si content and single Si(4Al) species shows longer lifetime and lower selectivity of propane (thus slow coking rate). The SAPO-34 synthesized at DICP is featured by the rapid crystallization, adjustable acidity, and on-purpose particle size.

In 1993 scientists at DICP synthesized SAPO-34 by employing a 1.0 M^3 autoclave in order to supply catalyst for the pilot-scale experiments. In 2005, DICP built a pilot plant for catalyst synthesis in which two 2 M^3 autoclaves were used.

2.2 Synthesis of fluidized bed MTO catalyst

The fluidized bed MTO catalyst with SAPO-34 as the active component is prepared by spray drying method, which has the following characteristics: spherical, particle size distribution similar to FCC catalyst, sufficient mechanical strength, and good catalytic performance. In 2005, roughly 25 tons of MTO catalyst was produced and supplied for the DMTO demonstration unit (16 kt/a MeOH feed). In 2008, a 2000 t/a MTO catalyst plant (containing molecular sieve synthesis and catalyst manufacturing) was constructed and put into stream. The manufacturing process is composed of mixing, milling, spraying, drying, calcination, packaging, and storage.

Establishment of quality controlling system and standard test procedure is critical to keep the excellent performance of MTO catalyst. Five quality control points were set for SAPO-34 synthesis process: raw materials, purity of the solid in the slurry, content of distilled amine solution, collected waste liquids, and SAPO-34 molecular sieves product. The latter includes crystal phase, composition, morphology, textural properties, solid content, and catalytic performance. Meanwhile, four quality controlling points were set for the catalyst particle manufacturing: raw materials, particle size distribution of the milled slurry, particle size distribution and morphology of the semiproduct, and calcined product. The parameters for final quality controlling of the calcined product includes solid content, composition, morphology, particle size distribution, attrition index, textural properties, bulk density, and catalytic performance.

3. DMTO technology

3.1 MTO reaction in fluidized bed reactor

The MTO reaction over SAPO-34 catalyst is highly exothermal, which will release about 196 kcal heat when 1 kg methanol feed is completely converted at 495 °C. A feature of MTO reaction over SAPO-34 is the rapid catalyst deactivation due to coke deposition. Figure 1 shows the methanol conversion and selectivity to ethylene and propylene in a fluidized bed reactor without external catalyst circulation. As can be seen, a rapid drop of methanol conversion occurs after the coke content on catalyst reaching a certain *critical* value. In order to maintain high activity of catalyst in the reactor, the in-line combustion of coke is required. Another feature of MTO reaction over SAPO-34 is that the selectivity to ethylene is very sensitive to the coke content on catalyst. At the *critical* coke content on catalyst, the selectivity to ethylene becomes the maximum. The selectivity to propylene basically keeps stable with the increasing coke content. This is very important for MTO reactor operation.

Figure 1: Methanol conversion and light olefins selecitivty

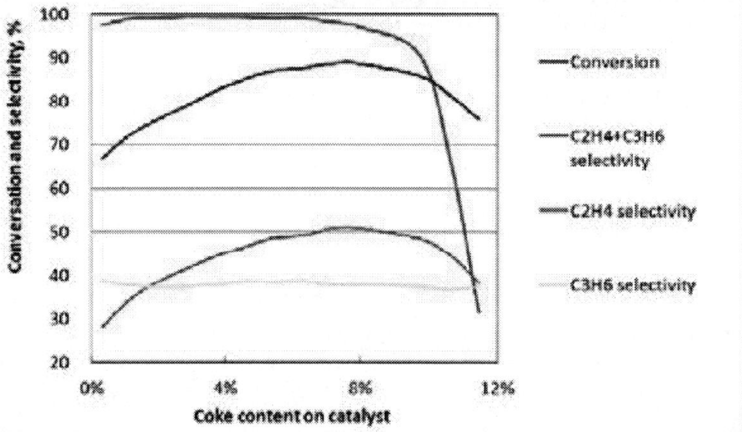

The particle size of DMTO catalyst is similar to that of fluid catalytic cracking (FCC) catalyst, which belongs to Geldart A type of particles. Despite the good fluidization performance, the great quantity of fine catalyst particles can provide large surface area and benefits the heat transfer in the fluidized bed. The temperature difference in the DMTO fluidized bed can be readily reduced to just several degrees via optimal operation.

3.2 MTO fluidized bed reactor scale-up

The scale up of fluidized bed reactor is one of big challenges in fluidized bed process development. The DMTO fluidized bed reactor was scaled up step by step from micro-scale fluidized bed to commercial scale. In the micro-scale reactor (tube diameter is roughly ~1 cm), catalytic performance of DMTO catalyst was evaluated, and the optimal operation conditions were identified. Note that the coke content on catalyst is a function of catalyst residence time (RT), an optimal catalyst RT in pilot-scale reactor (tube diameter is ~10 cm) can be established based on the time-on-stream (TOS) in the micro-scale fluidized bed reactor. In addition, the reaction kinetics of DMTO reaction has also been studied in the micro-scale facility. In the pilot-scale experimental facility, the reactor is connected to the regenerator via standpipes, thus a complete catalyst circulation between reactor and regenerator can be tested. In the pilot-scale reactor, the catalyst stripping was tested. The fluidization performance of DMTO catalyst was also checked in the pilot-scale reactor. The methanol conversion is above 99% for both micro-scale and pilot-scale reactors, while the typical selectivity (in nonaqueous products) to ethylene and propylene is above 80% in both micro-scale reactor and pilot-scale reactor. In 2006, a semi-industrial DMTO demonstration unit was constructed in Huaxian, Shanxi, which consists of fluidized bed reactor, fluidized bed regenerator, quench tower, and sour water stripper. The aims of the demonstration unit are to study the hydrodynamics and reaction performance of DMTO turbulent fluidized bed reactor and provide data for commercial reactor design. The diameter of the reactor in the DMTO demonstration unit is ~1 m, with a methanol feed rate of 2.5 t/h. The unit was running for 1200 hours, and an average selectivity to ethylene and propylene was 79.21%, and the methanol conversion was higher than 99%. Furthermore, the ethylene/propylene ratio could be adjusted in the range of 0.8 to 1.2 by varying operation conditions. The data from this demonstration unit was implemented in the design package of commercial DMTO units.

3.3 DMTO commercial units

Figure 2. The schematic diagram of DMTO process

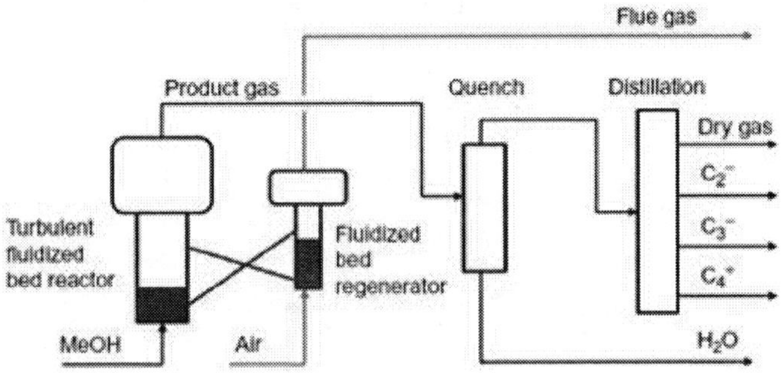

In May 2010, the world's first DMTO commercial unit was constructed in Baotou, North China by Shenhua group. The schematic diagram of DMTO unit is shown in Figure 2. The capacity of light olefins production is 0.6 Mt/a, where the reactor is a shallow fluidized bed with a diameter of 11 m and bed height of 3 m. The unit was successfully started up in August 8, 2010. The heat released by the MTO reaction was used to heat the reactor and regenerator during the start-up, which can on one hand accelerate the start-up process, and on the other hand reduce the energy consumption. Figure 3 illustrates the methanol conversion and selectivity to light olefins in the commercial unit. As can be seen, the methanol conversion reaches ~100%, and the selectivity is higher than ~80%. Typical production distribution in DMTO unit is shown in Table 1. In March 2011, a 72-hour calibration of the Shenhua Baotou DMTO unit was conducted, and the methanol consumption of 2.96 tons per ton product of ethylene and propylene had been obtained.

Table 1. Product distribution for DMTO reactor in commercial units

Product	Selectivity, %
H2	0.20
CO2	0.03
CO	0.12
CH4	1.74
C2H6	1.56
C2H4	41.40

C3H8	3.00
C3H6	39.61
C4	10.92
C5+	1.42
总计	100.00

In total 19 DMTO units have been licensed in China, with a light olefin product capacity of 11.26 Mt/a. And so far 8 DMTO units have been brought into stream. The majority of these DMTO units were constructed together with coal-to-methanol units for methanol supply. Several units built in the East China import methanol from overseas as their feedstock. Figure 4 shows the locations of these DMTO units. The smooth start-up and steady operation of 8 DMTO units indicate that the DMTO technology is becoming a main stream technology for light olefins production in China.

Figure 3. The conversion and selectivity to light olefins in commercial DMTO unit

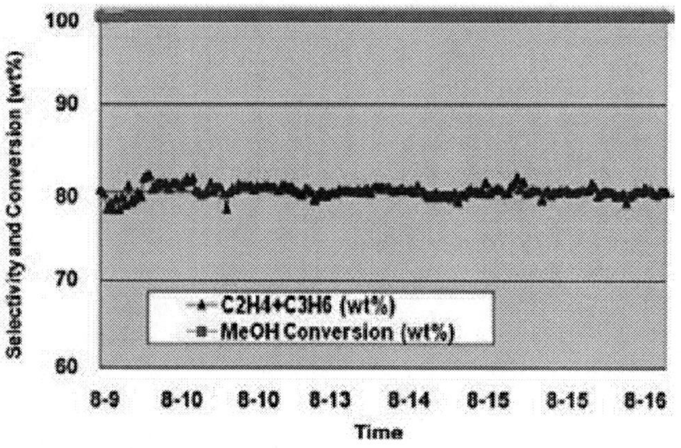

Figure 4. The location distribution of licensed DMTO units in China

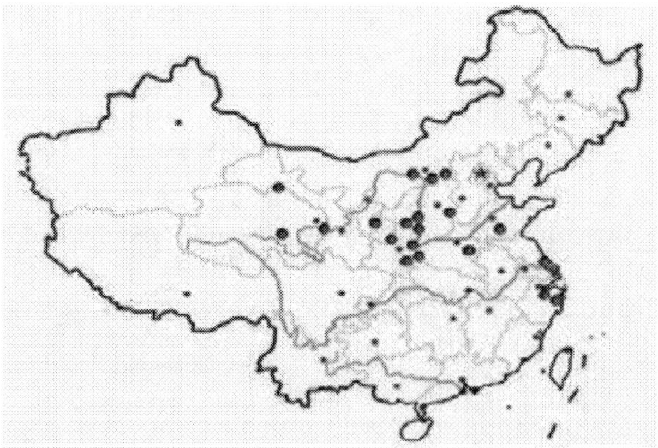

4. DMTO-II technology

4.1 DMTO-II process

DMTO-II is the second generation DMTO process. In this process the byproduct of C4+ was separated from the product and further converted to ethylene and propylene in a second fluidized bed reactor. Figure 5 shows the scheme of DMTO-II process. The advantage of the DMTO-II process is that a double-functional catalyst is used, which can catalyze both the MTO reaction and C4+ conversion reaction. Thus only one regenerator for catalyst regeneration is required.

In 2009, the DMTO demonstration unit was revamped by adding a second bubbling fluidized bed reactor for C4+ cracking. From 2009 to 2010, the DMTO-II process had been verified in the demonstration unit. The selectivity to ethylene and propylene increased from 79.21% in DMTO process to 85.68% in the DMTO-II process with 60 wt % of C4+ recycling, and the methanol consumption for producing one ton ethylene and propylene reduced from 2.96 to 2.67 tons.

Figure 5. The schematic diagram of DMTO-II process

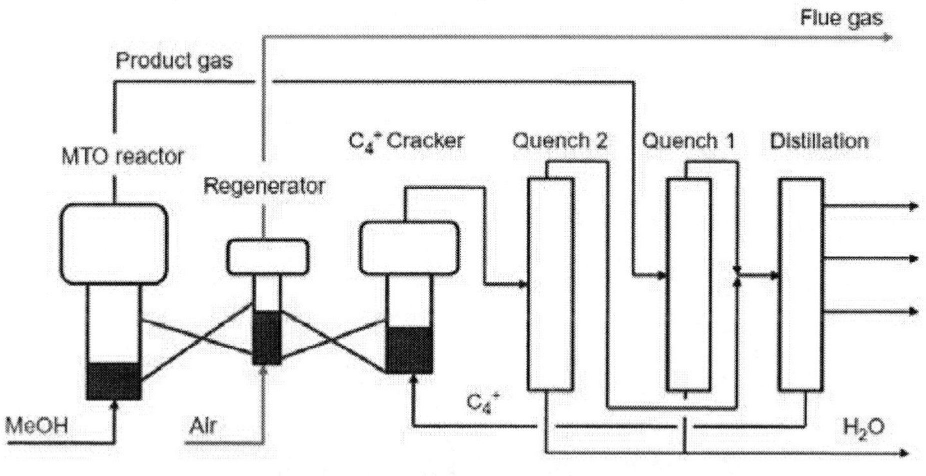

Figure 6. The commercial DMTO-II unit

Table 2. Typical results for C4+ cracker in DMTO-II commercial unit

Product selectivity,%	Condition 1	Condition 2
	Low catalyst circulation rate	High catalyst circulation rate
C2H4	15.74	20.00
C3H6	25.34	32.69
Conversion, %		
C4	66.48	75.63
C5+	65.29	78.76

4.2 DMTO-II commercial unit

In October 2010, DICP licensed the first DMTO-II unit (0.67 Mt/a polyethylene and polypropylene product) to Pucheng Clean Energy Chemical Co. Ltd. In December 2014, the first commercial DMTO-II unit was successfully started up. Table 2 shows the typical results for the commercial DMTO-II unit. Figure 6 shows the photo of the unit.

5. Sustainable development of DMTO technology

DMTO has been widely accepted as the advanced technology in ethylene and propylene production. Beyond that, however, the sustainable development of DMTO technology and its derivatives continues by many other industrial orientated technologies. In the following, several examples are introduced to illustrate DICP's strategy in sustainable development of chemical processes.

5.1 Fundamental research on catalysis

As one of the leading energy research institutes in China, DICP is always committed to develop new catalyst as well as processes for the sustainable, efficient, and clean usage of traditional fossil fuel like oil and coal. In the past years, coal-to-chemicals (CTC) was one of the focuses in DICP, and many breakthroughs have been achieved in our fundamental research. For example, a recent developed bifunctional catalyst that can convert syngas to light olefins with very high selectivity, which exceeds the mechanistic limitation of <60% conversion for Fischer-Tropsch catalysts. As to MTO process, a lot of fundamental research has been done to understand the reaction mechanism, such as the formation of first C-C bond, the induction period, the intermediate product in the reaction, the deactivation kinetics, and diffusion inside cages and channels in zeolites. Based on these research results, the on-purpose improvement of MTO process and optimization of catalyst can be achieved to satisfy different demands in the market.

5.2 DMTP process

As propylene is also highly demanded in market, the on-purpose production of propylene stimulates our research on methanol-to-propylene (DMTO) process development. In this process a catalyst with dual function which can convert methanol to light olefins (ethylene and propylene) and further transform the ethylene is used. And a special designed fluidized bed reactor is developed to maximize the performance of catalyst. The pilot experiments with the scale of 300 t/a methanol feed have been completed. Figure 7 shows the scheme of the DMTP process. Figure 8 shows the concentration of ethylene at the inlet and outlet of the reactor R1 in typical operation, which is calculated in the CH_2 basis. As can be seen, the concentration of ethylene is balanced during this operation, and this means that the net production of ethylene can be minimized.

Figure 7. The scheme of DMTP process

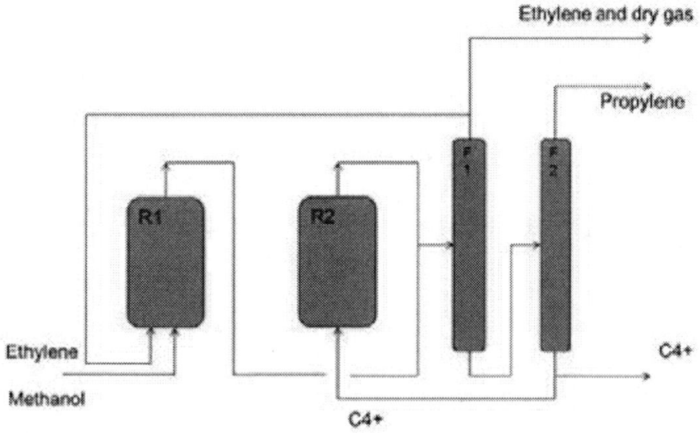

Figure 8. The concentration of ethylene at the inlet and outlet of the reactor R1 in typical operation

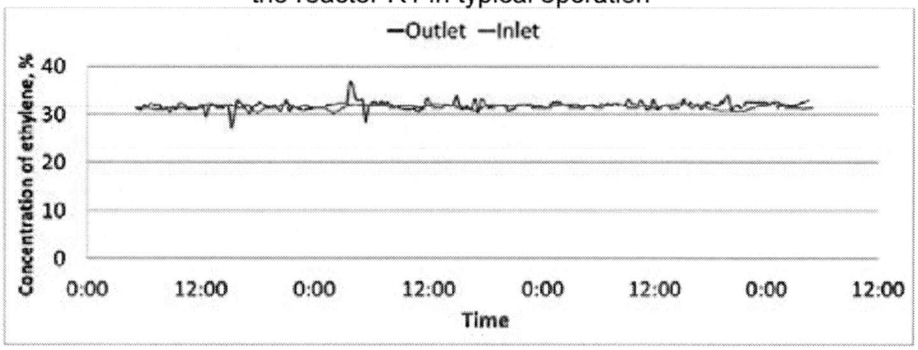

A process simulation has been carried out, and it has been shown that the overall selectivity to propylene of 75% can be reached. This technology is ready for commercialization in DICP.

5.3 Methanol-coupled-naphtha to light olefins

Naphtha steam cracking is still the main stream for light olefins production. Typically about 30% of ethylene and 15% of propylene can be obtained at a temperature of 800~900 ℃ by controlling the contact time of the reactants. Catalytic cracking of naphtha can take place at around 600°C, which is economically attractive in terms of energy consumption. But naphtha cracking reaction is still characterized by highly endothermic, and requires a continuous heat supply. A possible approach is to feed methanol to the naphtha cracking reactor. As the conversion of methanol to light olefin is exothermic, if coupled with methanol feed, the heat required by the naphtha cracking reaction can be partly compensated by heat of methanol conversion. It is known that both naphtha cracking and methanol conversion reactions could be catalyzed by solid acids, and zeolites have been proved to be most efficient catalysts, which provides a possibility to develop one zeolitic catalyst for the both reactions.

The mechanism of methanol-coupled naphtha cracking (MCNC) process is very complicated. The notable character of this coupling transformations is the presence of methanol can lower activation energy required for the alkenes cracking reactions, and promote the cracking reaction rate. This has been confirmed by the coupling conversion of n-hexane and methanol over HZSM-5 zeolite catalysts in a pulse reaction experiment in DICP (as shown in Figure 9). In addition to that, the exothermic conversion of methanol to light olefin can partly compensate the heat required by the naphtha cracking reaction.

Figure 9. Methanol-coupled n-hexane cracking reaction over HZSM-5 catalyst.

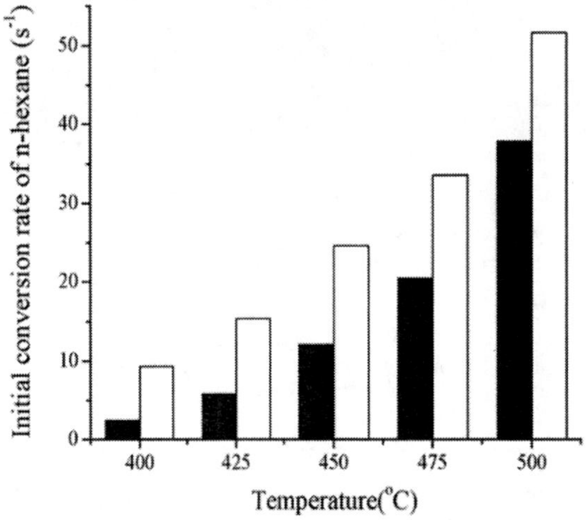

Table 3. The typical light olefins yield for different production processes.

	Naphtha Steam Cracking	Naphtha Catalytic Cracking	Methanol-coupled naphtha cracking		MTO
Reaction Temperature, °C	800-900	600-650	600-650		450-500
Feedstock, %					
Naphtha		83.33	75.00	50.00	0
Water		16.67	16.67	14.29	60
MeOH		0.00	8.33	35.71	40
Total		100.00	100.00	100.00	100.00
Product,%					
Ethylene + propylene	43.37	40.38	40.16	44.3	80.00
BTX	11.51	13.07	11.95	9.84	0

In Table 3, we show the typical yield data for four olefin production processes. The data for naphtha steam cracking was obtained from open literature. The data for naphtha catalytic cracking and methanol-coupled naphtha cracking process are based on lab-scale fluidized bed reactor experiments, which were carried out in DICP. It should be noted, however, that the yield distribution is quite rough and the attention was paid to the main components such as ethylene, propylene, and BTX. All the data are collected in a dry basis, and water is excluded in the analysis. Apparently, the advantages of the methanol-coupled naphtha cracking process include:

(a) heat required for cracking reaction is partially compensated by that released by MTO reaction;
(b) one catalyst for two coupled reaction;
(c) presence of methanol lowers the activation energy required for the alkanes cracking reactions;
(d) product including more aromatics;
(e) flexibility of feedstock.

In fact, methanol is a feedstock that can be readily obtained from either natural gas or coal. The wide availability of natural gas and coal makes methanol much cheaper than naphtha in the market. Commercialization of methanol-coupled naphtha cracking process can certainly improve the margin of olefin production plant.

6. Conclusion

This paper introduced the research and development (R&D) activities at DICP on DMTO and DMTO-II technologies, which covers the zeolite and catalyst synthesis, process development, reactor scale-up, and commercial operations. Beyond that, DICP's strategy of innovative catalyst and process research for on-purpose production of light olefins from methanol is also highlighted by two examples, i.e. the methanol-to-propylene (DMTP) and methanol-coupled naphtha cracking processes. The DMTO technologies, as well as these new processes, will certainly provide extra choices for light olefins production in the market.

7. References

[1] Tian Peng, Wei, Yingxu, Ye, Mao, and Liu, Zhongmin, "Methanol to olefins (MTO): from fundamentals to commercialization", *ACS Catalysis*, Volume 5, NO.3, 2015, pp.1922-1938

[2] Ye, Mao, Li, Hua, Zhao, Yinfeng, Zhang, Tao, and Liu, Zhongmin, "MTO processes development: the key of mesoscale studies", Advances in Chemical Engineering, Volume 47, NO.2, 2015, pp.79-335

[3] Liu, Zhongmin et al., *Methanol to olefins*. Science Press, Beijing, 2015 (inn Chinese)

[4] Jiao, Feng et al., "Selective conversion of syngas to light olefins", *Science*, Volume 351, NO. 6277, 2016, pp.1065-1068

AIChE PAPER 192b

BOOST ETHYLENE AND PROPYLENE PRODUCTION WITH AIR LIQUIDE'S MTP PROCESS AND TECHNIP'S ETHYLENE TECHNOLOGY

Jean-Francois Fournier
Supervising Process Engineer
Technip Stone & Webster Process Technology
Eric Wagner
Vice President, Technology Development
Technip Stone & Webster Process Technology
Jean-Paul Laugier
Vice President, Ethylene Product Line
Technip

Martin Gorny
Product Manager Methanol and MTP Technology
Air Liquide Global E&C Solutions
Thomas Wurzel
E & C Director, Downstream and Petrochemical Product Line
Air Liquide Global E&C Solutions
Stéphane Haag
Senior Research Scientist
Air Liquide Forschung und Entwicklung

Prepared for Presentation at the 2016 Spring National Meeting
Houston, Texas, April 11 – 14, 2016

BOOST ETHYLENE AND PROPYLENE PRODUCTION WITH AIR LIQUIDE'S MTP PROCESS AND TECHNIP'S ETHYLENE TECHNOLOGY

Jean-Francois Fournier

Supervising Process Engineer

Technip Stone & Webster Process Technology

Eric Wagner

Vice President, Technology Development

Technip Stone & Webster Process Technology

Jean-Paul Laugier

Vice President, Ethylene Product Line

Technip

Martin Gorny

Product Manager Methanol and MTP Technology

Air Liquide Global E&C Solutions

Thomas Wurzel

E & C Director, Downstream and Petrochemical Product Line

Air Liquide Global E&C Solutions

Stéphane Haag

Senior Research Scientist

Air Liquide Forschung und Entwicklung

Abstract: In recent years, the steam cracker industry has moved to lighter feedstocks. In particular, a significant number of steam crackers in the US have been revamped to switch to ethane feedstock, thereby reducing the propylene production. As a result, olefin producers must find new ways to restore the balance between the ethylene and propylene products.

Conversion of methanol to propylene via Lurgi MTP™ from Air Liquide is one of the proven technologies available for on-purpose propylene production. Up to now, Air Liquide MTP™ plants have been designed and installed as stand-alone units.

Air Liquide and Technip are currently proposing a new design approach, which combines an MTP plant and a steam cracker unit. Compared to a stand-alone MTP plant and a stand-alone ethane cracker, the integration of the two units boosts the overall ethylene and propylene production by 5 to 10%, depending of the site configuration. At the same time the integration reduces the specific CAPEX per ton of light olefin by a similar percentage. The proposed integration is flexible and can be applied to both existing steam crackers as well as grassroot plants.

BOOST ETHYLENE AND PROPYLENE PRODUCTION WITH AIR LIQUIDE'S MTP PROCESS AND TECHNIP'S ETHYLENE TECHNOLOGY

Jean-Francois Fournier and Eric Wagner, Technip Stone & Webster Process Technology
Jean-Paul Laugier, Technip
Martin Gorny, and Thomas Wurzel, Air Liquide Global E&C Solutions
Stéphane Haag, Air Liquide Forschung und Entwicklung

Introduction

In recent years, the steam cracker industry has moved to lighter feedstocks, in particular ethane, which offers a high olefin yield at a low feedstock price. In the USA, the shale gas development has boosted availability of ethane and a significant number of steam crackers in the USA have already been revamped to switch to ethane, thereby reducing the propylene production. As a result, olefin producers must find new ways to restore the balance between the ethylene and propylene products.

Conversion of methanol to propylene via Lurgi MTP™ process from Air Liquide is a well referenced technology for on-purpose propylene production. Integrated with natural gas based methanol production it offers the best in class technology, and low cash cost route to on-purpose propylene [1].

Up to now, Air Liquide MTP plants have been designed and installed as stand-alone units. Integration was only applied with up and downstream units, such as methanol and air separation units.

Air Liquide and Technip have jointly developed a new design approach, which combines an MTP unit and a Steam Cracker unit. In addition to sharing common equipment between the two plants, the MTP – Steam Cracker integration effort has also identified several synergies that boost the ethylene + propylene production using the same feedstock input as the corresponding stand-alone units.

In the following, different levels of integration are described and referenced to a grassroot concept for both Steam Cracker and MTP unit. The corresponding grassroot plants are used as a basis as they allow for simpler comparison and therefore help to illustrate better the benefits of different levels of integration.

Subsequently the potential for integration of an MTP unit as a "propylene booster" into an existing Steam Cracker is described. Since this integration is more site dependent, the effect of different levels of integration will focus on additional benefits that can be derived from the add-on MTP unit.

Assuming that the reader is quite familiar with Steam Cracking technology but not methanol to olefins processes, an introduction to the MTP process features is given upfront. A more detailed description of the Lurgi MTP™ process can be found here [2].

Lurgi MTP™ Technology

The Lurgi MTP™ process produces propylene from methanol employing a zeolite-based catalyst from Clariant and a fixed bed, multi-stage reactor system as shown in Figure 1.

Figure 1
MTP – Process Overview

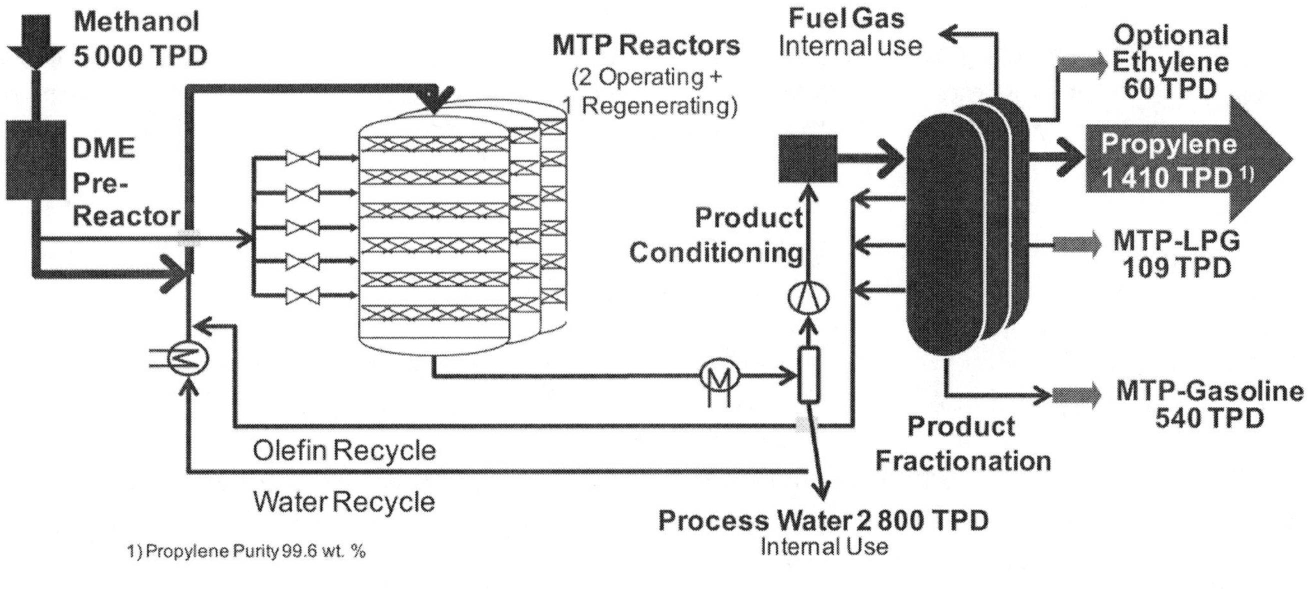

Vaporized methanol is converted to DME (Di-Methyl Ether) and water in an adiabatic DME pre-reactor on an alumina oxide catalyst according to the following equation:

$$2\ CH_3OH\ \text{(methanol)} \quad \text{->} \quad CH_3OCH_3\ \text{(DME)} + H_2O$$

The catalyst, which is robust, highly active and highly selective towards DME, nearly achieves thermodynamic equilibrium for this reaction step.

One part of the DME product stream is combined with process steam and a hydrocarbon recycle from the MTP product fractionation. The combined stream is heated up further in a gas-fired heater and routed to the first catalyst bed of the MTP reactor. The major portion of the remaining DME reactor product is cooled down prior to being fed to the subsequent catalyst beds of the MTP reactor.

The MTP reactor is designed as a multi-stage adiabatic fixed bed reactor. In the MTP reactor the DME/methanol – mixture is converted to olefins on a zeolite-based catalyst according to the following main reaction:

$$n\ CH_3OCH_3\ (DME) \quad -> \quad 2\ C_n H_{2n} + n\ H_2O \qquad n = 2 \ldots 8$$

The conversion of oxygenates to olefins is exothermic and other reaction pathways result in formation of various paraffins, naphtenes and aromatics. High selectivity towards propylene requires control of the reaction conditions. Dedicated feed distribution and process conditions for the MTP reactor stages maintain similar reaction conditions optimized for maximum overall propylene yield. For this the hydrocarbon recycle is very important as it allows for conversion of olefins to propylene and therefore substantially enhances overall propylene yield. The product of the MTP reactor is an olefin-rich hydrocarbon mixture with propylene as the predominant component.

The MTP reactors are installed in a 2+1 reactor concept (2 reactors in operation and 1 reactor in regeneration mode or standby) which allows regeneration as well as the option to exchange the catalyst while production continues. Each reactor represents 50% of the plant capacity.

After a cycle of about 600 hours of operation, the MTP catalyst can be regenerated in-situ by burning off the coke on the catalyst with a nitrogen/air mixture. The regeneration is carried out at similar temperatures as the reaction itself so that temperature stress within the catalyst structure is avoided. Because of the 2+1 reactor concept, regeneration of one reactor does not disturb operation of the MTP unit.

A nitrogen purge after completion of the regeneration ensures that oxygen cannot get in contact with the MTP synthesis. This reduces operation risk and minimizes the formation of oxygenates other than methanol and DME.

The MTP reactor effluent is cooled in a heat recovery system and finally in the quench system, where the hydrocarbon product is separated from the bulk of the byproduct water. The hydrocarbons leave the quench system as overhead vapor whereas the water is condensed and sent to the Methanol Recovery Column. In this column methanol and DME are recovered and subsequently fed as unconverted fresh feed to the DME reactor.

The gas separation and compression as well as the product separation and purification follow standard techniques, quite similar to the practices used in common steam cracker processes.

The compressed hydrocarbon vapour is routed to a product purification section together with the liquid hydrocarbons. Traces of water, carbon dioxide and DME are removed by standard techniques including adsorption and extraction. The hydrocarbon product is then further processed by distillation yielding polymer-grade propylene and optionally some ethylene.

Optional ethylene production is tailored to meet requirements of downstream propylene polymerization units to be able to manufacture Co-Polymer grades. Several olefin-containing streams are sent back to the main MTP synthesis loop. To avoid accumulation of inert or undesired materials in the loop, a small purge is required for light-ends used as fuel gas. Besides the main products - propylene and ethylene - MTP-LPG and MTP-Gasoline are produced as valuable co-products.

MTP – Steam Cracker Integration: Main Constraints

The integration of Steam Cracker and MTP unit is limited by certain constraints derived from the two processes.

The MTP reactor uses a hydrocarbon recycle rich in C4 - C6 olefins. As described above, the olefins in this Hydrocarbon Recycle increase the propylene yield of the MTP unit. Therefore any stream containing C4 – C6 olefins sent to the MTP reactor will increase propylene yield. At the same time such streams should be limited in aromatics, specifically benzene, as their conversion consumes methanol and increases coking, and in acetylenics and di-olefins, which promote catalyst fouling. In addition, it must be ensured that streams sent to the MTP unit are virtually free of poisons for zeolite catalysts such as alkaline metals and sulphur.

Contamination of the Steam Cracker feeds by oxygenates shall be kept to the minimum, i.e. at ppm levels, as it can increase corrosion and fouling issues. The impact of oxygenates in a Steam Cracker have been discussed in previous papers [3, 4].

The effluent from cracking furnaces contain relatively large amounts of di-olefins and benzene. For example, in our Grassroot Study Case (ethane steam cracking), butadiene accounts for 70% of the C4s produced by the Steam Cracker and benzene for 60% of the C6s.

The MTP reactor effluent contains significant amounts of oxygenates, mostly unreacted methanol and DME. It also contains trace amounts of other oxygenates, such as ketones. In the MTP stand-alone plant, the process scheme progressively and efficiently recovers and recycles the unreacted methanol and DME. The bulk of the Oxygenates Removal is actually combined with the MTP Product Fractionation generating the Hydrocarbon Recycle.

Based on the above considerations, Air Liquide and Technip have developed the integration of the MTP unit and the Steam Cracker unit on the following principles:

- For the MTP unit:
  - Segregation of the MTP reactor effluent from the cracking furnaces effluent until the Hydrocarbon Recycle to the MTP unit is generated.
  - Enhancement of the Oxygenates Removal scheme for MTP streams sent to the Steam Cracker.

- For the Steam Cracker:
  - Segregation of the cracking furnace effluent from the MTP reactor effluent until oxygenates are removed in the MTP unit.
  - Additional treatment (such as hydrogenation or extraction) of Cracker streams sent to the MTP reactor and containing acetylenics, di-olefins or aromatics.

MTP – Steam Cracker Integration: Overall Material Balance Improvements

Air Liquide and Technip understand that each plant configuration is unique and that production objectives and associated economics can vary widely from one plant location to another, and/or from one company to another.

As a result, Air Liquide and Technip have developed several flexible integration steps that can be implemented, or not, to improve the Overall Material Balance of a combined MTP – Steam Cracker plant. A few examples of such integration steps are presented in this paper to illustrate their benefits.

For each integration step, the improvements in the Overall Material Balance are quantified against the production of two stand-alone units and at constant fresh feedstock consumption, i.e. at constant methanol and ethane feed consumption.

For a stand-alone MTP unit, propylene production is usually favored over ethylene production due to the purpose of the stand-alone MTP unit and the MTP reactions. For a stand-alone Steam Cracker, ethylene production is usually the primary target and propylene production a secondary target, in particular for gas crackers.

In the context of the MTP – Steam Cracker integration, both the ethylene and propylene productions become primary targets. Improving the Overall Material Balance therefore translates into maximizing the combined ethylene + propylene production.

As a result, the MTP unit design no longer needs to maximize propylene production over ethylene production, but can rather target the maximum light olefins production when it is combined with a Steam Cracker.

MTP – Steam Cracker Integration: Grassroot Study Case

Three sequential degrees of integration will be applied to a Grassroot Study Case which combines two separate and independent grassroot units:

- Lurgi MTP™ unit with a name plate capacity of 1,410 TPD propylene based on 5,000 TPD methanol feed.

- Gas Steam Cracker unit, producing 1,000 kTA ethylene, ethane feed, Technip SMK™ cracking furnaces, ethane and propane recycled to extinction.

The Overall Material Balance for the two stand-alone units is shown in Table 1, considering middle of run catalysts conditions and an on-stream factor is 8,400 hr / year.

Table 1
Stand-alone Steam Cracker + Stand-alone MTP
Overall Material Balance

Hydrocarbons Products	**Cracker kTA**	**MTP kTA**	**Total kTA**	**wt%**
Ethylene	1,000	26.4	1,026	50.0%
Propylene	22.5	500	523	25.4%
Fuel Gas	197	13.2	210	10.2%
Raw C4 Cut / LPG	35.8	37.2	73.0	3.6%
Pygas / Gasoline	30.1	188	218	10.6%
Fuel Oil	3.9	-	3.9	0.2%
Total Hydrocarbons	**1,289**	**765**	**2,054**	**100%**

MTP – Steam Cracker Integration: Equipment Sharing

Sharing equipment in common is driven by the overall CAPEX reduction. However, equipment sharing shall not compromise:

- the process chemistry,
- the operability of the units,
- the overall energy efficiency, i.e. with significant higher OPEX.

The treatment of the C2s and lighter can advantageously be combined between the two units. While this process section is at the heart of the Steam Cracker, in the MTP unit many equipment are required to recover small quantities of ethylene.

The C2s and lighter from the MTP Deethanizer are routed to the Steam Cracker unit upstream of the Cracker Caustic Wash. This allows in the MTP unit the elimination of all equipment related to ethylene purification such as: CO2 Scrubber and Adsorber, Demethanizer, C2 Splitter and C2 Refrigeration. It as well eliminates all cryogenic heat exchangers.

In our Grassroot Case Study example, as per Table 1, the impact on the size of the Steam Cracker equipment is limited due to the fairly low quantity of C2s and lighter produced by the MTP unit compared with the Steam Cracker.

In the same way, the C3 Splitter system can also be combined between the two units.

For a configuration similar to our Grassroot Study Case, the Steam Cracker C3 Cut is routed directly to the MTP C3 Splitter, eliminating the need for a dedicated C3 Splitter in the Steam Cracker. As per Table 1, the impact on equipment size is limited due to the fairly low quantity of C3s produced by the Steam Cracker compared with the MTP unit.

For grassroot plants, the cooling required in the MTP unit at temperatures below cooling water can be done by a common C3 Refrigeration system shared by the two units.

For some of the steps of integration, additional equipment is required and described under the respective step.

Integration Step 1

The overall process scheme for this step is illustrated in Figure 2. In this step, the equipment sharing concepts previously developed are implemented:

- the MTP C2s and lighter are sent to the Steam Cracker upstream of the Caustic Wash,
- the Cracker C3 Cut is sent to the MTP C3 Splitter.

The propane from the MTP C3 Splitter, which now combines both the MTP propane and the Cracker propane, is mixed with the MTP C4s. This C3/C4 mixture, which has also been subjected to enhanced oxygenate recovery, is sent to the cracking furnaces.

With this configuration, the MTP C2s and light ends are recovered and fractionated in the Steam Cracker. In the end, the MTP ethane and the MTP LPG (C3/C4) are used as a supplemental feed to the cracking furnaces.

Figure 2
MTP - Steam Cracker Integration Step 1

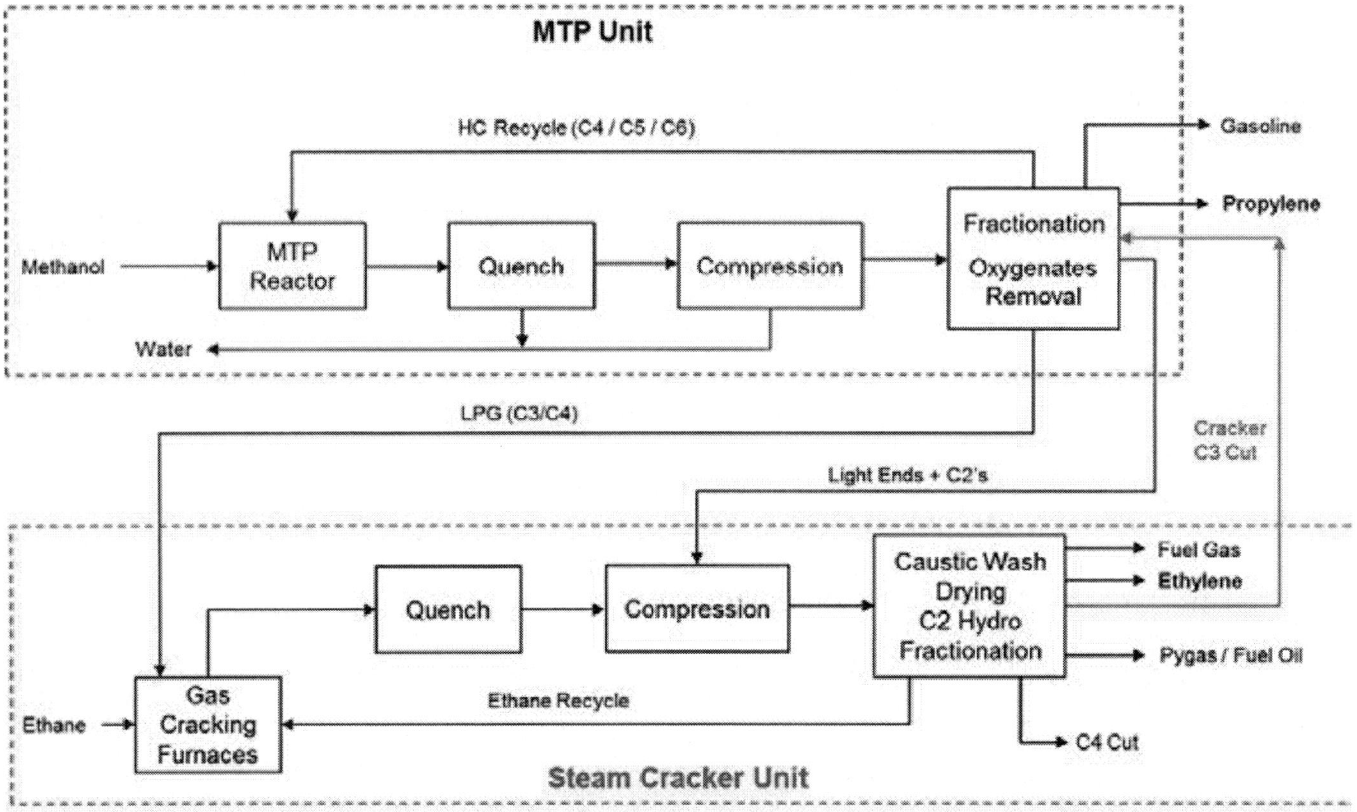

Integration Step 2

In a second integration step, the MTP Gasoline is sent to the Steam Cracker to feed a dedicated liquid cracking furnace, using Technip GK6® coil technology in this study case. Since the MTP Gasoline contains very low amounts of di-olefins, no partial or full hydrogenation of the stream is required prior to feeding it to the cracking furnace. Once-through steam cracking yields evaluated with Technip's SPYRO® software show that approximately 30% of the MTP Gasoline can be directly converted to light olefins. In a stand-alone MTP plant, the Gasoline product is normally used as low sulfur blend stock for gasoline pools.

The overall process scheme for this step is illustrated in Figure 3.

Figure 3
MTP - Steam Cracker Integration Step 2

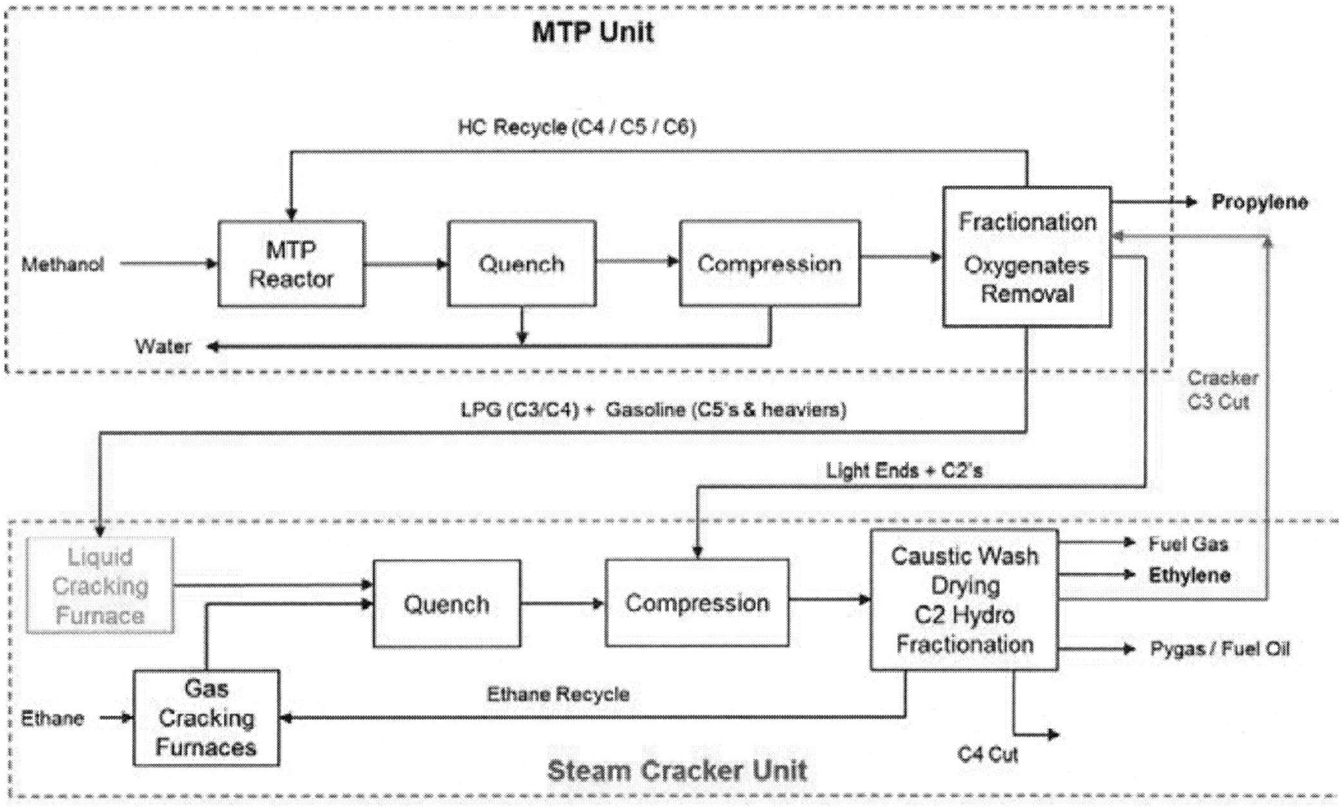

A further improvement of the gasoline conversion to light olefins could be achieved through hydrogenation of the MTP Gasoline before feeding it to the Steam Cracker. This hydrogenation step to convert olefins into paraffins would only be done to further increase the cracking yields to ethylene and propylene.

Integration Step 3

In a third integration step, the olefin-rich streams from the Steam Cracker are sent to the MTP reactor for conversion to propylene. The most obvious candidate is the C4 Cut from the Steam Cracker.

In general, conversion of the C4 Cut is in competition with other potential uses of the stream, such as extraction of butadiene or recycle of the C4 stream to the Cracker.
For the third integration step the recycle of the full C4 Cut after partial hydrogenation of the di-olefins is shown.

Any other olefin-rich stream from the Steam Cracker may be a good candidate for feed to the MTP reactor as well, as long as it does not contain di-olefins or aromatics.

The overall process scheme for this step is illustrated in Figure 4.

Figure 4
MTP - Steam Cracker Integration Step 3

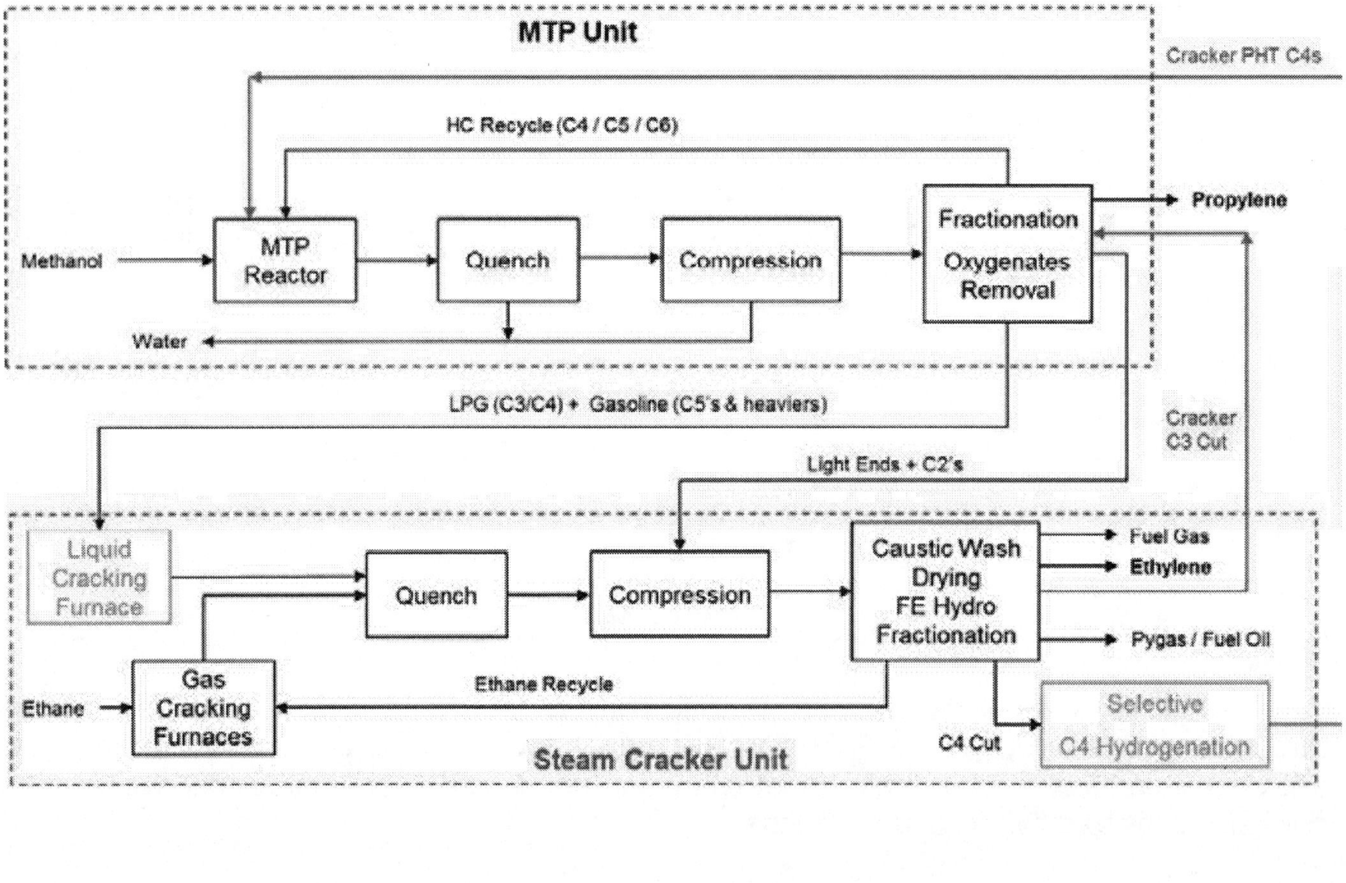

Comparison of the steps of integration and Conclusion

For the integration steps 1 and 2, the MTP by-products are converted to light olefins by pyrolysis in the Steam Cracker furnaces.

For the integration step 3, the Steam Cracker C4 Cut by-product is converted to light olefins in the MTP reactor.

Air Liquide and Technip have performed a cost analysis of each integration step for comparison with the Total Installed Cost of two stand-alone units using existing projects.

The benefits of each integration step is summarized in Table 2.

Table 2
MTP - Steam Cracker Integration
Grassroot Study Case
Integration Steps Summary

Production kTA	**Stand-alone 1 + 1**	**Step 1**	**Step 2**	**Step 3**
Ethylene	1,026	1,086	1,120	1,132
Propylene	523	507	533	556
Total Light Olefins	**1,549**	**1,593 + 3%**	**1,653 + 7%**	**1,688 + 9%**
Butadiene export	24.8	25.5	33.3	0.0
CAPEX	**Base**	**- 4.5%**	**- 1%**	**- 0%**
CAPEX / kTA Light Olefins	**Base**	**- 7%**	**- 7.5%**	**- 8.5%**

Compared to two stand-alone units, the first integration step leads to a slightly lower propylene production. While this may seem surprising, it is actually in line with our target of maximizing the light olefins production. It is a deliberate result from design simplifications made in the integrated MTP unit.

In the context of our Case Study based on an ethane cracker equipped with gas cracking furnaces, both the second and third integration steps require the addition of dedicated liquid cracking capabilities in the Steam Cracker to handle the MTP Gasoline.

This was selected as a "worst-case scenario" from a CAPEX stand-point. Nevertheless, the savings resulting from the equipment sharing between the two units balance the additional costs of the new liquid cracking capabilities.
The CAPEX benefits of the integration steps 2 and 3 would largely increase when combining an MTP unit with a Mixed Feed, Flexible Feed or Liquid Feed Steam Cracker already including some liquid cracking capabilities.

Sending the partially hydrogenated C4 Cut from the Steam Cracker to the MTP unit in order to increase the Propylene production can also be implemented without sending the MTP Gasoline to the Steam Cracker.

However, it shall be noted that cracking the MTP Gasoline increases the Butadiene / C4 Cut production from the Cracker, therefore it has a synergetic effect on the light olefins production when sending the treated C4 Cut back to the MTP reactor.

As mentioned previously, the Cracker C4 Cut may compete with other users. For example, a common practice is to recycle the C4 Cut to the cracking furnaces. In terms of yield to light olefins, sending the partially hydrogenated C4 Cut to the MTP unit is equivalent to recycling the fully hydrogenated C4 Cut to the cracking furnaces. With the MTP route, however, the yield to propylene is much higher than the yield to ethylene. This advantage offers an excellent opportunity to use the Cracker C4 Cut as a "propylene booster" in an integrated MTP – Steam Cracker plant.

The different steps of integration described above represent the range of options available. The optimum solution for an individual project requires all specific aspects to be analyzed.

For better overview, the impact of step 3 on the Overall Material Balance is quantified against the two stand-alone units in Table 3. The integration step 3 boosts the light olefins production (ethylene + propylene) by almost 10% while reducing the quantity of by-products by nearly 30%.

Table 3
MTP - Steam Cracker Integration
Grassroot Study Case
Overall Material Balance

Hydrocarbons Products kTA	Stand-alone Units 1 + 1	Integration Step 3
Ethylene	1,026	1,132
Propylene	523	556
Total Light Olefins	**1,549**	**1,688** **+ 9%**
Fuel Gas	210	245
Raw C4 Cut / LPG	73	0
Pygas / Gasoline	218	88
Fuel Oil	4	33
Total By-Products	**505**	**366** **- 28%**
Total Hydrocarbons	2,054	2,054

MTP – Steam Cracker Integration: Existing Plant Study Case

This Study Case has been developed to highlight the benefits of combining a new MTP unit to an actual existing Steam Cracker where the existing C3 Splitter is no longer used to its full capacity.

Such an integration has the potential for a substantially higher degree of use of common equipment as the Grassroot Study Case. The most obvious is to route the MTP C3 Cut directly to the existing Steam Cracker C3 Splitter to eliminate the need for a new dedicated C3 Splitter in the MTP unit, hence reducing CAPEX.

On the other hand the use of a common C3 Refrigerant system may be more difficult, as the existing refrigeration system may not have sufficient spare capacity. In such cases specific concepts must be developed to meet the MTP plants cooling needs (e.g. Absorption cooling).

The Study Case presented here is based on the following scenario. The existing Steam Cracker is a flexible feed unit equipped with cracking furnaces able to handle gas and liquid feeds, including naphtha. The C3 Splitter design capacity is around 250 kTA propylene. Due to a switch to lighter feedstock, the actual propylene production is down to 42 kTA while the ethylene production is 954 kTA. The raw C4 Cut is exported.

The capacity of the new MTP plant is set at 2,000 TPD methanol feed in order to fit the design capacity of the existing Cracker C3 Splitter. The MTP unit production is 200 kTA propylene and 10 kTA ethylene. For the two stand-alone plants the combined propylene production (MTP + Steam Cracker) is 242 kTA.

Since the existing Steam Cracker has already liquid cracking capabilities, the new MTP unit is combined as per the integration step 3 previously described, refer also to Figure 4. The main difference is that in this case the MTP C3 Cut is sent to the existing C3 Splitter in the steam cracker. The MTP LPG / Gasoline can be handled by a single existing liquid furnace.

No estimation of the overall investment cost is presented, as it depends of the actual spare capacities available in the cracker. However, with an existing cracker already equipped with a C3 Splitter and liquid cracking capabilities, most of the CAPEX will be spent for the new MTP unit. Air Liquide and Technip have therefore performed a cost analysis for the MTP unit alone to illustrate the potential in CAPEX reduction.

The benefits of the integration is summarized in Table 4.

Table 4
MTP – Steam Cracker Integration
Existing Plant Study Case

Production kTA	**Stand-alone Units 1+1**	**Integrated Units**
Ethylene	964	1,010 **+ 5%**
Propylene	242	269 **+ 11%**
Total Light Olefins	**1,206**	**1,279 + 6%**
Butadiene Export	28	0
MTP unit CAPEX	**Base**	**-30%**

Compared to our Grassroot Study Case, the impact of sending the Cracker C4 Cut to the MTP unit is more visible on the propylene production, due to the change in the relative sizes of the combined units.

As mentioned previously, a range of options can further improve the integration benefits. For an existing steam cracker, some of the options mentioned previously may not be applicable; however additional feedstock opportunities may be available on site.

For example, the partially hydrogenated C5 Cut from an existing pygas hydrogenation unit is typically recycled to the cracking furnaces. Instead, it can be routed as a feed to the new MTP unit to boost propylene production.

Some petrochemical sites include multiple steam crackers. The treated cracker C4 Cuts and/or C5 Cuts can be combined and send to a single new integrated MTP unit. As illustrated in Table 4, the relative size of the combined units can significantly boost the propylene production, therefore the crackers C4 / C5 by-products would in this example provide a very substantial amount of the final propylene production.

This example illustrates the strong advantages to integrate a new MTP unit as a "propylene booster" into an existing Steam Cracker through synergistic effects of exchanging streams and CAPEX reduction through use of spare capacities in existing equipment.

MTP – Steam Cracker Integration: Overall Conclusion

The cases of Grassroot Study Case show that the integration of a Steam Cracker with an MTP plant offers the potential to achieve a synergistic increase in light olefin yield of nearly 10% compared to independent systems without any increase in CAPEX. The integration scheme is flexible and can alternatively achieve a 4.5% reduction in CAPEX while still providing a 3% increase in light olefin production.

The integration of the two technologies can be done in different steps offering best adaptability to local conditions. The proposed integration offers the required flexibility to meet specific objectives and addresses the full range of Steam Cracker configurations, from ethane to naphtha crackers including mixed and flexible feed crackers.

The Existing Plant Study Case shows that integration of a new MTP plant into an existing steam cracker allows for CAPEX reduction of up to 30% for the new MTP unit and at the same time for a synergistic increase in light olefin production of 6%. This makes an MTP unit an ideal "propylene booster" for existing steam crackers.

With Technip's vast experience in Steam Cracking technologies and Air Liquide's expertise in MTP technology the integration can be tailored to best meet the needs of any specific application and its relevant markets.

References

1. Diercks Rainer(2014), "Petrochemicals The Heart of BASF's Verbund" BASF Investor Day Chemicals Segment May 22, 2014. *https://www.basf.com/documents/corp/en/investor-relations/calendar-and-publications/calendar/2014/basf_investor_day_chemicals/BASF_Investor-Day-Chemicals_Petrochemicals.pdf (accessed February 17 2016)*
2. Haag, Stéphane; Pohl, Sven; Gorny, Martin and Martin Rothaemel (2012), "Methanol to propylene: from development to commercialization" proceedings from DGMK Tagung 2012-3, p 115-122.
3. Nowowiejski, George and John A. Reid (2003), "An Overview of Oxygenates in Olefins Units in Relation to Corrosion, Fouling, Product Specifications, and Safety," proceedings from the 15th Ethylene Producers' Conference.
4. Hood, David and Steve Coleman (2007), "Minor Components, Major Issues: An Overview of Contaminants in Ethylene Plants" proceedings from the 19th Ethylene Producers' Conference.

AIChE Paper 192C

A NEW ERA IN ON-PURPOSE BUTADIENE TECHNOLOGY

Clifford Alan Maat

Senior Staff Engineer

TPC Group

John Senetar

Fellow

UOP LLC, a Honeywell Company

Christopher DiGiulio

Senior Engineer

UOP LLC, a Honeywell Company

Jeannie Blommel

Lead Design Engineer

UOP LLC, a Honeywell Company

Joseph Duff

Staff Research Chemist

TPC Group

Jillian Horn

Senior Research Scientist

TPC Group

Michael Nutt

Principal Research Engineer

TPC Group

Prepared for Presentation at the 2016 Spring National Meeting

EPC Fundamentals of Ethylene Technology Session

Houston, Texas, April 11 -14, 2016

AIChE Paper 192C

A NEW ERA IN ON-PURPOSE BUTADIENE TECHNOLOGY

Clifford Alan Maat

Senior Staff Engineer

TPC Group

John Senetar

Fellow

UOP LLC, a Honeywell Company

Christopher DiGiulio

Senior Engineer

UOP LLC, a Honeywell Company

Jeannie Blommel

Lead Design Engineer

UOP LLC, a Honeywell Company

Joseph Duff

Staff Research Chemist

TPC Group

Jillian Horn

Senior Research Scientist

TPC Group

Michael Nutt

Principal Research Engineer

TPC Group

Abstract: Butadiene is produced almost entirely as a byproduct of ethylene plants. As ethylene plants switch from naphtha feedstock to cheaper ethane, the amount of butadiene byproduct decreases. A supply shortage is anticipated. On-Purpose Butadiene will be needed to meet the demand. One way to make On-Purpose Butadiene is Oxidative Dehydrogenation of butenes, traditionally done by TPC Group (Texas Petrochemicals, Petro-Tex). TPC and UOP have jointly developed an improved process based on the TPC technology. The improvements are reviewed and the new technology compared with alternative routes for On-Purpose Butadiene.

Background

Butadiene was produced in moderate volumes from ethanol or acetylene prior to the Second World War.

Direct Dehydrogenation - early 1940's to mid 1960's

Interruption of global trade in natural rubber by the Second World War created an immediate and urgent demand for on-purpose butadiene. Most of the new production capacity used direct dehydrogenation processes, with n-butane or normal butenes as the feedstock. Further expansion of butadiene production capacity took place in the 1950's.

Oxidative Dehydrogenation - mid 1960's to late 1980's

A new process for butadiene production from butenes was introduced by Petro-Tex in 1965. Addition of oxygen to the feedstock resulted in a chemical equilibrium much more favorable for butadiene. The oxidative dehydrogenation process eventually displaced much of the direct dehydrogenation production.

Adequate Supply From Byproduct Butadiene - late 1980's to mid-2010's

A wave of construction of new ethylene plants in the late 1980's enlarged the supply of butadiene as a byproduct of ethylene manufacture. The low-cost byproduct butadiene displaced all other manufacturing routes, and large scale on-purpose butadiene production eventually stopped.

New Demand for On-Purpose Butadiene - mid-2010's...

The shale gas revolution has made low-cost ethane available to many ethylene plants in North America. As the ethylene plants have switched to feed more ethane, the amount of byproduct butadiene has decreased. On-purpose butadiene production capacity has been installed in China, and a need for new on-purpose butadiene capacity in other regions is anticipated in the near future.

Commercial History of the Oxidative Dehydrogenation Process

In 1965, Petro-Tex (predecessor of Texas Petrochemicals and TPC Group LLC) commercialized butadiene production via oxidative dehydrogenation of butenes in its TPC Group OXO-D® process [1]. The OXO-D technology provided significant savings in raw material and energy costs when compared to conventional dehydrogenation technologies, which are endothermic and equilibrium limited. The first OXO-D plant was operated by TPC Group in Houston, Texas, with an annual butadiene production capacity of 350 thousand metric tons/yr (kMTA). In the years that followed, six additional OXO-D plants were licensed with production capacities ranging from 75 to 350 kMTA. The OXO-D units successfully operated with a wide range of C4 feedstock compositions and downstream butadiene extractive distillation technologies. All of the OXO-D units had excellent safety records.

The original OXO-D plant in Houston remained in operation until 2004, when it was finally decommissioned after four decades of successful operation.

Process Chemistry/Catalysts

The reaction chemistry is summarized in Figure 1, with the heats of reaction reported per mol of 1-butene and oxygen consumed, respectively [2]. Butene is reacted with molecular oxygen to form butadiene and water, with air being the most readily-available oxygen source. The formation of 1,3-butadiene is greatly favored over 1,2-butadiene. Thus for convenience, the desired 1,3-butadiene product is herein simply referred to as butadiene. The conversion and selectivity to butadiene for the normal-butene isomers (1-butene, cis-2-butene, and trans-2-butene) are similar in the OXO-D process. The main byproducts are the combustion products CO and CO2. In this process, oxygen is consumed to completion.

FIGURE 1 – Oxidative Dehydrogenation Reactions

Stoichiometry	ΔH_{RXN} (kJ/mol_{BUTENE})	ΔH_{RXN} (kJ/mol_{OXYGEN})
$C_4H_8 + ½O_2 \rightarrow C_4H_6 + H_2O$ 1 or 2 Butenes, Oxygen, 1,3 Butadiene, Water	-131.9	-263.8
$C_4H_8 + 4O_2 \rightarrow 4CO + 4H_2O$ 1 or 2 Butenes, Oxygen, Carbon Monoxide, Water	-1409.3	-352.3
$C_4H_8 + 6O_2 \rightarrow 4CO_2 + 4H_2O$ 1 or 2 Butenes, Oxygen, Carbon Dioxide, Water	-2541.3	-423.6
$iC_4H_8 + 6O_2 \rightarrow 4CO_2 + 4H_2O$ Isobutene, Oxygen, Carbon Dioxide, Water	-2524.3	-420.7

Minor byproducts include ethyl and vinyl acetylenes, formaldehyde, acetaldehyde, acrolein, methacrolein, acetone, furan, isoprene, vinylcyclohexene, and aromatics (BTX, styrene, naphthalene). Even small amounts of acetylenes and some oxygenates can cause operating problems in downstream processing units, so there is a premium on minimizing the amounts of these chemicals in the butadiene product.

Process Description

Figure 2 provides a simplified process flow diagram for the TPC/UOP OXO-D process. Butenes, steam and air are mixed and fed into a two-stage reactor system. The reactors operate adiabatically between 0.1 to 0.3 MPa(g) , inlet temperatures from 300 to 400 °C, and outlet temperatures from 500 to 600 °C. Depending on the feed composition and targeted yield, the steam to butene molar ratio for the reactors can be between 9:1 to 14:1. Additional air is introduced in between Stage 1 and Stage 2 to maximize butene conversion. Splitting the air addition between the two stages has the added benefit of limiting the exotherm in each stage, which reduces the overall steam requirement for the process. The reactor effluent is sent to a Quench Tower, which serves to condense both the dilution steam added to the feed and the additional water vapor generated via the oxidative reactions. Condensed water is recovered and vaporized to provide all of the reactor steam requirements. The utilities for steam generation are minimized by using process heat recovery (not shown). Net water produced by the process is sent to water treatment. After bulk water removal, the net product vapor is compressed and sent to an Aldehyde Scrubber to remove trace aldehydes and other oxygenates. The rich solvent from the Aldehyde Scrubber is sent to the Aldehyde Stripper to generate a lean solvent for reuse and a vent stream containing the oxygenate contaminates. The contaminant-free vapor from the Aldehyde Scrubber is sent to the Product Absorber where C4+ hydrocarbons are absorbed in a circulating solvent. The rich solvent from the Product Absorber is sent to the Degasser to remove any absorbed light species, and then sent to the Product Stripper for recovery overhead of the crude butadiene product, which is typically in the range of 55 to 65 percent butadiene. Lean solvent from the Product Stripper is recycled to the Product Absorber. Solvent regeneration (not shown) is included to remove byproducts accumulated in the solvent. The crude butadiene is then sent to an extractive distillation unit to produce high-purity butadiene.

Figure 2

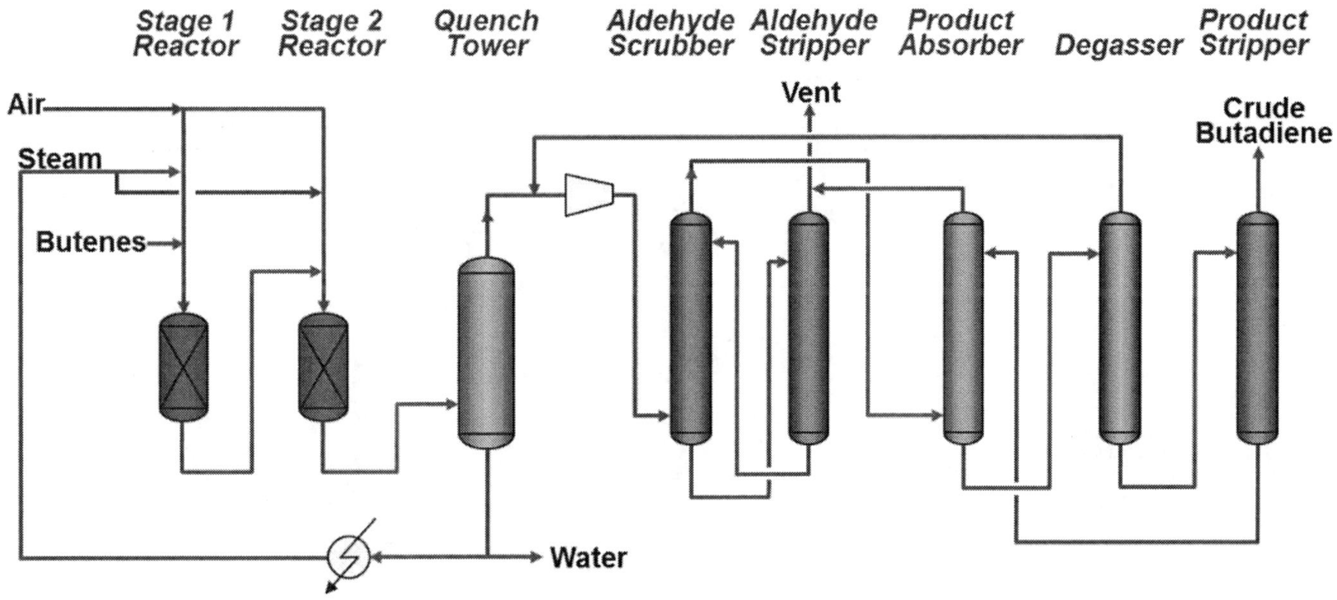

TPC/UOP OXO-D Process

Feedstocks

The unit design is flexible and can process a wide variety of feeds. Butanes pass through the reactor essentially unconverted. As indicated above, isobutene will largely convert to carbon dioxide. However, isobutene in the feed is acceptable and the suitable level is simply an economic consideration. Generally a raffinate-II with isobutene less than 2 percent is a preferred feedstock. Byproduct C4s from a methanol-to-olefins (MTO) process are an excellent feed, as MTO C4s are primarily 1- and 2-butenes with only a small amount of saturates and almost no isobutene.

Reactor Design and Operation

Conversion and Selectivity

Unlike the direct dehydrogenation processes, in which conversion is limited by the equilibrium concentrations of the products and feed, the oxidative dehydrogenation reaction proceeds essentially to completion. The conversion in each reactor can be adjusted simply by increasing or decreasing the amount of oxygen feed.

Selectivity, however, presents a practical limit to the feasible conversion per reactor pass. As incremental oxygen is added, selectivity decreases, producing relatively more CO and CO2. There is a point of maximum yield, beyond which further addition of oxygen results in lower butadiene production.

Heat and Temperature

Reactor temperature can also limit the feasible conversion per pass. Because both the main reaction and the significant side reactions to CO/CO2 are exothermic, the reactor effluent temperatures increase rapidly as the reactions progress. Typical temperature profiles are shown in Figure 3:

Figure 3

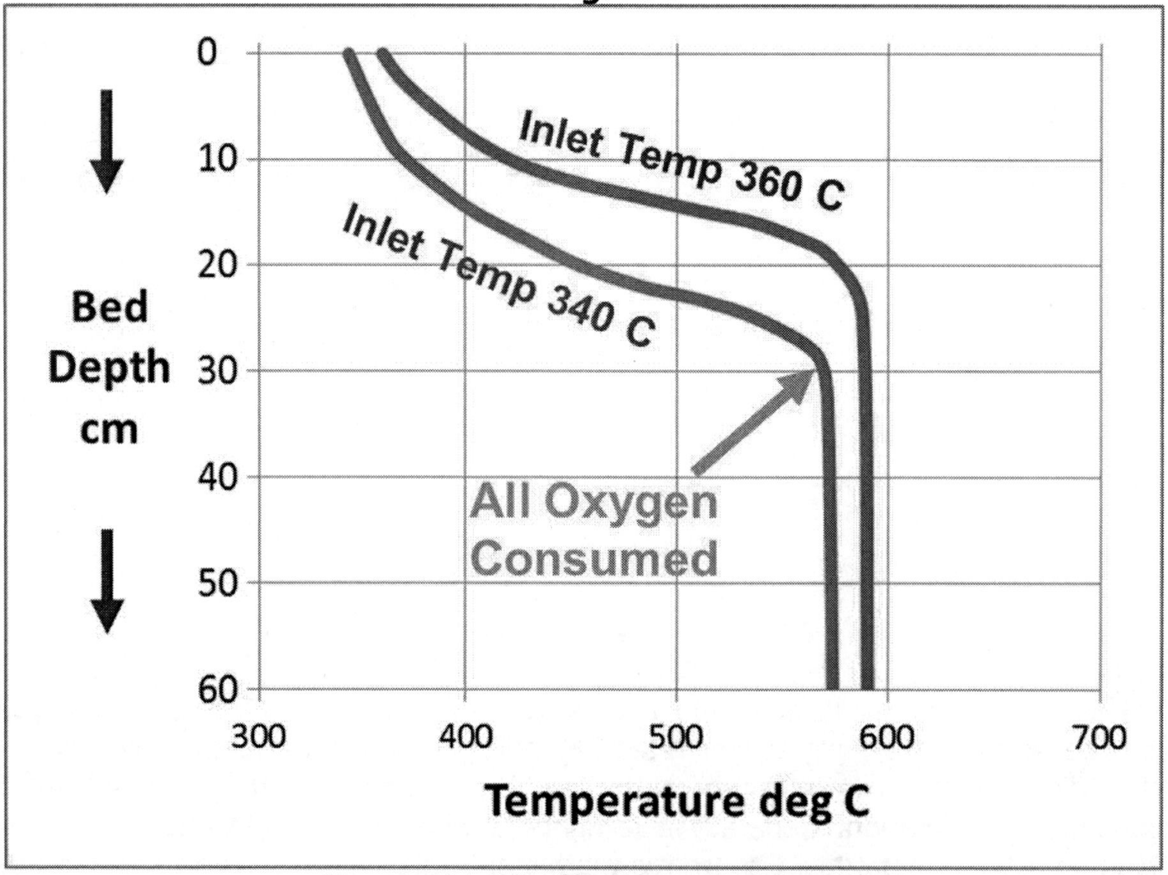

Temperature Profiles in an OXO-D Reactor

Steam feed serves partly to control the reactor outlet temperature; the steam absorbs a large part of the heat of reaction.

Single Stage versus Two Stage

The traditional commercial OXO-D process used a single stage reactor. While use of a second stage implies a loss of selectivity by feeding butadiene to the second stage, the two-stage reactor can be designed to minimize butadiene loss and, by virtue of much higher yield, result in higher overall plant efficiency. A qualitative comparison of reactor yields is shown in Figure 4:

Figure 4

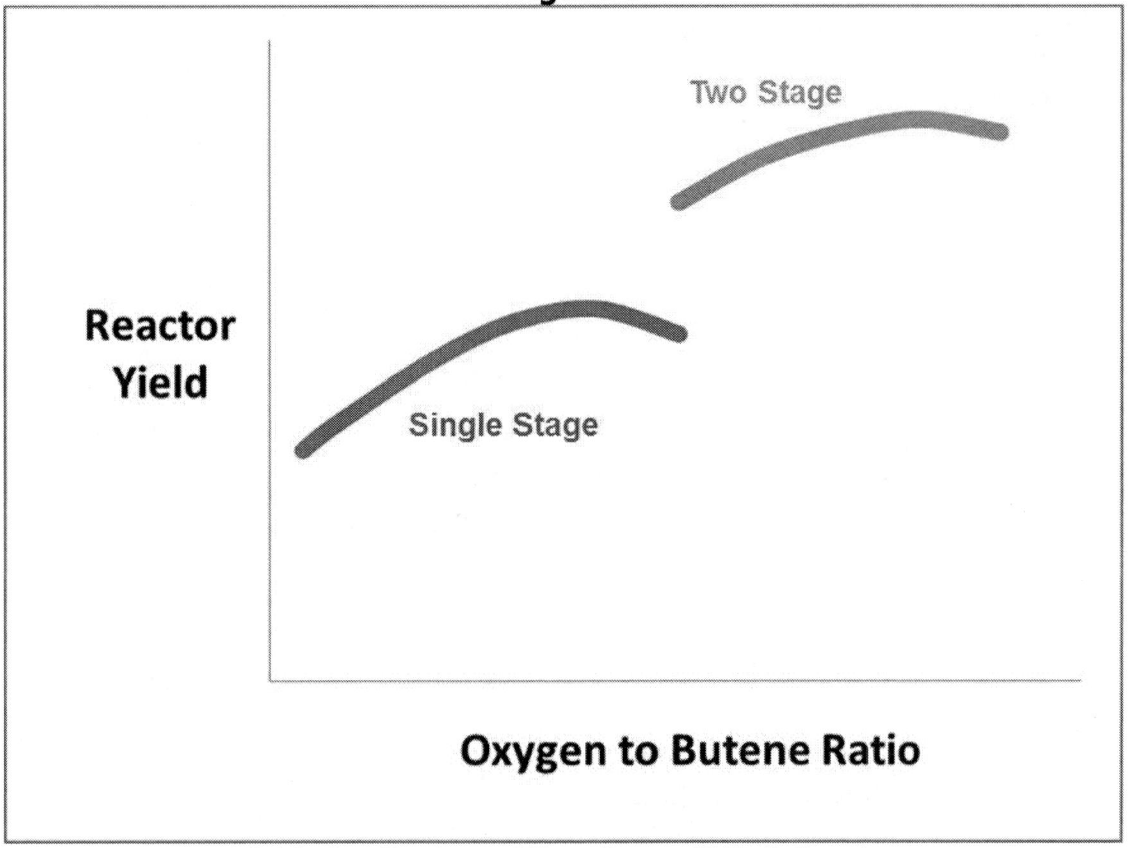

Yields for Single and Two Stage OXO-D Reactors

Byproduct Reduction

Oxygenated byproducts, as listed previously, are likely to have detrimental effects on downstream butadiene processing units. Some oxidative dehydrogenation processes, such as some with molybdenum-bismuth catalyst systems, use excess oxygen in the reactor and therefore have a propensity to form relatively large amounts of oxygenated byproducts. The OXO-D process is designed to minimize these byproducts as follows:

1. Butene is fed in excess, so all oxygen is consumed and the residence time of the oxygen is minimized.

2. The reactor catalyst system is designed such that any byproduct formation favors the relatively benign byproduct CO2. Figure 5 shows the difference between the OXO-D catalyst and an oxidative dehydrogenation catalyst designed only for high yield.

Figure 5

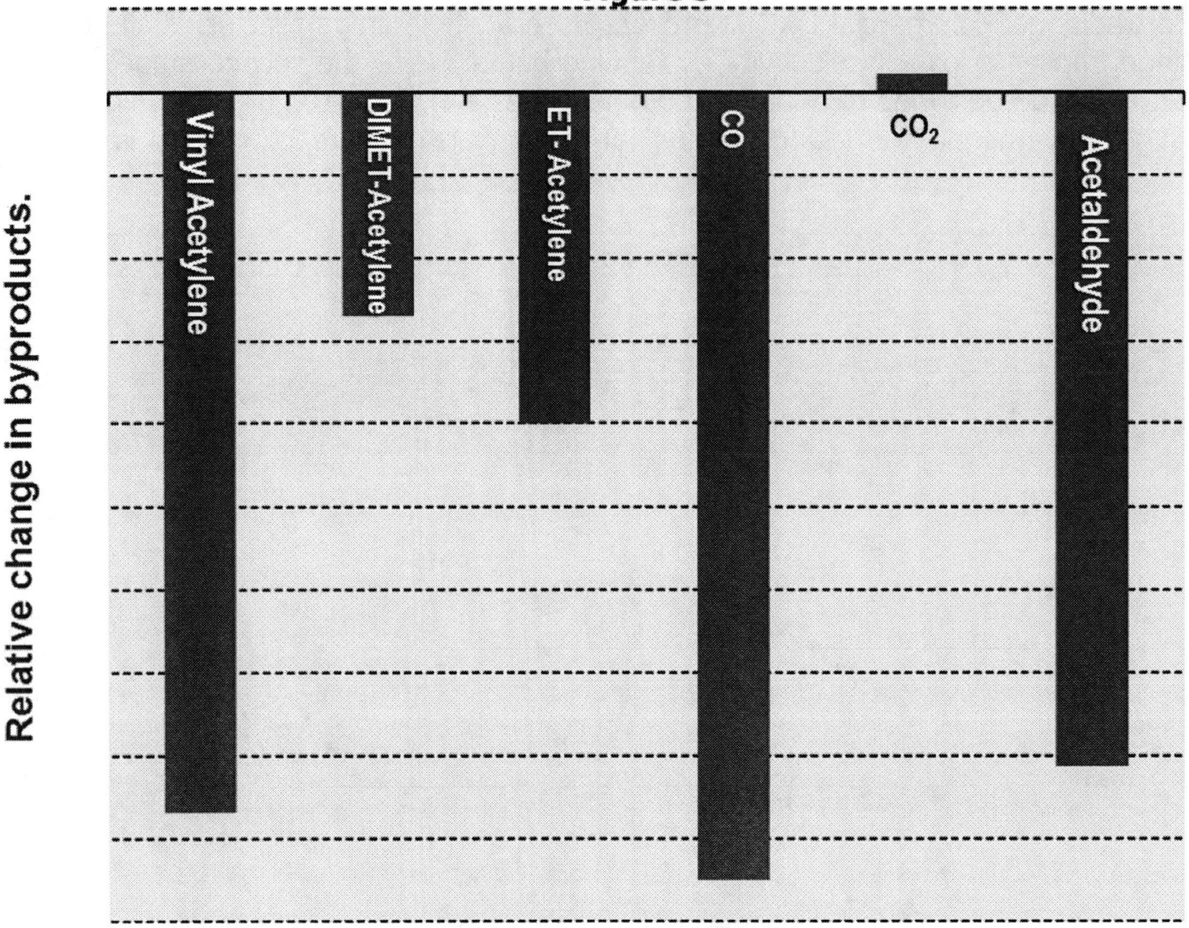

Relative Destruction Efficiency of TPC's Byproduct Reduction Catalyst

3. Oxygenated byproducts that do manage to form in the reactor are mostly removed in the Aldehyde Scrubber.

Process Safety

Reactor Temperature Control

The oxidative dehydrogenation reactions are exothermic and have a high initiation temperature (300 to 400 °C), so the reactor effluent temperature must not be allowed to get too hot. Some oxidative dehydrogenation processes and many other refining and chemical processes, hydrogenation reactors for example, also must maintain temperatures within a certain range while generating heat, and temperature control measures such as cooling sprays and emergency depressuring systems are common.

The OXO-D process temperature control is comparatively simple, because once initiated, the reactions proceed to completion. There is no need to remove heat from the catalyst bed or to try to control the temperature within the reactor. The temperature is controlled by

1. Inlet Temperature
2. Oxygen Feed
3. Steam Feed

All of these are easily instrumented and controlled upstream of the reactor.

Mixing Oxygen with Hydrocarbon

The possibility of unwanted combustion must always be considered when mixing oxygen and hydrocarbons, especially at high temperature. The OXO-D process has been operated safely as follows:

First, the air-steam-hydrocarbon mixture entering the reactor has too little oxygen to burn. This is illustrated in Figure 6, a triangular composition diagram and flammability envelope for 1-Butene/Oxygen/Nitrogen at ambient conditions (the true OXO-D flammability envelope is somewhat different due to operating conditions and other components, but Figure 6 illustrates the concept). A representative OXO-D inlet composition is the point labeled "Example Reactor Inlet Composition." All practical operating reactor inlet compositions would be in this region, outside the flammable envelope.

Second, mix the air with the steam first. No mixing process takes place instantaneously, so all possible proportions of two mixing feeds can be expected to be temporarily present during the mixing process. By adding steam to the air first, the oxygen concentration is diluted to a level lower than the minimum required to support combustion. Then, any amount of butene may be added safely. The commercial OXO-D process operated this way for over thirty years.

Figure 6

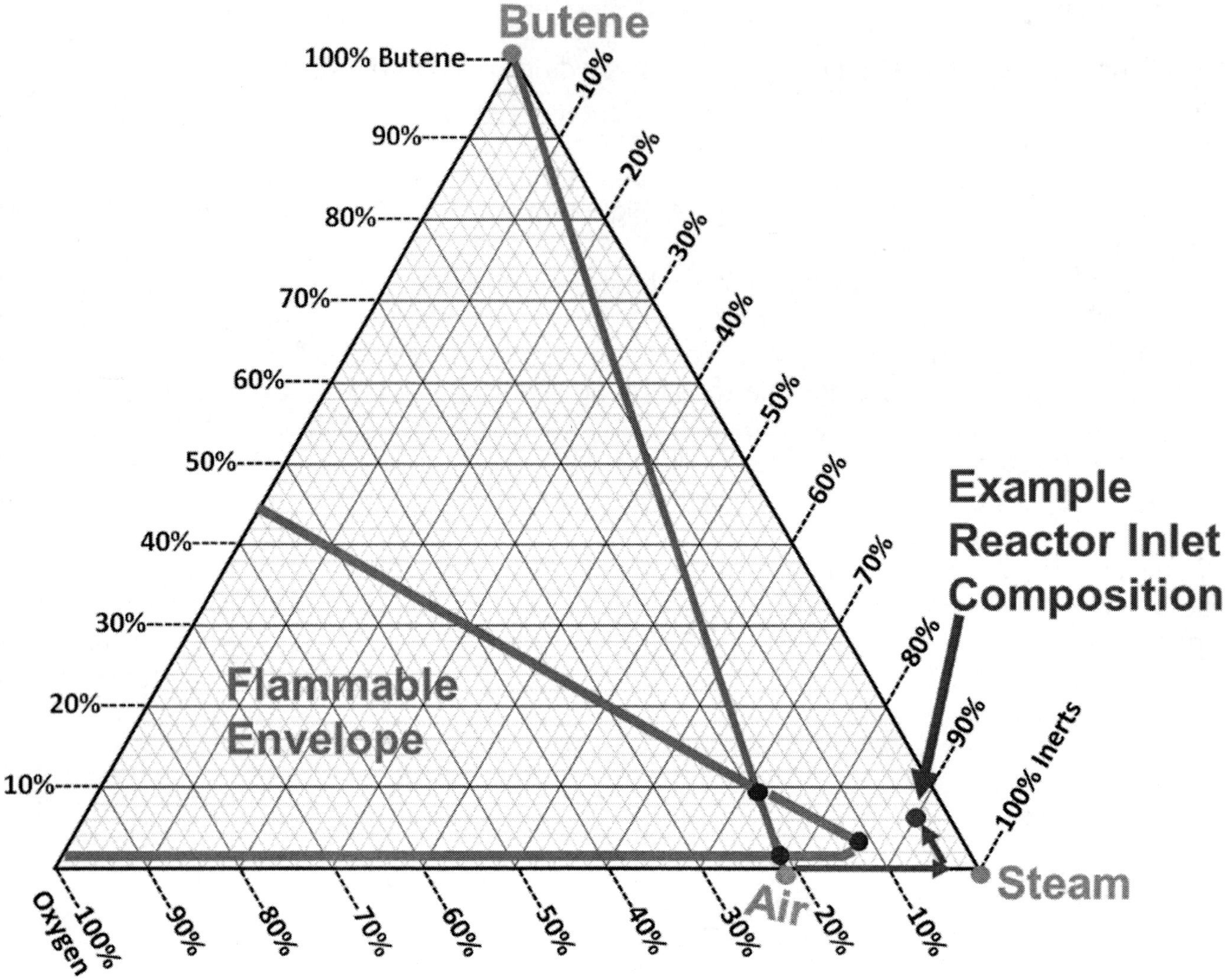

Flammability Envelope for 1-Butene/Oxygen/Nitrogen

Environmental Performance

Water Management

The steam feed to the OXO-D reactors is generated entirely from process condensate from the Quench Tower. Commercial experience from the TPC OXO-D unit as well as from the licensed plants enables proper design of process water treatment to minimize fouling and contamination problems. Since no fresh water is added to the process during normal operation, only a small purge stream of wastewater equal to the net production of water in the reactors must be treated for disposal.

Absorber Offgas

The Absorber offgas is mostly nitrogen with some CO2 and minor components. The heating value is too low for direct use as fuelgas, but it can be mixed with fuelgas and used in fired heaters or thermal oxidizers according to local environmental regulations.

Energy Efficiency

The hot reactor effluent contains high quality sensible heat that is recovered. The Quench Tower bottoms contains a large amount of heat; though it is at lower temperature and therefore more difficult to recover, opportunities should exist in most plants for heat integration. Further heat integration within the OXO-D unit has been incorporated into the TPC/UOP design.

Reliability and Campaign Length

All process plant designs involve some trade-off between capital cost and operating reliability and run-time. The commercial TPC plant operated with moderate campaign lengths. The TPC/UOP OXO-D process is designed for continuous four-year operation. Among the design features are

1. Spare Reactor – The catalyst does not last four years; consequently, one reactor can be taken offline for catalyst regeneration/replacement while full production rates are maintained.
2. Ceiling Temperatures – Streams containing large amounts of butadiene are kept cool enough to eliminate fouling problems caused by polymerization.
3. Spare Reboilers – While the reboilers in the butadiene recovery section do not foul rapidly, they cannot run continuously for four years.

If a four year run time is not needed, capital cost may be reduced by installing fewer spares and allowing a wider range of operating conditions.

Economics

Material Balance

An overall unit material balance is shown in Table 1 for a plant producing 100 kMTA of butadiene (as butadiene contained in the crude butadiene product) based on 8,000 operating hours per year. The feed in this case is assumed to be a raffinate-II feed containing 72wt% 1- and 2-butene and 0.5% isobutene. Butadiene yield is 71.6 wt% of combined 1- and 2-butenes in the fresh feed. Yields can range from 70wt% to 75wt%, depending on the feed purity and the desired operating severity.

Table 1 - Overall Unit Material Balance

	Inputs (kg/hr)
C4 Feed	24,263
1- and 2-Butene	17,469
Isobutene	121
Butanes	6,672
Air to Process	40,399
Deaerated Water	1,224
MP Steam to Process	1,631
Make-up Solvent	138
Total	67,656
	Outputs (kg/hr)
Crude Butadiene Product	21,464
Butadiene	12,500
1- and 2-Butene	2,299
Isobutene and Butanes	6,633
Other	33
Water	7,688
Vent	38,390
Heavies Purge	114
Total	67,656

Utility Consumption

Utility consumption for an OXO-D process unit producing 100 kMTA of butadiene molecules in the crude butadiene product is given in Table 2. This utility consumption does not include the utilities required for the downstream butadiene extractive distillation unit.

Table 2 – Utility Consumption

ISBL Utility Requirements		Time Averaged Consumption
Electricity (1)	kW	7,768
HP Steam (46 barg)	MT/hr	3.2
MP Steam (16 barg)	MT/hr	59.8
Boiler Feed Water	MT/hr	2.6
Steam Condensate (2)	MT/hr	(65.5)
Water Loss	MT/hr	0.10
Cooling Water Circulation	m3/hr	2,088
Net Fuel Import/Export	MM kcal/hr	5.55

1. Assumes electric compressor drivers
2. Steam condensate export to utility area

Summary

Butadiene industry conditions are changing such that new on-purpose butadiene production capacity will be needed. The commercially proven TPC OXO-D process is well suited to meet the new demand and has been further improved by TPC and UOP. Features of this advanced design include

- Safe handling and complete consumption of oxygen
- Feedstock flexibility
- High conversion and selectivity
- Minimization of acetylene and oxygenate byproducts
- Simple and robust controls
- Energy efficiency
- Reliable extended operation

References

1. Welch, L.M., Croce, L.J., and Christmann, H.F. (1978), "Butadiene Via Oxidative Dehydrogenation," *Hydrocarbon Processing,* pp. 131–136.

2. *CRC Handbook of Chemistry and Physics,* 95th Edition.

AIChE Paper Number 192d

Dow Fluidized Catalytic Dehydrogenation (FCDh)
The Future of On-Purpose Propylene Production

Matt Pretz

Sr. Research Scientist
The Dow Chemical Company

Barry Fish

Process Development Leader
The Dow Chemical Company

Lin Luo

Sr. Research Scientist
The Dow Chemical Company

Brien Stears

Associate Director
The Dow Chemical Company

Prepared for Presentation at the 2016 Spring National Meeting

Houston, TX, April 13, 2016

Dow Fluidized Catalytic Dehydrogenation (FCDh)

The Future of On-Purpose Propylene Production

Matt Pretz

Sr. Research Scientist
The Dow Chemical Company

Barry Fish

Process Development Leader
The Dow Chemical Company

Lin Luo

Sr. Research Scientist
The Dow Chemical Company

Brien Stears

Associate Director
The Dow Chemical Company

Abstract: Shale gas produced an imbalance in propylene supply and demand. This has driven the implementation of several projects using available propane dehydrogenation commercial technology. These technologies were adapted from existing processes. These adapted technologies are not tailored to efficiently manage the inherent challenges of catalyst activity, heat input, reaction equilibrium and unselective thermal reactions. Therefore, the Dow Chemical Company has developed a Catalytic Dehydrogenation Process that is designed from the standpoint of an operating company to efficiently solve the problems inherent in catalytic dehydrogenation processes.

The new technology is a circulating fluid bed technology based on existing Fluid Catalytic Cracking (FCC) technology. The Dow process utilizes a proprietary catalyst with a proprietary reactor and regenerator design to achieve 45% propane conversion at approximately 93 mol% selectivity to propylene. The simplicity of the system enables a capital savings greater than 20% versus other commercial offerings. In addition, the higher conversion at moderate pressure reduces the energy requirement per pound of propylene.

This new technology can be fully integrated with existing ethylene crackers in parallel with new or existing furnaces. In this way, a producer can increase production in an existing light hydrocarbon facility or build a new facility with exactly the amount of ethylene and propylene product that is desired based on ethane and propane feeds. Dow is currently planning a project to implement this technology in an existing cracker at commercial scale and would license the technology upon request.

Introduction

In the early 2000's, very few predicted the innovation that was going to change the landscape in the energy, refining and petrochemical sectors for the next several decades. But, innovators successfully combined hydraulic fracturing and horizontal drilling to produce unprecedented shale gas and shale oil. This innovation rapidly enabled petrochemical producers to begin producing ethylene from cheap ethane. Initially, the price of crude oil remained high which meant naphtha was rapidly phased out as the preferred feedstock where feasible in place of ethane. When the rate of naphtha feed was reduced, the co-product propylene that was produced also was reduced. This created an imbalance in the supply and demand for propylene. As this imbalance occurred, the economics for producing propylene from propane began to look very attractive. Fortunately, the Catofin® and Oleflex™ processes had been adapted from existing technologies for the production of propylene[11,12]. The licensors of these technologies were able to take advantage of market conditions and sold many units to help bring the markets back into balance.

Now there is another innovation ready to change the landscape of the catalytic dehydrogenation market place. Few are predicting the change that this technology will bring to the production of hydrocarbons in the petrochemical industry. In order to understand the significance of this new innovative technology, it is imperative to understand the fundamentals and evolution of catalytic cracking process technology.

Fundamentals and Evolution of Catalytic Cracking Process Technology

Fundamentals

There are several basic fundamentals that constrain potential process solutions when cracking or dehydrogenating hydrocarbons. In this paper, cracking and dehydrogenation will be referred to collectively as "cracking". The first is that the reactions are endothermic which means heat has to be added to the process to accomplish the reaction. Secondly, the chemical equilibrium of the dehydrogenation reaction favors olefins at low pressure and high temperature. The low partial pressure preference of the dehydrogenation reaction is at odds with the cryogenic separation requirement for light hydrocarbons; therefore, the reactor pressure needs to be optimized with the entire process.

$$C_3H_8 \rightleftharpoons C_3H_6 + H_2$$

Third, the high temperature requirement inherently results in the formation of coke on the catalyst which depresses catalyst activity. The resulting loss of catalyst activity requires regeneration to remove coke. Fourth, the thermal or gas phase reaction has a much lower selectivity to the desired product than the catalytic reaction, requiring the gas residence time at temperature to be minimized.

Each process that has been successful in the area of catalytic cracking of hydrocarbons has solved or managed each of these fundamental constraints. Table 1 illustrates 4 processes in this area including Fluidized Catalytic Cracking (FCC) for gasoline, Catofin® and Oleflex™ for propylene, and styrene technology.

Table 1. Fundamentals of Catalytic Cracking Process Technology

	Gasoline	Propylene	Propylene	Styrene
Process	Fluidized Catalytic Cracking (FCC)	CATOFIN® Dehydrogenation Process	OLEFLEX™ Dehydrogenation Process	
Heat Input	Catalyst[22]	Catalyst[6]	Feed[2,11]	Steam Diluent[5]
Pressure controls equilibrium	25 psia[22]	7 psia[1], [3]	16-24 psia[7]	6 psia[5]
Catalyst Activity Recovery	Burn Coke with air[22]	Burn Coke with air [1], [6]	Burn Coke off with air[2], reactivate with chlorine[3], inject sulfur & recycle H2 to inhibit coke[2], [4]	React coke off with steam and catalyst promoter[5]
Reactor Type controls residence time	Fluidized Bed[22]	Fixed Bed with cyclic operation[6]	4 Radial Adiabatic Moving Beds with inter-stage heaters [2]	2 or 3 Radial Adiabatic Fixed Beds with inter-stage reheat[5]

To put heat into the reaction, both FCC and Catofin®[12,22] primarily use the catalyst as a source of heat while Oleflex™ uses the superheated feed[11] and conventional styrene processes use a co-feed steam diluent[5]. Processes that don't require the feed to be preheated over the thermal reaction temperature have an inherent advantage in product selectivity.

To drive the reaction equilibrium, the styrene and Catofin® processes use vacuum[1,3,5]. For the styrene process, creating vacuum can be done efficiently since the diluent steam and product liquids can be condensed prior to compression of the vent gas which minimizes the compression duty. The FCC process is not as adversely impacted by higher pressure due to different chemistry; therefore, the process has tended to increase pressure over the life of the technology to reduce capital investment per ton of product. Finally, the Oleflex™ process tends to use more moderate pressures to minimize the energy and capital cost of the compression train. The Oleflex™ process does require a significant hydrogen recycle which is one reason for the moderate pressure selection[7].

The catalyst activity loss due to coke formation is mitigated primarily through burning the coke in the FCC, Catofin® and Oleflex™ processes[1,2,4,22]. The Oleflex™ process does require a relatively complex regeneration in that platinum must be redistributed with chlorine[3]. In addition, sulfur and hydrogen are intentionally injected to control coke formation[2,4]. Uniquely, the styrene process uses a novel solution by reacting the coke off of the catalyst in situ using steam and a catalyst promoter[5].

Finally, the residence time of the gas at temperature must be managed to minimize selectivity loss and that is dictated by the reactor type and heat input scheme. Both the conventional styrene process and the Oleflex™ process use radial adiabatic contacting reactors which are known for low residence times[2,5]. The Catofin® process uses fixed adiabatic reactors that operate in a cyclic mode[12]. Finally, FCC uses a circulating fluid bed and rapid gas-solid separation to control gas residence time[22].

All of these processes include designs and mitigations to manage the fundamental technical constraints, but some more than others have evolved and form the basis for the next generation of catalytic dehydrogenation processes. Technology ultimately will converge to the best overall economic solution as capitalism forces companies to find additional ways to increase profits. The question is not if .... but rather when.

Evolution

Table 2 illustrates the evolution of processes for the production of gasoline, styrene, and propylene. The gasoline or refining industry is large with over 1,200,000 kTa of gasoline produced each year[8]. With a large market, there are economic drivers to improve technologies that can reduce the cost of production. Because of this, there has been significant improvement in technologies to produce gasoline starting with the Houdry process in 1936 which utilized a batch reaction step followed by batch regeneration[21]. The Houdry process was the original technology that gave rise to the Catofin® technology[12]. Later in 1938, a moving bed technology with bucket conveyors enabled the process to transition to a continuous one[22]. This process was plagued with mechanical problems which drove the need for further innovation. Finally, in 1942, during the middle of the World War II, the Fluidized Catalytic Cracking technology (FCC) was born[22]. The initial process used acidized clay catalysts which were not as active and required bed cracking[22]. Finally, in the 1960's, zeolite catalysts were invented which had higher activity and ultimately resulted in riser cracking[22]. After the catalyst introduction, several hardware improvements followed such as the 2 stage regenerator and rapid separation riser termination devices[22]. But, the lesson here is that the technology evolved rapidly to the most economical solution which was a circulating fluid bed.

The styrene market is much smaller with only 27,500 kTa of styrene produced (2014)[9]. Therefore, the rate of innovation was slower over the years. In 1940, the United States required styrene for rubber production for World War 2; therefore, axial adiabatic reactors were rapidly built to produce styrene due to their ease of construction and speed of implementation, but the high temperatures and residence time resulted in low selectivity[23]. The heated case reactors were introduced in 1965 which essentially used a heat exchanger to improve the heat transfer rates which enabled lower operating temperature gradients and lower gas residence times[23]. The radial adiabatic reactors were introduced in 1972 which further decreased residence time and allowed very large single train plants to be built[23]. Finally, in 2005, Dow and Snamprogetti successfully demonstrated the first circulating fluidized bed process for the production of styrene without a steam diluent. Unfortunately, due to commercial decisions to exit the styrene business, the commercial implementation of this technology wasn't realized.

Finally, the propylene market is about 89,000 kTa but only 5,000 kTa of this propylene is on-purpose propylene[9]. Up until 1990, there was no on-purpose propylene production as all of the propylene was produced as an FCC byproduct or a steam cracking byproduct[11].

Therefore, there has been very little emphasis as an industry in regards to improving technology in this area. The first Catofin® system was commissioned in 1980's for C4 production and was later used for propane dehydrogenation in the 1990's[12]. In Thailand, UOP commissioned the first Oleflex™ unit in 1990 due to a unique market condition in which propane was available and propylene was needed[11]. Both of these technologies were initially developed due to unique circumstances which enabled both licensors to be ready for the unforeseen impacts of shale gas. But, the truth of technology evolution still remains. Technology will ultimately converge to the best economic solution which is a circulating fluid bed.

Table 2. Evolution of Catalytic Cracking Processes

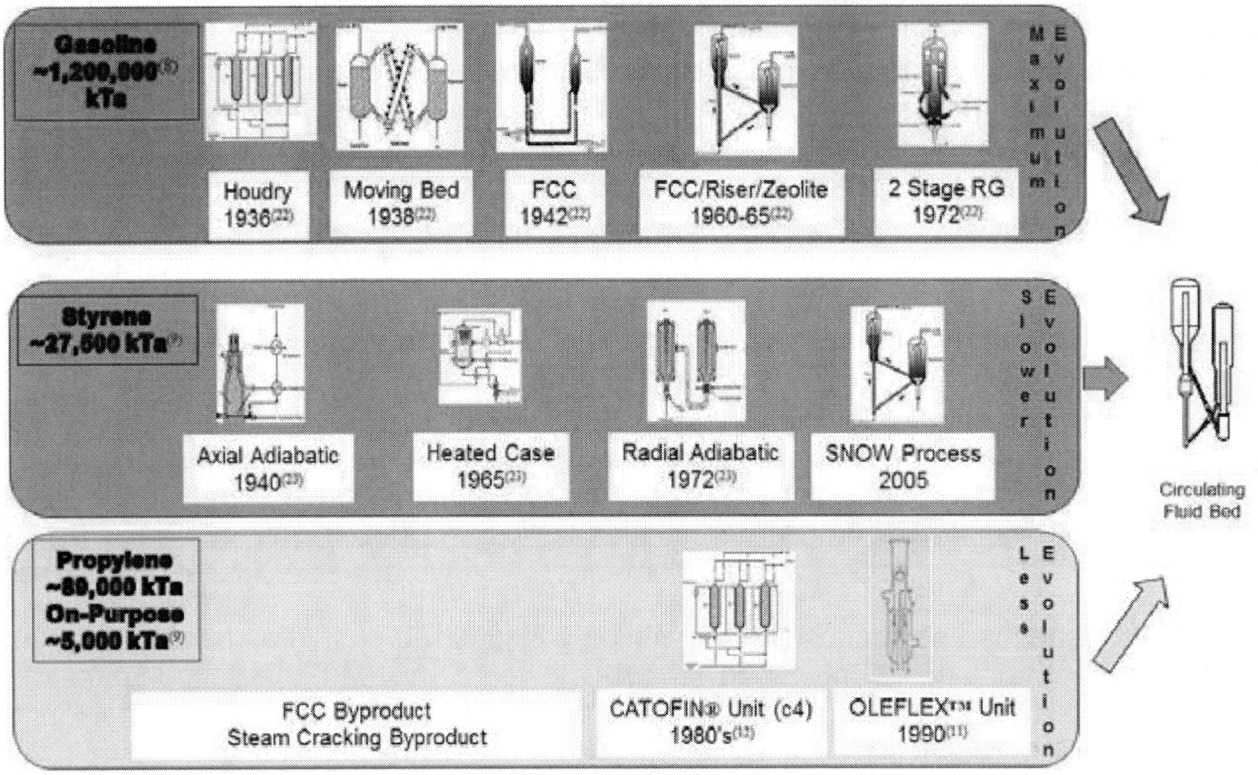

Introduction to Dow's Advanced PDH Technology

Process

The Dow Catalytic Dehydrogenation Reactor-Regenerator system is shown in Figure 1. The technology is a circulating fluid bed system with an alumina supported catalyst. Propane is introduced into a turbulent or fast fluidized reactor followed by a plug flow riser reactor which transports the catalyst to close coupled cyclones for rapid product and catalyst separation. The catalyst is then stripped with a stripping agent to recover any product entrained with the catalyst. The spent catalyst is transported to a combustor where it is reheated in a fluidized bed using external fuel for the required heat. The catalyst and flue gas then flow through a riser and a termination device. The remaining catalyst is then separated from the flue gas in cyclones and returned to a bubbling bed. The catalyst is then rejuvenated with air to recover the activity in an oxygen soak zone.

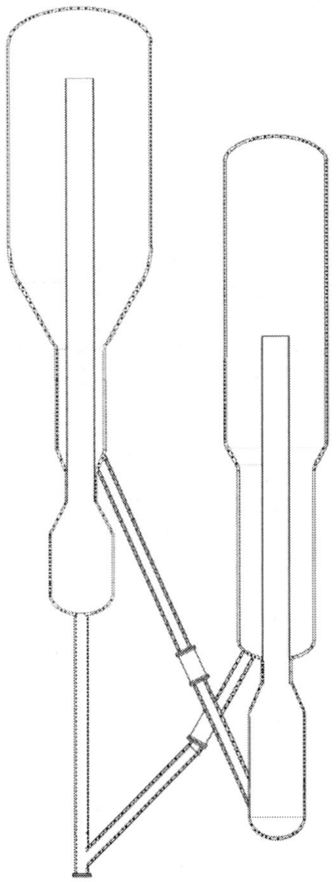

Figure 1. Dow Catalytic Dehydrogenation Reactor-Regenerator System

Catalyst

The catalyst shown in Figure 2 is produced utilizing a commercially available alumina support and impregnated with the active components which include gallium and low levels of platinum. The catalyst has been developed over many decades and has been shown to meet all the requirements of operating in a fluid bed dehydrogenation process. For example, the catalyst can perform the dehydrogenation of paraffins and alkyl aromatic compounds including: ethane, propane, butane, isobutane, ethylbenzene and others. The catalyst also promotes the combustion of supplemental fuels such as methane, hydrogen, ethane and others. As a catalyst operates in a circulating fluid bed system, it is desirable that any deactivating function that may occur in the reactor, reactor stripper or combustor be reversible. As such, the catalyst is capable of cyclic reactivation after any temporary activity loss is experienced. In addition, the catalyst can tolerate reasonable levels of impurities that are commonly found in commercially available propane streams. Finally, the catalyst has the appropriate mechanical properties to make it fluidizable (particle size and density) as well as attrition resistant.

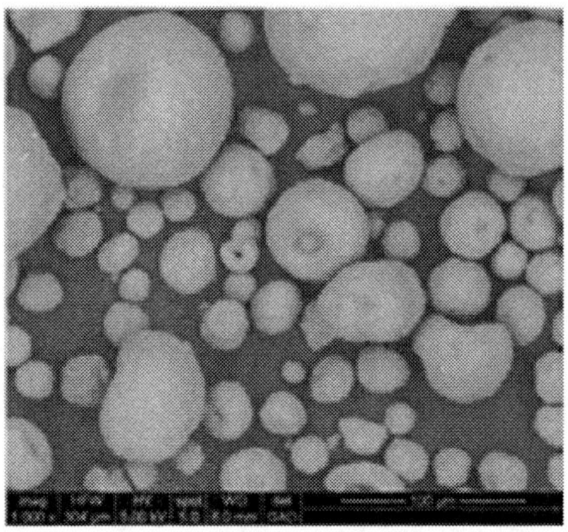

Figure 2. Dow Catalytic Dehydrogenation Catalyst

Dow Reactor Design

The reactor design (Figure 3) for the Dow circulating fluid bed process is an up flow reactor where the average velocity of both the catalyst and gas are in the upward direction. The bottom portion of the reactor operates as a turbulent or fast fluidized reactor and is followed by a conventional riser reactor. Upon completion of the catalytic dehydrogenation reaction, the catalyst and product gases are rapidly separated in a cyclonic separation device that is commonly used in Fluidized Catalytic Cracking processes. After the catalyst is removed, a stripping gas is injected in the annular space and the product gases are recovered prior to transporting the catalyst to the regenerator.

The advantage of the Dow reactor and catalyst process is that it can achieve high conversions such as 45% of the propane feed at moderate pressures. In addition, there is a minimal impact from feed impurities such as sulfur which eliminates the need for costly front-end capital investment. Finally, the selectivity to propylene is maximized due to the low gas residence time at temperature requirements and the low levels of coke that are formed on the catalyst during the reaction step.

Figure 3. Dow Reactor Design

Dow Regenerator Design

The Dow regenerator design as shown in Figure 4 consists of a lower combustor section and is followed by a catalyst removal section at the top and a catalyst reactivation section in the center. Due to the low coke generation, supplemental fuel and air are injected into the combustor which generates the heat for the exothermic dehydrogenation process. The combustor is a bubbling bed that operates with an average gas and catalyst velocity in the upward direction. The catalyst and flue gas are then transported through a riser where the catalyst is partially removed from the flue gas. The flue gas is then processed in a cyclone system which removes any remaining catalyst. The catalyst is then reactivated by a simple air treatment prior to transport back to the reaction system.

Advantages of the Dow regenerator process include the catalytic combustion of a fuel which minimizes any NOx formation versus competing alternatives which require a flame for combustion. In addition, the catalyst is continuously regenerated which eliminates the need for large high temperature valves operating at a high frequency which have been prone to failures in batch regenerator processes. In addition, there is no need for complex regeneration steps such as chlorine introduction.

Figure 4. Dow Regenerator Design

Comparison to Existing Technologies

As with any new technology, it is imperative to benchmark any new process with the current technologies to ensure a sufficient improvement in performance can be obtained. Table 3 summarizes the key differences between the Catofin®, Oleflex™, and Dow PDH process.

In reactor technology, both Catofin® and Oleflex™ require multiple reactors to process the required amount of propane to achieve PDH economies of scale. For Oleflex™, four reactors are used to achieve the target conversion due to multiple heat up steps being required[2]. The Catofin® process requires multiple reactors depending on capacity and each reactor spends a significant time operating in the regeneration step[12]. The Dow process utilizes a single reactor and operates in reaction mode continuously.

Heat introduction into the reactor is extremely important as it impacts the selectivity of the process as well as the capital intensity. The Catofin® process provides the heat by a combination of hot catalyst generated in the decoking cycle and hot propane[1,10,12]. The cyclic heat introduction by catalyst results in a high temperature initially during the start of a reaction cycle and then a low temperature just prior to regeneration. This cycle results in a non-optimum temperature for the reaction thereby reducing the overall selectivity of the process. In addition, by feeding the propane at temperatures above the thermal cracking temperatures non-selective products are formed. Alternatively, the Oleflex™ process preheats the propane and gas products before each reactor to temperatures above the thermal cracking temperature of propane[2,11]. This results in non-selective reactions due to the residence time in these sections as well as increased capital costs and NOx emissions from the fired heaters. The Dow process introduces propane below the thermal cracking temperature and utilizes the catalyst to rapidly heat up the propane to reaction temperature. The high surface area of the catalyst and rapid heat transfer is one of the most efficient ways to transfer heat to the propane vapor.

Each process utilizes a different catalyst which impacts the economics and handling requirements. For example, the Catofin® process uses a very large quantity of chromium catalyst which is relatively inexpensive[12]. The catalyst possesses environmental handling challenges which must be managed. Alternatively, the Oleflex™ process utilizes a catalyst which contains ~ 3000 ppm of platinum supported on alumina[13]. The Dow process uses an more sustainable catalyst that is significantly less expensive than the Oleflex catalyst as the platinum levels are significantly lower.

The regeneration of all the processes involves burning any coke build-up on the catalyst in either a continuous or batch process. The Oleflex™ process uses a complex cycle of coke combustion, drying, platinum re-dispersion with chlorine and oxygen, followed by a hydrogen reduction[14,15]. Both the Dow and Catofin® regeneration process are relatively simple and straight forward.

Table 3. Leading Commercial Technologies vs Dow PDH

	CATOFIN®	OLEFLEX™	Dow's Process
Reactor Design	(20) 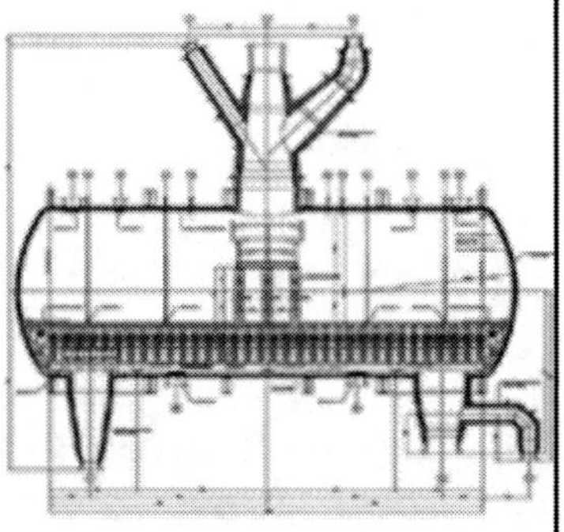	(11) 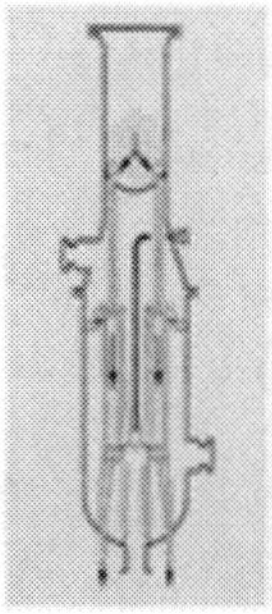	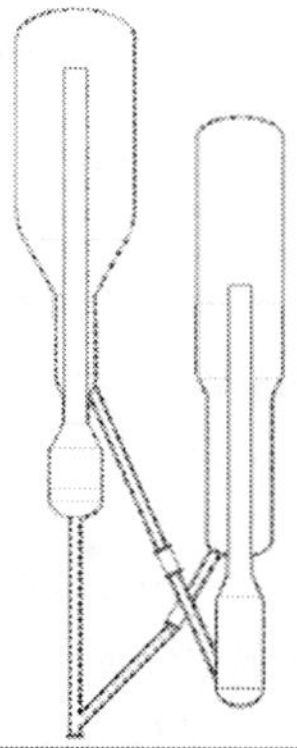
Reactor Technology	Fixed Bed Parallel Reactors [1]	4 Moving Bed Reactors 1 CCR Regenerator[2]	1 Fluidized Reactor 1 Fluidized Regenerator
Method of Heat Introduction for Reaction	Cyclic operation of reaction vs. coke combustion with air in regeneration step to reintroduce heat after reaction step cools catalyst bed[10] Propane Feed at 590-650 C[10], [1]	Direct-fired furnace heats process gas prior to each of 4 adiabatic reactors in series to 650 C[2], [11]	Heat carried to reactor via fluidized catalyst after heating by combustion in regeneration step. Propane fed at temperatures below thermal cracking
Catalyst type	Cr2O3 on alumina[12]	0.3 wt% Pt supported on alumina[3], [13]	Alumina with Pt and Ga
Catalyst Regeneration	Coke from reaction step burned with air.	Complex cycle of coke combustion, drying, Pt re-dispersion with chlorine and O2 and then H2 reduction. [14], [15]	FCC like cycling to perform combustion, decoke, and regeneration after reaction
RX Catalyst Residence Time	7-15 minutes [12]	~7,200-14,400 minutes (5-10 days) [16]	< 2 minutes
Pressure	7 psia[1], [3]	16-24 psia[7]	19-25 psia
Conversion	48 - 53%[6]	35(7) - 40% [19]	43-53%
Selectivity (mol%)	86+[6] @ XX% conv	89%(18)@ XX conv 84% @ 40% conv [19]	92-96% @ 43%-48% conv

The residence time of the catalyst in the reactor varies widely between the 3 processes and generally impacts the ability of the catalyst to withstand feed impurities and deactivation. For example, the Catofin® reactor undergoes a regeneration every 7-15 minutes which enables any deactivation that may have occurred from feed impurities to be removed each cycle[(12)]. Alternatively, the Oleflex™ process has a catalyst residence time of 7,200-14,400 minutes (5-10 days) which makes the catalyst much more susceptible to feed impurities and may require additional capital to mitigate[(16)]. The Dow PDH process has a catalyst residence time in the reactor of less than 2 minutes which makes the process more forgiving to feed impurities and potential impacts to the catalyst.

The pressure of the processes dictates the maximum potential conversion due to chemical equilibrium. For example, the Catofin® reactor operates at a low pressure of ~7 psia to drive higher conversions[(1,3)]. This increases compressor operating cost and capital. Alternatively, both the Dow PDH and Oleflex™ process operate between ~16 and ~25 psia which is slightly above atmospheric pressure[(7)]. The Oleflex™ process does require recycle hydrogen to suppress coke make due to the long catalyst residence times which additionally suppresses the maximum possible conversion[(2,4)].

Finally, the reactor conversion and selectivity ultimately dictate the economics of compression and separation sections of a commercial plant. After a thorough literature review, the conversion and selectivity for both the Catofin® and Oleflex™ processes are difficult to find reported together. The Catofin® process has published propane conversion numbers ranging from 48-53% with a separate selectivity performance of 86 mol% at an unknown conversion[(6)]. The Oleflex™ process has published propane conversions ranging from 35-40%[(7,19)]. The Oleflex™ selectivity has been reported to be 89 mol% at an unknown conversion and 84 mol% at 40% conversion in a UOP patent[(18,19)]. The Dow PDH process is expected to operate at 43-48% conversion with propane to propylene selectivities of 92-96 mol% which is significantly improved over the current commercialized technologies. The performance of the units is expected to change over time and as a function of plant scale which provides some challenges when comparing conversion and selectivity performance.

The Dow process delivers significant capital and operating cost advantages due its inherent simplicity and multi-functional nature. The capital outlay is expected to be at least 20% less than other existing commercialized technologies based on internal Dow studies. In addition, by performing the reaction at moderate pressure without hydrogen recycle, the process is expected to have 20% lower energy consumption. The low gas residence time at temperature design enables the process to achieve a 5% lower propane unit ratio than current processes. In addition, through catalytic combustion of process gas, the NOx emissions will be lower than alternative technologies. In conclusion, the technical advantages of Dow PDH technology offer significant economic advantages versus current technologies.

Dow's History of Technology Development and Status

Dow scientists identified a fluidized reaction process as the most economical for catalytic dehydrogenation more than 25 years ago. Since that time, a fleet of different lab reactors have been built and operated to test and develop improved catalysts as well as understand the chemistry of the process to enable design of a commercial unit. Fixed bed reactors, bubbling bed reactors, ACE fluidized lab reactors, attrition units, combustion units, and several others have been designed and operated. In addition, two pilot units such as a Davidson Circulating Reactor (DCR) and a large Process Demonstration Unit (PDU) were constructed to study the circulating aspects of the chemistry and better understand the fluid dynamics at high temperature. Finally, more than 3 large cold flow programs were chartered at PSRI to measure and quantify the catalyst hold-up as a function of superficial gas velocity and catalyst flux over a broad range of reactor design conditions. PSRI subject matter experts on fluidization consulted on the project and provided recommendations for multiple areas in the plant design.

In addition to the hardware and R&D programs, Fluidized Catalytic Cracking (FCC) subject matter experts with a broad range of experience were leveraged to design and scale-up the process. Many of the challenges of the development were synonymous with FCC such as catalyst circulation and catalyst retention. Dow modeling experts also generated extremely valuable scale-up tools in the area of catalyst attrition and retention, catalyst kinetics, reactor hydrodynamics, and catalyst – gas separation devices.

In conclusion, Dow brought its entire arsenal of R&D and process know-how (Figure 5) to the table as well as leveraged leading industry fluidized solids experts to bring this process to fruition. The next step in the development is the construction of a small commercial demonstration unit. Upon completion, the data and know-how generated from the commercial demonstration unit will enable the design and construction of world scale units and future technology proliferation to the industry.

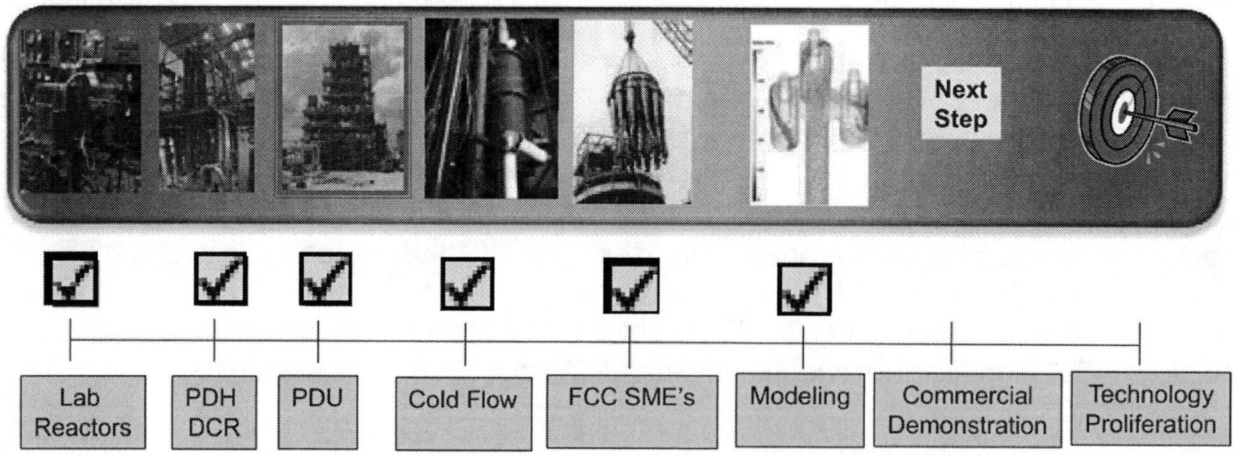

Figure 5. Dow Process Development Pathway

Commercial Application

The Fluidized Catalytic Dehydrogenation (FCDh) process is a revolutionary platform reactor technology. The potential applications for the process are numerous as shown in Figure 6. For example, in PDH, the reactor technology can plug into steam crackers that are new or be bolted on to an existing cracker. The reactors can also be used to build a new PDH unit or retrofit an existing PDH unit. Additionally, the reactors can be used to build new or retrofit existing ethylbenzene to styrene dehydrogenation facilities, butane to butene, or isobutane to isobutene plants. C4 dehydrogenation units can be integrated into refineries to upgrade butane or isobutane for use in the alkylation section to produce alkylate.

In the case of PDH plant integration with existing steam crackers, the Dow process is the only technology that can integrate economically with a cracker due to is comparable operating pressure, high propane conversion, lack of hydrogen recycle requirement and simple reactor section design. Because of the simple reactor-regenerator configuration, small plants still maintain attractive capital intensity when integrated with a steam cracker unlike competing technologies.

Figure 6. Platform Reactor Technology with Multiple Plug and Play Options

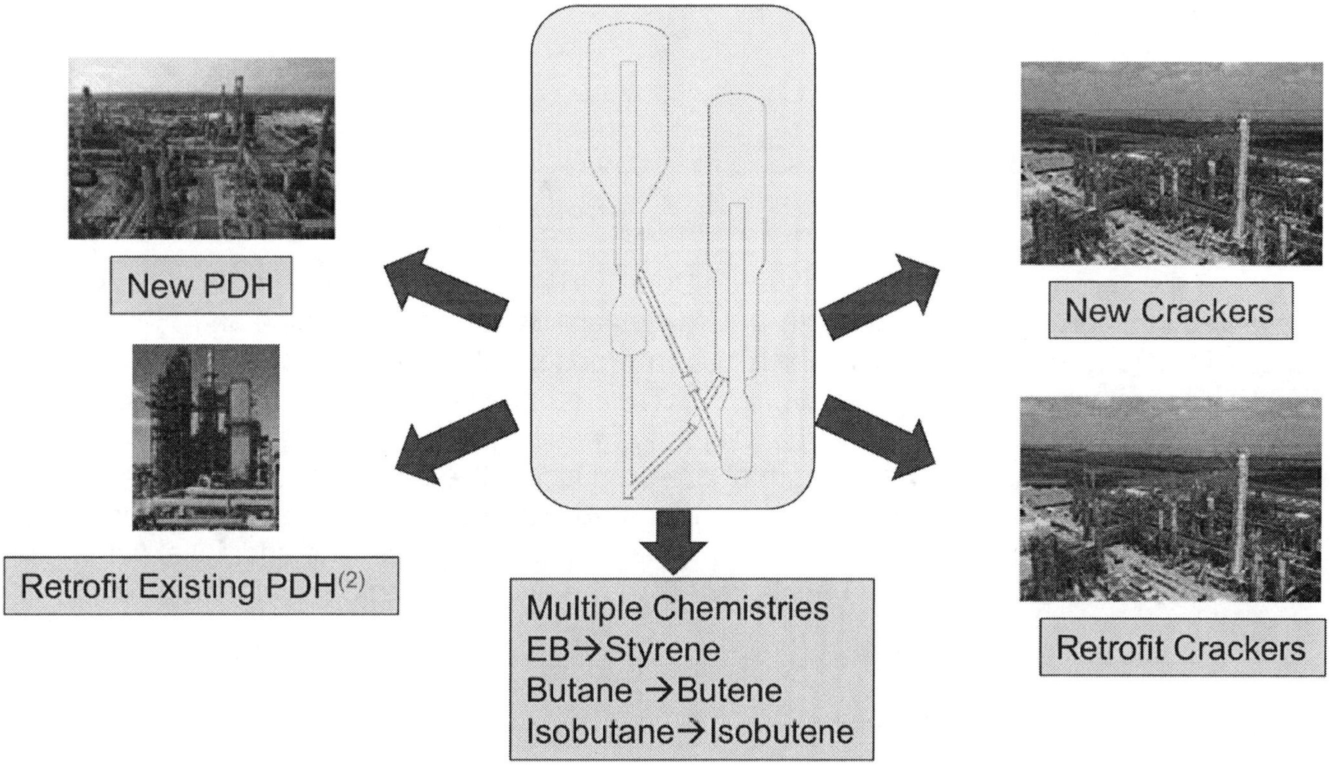

In order to show the advantage of an integrated Dow PDH reactor system and ethane cracker versus an ethane-propane cracker, two examples will be presented. The first example is an ethane-propane cracker and is shown in Figure 7. In this example, if a producer wanted to produce 1500 kTa of ethylene and 500 kTa of propylene, an EP cracker with approximately 8 furnaces could be envisioned. 2 furnaces would feed ethane and 5 would feed propane including the recycle streams. One furnace would be in decoke mode. The feedstock would be 250 kTa of ethane and 3000 kTa of propane. This cracker would produce 1250 kTa of other products.

Figure 7. 2000 kTa Ethane-Propane Cracker

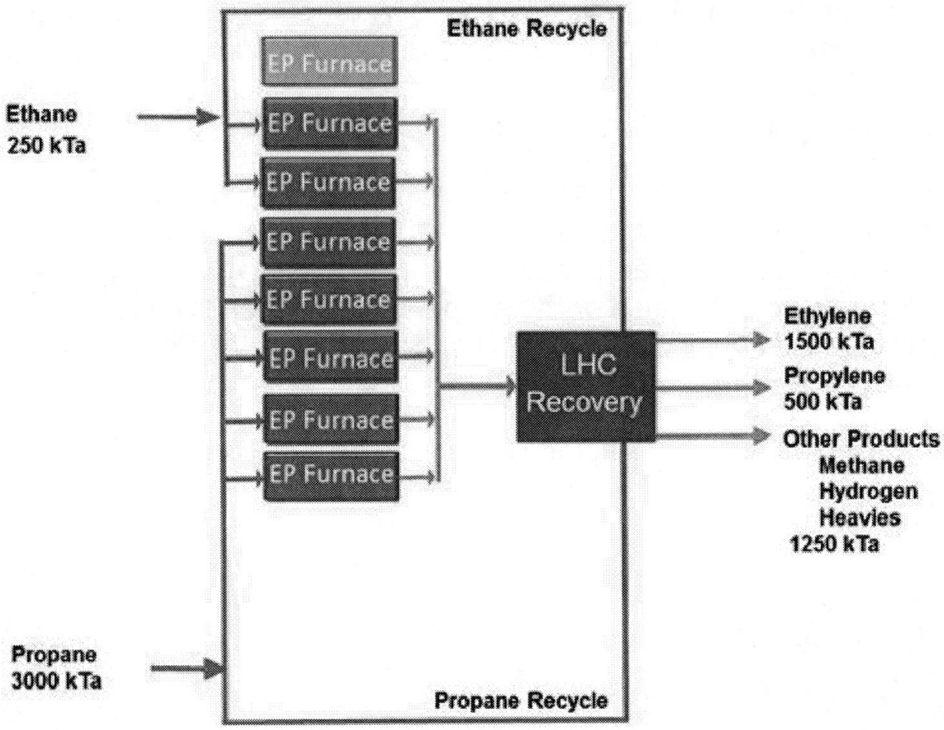

In the second example, an integrated Dow PDH reactor system with ethane furnaces is presented in Figure 8. In this example, 7 furnaces are used to crack ethane and 1 PDH reactor is used to process propane which in turn produces about 1500 kTa of ethylene and 500 kTa of propylene. In this example, 1950 kTa of ethane is required and 550 kTa of propane feedstock. There are only 500 kTa of byproducts produced. This is a ~750 kTa reduction in the amount of byproducts that are produced. The total predicted economic advantage of the integrated Dow PDH reactor and ethane cracker versus an ethane propane cracker that produces essentially the same product mix is $234 MM/yr of additional value due to the products versus the raw materials based on the average predicted price from 2016-2020[24].

Figure 8. Integrated Dow PDH Reactor System with Ethane Cracker

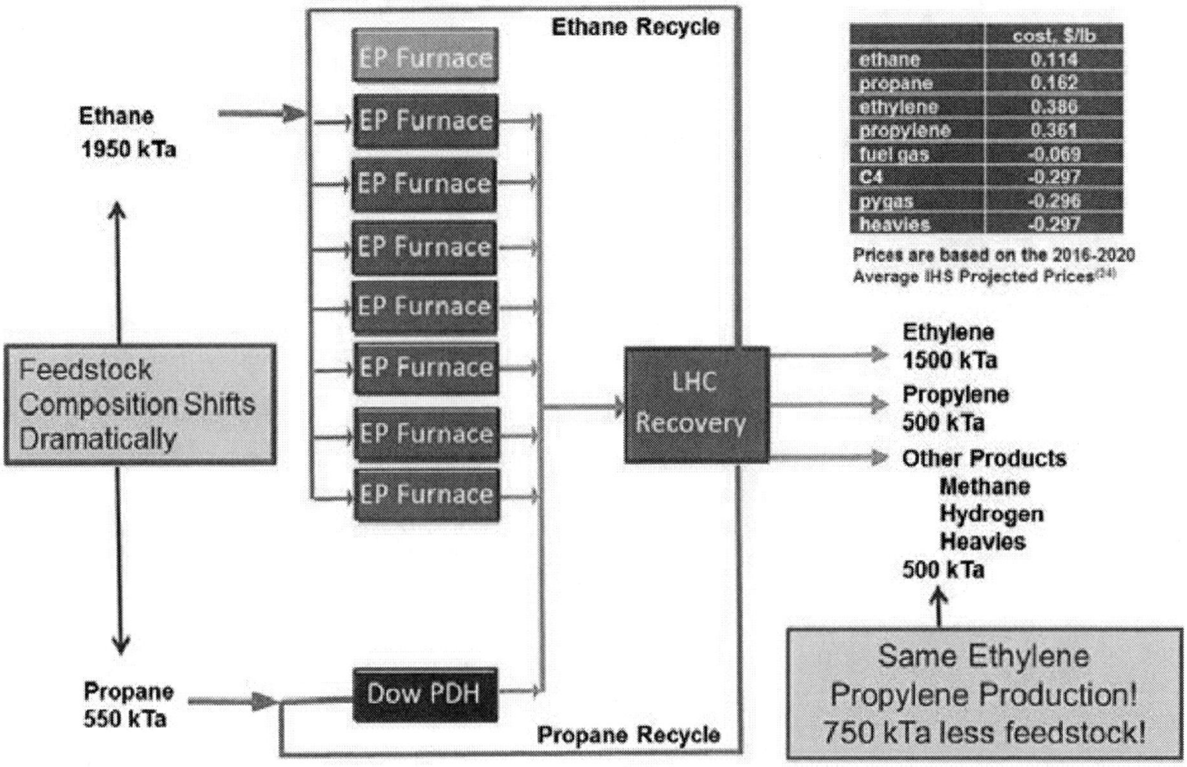

Conclusion

The Dow Catalytic Dehydrogenation reaction system is a breakthrough technology that provides significant advantage in the production of propylene from propane. Dow believes the technology represents the future of how on-purpose propylene will be produced. The process's elegant and simple design provides the most capital and energy efficient method to produce propylene from propane that is currently available. In addition, it provides for lower raw material usage and simplicity of operation versus other options.

This technology has the potential to change how hydrocarbons are produced in the chemical industry. To support this vision, Dow plans to implement a project at Dow and has plans for strategic expansion of the technology to other geographies and facilities. In addition, Dow would be willing to license the technology for applications in which catalytic dehydrogenation is the preferred solution.

References

1) Seung-Taek Seo, Wangyun Won, Kwang Soon Lee, Chansul Jung, and Seokho Lee, Repetitive control of CATOFIN process, Korean Journal of Chemical Engineering, Volume 24, Number 6 / November, 2007.

2) UOP Oleflex Process Brochure "UOP Oleflex™ Process for Propylene Production".

3) Weckhuysen, Sattler, Ruiz-Martinez, Santillan-Jimenez. Catalytic Dehydrogenation of Light Alkanes on Metals and Metal Oxides. ACS Publications. August 2014.

4) Heyse, Johnson, Mulaskey. "Dehydrogenation Processes, Equipment and Catalyst Loads Therefore." US Patent 5,406,014, April 11, 1995 to Chevron.

5) Styrene. Chen, Shiou-Shan. Raytheon Engineers and Constructors. Kirk-Othmer Encyclopedia of Chemical Technology. John Wily and Sons. Inc. December 4, 2000.

6) "CATOFIN® Dehydrogentation", Lummus Technology a CB&I Company, http://www.cbi.com/images/uploads/tech_sheets/Dehyrogenation.pdf, 2009.

7) Kogan, Herskowitz, Woerde, and van den Oosterkamp. "Article 16A: Select Dehydrogenation of Propane using an improved Dehydrogenation catalyst." 10th Ethylene Producers Conference, 1998.

8) World Crude Oil Production by Year. United States Energy Administration. 2014.

9) Global Annual Production of Styrene and Propylene Monomer. US Energy and Information Administration. 2015.

10) Gartside, Robert J., "Catalytic Hydrocarbon Dehydrogenation System with Prereaction", US Patent 6,392,113 B1, May 21, 2002 to ABB Lummus.

11) Chaiyavech, Pramote, "Commercialization of the World's First Oleflex Unit", The Journal of the Royal Institute of Thailand, Vol 27, No 3, Jul-Sept 2002.

12) Oviol, Bruns, Fridman, Merriam, Urbancic. "Mind the Tap", Hydrocarbon Engineering. Sept 2012.

13) Oleflex™ DeH-16 Catalyst Description, UOP Website, http://www.uop.com/objects/DeH16.pdf.

14) Cottrell, P. R., Fettis, M. E. "Process for the dehydrogenation of hydrocarbons", US 5,087,792 A to UOP.

15) Sechrist, P. A., Robinson, D. W., Schlueter, W. D., "*Method for controlling moisture in a catalyst regeneration process*", US 6,123,833 A to UOP.

16) Meyers, R.A., "Handbook of Petrochemicals Production Processes" McGraw-Hill Handbooks, **2005**.

17) Owen, Hartley, Mead, Belle, Zahner, John C., "Conversion of LPG Hydrocarbons into Distillate Fuels Using and Integral LPG Dehydrogenation-MOGD Process", Patent 4,675,461, June 23, 1987 to Mobil.

18) Houdek, J. Mark, Anderson, James. "On-Purpose" Propylene Technology Developments". UOP LLC. ARTC 8th Annual Meeting 4/29/2005 Kula Lumpur

19) Glover, B. K., Zarraga, J. A., Schultz, M. A., "Fluidized bed reactor with back-mixing for dehydrogenation of light paraffins", US patent application, US 2008/0161624 A1, to UOP.

20) "APPC PDH/PP Project – 8 Catofin Reactors – Jubail Saudi Arabia Engineering, Design, and Installation of Refractory Lining", Chimney and Refractory International SRL, Presentation, http://www.chimneysandrefractoriesinternational.com/attachment/APPC-SAMSUNG%20PDH-PP%20CATOFIN%20REACTORS%20PROJECT.pdf.

21) Letzsch, Warren. Hemler, Chuck. Nieskens, Mart. The 70 year History and Future of Cat Cracking. AFPM Conference: Cat Cracker Seminar. August 21-22, 2012.

22) Wilson, Joseph. Fluid Catalytic Cracking: Technology and Operation. PennWell Books. 1997.

23) Mitchell, J.E. Styrene: Its Polymers, Copolymers and Derivatives. Chapter 2: Manufacture of Styrene Monomer. Hafner Publishing Company. 1952

24) Chemical Price and Economics. IHS. January 5, 2016.

Catofin® is a registred trademark of CBI Lummus

Oleflex™ is a registered trademark of UOP

AIChE Paper Number 192E

C_2 Front-End Hydrogenation Improvements

Ernesto J Angueira

Technical Service Manager, Americas

CRI Catalyst Company LP

Pascal Coenen

Licensing Technical Manager

Shell Global Solutions International B.V.

Prepared for Presentation at the 2016 Spring National Meeting

Houston, Texas, April 12, 2016

C_2 Front-End Hydrogenation Improvements

Ernesto J Angueira

Technical Service Manager, Americas

CRI Catalyst Company LP

Pascal Coenen

Licensing Technical Manager

Shell Global Solutions International B.V.

Abstract: The presentation will address the parameters influencing selectivity and the impact over time. A side-by-side comparison of 5+ year operating data of the Shell/CRI' C_2 Front-End catalysts in both gas and liquid crackers gave new insides in excelling performance. The difference between the latest generations of catalyst looking at the key performance parameters for ethylene producers will be addressed. The selectivity comparison between the modeling (predicted selectivity) and the 5+ year real plant data (achieved selectivity) will be compared.

A business case demonstrates the value (margin gain) which has been achieved by reference plant which upgraded to the latest generation catalyst will be explained.

Ethylene continues to be one of the most important building blocks in the chemical industry; its production is a highly competitive business. The producers are continuously experiencing growing competition (e.g. low-cost feedstock producers), variation in feedstock costs, increasing energy costs and additional environmental regulations, requires plants to operate at maximum reliability and efficiency.

A few challenges in front-end hydrogenation units include: excess hydrogen, therefore a selective catalyst is very important to avoid runaways and to ensure high yield of ethylene. Another challenge for front-end hydrogenation operation is fluctuation of CO concentration in the feed. CO functions as an activity inhibitor to hydrogenation as it is adsorbed on the catalyst active sites. Temperature adjustments to CO activity inhibition and sudden CO content drop can trigger temperature runaway (Figure 1). As producers are generating hydrogenation feed with lower CO levels, depending on operation, feedstock and processes, a very stable catalyst is required.

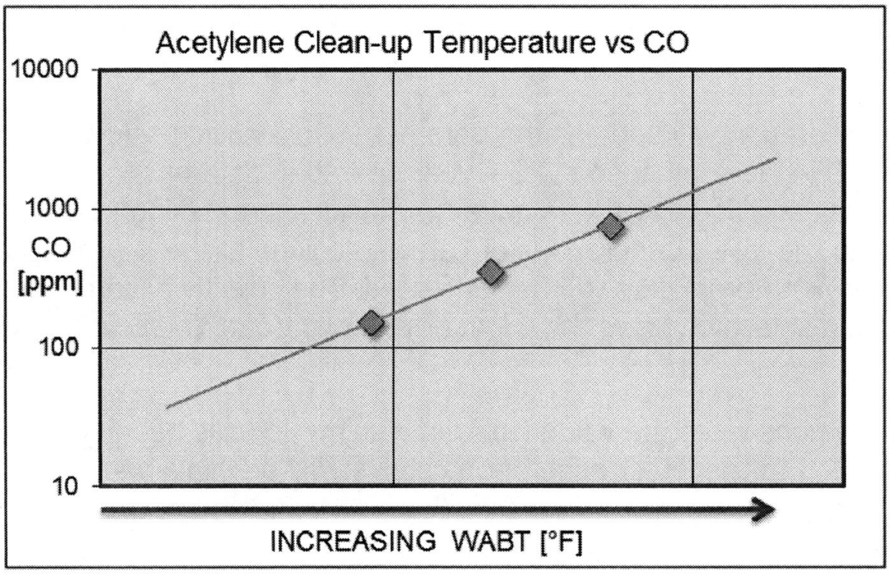

Figure 1. Temperature adjustments due to CO inhibition.

The selective hydrogenation of acetylenes is a very important unit operation within the ethylene process and catalyst selection can significantly overcome the different challenges while offering increased profitability and ease of operation to ethylene manufacturers.

This paper will discuss how new catalyst development can enhance the profitability of an ethylene plant by improvement in olefin yields, consistent delivery of on-specification product, and lower utility costs, resulting in significant improvements in overall plant operating economics.

Why new catalyst development?

Catalyst research, besides changing the formulation and palladium (Pd) impregnating techniques, are also looking into possibilities of improving catalyst performance by varying the size and shape of the catalyst as CRI developed in it KL7741B series.

One focus of catalyst shape optimization is the desired to reduce catalyst bed pressure drop which translates directly into energy savings as a result of reduced power consumption or higher throughput maintaining the same pressure drop.

A ring shape catalyst provides several advantages over the traditional tablet (or spherical) catalyst. The ring shape exposes more catalyst surface area to the process gas per catalyst volume than the tablet shaped particle. The catalyst is impregnated with Pd and since there is high geometric surface area, the Pd is better dispersed, resulting in high activity and poison resistance. Pressure drop across the catalyst bed is also reduced due to a higher void volume for the ring shape. Higher void volume results in low pressure drop, thus higher space velocity or increase in throughput (Figure 2). Lower density, therefore lower fill cost.

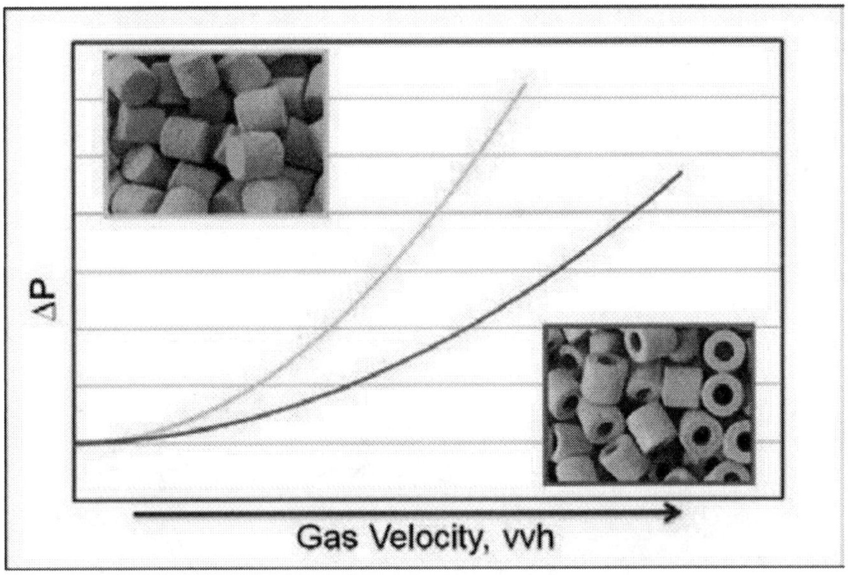

Figure 2. Pressure drop as function of GHSV.

The catalyst shape must achieve all of the above without compromising on catalyst activity or mechanical strength.

Performance

Commercial units that switched from tablets to rings shape catalyst experienced significant improvements. The ring shape catalyst features extreme stability maintaining a high level of ethylene selectivity for prolonged time as shown in Figure 3.

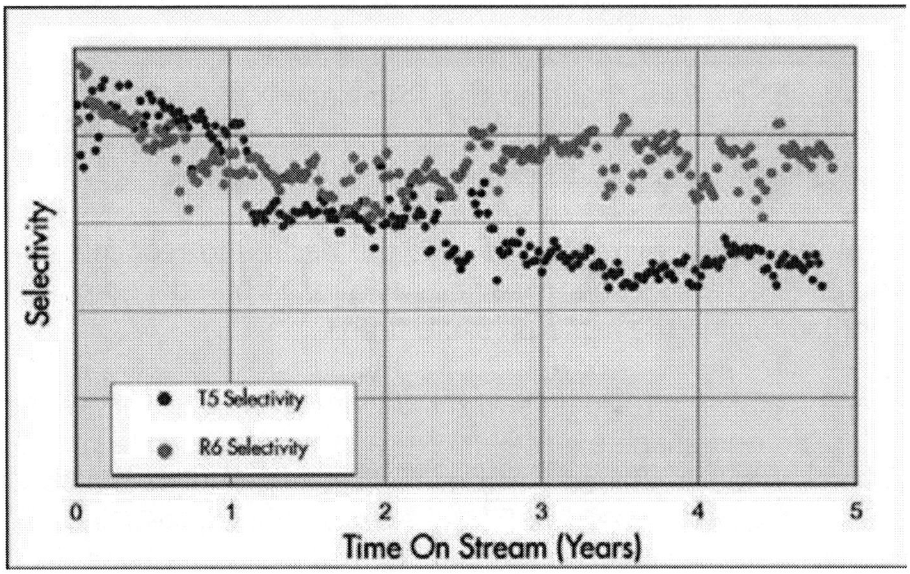

Figure 3. Rings vs Tablets shape – selectivity.

The stability of the catalyst resulted in higher selectivity during the cycle translating to extra margins therefore increased profitability.

In addition to the stability of the catalyst, its improved Pd dispersion provides a more active catalyst, therefore requiring lower operation temperature (Figure 4).

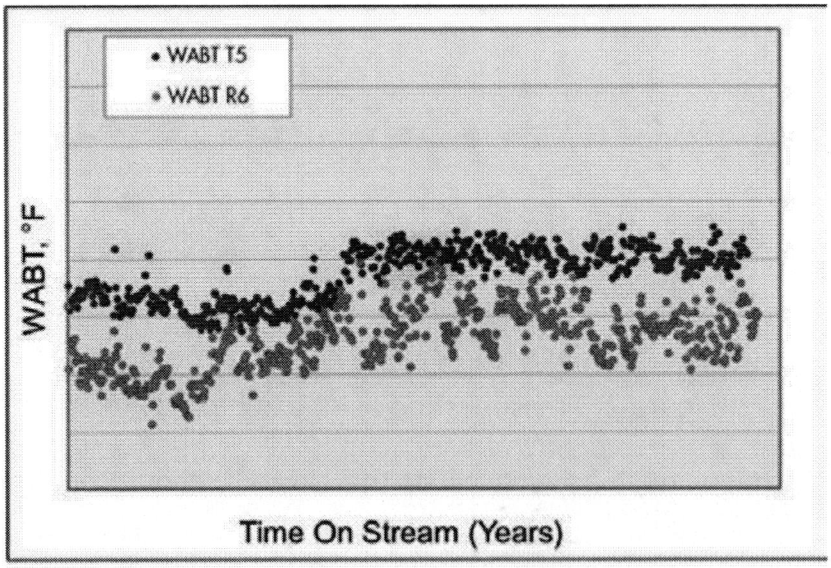

Figure 4. Rings vs Tablets shape – WABT.

KL7741B catalyst has proven to be very stable during CO swings, even under very low CO concentrations (100 ppm).

The ring shape catalyst, even with better metal dispersion exhibited robust operation during CO swings.

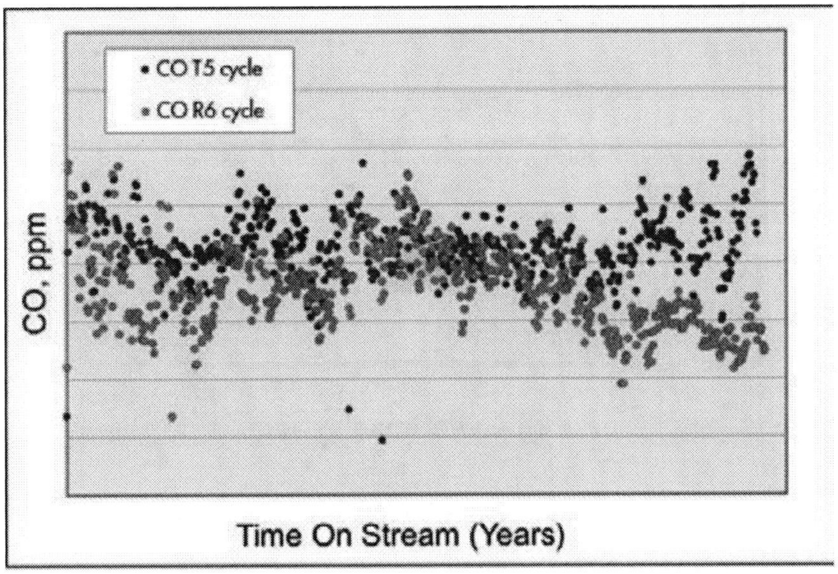

Figure 5. Rings vs Tablets shape – CO stability.

Modeling plant data

Using the data of multiple plants, an empirical equation was developed to model the aging of the catalyst.

ethylene selectivitytime = -A * (log (1 + [time]*B) – 1)

A = ethylene selectivity at SOR

B = C * D * E * constant

C = GHSV in h-1

D = "catalyst shape factor"

E = $[C_2H_2]^2$ in vol%

constant = y

Using several references gives the input to model the benefit as show in the business case.

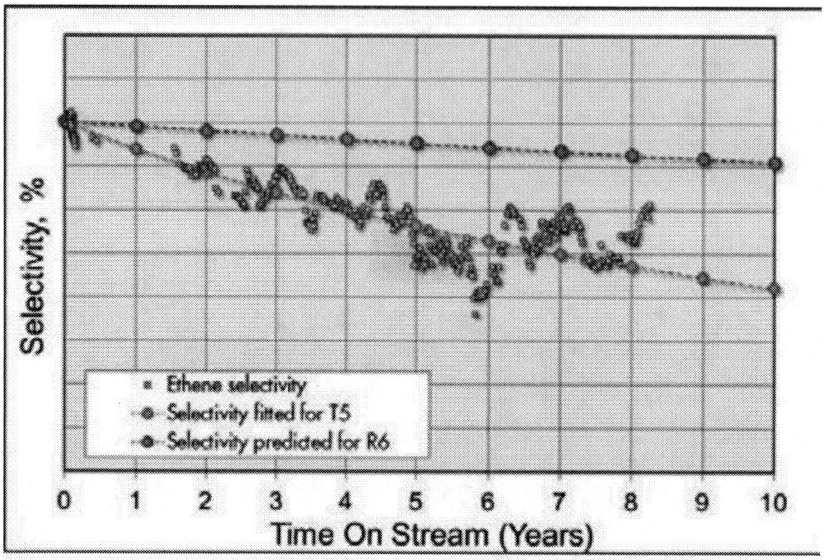

Figure 6. Predicted selectivity for rings.

Business Case Example

Based on the commercial data from several plants operating the tablet and ring shaped catalyst of CRI a great improvement (average over 5 year period) in ethylene yield has been observed. For this world scale steam cracker with the assumptions listed below leads to 2 million $ per year increase in margin as shown in Figure 7.

Steam Cracker - Gas		
Plant Capacity	kTA	World Scale
Ethylene margin	$/ton	400
On-stream factor	hr/yr	8,500

Yearly Benefit over 5 year cycle

Benefits		
Increased direct Ethylene Yield	$/yr	> 1,100,000
Increased benefit (more fresh feed instead of recycle)	$/yr	> 900,000
Estimated Total Margin Improvement	**$/yr**	**> 2,000,000**

Figure 7. Predicted selectivity for rings.

Conclusion

Ethylene producers expect the acetylene selective hydrogenation unit to operate in a safe and reliable manner. Due to high competitive market efficiency and profitability are key and therefore catalyst selection is critical.

Catalyst manufacturers continue the development of front-end hydrogenation catalysts, improving the performance with each new generation. Catalyst shape demonstrated to be an effective solution and KL7741B rings have been introduced to markets and commercially proven to successfully address the previously mentioned challenges.

The ring shape catalyst demonstrated enhanced stability, better selectivity, improved activity, and therefore lower operating temperature. Lower pressure drop resulted in economic benefits, from lower energy consumption to higher unit throughput. The largest benefit is achieved from the very low aging of the catalyst and high selectivity achieved over the cycle as shown in the example for a world scale cracker the benefit can be in excess of 2 million US$ per annum.

192F

FULL FURNACE SIMULATIONS AND OPTIMIZATION WITH COILSIM1D

Alexander Vervust
Laboratory for Chemical Technology
Ghent University

Ismaël Amghizar
Laboratory for Chemical Technology
Ghent University

Andrés E. Muñoz G.
R&D Manager
AVGI

Kevin M. Van Geem
Laboratory for Chemical Technology
Ghent University

Guy B. Marin
Laboratory for Chemical Technology
Ghent University

Prepared for Presentation at the 2016 Spring National Meeting
Houston, Texas, April, 13, 2016

FULL FURNACE SIMULATIONS AND OPTIMIZATION WITH COILSIM1D

Alexander Vervust

Laboratory for Chemical Technology

Ghent University

Ismaël Amghizar

Laboratory for Chemical Technology

Ghent University

Andrés E. Muñoz G.

R&D Manager

AVGI

Kevin M. Van Geem

Laboratory for Chemical Technology

Ghent University

Guy B. Marin

Laboratory for Chemical Technology

Ghent University

Abstract: COILSIM1D is a steam cracking simulation package that has been developed as the result of several decades of independent research and modelling activities at the Laboratory for Chemical Technology (LCT) of Ghent University. Using a friendly graphical user interface, it combines a detailed fundamental kinetic model with a one-dimensional reactor model, which provides high flexibility regarding reactor geometries, feeds and simulation strategies. Simulations for a very broad range of complex feeds up to (hydrotreated) VGO can be easily performed thanks to SIMCO, a molecular reconstruction module incorporated in the software based on commercial indices of the feed. Over time, and based on own and client initiatives, various other features have been incorporated into COILSIM1D, such as TLE and run length simulations, or a shooting method that allows to perform simulations using a variety of severity indicators as boundary conditions at the reactor outlet. Recently, the kinetic model has been expanded and fine-tuned based on data from a large ethylene producer, including oxygenates; a firebox and convection section model has been incorporated, to allow full cracker simulations (i.e. convection, firebox, reactor and TLE altogether) in a unique package, and with short simulation times. COILSIM1D has also been linked with Aspen and other process simulators. Illustrative simulations have been carried out to show the operation of these new features, which effectively take COILSIM1D to a new level, transforming it into a whole-furnace simulation and optimization package.

Introduction

COILSIM1D is a comprehensive steam cracking simulation package, developed as the natural result of several decades of independent research and modelling by the Laboratory for Chemical Technology (LCT) of Ghent University. AVGI, a spin-off company from Ghent University, recently acquired all technical and commercial rights to COILSIM1D. With this, further and faster development of the software can be carried out, whilst interacting on a business-to-business structure with clients spread over four continents, allowing for customization of the package, tailored to the specific needs of the client.

The first two AVGI-driven developments of COILSIM1D have been carried out within a framework of collaborative work with LCT, and comprise the expansion and fine-tuning of the kinetic network used to describe the chemistry of the process, as well as the incorporation of furnace simulation capabilities that allow users to carry out full cracker simulations in one software package, with an even higher level of accuracy at the reactor level.

COILSIM1D has been traditionally developed as a reactor simulator aimed at the prediction of yields at the reactor outlet, combining a fundamental single-event kinetic network with a one-dimensional plug-flow reactor model. Over time, and based on own insights and permanent interaction with the increasing number of licensees, new functionalities have been gradually added to the package, that are made available to all clients having annually renewable license agreements by means of regular updates. Clients using evaluation licenses also receive the most updated version of COILSIM1D.

All these functionalities have conferred large flexibility to COILSIM1D, making it able to handle a very broad range of reactor geometries, feeds and operating conditions. A feedstock reconstruction method (SIMCO) is integrated in the package, which allows to reconstruct the molecular composition of light mixtures to HVGO feeds, based on simple and easily available commercial indices of the feed. Also a module to carry out TLE simulations is incorporated, allowing to account for the effect of this equipment on the process yields, effluent temperature, and heat recovery.

Different simulation approaches can be applied with COILSIM1D, depending on the information that the user has available: heat-flux or temperature profiles along the reactor can be defined; outlet conditions based on frequently used severity indices can also be specified, such as the Coil Outlet Temperature (COT), the conversion of a given species, the ratio of propylene to ethylene in the effluent (P/E), or that of methane to propylene (M/P). Additionally, dedicated routines for yield maximization, sensitivity analysis and run length estimation are incorporated in COILSIM1D, enabling users to straightforwardly optimize their process while taking multiple targets into account.

Following the market's urge to count on a single tool that is able to accurately simulate steam crackers as a unit (i.e. firebox, convection section, reactor and TLE in one package), AVGI and LCT took on the challenge of taking COILSIM1D to the next level, successfully transforming it into a whole cracker simulation package.

This paper presents an overview of the main features of COILSIM1D, paying particular attention to the most recent developments regarding the kinetic network, as well as the firebox and convection section models. To demonstrate the operation of these new features and their harmonious integration in the already existing package, illustrative simulations combining the new and already existing models are presented. The user-friendliness of COILSIM1D and its ability to easily interact with external software packages are also discussed.

The COILSIM1D Simulation Package

SIMCO: Feedstock Reconstruction from Common Commercial Indices

To be able to take advantage of the flexibility and accuracy that a fundamental kinetic model can offer, knowledge of the detailed molecular composition of the used feed is crucial and will affect the accuracy of the obtained simulation results. [1] For liquid fractions, which can be made up of several thousands of different components, detailed feedstock compositions are not straightforward to determine; not even for naphtha fractions. Existing analytical state-of-the-art techniques such as GCxGC, can be used to determine the detailed feedstock composition, but they are costly and can be very time consuming. In industry, feed composition can change significantly on a daily or even hourly basis and hence these analyses are not feasible to perform.

A feasible alternative are feedstock reconstruction methods, that allow to compute a detailed molecular composition of the feedstock based on common commercial indices, e.g. the average molecular weight, information regarding boiling points, the specific density, the global PIONA weight fractions, etc. The ease to reconstruct different feeds makes SIMCO a powerful tool for rapid and accurate feed reconstruction. This numerical tool comes with a user friendly GUI, depicted in Figure 1, in which the different input parameters are filled in by the user in the units of choice.

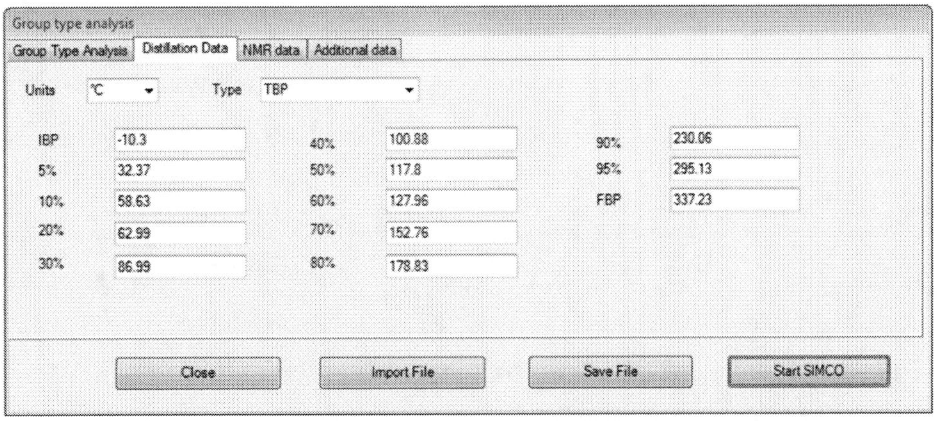

Figure 1: The GUI of SIMCO, where different parameters can be filled in as input for reconstruction of a feed.

Commercial indices can be determined by means of relatively fast and low cost standardized analytical procedures. In this respect, SIMCO, the feedstock reconstruction method available within the COILSIM1D package, is a powerful tool for industry to simulate the complex molecular composition as input for a fundamental kinetic model. SIMCO uses two methodologies, one based on the Maximisation of the Shannon Entropy (MSE) [2, 3], and another based on an artificial neural network (ANN). [4] MSE is the most general method and can be used for virtually every hydrocarbon feed. ANN, on the other hand, has shown to be more accurate for naphthas, provided that their characteristics fall within the application range of the network's training set. [4] The most suited method to reconstruct the feedstock is automatically selected by the software, offering the highest reliability on the reconstructed feed and the subsequent simulated product spectra. This enables users to take optimal advantage of feedstock flexibility. The feed reconstruction method also incorporates sulfur components. The reconstructed feed can be directly imported into a reactor simulation.

Reactor Definition

COILSIM1D incorporates an extensive built-in library that includes the reactor configurations most commonly used in industrial steam crackers. In this way only the length, internal diameter and wall thickness of each section have to be provided as input by the user. Since the reactor model of COILSIM1D can be applied to any type of tubular reactor, virtually any other reactor geometry that is not included in the built-in library can be easily implemented using the 'New coil geometry' tool in the GUI, as shown in Figure 2. The user can also choose from an assortment of reactor materials. Fins or other 3D geometries (e.g. straight, riffled, spiral) are also supported and accounted for. User defined reactor geometries can be saved and reused later in future simulations.

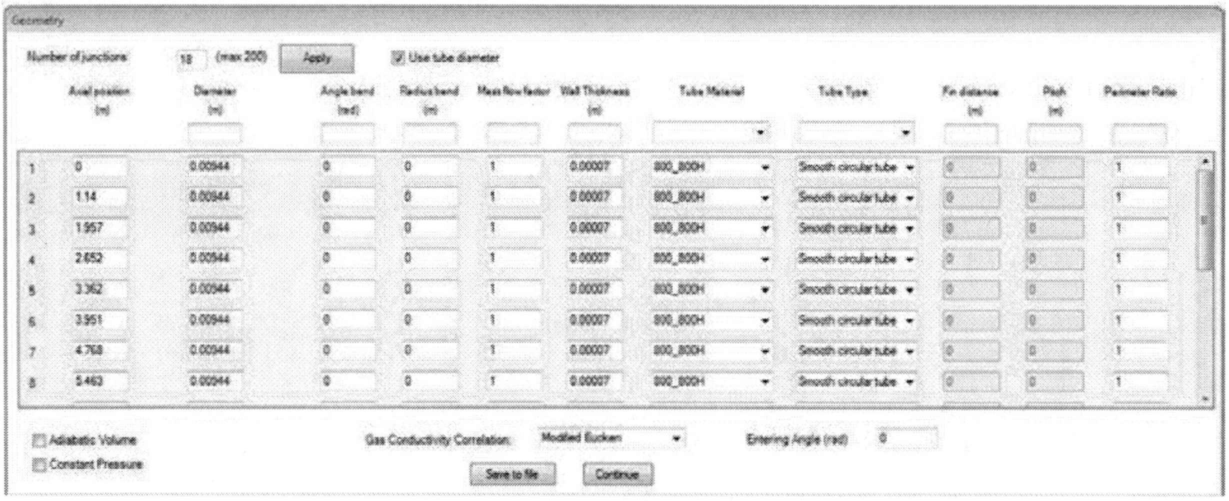

Figure 2. New coil geometry tool included in the GUI of COILSIM1D.

Transfer Line Exchanger

To explicitly account for the effect of the rapid quenching of the process gas leaving the reactor, COILSIM1D incorporates a TLE simulation model. [5] It takes into account the heat transfer from the process gas side to the steam/water mixture used for cooling. The high heat content of the process gas is usually used to produce high pressure steam and hence accurate simulations of this element are highly important for integral simulations of the furnace as a whole.

Different simulation strategies are possible with COILSIM1D, using either a specific water mass flow rate or a specific steam quality as boundary conditions. The model also allows the simulation of a full thermosiphon, which can be used to estimate the water recirculation rate as well as the steam production.

The TLE model is designed specifically to simulate double piped TLEs, like the one shown in Figure 3, which are the most common configuration in current commercial furnaces. In this type of TLE, the cracked gas flows through the inner tube and the water/steam mixture flows through the outer tube. Since the temperature of the cracked gas is still considerably high, the secondary reactions that can take place in the inner tube of the TLE are explicitly accounted for.

This module also includes a dedicated model to account for fouling in the TLE, caused mainly by the condensation of the heavy hydrocarbons that are often present in the reactor effluent. [6-8] Combined with the coke deposition models for gaseous and liquid feeds used in the reactor section, this feature provides the user with a very handy tool for the estimation of run lengths.

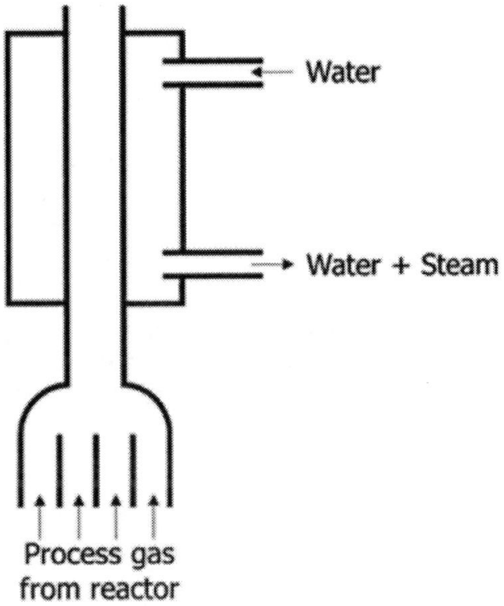

Figure 3. Schematic of a double piped TLE used to quench the effluent of reactor coils.

Run Length Simulations

Steam cracking is inevitably associated with the deposition of coke on the inner wall of the reactor, which largely affects the economics of the process. [9] Since decoking operations translate into loss of production capacity, accurate prediction of the run length of a furnace is highly interesting for the optimization of cracking furnaces. COILSIM1D offers the possibility to perform run length simulations that account for the effect of coke depositing inside the reactor over the course of a run, until the operational limitations are reached.

To accurately predict the formation of coke, two different models are incorporated: one for gaseous feeds [10] and one for liquid feeds [11]. Both account for the heterogeneous non-catalytic or so-called asymptotic coking regime, which is dominant over most of the run length. [12, 13] The effect of the brief catalytic coking regime can be accounted for in COILSIM1D, by giving the model an estimate of the expected duration of the catalytic coking regime.

The coking rate is computed using the selected model, and the increase in coke thickness during each specific time step is calculated and added to the already deposited layer, until reaching the maximum TMT or maximum pressure drop, as defined by the user in the GUI. An illustrative example of simulation results is shown in Figure 4, where the calculated TMT (accounting for faster catalytic coking at the beginning of the run) is plotted as a function of time. It is clear that the slope of the TMT rapidly drops to a constant value as a consequence of the shift of the dominant coke deposition mechanism.

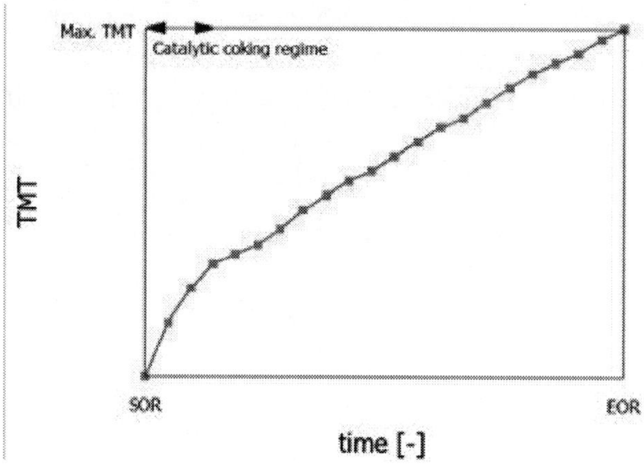

Figure 4: TMT as a function of time for a simulation case that accounts for a faster catalytic coking regime at the beginning of the run.

New Features of COILSIM1D

Kinetic Model for Steam Cracking of Hydrocarbons

The kinetic model of COILSIM1D is assuredly the most important feature of this software package. COILSIM1D comprises a very thorough mathematical model that consists of two parts: (i) A single event kinetic model called CRACKSIM, [1] which accurately describes the chemistry taking place in the reactor; and (ii) a one dimensional tubular reactor model describing the transport phenomena. The model equations are valid for any type of tubular reactor, allowing COILSIM1D to be used for a broad range of coil configurations. Both parts complement each other and help to structure a mathematical description of the problem to simulate. A dedicated fast and robust solving algorithm is applied to solve the mathematical problem. The required simulation time is usually less than 1 second allowing to simulate within a broad window of conditions and feeds in a limited time.

The accuracy and robustness that are characteristic of CRACKSIM are, to a great extent, the result of the fundamental approach that it uses, i.e. describing each of the occurring reactions as a single step, instead of simply accounting for global reactions. The corresponding kinetic and thermodynamic parameters are calculated using consistent and dedicated high level ab-initio calculations. [14-19] It is therefore no surprise that COILSIM1D incorporates the broadest and most accurate reaction network for steam cracking of hydrocarbons.

The kinetic model has been recently extended, to incorporate also common oxygenate contaminant species whose thermochemical decomposition yields products such as methanol and aldehydes, known for leading to several downstream process difficulties such as gum formation and off-spec products.

Currently more than 720 different molecular species and 43 radical species are incorporated, leading to a global network of more than 300,000 reactions. The model has been validated using an extensive set of data points originating from industrial furnaces, as well from the pilot plant and bench scale set steam cracking setups of LCT. The validation dataset includes a broad assortment of feeds, ranging from ethane to complex and heavy mixtures such as gasoil, waxes and even crude oil. Excellent agreement is observed between the measured and predicted yields of the main and minor products, [1, 20] as shown in the parity plots presented in Figure 5.

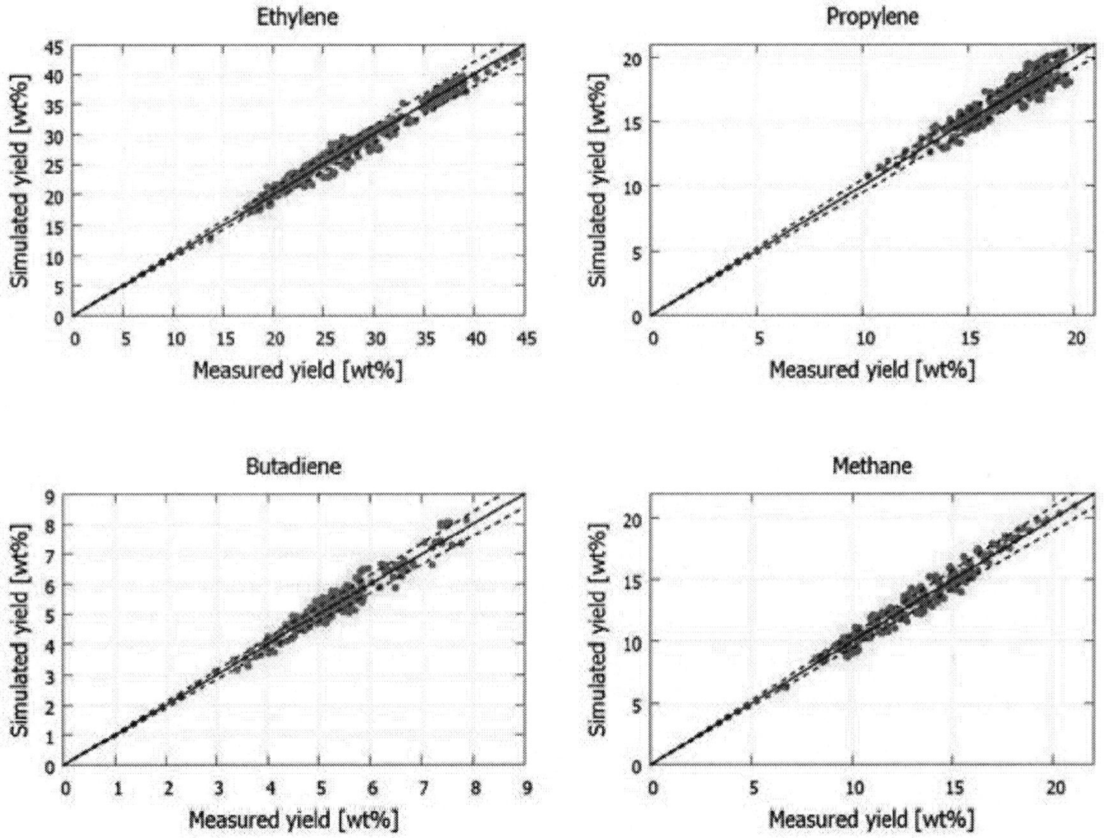

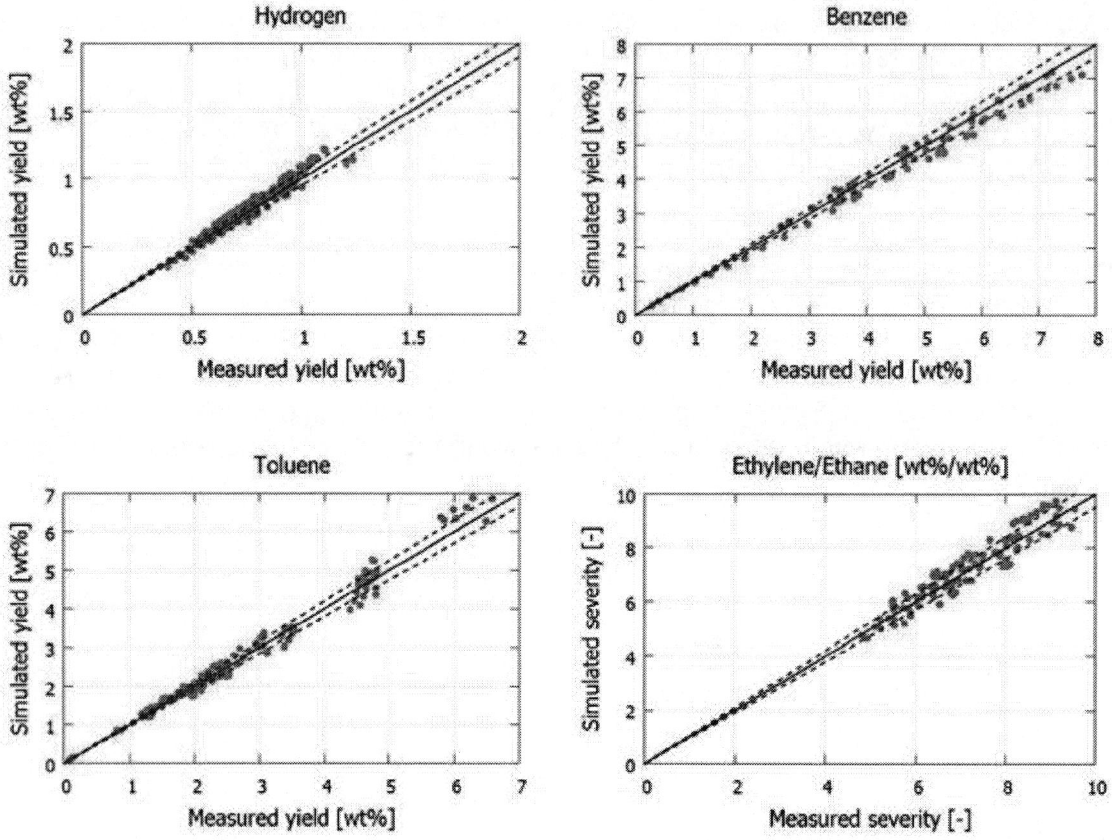

Figure 5: Parity plots comparing measured and simulated yields of various cracking products. Dashed lines indicate ±5% deviations.

Convection Section

The convection section is a new feature of COILSIM1D, and it is part of the furnace module of the GUI, which comprises two complementary models for simulations of firebox and convection section. The convection section simulations help to estimate different process parameters of interest such as flue gas temperature, utility steam production and total thermal power exchange based on limited user input.

When feeding liquids, evaporation also takes place in the convection section, which is an important aspect to consider for design and optimization, especially when dealing with heavy feeds that might not be completely evaporated and thus lead to fouling, in particular in the lower banks. Explicitly taking evaporation into account can help to locate hot spots and hence prevent extensive fouling of the convection banks. To meet this challenge, the a dedicated equilibrium based evaporation model has been developed.

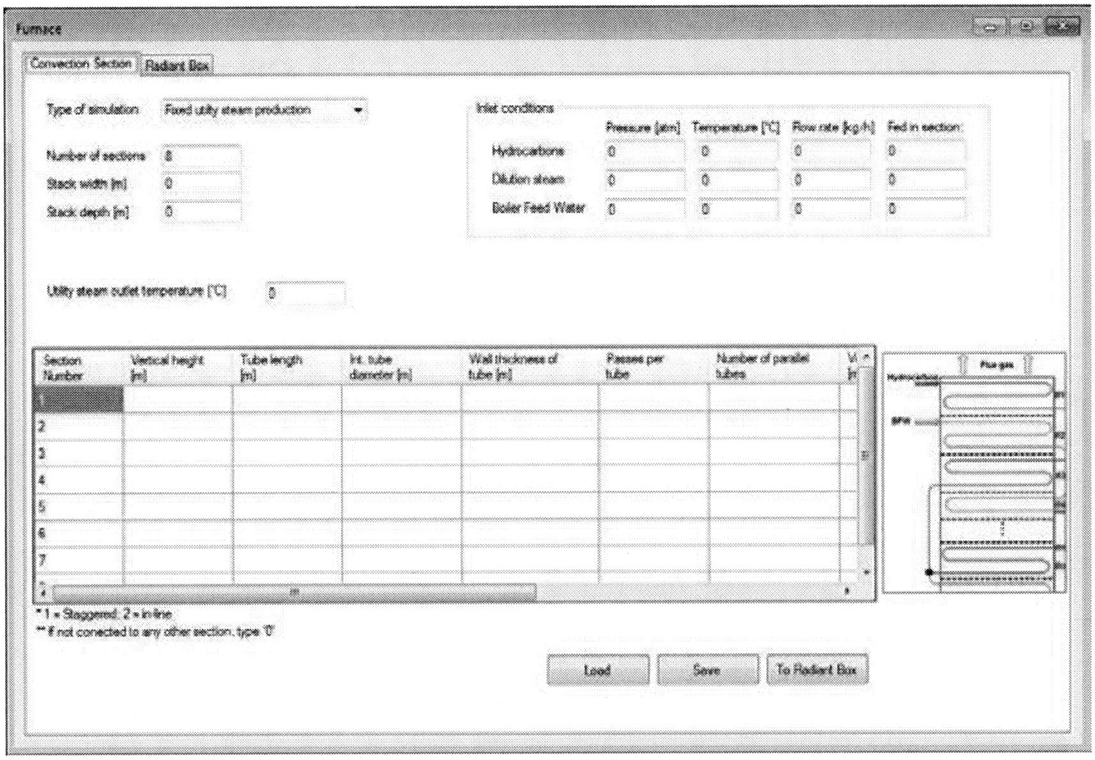

Figure 6: GUI of the convection section simulation module of COILISM1D.

The convection section is modelled as a finite number of tube banks in-series. As can be seen in Figure 6, the user can define the number of banks and the number of parallel tubes on each bank, which can make multiple passes. The feed entering each bank can be either a stream entering the convection section for the first time, a stream originating from another bank, or a mixture thereof. Furthermore, different tube arrangements (in-line or staggered) are supported, as well as different configurations (counter and concurrent flow), as shown in Figure 7. Additionally, different tube materials, with characteristic roughness and thermal properties can be chosen. The model can also account for the usage of finned tubes. In this way, virtually any type of convection section configuration can be easily simulated with COILSIM1D.

The model describes the duty pick-up of each bank from the hot flue gas to the process stream, based on different users input and process conditions (e.g. throughputs, composition, temperatures), parameters from of the firebox (thermal efficiency, combustion fuel rate, bridge wall temperature) and TLE (heat duty, steam flow rate). The tool contains an extensive library of pressure- and temperature-dependent physical properties for all the molecular species available in COILSIM1D.

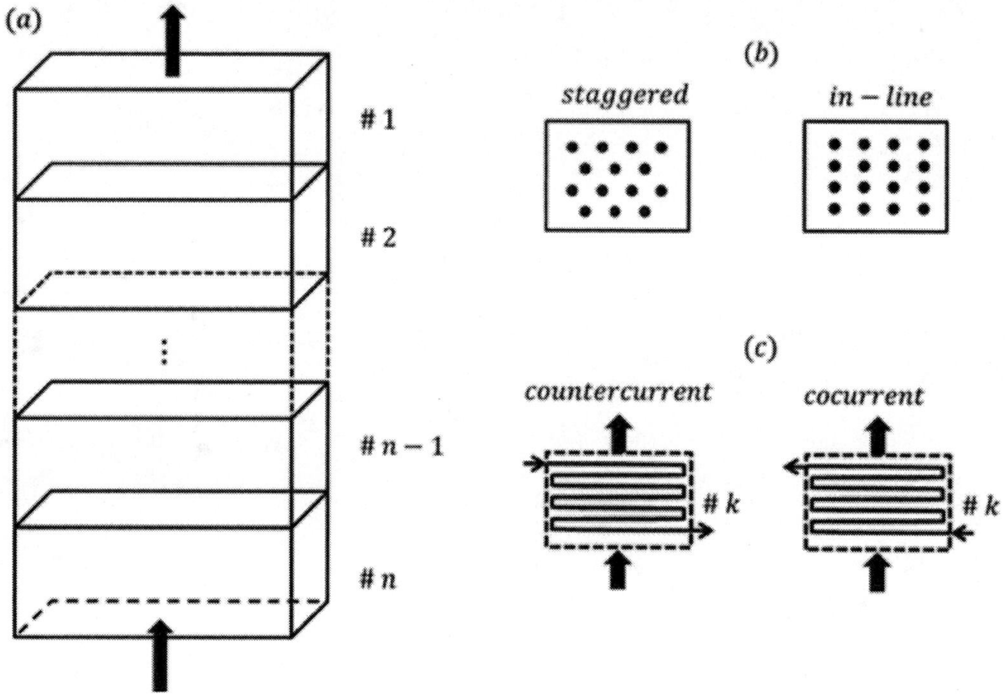

Figure 7. Schematic representation of the convection section in COILSIM1D: (a) tube banks in series; (b) supported tube arrangements; (c) supported flow pattern arrangements.

To illustrate the use of the convection section module, a demonstration case has been simulated for a typical arrangement, presented in Figure 8. The geometry of the convection section (i.e. number of passes, tube lengths, number of parallel tubes and pitch lengths) was defined via the GUI, as well as the mass flows of the input streams and the number of the bank in which they enter the convection section. Eight heat exchanging tube banks and six different inlet streams are defined. The naphtha is fed to the first bank, then mixed with dilution steam and taken to the third bank. The mixture is further heated in banks 3 and 4. The heated mixture is then mixed with superheated dilution steam (outlet stream from bank 7), and conducted to the lowest bank. The second bank is used to heat water at high pressure. High pressure steam is overheated in bank 5 and then mixed with desuperheating water. The resulting stream is then heated up in bank 6.

The hydrocarbon feed was reconstructed using SIMCO, based on a set of common commercial indices presented in Table 1. The reconstructed naphtha consists of 72 different components ranging from C_4 to C_{11}.

The calculated Coil Inlet Temperature (CIT) is 578.3°C and the pressure before the Venturi nozzles is 3.8 bara. These values are typical for a naphtha cracker, [21] and can be subsequently used as input for the reactor model to determine the product distribution.

Table 1. Commercial indices used for reconstructing the naphtha feed with SIMCO

Commercial indices of the naphtha feed	
Specific density [kg/m³]	708.8
ASTM D86 [°C]	
IBP	39
50%	99
FBP	165
PIONA [wt%]	
Paraffins	36.5
Iso-paraffins	32.8
Olefins	0.0
Naphtenes	21.4
Aromatics	9.20

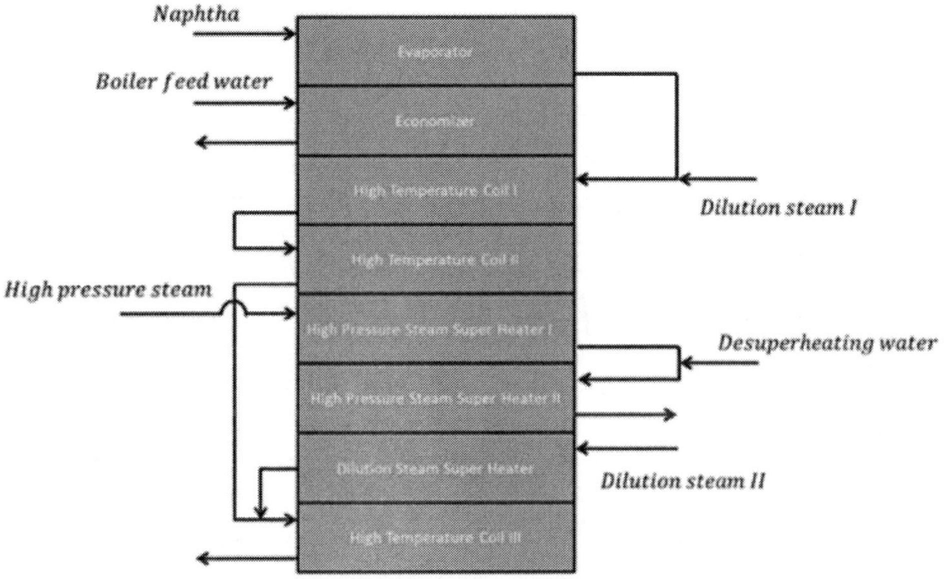

Figure 8. Schematic representation of the simulated convection section.

The calculated temperature profiles of the flue gas and process streams are shown in Figure 9. The flue gas leaves the stack at a temperature of 283°C, whilst the bridge wall temperature is estimated to be 1195°C. The vapor fraction is estimated to be 98.39% at a temperature of 301.4°C, indicating that the feed is almost completely evaporated in the first bank. It is worth noticing that, according to the ASTM D86 boiling points given in Table 1, the feed should be completely evaporated at such a high temperature, but due to the substantially high pressure in the evaporator (6 bar), the boiling points of the species are significantly higher than under atmospheric pressure. This shows the importance of accounting for pressure dependence when simulating the phase transition from liquid to vapor.

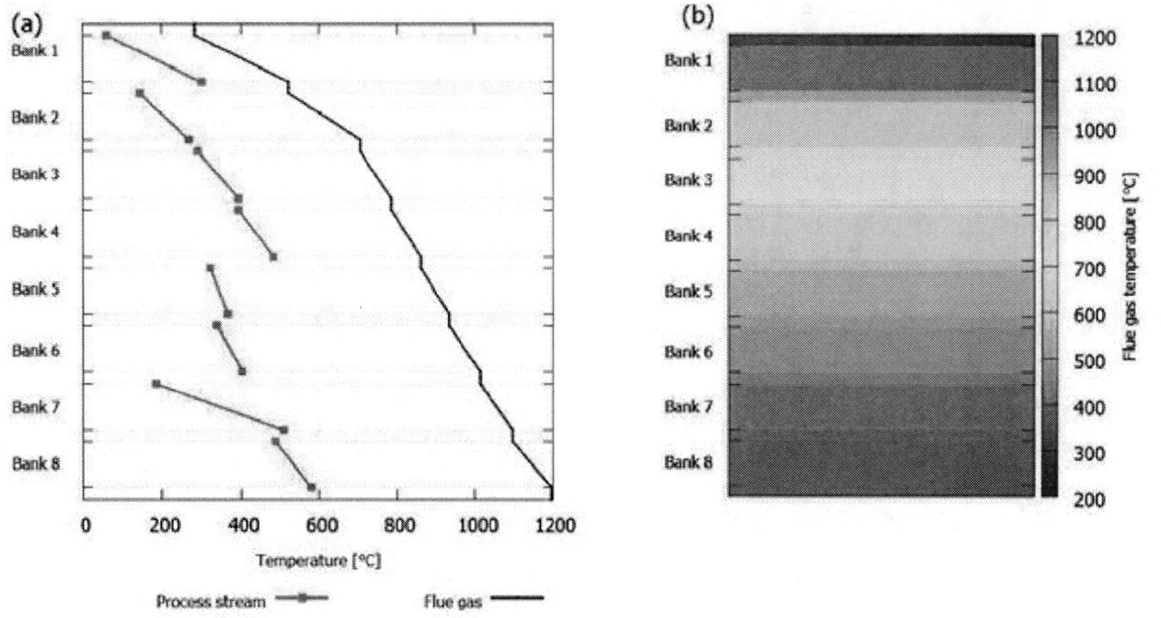

Figure 9. (a) Temperature profile of the process streams for all banks, (b) flue gas temperature profile along the convection section

Firebox Simulation Model

With the implementation of a firebox model, COILSIM1D is able to estimate additional process parameters of interest, such as the flue gas temperature profile inside the firebox, the bridge-wall temperature, the furnace efficiency and the heat flux profile to the reactor coils. Using the GUI of COILSIM1D, shown in Figure 10, the user can easily provide the geometry and operating conditions of the firebox, required as input for the simulation. The user is able to specify the furnace geometry, the type of burners (i.e. floor and/or wall burners) and, in the case of wall burners, the heights at which they are located. Also the operating conditions, such as the inlet temperatures and mass flow rates of fuel and oxidizer (e.g. air) per burner, and the operating pressure, are to be specified. This flexibility in user input guarantees the applicability of the model for a broad range of cracking furnaces.

The firebox model has been coupled to the reactor model, enabling the estimation of process yields, COT, COP, etc. of cracking furnaces, based on the geometry and inlet conditions of the furnace and tubular reactors, specified by the user.

An illustrative simulation of the firebox, reactor and TLE has been carried out with COILSIM1D for a Kellogg Millisecond furnace, shown in Figure 11 (left). The dimensions and operating conditions of the reactor and TLE are presented in Table 2. The simulation parameters are straightforwardly input by means of the GUI of COILSIM1D, as illustrated in Figure 10 for the geometry and operating conditions of the firebox. 53.9 kg/h of fuel (CH_4 + 5% H_2) are supplied to each long flame burner and burnt using air at ambient conditions. The software converts the three-dimensional furnace dimensions to their one-dimensional equivalent and discretizes the furnace along its height, as illustrated by the dotted lines in Figure 11 (right).

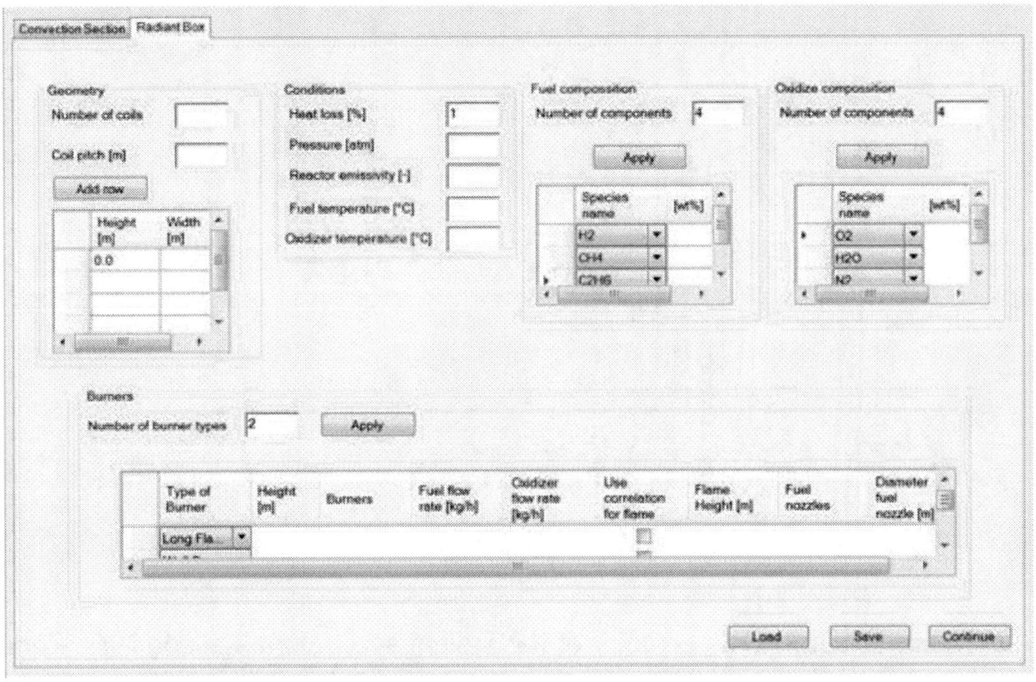

Figure 10. GUI of the firebox simulation module of COILISM1D.

Table 2. Dimensions and operating conditions of the reactor and TLE of the Kellogg Millisecond furnace

Reactor dimensions		**Feed composition**	[wt%]
Length [m]	10.556	C_3H_8	92.25
Internal diameter [m]	0.0348	C_2H_6	1.05
		C_3H_6	3.02
Operating conditions		i-C_4H_8	0.33
Feed mass flow rate [kg/h]	118.5	1-C_4H_8	0.27
Steam dilution [kg/kg]	0.3265	2-C_4H_8	0.08
Coil inlet temperature [°C]	630.5	i-C_4H_{10}	1.93
Coil inlet pressure [bara]	2.35	n-C_4H_{10}	0.45
		C_5^+	0.62
TLE dimensions			
Length [m]	9.369		
Diameter inner tube [m]	0.067		
Diameter outer tube [m]	0.119		
Number of joining coils [-]	4		

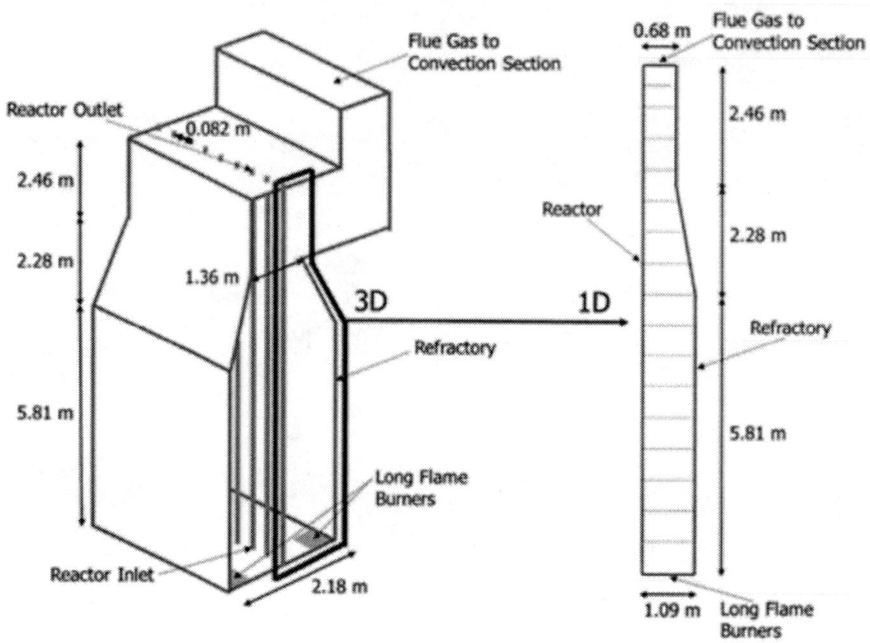

Figure 11. Three-dimensional geometry of the radiant section of a Kellogg Millisecond furnace (left) and its equivalent one-dimensional geometry used for the COILSIM1D simulation (right).

Table 3 shows the results of the COILSIM1D simulation, together with industrially measured values. The simulated values agree well with the available industrial data. The simulation results demonstrate not only the accuracy with which COILSIM1D predicts the process yields, but also its ability to estimate other parameters, such as the COT, furnace efficiency and bridge-wall temperature. Since the GUI facilitates setting up the simulation and considering that the computational time is less than one minute on one core of an Intel® Core™ i7-5600U CPU @ 2.60 GHz processor, this simulation case illustrates the ease and speed with which steam crackers can be simulated using COILSIM1D.

Table 3. COILSIM1D-simulated and industrial [22] data for the Kellogg Millisecond furnace.

	Simulated	**Industrial***
Product Yields [wt%]		
H_2	1.5	1.6
CH_4	19.9	19.3
C_2H_2	0.7	0.8
C_2H_4	35.1	35.1
C_2H_6	2.7	2.9
C_3H_4 (MA)	0.3	0.7
C_3H_4 (PD)	0.1	
C_3H_6	17.0	17.0
C_3H_8	15.8	15.7
1,3-C_4H_6	1.8	1.9
1-C_4H_8	0.8	1.2

2-C_4H_8	0.4	
i-C_4H_8	0.3	0.2
i-C_4H_{10}	0.2	0.1
n-C_4H_{10}	0.1	0.1
Other properties		
Reactor coil outlet temperature [°C]	878.9	
Temperature at exit TLE [°C]	445.9	435
Furnace efficiency [%]	43.2	
Bridge-wall temperature [°C]	1138	

*Industrial data are measured downstream from the transfer line exchanger (TLE).

Input/Results Files and Interaction with External Software

COILSIM1D is a text-based simulation tool. The user friendly GUI allows the user to easily set up a simulation case step by step. Operating parameters (and their units), simulation strategies and geometrical input can be typed in manually by the user, or can be loaded from previous simulation cases.

```
0
0       'fraction type(0,1,2,3 or 4)'
0       'reconstruction method (0,1 or 2)'
0.00    'Average molecular weight (g/mol)'
0.00    'H/C ration (kg/kg)'
0.7088  'Density (60 ° F) (g/ml)'
0       'Mass/volume indicator (0 or 1)'
0       '0:PIONA, 1:PNA, 2:SA'
36.50   'Total amount of paraffins (wt% or vol%)'
32.80   'Total amount of isoparaffins (wt% or vol%)'
0.00    'Total amount of olefins (wt% or vol%)'
21.40   'Total amount of napthenes (wt% or vol%)'
9.20    'Total amount of aromatics (wt% or vol%)'
0       'boiling point unit (0,1 or 2)'
0       'boiling point curve type (0,1 or 2)'
39.00   'Initial boiling point (° C,K or °F)'
0.00    '5% volume boiling point (° C,K or °F)'
0.00    '10% volume boiling point (° C,K or °F)'
0.00    '20% volume boiling point (° C,K or °F)'
0.00    '30% volume boiling point (° C,K or °F)'
0.00    '40% volume boiling point (° C,K or °F)'
99.00   '50% volume boiling point (° C,K or °F)'
0.00    '60% volume boiling point (° C,K or °F)'
0.00    '70% volume boiling point (° C,K or °F)'
0.00    '80% volume boiling point (° C,K or °F)'
0.00    '90% volume boiling point (° C,K or °F)'
0.00    '95% volume boiling point (° C,K or °F)'
165.50  'final boiling point (° C,K or °F)'
0       'NMR H (arom) (%)'
0       'NMR H alpha (%)'
0       'NMR H beta (%)'
Ln 18, Col 39
```

Figure 12: Example of an input file used for feedstock reconstruction with SIMCO.

All the input values are then used to create well-organized input files that are stored into project folders specifically created for each simulation. These input files are simple, self-explanatory, and have a fixed format. An example of an input file used for feedstock reconstruction is shown in Figure 12. The user manual of COILSIM1D explains the purpose and structure of all the input and output files.

The GUI incorporates a tool for quick visualization of the main results of a simulation, as shown in Figure 13. Furthermore, the GUI also displays the results (yields, temperatures, ...) in the form of graphs, which facilitates interpretation of the simulated results. This feature is illustrated in Figure 14. The files with all the simulation results are also stored in the project folder, and can be imported into spreadsheets for further analysis or post-processing.

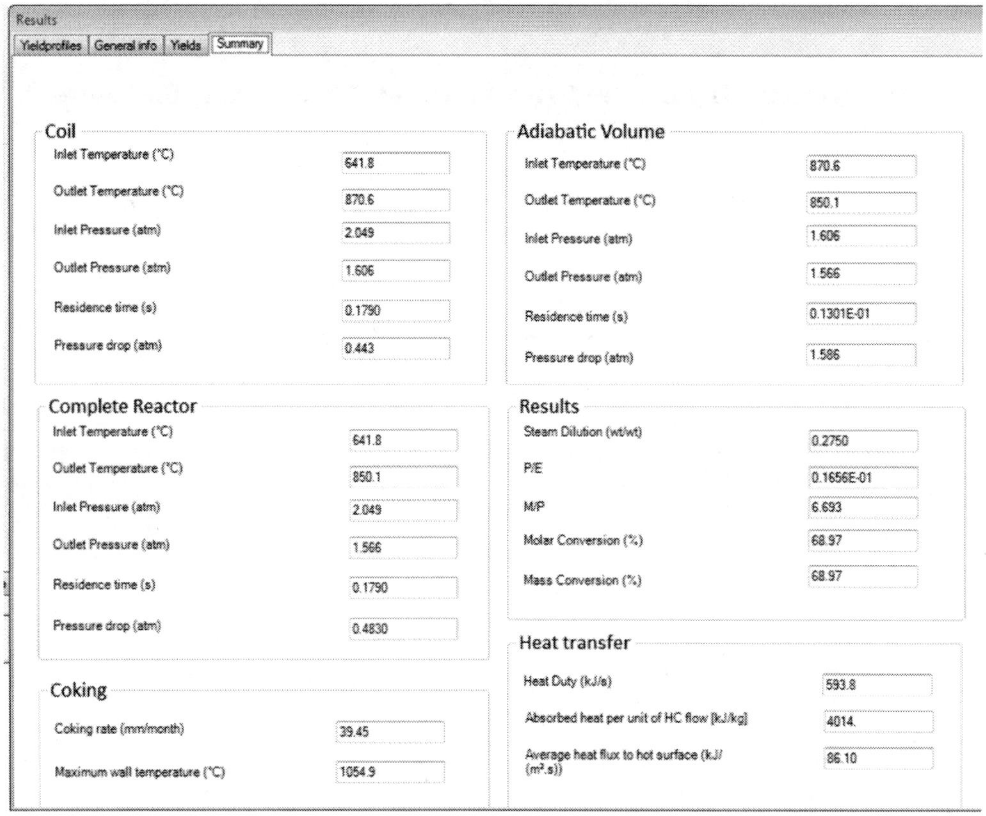

Figure 13: Results tab of the GUI, showing the main results of a simulation

Since the input/output files have a fixed format, they can be easily created and modified without using the GUI, thus allowing users to make use of scripts in the programming language of their preference to take care of these tasks. COILSIM1D includes also an executable file, which allows to carry out simulations bypassing the GUI. This enables users to write scripts for automated creation and execution of large batches of simulations without the need of constant human input.

The ability of COILSIM1D to work using text input/output files has the additional benefit of enabling easy interaction with external software packages, e.g. Aspen Plus®, HYSYS® or gPROMS. This way, whole plant simulations can be made taking into account the constraints of all the units involved. The interaction is simple and based on an interfacing script that reversibly converts the species and process conditions representation of both software tools. Integration with unit operation software packages or in-house optimization tools has been successfully performed with several licensees, considering their specific requirements. Integration of COILSIM1D in real time optimization has also been successfully carried out in a demonstration unit in Asia, showing great potential for industrial scale implementation.

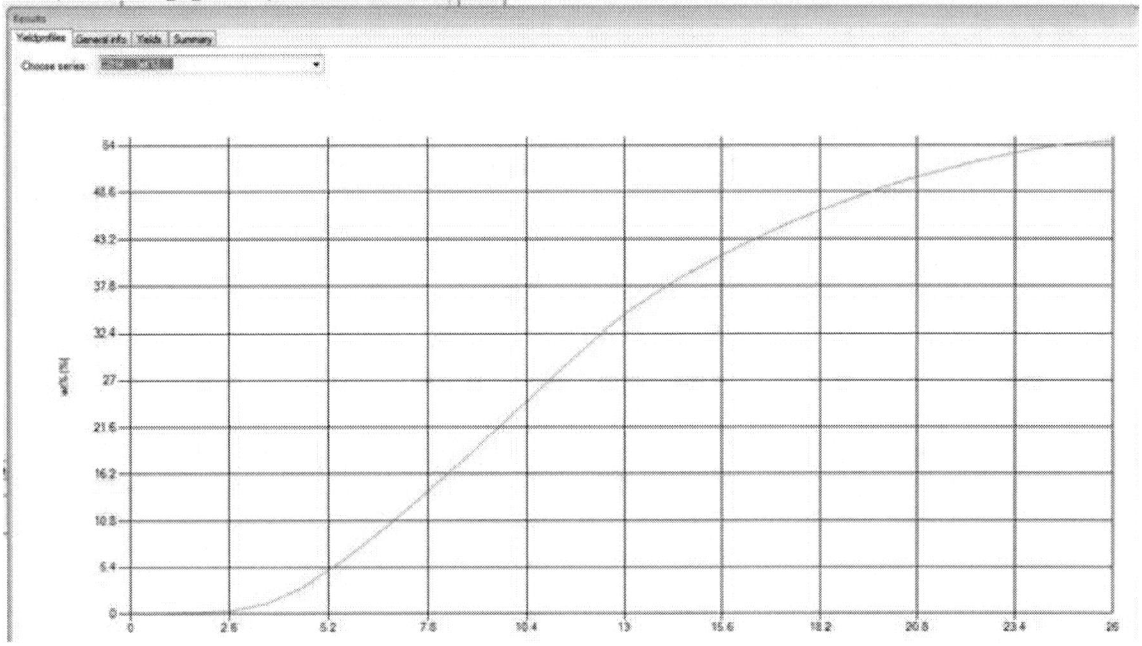

Figure 14: Graphic representation of yield profiles in COILSIM1D

Conclusions

COILSIM1D is a comprehensive simulation package for steam cracking processes. Its main strength lies in its fundamental kinetic model, the broadest currently available. Various features and simulation strategies have been developed and implemented in COILSIM1D: SIMCO, a molecular feedstock reconstruction method; a module to simulate the TLE; a module to estimate reactor run lengths using dedicated coking models; and a shooting method, that allows to carry out simulations using severity-related indices as boundary conditions. To take COILSIM1D to the next level and transform it into a whole furnace simulation package, new features have been incorporated and the kinetic model has been fine-tuned and further extended to account for oxygenate species.

A model to simulate the convection section of a steam cracker has been developed. It allows the user to evaluate the impact of a broad range of different process conditions and feeds in the process. Despite requiring a limited amount of input data, this model helps to easily evaluate the effect of a broad range of different process conditions and feeds. This is a very helpful tool for design and optimization of new and existing configurations. The model accurately accounts for the evaporation of the feed and can be used to prevent fouling in the convection section, which has significant technologic and economic impact on the process. An example simulation was defined and executed yielding satisfactory results.

Another new feature is the firebox model, which allows to reliably determine heat flux profiles for the reactor, and various other parameters of interest, such as the bridge wall temperature and thermal efficiencies. The model can be applied to a variety of firebox geometries, burner arrangements and fuel compositions. It has been successfully used to simulate a propane Millisecond reactor with its corresponding TLE, showing the accuracy and ease of using COILSIM1D when dealing with simulations that explicitly link various aspects of the process altogether.

COILSIM1D comes with a user-friendly GUI, which makes it extremely easy to create and execute all types of simulations. The input and output files of the simulations are stored in Excel readable files, that can be easily imported into spreadsheets for further analysis and processing. This has the additional advantage of enabling easy interaction of COILSIM1D with user-created scripts for large batches of simulations, or to couple COILSIM1D to external software, e.g. to carry out whole-plant simulation and optimization.

The combination of all these new and already existing features enables users to perform accurate simulations of the whole furnace with limited number of inputs, reasonable computational effort and unprecedented accuracy, all with one click.

References

1. Van Geem, K. M.; Reyniers, M. F.; Marin, G. B., Challenges of modeling steam cracking of heavy feedstocks. *Oil & Gas Science and Technology-Revue D Ifp Energies Nouvelles* 2008, 63, (1), 79-94.
2. Hudebine, D.; Verstraete, J. J., Reconstruction of petroleum feedstocks by entropy maximization. Application to FCC gasolines. *Oil & Gas Science and Technology–Revue d'IFP Energies nouvelles* 2011, 66, (3), 437-460.
3. Van Geem, K. M.; Hudebine, D.; Reyniers, M. F.; Wahl, F.; Verstraete, J. J.; Marin, G. B., Molecular reconstruction of naphtha steam cracking feedstocks based on commercial indices. *Computers & Chemical Engineering* 2007, 31, (9), 1020-1034.
4. Pyl, S. P.; Van Geem, K. M.; Reyniers, M. F.; Marin, G. B., Molecular reconstruction of complex hydrocarbon mixtures: An application of principal component analysis. *AIChE journal* 2010, 56, (12), 3174-3188.
5. Dijkmans, T. Steam cracking: from feedstock analysis to plant optimization. Ghent University, 2014.
6. Lahaye, J.; Badie, P.; Ducret, J., Mechanism of carbon formation during steamcracking of hydrocarbons. *Carbon* 1977, 15, (2), 87-93.
7. Albright, L. F.; Marek, J. C., Mechanistic model for formation of coke in pyrolysis units producing ethylene. *Industrial & Engineering Chemistry Research* 1988, 27, (5), 755-759.
8. Kopinke, F. D.; Bach, G.; Zimmermann, G., New results about the mechanism of TLE fouling in steam crackers. *Journal of Analytical and Applied Pyrolysis* 1993, 27, (1), 45-55.
9. Muñoz Gandarillas, A. E.; Van Geem, K. M.; Reyniers, M. F.; Marin, G. B., Coking resistance of specialized coil materials during steam cracking of sulfur-free naphtha. *Industrial & Engineering Chemistry Research* 2014, 53, (35), 13644-13655.
10. Plehiers, P. M.; Reyniers, G. C.; Froment, G. F., Simulation of the run length of an ethane cracking Furnace. *Industrial & Engineering Chemistry Research* 1990, 29, (4), 636-641.
11. Reyniers, G. C.; Froment, G. F.; Kopinke, F.-D.; Zimmermann, G., Coke formation in the thermal cracking of hydrocarbons. 4. Modeling of coke formation in naphtha cracking. *Industrial & engineering chemistry research* 1994, 33, (11), 2584-2590.
12. Reyniers, M.-F. S.; Froment, G. F., Influence of metal surface and sulfur addition on coke deposition in the thermal cracking of hydrocarbons. *Industrial & engineering chemistry research* 1995, 34, (3), 773-785.
13. Muñoz Gandarillas, A. E.; Van Geem, K. M.; Reyniers, M. F.; Marin, G. B., Influence of the reactor material composition on coke formation during ethane steam cracking. *Industrial & Engineering Chemistry Research* 2014, 53, (15), 6358-6371.
14. Sabbe, M. K.; De Vleeschouwer, F.; Reyniers, M.-F.; Waroquier, M.; Marin, G. B., First principles based group additive values for the gas phase standard entropy and heat capacity of hydrocarbons and hydrocarbon radicals. *The Journal of Physical Chemistry A* 2008, 112, (47), 12235-12251.
15. Sabbe, M. K.; Reyniers, M. F.; Van Speybroeck, V.; Waroquier, M.; Marin, G. B., Carbon-Centered Radical Addition and β-Scission Reactions: Modeling of Activation Energies and Pre-exponential Factors. *ChemPhysChem* 2008, 9, (1), 124-140.

16. Sabbe, M. K.; Reyniers, M. F.; Waroquier, M.; Marin, G. B., Hydrogen Radical Additions to Unsaturated Hydrocarbons and the Reverse β-Scission Reactions: Modeling of Activation Energies and Pre-Exponential Factors. *ChemPhysChem* 2010, 11, (1), 195-210.
17. Sabbe, M. K.; Vandeputte, A. G.; Reyniers, M.-F.; Waroquier, M.; Marin, G. B., Modeling the influence of resonance stabilization on the kinetics of hydrogen abstractions. *Physical Chemistry Chemical Physics* 2010, 12, (6), 1278-1298.
18. Saeys, M.; Reyniers, M. F.; Marin, G. B.; Van Speybroeck, V.; Waroquier, M., Ab initio group contribution method for activation energies for radical additions. *AIChE journal* 2004, 50, (2), 426-444.
19. Saeys, M.; Reyniers, M. F.; Van Speybroeck, V.; Waroquier, M.; Marin, G. B., Ab initio group contribution method for activation energies of hydrogen abstraction reactions. *ChemPhysChem* 2006, 7, (1), 188-199.
20. Van Geem, K. M.; Reyniers, M.-F.; Marin, G. B. In *Taking optimal advantage of feedstock flexibility with COILSIM1D*, AIChE Spring Meeting: Ethylene producers conference, New Orleans, LA, USA, 2008; New Orleans, LA, USA, 2008.
21. Zimmermann, H.; Walzl, R., Ethylene. *Ullmann's Encyclopedia of Industrial Chemistry* 2009.
22. Heynderickx, G. J. Modellering en Simulatie van Huidige en Nieuwe Technologieën voor de Thermische Kraking van Koolwaterstoffen. University Ghent, Ghent, 1993.

442072 Real-Time Analytic Dashboard Visualization

Abstract:

This session will focus on effective dashboard visualization to ensure real-time analytics projects achieve success. We will highlight examples of dashboard designs along the rationale for their use, and will discuss strategies for organizing and managing projects like these. We will also explore best practices for harnessing management support and funding, along with ensuring the project achieves its goals. Finally we will address implementation tips for developing organizational habits to maximize the effectiveness of the system.

444865 Put Big Data Technology in the Hands of the Engineer

Abstract:

Why do Process Engineers still rely heavily on Excel for analysing the data that has been generated for decades? Why haven't we copied the success of retail, satnav or medical Big Data solutions in the process industry? The first big reason is that the location of our data, the historian database, is not built for these types of questions and requires specific technologies that are usually expensive and require a heavy installation process before they can be used. Secondly, these technologies are built for data scientists,not for the engineer or operator. This creates a knowledge gap and discourages flexibility. TrendMiner is a Belgian software startup with many years of research experience in Process Industry analytics. This solution is to be installed and ready to use within one working day without the need of new datacenter infrastrcuture. In this presentation you'll gain insight into the technology and the cases it has already solved to date.

445137 Data Science in Process Automation: Evaluating Economic Advantage

Abstract:

Abstract

Each day businesses make decisions and take action as best they can based on available information and in the face of uncertainty. The goal of data science and analytics can be seen in terms of improving the amount and interpretation of information such that uncertainty can be reduced and quantified and better decisions and actions may be taken.

This goal – while agreeable – often raises new questions: How do we select the right data science problem that will return value for our efforts? How do we determine the performance required of a data analytics algorithm for it to be economically beneficial? How can data science techniques complement and augment traditional engineering techniques – and avoid having machine learning algorithm return simple mass-energy balance or automation configuration rules?

Honeywell provides an integrated suite of tools to monitor and automate many aspects of plants in the process industry – including refining, oil and gas, pulp and paper, chemicals, power generation, and mining.

In this presentation we will describe an approach in which the economic costs and benefits to the business can be evaluated for given data analytic solution. This economic framework can be introduced early in the data science analysis process and has proven to be an effective tool in guiding and engaging developers and stakeholders throughout all phases of the project – setting the question, exploratory analysis, model building, communication, and decision making.

16AIChE Spring Meeting, Houston, TX

12TH GLOBAL CONGRESS ON PROCESS SAFETY

Big Data Analytics Applied to Real-World Control Systems

From Instruments to Advanced Controls

16AIChE Spring Meeting, Houston, TX

12TH GLOBAL CONGRESS ON PROCESS SAFETY

Agenda

- Real-World Data from Control Systems
- Data to Information to Action
- Analysis of a Typical Plant
- Case Study
- Practical Suggestions

16AIChE Spring Meeting, Houston, TX

12TH GLOBAL CONGRESS ON PROCESS SAFETY

Real-World Data from Control Systems

Good News! You already have the data!

- Thousands of Data Points
  - Controllers (PV, SP, OP, MODE)
- Organized (for Control)
- Somewhat Organized (for Process)
- Historized
  - With Compression?

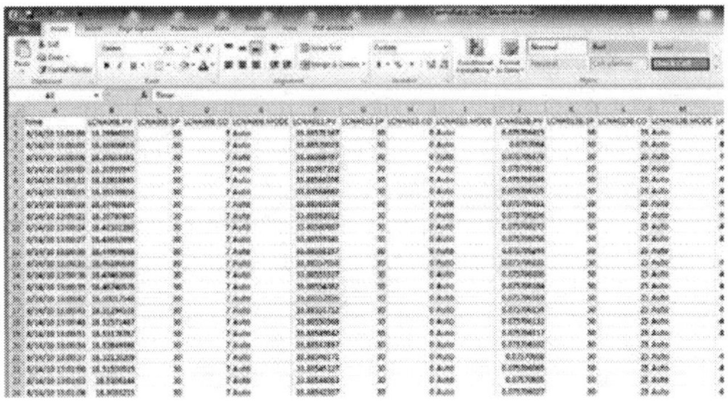

3

16AIChE Spring Meeting, Houston, TX

12TH GLOBAL CONGRESS ON PROCESS SAFETY

Data to Information to Action

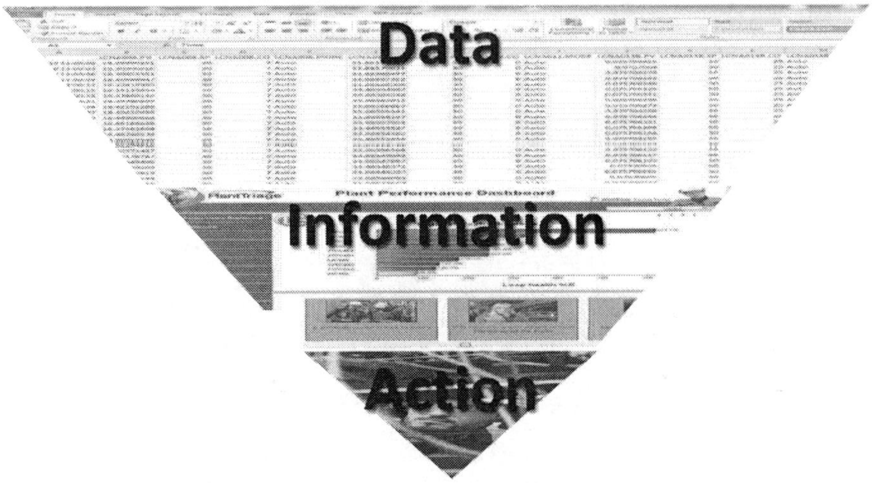

4

16AIChE Spring Meeting, Houston, TX

12TH GLOBAL CONGRESS ON PROCESS SAFETY

Data to Information

DATA

INFORMATION

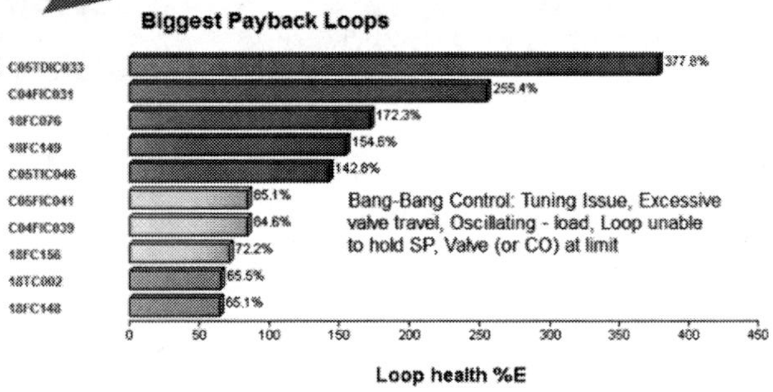

5

16AIChE Spring Meeting, Houston, TX

12TH GLOBAL CONGRESS ON PROCESS SAFETY

Simple, Yet Powerful Information from Data

Control Performance Statistics

- % of Time Loop in Manual
- % of Time Valve at Limits
- Statistics
  - Min / Max
  - Avg
  - Std. Deviation
- PV max = PV min

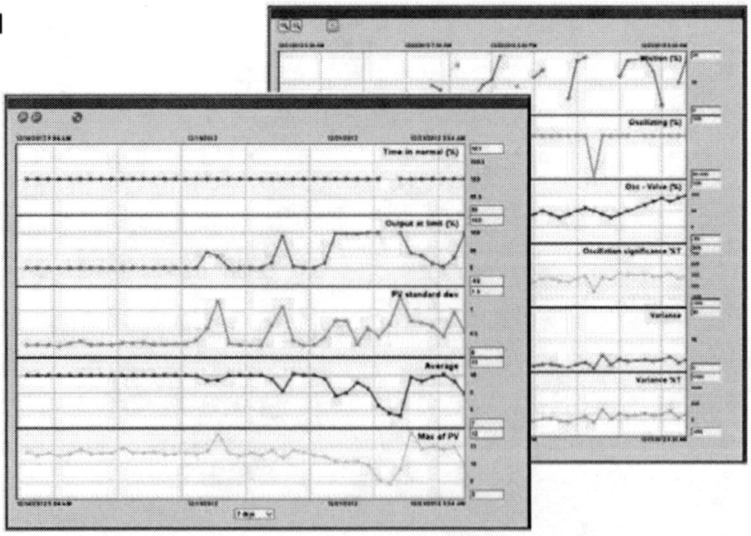

6

16AIChE Spring Meeting, Houston, TX

12TH GLOBAL CONGRESS ON PROCESS SAFETY

Automated Data to Information

Control Performance Monitoring

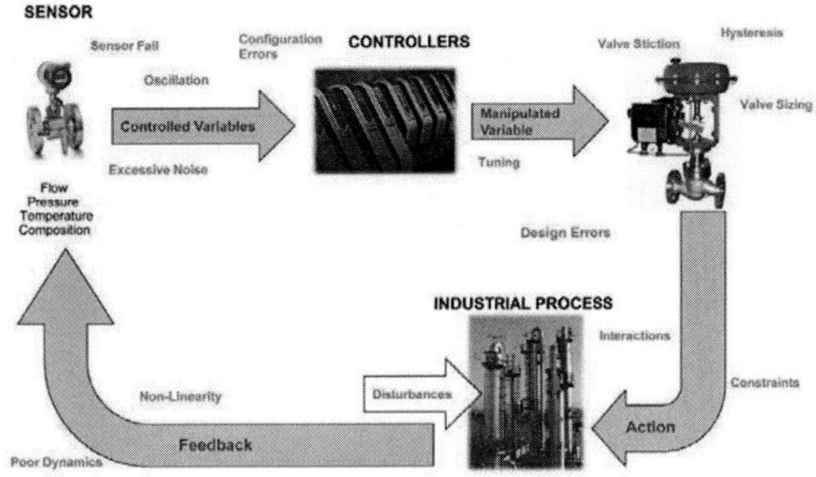

7

16AIChE Spring Meeting, Houston, TX

12TH GLOBAL CONGRESS ON PROCESS SAFETY

Example: Data to Information

Measuring Valve Stiction

- Is loop in AUTO?
- Is it Oscillating?
- Does it have a Valve?
- PV = Square Wave
- CO = Saw Tooth
- Peak-to-Trough of CO

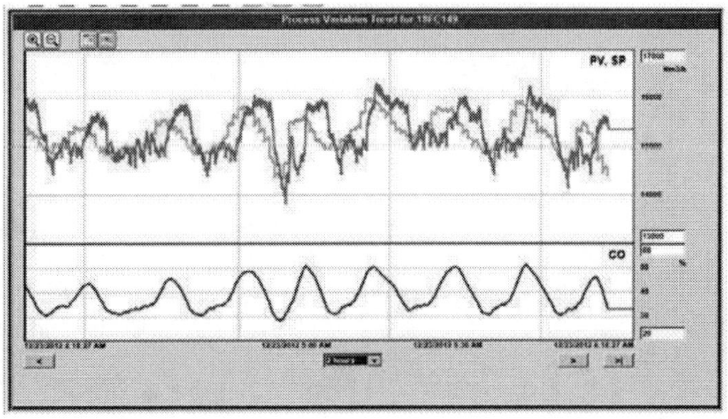

Existing Control Performance

- Control Performance – The Facts
  - 10%-35% of control loops are in **manual**
  - 1.5M to 5.3M in under utilized assets by running in manual in a 1000 loop site
  - 30% of control valves have problems
  - Are you repairing the correct valves?
  - 30% of control loops are tuned incorrectly, increasing variability in the process.
  - A tremendous amount of money can be saved by understanding the control loop interactions and implementing corrective action.

Information to Action

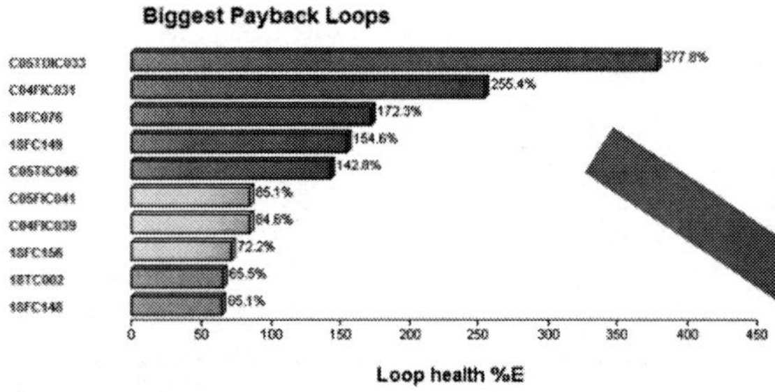

INFORMATION

ACTION

10

Analysis of a Typical Plant

One-Time Analysis to Identify Opportunity

260 Loops			
% Loops	Issue	Typical	World-Class
46	Loops Unable to Hold SP	1 - 5%	< 1%
89	Loops Not in Normal mode	10% - 25%	< 10%
35	Oscillating	10% - 40%	< 10%
40	Excessive Valve Travel	5% - 15%	< 5%
89	Valve or CO at Limits	5% - 15%	< 5%
53	Valve Stiction	10% - 20%	< 10%
15	Sluggish Tuning	3% - 5%	< 3%
3	Sensor Problems	1%- 3%	< 1%

11

Plenty of Data

Distillation column swings

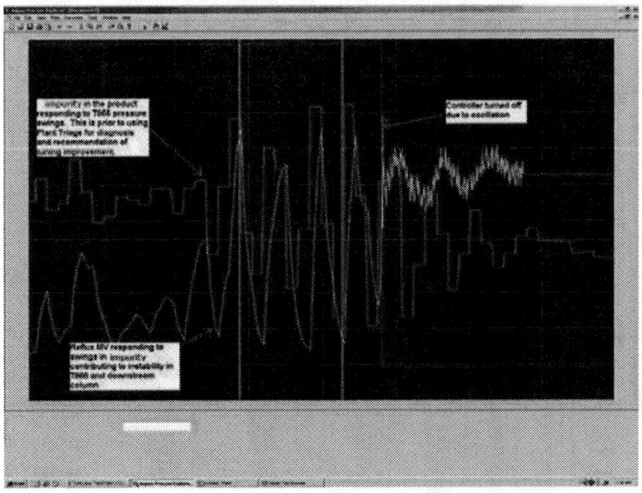

12

16AIChE Spring Meeting, Houston, TX
12TH GLOBAL CONGRESS ON PROCESS SAFETY

Use the Data to Obtain Information

Process mass amounts of data into a root cause

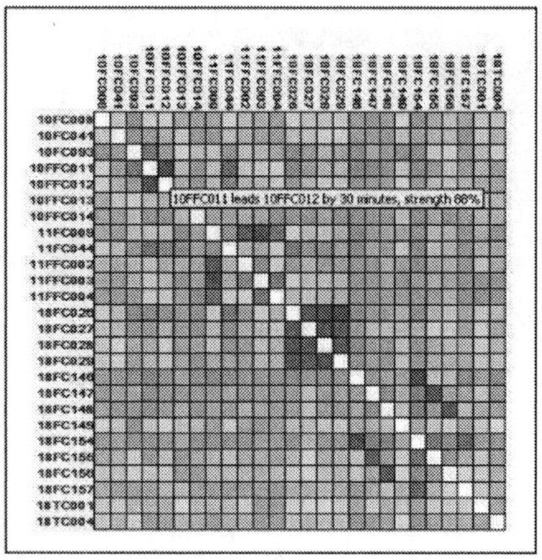

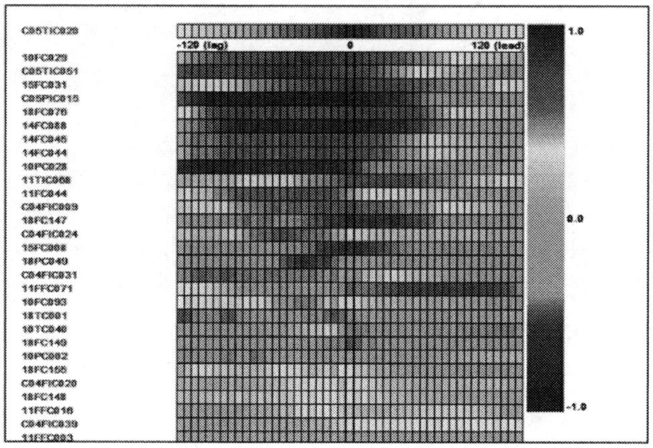

13

16AIChE Spring Meeting, Houston, TX
12TH GLOBAL CONGRESS ON PROCESS SAFETY

Take Action

Tune the Pressure controller -use "data" to obtain a process model

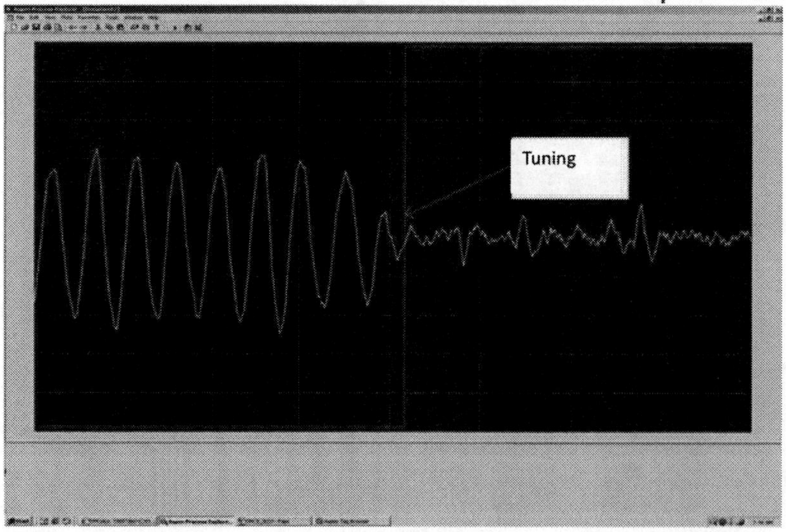

14

16AIChE Spring Meeting, Houston, TX

12TH GLOBAL CONGRESS ON PROCESS SAFETY

See Results

More stable and energy savings

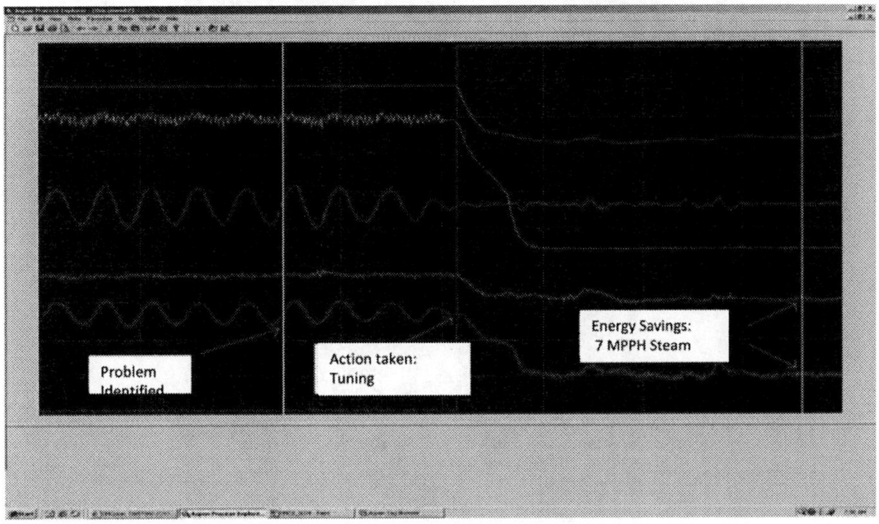

15

16AIChE Spring Meeting, Houston, TX

12TH GLOBAL CONGRESS ON PROCESS SAFETY

Example: Identify Tuning Opportunities

Active Model Capture

Loop	Description	Harris (normalized) (0-1)	Relative response time (sec)	Model quality	Process dead time (sec)	Process gain	Process lag 1 (sec)	Process lag 2 (sec)
Sort	Sort	Sort				Sort	Sort	Sort
38FC-6846	Темп.разбавитель	0.9982*				0.26†*	84†	0†
2220-LC-3796	L БАК 1ст.СОРТИРОВ./L TANK 1st.SORTIROV	0.9874*				7.2†*	43.2†	0†
2210-TC-3821	Пар на конд.перв.пар/Pair of kond.perv.par	0.9865*	830†	3†	63.01†	6.8†*	462†	0†
2210-TC-3220	Темп. после нагрева/Temp. after heating	0.9859*	1008†	3†	61.29†	3.4†*	9000†	0†
2220-LV-3995	БАК ФИЛЬТРАТА 233.11/LHC LEACHATE 233.11	0.9847*	2†	1†	1†	1†	1E-08†	
38FC-6127	Поток на входе пресса/Stream entering the press	0.9886*	210†	1†	13.28†	1.2†	0†	
38FC-6168	Поток на входе пресса/Stream entering the press	0.9867*	290†	2†	35.93†	1.9†	0†	
38FC-6087	Поток разбав.выгрузк. бака/Feed razbav.vydruzk. bak	0.9856*	46†	3†	15.1†	0.54†	1.3†	
2210-TCZ-3086	Пар КД в низ букв./OD pairs in the bottom of his e	0.9859*	620†	2†	96.26†	0.17†*	54.6†	

Models for 77 Loops

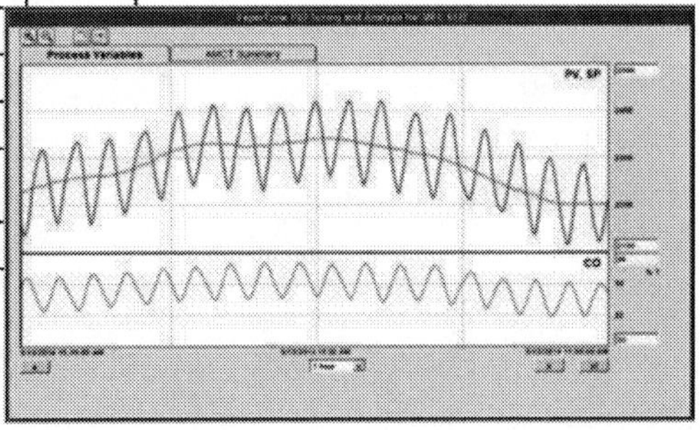

16

16AIChE Spring Meeting, Houston, TX

12TH GLOBAL CONGRESS ON PROCESS SAFETY

Practical Suggestions

An Approach to Control Performance

PLATFORM
- Software
- Hardware
- Installation
- Training
- Site Evaluation

SURGE
- Detailed Unit Evaluation
- Diagnosis
- Corrective Actions
- Value Documentation

SUSTAIN
- Continuous Improvement
- Diagnose & Correct New Issues
- Identify Additional Surges

17

16AIChE Spring Meeting, Houston, TX

12TH GLOBAL CONGRESS ON PROCESS SAFETY

More Practical Suggestions

- Do a Quick Survey
- Blanket an Area
- Don't be Overly Selective – i.e. Critical Loops Only
- Start with the Basics (Instrument, Valve, Tuning)
- Boost Data Collection Frequencies
- Follow Up is Key
- Learn from Others
- Document the Value

18

Questions & Answers

19

AUTHOR INDEX

AUTHOR INDEX

AUTHOR INDEX